SMART PHARMACEUTICAL NANOCARRIERS

SMART PHARMACEUTICAL NANOCARRIERS

Editor

Vladimir Torchilin

Northeastern University, USA

ICP Imperial College Press

Published by

Imperial College Press
57 Shelton Street
Covent Garden
London WC2H 9HE

Distributed by

World Scientific Publishing Co. Pte. Ltd.
5 Toh Tuck Link, Singapore 596224
USA office: 27 Warren Street, Suite 401-402, Hackensack, NJ 07601
UK office: 57 Shelton Street, Covent Garden, London WC2H 9HE

Library of Congress Cataloging-in-Publication Data
Smart pharmaceutical nanocarriers / [edited by] Vladimir Torchilin.
p. ; cm.
ISBN 978-1-78326-722-4 (alk. paper)
I. Torchilin, V. P., editor.
[DNLM: 1. Drug Carriers. 2. Nanoparticles. 3. Drug Delivery Systems. 4. Technology, Pharmaceutical. QV 785]
RS199.5
615'.6--dc23
2015021500

British Library Cataloguing-in-Publication Data
A catalogue record for this book is available from the British Library.

In-house Editors: Catharina Weijman/Chandrima Maitra

Typeset by Stallion Press
Email: enquiries@stallionpress.com

Printed in Singapore by Mainland Press Pte Ltd.

Preface

The use of pharmaceutical nanocarriers to enhance the *in vivo* efficacy of many drugs established itself well over the past couple of decades. Nanocarriers have revolutionized drug delivery practices, primarily due to the advances in materials chemistry, nanotechnology, and nanomedicine. This in turn, has spurred the development of a number of novel nanocarrier-based platforms and treatment strategies. Present day drug delivery strategies progressively lean towards "multi-pronged" combination approaches to make treatments more effective. To that end, nanocarriers that simultaneously incorporate multiple drugs to affect different pathways and act through different mechanisms, or combinations of drugs with biological therapeutics like genes, antibodies, proteins, or short interfering ribonucleic acids (siRNAs), have been the focus of recent active research. Furthermore, nanocarriers that respond to different types of stimuli, such as internal ones, which are intrinsic for the pathological area (e.g. changes in pH, temperature, redox condition, activity of certain enzymes), or external ones, which are artificially and locally applied (e.g. magnetic field, ultrasound, various irradiations) to trigger site-specific drug release, have been developed and represent an important and continuously growing area of research. Certain stimuli could also be considered both intrinsic and external. Thus, many pathological areas, such as inflammations and tumors, show a hyperthermia, which could be utilized as a possible stimulus to act on stimuli-sensitive nanopreparations. On the other hand, the required site could be heated from the outside to increase microvascular pore size and blood flow, which results in an increased extravasation of drug-loaded nanocarriers.

Stimuli-sensitive drug delivery, or smart delivery, is a very popular form of active targeting that has received a lot of attention, since it delivers and liberates the biologically active cargo at the desired site and at the required time. These systems target local stimuli or pathological changes occurring in affected tissues and use them as triggers for "switching on" certain desired functions (in addition to drug release, the function could be penetration inside pathological cells or generation of a diagnostic signal). The approach is most frequently applied to drug delivery into solid tumorsso far, since tumor tissues and associated inflammation and hypoxia demonstrate a lowered pH and a higher temperature compared to most healthy tissues. These tissues also exhibit a greater redox potential and the presence of peculiar enzyme systems, which may be present in small quantities or not be present at all in normal tissues. Still, interesting attempts to apply these approaches to treat diseases other than cancer also exist.

Stimuli sensitivity can also be used to allow certain functions of multifunctional nanopreparation to work only in required areas. Thus, although the longevity, targeting ligands, and stimuli-responsive moieties of nanomedicines significantly increased the drug disposition in the required area, insufficient cellular internalization of nanomedicines can be another barrier, especially for macromolecular drugs. To enhance the intracellular uptake, moieties such as cell-penetrating proteins/peptides could be used to modify the nanomedicines, since they had been demonstrated to be efficient in the translocation across the plasma membrane via either endocytosis or macropinocytosis. The direct electrostatic interaction between the positive residues of cell-penetrating peptides and the negative residues of the cell surface proteoglycans or glycosaminoglycans is required for internalization, regardless of the mechanism of cellular uptake. To minimize the nonspecific cell internalization of the modified nanocarriers, polyethylene glycol (PEG)ylation can be used to "shield" cell-penetrating function and prolong the circulation time of the nanocarrier, which provides sufficient time for nanocarriers to find their targets. In the target area, protecting the PEG coat could be eliminated by certain local stimuli (for example, if PEG was attached to a nanocarrier via a pH-sensitive bond degrading at the lowered pH in the tumor) and the exposed cell-penetrating function would bring the system inside the target pathological cell.

Thus, "smart" or stimuli-sensitive nanopreparations are designed to behave dynamically in response to various internal cues in the microenvironment or to certain external cues to release their therapeutic payloads, and offer exceptional spatial and temporal control over their properties, including the release of their cargoes.

In this book, an attempt was made to focus specifically on nanocarriers that exhibit stimuli sensitivity and incorporate various combinations of conventional small molecule therapeutic agents, biologics, and/or diagnostic moieties. In its fifteen chapters, the book provides an overview of the different internal and external stimuli relevant to various diseases, and discusses selected examples of stimuli-sensitive, combination nanopreparations from the recent literature with respect to each stimulus. We focus separately on each stimulus, internal and external, and discuss relevant examples of nanopreparations for combination therapy under each category.

The editor of this book is sincerely thankful to all outstanding scientists who agreed to contribute to this volume and hopes that the potential readers will find the information presented useful for their work in the important area of drug delivery.

Vladimir Torchilin
Boston, MA

Contents

Chapter 1

Stimuli-Sensitive Nanopreparations: Overview

Aditi Jhaveri[1], *Vitaly Shvets*[2] *and Vladimir Torchilin*[1]

[1]*Center for Pharmaceutical Biotechnology and Nanomedicine, Northeastern University, Boston, MA, USA*
[2]*Lomonosov Moscow University of Fine Chemical Technologies, Moscow, and Institute of General Pathology and Pathophysiology, Moscow, Russia*

1.1 Introduction

Free therapeutic molecules often exhibit various limitations, including low stability and fast inactivation or degradation *in vivo*, poor solubility, unfavorable pharmacokinetics, non-specific toxicity, and poor biodistribution, which may significantly complicate their use for effective therapy. The use of nanoparticulate pharmaceutical carriers (drug delivery systems) can overcome many of these limitations since these nanocarriers may solubilize hydrophilic, as well as hydrophobic drugs, increase their stability and longevity *in vivo*, provide controllable drug delivery and drug release, modify drug pharmacokinetics in a desirable way (since the drug's own pharmacokinetics is now substituted with the controllable pharmacokinetics of the delivery system), allow for targeted delivery of drugs to specific organs and tissues, or even to individual cells and cell organelles, and minimize undesirable side effects [1–3]. Advances in nanotechnology and nanomedicine have resulted in the development of nanocarrier platforms for therapeutics, which are expected to play a key role in improving disease management and treatment effectiveness. A number of nanoparticulate carriers including liposomes, polymeric micelles PMs, niosomes, solid lipid nanoparticles, polymeric nanoparticles, dendrimers, nanoemulsions, and quite a few others, are currently under investigation [4–6, 1, 7].

Moreover, these nanoparticulate drug delivery systems can be engineered in such a way as to selectively change their properties/functions (for example, facilitate drug release or cellular uptake) in response to specific internal or external stimuli/triggering mechanisms, i.e. behave as smart stimuli-sensitive preparations [8, 4, 9]. Such nanopreparations are designed to behave dynamically in response to various internal cues in the microenvironment of the pathological area or to certain external stimuli [10, 11]. Internal stimuli that are characteristic for the pathological areas, such as tumors, infarcts, sites of infection, etc., include local changes (compared to normal physiological values) in pH [12, 13], temperature (local hyperthermia that accompanies inflammation) [14–16], redox conditions (such as high intracellular glutathione levels) [17, 18], and the expression of certain molecules, including those with enzymatic activity [19, 20]. External stimuli or stimuli that could be artificially applied from outside of the body include heat, magnetic fields, light, and ultrasound, and can be employed to facilitate "on-demand" changes of certain functions of nanomedicines [8, 11]. See the schematic representation of various stimuli in Fig. 1.1.

A better understanding of local changes taking place in pathological areas, including tumors, has resulted in novel materials and engineering principles for building systems capable of responding to specific features of pathological tissues [21]. Many lipids and polymers undergo certain physicochemical changes as a response to these stimuli and can be used to build smart drug delivery systems [22]. Upon exposure to either extrinsic or intrinsic stimuli, such systems undergo the changes that are relatively sharp and abrupt and act as triggers to switch on certain functions of the system: drug release or exposure of hidden functions, such as intracellular penetration [4]. Drug release can proceed via the structural changes in delivery systems in the case of a physical entrapment of bioactive substances into a nanocarrier (cleavage of shell, charging of functional groups, carrier degradation), or by degrading certain chemical bonds if the bioactive is chemically attached to the carrier. The unique biochemistry of different pathological areas helps to engineer the responsive effects of stimuli-sensitive nanopreparations (SSNs) and can eventually allow for precise temporal and spatial delivery of the payload, improving its effectiveness and minimizing off-target effects [23, 4].

Here, we will provide the overview of various SSNs used for the delivery of pharmaceutical agents (see the general scheme in Fig. 1.1).

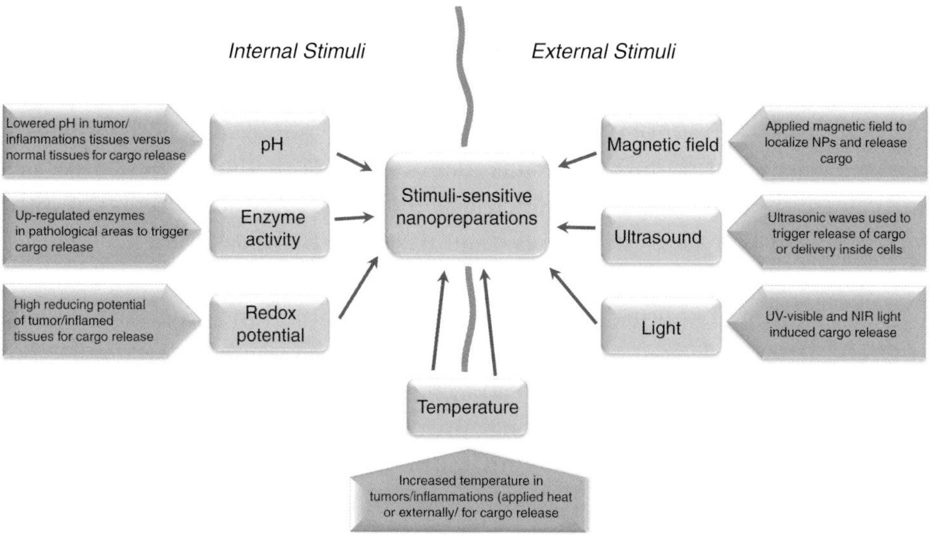

Figure 1.1. Schematic of SSNs. NP = nanoparticle; NIR = near-infrared; UV = ultraviolet.

1.2 Stimuli Characteristics for Pathological Tissues (see Table 1.1)

1.2.1 *pH*

The nutritional and metabolic environment of pathological/inflamed tissues differs significantly from that of normal cells. Thus, anaerobic conditions (hypoxia) are typical for many inflamed areas, including tumors, which lead to the production of lactic acid via the glycolytic pathway and are responsible for the lowered pH discovered in these hypoxic tissues [13]. The low extracellular/interstitial pH in tumors also results from additional mechanisms, like glutaminolysis, carbon dioxide production, and biocarbonate depletion [24]. The extracellular pH in solid tumors is lowered to ~6.5 compared to the normal pH of 7.4 [12, 25]. The difference in pH is, however, not uniform throughout the tumor tissue and can also fluctuate depending upon the cancer type [4]. The pH values in the intracellular organelles are further lowered below that of the tumor pH: for early endosomes, the pH is between 5–6, whereas lysosomes are even more acidic with a pH of about 4–5 [26]. This pH gradient can be successfully used for the stimuli(pH)-controlled drug delivery by engineering nanocarriers that can selectively release their therapeutic payloads at specific sites by responding to local changes in the pH of the environment [27].

The pH gradient is the most widely used stimulus for designing smart nanopreparations [27, 8]. The majority of the initial work on pH-sensitive systems was focused on building nanocarriers including liposomes with pH-responsive phospholipid components, such as phosphatidylethanolamine variants with unsaturated acyl chains that destabilize the liposomes at low pH and release the drug/deoxyribonucleic acid (DNA) load by protonation or by forming non-bilayered structures that destabilize the nanocarrier and/or endosomes [8, 28, 29]. Thus, one possible strategy to impart the pH-sensitivity to nanopreparations is to incorporate ionizable functional groups, like amines, carboxylic acids, and phosphoric acids, into polymers that are used to build or modify nanocarriers that can ionize at certain pH values [30, 31]. In some cases, the ionizable functional groups can reversibly change polymer solubility upon ionization [31]. Thus, carboxylic acids carry a negative charge and are deprotonated at neutral pH values, but lose the charge and become undissociated at a lower pH. Amines, in turn, are uncharged and hydrophobic when in neutral and alkaline solutions, but protonated and hydrophilic at mildly acidic pHs [10]. Protonation/deprotonation polymer transition can transform the polymer from hydrophobic to hydrophilic and *vice versa*, which can provoke its swelling or increase its solubility, thus allowing for drug release from particles made of such polymers [10]. In this way, for example, pH-responsive doxorybicin-loaded nanogel was prepared from glycol chitosan (GCS) grafted with functional 3-diethylaminopropyl (DEAP) groups (GCS-g-DEAP) [32]. At physiological pH, amphiphilic GCS-g-DEAP self-assembled, entrapping the drug, but the decrease in pH (in the tumor) led to the protonation of DEAP and a decrease in its hydrophobicity, resulting in nanogel destabilization and drug release. Another possible strategy to design pH-sensitive nanocarriers is the use of acid-labile bonds/functional groups, which undergo cleavage at a specific pH. A set of tested pH-labile linkers include benzoic imine [33], hydrazone [34–38], oxime [39], acetal [40], silyl ether [41], ortho ester [42, 43], Schiff base [44], and vinyl ether [45]. These pH-sensitive linkers can be a part of the polymeric components used to produce pH-responsive drug delivery systems, which are expected to be stable in normal tissues and in the blood, but can disintegrate and release the entrapped drugs at lowered pH values [46, 8]. These linkers/bonds may be present in the polymer main chain, side chains, or as a linkage between the hydrophilic and hydrophobic blocks in amphiphilic polymers [31]. Frequently, the use of such bonds was suggested for removing PEG from the surface of PEGylated long-circulating nanopreparations. Another important application of pH-cleavable linkers is in the design of long circulating nanoparticles. PEGylation allows for increased stability and extended circulation time of nanoparticles in the body, but could inhibit the interaction of the nanocarrier with the cell membrane and make the release of the drug from the nanocarrier difficult. To solve this problem, a pH-sensitive linker between PEG and the nanoparticle surface can be introduced, which remains stable in the circulation and allows for the nanopreparation to accumulate in the target (tumor); however, inside pathological tissues with the lowered pH, such linkers are rapidly hydrolyzed and eliminate the PEG layer [47, 27, 48]. pH-sensitivity of the PEG coats on nanoparticles also allows for designing multifunctional drug delivery systems; the PEG layer can shield certain functions while the nanocarrier is in circulation, but those functions (for example, cell-penetrating moieties) could become exposed after the PEG is eliminated by the lowered tumor pH [37]. Alternatively, drug moieties can be conjugated via such acid-labile linkers to nanoparticles to allow for the specific drug release in tumors [49]. A pH-responsive micellar system for the delivery of plasmid DNA (pDNA) was reported where phospholipids, such as phosphatidyl ethanolamine (PE), were conjugated to low molecular weight (MW) polyethylene imine (PEI) and such PEI–PE conjugates effectively complexed DNA (PEI–PE/DNA) [36]. The PEI–PE/DNA complexes, when mixed with low pH degradable PEG–hydrazone–PE (PEG–Hz–PE), mixed with high transfection efficacy at lowered pH values because of PEG elimination and easier uptake by DNA complexes by cells. See some examples of pH-sensitive crosslinkers in Fig. 1.2.

Other strategies for pH-sensitive drug delivery make use of charge-conversional or charge-reversal nanoparticles [50, 51], which are usually negatively charged at neutral pH and are capable of converting their surface charge into a positive one in response to the lowered extracellular pH (such as in tumors). Calcium phosphate (CaP)/charge-conversional polymer hybrid nanoparticles have been reported carrying vascular endothelial growth factor (VEGF) siRNA against pancreatic cancer [51]. These nanoparticles utilized a block copolymer of PEG and an endosome-disrupting polyanion, poly(*N*′-{*N*′-[(*N*-*cis*-aconityl)-2-aminoethyl]-2-aminoethyl}aspartamide) (PAsp), which is stable at the extracellular pH (7.4), but degrades inside the lysosome at pH 5.5 due to cleavage of the *cis*-aconitic amide linkages, reverting back to the parent polycation, PAsp. Nanoparticles showed high gene silencing efficiency in BxPC3 (pancreatic cancer) tumor model.

To increase the accumulation of nanoparticles in tumors, it was suggested to utilize the aggregation of nanoparticles stimulated by the tumor microenvironment. After extravasation into the tumor tissue, stimuli such as a reduced pH can cause the aggregation of nanoparticles to trap them at the tumor site due to their increased size, and lead to increased nanoparticle

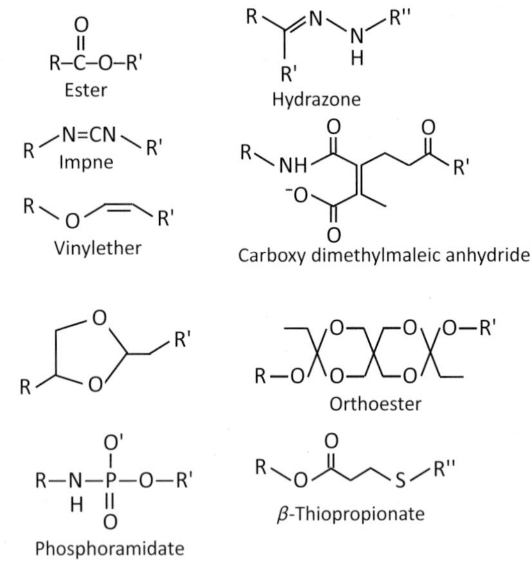

Figure 1.2. Commonly-used pH-sensitive cross-linkers and their chemical structure.

retention in tumors [52]. The aggregates eventually sediment at the tumor site and their uptake by the cancer cells increases [53, 54].

1.2.2 *Reduction–oxidation (redox) potential*

Another important physiological stimulus is the redox potential in the pathological tissues or in the intracellular compartments of cells, which is more reducing compared to the extracellular compartment. This reducing environment is due to the presence of higher levels of the glutathione tripeptide (γ-glutamyl-cysteinyl-glycine) (GSH) in the cytosol and subcellular compartments (approximately 2–10 mM), which are about 100–1,000 times higher than found in the extracellular compartment (approximately 2–20 μM) [55], where GSH is rapidly degraded by plasma enzymes. It has also been reported that levels of GSH are much higher in the tumor cytosol as compared to the normal cell cytosol, a difference that has been exploited to a great extent for redox-sensitive, tumor-specific delivery of therapeutics [56–58, 17]. Endosomes and lysosomes also have own redox potential that can be used to promote endosomal and subsequent lysosomal escape [59]. The inter-conversion of thiols and disulfides is one of the important biological processes that helps maintain the structure of native proteins in cells. Under reductive conditions, the disulfide bonds can be converted to thiols by reducing agents or can even undergo disulfide exchange in the presence of other thiols [60]. Generally, redox and thiol responsive materials have disulfide linkages in their structures that are stable at oxidizing extracellular conditions, but undergo rapid breakdown or thiol–disulfide exchange in the presence of reducing agents like GSH in the intracellular compartment [61]. The difference in redox potential between intra- and extracellular environments, as well as that between tumor cell surrounding and normal tissues, provides an excellent opportunity to design nanocarriers that can selectively release their cargo in response to this altered GSH concentration [62, 60]. The thiol–disulfide interconversion chemistry, which is a critical step in many biological processes, has been utilized synthetically for a number of bioconjugation protocols [18]. Typically, redox-responsive nanocarriers contain disulfide bonds, which hold the cargo (DNA, siRNA, drugs, protein) within the nanocarrier under normal conditions. However, upon encountering reducing conditions within the cell, the carrier is destabilized to release the entrapped cargo and facilitate intracellular accumulation [8, 63].

Many polymers have been designed by incorporating a thiol-reducible bond in their structure or by cross-linking reducible bonds to form biocompatible systems that have site-specific delivery in the areas with a higher redox potential, like the cancer cell cytosol [60]. To produce redox-responsive nanoparticles, various approaches have been utilized, including introduction of reduction-responsive linkers at the hydrophilic–hydrophobic junction in amphiphilic polymers (cleavable backbone), pendant groups, or side chains, and redox-sensitive bonds as cross-linkers [31, 63]. The disulfide bonds are stable under normal reductive conditions found in the extracellular compartment, but undergo reversible conversion to free thiols under the reductive environment intracellularly [64]. These reduction-triggered systems deliver the payload following a burst in response to the high redox potential by destruction of the carrier. In the presence of reducing agents like GSH, nanocarriers containing disulfide bonds in their structures lose their integrity and form free thiols to deliver drugs and DNA [65, 66]. Positively charged polymers have been used to complex DNA, with the inclusion of disulfide bonds in these complexes to form bridges that are broken down to thiols, resulting in DNA release [8]. Thiolated nanoparticles, thiolated PEIs, and redox-sensitive-lipid incorporated liposomes have also been explored for redox-responsive delivery [8, 4].

Nanopreparations that are sensitive to hypoxic tumor conditions are also being developed. Thus, azobenzene was used as a hypoxia-responsive bioreductive linker for hypoxia-specific release of siRNA from PEGylated nanopreparations [67]. Redox-sensitive preparations have also been used for co-delivery of paclitaxel (PXL) and anti-survivin siRNA to down-regulate the survivin protein, which is over-expressed in cancer cells compared to normal cells and to sensitize the PXL-resistant cells [68]. The authors utilized PEG_{2000}–PE PM, with the survivin siRNA conjugated to these micelles via a disulfide bond attached to the PE (siRNA–S–S–PE), and PXL was loaded into the core of such micelles. The pre-treatment of cells with siRNA–S–S–PE PM led to the release of siRNA under the reducing conditions of tumor cells and sensitized these cells to PXL treatment.

Redox-responsive micellar nanocarriers have been prepared from a variety of amphiphilic copolymers including linear [69, 70], star [71], graft [72, 61], and hyperbranched copolymers [73]. Thambi *et al.* reported PMs composed of PEG and poly(γ-benzyl-L-glutamate) (PBLG) bearing a disulfide bond (PEG–SS–PBLG) for the delivery of camptothecin (CPT) [74]. While only 40% of CPT was released in the absence of GSH, complete release of CPT occurred in 20 h in the presence of 10 mM GSH. PEG–SS–PBLG micelles were more cytotoxic to squamous carcinoma cells (SCC7) *in vitro* compared with PEG–*b*–PBLG and efficiently delivered CPT into the nuclei of these cells. Redox-sensitive cross-linked micelles have been developed for the delivery of drugs as well as nucleic acids, which are core-cross-linked, shell-cross-linked, or cross-linked at the core–shell interface [75–79].

The development of novel, biodegradable redox-sensitive polymers containing trimethyl-locked benzoquinone (TMBQ) as pendant groups was recently reported [80]. The polymer was designed to release lactone and unmask free amino groups when reduced. Nanoparticles made of this polymer released paclitazel upon conversion of the polymer from hydrophobic to hydrophilic with subsequent nanoparticle disassembly (swelling) resulting from protonation of the free amino groups. Reduction responsive PEG-poly(amino acid) synthesized from L-phenylalanine *N*-carboxyanhydride, and L-cysteine, with an amino group-terminated PEG monomethyl ether ($mPEG–NH_2$) as the macroinitiator, was used to prepare doxorubicin-loaded nanogels capable of releasing the drug under the action of GSH [81].

Redox-sensitive nanoparticles also provide a promising strategy for intracellular delivery of siRNA and genes. With this in mind, bio-reducible cationic lipids or polymers containing disulfide bonds, such as cholesterol disulfide cationic lipids [82], branched poly(disulfide amine) [83], hyperbranched poly(amido amine) (PAMAM) [84, 85], and disulfide-based polyethyleneimine (SS–PEI) [86], have been synthesized. Redox-sensitive micelles have been successfully used for the delivery of genes, anti-sense oligonucleotides, and siRNA [76, 87, 88]. Bioreducible, non-viral carriers were suggested for the intracellular delivery of siRNA against human telomerase reverse transcriptase [86]. The authors used a biodegradable PEI of ~800 Da, containing multiple disulfide bonds (SS–PEI), which could strongly bind siRNA to form nano-sized positively charged complexes that readily destabilize to release siRNA in a reducing environment both *in vitro* and *in vivo*. Lipid-based micelles and micelle-like nanoparticles have also been explored for redox-sensitive delivery of genes and siRNA. Musacchio *et al.* reported a non-toxic, redox-sensitive micellar system to stabilize and deliver siRNA *in vitro* [89]. Double stranded green fluorescent protein (GFP)–siRNA was reversibly modified with phosphothioethanol via a reducible disulfide bond to form siRNA–S–S–PE, and this siRNA derivative was incorporated into PEG–PE micelles to form nanosized siRNA–S–S–PE/PEG–PE mixed micelles. The mixed micelles protected siRNA against nuclease-mediated degradation and enabled the release of free siRNA under intracellularly relevant GSH concentrations. See some examples of the redox-sensitive cross-linkers in Fig. 1.3.

In addition to disulfide linkages, diselenide bonds (Se–Se) have also been investigated for redox-responsive gene delivery [90]. The authors have used diselenide bonds cross-linked to branched oligoethylenimine to develop novel reduction-sensitive gene carriers.

1.2.3 *Locally overexpressed enzymes*

The expression of enzymes and their composition varies in different tissues and may be specific to a particular tissue or tissue status. The expression of certain enzymes, like proteases, phospholipases, or glycosidases, in pathological tissues suffering inflammation or cancer is often greater compared to their expression in normal tissues [11, 91–94]. Thus, extracellular matrix (ECM) remodeling proteinases, such as the matrix metalloproteinases (MMPs), are found to be the chief mediators of changes in the tumor microenvironment associated with the progression of tumors [93]. The MMPs and related enzymes greatly influence various properties of tumors, since they are involved in the process of breaking down cell adhesion molecules, which in turn affects cell–cell and cell–ECM interactions [95–98]. Other enzymes specifically dysregulated in cancer, like gelatinase [99], hyaluronidase-I [100, 101], esterase [102], and phospholipase A2 [103, 104], have also been successfully exploited for enzyme-sensitive cancer therapy. Proteases are associated with the exterior of the tumor tissues and at the sites of angiogenesis, which are easily accessible to passively targeted nanoparticles via the enhanced permeability

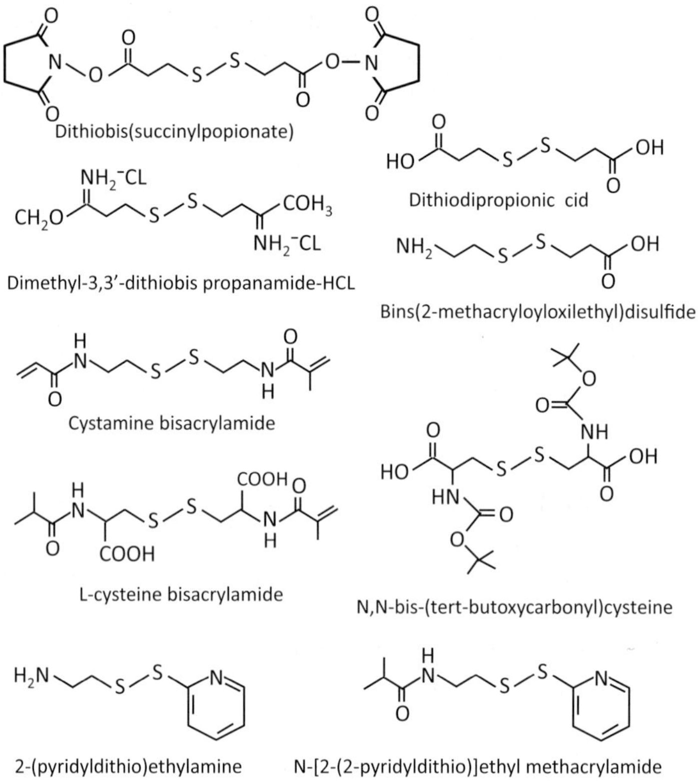

Figure 1.3. Commonly-used redox-responsive cross-linkers and their chemical structure.

and retention (EPR) effect, making them excellent candidates for enzyme-responsive delivery [105].

Designing nanoparticles that can respond to the biocatalytic actions of enzymes is an emerging strategy in the field of bioresponsive delivery of therapeutics [106, 19, 20]. Enzyme-responsive delivery from nanoparticles can be brought about by incorporating enzyme reactive moieties in the main chain or side chain of the polymeric material, which is recognized by the enzyme, or a product of the enzymatic reaction, both of which result in enzymatic transformation of the nanoparticles [19]. Alternately, when the nanomaterial itself is not sensitive to biocatalysis, its surface may be modified with molecules that elicit a change in the physical properties of the nanoparticle solution (solution-to-gel and gel-to-solution transition) upon enzymatic transformation [19, 106]. Enzyme catalyzed reactions that cause the transformation of polymers can involve multiple non-covalent interactions (e.g. van der Waals forces, hydrogen bonding, hydrophobic and electrostatic interactions), which may result in self-assembly, disintegration, and morphological transition of polymers or polymeric assemblies [106].

Enzyme-sensitive nanocarriers are built using the concept of cleavage of esters or short peptide sequences in the presence of proteases or esterases to facilitate drug release [4]. One such example of a commonly used material for building these systems is a polymer, N-(2-hydroxypropyl) methacrylamide (HPMA) as the backbone, attached with a commonly used tetrapetide Gly–Phe–Leu–Gly as a linker for attachment of active molecules. This tetrapeptide is a substrate of the lysosomal enzyme, cathepsin B, over-expressed in tumors, and is cleaved off in the tumor environment to release the therapeutic payload [4, 107–109]. Similarly, many other substrates of MMPs, like proteins, peptides, and polymers, are being used to design nanocarriers for tumor-specific delivery [105, 110–116].

Based on the MMP-2 sensitive enzyme-triggered strategy, many nanopreparations were built to deliver drug, DNA, or imaging agents to the tumor tissues or in combinations, for therapy or co-delivery. One such system allows for enzyme-sensitive combination therapy [117]. It combines siRNA against VEGF and doxorubicin in nanoparticles built of dendrigraft poly-L-lysine with cell-penetrating peptide (CPP) conjugated to it via malemidyl-ω-N-hydroxysuccinimidyl polyethylene glycol. The CPP was masked using a covalently attached pH-sensitive peptide, with a MMP2-sensitive linker. In circulation, the masking peptide was thought to protect the CPP and expose it only upon exposure to lower pH conditions and elevated MMP2 levels to activate the CPP. These nanoparticles were tested on glioma models both *in vitro* and *in vivo* and promoted apoptosis and anti-angiogenesis simultaneously.

Among the MMPs, MMP2 and 9 are the most important cancer-related MMPs [118]. Polymeric nanoparticles can be covalently modified with drugs through MMP cleavable linkers that can trigger drug delivery upon exposure to these enzymes in the target tissue. MMP-cleavable substrates (typically, cleavable peptides digested by proteases) have been used for the delivery of drugs and imaging agents [119, 120, 118]. A functionalized gold nanoparticle system (AuNP) for tumor imaging and targeting was recently reported, where doxorubicin was conjugated to the AuNPs by a thiol–Au bond with a MMP-2-cleavable peptide substrate, CPLGLAGG [120]. The nanoparticles were decorated with PEG to allow for extended circulation. The presence of a thiol–Au bond led to accelerated drug release once the particles were internalized via endocytosis. Injection of the AuNP into SCC-7 tumor-bearing mice caused rapid release of the drug, which resulted in efficient tumor growth inhibition.

Among other enzymes, the cysteine protease cathepsin B is implicated in tumor invasion and metastasis, and is overexpressed in many cancers at the messenger ribonucleic acid (mRNA), protein, and activity levels [109, 121]. Cathepsin B-responsive polymerosomes were prepared from a biodegradable block copolymer of mPEG and poly (d,l-lactide) (PDLLA) by incorporating a cathepsin cleavable peptide sequence glycine (Gly)–phenylalanine (Phe)–leucine (Leu)–Gly–Phe (GFLGF) between the two blocks (mPEG–pep–PDLLA) [122]. These polymerosomes destabilized at pH 5.5 in the presence of cathepsin B and showed rapid release of the model drug acridine orange under these conditions. Proteases like plasmin have been also explored for the delivery of drugs (proteins) using enzyme-responsive nanocapsules [123]; gelatinase responsive copper sulphide nanoparticles were reported for combined photoacoustic imaging, enzyme-responsive drug release and photothermal therapy [99], and liposomal formulations of lipid pro-drugs of retinoids sensitive to degradation by secretory phospholipase A_2 [124] have also been developed.

1.2.4 *Temperature*

Since changes in local tissue temperature could be either intrinsic or caused from the outside of the body, we will consider temperature-sensitive systems at the end of natural stimuli and close to external stimuli. Elevated temperature is commonly encountered in disease states, such as fever or inflammation, and can be applied from outside the body by local ultrasound or high frequency alternating magnetic fields (AMFs), which can cause oscillation of magnetic material accumulated at the target site with subsequent heat release [8]. The higher than normal temperature in inflamed or cancerous tissues is characteristic for such tissues, but does not differ markedly from normal tissues. Still, even not very big changes in temperature are accompanied by changes in the physiology of the corresponding tissues as well. Higher temperatures (hyperthermia) in the range of 40–45°C induces a greater blood flow in the tissue and increases the vasculature permeability as compared to that in normal tissues (37°C) [62, 125]. Hyperthermia is also responsible for decreased DNA synthesis, induction of heat shock proteins, altered protein synthesis, disruption of microtubule organizing centers, changes in the expression of receptors and binding of growth factors, and ultimately, changes in the cellular morphology [126]. Since the understanding of the behavior of the vasculature in the presence of hyperthermia has widened, it has become a popular rationale to use higher local temperatures in combination with thermally non-labile chemotherapeutic drugs. Non-temperature-sensitive polymers have also been found to accumulate in areas of hyperthermia due to increased permeability in these areas. Frequently, external heating sources are applied to create the temperature contrast between normal and pathological tissues. These sources could be electromagnetic, laser, water bath, or even high intensity focused ultrasound (HIFU) [62]. Nanopreparations designed using thermo-sensitive materials that exploit inherent or induced hyperthermia have become popular delivery systems. External control of local body temperature by physical means affords a more consistent signal and can be used as an extensive application for thermo-sensitive nanocarriers [127].

Thermo-responsive materials help to carry out spatial delivery of the payload resulting in efficient cancer therapy. Lipids with appropriate phase transition temperatures have been used to formulate thermo-sensitive liposomes that result in the release of the drug at elevated temperatures [62, 125]. One of the most commonly used and studied lipids for use in hyperthermia is dipalmitoylphosphatidylcholine (DPPC), which has an ideal gel-to-liquid crystalline transition temperature (Tg) of 41°C [125, 62]. Many studies have indicated that DPPC undergoes a phase transition and conformational changes compromising the integrity of the liposomes, leading to the release of its payload in response to hyperthermia [128]. Temperature-sensitive liposomes containing DPPC, in combination with hyperthermia produced by ultrasound or magnetic fluid hyperthermia, have been used for stimuli-responsive release of drugs [129–131].

Thermo-responsive proteins and polymers also tend to undergo a phase transition (usually at >40°C), which is termed as the critical point of these polymers. The temperature at which the transition occurs is termed as the lower critical solution temperature (LCST) [132]. Below the LCST, the polymers exist in a water-soluble hydrophilic form due to extensive hydrogen bond formation between the polymer and water molecules. This hydrogen-bond network collapses at temperatures above the LCST, resulting in the exclusion of water molecules from the polymer structure, making the polymer hydrophobic and insoluble and causing it to aggregate and precipitate in the hyperthermic tissues [133]. Polymers can be designed in a manner so that their LCST is ideal to build a nanopreparation that is soluble when administered by injection *in vivo*, but becomes insoluble and aggregated in the tumor regions heated above its LCST, leading to delivery by local targeting. Below the LCST, the hydrated polymers attached to the surface of a nanocarrier prevent the interaction of proteins with its surface; however, above the LCST, the dehydrated polymer exposes the nanocarrier surface to facilitate interaction [125]. One of the most extensively studied thermo-responsive polymers is poly (N-isopropylacrylamide) (PNIPAm), which has a LCST of about 32°C [134–136]. Often, a hydrophilic monomer is added to the polymer to raise its LCST for better thermo-selectivity [137]. A number of copolymers of pNIPAAm with various hydrophilic polymers have been developed to fine tune its LCST closer to the body temperature for biomedical applications. Polymers like *N*-vinyl-2-pyrrolidone [138], *N,N*-dimethylacrylamide [139, 140], acrylamide (AAm) [141], and PEG [142–144] among others, have been utilized to that end.

Thermo-sensitive comb-like copolymers of methoxy PEG (mPEG) blocks and hydrophobic polyacrylate (PA) backbones with thermo-sensitive PNIPAM graft chains (mPEG–b–PA–g–PNIPAM) have been reported and used to prepare micelles loaded with camptothecin, which released the drug at temperatures above 40°C, the rate of release being dependent on copolymer composition and the chain length of mPEG blocks [145]. Thermo-sensitive polyelectrolyte complex micelles have also been suggested for delivery of 5-fluorouracil (5-FU) using two biocompatible graft copolymers, chitosan-g-poly(N-isopropylacrylamide) (CS–g–PNIPAM) and carboxymethyl cellulose-g-poly(*N*-isopropylacrylamide) (CMC–g–PNIPAM), which self-assembled readily to form nano-sized core-shell micelles [146]. PNIPAM formed the thermo-sensitive shell whereas chitosan (positively charged) and CMC (negatively charged) formed the core of the micelles. The LCST of these micelles was near 37°C, and *in vitro*, with Henrietta Lacks (HeLa) cells, these micelles exhibited excellent temperature- and pH-controlled drug release properties [146] pNIPAAm copolymer (pNIPAAm–co–Am–SH) was used to functionalize AuNPs [147]. Resulting nanoparticles had a transition temperature within the range of 37–40°C, and the increase in temperature from 34°C to 40°C significantly increased the preparation uptake by human breast adenocarcinoma (MCF-7) cells.

Elastin-like polypeptides (ELPs) have been extensively investigated as novel vehicles for thermo-responsive delivery [148–150]. ELPs are genetically encoded biopolymers which possess repeats of a pentameric peptide sequence (valine (Val)–proline (Pro)–Gly–Xaa–Gly) derived from tropoelastin, where X can be any amino acid except Pro [151, 152]. ELPs show a phase transition temperature similar to the LCST observed with some polymers, wherein they are soluble at low temperatures, but can phase-separate into a gel-like phase above a critical transition temperature [148]. The transition temperature can be adjusted by changing the composition of the Xaa residue, although there are also other factors determining it, including the length of polypeptide, ionic strength and polarity of the solvent, and solution concentration [153].

1.3 Stimuli Acting from Outside of the Body (see Table 1.1)

1.3.1 *Magnetic field*

In magnetic drug targeting (MDT), the magnetically susceptible delivery vehicles loaded with drugs are guided to, and concentrated at the pathological site (tumors) using an external magnet and drugs are released from these carriers at the site. Magnetic field-responsive nanopreparations typically incorporate the therapeutic payload(s) along with a magnetically active component in a pharmaceutically stable formulation [8]. The magnetic sensitivity is most often imparted to nanopreparations by iron oxide nanoparticles (4–10 nm), which may be either magnetite (Fe_3O_4) or maghemite (Fe_2O_3) [200]. These iron oxide nanoparticles are attracted to a magnetic field, but do not retain the residual magnetism upon removal of the field, a property known as superparamagnetism [201]. Thus, they are referred to as superparamagnetic iron oxide nanoparticles (SPIONs). SPIONs are frequently utilized for biomedical applications like drug targeting, bioseparations, magnetic resonance imaging (MRI), magnetofection, and magnetic hyperthermia [8]. SPIONs are hydrophobic and are thus prone to aggregation, unless their surface is coated with materials like dextran, PEG, PEI, poly(vinyl alcohol), and citric acid, among others, to improve their biocompatibility [202]. SPIONs can be engineered to perform multiple functions, such as carriers for drugs, contrast enhancers for *in vivo* tumor imaging, hyperthermia, and MDT [202].

Magnetic cationic liposomes (MCLs) loaded with the anti-cancer drug vinblastine significantly improved vascular uptake of MCLs in a murine melanoma model in the presence of an external magnet, which resulted in reduced tumor nodules in metastatic sites compared to the no-magnet treatment group [203]. Novel GoldMag (Fe_3O_4/Au) nanoparticles with iron (Fe) oxide as the core and a layer of gold (Au) deposited as the shell were modified with PEG and loaded with doxorubicin. Biodistribution studies revealed a significantly higher doxorubicin concentration in the liver exposed to magnetic field versus no applied magnetic field [204]. Magnetic lipoplexes (ternary complexes of cationic lipids with plasmid DNA associated with SPIONs) can be concentrated onto target cells using magnetic fields. Wang *et al.* reported magnetic lipoplex delivery of short hairpin RNA (shRNA) to suppress Type 1 insulin-like growth factor receptor (IGF-1R) overexpression in lung cancer *in vitro* and *in vivo* [205]. Cui *et al.* reported the development of transferrin (Tf) conjugated magnetic silica poly(d,l-lactic-co-glycolic acid) (PLGA) nanoparticles co-loaded with doxorubicin and PXL to enable their effective delivery across the blood–brain barrier [206]. Tf was used specifically to target the glioma cells, which over-express Tf receptors.

Table 1.1. Examples of SSNs.

Nanopreparations	Sensitive components	Acting stimulus	Load	Property, action	Reference
Amphiphilic linear-dendritic prodrugs	Methoxy poly(*ethylene glycol*) (MPEG)-b-poly(amidoamine) (PAMAM)	pH	Doxorubicin/CPT	pH-dependent release of both drugs accompanied by effective uptake and suppressed cancer growth.	[154]
Hollow mesoporous silica nanoparticle (HMSNP)	MSNPs coated with folic acid conjugated to PEI (PEI-FA)	pH	Doxorubicin/siRNA	pH-responsive intracellular drug/siRNA release which minimized pre-release and reduced off-target effects, greatly suppressed Bcl-2(anti-apoptotic) protein expression.	
Micelleplexes (Cationic micellar nanoparticles)	Dual pH-responsive PDMA-b-PDPA di-block copolymers complexed with siRNA	pH	Bcl-2 siRNA/PXL	Down-regulation of anti-apoptosis gene of Bcl-2, good tool for overcoming MDR.	[155]
PMs	Doxorubicin conjugated SMAderivative with adipic dihydrazide (ADH) via an acid-cleavable hydrazone bond	pH	Doxorubicin/Disulfiram (DSF)	Temporal release of both drugs; combination effective in enhancing cytotoxicity, promoting apoptotic response and effective inhibitory effect on the growth in resistant breast cancer models.	[156]

Prodrug nanoparticles	Poly (lactic acid) conjugated — cisplatin prodrug modified-D-(TPGS), carboxy group modified co-polymer (TCPNPs)	pH	Cisplatin/ Docetaxel	pH-sensitive release of drugs; TCPNPs showed better therapeutic efficacy than individual drugs, TCPNPs of high docetaxel versus cisplatin ratio demonstrated highest efficacy, lowered IC50s observed in HER-2 over-expressing cells.	[157]
Polyion complex micelles (NPIC)	PEI-PEG copolymer, cationic PEI complexation of DNA with Doxorubicin loaded-TAT-modified NPIC with electrostatic adsorption of sulfamerazine (SA)-PEG-NGR	pH	Doxorubicin/ DNA	Enhanced cellular uptake and transfection efficiency.	[158]
Composite micelles	Biodegradable polymer methoxyl-poly-(ethylene glycol)-b-poly(lactide-co-2-methyl-2-carboxyl-propylene carbonate) with pendant carboxyl groups	Redox potential	Hydrophilic Oxaliplatin pro-drug (Oxa(IV)–COOH) Hydrophobic Daunorubicin (DRB)	Reduced systematic toxicity and greater synergistic effect of composite micelles than combination of anticancer drugs both *in vitro* and *in vivo*.	[159]

(*Continued*)

Table 1.1. (*Continued*)

Nanopreparations	Sensitive components	Acting stimulus	Load	Property, action	Reference
Nanoparticles	Bioreducible poly (b-amino esters) (PAEs), poly[bis(2-hydroxylethyl)-disulfide-diacrylate tetraethylenepentamine] (PAP)	Redox potential	iMdr-1-shRNA/ iSurvivin-shRNA	Low IC50 of Doxorubicin in MDR cells, down-regulation of P-gp and survivin, decreased tumor volumes *in vivo*.	[160]
Polymeric micelle	Arginine-grafted poly (cystaminebisacrylamide-diaminohexane) (ABP)	Redox potential	PXL/DNA	Increased cellular uptake efficiency, low cytotoxicity in non-reductive conditions.	[161]
Mesoporous silica nanoparticle (MSNP)	Amino-terminated alkyl chains with disulfide bonds functionalized on nanoparticle surface by the reaction of SATH and MSNPSH	Redox potential	Negatively charged ssDNA/ Doxorubicin	Enhanced cellular internalization and significant apoptosis induction.	[162]
Polyplex	Supramolecular self-assembled inclusion complex prepared from PXL, star-shaped cationic polymer containing g-CD, OEI arms with FA conjugated via a disulfide linker.	Redox potential	PXL/plasmid DNA encoding luciferase or p53 gene	Enhanced gene transfection, significant cell apoptosis.	[163]

Theranostic nanohybrids	Graphene oxide (GO) grafted PEI–PEG via MMP-2 cleavable PLGLAG peptide linkage	Enzyme MMP2 enzyme	Doxorubicin/ DNA	MMP2 responsive peptide cleavage controlled payload delivery, enhanced drug efficacies *in vitro*, efficient transfection comparable to that of PEI_{25k}.	[164]
Dendrimers	Dendritic thiolated hyperbranched polyglycerol	Enzyme cathepsin B	Maleimide-bearing Doxorubicin prodrug/ Maleimide-bearing Methotrexate prodrug	Drug release in the presence of cathepsin B, cytotoxicity against human tumor cell lines due to drug activity.	[165]
PMs	MMP-2 sensitive (PEG–pp–PEI–PE) co-polymer	Enzyme MMP2 enzyme	PXL/siRNA (anti-GFP or anti-survivin)	Improved tumor targeting, tumor cell internalization and synergistic anti-tumor activity of co-loaded PXL and siRNA.	[166]
Micelle	MMP2 sensitive GPLGIAGQ peptide	Enzyme MMP2 enzyme	PXL	Enhanced antitumor activity.	[110]
Copper sulphide Nanaparticles	Gelatine	Enzyme gelatinase	Doxorubicin	Enhanced cytotoxicity.	[99]
Liposome	Lipid C6/Phospholipase A(2)	Enzyme phospholipase A(2)	Retinoid	Enhanced cytotoxicity.	[124]
Core-shell Nanaparticles	Benzyl ester bond/ Esterase	Enzyme esterase	PXL and photosensitizer	Enhanced antitumor activity.	[102]
Hydrogel	Phe–Pip–Arg–Ser/ Thrombin\	Enzyme thrombin	Heparin	Autoregulation of heparin release.	[167]

(*Continued*)

Table 1.1. (*Continued*)

Nanopreparations	Sensitive components	Acting stimulus	Load	Property, action	Ref
Hydrogel nanocapsules	KNRVK/Plasmin	Enzyme protease	VEGF*	Blood vessel formation induced by enzyme-responsive release of VEGF.	[123]
MCLs	DPPC, DC-Chol, DOAB, Cholesterol and magnetic fluid Fe_3O_4	Temperature	Doxorubicin/SATB1-shRNA	Co-delivery of Doxorubicin and SATB1 shRNA enhanced inhibition of cell growth in a gastric cancer model *in vitro* and *in vivo*, compared to either bioactive agent alone.	[168]
Hydrogel	Linoleic acid-coupled poloxamer thermo-sensitive hydrogel	Temperature	PXL/Akt1 shRNA	Co-delivery of Doxorubicin and SATB1 shRNA enhanced inhibition of cell growth in a gastric cancer model *in vitro* and *in vivo*, compared to either bioactive agent alone.	[169]
Nanogel	Fibrinogen-graft-PNIPAAm	Temperature	5-Floururacil/ Megestrol acetate	Triggering of apoptosis and enhanced toxicity *in vitro*.	[170]
Thermo-responsive barrier gel	Injectable thermo-gelling PDEAEM-Pluronic F127 (PL)-PDEAEM pentablock copolymer (PB)	Temperature	PXL/ DNA	Enhanced transfection efficiency and anti-cancer effects of PXL *in vitro*.	[171]
Nanoparticles	Magnetic mesoporous silica nanoparticles (MMSNs)	Magnetic field	Doxorubicin/PXL or Rapamycin	Better internalization in A549 cells, enhanced apoptosis and growth inhibition over single-drug MMSNs, improved anticancer efficacy.	[172]

Nanoparticles	PLGA-magnetic NPs conjugated to Herceptin	Magnetic field	PXL/Carboplatin or Rapamycin	PXL+Rapa more effective than PXL+Carbo, increased cellular uptake and enhanced synergistic effect with HER2 conjugation.	[173]
Double emulsion (W/O/W) nanocapsules (DEC)	PVA (MW 16k) shell, iron oxide (IO) NP localized in the PVA shell, IVO24, a peptide targeting cancer cells, attached to DEC	Magnetic field	PXL/Doxorubicin	Magnetic field (50 kHz) induced controllable drug release, enhanced dual magneto-chemotherapy and magneto-hyperthermia in vitro and in vivo with multiple functionalities.	[174]
Nanoparticles	Glycerol monooleate (GMO) iron oxide magnetic NPs conjugated to HER2	Magnetic field	PXL/Rapamycin	High encapsulation efficiency(~95%) and sustained release of drugs, synergistic effect of co-loaded drugs enhanced by targeting.	[175]
Paramagnetic nanobeads	Magnetically sensitive nanoparticles	Magnetic field	Mixed-metal oxo cluster, $Mn_8Fe_4O_{12}(O_2CCH_3)_{16}(H_2O)_4$ or Mn_8Fe_4 and styrene	T2-weighted MRI.	[176]
PEGylated fullerene/ iron oxide nanocomposites	Magnetically sensitive nanoparticles	Magnetic field	Iron oxide	T2-weighted magnetic resonance imaging and photodynamic therapy, targeted drug delivery.	[177]

(Continued)

Table 1.1. (*Continued*)

Nanopreparations	Sensitive components	Acting stimulus	Load	Property, action	Ref
Synthetic melanin-like nanoparticles	Magnetically sensitive nanoparticles	Magnetic field	Melanins and Fe3+	T1 -weighted MRI.	[178]
Oil-in-water magnetic emulsion	Magnetically sensitive nanoparticles	Magnetic field	Iron oxide	T2-weighted MRI.	[179]
Silica nanoparticles	Magnetically sensitive nanoparticles	Magnetic field	Mn(2+) ions	T1-weighted MRI.	[180]
Gadolinium oxide (Gd_2O_3) nanoplates	Magnetically sensitive nanoparticles	Magnetic field	Gd_2O_3	Multimodal imaging in biomedical applications.	[181]
long circulating magnetoliposomes	Magnetically sensitive nanoparticles	Magnetic field	Superparamagnetic iron oxide nanoparticles	T2-weighted MRI.	[182]
Nanoparticles	Core-shell NP with iron oxide cores and mesoporous silica shells	Light	CPT Capping cytotoxic CdS NP	Cooperative anti-cancer activity of CPT (released by UV) and CdS NP.	[183]
Nanoparticles	PEO–PCL	Light	Docetaxel Zinc phthalocyanine (photosensitizer) ZnPc	Superior *in vitro* and *in vivo* activity of dual drug-loaded NP compared to Docetaxel NP alone, photo-dynamic therapy (PDT) by irradiation at 610 nm, singlet oxygen produced without release of ZnPc in the medium, sustained release of Docetaxel.	[184]

Nanoparticles	Heparin-folate-retinoic acid (HFR)	Light	Retinoic acid/ Pheophorbide (photosensitizer)	Dual chemotherapy and PDT, selective increase in phototoxicity of PhA upon 670 nm light irradiation with minimal dark toxicity, improved and targeted anti-cancer activity of RA *in vitro* in FR+ cells.	[185]
Polyion complex micelles (PICs)	Anionic dendrimer phthalocyanine (DPc) and PEG–PLL block copolymer to form DPc encapsulated PIC micelles (DPc/m)	Light	Doxorubicin/DPc (PS)	NP mediated dual PDT/PCI effects, Doxorubicin released from endo-lysosomes to nuclei after photoirradiation, increased efficacy of PDT with Doxorubicin *in vitro* and *in vivo*.	[186]
Nanoparticles	Galactose-functionalized mesoporous silica NPs	Light	CPT/Porphyrin	Synergistic anti-cancer effect of Porphirin with CPT *in vitro* (PDT: laser irradiation at 630–680 nm, 40 min).	[187]
Mesoporous silica nanoparticles	Theo-nitrobenzyl	Light 980 nm NIR irradiation	Doxorubicin	Controllable drug delivery and fluorescence imaging.	[188]
Mesoporous silica nanoparticles	Nitroveratryl carbamate-protected aminopropyl-functionality	Light, 350 nm UV irradiation	Doxorubicin	Light- and pH-Responsive release of Doxorubicin.	[189]
Dendritic micellar nanocarrier	Diazonaphthoquinone	Light 808 nm NIR irradition; 365 nm UV irradition	Doxorubicin	Light- sensitive release of Doxorubicin.	[190]

(*Continued*)

Table 1.1. (*Continued*)

Nanopreparations	Sensitive components	Acting stimulus	Load	Property, action	Reference
Micelle	Meta-tetra (hydroxyphenyl) chlorine	Light Light emitting diodes with the peak intensity at about 660 nm	Meta-tetra (hydroxyphenyl) chlorine	Photodynamic therapy.	[191]
Micelle	Coumarin groups	Light 794 nm NIR irradiation	Nile red	Controllable drug delivery.	[192]
Single-walled carbon nanotubes	Single-walled carbon nanotubes	Light 808 nm NIR irradiation		Cancer cell detection and NIR absorbance for selective photothermal ablation (PTA) of tumors.	[193]
Nanoemulsions	Perfluoropentane or Perfluoro-15-crown-5-ether	Ultrasound	PXL	Ultrasonic tumor imaging and targeted chemotherapy (1-MHz or 5-MHz).	[194, 195]
Liposome	Perfluoropropane gas	Ultrasound	pDNA	Ultrasound imaging and gene delivery (Frequency: 1 MHz, Duty: 50%, Burst rate: 2.0 Hz, Intensity: 1.0 W/cm^2, Time: 2 min).	[196]

Hetero-assembly of siRNA-loaded micelles and gas-cored liposomes	Octafluoropropane	Ultrasound	siRNA	Tumor gene therapy (Frequency of 1 MHz, pulse repetition frequency of 1 kHz with 50% a duty cycle, intensity of 1.0 W/cm2 and exposure time for 1 min).	[197]
Liposome-microbubble complexes	Perfluoropropane	Ultrasound	PXL	Ultrasound-triggered tumor-targeted chemotherapy (10-ms burst length, 1% duty cycle, 1 Hz pulse repetition frequency and 10-min sonication duration).	[198]
Micelle	Perfluoropentane or Perfluoro-15-crown-5-ether	Ultrasound	Doxorubicin	Ultrasound-Mediated Intracellular and Nuclear Trafficking (MHz continuous wave or pulsed ultrasound with 33% duty cycle at 3.4 W/cm^2 nominal power density).	[199]

SPIONs have been used extensively as MRI contrast agents for simultaneous imaging and stimuli-responsive drug delivery [207, 177]. PMs encapsulating SPIONs and conjugated to antibodies have been used to facilitate precise tumor targeting and improve MRI contrast (T_2) [208, 209]. PEG–PCL nanomicelles incorporating both SPIONs and doxorubicin and were targeted with s.c. anti-EGFR mAb (Cetuximab). Cetuximab immunomicelles were bound to EGFR-overexpressing cells and delivered higher DOX and SPIONs into the cells compared to non-targeted counterparts [208]. Gultepe *et al.* demonstrated that MCLs loaded with SPIONs could be guided to the tumor site efficiently using an external magnet and their accumulation could be monitored non-invasively using MRI, as well as biodistribution analysis. Quantitative MRI analysis and biodistribution studies in melanoma-bearing mice showed up to two-fold greater accumulation of the administered MCLs in tumor masses by magnetic targeting [210]. A PEGylated, fullerene (C_{60})-based multifunctional magnetic nanocomposite with surface decorated iron oxide nanoparticles and a new PDT anti-cancer drug, hematoporphyrin monomethyl ether (HMME), adsorbed to it was recently reported [177]. This system was capable of excellent magnetic drug targeting, powerful PDT, and MRI both *in vitro* and *in vivo*, thus showing excellent potential as a theranostic agent.

Another interesting application of magnetic nanoparticles is their ability to generate magnetic fluid hyperthermia. Once the magnetic nanoparticles localize in tumors, the application of an AMF of sufficient strength can cause the magnetic moment inside these particles to oscillate. This magnetic energy is then converted to heat through hysteresis losses or Néel relaxation and transferred to the surrounding environment. The heat produced can be tuned by varying the strength and duration of exposure to the AMF [211]. When a temperature of 41–43°C is reached, the tissue damage to normal tissues is reversible, while tumor cells are irreversibly damaged [212]. Local hyperthermia, in addition to its direct effects on cells, can cause an increase in perfusion of systemically administered drugs into the core of the tumor, and also lead to increased permeability [213]. Cationic albumin-conjugated magnetite nanoparticles have been recently reported as novel candidates for hyperthermia in cancer therapy [214]. Other nanoparticles capable of magnetic hyperthermia have also been reported [215, 216].

1.3.2 *Light*

Light is an attractive stimulus because of the flexibility it offers in terms of the spatio-temporal control over the release of cargo from nanocarriers [4, 217, 218]. Moreover, the response to light can be tuned by varying the wavelength and intensity of applied light [219]. A number of photo-responsive systems that respond to illumination by light of a specific wavelength in the ultraviolet (UV), visible, or near-infrared (NIR) regions have been engineered to trigger the release of various cargoes. Whereas UV light is unable to penetrate deeply in the body due to its absorption by the skin, blood, and tissues, NIR light can penetrate much deeper (10 cm) and is non-damaging to the cells [220, 218]. The release of cargoes from light-sensitive nanopreparations may be either a one-time or a repeatable "on–off" release. The photo-responsive groups or chromophores are typically encapsulated within, or conjugated to the surface of the nanocarriers [220]. This change in the structure of the nanocarrier is due to structural transformations of specific functional groups either in the backbone or in the side chains of the polymer forming the nanoparticle [60, 221, 218]. A wide range of mechanisms can induce light-triggered release of cargoes from the nanopreparations including photo-isomerization, photo-cross-linking, photosensitization-induced oxidation, light-triggered reversible or irreversible switch in polarity, photo-decross-linking, or photo-degradation of the polymer backbone [222, 221]. PMs and liposomes have been the two most widely utilized nanocarriers for the development of light-responsive vehicles [223, 224, 218]. However, other nanocarriers like dendrimers [225], silica nanoparticles [226], gold nanoparticles [227], and nanogels [228] have also been employed.

Azobenzene and its derivatives account for the most widely studied reversible photo-responsive groups. Upon UV irradiation, azobenzenes undergo a reversible *trans* to *cis* photo-isomerization of their nitrogen double bond (N=N). This converts the non-polar *trans* isomer into the polar *cis* isomer, which can be reconverted to the *trans* isomer by visible light [229, 221, 218]. This isomerization causes the nanocarriers to disrupt and release the cargo. The reconversion to *trans* isomer is a useful feature for on-demand drug delivery [222]. A number of photo-responsive nanocarriers bearing azobenzene groups have been reported with the greater emphasis being on PMs [230–234]. Cinnamic acid and its derivatives also exhibit a similar *trans–cis* photo-isomerization [229]. Recently, liposomes composed of dioleoylphosphatidylethanolamine (DOPE) and 2-(hexadecyloxy) cinnamic acid (HOCA) were reported, where photo-dimerization of the cinnamic acid residue of HOCA destabilized the liposomal membrane and enabled significant release of the loaded cargo [235]. Photo-switchable drug release is also obtained with spiropyran derivatives, which convert from a colorless spiropyran to a colored zwitterionic merocyanine upon UV light exposure. Such photo-switching spiropyran-containing nanoparticles were developed both for on-demand drug delivery and reversible changes in particle volume to enhance tissue penetration [236]. This photo-isomerization affects the polarity, which in turn, can alter the microenvironments within polymers [236]. Some chromophores, like 2-diazo-1,2-napthoquinone, undergo irreversible photo-cleavage after NIR irradiation, alter the polarity of the nanocarriers, and destabilize it to release the incorporated drugs [237, 238, 217, 220]. NIR light-sensitive PMs have been used for the delivery of doxorubicin [237]. Other chromophores, like coumarin, pyrenyl methyl esters, stilbene, dithienylethene (DTE), and o-nitrobenzyl, respond to UV/visible light or NIR irradiation and undergo either structural or phase changes to cause drug release. DTE and stilbene also undergo reversible photo-isomerization upon UV or visible light irradiation, which causes structural deformation of nanocarriers [239, 240].

Another mechanism, photo-polymerization or photo-cross-linking, involves the use of a polymerizable double bond, which is irradiated directly or in the presence of an initiator. The photo-polymerization of double bonds in the hydrophobic portion of the bilayer causes a shrinkage of some parts of the bilayer, disrupting its uniform molecular packaging and creates pores in the bilayer to allow for the release of payloads [222, 241]. A novel class of photo-triggerable liposomes made from diplamitoyl phosphatidylcholine (DPPC), a photopolymerizable diacetylene phospholipid (1,2 bis(tricosa-10,12-diynoyl)-*sn*-glycero-3-phosphocholine), and DSPE-PEG2000 was reported in the reference [241]. Doxorubicin was loaded into these liposomes and its release was triggered by irradiating them with visible light which provided better drug release compared to non-irradiated samples.

Gold nanostructures have a high surface plasmon resonance that allows them to absorb NIR radiation and convert the photons into heat, which is either used to trigger the release of chemotherapeutics from NIR responsive nanopreparations or for PTA, by increasing the temperature of the surrounding tissues [222]. Doxorubicin-loaded NIR light-sensitive liposomes have been prepared modified with hollow gold nanospheres attached to the liposome membrane [242]. Rapid drug release was observed form the liposomes upon NIR irradiation enhancing preparation cytotoxicity against cancer cells both *in vitro* and *in vivo*.

1.3.3 *Ultrasound*

Ultrasound can exert either thermal or non-thermal (mechanical) effects on biological tissues [243]. Thermal effects are due to the absorption of acoustic energy by fluids or tissues, while non-thermal effects are typically associated with oscillating or cavitating bubbles [244]. Generation of thermal energy, perturbation of cell membranes due to cavitation and enhanced permeability of blood capillaries constitute the main mechanisms of the biological effects of ultrasound [245]. For the purpose of drug delivery, ultrasound-responsive systems accumulate in desired areas and are made leaky by locally applied ultrasound to liberate incorporated drugs and genes. Ultrasound-triggered drug release depends on the time of ultrasound application, frequency of ultrasound, power density, use of pulsed versus continuous wave ultrasound, and drug lipophilicity [246, 247]. Low frequency ultrasound (20–100 kHz) can release drug from nanocarriers more effectively and can penetrate deeper into the tissue than high-frequency ultrasound (1–3 MHz) [248, 249]. However, compared to high-frequency ultrasound, it does not allow for sharp focusing [248]. For *in vivo* applications, low-frequency ultrasound is more appropriate for larger and deep-seated pathological areas (tumors), while high-frequency ultrasound is used for small and superficial lesions (tumors) [248, 249].

Ultrasound-triggered drug release has been studied extensively with Pluronic-based PMs [250–252, 245, 253]. P105 micelles with a folate targeting moiety were loaded with doxorubicin and low frequency ultrasound (70 kHz) was used as a stimulus to enable drug release, which increased with the increase in the power intensity of ultrasound [253]. HIFU was used to disassemble amphiphilic block copolymer micelles in aqueous solution via local conversion of amphiphilic unimers into hydrophilic residues to ultrasound-provoked conversion of the hydrophilization of the gydrophobic block [254]. HIFU-triggered drug release was also reported from poly(ethylene glycol)-*b*-poly[N-(2-hydroxypropyl) methacrylamide-lactate (mPEG-b-p(HPMAm-Lac_n) micelles [255].

Microbubbles have been used as ultrasound contrast agents (UCAs) in the clinic for a number of years. They usually incorporate a shell that is made up of protein, lipid, or polymer, which encloses a gas within it. The gas could be air, sulfur hexafluoride, or most commonly, a perfluorocarbon (PFC) gas [256]. Stable cavitation and inertial cavitation of microbubbles in an ultrasonic field can cause transient, reversible cell membrane permeabilization by localized amplification of the effects of ultrasound on cellular targets [257]. Microbubbles have also been investigated widely as drug carriers and enhancers of drug and gene delivery, as they perform the dual functions of target-specific delivery by responding to ultrasound and can deliver their payload intracellularly, resulting from transient membrane permeabilization [258, 256, 259–261]. However, microbubbles have large sizes (2–10 μm) and short circulation times (minutes), both of which preclude effective extravasation into tumors [260, 257]. The use of "nanobubbles" was suggested to overcome this limitation [262, 260, 263]. Alternately, nanobubbles or "bubble liposomes" combining benefits of liposomes with ultrasound-responsiveness have been prepared by sonicating liposomes in the presence of fluorinated gases, and they have been successfully utilized for drug and gene delivery [264, 265]. Novel nanocarriers known as eLiposomes have been described, which are essentially liposomes encapsulating an emulsion nanodroplet capable of phase change, as well as a therapeutic agent [266, 267]. Typically, PFC liquids are used as nanoemulsions within the liposomes and are stabilized using phospholipids. Application of ultrasound can cause the emulsion droplet to change from liquid to gas, causing an increase in the volume within the vesicle, which results in rupture and release of the drug from the liposomes.

1.4 Concluding Remarks

Advances in biomedical nanotechnology have enabled the development of pharmaceutical nanocarriers with specialized features and tunable surface chemistries to tailor them for delivery of a variety of different payloads from small molecules to complex biologics, like DNA, siRNA, and proteins. From passively targeted nanocarriers to actively targeted ones bearing various functionalities, then to nanocarriers with the ability to selectively internalize into intracellular structures, and finally to smart nanocarriers, which can respond to subtle changes in the environment, the levels of technological sophistication of these nanocarriers have increased. Evidently, stimuli-responsive nanoparticles show a better control over the temporal and spatial release of drugs compared to conventional drug delivery systems. Still, they are not devoid of off-target effects, which could arise from non-specifically located stimuli. Thus, a careful understanding of not only the stimulus, but also the extent of stimuli-sensitivity of the designed nanomaterial is desired. Combining unique targeting ligands, along with the stimuli-sensitivity feature, can improve the capability of these systems for minimizing undesirable effects and offer better control over their distribution and localization in the body. It looks like the main challenge will be to develop nanocarriers that are responsive to biochemical signals, or biomarkers typically present in nanomolar concentration ranges or even lower, and engineer nanocarriers incorporating signal amplification mechanisms. The current trend is towards incorporating multiple functions into nanocarriers allowing for integrating detection, diagnosis, and therapy in a single nanoparticle. We are at the beginning of a long road and this chapter provides just a few examples of multiple interesting possibilities that exist for the development of stimuli-responsive nanocarriers.

References

1. Torchilin, V. P. (2006). Multifunctional nanocarriers. *Adv. Drug Deliv. Rev.,* **58**, 1532–1555.
2. Peer, D., Karp, J. M., Hong, S., Farokhzad, O. C., Margalit, R. and Langer, R. (2007). Nanocarriers as an emerging platform for cancer therapy. *Nature Nanotechnology,* **2**, 751–760.
3. Farokhzad, O. C. and Langer, R. 2009. Impact of nanotechnology on drug delivery. *ACS Nano.,* **3**, 16–20.
4. Fleige, E., Quadir, M. A. and Haag, R. (2012). Stimuli-responsive polymeric nanocarriers for the controlled transport of active compounds: Concepts and applications. *Adv. Drug Deliv. Rev.,* **64**, 866–884.
5. Wang, A. Z., Langer, R. and Farokhzad, O. C. (2012). Nanoparticle delivery of cancer drugs. *Annu. Rev. Med.,* **63**, 185–198.
6. Ferrari, M. (2005). Cancer nanotechnology: Opportunities and challenges. *Nat. Rev. Cancer,* **5**, 161–171.
7. Torchilin, V. P. (2005). Recent advances with liposomes as pharmaceutical carriers. *Nat. Rev. Drug Discov.,* **4**, 145–160.
8. Torchilin, V. (2009). Multifunctional and stimuli-sensitive pharmaceutical nanocarriers. *Eur. J. Pharm. Biopharm.,* **71**, 431–444.
9. Shim, M. S. and Kwon, Y. J. (2012). Stimuli-responsive polymers and nanomaterials for gene delivery and imaging applications. *Adv. Drug Deliv. Rev.,* **64**, 1046–1059.
10. Colson, Y. L. and Grinstaff, M. W. (2012). Biologically responsive polymeric nanoparticles for drug delivery. *Adv. Mater.,* **24**, 3878–3886.
11. Mura, S., Nicolas, J. and Couvreur, P. (2013). Stimuli-responsive nanocarriers for drug delivery. *Nature materials,* **12**, 991–1003.
12. Gerweck, L. E. and Seetharaman, K. (1996). Cellular pH gradient in tumor versus normal tissue: Potential exploitation for the treatment of cancer. *Cancer Res.,* **56**, 1194–1198.
13. Tannock, I. F. and Rotin, D. (1989). Acid pH in tumors and its potential for therapeutic exploitation. *Cancer Res.,* **49**, 4373–4384.
14. Colotta, F., Allavena, P., Sica, A., Garlanda, C. and Mantovani, A. (2009). Cancer-related inflammation, the seventh hallmark of cancer: Links to genetic instability. *Carcinogenesis,* **30**, 1073–1081.
15. Gromkowski, S. H., Yagi, J. and Janeway, C. A., Jr. (1989). Elevated temperature regulates tumor necrosis factor-mediated immune killing. *Eur. J. Immunol.,* **19**, 1709–1714.
16. Stefanadis, C., Chrysochoou, C., Markou, D., Petraki, K., Panagiotakos, D. B., Fasoulakis, C., Kyriakidis, A., Papadimitriou, C. and Toutouzas, P. K. (2001). Increased temperature of malignant urinary bladder tumors *in vivo*: The application of a new method based on a catheter technique. *J. Clin. Oncol.,* **19**, 676–681.
17. Russo, A., Degraff, W., Friedman, N. and Mitchell, J. B. (1986). Selective modulation of glutathione levels in human normal versus tumor cells and subsequent differential response to chemotherapy drugs. *Cancer Res.,* **46**, 2845–2848.

18. Saito, G., Swanson, J. A. and Lee, K. D. (2003). Drug delivery strategy utilizing conjugation via reversible disulfide linkages: Role and site of cellular reducing activities. *Adv. Drug Deliv. Rev.,* **55**, 199–215.
19. De La Rica, R., Aili, D. and Stevens, M. M. (2012). Enzyme-responsive nanoparticles for drug release and diagnostics. *Adv. Drug Deliv. Rev.,* **64**, 967–978.
20. Ghadiali, J. E. and Stevens, M. M. (2008). Enzyme-responsive nanoparticle systems. *Adv. Mater.,* **20**, 4359–4363.
21. Kamaly, N., Xiao, Z., Valencia, P. M., Radovic-Moreno, A. F. and Farokhzad, O. C. 2012. Targeted polymeric therapeutic nanoparticles: design, development and clinical translation. *Chem. Soc. Rev.,* **41**, 2971–3010.
22. Zhuang, J., Gordon, M. R., Ventura, J., Li, L. and Thayumanavan, S. (2013). Multistimuli responsive macromolecules and their assemblies. *Chem. Soc. Rev.,* **42**, 7421–7435.
23. Cheng, R., Meng, F., Deng, C., Klok, H. A. and Zhong, Z. (2013). Dual and multistimuli responsive polymeric nanoparticles for programmed site-specific drug delivery. *Biomaterials,* **34**, 3647–3657.
24. Helmlinger, G., Sckell, A., Dellian, M., Forbes, N. S. and Jain, R. K. (2002). Acid production in glycolysis-impaired tumors provides new insights into tumor metabolism. *Clin. Cancer Res.,* **8**, 1284–1291.
25. Vaupel, P., Kallinowski, F. and Okunieff, P. (1989). Blood flow, oxygen and nutrient supply, and metabolic microenvironment of human tumors: A review. *Cancer. Res.,* **49**, 6449–6465.
26. Casey, J. R., Grinstein, S. and Orlowski, J. (2010). Sensors and regulators of intracellular pH. *Nat. Rev. Mol. Cell Biol.,* **11**, 50–61.
27. Gao, W., Chan, J. M. and Farokhzad, O. C. (2010). pH-responsive nanoparticles for drug delivery. *Mol. Pharm.,* 7, 1913–1920.
28. Wang, C. Y. and Huang, L. (1989). Highly efficient DNA delivery mediated by pH-sensitive immunoliposomes. *Biochemistry,* **28**, 9508–9514.
29. Connor, J. and Huang, L. (1986). pH-sensitive immunoliposomes as an efficient and target-specific carrier for antitumor drugs. *Cancer Res.,* **46**, 3431–3435.
30. Schmaljohann, D. (2006). Thermo- and pH-responsive polymers in drug delivery. *Adv. Drug Deliv. Rev.,* **58**, 1655–1670.
31. Li, Y., Gao, G. H. and Lee, D. S. (2013). Stimulus-sensitive polymeric nanoparticles and their applications as drug and gene carriers. *Advanced healthcare materials,* **2**, 388–417.
32. Oh, N. M., Oh, K. T., Baik, H. J., Lee, B. R., Lee, A. H., Youn, Y. S. and Lee, E. S. (2010). A self-organized 3-diethylaminopropyl-bearing glycol chitosan nanogel for tumor acidic pH targeting: *In vitro* evaluation. *Colloids Surf. B. Biointerfaces,* **78**, 120–126.
33. Guo, J., Cheng, W. P., Gu, J., Ding, C., Qu, X., Yang, Z. and O'driscoll, C. (2012). Systemic delivery of therapeutic small interfering RNA using a pH-triggered amphiphilic poly-L-lysine nanocarrier to suppress prostate cancer growth in mice. *Eur. J. Pharm. Sci.,* **45**, 521–532.
34. Dong, D. W., Xiang, B., Gao, W., Yang, Z. Z., Li, J. Q. and Qi, X. R. (2013). pH-responsive complexes using prefunctionalized polymers for synchronous delivery of doxorubicin and siRNA to cancer cells. *Biomaterials,* **34**, 4849–4859.

35. Fang, C., Kievit, F. M., Cho, Y. C., Mok, H., Press, O. W. and Zhang, M. (2012). Effect of cationic side-chains on intracellular delivery and cytotoxicity of pH sensitive polymer-doxorubicin nanocarriers. *Nanoscale,* **4**, 7012–7020.
36. Sawant, R. R., Sriraman, S. K., Navarro, G., Biswas, S., Dalvi, R. A. and Torchilin, V. P. (2012). Polyethyleneimine-lipid conjugate-based pH-sensitive micellar carrier for gene delivery. *Biomaterials,* **33**, 3942–3951.
37. Koren, E., Apte, A., Jani, A. and Torchilin, V. P. (2012). Multifunctional PEGylated 2C5-immunoliposomes containing pH-sensitive bonds and TAT peptide for enhanced tumor cell internalization and cytotoxicity. *J. Control. Release,* **160**, 264–273.
38. Huan, M., Zhang, B., Teng, Z., Cui, H., Wang, J., Liu, X., Xia, H., Zhou, S. and Mei, Q. (2012). In vitro and in vivo antitumor activity of a novel pH-activated polymeric drug delivery system for doxorubicin. *PLoS One,* **7**, e44116.
39. Jin, Y., Song, L., Su, Y., Zhu, L., Pang, Y., Qiu, F., Tong, G., Yan, D., Zhu, B. and Zhu, X. (2011). Oxime linkage: A robust tool for the design of pH-sensitive polymeric drug carriers. *Biomacromolecules,* **12**, 3460–3468.
40. Gu, Y., Zhong, Y., Meng, F., Cheng, R., Deng, C. and Zhong, Z. (2013). Acetal-linked paclitaxel prodrug micellar nanoparticles as a versatile and potent platform for cancer therapy. *Biomacromolecules,* **14**, 2772–2780.
41. Parrott, M. C., Luft, J. C., Byrne, J. D., Fain, J. H., Napier, M. E. and Desimone, J. M. (2010). Tunable bifunctional silyl ether cross-linkers for the design of acid-sensitive biomaterials. *J. Am. Chem. Soc.,* **132**, 17928–17932.
42. Cheng, J., Ji, R., Gao, S. J., Du, F. S. and Li, Z. C. (2012). Facile synthesis of acid-labile polymers with pendent ortho esters. *Biomacromolecules,* **13**, 173–179.
43. Bruyere, H., Westwell, A. D. and Jones, A. T. (2010). Tuning the pH sensitivities of orthoester based compounds for drug delivery applications by simple chemical modification. *Bioorg. Med. Chem. Lett.,* **20**, 2200–2203.
44. Wang, B., Xu, C., Xie, J., Yang, Z. and Sun, S. (2008). pH controlled release of chromone from chromone-Fe3O4 nanoparticles. *J. Am. Chem. Soc.,* **130**, 14436–14437.
45. Xu, Z., Gu, W., Chen, L., Gao, Y., Zhang, Z. and Li, Y. (2008). A smart nanoassembly consisting of acid-labile vinyl ether PEG-DOPE and protamine for gene delivery: Preparation and in vitro transfection. *Biomacromolecules,* **9**, 3119–3126.
46. Bertrand, N., Simard, P. and Leroux, J. C. (2010). Serum-stable, long-circulating, pH-sensitive PEGylated liposomes. *Methods Mol. Biol.,* **605**, 545–558.
47. Tiernan, J. P., Perry, S. L., Verghese, E. T., West, N. P., Yeluri, S., Jayne, D. G. and Hughes, T. A. (2013). Carcinoembryonic antigen is the preferred biomarker for *in vivo* colorectal cancer targeting. *Br. J. Cancer,* **108**, 662–667.
48. Thambi, T., Deepagan, V. G., Yoo, C. K. and Park, J. H. (2011). Synthesis and physicochemical characterization of amphiphilic block copolymers bearing acid-sensitive orthoester linkage as the drug carrier. *Polymer,* **52**, 4753–4759.
49. Kaminskas, L. M., Kelly, B. D., Mcleod, V. M., Sberna, G., Owen, D. J., Boyd, B. J. and Porter, C. J. (2011). Characterisation and tumour targeting of PEGylated polylysine dendrimers bearing doxorubicin via a pH labile linker. *J. Control. Release,* **152**, 241–248.
50. Lee, Y., Ishii, T., Cabral, H., Kim, H. J., Seo, J. H., Nishiyama, N., Oshima, H., Osada, K. and Kataoka, K. (2009). Charge-conversional polyionic complex micelles-efficient nanocarriers for protein delivery into cytoplasm. *Angew. Chem. Int. Ed. Engl.,* **48**, 5309–5312.

51. Pittella, F., Miyata, K., Maeda, Y., Suma, T., Watanabe, S., Chen, Q., Christie, R. J., Osada, K., Nishiyama, N. and Kataoka, K. (2012). Pancreatic cancer therapy by systemic administration of VEGF siRNA contained in calcium phosphate/charge-conversional polymer hybrid nanoparticles. *J. Control. Release,* **161**, 868–874.
52. Liu, X., Chen, Y., Li, H., Huang, N., Jin, Q., Ren, K. and Ji, J. (2013). Enhanced retention and cellular uptake of nanoparticles in tumors by controlling their aggregation behavior. *ACS nano,* 7, 6244–6257.
53. Albanese, A. and Chan, W. C. (2011). Effect of gold nanoparticle aggregation on cell uptake and toxicity. *ACS nano,* **5**, 5478–5489.
54. Cho, E. C., Zhang, Q. and Xia, Y. (2011). The effect of sedimentation and diffusion on cellular uptake of gold nanoparticles. *Nature nanotechnology,* **6**, 385–391.
55. Schafer, F. Q. and Buettner, G. R. (2001). Redox environment of the cell as viewed through the redox state of the glutathione disulfide/glutathione couple. *Free Radic. Biol. Med.,* **30**, 1191–1212.
56. Estrela, J. M., Ortega, A. and Obrador, E. (2006). Glutathione in cancer biology and therapy. *Crit. Rev. Clin. Lab. Sci.,* **43**, 143–181.
57. Balendiran, G. K., Dabur, R. and Fraser, D. (2004). The role of glutathione in cancer. *Cell Biochem. Funct.,* **22**, 343–352.
58. Ballatori, N., Krance, S. M., Notenboom, S., Shi, S., Tieu, K. and Hammond, C. L. (2009). Glutathione dysregulation and the etiology and progression of human diseases. *Biol. Chem.,* **390**, 191–214.
59. Cheng, R., Feng, F., Meng, F., Deng, C., Feijen, J. and Zhong, Z. (2011). Glutathione-responsive nano-vehicles as a promising platform for targeted intracellular drug and gene delivery. *J. Control. Release,* **152**, 2–12.
60. Roy, D., Cambre, J. N. and Sumerlin, B. S. (2010). Future perspectives and recent advances in stimuli-responsive materials. *Prog. Polym. Sci.,* **35**, 278–301.
61. Li, J., Huo, M., Wang, J., Zhou, J., Mohammad, J. M., Zhang, Y., Zhu, Q., Waddad, A. Y. and Zhang, Q. (2012). Redox-sensitive micelles self-assembled from amphiphilic hyaluronic acid-deoxycholic acid conjugates for targeted intracellular delivery of paclitaxel. *Biomaterials,* **33**, 2310–2320.
62. Zhu, L. and Torchilin, V. P. (2013). Stimulus-responsive nanopreparations for tumor targeting. *Integr. Biol. (Camb.),* **5**, 96–107.
63. Wei, H., Zhuo, R. X. and Zhang, X. Z. (2012). Design and development of polymeric micelles with cleavable links for intracellular drug delivery. *Prog. Polym. Sci.,* **38**, 503–535.
64. Sun, P., Zhou, D. and Gan, Z. (2011). Novel reduction-sensitive micelles for triggered intracellular drug release. *J. Control. Release,* **155**, 96–103.
65. Goldenbogen, B., Brodersen, N., Gramatica, A., Loew, M., Liebscher, J., Herrmann, A., Egger, H., Budde, B. and Arbuzova, A. (2011). Reduction-sensitive liposomes from a multifunctional lipid conjugate and natural phospholipids: Reduction and release kinetics and cellular uptake. *Langmuir,* **27**, 10820–10829.
66. Dunn, S. S., Tian, S., Blake, S., Wang, J., Galloway, A. L., Murphy, A., Pohlhaus, P. D., Rolland, J. P., Napier, M. E. and Desimone, J. M. (2012). Reductively responsive siRNA-conjugated hydrogel nanoparticles for gene silencing. *J. Am. Chem. Soc.,* **134**, 7423–7430.

67. Perche, F., Biswas, S., Wang, T., Zhu, L. and Torchilin, V. P. (2014). Hypoxia-targeted siRNA delivery. *Angew. Chem. Int. Ed. Engl.*, **53**, 3362–3366.
68. Salzano, G., Riehle, R., Navarro, G., Perche, F., de Rosa, G. and Torchilin, V. P. (2014). Polymeric micelles containing reversibly phospholipid-modified anti-survivin siRNA: A promising strategy to overcome drug resistance in cancer. *Cancer Lett.*, **343**, 224–231.
69. Li, X. Q., Wen, H. Y., Dong, H. Q., Xue, W. M., Pauletti, G. M., Cai, X. J., Xia, W. J., Shi, D. and Li, Y. Y. (2011). Self-assembling nanomicelles of a novel camptothecin prodrug engineered with a redox-responsive release mechanism. *Chem. Commun. (Camb.)*, **47**, 8647–8649.
70. Khorsand Sourkohi, B., Cunningham, A., Zhang, Q. and Oh, J. K. (2011). Biodegradable block copolymer micelles with thiol-responsive sheddable coronas. *Biomacromolecules*, **12**, 3819–3825.
71. Ren, T. B., Feng, Y., Zhang, Z. H., Li, L. and Li, Y. Y. (2011). Shell-sheddable micelles based on star-shaped poly(ε-caprolactone)-SS-poly(ethyl glycol) copolymer for intracellular drug release. *Soft matter*, 7, 2329–2331.
72. Zhang, X., Du, F., Huang, J., Lu, W., Liu, S. and Yu, J. (2012). Fabrication of biodegradable micelles with reduction-triggered release of 6-mercaptopurine profile based on disulfide-linked graft copolymer conjugate. *Colloids Surf. B. Biointerfaces*, **100**, 155–162.
73. Liu, J., Pang, Y., Huang, W., Huang, X., Meng, L., Zhu, X., Zhou, Y. and Yan, D. (2011). Bioreducible micelles self-assembled from amphiphilic hyperbranched multiarm copolymer for glutathione-mediated intracellular drug delivery. *Biomacromolecules*, **12**, 1567–1577.
74. Thambi, T., Yoon, H. Y., Kim, K., Kwon, I. C., Yoo, C. K. and Park, J. H. (2011). Bioreducible block copolymers based on poly(ethylene glycol) and poly(gamma-benzyl L-glutamate) for intracellular delivery of camptothecin. *Bioconjug. Chem.*, **22**, 1924–1931.
75. Cajot, S., Lautram, N., Passirani, C. and Jerome, C. (2011). Design of reversibly core cross-linked micelles sensitive to reductive environment. *J. Control. Release*, **152**, 30–36.
76. Matsumoto, S., Christie, R. J., Nishiyama, N., Miyata, K., Ishii, A., Oba, M., Koyama, H., Yamasaki, Y. and Kataoka, K. (2009). Environment-responsive block copolymer micelles with a disulfide cross-linked core for enhanced siRNA delivery. *Biomacromolecules*, **10**, 119–127.
77. Zhang, A., Zhang, Z., Shi, F., Xiao, C., Ding, J., Zhuang, X., He, C., Chen, L. and Chen, X. (2013). Redox-sensitive shell-cross-linked polypeptide-block-polysaccharide micelles for efficient intracellular anticancer drug delivery. *Macromol. Biosci.*, **13**, 1249–1258.
78. Koo, A. N., Min, K. H., Lee, H. J., Lee, S. U., Kim, K., Kwon, I. C., Cho, S. H., Jeong, S. Y. and Lee, S. C. (2012). Tumor accumulation and antitumor efficacy of docetaxel-loaded core-shell-corona micelles with shell-specific redox-responsive cross-links. *Biomaterials*, **33**, 1489–1499.
79. Kim, J. E., Cha, E. J. and Ahn, C. H. (2010). Reduction-sensitive self-aggregates as a novel delivery system. *Macromol. Chem. Phys.*, **211**, 956–961.

80. Cho, H., Bae, J., Garripelli, V. K., Anderson, J. M., Jun, H. W. and Jo, S. (2012). Redox-sensitive polymeric nanoparticles for drug delivery. *Chem. Commun. (Camb.)*, **48**, 6043–6045.
81. Ding, J., Shi, F., Xiao, C., Lin, L., Chen, L., He, C., Zhuang, X. and Chen, X. (2011). One-step preparation of reduction-responsive poly(ethylene glycol)-poly(amino acid)s nanogels as efficient intracellular drug delivery platforms. *Polym. Chem.*, **2**, 2857–2864.
82. Sheng, R., Luo, T., Zhu, Y., Li, H., Sun, J., Chen, S., Sun, W. and Cao, A. (2011). The intracellular plasmid DNA localization of cationic reducible cholesterol-disulfide lipids. *Biomaterials,* **32**, 3507–3519.
83. Shim, M. S., Bhang, S. H., Yoon, K., Choi, K. and Xia, Y. (2012). A bioreducible polymer for efficient delivery of Fas-silencing siRNA into stem cell spheroids and enhanced therapeutic angiogenesis. *Angew. Chem. Int. Ed. Engl.,* **51**, 11899–11903.
84. Rahbek, U. L., Nielsen, A. F., Dong, M., You, Y., Chauchereau, A., Oupicky, D., Besenbacher, F., Kjems, J. and Howard, K. A. (2010). Bioresponsive hyperbranched polymers for siRNA and miRNA delivery. *J. Drug Target.,* **18**, 812–820.
85. Yu, Z. Q., Yan, J. J., You, Y. Z. and Zhou, Q. H. (2012). Bioreducible and acid-labile poly(amido amine)s for efficient gene delivery. *Int. J. Nanomedicine,* 7, 5819–5832.
86. Xia, W., Wang, P., Lin, C., Li, Z., Gao, X., Wang, G. and Zhao, X. (2012). Bioreducible polyethylenimine-delivered siRNA targeting human telomerase reverse transcriptase inhibits HepG2 cell growth in vitro and in vivo. *J. Control. Release,* **157**, 427–436.
87. Oishi, M., Hayama, T., Akiyama, Y., Takae, S., Harada, A., Yamasaki, Y., Nagatsugi, F., Sasaki, S., Nagasaki, Y. and Kataoka, K. (2005). Supramolecular assemblies for the cytoplasmic delivery of antisense oligodeoxynucleotide: Polyion complex (PIC) micelles based on poly(ethylene glycol)-SS-oligodeoxynucleotide conjugate. *Biomacromolecules,* **6**, 2449–2454.
88. Kumagai, M., Shimoda, S., Wakabayashi, R., Kunisawa, Y., Ishii, T., Osada, K., Itaka, K., Nishiyama, N., Kataoka, K. and Nakano, K. (2012). Effective transgene expression without toxicity by intraperitoneal administration of PEG-detachable polyplex micelles in mice with peritoneal dissemination. *J. Control. Release,* **160**, 542–551.
89. Musacchio, T., Vaze, O., D'souza, G. and Torchilin, V. P. (2010). Effective stabilization and delivery of siRNA: Reversible siRNA-phospholipid conjugate in nanosized mixed polymeric micelles. *Bioconjug. Chem.,* **21**, 1530–1536.
90. Cheng, G., He, Y., Xie, L., Nie, Y., He, B., Zhang, Z. and Gu, Z. (2012). Development of a reduction-sensitive diselenide-conjugated oligoethylenimine nanoparticulate system as a gene carrier. *Int. J. Nanomedicine,* 7, 3991–4006.
91. Roy, R., Yang, J. and Moses, M. A. (2009). Matrix metalloproteinases as novel biomarkers and potential therapeutic targets in human cancer. *J. Clin. Oncol.,* **27**, 5287–5297.
92. Kessenbrock, K., Plaks, V. and Werb, Z. (2010). Matrix metalloproteinases: Regulators of the tumor microenvironment. *Cell,* **141**, 52–67.
93. Gialeli, C., Theocharis, A. D. and Karamanos, N. K. (2011). Roles of matrix metalloproteinases in cancer progression and their pharmacological targeting. *FEBS J.,* **278**, 16–27.
94. Park, J. B., Lee, C. S., Jang, J. H., Ghim, J., Kim, Y. J., You, S., Hwang, D., Suh, P. G. and Ryu, S. H. (2012). Phospholipase signalling networks in cancer. *Nat. Rev. Cancer,* **12**, 782–792.

95. Loechel, F., Fox, J. W., Murphy, G., Albrechtsen, R. and Wewer, U. M. (2000). ADAM 12-S cleaves IGFBP-3 and IGFBP-5 and is inhibited by TIMP-3. *Biochem. Biophys. Res. Commun.*, **278**, 511–515.
96. Murphy, G. (2008). The ADAMs: Signalling scissors in the tumour microenvironment. *Nat. Rev. Cancer,* **8**, 929–941.
97. Rocks, N., Paulissen, G., El Hour, M., Quesada, F., Crahay, C., Gueders, M., Foidart, J. M., Noel, A. and Cataldo, D. (2008). Emerging roles of ADAM and ADAMTS metalloproteinases in cancer. *Biochimie,* **90**, 369–379.
98. El Hour, M., Moncada-Pazos, A., Blacher, S., Masset, A., Cal, S., Berndt, S., Detilleux, J., Host, L., Obaya, A. J., Maillard, C., Foidart, J. M., Ectors, F., Noel, A. and Lopez-Otin, C. (2010). Higher sensitivity of Adamts12-deficient mice to tumor growth and angiogenesis. *Oncogene,* **29**, 3025–3032.
99. Zha, Z., Zhang, S., Deng, Z., Li, Y., Li, C. and Dai, Z. (2013). Enzyme-responsive copper sulphide nanoparticles for combined photoacoustic imaging, tumor-selective chemotherapy and photothermal therapy. *Chem. Commun. (Camb.),* **49**, 3455–3457.
100. Chen, Z., Li, Z., Lin, Y., Yin, M., Ren, J. and Qu, X. (2013). Bioresponsive hyaluronic acid-capped mesoporous silica nanoparticles for targeted drug delivery. *Chemistry,* **19**, 1778–1783.
101. Vangara, K. K., Liu, J. L. and Palakurthi, S. (2013). Hyaluronic acid-decorated PLGA-PEG nanoparticles for targeted delivery of SN-38 to ovarian cancer. *Anticancer Res.,* **33**, 2425–2434.
102. Lee, C. S., Park, W., Park, S. J. and Na, K. (2013). Endolysosomal environment-responsive photodynamic nanocarrier to enhance cytosolic drug delivery via photosensitizer-mediated membrane disruption. *Biomaterials,* **34**, 9227–9236.
103. Foged, C., Nielsen, H. M. and Frokjaer, S. (2007). Phospholipase A2 sensitive liposomes for delivery of small interfering RNA (siRNA). *J. Liposome Res.,* **17**, 191–196.
104. Kaasgaard, T., Andresen, T. L., Jensen, S. S., Holte, R. O., Jensen, L. T. and Jorgensen, K. (2009). Liposomes containing alkylated methotrexate analogues for phospholipase A(2) mediated tumor targeted drug delivery. *Chem. Phys. Lipids,* **157**, 94–103.
105. Harris, T. J., Von Maltzahn, G., Lord, M. E., Park, J. H., Agrawal, A., Min, D. H., Sailor, M. J. and Bhatia, S. N. (2008). Protease-triggered unveiling of bioactive nanoparticles. *Small,* **4**, 1307–1312.
106. Hu, J., Zhang, G. and Liu, S. (2012). Enzyme-responsive polymeric assemblies, nanoparticles and hydrogels. *Chem. Soc. Rev.,* **41**, 5933–5949.
107. Jedeszko, C. and Sloane, B. F. (2004). Cysteine cathepsins in human cancer. *Biol. Chem.,* **385**, 1017–1027.
108. Duncan, R. (2009). Development of HPMA copolymer-anticancer conjugates: clinical experience and lessons learnt. *Adv. Drug Deliv. Rev.,* **61**, 1131–1148.
109. Podgorski, I. and Sloane, B. F. (2003). Cathepsin B and its role(s) in cancer progression. *Biochem. Soc. Symp.*, **70**, 263–276.
110. Zhu, L., Wang, T., Perche, F., Taigind, A. and Torchilin, V. P. (2013). Enhanced anticancer activity of nanopreparation containing an MMP2-sensitive PEG-drug conjugate and cell-penetrating moiety. *Proc. Natl. Acad. Sci. USA.,* **110**, 17047–17052.
111. Gao, W., Xiang, B., Meng, T. T., Liu, F. and Qi, X. R. (2013). Chemotherapeutic drug delivery to cancer cells using a combination of folate targeting and tumor microenvironment-sensitive polypeptides. *Biomaterials,* **34**, 4137–4149.

112. Shi, N. Q., Gao, W., Xiang, B. and Qi, X. R. (2012). Enhancing cellular uptake of activable cell-penetrating peptide-doxorubicin conjugate by enzymatic cleavage. *Int. J. Nanomedicine,* 7, 1613–1621.
113. Garripelli, V. K., Kim, J. K., Son, S., Kim, W. J., Repka, M. A. and Jo, S. (2011). Matrix metalloproteinase-sensitive thermogelling polymer for bioresponsive local drug delivery. *Acta. Biomater.,* 7, 1984–1992.
114. Tokatlian, T., Shrum, C. T., Kadoya, W. M. and Segura, T. (2010). Protease degradable tethers for controlled and cell-mediated release of nanoparticles in 2- and 3-dimensions. *Biomaterials,* **31**, 8072–8080.
115. Hatakeyama, H., Ito, E., Akita, H., Oishi, M., Nagasaki, Y., Futaki, S. and Harashima, H. (2009). A pH-sensitive fusogenic peptide facilitates endosomal escape and greatly enhances the gene silencing of siRNA-containing nanoparticles in vitro and in vivo. *J. Control. Release,* **139**, 127–132.
116. Andresen, T. L., Thompson, D. H. and Kaasgaard, T. (2010). Enzyme-triggered nanomedicine: drug release strategies in cancer therapy. *Mol. Membr. Biol.,* **27**, 353–363.
117. Huang, S., Shao, K., Liu, Y., Kuang, Y., Li, J., An, S., Guo, Y., Ma, H. and Jiang, C. (2013). Tumor-targeting and microenvironment-responsive smart nanoparticles for combination therapy of antiangiogenesis and apoptosis. *ACS nano,* 7, 2860–2871.
118. Li, R., Wu, W., Liu, Q., Wu, P., Xie, L., Zhu, Z., Yang, M., Qian, X., Ding, Y., Yu, L., Jiang, X., Guan, W. and Liu, B. (2013). Intelligently targeted drug delivery and enhanced antitumor effect by gelatinase-responsive nanoparticles. *PLoS One,* **8**, e69643.
119. Zhu, L., Kate, P. and Torchilin, V. P. (2012). Matrix metalloprotease 2-responsive multifunctional liposomal nanocarrier for enhanced tumor targeting. *ACS nano,* **6**, 3491–3498.
120. Chen, W. H., Xu, X. D., Jia, H. Z., Lei, Q., Luo, G. F., Cheng, S. X., Zhuo, R. X. and Zhang, X. Z. (2013). Therapeutic nanomedicine based on dual-intelligent functionalized gold nanoparticles for cancer imaging and therapy *in vivo*. *Biomaterials,* **34**, 8798–8807.
121. Elliott, E. and Sloane, B. (1996). The cysteine protease cathepsin B in cancer. *Perspect. Drug Discov. Des.,* **6**, 12–32.
122. Lee, J. S., Groothuis, T., Cusan, C., Mink, D. and Feijen, J. (2011). Lysosomally cleavable peptide-containing polymersomes modified with anti-EGFR antibody for systemic cancer chemotherapy. *Biomaterials,* **32**, 9144–9153.
123. Wen, J., Anderson, S. M., Du, J., Yan, M., Wang, J., Shen, M., Lu, Y. and Segura, T. (2011). Controlled protein delivery based on enzyme-responsive nanocapsules. *Adv. Mater.,* **23**, 4549–4553.
124. Pedersen, P. J., Adolph, S. K., Subramanian, A. K., Arouri, A., Andresen, T. L., Mouritsen, O. G., Madsen, R., Madsen, M. W., Peters, G. H. and Clausen, M. H. (2010). Liposomal formulation of retinoids designed for enzyme triggered release. *J. Med. Chem.,* **53**, 3782–3792.
125. Kono, K. (2001). Thermosensitive polymer-modified liposomes. *Adv. Drug Deliv. Rev.,* **53**, 307–319.
126. Chilkoti, A., Dreher, M. R., Meyer, D. E. and Raucher, D. (2002). Targeted drug delivery by thermally responsive polymers. *Adv. Drug Deliv. Rev.,* **54**, 613–630.
127. Kang, H., Lee, E., Na, K. and Bae, Y. (2008). 'Stimuli-sensitive nanosystems: For drug and gene delivery', in Torchilin, V. (ed.), *Multifunctional Pharmaceutical Nanocarriers,* Springer, New York, pp. 161–199.

128. Bikram, M. and West, J. L. (2008). Thermo-responsive systems for controlled drug delivery. *Expert opinion on drug delivery,* **5**, 1077–1091.
129. Al-Jamal, W. T., Al-Ahmady, Z. S. and Kostarelos, K. (2012). Pharmacokinetics and tissue distribution of temperature-sensitive liposomal doxorubicin in tumor-bearing mice triggered with mild hyperthermia. *Biomaterials,* **33**, 4608–4617.
130. Manzoor, A. A., Lindner, L. H., Landon, C. D., Park, J. Y., Simnick, A. J., Dreher, M. R., Das, S., Hanna, G., Park, W., Chilkoti, A., Koning, G. A., Ten Hagen, T. L., Needham, D. and Dewhirst, M. W. (2012). Overcoming limitations in nanoparticle drug delivery: Triggered, intravascular release to improve drug penetration into tumors. *Cancer Res.,* **72**, 5566–5575.
131. Lapointe, J. and Martel, S. (2009). Thermoresponsive hydrogel with embedded magnetic nanoparticles for the implementation of shrinkable medical microrobots and for targeting and drug delivery applications. *Conf. Proc. IEEE Eng. Med. Biol. Soc.,* **2009**, 4246–4249.
132. Schwerdt, A., Zintchenko, A., Concia, M., Roesen, N., Fisher, K., Lindner, L. H., Issels, R., Wagner, E. and Ogris, M. (2008). Hyperthermia-induced targeting of thermosensitive gene carriers to tumors. *Hum. Gene Ther.,* **19**, 1283–1292.
133. Yang, M., Ding, Y., Zhang, L., Qian, X., Jiang, X. and Liu, B. (2007). Novel thermosensitive polymeric micelles for docetaxel delivery. *J. Biomed. Mater. Res. A,* **81**, 847–857.
134. Sahoo, B., Devi, K. S., Banerjee, R., Maiti, T. K., Pramanik, P. and Dhara, D. (2013). Thermal and pH responsive polymer-tethered multifunctional magnetic nanoparticles for targeted delivery of anticancer drug. *ACS applied materials and interfaces,* **5**, 3884–3893.
135. XIong, W., Wang, W., Wang, Y., Zhao, Y., Chen, H., Xu, H. and Yang, X. (2011). Dual temperature/pH-sensitive drug delivery of poly(N-isopropylacrylamide-co-acrylic acid) nanogels conjugated with doxorubicin for potential application in tumor hyperthermia therapy. *Colloids Surf. B. Biointerfaces,* **84**, 447–453.
136. Ta, T., Convertine, A. J., Reyes, C. R., Stayton, P. S. and Porter, T. M. (2010). Thermosensitive liposomes modified with poly(N-isopropylacrylamide-co-propylacrylic acid) copolymers for triggered release of doxorubicin. *Biomacromolecules,* **11**, 1915–1920.
137. CHeng, C., Wei, H., Zhang, X. Z., Cheng, S. X. and Zhuo, R. X. (2009). Thermo-triggered and biotinylated biotin-P(NIPAAm-co-HMAAm)-b-PMMA micelles for controlled drug release. *J. Biomed. Mater. Res. A,* **88A**, 814–822.
138. DIncer, S., Rzaev, Z. O. and Piskin, E. (2006). Synthesis and characterization of stimuli-responsive poly(N-isopropylacrylamide-co-N-vinyl-2-pyrrolidone). *Journal of Polymer Research,* **13**, 121–131.
139. Qiu, L. Y., Wu, X. L. and Jin, Y. (2009). Doxorubicin-loaded polymeric micelles based on amphiphilic polyphosphazenes with poly(N-isopropylacrylamide-co-N,N-dimethylacrylamide) and ethyl glycinate as side groups: Synthesis, preparation and in vitro evaluation. *Pharm. Res.,* **26**, 946–957.
140. Akimoto, J., Nakayama, M., Sakai, K. and Okano, T. (2010). Thermally controlled intracellular uptake system of polymeric micelles possessing poly(N-isopropylacrylamide)-based outer coronas. *Mol. Pharm.,* 7, 926–935.
141. Yavuz, M. S., Cheng, Y., Chen, J., Cobley, C. M., Zhang, Q., Rycenga, M., Xie, J., Kim, C., Song, K. H., Schwartz, A. G., Wang, L. V. and Xia, Y. (2009). Gold nanocages

covered by smart polymers for controlled release with near-infrared light. *Nature materials,* **8**, 935–939.

142. Neradovic, D., Soga, O., Van Nostrum, C. F. and Hennink, W. E. (2004). The effect of the processing and formulation parameters on the size of nanoparticles based on block copolymers of poly(ethylene glycol) and poly(N-isopropylacrylamide) with and without hydrolytically sensitive groups. *Biomaterials,* **25**, 2409–2418.
143. Motokawa, R., Morishita, K., Koizumi, S., Nakahira, T. and Annaka, M. (2005). Thermosensitive diblock copolymer of poly(N-isopropylacrylamide) and poly(ethylene glycol) in water: Polymer preparation and solution behavior. *Macromolecules,* **38**, 5748–5760.
144. Zhang, W., Shi, L., Wu, K. and An, Y. (2005). Thermoresponsive micellization of poly(ethylene glycol)-b-poly(N-isopropylacrylamide) in water. *Macromolecules,* **38**, 5743–5747.
145. Yang, X. L., Luo, Y. L., Xu, F. and Chen, Y. S. (2014). Thermosensitive mPEG-b-PA-g-PNIPAM comb block copolymer micelles: Effect of hydrophilic chain length and camptothecin release behavior. *Pharm. Res.,* **31**, 291–304.
146. Li, G., Meng, Y., Guo, L., Zhang, T. and Liu, J. (2014). Formation of thermo-sensitive polyelectrolyte complex micelles from two biocompatible graft copolymers for drug delivery. *J. Biomed. Mater. Res. A,* **102**, 2163–2172.
147. Salmaso, S., Caliceti, P., Amendola, V., Meneghetti, M., Magnusson, J. P., Pasparakis, G. and Alexander, C. (2009). Cell up-take control of gold nanoparticles functionalized with a thermoresponsive polymer. *J. Mater. Chem.,* **19**, 1608–1615.
148. Mcdaniel, J. R., Dewhirst, M. W. and Chilkoti, A. 2013. Actively targeting solid tumours with thermoresponsive drug delivery systems that respond to mild hyperthermia. *Int. J. Hyperthermia,* **29**, 501–510.
149. Mcdaniel, J. R., Bhattacharyya, J., Vargo, K. B., Hassouneh, W., Hammer, D. A. and Chilkoti, A. 2013. Self-assembly of thermally responsive nanoparticles of a genetically encoded peptide polymer by drug conjugation. *Angew. Chem. Int. Ed. Engl.,* **52**, 1683–1687.
150. Mcdaniel, J. R., Macewan, S. R., Dewhirst, M. and Chilkoti, A. (2012). Doxorubicin-conjugated chimeric polypeptide nanoparticles that respond to mild hyperthermia. *J. Control. Release,* **159**, 362–367.
151. Gray, W. R., Sandberg, L. B. and Foster, J. A. (1973). Molecular model for elastin structure and function. *Nature,* **246**, 461–466.
152. Tatham, A. S. and Shewry, P. R. (2000). Elastomeric proteins: Biological roles, structures and mechanisms. *Trends Biochem. Sci.,* **25**, 567–571.
153. Meyer, D. E. and Chilkoti, A. (2004). Quantification of the effects of chain length and concentration on the thermal behavior of elastin-like polypeptides. *Biomacromolecules,* **5**, 846–851.
154. Zhang, Y., Xiao, C., Li, M., Chen, J., Ding, J., He, C., Zhuang, X. and Chen, X. (2013). Co-delivery of 10-hydroxycamptothecin with doxorubicin conjugated prodrugs for enhanced anticancer efficacy. *Macromol. Biosci.,* **13**, 584–594.
155. Yu, H., Xu, Z., Chen, X., Xu, L., Yin, Q., Zhang, Z. and Li, Y. (2014). Reversal of lung cancer multidrug resistance by pH-responsive micelleplexes mediating co-delivery of siRNA and paclitaxel. *Macromol. Biosci.,* **14**, 100–109.

156. Duan, X., Xiao, J., Yin, Q., Zhang, Z., Yu, H., Mao, S. and Li, Y. (2013). Smart pH-sensitive and temporal-controlled polymeric micelles for effective combination therapy of doxorubicin and disulfiram. *ACS nano,* 7, 5858–5869.
157. Mi, Y., Zhao, J. and Feng, S. S. (2013). Targeted co-delivery of docetaxel, cisplatin and herceptin by vitamin E TPGS-cisplatin prodrug nanoparticles for multimodality treatment of cancer. *J. Control. Release,* **169**, 185–192.
158. Liu, C., Liu, F., Feng, L., Li, M., Zhang, J. and Zhang, N. (2013). The targeted co-delivery of DNA and doxorubicin to tumor cells via multifunctional PEI-PEG based nanoparticles. *Biomaterials,* **34**, 2547–2564.
159. Xiao, H., Li, W., Qi, R., Yan, L., Wang, R., Liu, S., Zheng, Y., Xie, Z., Huang, Y. and Jing, X. (2012). Co-delivery of daunomycin and oxaliplatin by biodegradable polymers for safer and more efficacious combination therapy. *J. Control. Release,* **163**, 304–314.
160. Yin, Q., Shen, J., Chen, L., Zhang, Z., Gu, W. and Li, Y. (2012). Overcoming multidrug resistance by co-delivery of Mdr-1 and survivin-targeting RNA with reduction-responsible cationic poly(beta-amino esters). *Biomaterials,* **33**, 6495–6506.
161. Nam, K., Nam, H. Y., Kim, P. H. and Kim, S. W. (2012). Paclitaxel-conjugated PEG and arginine-grafted bioreducible poly (disulfide amine) micelles for co-delivery of drug and gene. *Biomaterials,* **33**, 8122–8130.
162. Kagaya, H., Oba, M., Miura, Y., Koyama, H., Ishii, T., Shimada, T., Takato, T., Kataoka, K. and Miyata, T. (2012). Impact of polyplex micelles installed with cyclic RGD peptide as ligand on gene delivery to vascular lesions. *Gene Ther.,* **19**, 61–69.
163. Zhao, F., Yin, H. and Li, J. (2014). Supramolecular self-assembly forming a multifunctional synergistic system for targeted co-delivery of gene and drug. *Biomaterials,* **35**, 1050–1062.
164. Qin, S. Y., Feng, J., Rong, L., Jia, H. Z., Chen, S., Liu, X. J., Luo, G. F., Zhuo, R. X. and Zhang, X. Z. (2014). Theranostic GO-based nanohybrid for tumor induced imaging and potential combinational tumor therapy. *Small,* **10**, 599–608.
165. Calderon, M., Graeser, R., Kratz, F. and Haag, R. (2009). Development of enzymatically cleavable prodrugs derived from dendritic polyglycerol. *Bioorg. Med. Chem. Lett.,* **19**, 3725–3728.
166. Zhu, L., Perche, F., Wang, T. and Torchilin, V. P. (2014). Matrix metalloproteinase 2-sensitive multifunctional polymeric micelles for tumor-specific co-delivery of siRNA and hydrophobic drugs. *Biomaterials,* **35**, 4213–4222.
167. Maitz, M. F., Freudenberg, U., Tsurkan, M. V., Fischer, M., Beyrich, T. and Werner, C. (2013). Bio-responsive polymer hydrogels homeostatically regulate blood coagulation. *Nat Commun,* **4**, 2168.
168. Peng, Z., Wang, C., Fang, E., Lu, X., Wang, G. and Tong, Q. (2014). Co-delivery of doxorubicin and SATB1 shRNA by thermosensitive magnetic cationic liposomes for gastric cancer therapy. *PLoS One,* **9**, e92924.
169. Guo, D. D., Hong, S. H., Jiang, H. L., Kim, J. H., Minai-Tehrani, A., Kim, J. E., Shin, J. Y., Jiang, T., Kim, Y. K., Choi, Y. J., Cho, C. S. and Cho, M. H. (2012). Synergistic effects of Akt1 shRNA and paclitaxel-incorporated conjugated linoleic acid-coupled poloxamer thermosensitive hydrogel on breast cancer. *Biomaterials,* **33**, 2272–2281.
170. Rejinold, N. S., Baby, T., Chennazhi, K. P. and Jayakumar, R. (2014). Dual drug encapsulated thermo-sensitive fibrinogen-graft-poly (N-isopropyl acrylamide) nanogels for breast cancer therapy. *Colloids Surf. B. Biointerfaces,* **114**, 209–217.

171. Zhang, B., Jia, F., Fleming, M. Q. and Mallapragada, S. K. (2012). Injectable self-assembled block copolymers for sustained gene and drug co-delivery: An in vitro study. *Int. J. Pharm.*, **427**, 88–96.
172. Liu, Q., Zhang, J., Sun, W., Xie, Q. R., Xia, W. and Gu, H. (2012). Delivering hydrophilic and hydrophobic chemotherapeutics simultaneously by magnetic mesoporous silica nanoparticles to inhibit cancer cells. *Int. J. Nanomedicine,* 7, 999–1013.
173. Singh, A., Dilnawaz, F., Mewar, S., Sharma, U., Jagannathan, N. R. and Sahoo, S. K. (2011). Composite polymeric magnetic nanoparticles for co-delivery of hydrophobic and hydrophilic anticancer drugs and MRI imaging for cancer therapy. *ACS applied materials and interfaces,* **3**, 842–856.
174. Hu, S. H., Liao, B. J., Chiang, C. S., Chen, P. J., Chen, I. W. and Chen, S. Y. (2012). Core-shell nanocapsules stabilized by single-component polymer and nanoparticles for magneto-chemotherapy/hyperthermia with multiple drugs. *Adv. Mater.,* **24**, 3627–3632.
175. Dilnawaz, F., Singh, A., Mohanty, C. and Sahoo, S. K. (2010). Dual drug loaded superparamagnetic iron oxide nanoparticles for targeted cancer therapy. *Biomaterials,* **31**, 3694–3706.
176. Pablico-Lansigan, M. H., Hickling, W. J., Japp, E. A., Rodriguez, O. C., Ghosh, A., Albanese, C., Nishida, M., Van Keuren, E., Fricke, S., Dollahon, N. and Stoll, S. L. (2013). Magnetic nanobeads as potential contrast agents for magnetic resonance imaging. *ACS nano,* 7, 9040–9048.
177. Shi, J., Yu, X., Wang, L., Liu, Y., Gao, J., Zhang, J., Ma, R., Liu, R. and Zhang, Z. (2013). PEGylated fullerene/iron oxide nanocomposites for photodynamic therapy, targeted drug delivery and MR imaging. *Biomaterials,* **34**, 9666–9677.
178. Ju, K. Y., Lee, J. W., Im, G. H., Lee, S., Pyo, J., Park, S. B., Lee, J. H. and Lee, J. K. (2013). Bio-inspired, melanin-like nanoparticles as a highly efficient contrast agent for T1-weighted magnetic resonance imaging. *Biomacromolecules,* **14**, 3491–3497.
179. Ahmed, N., Jaafar-Maalej, C., Eissa, M. M., Fessi, H. and Elaissari, A. (2013). New oil-in-water magnetic emulsion as contrast agent for in vivo magnetic resonance imaging (MRI). *Journal of biomedical nanotechnology,* **9**, 1579–1585.
180. Kim, S. M., Im, G. H., Lee, D. G., Lee, J. H., Lee, W. J. and Lee, I. S. (2013). Mn(2+)-doped silica nanoparticles for hepatocyte-targeted detection of liver cancer in T1-weighted MRI. *Biomaterials,* **34**, 8941–8948.
181. Paik, T., Gordon, T. R., Prantner, A. M., Yun, H. and Murray, C. B. (2013). Designing tripodal and triangular gadolinium oxide nanoplates and self-assembled nanofibrils as potential multimodal bioimaging probes. *ACS nano,* 7, 2850–2859.
182. Martins, M. B., Corvo, M. L., Marcelino, P., Marinho, H. S., Feio, G. and Carvalho, A. (2013). New long circulating magnetoliposomes as contrast agents for detection of ischemia-reperfusion injuries by MRI. *Nanomedicine: Nanotechnology, biology, and medicine*, **10**, 207–214.
183. Knezevic, N. Z. and Lin, V. S. (2013). A magnetic mesoporous silica nanoparticle-based drug delivery system for photosensitive cooperative treatment of cancer with a mesopore-capping agent and mesopore-loaded drug. *Nanoscale,* **5**, 1544–1551.
184. Conte, C., Ungaro, F., Maglio, G., Tirino, P., Siracusano, G., Sciortino, M. T., Leone, N., Palma, G., Barbieri, A., Arra, C., Mazzaglia, A. and Quaglia, F. (2013). Biodegradable core-shell nanoassemblies for the delivery of docetaxel and Zn(II)-phthalocyanine inspired by combination therapy for cancer. *J. Control. Release,* **167**, 40–52.

185. Tran, T. H., Bae, B. C., Lee, Y. K., Na, K. and Huh, K. M. (2013). Heparin-folate-retinoic acid bioconjugates for targeted delivery of hydrophobic photosensitizers. *Carbohydr. Polym.*, **92**, 1615–1624.
186. Lu, H. L., Syu, W. J., Nishiyama, N., Kataoka, K. and Lai, P. S. (2011). Dendrimer phthalocyanine-encapsulated polymeric micelle-mediated photochemical internalization extends the efficacy of photodynamic therapy and overcomes drug-resistance in vivo. *J. Control. Release,* **155**, 458–464.
187. Gary-Bobo, M., Hocine, O., Brevet, D., Maynadier, M., Raehm, L., Richeter, S., Charasson, V., Loock, B., Morere, A., Maillard, P., Garcia, M. and Durand, J. O. (2012). Cancer therapy improvement with mesoporous silica nanoparticles combining targeting, drug delivery and PDT. *Int. J. Pharm.*, **423**, 509–515.
188. Yang, Y., Velmurugan, B., Liu, X. and Xing, B. (2013). NIR photoresponsive cross-linked upconverting nanocarriers toward selective intracellular drug release. *Small,* **9**, 2937–2944.
189. Knezevic, N. Z., Trewyn, B. G. and Lin, V. S. (2011). Light- and pH-responsive release of doxorubicin from a mesoporous silica-based nanocarrier. *Chemistry,* **17**, 3338–3342.
190. Sun, L., Yang, Y., Dong, C. M. and Wei, Y. (2011). Two-photon-sensitive and sugar-targeted nanocarriers from degradable and dendritic amphiphiles. *Small,* 7, 401–406.
191. Shieh, M. J., Peng, C. L., Chiang, W. L., Wang, C. H., Hsu, C. Y., Wang, S. J. and Lai, P. S. (2010). Reduced skin photosensitivity with meta-tetra(hydroxyphenyl)chlorin-loaded micelles based on a poly(2-ethyl-2-oxazoline)-b-poly(d,l-lactide) diblock copolymer in vivo. *Mol. Pharm.*, 7, 1244–1253.
192. Babin, J., Pelletier, M., Lepage, M., Allard, J. F., Morris, D. and Zhao, Y. (2009). A new two-photon-sensitive block copolymer nanocarrier. *Angew. Chem. Int. Ed. Engl.*, **48**, 3329–3332.
193. Xiao, Y., Gao, X., Taratula, O., Treado, S., Urbas, A., Holbrook, R. D., Cavicchi, R. E., Avedisian, C. T., Mitra, S., Savla, R., Wagner, P. D., Srivastava, S. and He, H. (2009). Anti-HER2 IgY antibody-functionalized single-walled carbon nanotubes for detection and selective destruction of breast cancer cells. *BMC Cancer,* **9**, 351.
194. Rapoport, N., Gao, Z. and Kennedy, A. (2007). Multifunctional nanoparticles for combining ultrasonic tumor imaging and targeted chemotherapy. *J. Natl. Cancer Inst.*, **99**, 1095–1106.
195. Rapoport, N. Y., Kennedy, A. M., Shea, J. E., Scaife, C. L. and Nam, K. H. (2009). Controlled and targeted tumor chemotherapy by ultrasound-activated nanoemulsions/microbubbles. *J. Control. Release,* **138**, 268–276.
196. Endo-Takahashi, Y., Negishi, Y., Nakamura, A., Suzuki, D., Ukai, S., Sugimoto, K., Moriyasu, F., Takagi, N., Suzuki, R., Maruyama, K. and Aramaki, Y. (2013). pDNA-loaded Bubble liposomes as potential ultrasound imaging and gene delivery agents. *Biomaterials,* **34**, 2807–2813.
197. Yin, T., Wang, P., Li, J., Zheng, R., Zheng, B., Cheng, D., Li, R., Lai, J. and Shuai, X. (2013). Ultrasound-sensitive siRNA-loaded nanobubbles formed by hetero-assembly of polymeric micelles and liposomes and their therapeutic effect in gliomas. *Biomaterials,* **34**, 4532–4543.
198. Yan, F., Li, L., Deng, Z., Jin, Q., Chen, J., Yang, W., Yeh, C. K., Wu, J., Shandas, R., Liu, X. and Zheng, H. (2013). Paclitaxel-liposome-microbubble complexes as ultrasound-triggered therapeutic drug delivery carriers. *J. Control. Release,* **166**, 246–255.

199. Mohan, P. and Rapoport, N. (2010). Doxorubicin as a molecular nanotheranostic agent: effect of doxorubicin encapsulation in micelles or nanoemulsions on the ultrasound-mediated intracellular delivery and nuclear trafficking. *Mol. Pharm.*, 7, 1959–1973.
200. Tartaj, P., Morales, M. D. P., Veintemillas-Verdaguer, S., González-Carreño, T. and Serna, C. J. (2003). The preparation of magnetic nanoparticles for applications in biomedicine. *J. Phys. D: Appl. Phys.*, **36**, R182–R197.
201. Šafařík, I. and Šafaříková, M. (2002). 'Magnetic Nanoparticles and Biosciences', in Hofmann, H., Rahman, Z. and Schubert, U. (eds.), *Nanostruct. Mater.*, Springer, Vienna, pp. 1–23.
202. Santhosh, P. B. and Ulrih, N. P. (2013). Multifunctional superparamagnetic iron oxide nanoparticles: Promising tools in cancer theranostics. *Cancer Lett.*, **336**, 8–17.
203. Dandamudi, S., Patil, V., Fowle, W., Khaw, B. A. and Campbell, R. B. (2009). External magnet improves antitumor effect of vinblastine and the suppression of metastasis. *Cancer Sci.*, **100**, 1537–1543.
204. Chao, X., Guo, L., Zhao, Y., Hua, K., Peng, M., Chen, C. and Cui, Y. (2011). PEG-modified GoldMag nanoparticles (PGMNs) combined with the magnetic field for local drug delivery. *J. Drug Target.*, **19**, 161–170.
205. Wang, C., Ding, C., Kong, M., Dong, A., Qian, J., Jiang, D. and Shen, Z. (2011). Tumor-targeting magnetic lipoplex delivery of short hairpin RNA suppresses IGF-1R overexpression of lung adenocarcinoma A549 cells in vitro and in vivo. *Biochem. Biophys. Res. Commun.*, **410**, 537–542.
206. Cui, Y., Xu, Q., Chow, P. K., Wang, D. and Wang, C. H. (2013). Transferrin-conjugated magnetic silica PLGA nanoparticles loaded with doxorubicin and paclitaxel for brain glioma treatment. *Biomaterials*, **34**, 8511–8520.
207. Singh, R. K., Patel, K. D., Kim, J. J., Kim, T. H., Kim, J. H., Shin, U. S., Lee, E. J., Knowles, J. C. and Kim, H. W. (2014). Multifunctional hybrid nanocarrier: Magnetic CNTs ensheathed with mesoporous silica for drug delivery and imaging system. *ACS applied materials and interfaces*, **6**, 2201–2208.
208. Liao, C., Sun, Q., Liang, B., Shen, J. and Shuai, X. (2011). Targeting EGFR-overexpressing tumor cells using Cetuximab-immunomicelles loaded with doxorubicin and superparamagnetic iron oxide. *Eur. J. Radiol.*, **80**, 699–705.
209. Sawant, R. M., Sawant, R. R., Gultepe, E., Nagesha, D., Papahadjopoulos-Sternberg, B., Sridhar, S. and Torchilin, V. P. (2009). Nanosized cancer cell-targeted polymeric immunomicelles loaded with superparamagnetic iron oxide nanoparticles. *J. Nanopart. Res.*, **11**, 1777–1785.
210. Gultepe, E., Reynoso, F. J., Jhaveri, A., Kulkarni, P., Nagesha, D., Ferris, C., Harisinghani, M., Campbell, R. B. and Sridhar, S. (2010). Monitoring of magnetic targeting to tumor vasculature through MRI and biodistribution. *Nanomedicine (Lond)*, **5**, 1173–1182.
211. Laurent, S., Dutz, S., Hafeli, U. O. and Mahmoudi, M. (2011). Magnetic fluid hyperthermia: Focus on superparamagnetic iron oxide nanoparticles. *Adv. Colloid Interface Sci.*, **166**, 8–23.
212. Neuberger, T., Schöpf, B., Hofmann, H., Hofmann, M. and Von Rechenberg, B. (2005). Superparamagnetic nanoparticles for biomedical applications: Possibilities and limitations of a new drug delivery system. *J. Magn. Magn. Mater.*, **293**, 483–496.
213. Kong, G. and Dewhirst, M. W. (1999). Hyperthermia and liposomes. *Int. J. Hyperthermia*, **15**, 345–370.

214. Cheraghipour, E. and Javadpour, S. (2013). Cationic albumin-conjugated magnetite nanoparticles, novel candidate for hyperthermia cancer therapy. *Int. J. Hyperthermia,* **29**, 511–519.
215. Bealle, G., Di Corato, R., Kolosnjaj-Tabi, J., Dupuis, V., Clement, O., Gazeau, F., Wilhelm, C. and Menager, C. (2012). Ultra magnetic liposomes for MR imaging, targeting, and hyperthermia. *Langmuir,* **28**, 11834–11842.
216. Yoshida, M., Watanabe, Y., Sato, M., Maehara, T., Aono, H., Naohara, T., Hirazawa, H., Horiuchi, A., Yukumi, S., Sato, K., Nakagawa, H., Yamamoto, Y., Sugishita, H. and Kawachi, K. (2010). Feasibility of chemohyperthermia with docetaxel-embedded magnetoliposomes as minimally invasive local treatment for cancer. *Int. J. Cancer,* **126**, 1955–1965.
217. Gohy, J. F. and Zhao, Y. (2013). Photo-responsive block copolymer micelles: Design and behavior. *Chem. Soc. Rev.*, **42**, 7117–7129.
218. Zhao, Y. (2007). Rational design of light-controllable polymer micelles. *Chem. Rec.,* **7**, 286–294.
219. Schumers, J. M., Fustin, C. A. and Gohy, J. F. (2010). Light-responsive block copolymers. *Macromol. Rapid Commun.*, **31**, 1588–1607.
220. Oerlemans, C., Bult, W., Bos, M., Storm, G., Nijsen, J. F. and Hennink, W. E. (2010). Polymeric micelles in anticancer therapy: Targeting, imaging and triggered release. *Pharm. Res.,* **27**, 2569–2589.
221. Yan, Q., Han, D. and Zhao, Y. (2013). Main-chain photoresponsive polymers with controlled location of light-cleavable units: From synthetic strategies to structural engineering. *Polym. Chem.,* **4**, 5026–5037.
222. Fomina, N., Sankaranarayanan, J. and Almutairi, A. (2012). Photochemical mechanisms of light-triggered release from nanocarriers. *Adv. Drug Deliv. Rev.,* **64**, 1005–1020.
223. Alvarez-Lorenzo, C., Bromberg, L. and Concheiro, A. (2009). Light-sensitive intelligent drug delivery systems. *Photochem. Photobiol.,* **85**, 848–860.
224. Yavlovich, A., Smith, B., Gupta, K., Blumenthal, R. and Puri, A. (2010). Light-sensitive lipid-based nanoparticles for drug delivery: Design principles and future considerations for biological applications. *Mol. Membr. Biol.,* **27**, 364–381.
225. Sun, L., Ma, X., Dong, C. M., Zhu, B. and Zhu, X. (2012). NIR-responsive and lectin-binding doxorubicin-loaded nanomedicine from janus-type dendritic PAMAM amphiphiles. *Biomacromolecules,* **13**, 3581–3591.
226. Chang, Y. T., Liao, P. Y., Sheu, H. S., Tseng, Y. J., Cheng, F. Y. and Yeh, C. S. (2012). Near-infrared light-responsive intracellular drug and siRNA release using Au nanoensembles with oligonucleotide-capped silica shell. *Adv. Mater.,* **24**, 3309–3314.
227. Niikura, K., Iyo, N., Matsuo, Y., Mitomo, H. and Ijiro, K. (2013). Sub-100 nm gold nanoparticle vesicles as a drug delivery carrier enabling rapid drug release upon light irradiation. *ACS applied materials and interfaces,* **5**, 3900–3907.
228. Wang, C., Mallela, J., Garapati, U. S., Ravi, S., Chinnasamy, V., Girard, Y., Howell, M. and Mohapatra, S. (2013). A chitosan-modified graphene nanogel for noninvasive controlled drug release. *Nanomedicine,* **9**, 903–911.
229. Rijcken, C. J., Soga, O., Hennink, W. E. and Van Nostrum, C. F. (2007). Triggered destabilisation of polymeric micelles and vesicles by changing polymers polarity: An attractive tool for drug delivery. *J. Control. Release,* **120**, 131–148.
230. Tang, X., Gao, L., Fan, X. J., Liang, X. and Zhou, Q. (2009). Self-assembly and photoresponsivity property of amphiphilic ABA triblock copolymers containing azobenzene moieties in dilute solution. *Macromol. Chem. Phys.,* **210**, 1556–1562.

231. Boissiere, O., Han, D., Tremblay, L. and Zhao, Y. (2011). Flower micelles of poly(N-isopropylacrylamide) with azobenzene moieties regularly inserted into the main chain. *Soft matter,* 7, 9410–9415.
232. Jochum, F. D. and Theato, P. (2010). Thermo- and light responsive micellation of azobenzene containing block copolymers. *Chem. Commun. (Camb.),* **46**, 6717–6719.
233. Yuan, Q., Zhang, Y., Chen, T., Lu, D., Zhao, Z., Zhang, X., Li, Z., Yan, C. H. and Tan, W. (2012). Photon-manipulated drug release from a mesoporous nanocontainer controlled by azobenzene-modified nucleic acid. *ACS nano,* 6, 6337–6344.
234. Xing, M. Y., Guo, W. H., Cheng, Z. P., Zhang, Z. B., Zhu, J. and Zhu, X. L. (2011). Synthesis of azobenzene functionalized dendritic block copolymer based on hyper-branched PDMAEMA and investigation of its drug release properties. *J. Control. Release,* **152**, E104–E105.
235. Seo, H. J. and Kim, J. C. (2013). Liposomes composed of dioleoylphosphatidylethanolamine and 2-(hexadecyloxy)cinnamic acid: effects of UV irradiation and pH value on release. *Journal of nanoscience and nanotechnology,* **13**, 1727–1732.
236. Tong, R., Hemmati, H. D., Langer, R. and Kohane, D. S. (2012). Photoswitchable nanoparticles for triggered tissue penetration and drug delivery. *J. Am. Chem. Soc.,* **134**, 8848–8855.
237. Liu, G. Y., Chen, C. J., Li, D. D., Wang, S. S. and Ji, J. (2012). Near-infrared light-sensitive micelles for enhanced intracellular drug delivery. *J. Mater. Chem.,* **22**, 16865–16871.
238. Katz, J. S. and Burdick, J. A. (2010). Light-responsive biomaterials: Development and applications. *Macromol. Biosci.,* **10**, 339–348.
239. Chen, Z., He, Y., Wang, Y. and Wang, X. (2011). Amphiphilic diblock copolymer with dithienylethene pendants: Synthesis and photo-modulated self-assembly. *Macromol. Rapid Commun.,* **32**, 977–982.
240. Menon, S., Thekkayil, R., Varghese, S. and Das, S. (2011). Photoresponsive soft materials: Synthesis and photophysical studies of a stilbene-based diblock copolymer. *J. Polym. Sci. Part A: Polym. Chem.,* **49**, 5063–5073.
241. Yavlovich, A., Singh, A., Blumenthal, R. and Puri, A. (2011). A novel class of photo-triggerable liposomes containing DPPC:DC(8,9)PC as vehicles for delivery of doxorubcin to cells. *Biochim. Biophys. Acta,* **1808**, 117–126.
242. You, J., Zhang, P., Hu, F., Du, Y., Yuan, H., Zhu, J., Wang, Z., Zhou, J. and Li, C. (2014). Near-infrared light-sensitive liposomes for the enhanced photothermal tumor treatment by the combination with chemotherapy. *Pharm. Res.,* **31**, 554–565.
243. Rapoport, N. (2007). Physical stimuli-responsive polymeric micelles for anti-cancer drug delivery. *Prog. Polym. Sci.,* **32**, 962–990.
244. Husseini, G. A. and Pitt, W. G. (2008). Micelles and nanoparticles for ultrasonic drug and gene delivery. *Adv. Drug Deliv. Rev.,* **60**, 1137–1152.
245. Gao, Z. G., Fain, H. D. and Rapoport, N. (2005). Controlled and targeted tumor chemotherapy by micellar-encapsulated drug and ultrasound. *J. Control. Release,* **102**, 203–222.
246. Husseini, G. A., Myrup, G. D., Pitt, W. G., Christensen, D. A. and Rapoport, N. Y. (2000). Factors affecting acoustically triggered release of drugs from polymeric micelles. *J. Control. Release,* **69**, 43–52.

247. Marin, A., Muniruzzaman, M. and Rapoport, N. (2001). Acoustic activation of drug delivery from polymeric micelles: Effect of pulsed ultrasound. *J. Control. Release,* **71**, 239–249.
248. Rapoport, N. (2012). Ultrasound-mediated micellar drug delivery. *Int. J. Hyperthermia,* **28**, 374–385.
249. Marin, A., Sun, H., Husseini, G. A., Pitt, W. G., Christensen, D. A. and Rapoport, N. Y. (2002). Drug delivery in pluronic micelles: Effect of high-frequency ultrasound on drug release from micelles and intracellular uptake. *J. Control. Release,* **84**, 39–47.
250. Husseini, G. A., Diaz de la Rosa, M. A., Gabuji, T., Zeng, Y., Christensen, D. A. and Pitt, W. G. (2007). Release of doxorubicin from unstabilized and stabilized micelles under the action of ultrasound. *Journal of nanoscience and nanotechnology,* 7, 1028–1033.
251. Ugarenko, M., Chan, C. K., Nudelman, A., Rephaeli, A., Cutts, S. M. and Phillips, D. R. (2009). Development of pluronic micelle-encapsulated doxorubicin and formaldehyde-releasing prodrugs for localized anticancer chemotherapy. *Oncol. Res.,* **17**, 283–299.
252. Husseini, G. A., Stevenson-Abouelnasr, D., Pitt, W. G., Assaleh, K. T., Farahat, L. O. and Fahadi, J. (2010). Kinetics and thermodynamics of acoustic release of doxorubicin from non-stabilized polymeric micelles. *Colloids and surfaces. A, Physicochemical and engineering aspects,* **359**, 18–24.
253. Husseini, G. A., Velluto, D., Kherbeck, L., Pitt, W. G., Hubbell, J. A. and Christensen, D. A. (2013). Investigating the acoustic release of doxorubicin from targeted micelles. *Colloids Surf. B. Biointerfaces,* **101**, 153–155.
254. Xuan, J., Boissiere, O., Zhao, Y., Yan, B., Tremblay, L., Lacelle, S. and Xia, H. (2012). Ultrasound-responsive block copolymer micelles based on a new amplification mechanism. *Langmuir,* **28**, 16463–16468.
255. Deckers, R., Paradissis, A., Oerlemans, C., Talelli, M., Storm, G., Hennink, W. E. and Nijsen, J. F. (2013). New insights into the HIFU-triggered release from polymeric micelles. *Langmuir,* **29**, 9483–9490.
256. Nomikou, N. and Mchale, A. P. (2010). Exploiting ultrasound-mediated effects in delivering targeted, site-specific cancer therapy. *Cancer Lett.,* **296**, 133–143.
257. Geers, B., Dewitte, H., De Smedt, S. C. and Lentacker, I. (2012). Crucial factors and emerging concepts in ultrasound-triggered drug delivery. *J. Control. Release,* **164**, 248–255.
258. Lentacker, I., Geers, B., Demeester, J., De Smedt, S. C. and Sanders, N. N. (2010). Design and evaluation of doxorubicin-containing microbubbles for ultrasound-triggered doxorubicin delivery: Cytotoxicity and mechanisms involved. *Mol. Ther.,* **18**, 101–108.
259. Fokong, S., Theek, B., Wu, Z., Koczera, P., Appold, L., Jorge, S., Resch-Genger, U., Van Zandvoort, M., Storm, G., Kiessling, F. and Lammers, T. (2012). Image-guided, targeted and triggered drug delivery to tumors using polymer-based microbubbles. *J. Control. Release,* **163**, 75–81.
260. Rapoport, N., Nam, K. H., Gupta, R., Gao, Z., Mohan, P., Payne, A., Todd, N., Liu, X., Kim, T., Shea, J., Scaife, C., Parker, D. L., Jeong, E. K. and Kennedy, A. M. (2011). Ultrasound-mediated tumor imaging and nanotherapy using drug loaded, block copolymer stabilized perfluorocarbon nanoemulsions. *J. Control. Release,* **153**, 4–15.

261. Gao, Z., Kennedy, A. M., Christensen, D. A. and Rapoport, N. Y. (2008). Drug-loaded nano/microbubbles for combining ultrasonography and targeted chemotherapy. *Ultrasonics,* **48**, 260–270.
262. Huang, H. Y., Hu, S. H., Hung, S. Y., Chiang, C. S., Liu, H. L., Chiu, T. L., Lai, H. Y., Chen, Y. Y. and Chen, S. Y. (2013). SPIO nanoparticle-stabilized PAA-F127 thermosensitive nanobubbles with MR/US dual-modality imaging and HIFU-triggered drug release for magnetically guided in vivo tumor therapy. *J. Control. Release,* **172**, 118–127.
263. Perera, R. H., Solorio, L., Wu, H., Gangolli, M., Silverman, E., Hernandez, C., Peiris, P. M., Broome, A. M. and Exner, A. A. (2014). Nanobubble ultrasound contrast agents for enhanced delivery of thermal sensitizer to tumors undergoing radiofrequency ablation. *Pharm. Res.,* **31**, 1407–1417.
264. Wang, Y., Li, X., Zhou, Y., Huang, P. and Xu, Y. (2010). Preparation of nanobubbles for ultrasound imaging and intracelluar drug delivery. *Int. J. Pharm.,* **384**, 148–153.
265. Suzuki, R., Namai, E., Oda, Y., Nishiie, N., Otake, S., Koshima, R., Hirata, K., Taira, Y., Utoguchi, N., Negishi, Y., Nakagawa, S. and Maruyama, K. (2010). Cancer gene therapy by IL-12 gene delivery using liposomal bubbles and tumoral ultrasound exposure. *J. Control. Release,* **142**, 245–250.
266. Javadi, M., Pitt, W. G., Belnap, D. M., Tsosie, N. H. and Hartley, J. M. (2012). Encapsulating nanoemulsions inside eLiposomes for ultrasonic drug delivery. *Langmuir,* **28**, 14720–14729.
267. Lin, C. Y., Javadi, M., Belnap, D. M., Barrow, J. R. and Pitt, W. G. (2014). Ultrasound sensitive eLiposomes containing doxorubicin for drug targeting therapy. *Nanomedicine,* **10**, 67–76.

Chapter 2

pH-Sensitive Nanosystems

Jun Hu and You Han Bae

Department of Pharmaceutics and Pharmaceutical Chemistry, University of Utah, Salt Lake City, UT, USA

2.1 Introduction

Targeted drug delivery was introduced as a new strategy for fighting against cancer, which alters the biodistribution of therapeutic agents, improves drug availability to diseased sites, and decreases availability to normal tissues. Nanosystems or nano-sized drug carriers (nanocarriers) have been considered to be one of the approaches to achieve targeted delivery for two major reasons: (1) the leaky nature of microvasculatures in diseased areas, such as solid tumor and inflamed areas, which allows for translocalization of the nanocarriers from the blood to the diseased compartments, and (2) design flexibility and functionality in selecting materials. Amongst, incorporating stimuli-responsive characteristics to the nanosystems has attracted investigators for developing better delivery systems. Accordingly, a plethora of nanosystems that respond to stimuli, such as light, electric potential or current, temperature, magnetic field, ionic species, ultrasound, pH, enzymes, glucose, metals, and metabolites, have been synthesized and tested, in particular, for cancer therapy [1–3]. Recently, pH-sensitive nanosystems have obtained popularity and have been appearing in literatures. As shown in Fig. 2.1, the search using "pH-sensitive", "pH-responsive", or " 'pH-activatable" in titles and abstracts from the PubMed database shows an exponential increase in publications from 2000 onwards.

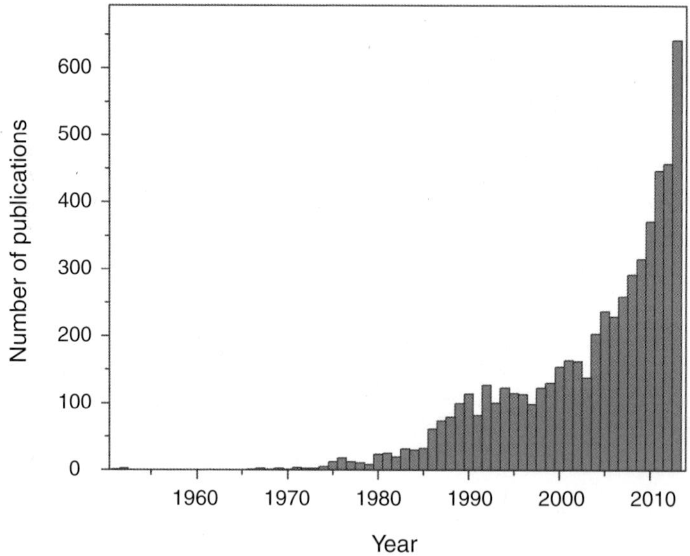

Figure 2.1. The number of publications in the PubMed database that contain either pH-sensitive, pH-responsive, or pH-activatable in their titles and abstracts since 1950.

The gastrointestinal (GI) tract is one possible site that can take advantage of pH-sensitive nanosystems. The pH variation in the GI tract has long been known to range from acidic pH (1–3) in the lumen of the stomach [4] to neutral and alkaline pH (6.6–7.5) in the duodenum and ileum [5]. Solid tumors exhibit slightly acidic pH (6.5–7.0) in the extracellular space [6], while healthy tissues maintain their pH at 7.4. One of the factors influencing tumor pH is relevant to abnormal blood vessels [7] in growing solid tumors, which have branching patterns that feature excessive loops and arteriolar–venous shunts [8]. This often results in the insufficient supply of nutrients and oxygen to the cancer cells. In addition, the cancer cells adopt high rate of glycolysis in both aerobic and anaerobic conditions, producing lactate, carbon dioxide (CO_2), and carbonic acid, for a strategy of survival and high proliferation [9, 7]. Fewer functional lymphatic vessels in the tumor [10] slow the clearance rates of the acidic by-products, resulting in an acidic tumor microenvironment. The greater pH differences found in subcellular organelles, ranging from neutral in the endoplasmic reticulum, to mildly acidic in the early endosome, and to highly acidic in the late endosome/lysosome (pH 5.4–6.2 in endosomes and ~4.8 in lysosomes) [11, 12].

This chapter briefly summarizes the mechanism of pH responsiveness, pH-sensitive nanosystems, and their applications for drug delivery and others.

2.2 Mechanism of pH-Sensitivity

2.2.1 *Protonation/deprotonation*

pH-sensitive polymers respond to pH variation by changing their hydrodynamic diameter, molecular state and/or chemical reactions, such as bond cleavage. Ionizable polymers donate or accept protons, depending on their environmental pH. This in turn influences their charge density, water solubility, and conformation, etc. The polymers can be classified as either anionic or cationic by parent ionizable groups. The chemical structures and acid dissociation constant (pK_a) values of the ionizable groups from polymer backbones or side chains, often being employed for the construction of pH-sensitive nanosystems, are summarized in Table 2.1.

Table 2.1. Chemical structures and pKa of some polymers showing pH-sensitivity with protonation/deprotonation.

Name	Typical structure	pKa	Reference
Poly (tertiary amine ethyl methacrylates)	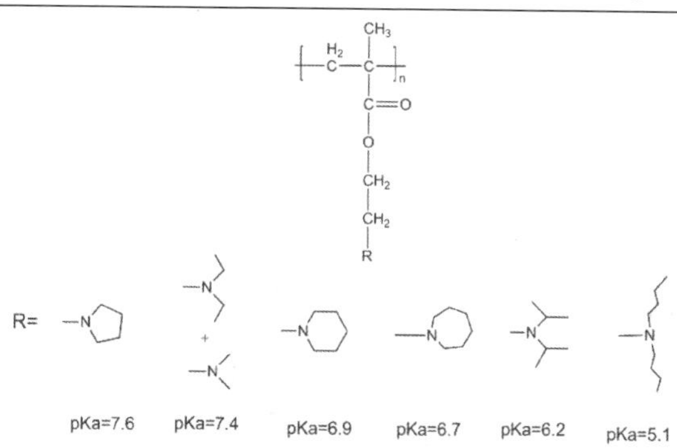		[13]
Poly (β-amino ester)	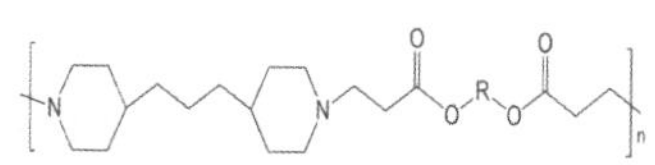	5.6–7.2	[17, 18] [24]
	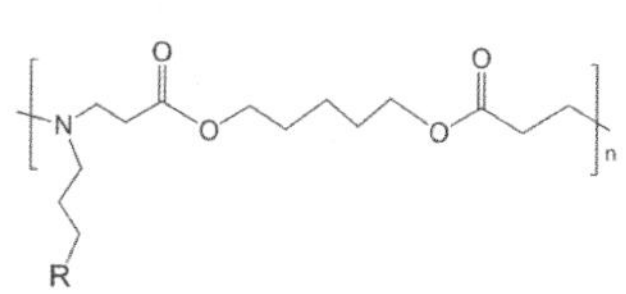	~6.0	
Poly (L-histidine)	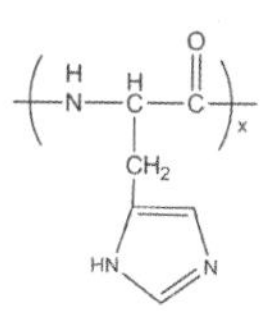	~6.0	[26]

(Continued)

Table 2.1. (*Continued*)

Name	Typical structure	pKa	Reference
Poly (vinylpyridine)	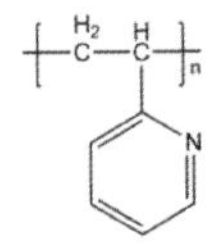	~5	[31]
	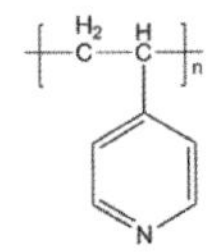	~5.6	
Polysulfonamide	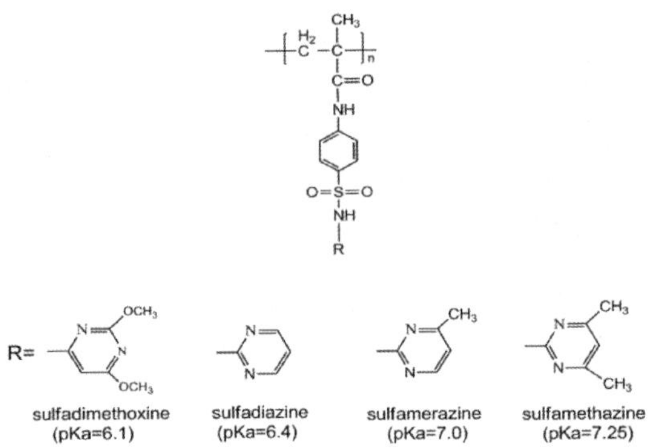		[37]
Poly (alkyl acrylic acid)	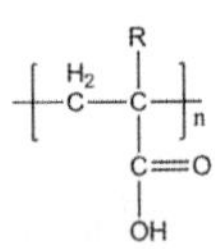	4–6	[39]

Incationic (basic) polymers, the tertiary amino groups from poly (tertiary amine alkyl methacrylates) and poly(β-amino ester), the imidazole group from poly(L-histidine) (PHis), and the pyridine group from polyvinylpyridine are ionized by binding with protons under acidic conditions and releasing protons at a higher pH, losing charges. Various nanosystems have been developed using the basic polymers. Gao *et al.* reported a series of tunable, pH-activatable micellar nanoparticles based on the self-assembly of methyl ether poly(ethylene glycol) (mPEG)-b-poly(tertiary amine ethyl methacrylate) [13–16]. Poly (tertiary amine alkyl methacrylates) undergo a hydrophobic–hydrophilic phase transition as the pH drops and are rapidly solubilized at pH values below their pK_a, which is tuned by the alkyl substituent to the tertiary amine. Similarly, poly (β-amino ester) (PAE) responds to the pH of milieu by changing the water solubility from insoluble solid in a pH range of 7.0 to 7.4 to soluble molecules after protonation at acidic conditions [17, 18]. PAE is prepared from bis(secondary amines) or primary amines with bis(acrylate ester) groups. Various PAE-based nanosystems have been developed, including polymeric micelles [18–20], hydrogels [21, 22], polymersomes [23, 24], and nanocapsules [25]. PHis is also acationic polymer which contains a weak basic group of an imidazole ring with a pKa of ~6.5 [26]. After protonation, the imidazole ring bears two nitrogen–hydrogen (NH) bonds with a positive charge that is equally distributed between two nitrogen atoms. The pK_a-and pH-dependent solubility of PHis are influenced by its molecular weight. Ring-opening polymerization of His-*N*-carboxy anhydride is one of the few feasible methods of PHis synthesis reported in 1957 [27] and prospered from 2003 [28]. Ascribing to its proton-absorbing capacity and fusogenic property, PHis has been used as a gene carrier component [26]. Micellar nanosystems from the copolymers of hydrophilic and PHis blocks, such as poly(ethylene glycol) (PEG)-*b*-PHis [28, 29] and polyethyleneimine (PEI)-*b*-Phis [30], have been extensively explored as a pH-sensitive drug carrier. Polyvinylpyridines are pH-sensitive cationic polymers with pyridine groups, including poly(2-vinylpyridine) and poly(4-vinylpyridine) [31]. The water-solubility of polyvinylpyridines relies on the extent of protonation of pyridine groups. When 35% of the pyridine groups are protonated, polyvinylpyridines change the water-solubility from insoluble to soluble [32]. The pH-sensitive nanosystems constructed from polyvinylpyridines includes micellar systems [33], hydrogels [34], and other forms [35].

Representative ionizable groups responsible for pH-sensitive anionic (acidic) polymers are sulfonamide and carboxylate. Under an acidic environment, the polymers are protonated and relatively hydrophobic, while at neutral or higher pH, the polymers become deprotonated, ionized, and hydrophilic. The sulfonamide group is readily ionized at high pH because the strong electronegativity of the oxygen atoms from the sulfonyl group draws electrons from the sulfur atom, which in turn, pulls the electrons from the neighboring nitrogen atom, resulting in deprotonation with negative charge and hydrophilicity [36]. The pK_a of sulfonamides is controlled by the pendent group (R) [37]. Polysulfonamide or oligosulfonamide shows pH-induced solubility transition. The first sulfonamide-based pH-sensitive polymer was reported by the copolymerization of modified sulfamethazine with a polymerizable acryloyl moiety and *N*, *N*-dimethylacrylamide monomer. The linear copolymers and hydrogel thereof showed a sharp transition in solubility and swelling at a pH range within 0.2 pH units, which was tuned by the copolymer composition [38].

Another series of anionic polymers are those functionalized with carboxylic groups, including poly(alkyl acrylic acid)s and poly(α-amino acid)s. For poly(alkyl acrylic acid)s, the pK_a of the homopolymers is determined by the length of the alkyl chain in the monomeric units; a longer alkyl segment raises the pK_a of the carboxylate [39]. Their pH-sensitivity can be further optimized by copolymerization with other monomers [40]. Because the transition pH of polycarboxylates is around 4–6, and their conformational change from coil to globule with a decrease in pH gives membrane disruption capability, poly(alkyl acrylic acid)s have been extensively used to free the delivery systems with therapeutic agents from endosomes before they reach the harsh and digestive lysosomal environment [41–43, 39, 44].

Nucleic acids also present pH-sensitivity with protonation. For example, i-motif deoxyribonucleic acid (DNA) is a four-stranded DNA structure with the stretched cytosine base (C). At low pH, the C residues are partially protonated (C^+) and the DNA folds into a closed structure. When the pH increases to basic values, the C^+ residues are deprotonated and the DNA unfolds to a single-strand form [45].

There are a series of small molecules named fluorophores, which switch off fluorescence at neutral pH because of the quenching of excited-state energy by the PET process, but switch on at acidic conditions by blocking the electron transfer process after protonation [46, 47].

Inorganic materials, metal oxides, and bases are relatively stable at physiological pH, but dissolved or hydrolyzed in acidic environments. Recently, the bicarbonate ion was encapsulated in the liposome, which was stable at pH 7.4, but produced CO_2 gas at pH 5.0, causing the remarkably enhanced permeability of the liposomes for drug release [48].

2.2.2 *Acid-labile bonds*

Another tool for endowing pH-sensitivity for drug delivery applications is acid-labile bonding between the therapeutic agents and nanosystems or nanocarriers. The bonds are relatively stable at a neutral to alkaline pH, but are cleaved or hydrolyzed readily in an acidic environment. The most widely used bonds as linkers are amide, hydrazone (HZ), vinyl ether, boronate ester, and Schiff base. The bond structures and their cleavage reactions are illustrated in Fig. 2.2.

A series of amide bonds are formed between primary amines and unsaturated anhydrides, including 2, 3-dimethylmaleic anhydride, citraconic anhydride, and *cis*-aconitic anhydride. The bond cleavage occurs in an acidic pH region (e.g. pH 6.8 for dimethylmaleic amide [49] and pH 5.0 for citraconic amide [50] and *cis*-aconitic amide [51]). Besides, the bond shows a reversible change in charge density, which potentially enhances the pH-sensitivity. The HZ bond contains a carbon–nitrogen double bond with the nitrogen connected

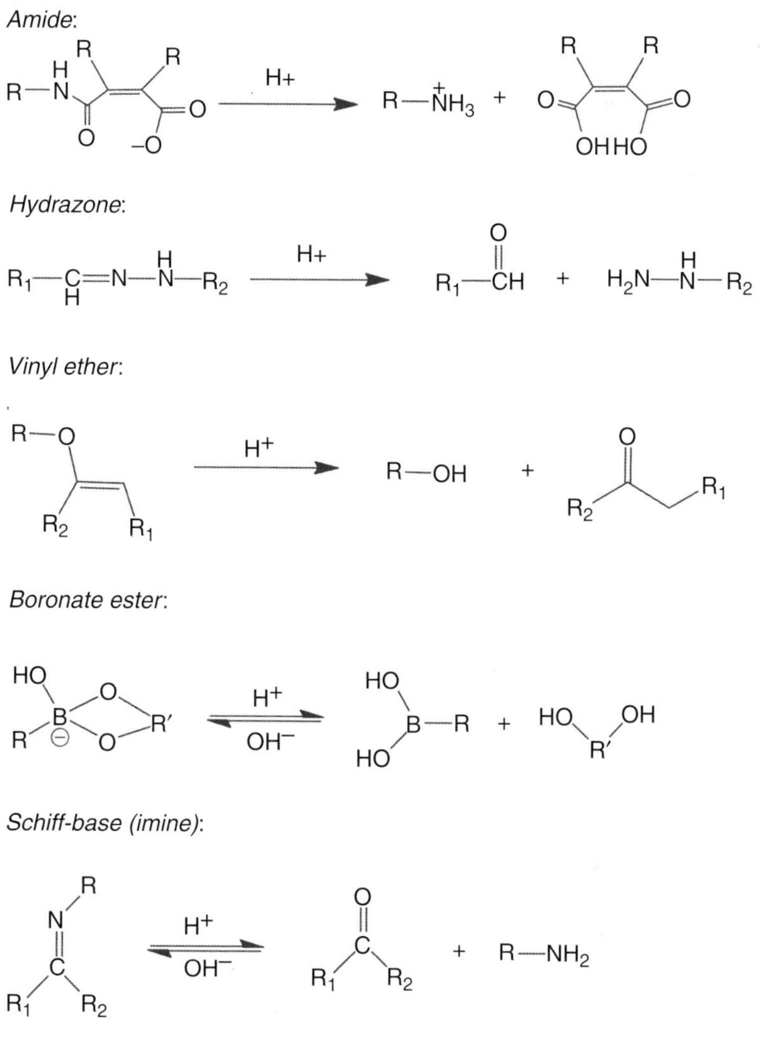

Figure 2.2. Schematic illustration of different acid-labile bonds with pH responsiveness. R = side chain; N = nitrogen; O = oxygen; H = hydrogen; OH = hydroxyl; B = boron; C = carbon.

to an amine group. The HZ bond is usually formed by the interaction of hydrazine on a ketone or aldehyde and is stable at pH 7.4, but is readily cleaved in an acidic environment (e.g. within minutes at pH 6 [52]). Vinyl ether groups contain a carbon–carbon double bond with one oxygen atom attached to one of carbon atoms. In acidic conditions, such as pH 4.5, a proton transfers to the β-carbon of the vinyl ether followed by hydration and decomposition to form an aldehyde and alcohol [53]. The hydrolysis rate of vinyl ether bonds can be influenced by the electronic characteristics of the α-substituent. Generally, the hydrolysis rate increases with the electron-donating ability of α-substituent [54]. Some systems have succeeded in releasing loaded cargoes in early endosomes using vinyl ether [55, 56].

Boronate esters and Schiff bases are reversible acid-labile bonds. Boronate esters originate from the interaction between boronic acid and a diol. At neutral or alkaline pH, stable boronate esters are formed, but in a low pH environment, the ester conjugation dissociates readily [57, 58]. The specific pH range at which reversible reaction occurs is controlled by using different kinds of boronic acid [59], diols, and polymer backbones conjugated with boronic acids or diols. Due to the simplicity in forming such bonds and the abundant availability of boric acids and diols in natural resources, designing pH-sensitive nanosystems based on this reversible chemical bond provides a versatile, chemoselective approach for targeted delivery [60]. Schiff bases, also known as imine groups, are formed by the reaction between carbonyl and primary amine groups. Same as other acid-labile bonds, Schiff bases are also relatively stable in alkaline solution, but are cleaved in a low pH environment, such as pH 6.5. Among different imines, the benzonic imine linker from the aromatic carbonyl is preferable in terms of preparation, stability, and reversibility [61–63].

2.3 Representative pH-Sensitive Nanosystems

In this section, different types of pH-sensitive nanosystems are described. Some of the relevant citations of pH-sensitive nanosystems with their functionalities and pH targets are summarized in Table 2.2.

2.3.1 *Micelle*

Micelles in aqueous solution are the aggregates of surfactant or amphiphilic copolymers composed of a hydrophilic shell and hydrophobic core. The most common pH-sensitive micelles are those originated from the self-assembly of pH-sensitive amphiphilic copolymers, such as PEG-*b*-PHis [64], PEG-b-PAE [65], PEG-*b*-poly(tertiary amine)s [66], etc. [67]. In these cases, the reversible change of solubility due to pH promotes the destabilization of micelles. Mainly, there are three advantages of these pH-sensitive polymeric micelles over other types of nanosystems. The first is the ability to incorporate hydrophobic agents into micelles by hydrophobic interactions with simple mixing [68], and the release rates and targeting sites can be optimized by the pK_a of hydrophobic parts [69]. Secondly, surface modification is easy and endows micelles with different kinds of functions. For example, PEG can be used as the shell to reduce opsonization by blood proteins and uptake by macrophages

Table 2.2. Summary of some pH-sensitive nanosystems with their functionalities, targets and forms.

Year	Bioactive agent	pH-sensitivity	Target	Carrier	Other responsiveness	Reference
2003	Doxorubicin and (DOX)	Poly(L-histidine)	Extracellular	Micelle		[93]
2005	DOX	Acid-labile bond (hydrazone)	Endosome	Micelle		[76]
2005	DOX	Poly(L-histidine)	Endosome	Micelle		[74]
2005	DOX	Poly(L-histidine)	Endosome and Extracellular	Micelle		[73]
2006	PTX	Sulfonamide		Micelle		[92]
2007	DOX	Poly(β-amino ester)	Extracellular	Micelle		[65]
2007	siRNA	Acid-labile bond (Carboxy dimethylmaleic amide)	Hepatocyte	Polyconjugates	-S–S-	[91]
2007	Protein	Acid-labile bond (citraconic amide)	Endosome	Micelle		[50]
2008	DOX	Poly(L-histidine)	Endosome and Extracellular	Micelle		[75]
2010	pDNA	Acid-labile bond (hydrazone)	Endosome	Polyplex		[88]
2010	Imaging	Poly(β-amino ester)	Tumoral pH	Micelle		[18,108]
2011	pDNA	Poly(β-amino ester)	Endosome	Nanocapsules		[25]
2011		Acid-labile bond (hydrazone)	Extracellular	Liposome		[83]
2011		iminobiotin-neutravidin interaction	Extracellular	LBL nanoparticles		[90]

(*Continued*)

Table 2.1. (*Continued*)

Year	Bioactive agent	pH-sensitivity	Target	Carrier	Other responsiveness	Reference
2011	DOX	Acid-labile bond (hydrazone) & (dimethylmaleic amide)	Endosome & Extracellular	Micelle		[77]
2012	pDNA	Acid-labile bond (Carboxy dimethylmaleic amide)	Endosome	Polyplex	-S–S-	[102]
2012		Hydrazone	Extracellular	Liposome		[97]
2013	PTX	Poly(L-histidine) Sulfonamide Oligomer	Endosome & Extracellular	Micelle		[30]
2013	Imaging	fluorophore	Tumoral pH	Micelle		[78]
2013	pH measurment	tertiary amine	Tumoral pH	Micelle		[66]
2014	Protein	Poly(acrylic acid)	intestines pH	Nanocargo		[89]
2014	imaging	Poly(L-histidine)	Tumoral pH	Micelle		[110]

of the mononuclear phagocyte system [70], and PEI can be used as the shell for the enhanced cell internalization and deeper penetration of micelles [30]. Also, different kinds of ligands can be conjugated at the surface of micelles for enhanced selection of sites and cell internalization [71–75, 29]. Lastly, pH-sensitive micelles can be used in different fields [65]. There are other means to prepare pH-sensitive micelles besides the self-assembly of pH-sensitive amphiphilic copolymers. The micelles that are conjugated with therapeutic agents by acid-labile bonds [76,77], or loaded with pH-sensitive molecules, are also defined as pH-sensitive micellar nanosystems [78].

2.3.2 *Liposome*

A liposome is an artificially prepared vesicle composed of amphiphilic bilayers surrounding equal numbers of aqueous compartments. Compared with other carriers, liposomes are suitable for the loading of hydrophobic, amphipathic, and hydrophilic agents. pH-sensitive liposomes have been confirmed as a means to increase the cytoplasmic delivery of therapeutic agents by destabilization in acidic endosomes after endocytosis. The common method to achieve pH-sensitive liposomes is by constructing liposomes from pH-sensitive components and, nowadays, the most studied pH-sensitive liposomal formulations are those consisting of unsaturated phosphatidylethanolamine (PE) combined with mildly acidic amphiphiles [79]. As liposomes are eliminated rapidly from the blood and captured by the cells of the reticulo-endothelial system [80], one of the focuses when developing pH-sensitive liposomes is the way to increase the stability in serum and blood circulation time. Introducing PEG, pH-sensitive terminally alkylated co-polymer of *N*-isopropylacrylamide (NIPAM), and methacrylic acid to PEG–PE-containing liposomes triggered pH-sensitivity and serum-stability with long-circulation [81]. Recently, besides pH-sensitivity and longevity, developing a ligand-mediated targeting has become a new research focus [82, 83].

2.3.3 *Mesoporous silica nanoparticles*

Besides the approaches based on the destabilization of carriers or the cleavage of linkers between therapeutic agents, an alternative method has been reported, as shown in Fig. 2.3, using mesoporous silica nanoparticles (MSN) as nanocarriers, which control the release of loaded cargoes by opening/closing the pore responding to the pH change [84, 85]. Nucleic acids show pH-sensitive conformational changes as described before; when i-motif DNA was introduced as a pore blocker, the pore was blocked at low pH, but opened at

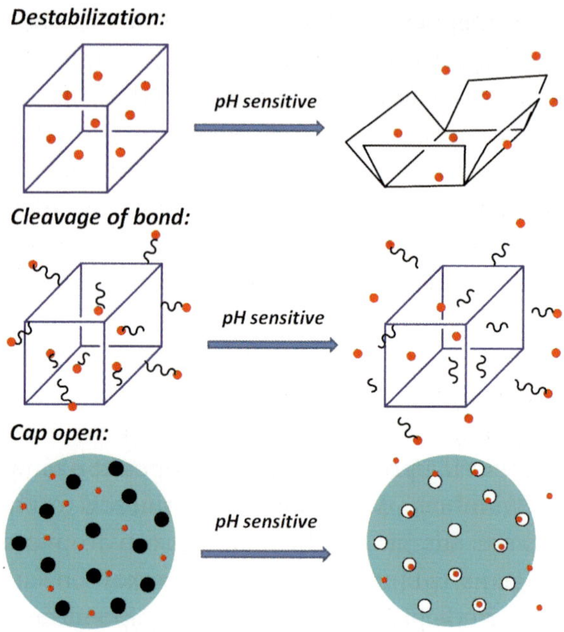

Figure 2.3. Schematic illustration of different releasing pathways.

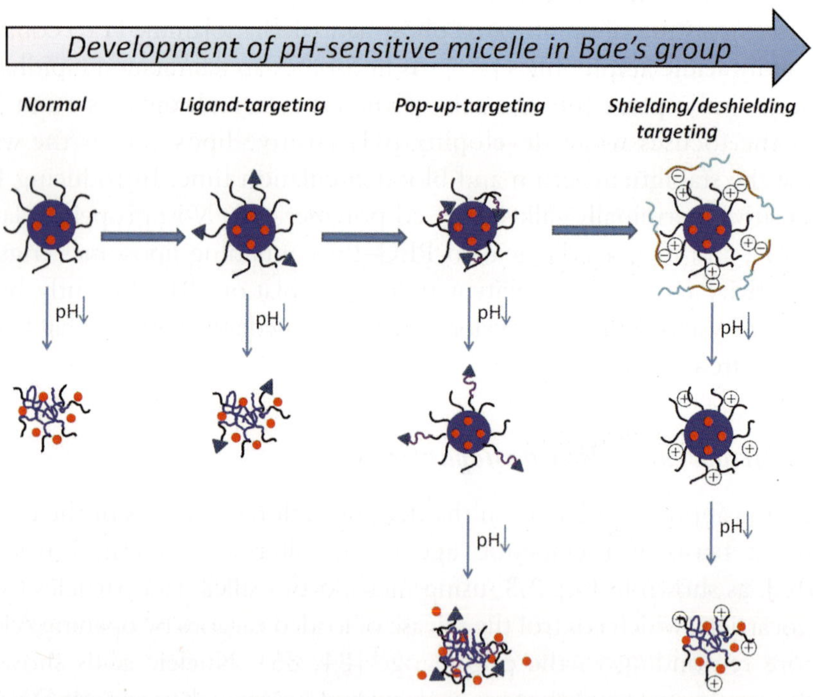

Figure 2.4. Schematic illustration of the overall strategy of pH-responsive drug delivery using block copolymers containing poly(L-histidine) (PHis) from Dr. Bae's group.

high pH, which resulted in the release of loaded rhodamine B, with 91% after 24 h at pH 8.0 and only negligible release at pH 5.0 [45]. In contrast to high pH triggering release, when natural gelatin was used as a pore blocker, the blocked pore was uncapped in slightly acidic environments because of the electrostatic repulsion between the gelation and MSN. In this system, DOX release was less than 4% over 440 min at pH 7.4, while 18%, 44%, 54%, and 83% were released within 440 min at pH 6.0, 5.0, 4.0, and 2.0, respectively [86]. Besides using pH-sensitive materials as a blocker, the introduction of ionizable groups to the pore surface also can control the release of ionic molecules with pH control [87]. Even though therapeutic agents can be loaded into the pores of MSN simply by mixing in aqueous solutions, there is no report evidencing loading of hydrophobic molecules in the MSN, which maybe the biggest challenge for this system in cancer therapy.

2.3.4 *Others*

Other types of nanosystems include polyplex [88], nanocargoes [89], nanocapsules [25], layer-by-layer (LBL) nanoparticles [90], polyconjugates [91], etc. These will not be discussed individually here, but there are some common points among these nanosystems: (1) The systems are required to be stable in the blood during circulation, (2) pH-sensitive ligand-mediated targeting and cell internalization are preferable in minimizing impairment of delivery efficacy, and (3) pH-sensitive cytoplasmic delivery of therapeutic agents is the best choice for higher therapeutic efficiency.

2.4 pH-Sensitive Nanosystems for Drug Delivery

pH-sensitive nanosystems have been extensively developed in various fields, including drug delivery, imaging, pH measurement, etc. This section summarizes pH-sensitive nanosystems used for the delivery purposes of small molecule drugs, genes, and proteins.

2.4.1 *Delivery of small molecule drugs*

Among pH-sensitive drug carriers, the polymeric micelle system is one of the most investigated pH-sensitive drug carriers, because the micelles can be tuned easily with pH-dependent polymers that have versatile chemical properties.

To set up pH-sensitive polymeric nanosystems for controlled drug release, approaches based on introducing a pH-sensitive polymer block as a part of amphiphilic copolymers that self-assemble in micelles or conjugating polymers with drugs via acid-labile bonds have been employed. mPEG-*b*-poly (d,l-lactide) (PDLLA)-b-oligosulfadimethazine(OSM), because of the ionization–deionization capability of OSM within a relatively small pH range (7.2–8.4), showed reversible micelle–unimer transition, and the hydrophobic drug paclitaxel (PTX) was incorporated successfully into the micelle with high loading efficiency when the pH of the aqueous solution of PTX and the polymer was adjusted from 9.0 to 7.4 without using any organic solvents. Compared with the micelles prepared by using the solid dispersion method, the release rate of PTX from the micelles prepared by the pH induced micellization method was a little slower, and the initial release amount (burst release) of PTX was lower as well [92]. After conjugating doxorubicin (DOX) to the polyaspartate block from PEG-b-polyaspartate using an HZ bond, the resulting block copolymer assembled into a micelle. The micelle showed endosomal and lysosomal pH (pH_i)-triggered drug release and effective antitumor activity with low toxicity [76].

An overall strategy of pH-responsive drug delivery using block copolymers containing PHis is illustrated in Fig. 2.4. An amphiphilic diblock copolymer, PHis(5K)-b-PEG(2K), was prepared firstly and self-assembled into micelles with a diameter of ~110 nm at pH 8.0, which gradually destabilized when hydrophobic PHis changed to hydrophilic at lower pH (pH<7.4) because of the protonation of imidazole groups in PHis [64]. After that, the mixed micelles from PHis(5K)-b-PEG(2K) and poly(L-lactic acid) (PLLA)(3K)-b-PEG(2K) block copolymers were prepared and investigated for pH-triggered drug release. The drug triggering pH of the micelles was modulated from 7.4 to 7.2~6.6 by varying the blending composition of two block copolymers. Typically, when the mixed micelles containing 25 wt% PLLA-b-PEG, the released amount of loaded DOX increased from 30 wt% at pH 8.0 to 32 wt% at pH 7.0, 70 wt% at pH 6.8 and 82 wt% at pH 5.0 [93]. Another self-assembled pH-sensitive micelle was reported by Ko *et al.* using mPEG-b-PAE, which was prepared by a Michael-type step polymerization. After loading DOX with 74.5% loading efficiency, the micelles did not release DOX at pH 7.4, while more than 60% of the DOX was released over 6 h at pH 6.4. In the evaluation of *in vivo* anti-tumor activity using B16F10 tumor-bearing mice, the pH-sensitive micelles loading DOX at 1 and 2 mg DOX/kg doses inhibited tumor sizes by 52.74% and 72.69%, comparing 32.4% and 45.75% from free DOX at 1 and 2 mg/kg, respectively. The survival rates of mice treated with 1 and 2 mg/kg of free DOX were 20% and 0% at 3 weeks, respectively, while that from the pH-sensitive micelles loading DOX were 60% and 40% at 1 and 2 mg DOX/kg after 3 weeks, respectively [65].

Although simple pH-sensitive nanosystems showed improved anti-tumor effects for wild-type sensitive tumors [94], their passive internalization is not recommended for resistant cancer cells [7]. In order to enhance the cell uptake of the systems, hence more drugs can be released inside the cancer cells instead of extracellularly, pH-sensitive nanosystems with an active endocytic mechanism, corresponding to the "Ligand-targeting" in Fig. 2.4, have been developed. The role of ligands in promoting the interaction between cells and nanosystems was demonstrated by Bae *et al.* using the PEG-b-poly (aspartate) micelles conjugated with folate [71, 72]. The original pH-sensitive PEG-b-poly(aspartate) micelles with the HZ bond released DOX by cleavage of the HZ bond at the intracellular pH (pH_i) in a relatively long exposure time for inhibiting cell growth (>10 h), but after conjugating with folate, the micelles showed efficient cytotoxicity within a short exposure time of 3h. It is of significance that, even though the folate conjugation did not significantly improve the tumor accumulation of the micelles, the folate-conjugated micelles showed higher cancer treatment efficiency than free drugs or the micelles without folate conjugation. The introduction of ligands into nanosystems to overcome multi drug resistance (MDR) was reported by Bae's group. After functionalizing the DOX loaded with mixed micelles of PH is (5K)-b-PEG(2K) and PLLA(3K)-b-PEG(2K) with folate, the cytotoxicity of the loaded DOX was much higher than that of free DOX (20% cell viability for micellar DOX at 10 μg/mL versus 85–90% for free DOX) when treated with MDR tumor cells. In the MCF-7/DOX^R (DOX resistant) xenograft model, the tumor accumulation level of DOX from the micelles with folate was 20 times higher than that of the free DOX, and three times higher than that of the micelles without a folate group [74].

Cell-specific targeting strategies have shown many promising outcomes by preferable over-expression of target receptors on cancer cells to non-target cells; however, there are absolutely no cell-specific receptors distinguishing between cancer and normal cells [95, 96]. In order to overcome tumor heterogeneity in cancer cell population, a proposed idea was hiding non-specific cell penetrating moieties during the systemic circulation, while exposing them for cell internalization after accumulation. As shown in Fig. 2.4, corresponding to the "pop-up targeting", the micelles presented cell-interacting moieties (e.g. biotin [73] or cell penetrating peptides (trans-activating transcriptional activator: TATp)) [75] on their surfaces in response to small pH changes, resulting in the non-specific cellular interactions only in acidic tumor environments.

These micelles were constructed by mixing PHis(long block)-b-PEG and PLLA-*b*-PEG-*b*-PHis(short block)-ligand (biotin or TATp). During the blood circulation (above pH 7.2), the ligands were buried by a PEG shell to avoid non-specific interactions with cells. After reaching the tumor site, the extracellular pH (pH_e) caused the ionization of the short PHis block located at the micelle core/shell interface to expose the ligands to the micelle surface for cell internalization. The long block of PHis in the micelle core was ionized at lower pH to destabilize the micelle, which in turn, accelerated the release of loaded DOX, and at the same time, the protonated PHis helped break the endosomal membrane, hence increasing drug dose concentrations in the cytoplasm in a relatively short time period. In the drug-resistant human ovarian tumor cell A2780/AD xenograft model, tumor volume from the group of pop-up systems loading DOX using TATp as a ligand was approximately 18 times smaller than that observed with the free DOX treatment under the same condition; more impressively, the tumor volume from the pop-up system was 6 times smaller than that of mixed micelles with TATp exposure at all pH values, which confirmed the effectiveness of this strategy. With a similar strategy, as shown in Fig. 2.5A, Torchilin's group reported PEG shedding systems initiated by the tumor pH. When employing different chain lengths of PEG, the activity of TAT attached to shortened PEG was interfered at normal pH due to the steric hindrance of longer PEG, allowing TATp to be exposed for enhanced cellular uptake after the degradation of the HZ bond to remove the long PEG chains when the system was exposed to lowered pH [97]. Figure 2.5B illustrates another method, where the targeting ligand was attached to the distal end of the PEG chain to facilitate easy access to the target site of interaction. The targeting ligands were cleaved off at lower pH values to expose TATp for intracellular penetration [83].

A new approach was used for designing novel pH-sensitive nanosystems, which displayed: (1) effective intracellular penetration by replacing expensive and non-specific ligands, such as TATp, into cationic surfaces that had better interactions with negatively charged cell membrane; (2) lower cell cytotoxicity and less aggregation with serum by shielding cationic surface during circulation; (3) subsequent deshielding of the cationic surface at pH_e, which are illustrated in Fig. 2.4 under "shielding/deshielding targeting". In detail, new cationic micelles (unshielded micelles, US-M), self-assembled from PHis-*b*-short branched polyethyleneimine (PHis-*b*-sbPEI), were confirmed to be powerful in being internalized into cells and escaping from endosomes. In order to increase the stability of cationic micelles and to minimize their non-specific interactions during circulation, negatively charged mPEG-*b*-poly (methacryloyl sulfadimethoxine) (mPEG-*b*-PSDM) was used to shield the

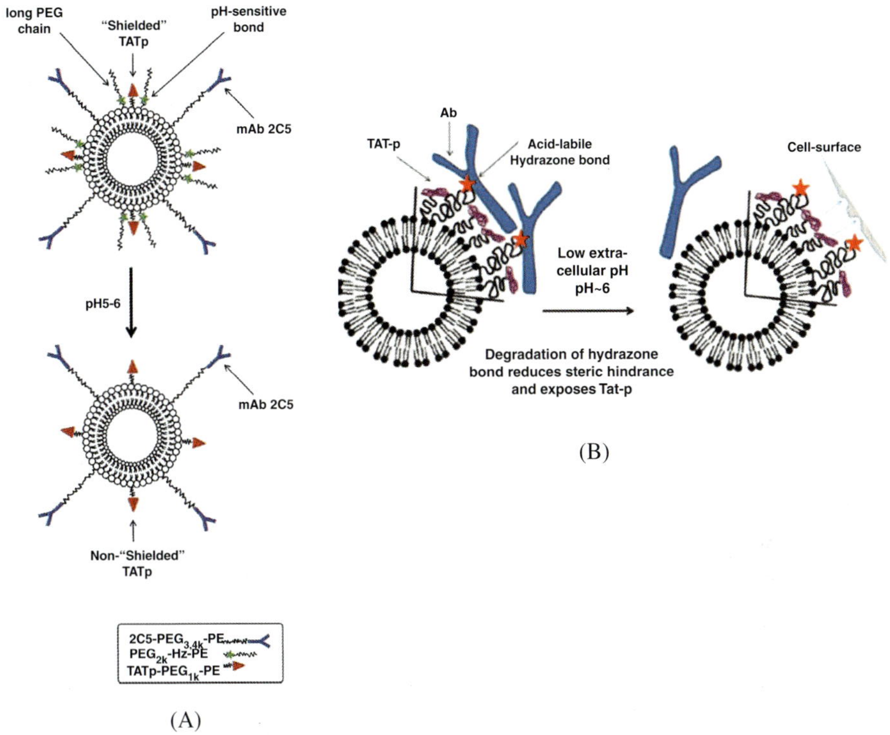

Figure 2.5. (A) Schematic representation of the low pH effect on TATp-modified pH-sensitive immunoliposomes composed of a pH-degradable PEG2k-Hz-PE with long PEG blocks, TATp-PEG1k-PE with short PEG blocks, and mAb2C5-PEG3.4k-PE. This research was originally published in *J. Control. Rel.* Koren, E.; Apte, A.; Jani, A.; Torchilin, V. P., 2012, Multifunctional PEGylated 2C5-immunoliposomes containing pH-sensitive bonds and TATp peptide for enhanced tumor cell internalization and cytotoxicity. **160**(2), 264–273. Copyright © 2012, Elsevier. (B) Schematic representation of the low pH effect on degradation of hydrazone bond reducing steric hindrance and exposing Tatp. Reprinted with permission from *Bioconjugate Chem.* Biswas, S.; Dodwadkar, N. S.; Sawant, R. R.; Torchilin, V. P., 2011, Development of the novel PEG-PE-based polymer for the reversible attachment of specific ligands to liposomes: Synthesis and *in vitro* characterization. **22**(10), 2005–2013. Copyright ©2011, American Chemical Society. PEG = polyethylene glycol, TATp = trans-activating transcriptional activator peptide, mAb = monoclonal antibody, Hz = hydrazone.

cationic surface by electrostatic interactions at pH 7.4 resulting in shielded micelles (S-M). When the S-Maccumulated in acidic tumor sites (pH 6.6), pH-sensitive sulfonamide groups quickly lost their charges and dissociated from the micelles, which resulted in deshielded micelles (DS-M) with a positive surface for non-specific cellular uptake to overcome the heterogeneity of solid tumors. After loading PTX, the size of micelles increased from 40 nm for US-M to 65 nm for S-M, and the median inhibitory concentration (IC_{50}) of loaded PTX in S-M changed from 295 ng/mL at pH 7.4 to 37 ng/mL at pH 6.6. Over a 29-day period, the *in vivo* anti-tumor activity evaluation with the human breast tumor MCF-7 xenograft model revealed that the tumor volume from the S-M group was 95 ± 33 mm^3 (n=7), comparing with 210±33 mm^3 in the free PTX group and 299 ± 30 mm^3 in the US-M group. It is worthy to note that, in the evaluation of intratumoral distribution of the fluorescence labeled shielding/deshielding system, the peak fluorescene intensity from the S-M was observed to be about 15 μm from the blood vessels and penetrated more than 60 μm, which confirmed the great ability of this system to penetrate deeply into tumor tissues after deshielding [30]. Another example for this strategy is pH-sheddable LBL nanoparticles. There was a trilayer architecture of poly-L-lysine (PLL) modified with iminobiotin, followed by the linker protein neutravidin and biotin end-functionalized poly(ethylene glycol) (PEG). With the PEG outer stealth layer, the nanoparticles could target and be retained in tumor regions by the EPR effect. After accumulation in the acidic tumor environment, the PEG layer was removed because of the decrease of iminobiotin–neutravidin interactions, which exposed the cationic polylysine layer to facilitate cellular uptake [90].

2.4.2 *Gene delivery*

Gene therapy is an approach for the treatment of various diseases arising from genetic abnormalities, including cancer, viral infections, and inflammatory disorders. Because of the rapid degradation of genes by nucleases, poor cellular uptake due to negative charge and large size, and the inability to reach the nucleus for action, a safe and efficient carrier is essential for gene therapy. Non-viral gene carriers have addressed these problems by forming nanosized polyelectrolyte complexes via ionic interactions between a cationic carrier and negatively charged genes. Since these complexes are usually generated as cationic nanoparticles, the positive charge induces cell toxicity and aggregation with serum components, in which further modification is required to enhance their transfection efficiency. One commonly used method is PEG shielding (PEGylation). A pentablock copolymer consisting of a central triblock copolymer of poly (ε-caprolactone)-*b*-PEG-poly(ε-caprolactone) (PCL-*b*-PEG-*b*-PCL) and two identical blocks of pH-sensitive PAE at the both ends of the central copolymer was synthesized and used as a nontoxic gene carrier. At low pH (<6.0), the copolymer was protonated and interacted with negatively charged plasmid (pDNA) to form polyplexes. At physiological pH, the PAE blocks were deprotonated and formed an insoluble skin, resulting in the formation of nanocapsules that encapsulated pDNA. The PEG surface increased the solubility and blood–serum compatibility, and the unprotonated tertiary amines from the PAE blocks facilitated the endosomal escape of the nanocapsules by adsorbingprotons (H^+) in the endosome [25]. Moreover, because PEGylation strongly inhibits the cellular uptake and endosomal escape [98], pH-detachable PEG is preferable for the enhanced transfection [91, 82, 99, 100]. For example, cationic diblock copolymer, poly(aspartate-hydrazide-PEG)-block-poly(aspartate-diaminoethane), complexed with pDNA resulted in polyplexes with a size range of 70–90 nm, whose surface charge was effectively shielded by the PEG layer. The transfection efficiency of the reversibly PEGylated polyplexes was comparable to that of non-PEGylated counterparts, but was 1,000 times higher than that of the irreversibly PEGylated polyplexes [88].

Except PEG, pH-sensitive anionic materials that can stabilize the cationic polyplexes during circulation and detach from the polyplexes in the intratumoral pH have been used more and more for shielding. Bae and his coworkers demonstrated the enhanced transfection efficiency of a tumor acidity-responsive sulfonamide/PEI/pDNA delivery *in vitro*. In detail, plasmid DNA was complexed with PEI and further with a pH-sensitive diblock copolymer, poly (methacryloylsulfadimethoxine)-b-PEG (PSD-*b*-PEG), to get nanoparticles of about 300 nm. At pH 7.4, these nanoparticles with negative surface charge showed 60% reduced cytotoxicity because of the decreased interaction between cells and the positively charged complex of PEI/pDNA after shielding. After decreasing the pH to 6.6, the coating copolymer was detached and the complex of PEI/pDNA recovered high cytotoxicity and transfection [36]. In a similar manner, PEG-*b*-poly (propyl methacrylate-*co*-methacrylic acid) (PEG-*b*-P(PrMA-*co*-MAA) complexed with poly (amino amine) (PAMAM) dendrimers and nucleic acids to form pH-responsive polyanion complex micelles. These micelles had the ability to lose their shell and release the PAMAM/nucleic acid core under mildly acidic conditions, such as those encountered in the endosomal compartment [101]. A pH-sensitive block copolymer of PEG and charge-reversional polymer, PEG-*b*-poly(aspartamide(DET-aco))(PEG-PAsp(DET-Aco), comprising two repeating units of aminoethylene in PAsp side chains and primary amines modified with anionic *cis*-aconitic groups, was used to coat the polyplex of pDNA with poly(L-lysine) containing thiol groups [102] or short interfering ribonucleic acid (siRNA) with calcium phosphate (CaP) [103] to produce ternary polyplexes. PEG-PAsp(DET-Aco) degraded at acidic pH to generate the parent, PEG-PAsp(DET), which was active towards late endosomal/lysosomal membranes and, thus, assisted in the endosomal escape of pDNA following endocytosis. When pH-sensitive carboxymethyl PHis was used to coat PEI/DNA [104] and PAE/DNA [105] complexes, gene transfection efficiencies were enhanced in both cases. Besides the contribution from the shielding/deshielding method, the pH-sensitivity of imidazole groups was considered to be another contributor to the enhanced endosomal escape.

The delivery of genes to the nucleus encounters another challenge, as the endosomal/lysosomal barrier generally hinders it. Incorporating pH-sensitive synthetic polymers, such as poly(methacrylic acid) (PMAA) copolymers [40, 106], poly(propylacrylic acid)(PPAA) [42], sulfonamide oligomers [37], or endosomolytic melittin peptide [107], into polyplexes can overcome this barrier by destabilizing membrane bilayers at pH values typically found in the endosomes.

2.4.3 *Protein/peptide delivery*

Proteins are the main components of many basic functions for cell life. Among various vehicles, protein-encapsulated nanocomplexes are considered to be potential and promising ones that can carry proteins to cells. The mechanism of pH-sensitive nanocomplexes releasing proteins in a controllable manner is often based on the change of electrical interaction between proteins and carriers by pH changes. A typical example of the delivery of cationic proteins in an acidic environment was reported by Kataoka *et al.* Negatively charged PEG-poly[(N′-citraconyl-2-aminoehtyl) aspartamide complexed with a cationic model protein lysozyme to obtain 130 nm micelles at neutral pH by the electrical interaction. At the endosomal pH, the degradation of citraconic amides, which is directly related to the charge conversion from negative to positive, promoted the destabilization of micelles and released lysozymes by the charge repulsion. There was no apparent activity at pH 7.4 for over 5h, while at pH 5.5, the micelles showed lysozyme activity after 2h, and reached 97% of that of free lysozymes after 5h [50]. As anionic carriers offer a potential to form complexes with cationic proteins, cationic carriers can be used to complex with anionic proteins. mPEG-*b*-PAE with piperidine and imidazole groups dissolved completely in water at low pH (<6.9) and the cationic PAE segment complexed with negatively charged albumin by the electrostatic interactions. As the pH of solution increased, the PAE segment was fully deionized and became hydrophobic, resulting in the stable protein-encapsulated micelles shielded by the PEG as a shell with a size range of 50–70 nm. When the pH was lower than 7.0, the charge of albumin-encapsulated polymeric micelles increased gradually to net positive charge due to the ionization of the amino groups in the PAE chain. As a result, the positive charge enhanced the cell uptake to the acidic pathological area, and subsequently, the polymers were dissolved, leaving the loaded albumin in the area [19]. The pH-sensitivity of carriers can also be applied for the controlled release of proteins in alkaline conditions. Poly(acrylic acid)-grafted grapheme oxide (GO-PAA) was used for loading bovine serum albumin (BSA) through noncovalent interactions. At lower pH, the PAA chain was protonated and shrank, preventing protein release, while at intestinal pH, the PAA was deprotonated and swelled by electrostatic repulsions to release protein. As a matter of fact, there was no release of BSA at pH 3.0 (similar to the pH in a fasting stomach) after 3 days, while the release was ~20% at pH 7.4 (similar to the pH of body fluids) and ~79% at pH 8 (intestinal pH) [89].

2.5 Other Applications of pH-Sensitive Nanosystems

2.5.1 *Imaging*

In cancer therapy, non-invasive imaging of acidic pH environments plays an important role in the diagnosis and tracking of therapeutic systems. Among various imaging probes, pH-sensitive micelles loading optical imaging agents have attracted much attention because of their narrow and specific pH-responsiveness. The main idea of this strategy is that at neutral pH, the hydrophobic imaging agents are silenced inside of the core, while they are activated in the acidic environment where the carrier destabilizes. After encapsulation of fluorescent dye tetramethylrhodamine isothiocyanate (TRITC) and black hole quencher (BHQ) into mPEG-*b*-PAE micelles, as shown in Fig. 2.6A, the fluorescence signal from TRITC was quenchedinside of the micellar cores and was restored after the disassembly of the micelles when it encountered pH changes in the tumor microenvironment. Under *in vivo* conditions, these quenched nanosized micelles exhibited specific accumulation at tumor tissue and gave a fluorescence signal after demicellization [18]. The hydrophobic photosensitizer protoporphyrin (PpIX) was also encapsulated into mPEG-*b*-PAE micelles. The PpIX was released by demicellization and produced fluorescence and singlet oxygen for simultaneous diagnosis and therapy. Compared with the major accumulation of free PpIX in the liver, PpIX loaded in the micelles showed strong fluorescence intensity at tumors, and the total photon counts in tumor tissue at 48 h post-injection were ten-fold higher than that of free PpIX, which confirmed the tumor targeted delivery of PpIX [108]. Besides optical imaging agents, contrast imaging agents also have been loaded in pH-sensitive micelles for non-invasive imaging of acidic pH environments. Magnetite nanoparticles, iron oxide (Fe_3O_4), were loaded in the pH-sensitive mPEG-*b*-PAE micelles by the hydrophobic interaction. The micelles endowed the controlled release of Fe_3O_4 in the acidic environment for magnetic resonance imaging (MRI) from the micellization/demicellization transition [68, 109]. Recently, Kim *et al.* loaded contrast agent gadolinium (Gd^{3+}) in the pH-sensitive polymeric micelles, made of mPEG-*b*-PLA-diethylenetriaminopenta acetic acid dianhydride (DTPA)-gadolinium chelate (Gd) and PEG-*b*-PHis. The micelles were stable at physiological pH, but destabilized in the acidic tumoral environment due to the protonation of the imidazole of PHis blocks, enhancing the effective magnetic resonance (MR) contrast in the tumor region [110].

pH-sensitive micelles loading imaging agents is one way of targeting the acidic pH region for imaging. Another way is to load pH-sensitive molecules, often fluorophores, in micellar systems. Depending on the OFF/ON switch of fluorescence from fluorophores, lower pH tumor sites or acidic organelles can be distinguished, and because of the low bioactivity of fluorophores themselves, combining them with nanoparticles is beneficial. For the chromophore of the fluorephore, N,N′-di-n-dodecyl-2,6-di(4-methyl-piperazin-1-yl) naphthalene-1,4,5,8-tetracarboxylic acid diimide (NDI), as shown in Fig. 2.6B, photo-induced electron transfer (PET) from the methyl-substituted nitrogen atoms to naphthalene diimide chromophore quenched the fluorescence (OFF), which was blocked by the protonation, resulting in the specific pH response from fluorescence OFF (NDI) to fluorescence ON (NDI-H^+). After being loaded in the poly(styrene-block-acrylic acid) (PS-b-PAA) micelles, NDI fluorescence intensity was enhanced by 55-fold upon changing from neutral to acidic pH (pH 3.4) [78].

2.5.2 *pH measurement*

Intracellular pH plays important roles in cellular functioning, where the pH detection is essential to understand the cellular processes. At the same time, a rapid development of pH-sensitive nanotechnology, especially in the chemotherapy, urges us to quantify the pH fluctuations to understand and optimize different delivery approaches.

Conventional methods have been using proton (^1H)-MRI contrast agents [111, 112] or ^1H-nanoprobes to measure intracellular pH. Comparing with ^1H probes, fluorine (^{19}F) is advantageous because many ^{19}F reporter molecules have diverse chemical shifts and narrow peak widths that can be easily differentiated. In designing pH activatable ^{19}F-nanoprobes, ^{19}F reporter molecules should be highly responsive to small pH transitions, and each pH

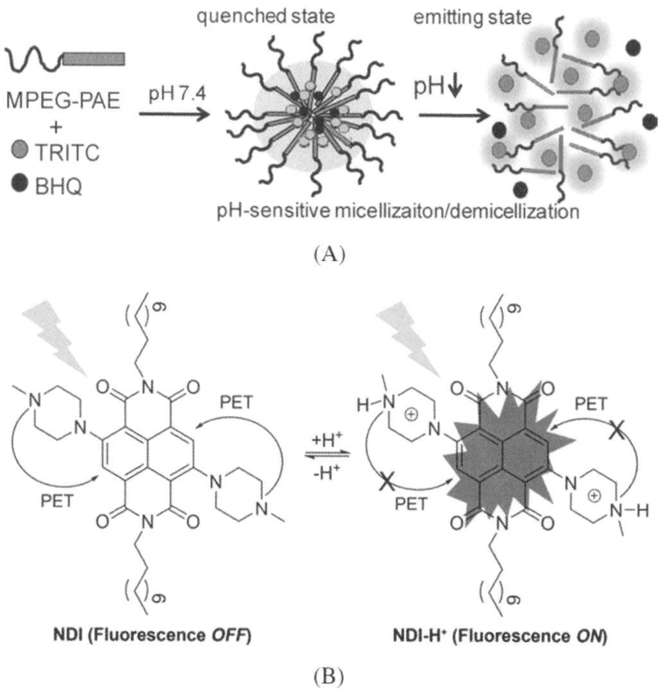

Figure 2.6. (A) Schematic illustration of the pH-dependent fluorescence recovery system under acidic conditions. The fluorescent dye (TRITC) and BHQ are encapsulated by self-assembly of MPEG-PAE block copolymer, called a quenched state. Upon lowering the pH, the strong fluorescence intensity is recovered by disassembly of MPEG-PAE block copolymer, called an emitting state. This research was originally published in *Small.* Ko, J. Y.; Park, S.; Lee, H.; Koo, H.; Kim, M. S.; Choi, K.; Kwon, I. C.; Jeong, S. Y.; Kim, K.; Lee, D. S. 2010, pH-Sensitive nanoflash for tumoral acidic pH imaging in live animals. **6**(22), 2539–2544. Copyright © 2010 WilEY-VCH Verlag Gmbh & Co. KGaA, Weinheim. (B) pH response of NDI based on PEG mechanism which can be blocked by the protonation of piperazines. This research was originally published in *Biomaterials.* Chang, S.; Wu, X.; Li, Y.; Niu, D.; Gao, Y.; Ma, Z.; Gu, J.; Zhao, W.; Zhu, W.; Tian, H.; Shi, J. 2013, A pH-responsive hybrid fluorescent nanoprober for real time cell labeling and endocytosis tracking. **34**(38), 10182–10190. Copyright © 2013, Elsevier.

transition should be encoded with a ^{19}F-specific signature. In these regards, Dr. Gao and co-workers prepared multi-chromatic pH-activatable MRI nanoprobes using pH-sensitive polymeric micelles. As shown in Fig. 2.7A, different ^{19}F reporters were conjugated to hydrophobic parts of the amphiphilic copolymers consisting of the hydrophilic block PEG and tertiary amine/ammonium segment with different pK_a values. When optimized with appropriate molar ratios of the ^{19}F reporter, each copolymer self-assembled into micelles at pH>pK_a; at this condition, the ^{19}F signal was suppressed because of the restriction of chain motion. When pH<pK_a, the micelles destabilized to unimers and restored the ^{19}F signal. More impressively, these nanoprobes responded within 0.25 pH units and could be tuned for specific pH transition by the pK_a of copolymers. After confirming the ON/OFF pH-activatable MR imaging capability of each nanoprobe, three different ones with pK_a 7.0, 6.1, and 5.0 were mixed together. As shown in Fig. 2.7B, the mixture showed different nuclear magnetic resonance (NMR)^{19}F spectra and ^{19}F MR images

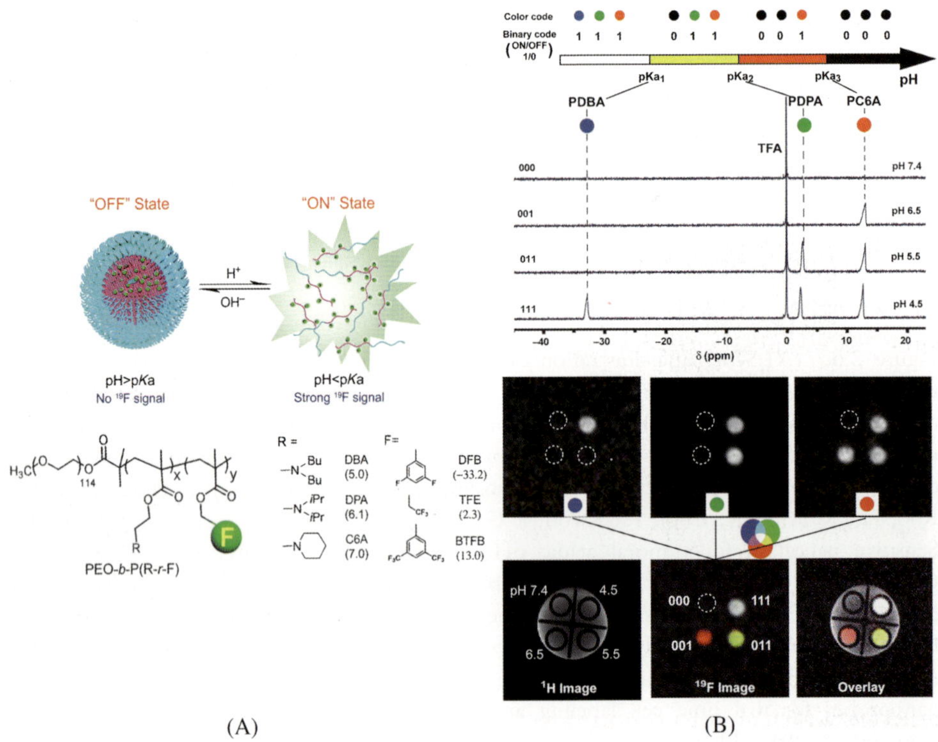

Figure 2.7. (A) pH-activatable ON/OFF ^{19}F-MRI nanoprobes from ionizable diblock copolymer and the structure formula of three representative diblock copolymers containing different pH responsive segments and ^{19}F reporter moieties with their pKa values and ^{19}F chemical shifts. (B) Schematic illustration of the activation barcode concept for direct readout of pH values between adjacent pKa values, ^{19}F MR imaging of different pH values and ^{19}F MR imaging of the same nanoprobe mixture in solutions with different pH values. This research was originally published in *Angew. Chem. Int. Ed.* Huang, X.; Huang, G.; Zhang, S.; Sagiyama, K.; Togao, O.; Ma, X.; Wang, Y.; Li, Y.; Soesbe, T. C.; Sumer, B. D.; Takahashi, M.; Sherry, A. D.; Gao, J., 2013, Multi-chromatic pH-activatable ^{19}F-MRI nanoprobes with binary ON/OFF pH transitions and chemical-shift barcodes. **52**(31), 8074–8078. Copyright ©2013 WilEY-VCH Verlag Gmbh & Co. KGaA, Weinheim.

when the mixed solution was exposed to different pH values (pH 7.4, 6.5, 5.5, and 4.5), which confirmed the feasibility of this system to discriminate the pH difference in the microenvironment [66].

2.5.3 *Miscellaneous*

pH-sensitive nanosystems were reported as pickering emulsifiers [113, 114]. Instead of using conventional surfactants to stabilize emulsion, pickering emulsion uses colloidal particles for stabilization [115]. After introducing pH-sensitive pickering emulsifiers, the emulsions break readily within seconds after the emulsifier is spontaneously desorbed from the emulsion droplet interface on lowering the solution pH to below the emulsifier's pK_a. More interestingly, the reversible pH-sensitivity of latex endows the reversible change between emulsification and demulsification by inducing H^+ and salt or CO_2/nitrogen (N_2) to regulate the aqueous pH [116].

2.6 Conclusion and Perspective

pH differences in pathological regions, particularly at tumors and inflamed sites, is a natural and unique signal to induce the chemical reactions or changes in physicochemical properties. Unlike experimental settings, the pH depth profile from the blood vessels in clinical tumors is heterogeneous and there is significant variation. One positive aspect is that the tumor pH can be further acidified by bolus glucose administration due to the high metabolic rate, energy consumption, and glycolysis activity of cancer cells. However, the tumor P He ranges from 6.8 to 7.2 and hardly drops below pH 6.5, even at semi-necrotic areas (the distal region from the blood vessels). This suggests that pH-sensitive nanosystems should be able to penetrate deeply into the tumor tissue to meet the acidic regions before they take advantage of acidity and should be able to discriminate between pH values between 7.4 (physiological) and ~6.8–7.2 (pathological). Although there are functional groups that are sensitive enough to respond to the small pH drops, the pH-sensitive nanosystems share the issues with all nanosized drug carriers experiencing transport problems in solid tumors associated with high interstitial tumor pressure, which interferes with convective transport, diffusivity, and interactions with local biological components. In order to have high potential to be translated into actual clinical use, the chemistry involved in the technology needs to be simple and safe without leaving any reactive groups behind with scalability. More importantly, the technology should be validated in a clinical setting to further encourage the research in the area.

Acknowledgments

The works performed by the Bae group have been in part supported by NIH CA101850 and CA122356. The authors acknowledge Sun Jin Kim for her proof reading.

References

1. Muthu, M. S., Rajesh, C. V., Mishra, A. and Singh, S. (2009). Stimulus-responsive targeted nanomicelles for effective cancer therapy. *Nanomedicine,* **4**, 657–667.
2. Kang, S. and Bae, Y. H. (2003). A sulfonamide-based glucose-responsive hydrogel with covalently immobilized glucose oxidase and catalase. *J. Control. Release,* **86**, 115–121.
3. Ulijn, R. V. (2006). Enzyme-responsive materials: A new class of smart biomaterials. *J. Mater. Chem.,* **16**, 2217–2225.
4. Dressman, J. B., Berardi, R. R., Dermentzoglou, L. C., Russell, T. L., Schmaltz, S. P., Barnett, J. L. and Jarvenpaa, K. M. (1990). Upper gastrointestinal (GI) pH in young, healthy men and women. *Pharm. Res.,* 7, 756–761.
5. Kararli, T. T. (1995). Comparison of the gastrointestinal anatomy, physiology, and biochemistry of humans and commonly used laboratory animals. *Biopharm. Drug Dispos.,* **16**, 351–380.
6. Stubbs, M., Mcsheehy, P. M., Griffiths, J. R. and Bashford, C. L. (2000). Causes and consequences of tumor acidity and implications for treatment. *Mol. Med. Today,* **6**, 15–19.
7. Trédan, O., Galmarini, C. M., Patel, K. and Tannock, I. F. (2007). Drug resistance and the solid tumor microenvironment. *J. Natl. Cancer Inst.,* **99**, 1441–1454.
8. Jain, R. K. (1988). Determinants of tumor blood flow: A review. *Cancer Res.,* **48**, 2641–2658.
9. Warburg, O. (1956). On the origin of cancer cells. *Science,* **123**, 309–314.
10. Leu, A. J., Berk, D. A., Lymboussaki, A., Alitalo, K. and Jain, R. K. (2000). Absence of functional lymphatics within a murine sarcoma: A molecular and functional evaluation. *Cancer Res.,* **60**, 4324–4327.
11. Nishi, T. and Forgac, M. (2002). The vacuolar (H+)-ATPases-nature's most versatile proton pumps. *Nat. Rev. Mol. Cell Biol.,* **3**, 94–103.
12. Lafourcade, C., Sobo, K., Kieffer-Jaquinod, S., Garin, J. and van der Goot, F. G. (2008). Regulation of the V-ATPase along the endocytic pathway occurs through reversible subunit association and membrane localization. *PLoS One,* **3**, e2758–e2772.
13. Zhou, K., Wang, Y., Huang, X., Luby-Phelps, K., Sumer, B. D. and Gao, J. (2011). Tunable, ultrasensitive pH-responsive nanoparticles targeting specific endocytic organelles in living cells. *Angew. Chem. Int. Ed.,* **50**, 6109–6114.
14. Zhou, K., Liu, H., Zhang, S., Huang, X., Wang, Y., Huang, G., Sumer, B. D. and Gao, J. (2012). Multicolored pH-tunable and activatable fluorescence nanoplatform responsive to physiologic pH stimuli. *J. Am. Chem. Soc.,* **134**, 7803–7811.
15. Zhang, S., Zhou, K., Huang, G., Takahashi, M., Sherry, A. D. and Gao, J. (2013). A novel class of polymeric pH-responsive MRI CEST agents. *Chem. Commun.,* **49**, 6418–6420.
16. Wang, Y., Zhou, K., Huang, G., Hensley, C., Huang, X., Ma, X., Zhao, T., Sumer, B. D., Deberardinis, R. J. and Gao, J. (2014). A nanoparticle-based strategy for the imaging

of a broad range of tumours by nonlinear amplification of microenvironment signals. *Nat. Mater.,* **13**, 204–212.

17. Lynn, D. M., Amiji, M. M. and Langer, R. (2001). pH-responsive polymer microspheres: Rapid release of encapsulated material within the range of intracellular pH. *Angew. Chem. Int. Ed.,* **40**, 1707–1710.
18. Ko, J. Y., Park, S., Lee, H., Koo, H., Kim, M. S., Choi, K., Kwon, I. C., Jeong, S. Y., Kim, K. and Lee, D. S. (2010). pH-sensitive nanoflash for tumoral acidic pH imaging in live animals. *Small,* **6**, 2539–2544.
19. Gao, G. H., Park, M. J., Li, Y., Im, G. H., Kim, J. H., Kim, H. N., Lee, J. W., Jeon, P., Bang, O. Y., Lee, J. H. and Lee, D. S. (2012). The use of pH-sensitive positively charged polymeric micelles for protein delivery. *Biomaterials,* **33**, 9157–9164.
20. Kim, M. S., Gao, G. H., Kang, S. W. and Lee, D. S. (2011). Evaluation of pH-sensitive poly(beta-amino ester)-graft-poly(ethylene glycol) and its usefulness as a pH-sensor and protein carrier. *Macromol. Biosci.,* **11**, 946–951.
21. Huynh, D. P., Nguyen, M. K., Pi, B. S., Kim, M. S., Chae, S. Y., Lee, K. C., Kim, B. S., Kim, S. W. and Lee, D. S. (2008). Functionalized injectable hydrogels for controlled insulin delivery. *Biomaterials,* **29**, 2527–2534.
22. Manokruang, K. and Lee, D. S. (2013). Albumin-conjugated pH/thermo responsive poly(amino urethane) multiblock copolymer as an injectable hydrogel for protein delivery. *Macromol. Biosci.,* **13**, 1195–1203.
23. Kim, M. S. and Lee, D. S. (2010). Biodegradable and pH-sensitive polymersome with tuning permeable membrane for drug delivery carrier. *Chem. Commun.,* **46**, 4481–4483.
24. Jeong, I. K., Gao, G. H., Li, Y., Kang, S. W. and Lee, D. S. (2013). A biodegradable polymersome with pH-tuning on-off membrane based on poly(beta-amino ester) for drug delivery. *Macromol. Biosci.,* **13**, 946–953.
25. Lee, M. S., Jang, Y. L., Huynh, D. P., Huynh, C. T., Lee, Y., Chae, S. Y., Kim, S. H., Park, T. G., Lee, D. S. and Jeong, J. H. (2011). pH-sensitive pentablock copolymer nanocapsules as nontoxic and efficient gene carriers. *Macromol. Biosci.,* **11**, 789–796.
26. Pack, D. W., Putnam, D. and Langer, R. (2000). Design of imidazole-containing endosomolytic biopolymers for gene delivery. *Biotechnol. Bioeng.,* **67**, 217–223.
27. Patchornik, A., Berger, A. and Katchalski, E. (1957). Poly-L-histidine. *J. Am. Chem. Soc.,* **79**, 5227–5230.
28. Lee, E. S., Shin, H. J., Na, K. and Bae, Y. H. (2003). Poly(L-histidine)-PEG block copolymer micelles and pH-induced destabilization. *J. Control. Release,* **90**, 363–374.
29. Wu, H., Zhu, L. and Torchilin, V. P. (2013). pH-sensitive poly(histidine)-PEG/DSPE-PEG co-polymer micelles for cytosolic drug delivery. *Biomaterials,* **34**, 1213–1222.
30. Hu, J., Miura, S. J., Na, K. and Bae, Y. H. (2013). pH-responsive and charge shielded cationic micelle of poly(L-histidine)-block-short branched PEI for acidic cancer treatment. *J. Control. Release,* **172**, 69–76.
31. Andon, R. J. L., Cox, J. D. and Herington, E. F. G. (1954). The ultra-violet absorption spectra and dissociation constants of certain pyridine bases in aqueous solution. *Trans. Faraday Soc.,* **50**, 918–927.
32. Satoh, M., Yoda, E. J., Hayashi, H. and Komiyama, J. (1989). Potentiometric titration of poly(vinylpyridines) and hydrophobic interaction in the counterion binding. *Macromolecules,* **22**, 1808–1812.

33. van Butsele, K., Fustin, C. A., Gohy, J. F., Jerome, R. and Jerome, C. (2009). Self-assembly and pH-responsiveness of ABC miktoarm star terpolymers. *Langmuir,* **25**, 107–111.
34. Risbud, M. V., Hardikar, A. A., Bhat, S. V. and Bhonde, R. R. (2000). pH-sensitive freeze-dried chitosan-polyvinyl pyrrolidone hydrogels as controlled release system for antibiotic delivery. *J. Control. Release,* **68**, 23–30.
35. Basset, C., Harder, C., Vidaud, C. and Dejugnat, C. (2010). Design of double stimuli-responsive polyelectrolyte microcontainers for protein soft encapsulation. *Biomacromolecules,* **11**, 806–814.
36. Sethuraman, V. A., Na, K. and Bae, Y. H. (2006). pH-responsive sulfonamide/PEI system for tumor specific gene delivery: An *in vitro* study. *Biomacromolecules,* 7, 64–70.
37. Kang, H. C. and Bae, Y. H. (2007). pH-tunable endosomolytic oligomers for enhanced nucleic acid delivery. *Adv. Funct. Mater.,* **17**, 1263–1272.
38. Park, S. Y. and Bae, Y. H. (1999). Novel pH-sensitive polymers containing sulfonamide groups. *Macromol. Rapid Commun.,* **20**, 269–273.
39. Murthy, N., Robichaud, J. R., Tirrell, D. A., Stayton, P. S. and Hoffman, A. S. (1999). The design and synthesis of polymers for eukaryotic membrane disruption. *J. Control. Release,* **61**, 137–143.
40. Yessine, M. A. and Leroux, J. C. (2004). Membrane-destabilizing polyanions: Interaction with lipid bilayers and endosomal escape of biomacromolecules. *Adv. Drug Deliv. Rev.,* **56**, 999–1021.
41. Flanary, S., Hoffman, A. S. and Stayton, P. S. (2009). Antigen delivery with poly(propylacrylic acid) conjugation enhances MHC-1 presentation and T-cell activation. *Bioconjugate Chem.,* **20**, 241–248.
42. Cheung, C. Y., Murthy, N., Stayton, P. S. and Hoffman, A. S. (2001). A pH-sensitive polymer that enhances cationic lipid-mediated gene transfer. *Bioconjugate Chem.,* **12**, 906–910.
43. Kyriakides, T. R., Cheung, C. Y., Murthy, N., Bornstein, P., Stayton, P. S. and Hoffman, A. S. (2002). pH-Sensitive polymers that enhance intracellular drug delivery *in vivo. J. Control. Release,* **78**, 295–303.
44. Felber, A. E., Dufresne, M. H. and Leroux, J. C. (2012). pH-Sensitive vesicles, polymeric micelles, and nanospheres prepared with polycarboxylates. *Adv. Drug Deliv. Rev.,* **64**, 979–992.
45. Chen, C., Pu, F., Huang, Z., Liu, Z., Ren, J. and Qu, X. (2011). Stimuli-responsive controlled-release system using quadruplex DNA-capped silica nanocontainers. *Nucleic Acids Res.,* **39**, 1638–1644.
46. Tian, J. W., Ding, L., Xu, H. J., Shen, Z., Ju, H. X., Jia, L., Bao, L. and Yu, J. S. (2013). Cell-specific and pH-activatable rubyrin-loaded nanoparticles for highly selective near-infrared photodynamic therapy against cancer. *J. Am. Chem. Soc.,* **135**, 18850–18858.
47. Zhu, X. C., Lu, W. T., Zhang, Y. Z., Reed, A., Newton, B., Fan, Z., Yu, H. T., Ray, P. C. and Gao, R. M. (2011). Imidazole-modified porphyrin as a pH-responsive sensitizer for cancer photodynamic therapy. *Chem. Commun.,* **47**, 10311–10313.
48. Liu, J., Ma, H. L., Wei, T. and Liang, X. J. (2012). CO_2 gas induced drug release from pH-sensitive liposome to circumvent doxorubicin resistant cells. *Chem. Commun.,* **48**, 4869–4871.
49. Du, J. Z., Sun, T. M., Song, W. J., Wu, J. and Wang, J. (2010). A tumor-acidity-activated charge-conversional nanogel as an intelligent vehicle for promoted tumoral-cell uptake and drug delivery. *Angew. Chem. Int. ED.,* **49**, 3621–3626.

50. Lee, Y., Fukushima, S., Bae, Y., Hiki, S., Ishii, T. and Kataoka, K. (2007). A protein nanocarrier from charge-conversion polymer in response to endosomal pH. *J. Am. Chem. Soc.,* **129**, 5362–5363.
51. Sanjoh, M., Hiki, S., Lee, Y., Oba, M., Miyata, K., Ishii, T. and Kataoka, K. (2010). pDNA/poly(L-lysine) polyplexes functionalized with a pH-sensitive charge-conversional poly(aspartamide) derivative for controlled gene delivery to human umbilical vein endothelial cells. *Macromol. Rapid Commun.,* **31**, 1181–1186.
52. Nguyen, R. and Huc, I. (2003). Optimizing the reversibility of hydrazone formation for dynamic combinatorial chemistry. *Chem. Commun.,* **8**, 942–943.
53. Kresge, A. J., Sagatys, D. S. and Chen, H. L. (1977). Vinyl ether hydrolysis. 9. isotope effects on proton transfer from the hydronium ion. *J. Am. Chem. Soc.,* **99**, 7228–7233.
54. Kim, H. K., van den Bossche, J., Hyun, S. H. and Thompson, D. H. (2012). Acid-triggered release via dePEGylation of fusogenic liposomes mediated by heterobifunctional phenyl-substituted vinyl ethers with tunable pH-sensitivity. *Bioconjugate Chem.,* **23**, 2071–2077.
55. Boomer, J. A., Inerowicz, H. D., Zhang, Z. Y., Bergstrand, N., Edwards, K., Kim, J. M. and Thompson, D. H. 2003. Acid-triggered release from sterically stabilized fusogenic liposomes via a hydrolytic dePEGylation strategy. *Langmuir,* **19**, 6408–6415.
56. Shin, J., Shum, P., Grey, J., Fujiwara, S., Malhotra, G. S., Gonzalez-Bonet, A., Hyun, S. H., Moase, E., Allen, T. M. and Thompson, D. H. (2012). Acid-labile mPEG-vinyl ether-1,2-dioleylglycerol lipids with tunable pH sensitivity: Synthesis and structural effects on hydrolysis rates, DOPE liposome release performance, and pharmacokinetics. *Mol. Pharm.,* **9**, 3266–3276.
57. Roberts, M. C., Hanson, M. C., Massey, A. P., Karren, E. A. and Kiser, P. F. (2007). Dynamically restructuring hydrogel networks formed with reversible covalent crosslinks. *Adv. Mater.,* **19**, 2503–2507.
58. Zhou, C. Y., Gao, W. X., Yang, K. W., Xu, L., Ding, J. C., Chen, J. X., Liu, M. C., Huang, X. B., Wang, S. and Wu, H. Y. (2013). A novel glucose/pH responsive low-molecular-weight organogel of easy recycling. *Langmuir,* **29**, 13568–13575.
59. Yan, J., Springsteen, G., Deeter, S. and Wang, B. H. (2004). The relationship among pKa, pH, and binding constants in the interactions between boronic acids and diols–it is not as simple as it appears. *Tetrahedron,* **60**, 11205–11209.
60. Su, J., Chen, F., Cryns, V. L. and Messersmith, P. B. (2011). Catechol polymers for pH-responsive, Targeted drug delivery to cancer cells. *J. Am. Chem. Soc.,* **133**, 11850–11853.
61. Wang, C., Wang, G., Wang, Z. and Zhang, X. (2011). A pH-responsive superamphiphile based on dynamic covalent bonds. *Chemistry,* **17**, 3322–3325.
62. Gu, J., Cheng, W. P., Liu, J., Lo, S. Y., Smith, D., Qu, X. and Yang, Z. (2008). pH-Triggered reversible "stealth" polycationic micelles. *Biomacromolecules,* **9**, 255–262.
63. Xin, Y. and Yuan, J. J. (2013). Schiff's base as a stimuli-responsive linker in polymer chemistry. *Polym. Chem.,* **3**, 3045–3055.
64. Lee, E. S., Shin, H. J., Na, K. and Bae, Y. H. (2003). Poly(l-histidine) — PEG block copolymer micelles and pH-induced destabilization. *J. Control. Release,* **90**, 363–374.
65. Ko, J., Park, K., Kim, Y. S., Kim, M. S., Han, J. K., Kim, K., Park, R. W., Kim, I. S., Song, H. K., Lee, D. S. and Kwon, I. C. (2007). Tumoral acidic extracellular pH targeting of pH-responsive MPEG-poly(beta-amino ester) block copolymer micelles for cancer therapy. *J. Control. Release,* **123**, 109–115.

66. Huang, X., Huang, G., Zhang, S., Sagiyama, K., Togao, O., Ma, X., Wang, Y., Li, Y., Soesbe, T. C., Sumer, B. D., Takahashi, M., Sherry, A. D. and Gao, J. (2013). Multichromatic pH-activatable 19F-MRI nanoprobes with binary ON/OFF pH transitions and chemical-shift barcodes. *Angew. Chem. Int. Ed.,* **52**, 8074–8078.
67. Bontha, S., Kabanov, A. V. and Bronich, T. K. (2006). Polymer micelles with crosslinked ionic cores for delivery of anticancer drugs. *J. Control. Release,* **114**, 163–174.
68. Gao, G. H., Im, G. H., Kim, M. S., Lee, J. W., Yang, J., Jeon, H., Lee, J. H. and Lee, D. S.(2010). Magnetite-nanoparticle-encapsulated pH-responsive polymeric micelle as an MRI probe for detecting acidic pathologic areas. *Small,* **6**, 1201–1204.
69. Kim, D., Lee, E. S., Oh, K. T., Gao, Z. G. and Bae, Y. H. (2008). Doxorubicin-loaded polymeric micelle overcomes multidrug resistance of cancer by double-targeting folate receptor and early endosomal pH. *Small,* **4**, 2043–2350.
70. Romberg, B., Hennink, W. E. and Storm, G. (2008). Sheddable coatings for long-circulating nanoparticles. *Pharm. Res.,* **25**, 55–71.
71. Bae, Y., Jang, W. D., Nishiyama, N., Fukushima, S. and Kataoka, K. (2005). Multifunctional polymeric micelles with folate-mediated cancer cell targeting and pH-triggered drug releasing properties for active intracellular drug delivery. *Mol. BioSyst.,* **1**, 242–250.
72. Bae, Y., Nishiyama, N. and Kataoka, K. (2007). In vivo antitumor activity of the folate-conjugated pH-sensitive polymeric micelle selectively releasing adriamycin in the intracellular acidic compartments. *Bioconjugate Chem.,* **18**, 1131–1139.
73. Lee, E. S., Na, K. and Bae, Y. H. (2005). Super pH-sensitive multifunctional polymeric micelle. *Nano Lett.,* **5**, 325–329.
74. Lee, E. S., Na, K. and Bae, Y. H. (2005). Doxorubicin loaded pH-sensitive polymeric micelles for reversal of resistant MCF-7 tumor. *J. Control. Release,* **103**, 405–418.
75. Lee, E. S., Gao, Z., Kim, D., Park, K., Kwon, I. C. and Bae, Y. H.(2008). Super pH-sensitive multifunctional polymeric micelle for tumor pH(e) specific TAT exposure and multidrug resistance. *J. Control. Release,* **129**, 228–236.
76. Bae, Y., Nishiyama, N., Fukushima, S., Koyama, H., Yasuhiro, M. and Kataoka, K.(2005). Preparation and biological characterization of polymeric micelle drug carriers with intracellular pH-triggered drug release property: Tumor permeability, controlled subcellular drug distribution, and enhanced *in vivo* antitumor efficacy. *Bioconjugate Chem.,* **16**, 122–130.
77. Du, J. Z., Du, X. J., Mao, C. Q. and Wang, J. (2011). Tailor-made dual pH-sensitive polymer-doxorubicin nanoparticles for efficient anticancer drug delivery. *J. Am. Chem. Soc.,* **133**, 17560–17563.
78. Chang, S., Wu, X., Li, Y., Niu, D., Gao, Y., Ma, Z., Gu, J., Zhao, W., Zhu, W., Tian, H. and Shi, J.(2013). A pH-responsive hybrid fluorescent nanoprober for real time cell labeling and endocytosis tracking. *Biomaterials,* **34**, 10182–10190.
79. Connor, J., Yatvin, M. B. and Huang, L. (1984). pH-sensitive liposomes: Acid-induced liposome fusion. *Proc. Natl. Acad. Sci. U S A,* **81**, 1715–1718.
80. Torchilin, V. P. (2005). Recent advances with liposomes as pharmaceutical carriers. *Nat. Rev. Drug Discov.,* **4**, 145–160.
81. Roux, E., Passirani, C., Scheffold, S., Benoit, J. P. and Leroux, J. C. (2004). Serum-stable and long-circulating, PEGylated, pH-sensitive liposomes. *J. Control. Release,* **94**, 447–451.

82. Kale, A. A. and Torchilin, V. P. (2007). Enhanced transfection of tumor cells *in vivo* using "Smart" pH-sensitive TAT-modified pegylated liposomes. *J. Drug Target.*, **15**, 538–545.
83. Biswas, S., Dodwadkar, N. S., Sawant, R. R. and Torchilin, V. P. (2011). Development of the novel PEG-PE-based polymer for the reversible attachment of specific ligands to liposomes: Synthesis and *in vitro* characterization. *Bioconjugate Chem.*, **22**, 2005–2013.
84. Rim, H. P., Min, K. H., Lee, H. J., Jeong, S. Y. and Lee, S. C. (2011). pH-Tunable calcium phosphate covered mesoporous silica nanocontainers for intracellular controlled release of guest drugs. *Angew. Chem. Int. ED.*, **50**, 8853–8857.
85. Xing, L., Zheng, H., Cao, Y. and Che, S. (2012). Coordination polymer coated mesoporous silica nanoparticles for pH-responsive drug release. *Adv. Mater.*, **24**, 6433–6437.
86. Zou, Z., He, D. G., He, X. X., Wang, K. M., Yang, X., Qing, Z. H. and Zhou, Q. (2013). Natural gelatin capped mesoporous silica nanoparticles for intracellular acid-triggered drug delivery. *Langmuir*, **29**, 12804–12810.
87. Dergunov, S. A., Durbin, J., Pattanaik, S. and Pinkhassik, E. (2014). pH-Mediated catch and release of charged molecules with porous hollow nanocapsules. *J. Am. Chem. Soc.*, **136**, 2212–2215.
88. Lai, T. C., Bae, Y., Yoshida, T., Kataoka, K. and Kwon, G. S. (2010). pH-sensitive multi-PEGylated block copolymer as a bioresponsive pDNA delivery vector. *Pharm. Res.*, **27**, 2260–2273.
89. Kavitha, T., Kang, I. K. and Park, S. Y. (2014). Poly(acrylic acid)-grafted graphene oxide as an intracellular protein carrier. *Langmuir*, **30**, 402–409.
90. Poon, Z., Chang, D., Zhao, X. and Hammond, P. T. (2011). Layer-by-layer nanoparticles with a pH-sheddable layer for *in vivo* targeting of tumor hypoxia. *ACS Nano*, **5**, 4284–4292.
91. Rozema, D. B., Lewis, D. L., Wakefield, D. H., Wong, S. C., Klein, J. J., Roesch, P. L., Bertin, S. L., Reppen, T. W., Chu, Q., Blokhin, A. V., Hagstrom, J. E. and Wolff, J. A. (2007). Dynamic PolyConjugates for targeted *in vivo* delivery of siRNA to hepatocytes. *Proc. Natl. Acad. Sci.*, **104**, 12982–12987.
92. Shim, W. S., Kim, S. W., Choi, E. K., Park, H. J., Kim, J. S. and Lee, D. S. (2006). Novel pH sensitive block copolymer micelles for solvent free drug loading. *Macromol. Biosci.*, **6**, 179–186.
93. Lee, E. S., Na, K. and Bae, Y. H. (2003). Polymeric micelle for tumor pH and folate-mediated targeting. *J. Control. Release*, **91**, 103–113.
94. Tomlinson, R., Heller, J., Brocchini, S. and Duncan, R. (2003). Polyacetal-doxorubicin conjugates designed for pH-dependent degradation. *Bioconjugate Chem.*, **14**, 1096–1106.
95. Parker, N., Turk, M. J., Westrick, E., Lewis, J. D., Low, P. S. and Leamon, C. P. (2005). Folate receptor expression in carcinomas and normal tissues determined by a quantitative radioligand binding assay. *Anal. Biochem.*, **338**, 284–293.
96. Cheng, Z., Al Zaki, A., Hui, J. Z., Muzykantov, V. R. and Tsourkas, A. (2012). Multifunctional nanoparticles: Cost versus benefit of adding targeting and imaging capabilities. *Science*, **338**, 903–910.
97. Koren, E., Apte, A., Jani, A. and Torchilin, V. P. (2012). Multifunctional PEGylated 2C5-immunoliposomes containing pH-sensitive bonds and TAT peptide for enhanced tumor cell internalization and cytotoxicity. *J. Control. Release*, **160**, 264–273.

98. Hatakeyama, H., Akita, H. and Harashima, H. (2013). The polyethyleneglycol dilemma: Advantage and disadvantage of PEGylation of liposomes for systemic genes and nucleic acids delivery to tumors. *Biol. Pharm. Bull.,* **36**, 892–899.
99. Li, W., Huang, Z., Mackay, J. A., Grube, S. and Szoka, F. C., Jr. (2005). Low-pH-sensitive poly(ethylene glycol) (PEG)-stabilized plasmid nanolipoparticles: Effects of PEG chain length, lipid composition and assembly conditions on gene delivery. *J. Gene Med.,* 7, 67–79.
100. Masson, C., Garinot, M., Mignet, N., Wetzer, B., Mailhe, P., Scherman, D. and Bessodes, M. (2004). pH-Sensitive PEG lipids containing orthoester linkers: New potential tools for nonviral gene delivery. *J. Control. Release,* **99**, 423–434.
101. Felber, A. E., Castagner, B., Elsabahy, M., Deleavey, G. F., Damha, M. J. and Leroux, J. C. (2011). siRNA nanocarriers based on methacrylic acid copolymers. *J. Control. Release,* **152**, 159–167.
102. Sanjoh, M., Miyata, K., Christie, R. J., Ishii, T., Maeda, Y., Pittella, F., Hiki, S., Nishiyama, N. and Kataoka, K. (2012). Dual environment-responsive polyplex carriers for enhanced intracellular delivery of plasmid DNA. *Biomacromolecules,* **13**, 3641–3649.
103. Pittella, F., Miyata, K., Maeda, Y., Suma, T., Watanabe, S., Chen, Q., Christie, R. J., Osada, K., Nishiyama, N. and Kataoka, K. (2012). Pancreatic cancer therapy by systemic administration of VEGF siRNA contained in calcium phosphate/charge-conversional polymer hybrid nanoparticles. *J. Control. Release,* **161**, 868–874.
104. Asayama, S., Sudo, M., Nagaoka, S. and Kawakami, H. (2008). Carboxymethyl poly(L-histidine) as a new pH-sensitive polypeptide to enhance polyplex gene delivery. *Mol. Pharm.,* **5**, 898–901.
105. Gu, J. J., Wang, X., Jiang, X. Y., Chen, Y. Z., Chen, L. C., Fang, X. L. and Sha, X. Y. (2012). Self-assembled carboxymethyl poly (L-histidine) coated poly (b-amino ester)/DNA complexes for gene transfection. *Biomaterials,* **33**, 644–658.
106. Yessine, M. A., Meier, C., Petereit, H. U. and Leroux, J. C. (2006). On the role of methacrylic acid copolymers in the intracellular delivery of antisense oligonucleotides. *Eur. J. Pharm. Biopharm.,* **63**, 1–10.
107. Meyer, M., Philipp, A., Oskuee, R., Schmidt, C. and Wagner, E. (2008). Breathing life into polycations: Functionalization with pH-responsive endosomolytic peptides and polyethylene glycol enables siRNA delivery. *J. Am. Chem. Soc.,* **130**, 3272–3273.
108. Koo, H., Lee, H., Lee, S., Min, K. H., Kim, M. S., Lee, D. S., Choi, Y., Kwon, I. C., Kim, K. and Jeong, S. Y. (2010). *In vivo* tumor diagnosis and photodynamic therapy via tumoral pH-responsive polymeric micelles. *Chem. Commun.,* **46**, 5668–5670.
109. Gao, G. H., Lee, J. W., Nguyen, M. K., Im, G. H., Yang, J., Heo, H., Jeon, P., Park, T. G., Lee, J. H. and Lee, D. S. (2011). pH-responsive polymeric micelle based on PEG-poly(beta-amino ester)/(amido amine) as intelligent vehicle for magnetic resonance imaging in detection of cerebral ischemic area. *J. Control. Release,* **155**, 11–17.
110. Kim, K. S., Park, W., Hu, J., Bae, Y. H. and Na, K. (2014). A cancer-recognizable MRI contrast agents using pH-responsive polymeric micelle. *Biomaterials,* **35**, 337–343.
111. Wu, Y., Soesbe, T. C., Kiefer, G. E., Zhao, P. and Sherry, A. D. (2010). A responsive europium(III) chelate that provides a direct readout of pH by MRI. *J. Am. Chem. Soc.,* **132**, 14002–14003.
112. Aime, S., Delli Castelli, D. and Terreno, E. (2002). Novel pH-reporter MRI contrast agents. *Angew. Chem. Int. Ed.,* **41**, 4334–4336.

113. Fujii, S., Read, E. S., Binks, B. P. and Armes, S. P. (2005). Stimulus-responsive emulsifiers based on nanocomposite microgel particles. *Adv. Mater.,* **17**, 1014–1018.
114. Dupin, D., Armes, S. P., Connan, C., Reeve, P. and Baxter, S. M. (2007). How does the nature of the steric stabilizer affect the pickering emulsifier performance of lightly cross-linked, acid-swellable poly(2-vinylpyridine) latexes? *Langmuir,* **23**, 6903–6910.
115. Finkle, P., Draper, H. D. and Hildebrand, J. H. (1923). The theory of emulsification. *J. Am. Chem. Soc.,* **45**, 2780–2788.
116. Morse, A. J., Armes, S. P., Thompson, K. L., Dupin, D., Fielding, L. A., Mills, P. and Swart, R. (2013). Novel pickering emulsifiers based on PH-responsive poly(2-(diethylamino)ethyl methacrylate) latexes. *Langmuir,* **29**, 5466–5475.

Chapter 3

Matrix Metalloproteinase-Sensitive Nanocarriers

Ying Tu and Lin Zhu

Department of Pharmaceutical Sciences,
Irma Lerma Rangel College of Pharmacy
Texas A&M University Health Science Center,
Kingsville, TX, USA

3.1 Introduction

Nanotechnology has brought deep impacts and new possibilities to almost all scientific fields including drug discovery and development [1, 2]. To improve unfriendly physicochemical or pharmacokinetic properties, as well as impart the new/desired characteristics of medicine, nanostructured drug carriers are indispensable. The most commonly used diagnostic/therapeutic nanocarriers include particulate delivery systems, such as liposomes, micelles, dendrimers, other polymeric nanoparticles, and inorganic nanoparticles (e.g. gold nanoparticles, carbon nanotubes, and quantum dots) and macromolecular delivery systems such as antibody-drug conjugates (ADC) and drug–polymer conjugates (e.g. polyethylene glycol (PEG)–protein conjugates and PEG–short interfering ribonucleic acid (siRNA) conjugates) [3–7].

Therapeutic or imaging molecules can be encapsulated into the inner layer or core of nanocarriers, or adsorbed on the surface or outer layer of nanocarriers, via physical forces or covalent bonds. Upon loading, the properties of the loaded molecules can be significantly altered in terms of their physical and chemical properties and their interaction with biological systems, since they are negligible in quantity and surrounded by carrier matrices [8]. In such a way, the therapeutic outcomes of drug-loaded nanocarriers/nanomedicine are expected to be improved. In addition to the tiny size, the components of nanocarriers provide an opportunity for further engineering and modification, especially on their surface, so as to finely tune their biological properties and *in vivo* behaviors. These result in the improvement of drug stability, blood circulation half-life, bioavailability, and targetability, as well as minimization of the non-specific protein adsorption, immunogenicity, undesirable biodistribution, and other side effects [9]. Over the last few decades since the first use of liposomes as a carrier in biomedical sciences, the drug-loaded nanocarriers have demonstrated various advantages over free therapeutic molecules.

In preclinical and even clinical applications, many drug delivery strategies have been developed and investigated, such as the prolonged circulation time in blood due to the hydrophilic polymer shell, the use of the enhanced permeability and retention (EPR) effect for tumor targeting, the surface modification of nanocarriers with targeting ligands, the on-demand drug release/delivery via the stimulus-sensitivity, the cell-penetrating protein and peptide (CPP)-induced intracellular drug delivery, and so on.

For targeted drug delivery, the main strategies are based on antibodies against antigens or ligands for receptors. The elaborate engineering for minimizing immunogenicity [10, 11] and limited penetration of solid tumors [12] impairs the clinical benefits of antibody-based strategies, although monoclonal antibodies (mAbs) have been successful in cancer therapy [13]. Small molecule ligands such as folic acid and galactose have been preliminarily explored, but their specificity/targetability is low compared to that of mAbs [14]. The lack of amplification of simple antibody or ligand binding (usually low affinity or monovalent), however, is another concern [14, 15].

The idea of a stimulus-sensitive drug delivery system is based on the fact of abnormalities during disease initiation and progression. For example, in the tumor microenvironment, the significant abnormalities include acidic pH, altered redox potential, and up-regulated proteins. These internal conditions, as well as external stimuli such as hyperthermia, magnetic field, and ultrasound, can be used to design a stimulus-sensitive drug delivery system for on-demand drug delivery or/and drug release [16].

Unlike pH and redox potential, the abnormal changes of certain proteins like matrix metalloproteinases (MMPs) are usually more significant and robust. The important role of MMPs and the fact of the up-regulation of MMPs in various malignancies, such as cancer and rheumatoid arthritis, have them well characterized and widely used as the biomarkers for the purpose of diagnosis, prognosis, and as the therapeutic targets in clinics as well. Therefore, the use of MMPs as the stimuli ensures a better stimulus–response for the stimulus-sensitive drug delivery. Besides, more and more emerging engineering and biological technologies make MMP-sensitive systems more reliable and efficient for various biomedical applications.

3.2 MMPs and their Physiological Functions

MMPs also called matrixins, are a family of extracellular matrix (ECM)-degrading endopeptidases that share common features in their structures, functions, and activation mechanisms. So far, more than 23 human MMPs have been discovered. A typical MMP consists of several common functional domains including a pro-peptide domain of about 80 amino acids, a highly conserved zinc binding domain of about 170 amino acids, a linker peptide of variable lengths (also called the "hinge region"), and a hemopexin (Hpx) domain of about 200 amino acids [17]. MMPs are historically named as collagenases, gelatinases, stromelysins, or matrilysins, based on their specificity for ECM components, or named currently by a sequential numbering system. Most MMPs are secreted proteins or membrane-type MMPs (MT-MMPs) [18]. To regulate cell behaviors, these MMPs usually reside at or near the cell membrane. MT-MMPs are linked to the cell membrane via covalent bonds, while secreted MMPs can also be attached to the cell surface by physical binding to membrane receptors/proteins, such as integrin $\alpha v\beta 3$, CD44, or other cell-surface associated biological molecules [18]. However, recently, some MMPs, such as MMP-1, –2, and –11 have also been found in subcellular compartments and may function as a subcellular proteases [17].

The main functions of MMPs are the degradation and removal of ECM molecules from the tissue. However, a number of non-matrix molecules are also their substrates. MMPs may affect cell behaviors, including migration, proliferation, apoptosis, and morphogenesis, modulate the activity of biological molecules, and alter the balance of protease activity [19]. Therefore, MMPs play critical roles in many physiological processes, such as blastocyst implantation, embryonic development, organ morphogenesis, nerve growth, ovulation, wound healing, angiogenesis, tissue repair and remodeling, and pathological processes such as arthritis, cancer, cardiovascular disease, nephritis, neurological disease, ulcers, liver and lung fibrosis, and emphysema.

Under normal physiological conditions, these multi-domain proteins with very low or negligible activity are precisely regulated via the activation of the precursor zymogens and the inhibition by endogenous inhibitors, the tissue inhibitors of metalloproteinases (TIMPs), and also transcriptionally controlled by inflammatory cytokines, growth factors, hormones, and cell–cell and cell–matrix interactions. But under certain disease conditions, such as cancer and inflammation, the expression and activity of MMPs are dysregulated.

3.3 Up-Regulation of MMPs in Pathological Conditions

In the tumor, to degrade basement membranes and invade surrounding connective tissues, tumor cells overexpress MMPs and other proteases and induce expression of these enzymes in neighboring stromal cells as well. The presence or up-regulation of many MMPs in both primary tumors and/or metastases is highly associated with tumor growth and angiogenesis. The elevated proteolytic activity is also required for a cancer cell to perform intravasation and extravasation during tumor cell migration [20].

After MMP-induced cleavage, the ECM degradation products exhibit biological properties and usually serve as cellular and/or extracellular signals in biological/pathological events, such as cell adhesion, migration, and epithelial to mesenchymal transition, a hallmark of cancer progression [20, 21]. The degradation of non-ECM proteins, such as growth factor precursors, cell surface adhesion molecules, angiogenic inhibitor precursors, and other surface receptors, also benefit tumor growth and metastasis [20]. In addition, MMPs are involved in the cytokine and chemokine processing, tumor-associated angiogenesis, and so on [20].

Because of many human cancers associating with abnormal levels of MMPs, MMPs have received great attention in recent years as putative cancer biomarkers for the purposes of diagnostics and prognostics. In breast cancer patients, MMP-2 and -9 levels have been found higher in the serum [22], while, in the tumor tissue, the elevated MMP-9 level is significant [23]. In pancreatic cancer, the activation ratio of MMP-2 [active form versus total MMP-2] in pancreatic juice is about ten times higher than that in chronic pancreatitis and about 20 times higher than that in the healthy controls [24]. Compared to the healthy population and patients with chronic pancreatitis, the patients with pancreatic ductal adenocarcinoma have higher levels of MMP-7 [25] and MMP-9 [26] in their serum and tissue. The elevated levels of MMP-9 were found in the serum, but not in the bronchial lavage fluid of the patients with advanced small-cell lung cancer or non-small-cell lung cancer [27]. In non-small-cell lung cancer, MMP-2 [28] and MMP-7 [29] have also been found up-regulated in the tumor tissue. In prostate cancer, both the activities and expression of plasma and tissue MMP-2 and -9 are increased significantly during tumor progression and metastasis [30]. Other MMPs, like MMP-10 and -26, are also overexpressed in prostate cancer [31]. Both up-regulated MMP-2 and -9 have been found in the tumor tissue, cerebrospinal fluid, and urine of the patients with the malignant brain cancer [32]. It has also been reported that MMP-2 overexpression was highly associated with shorter survival rates in the patients with the malignant brain tumor [33]. Other types of cancer like ovarian cancer, bladder cancer, colorectal cancer, melanoma, etc. have also showed higher levels of MMPs in tumor tissues and/or serum [18, 20].

Given the important roles of MMPs in tumor proliferation, angiogenesis, and metastasis, MMPs have been considered as important therapeutic targets for cancer therapy. The design of matrix metalloproteinase inhibitors (MMPIs) is the major strategy and several promising MMPIs have even entered into clinical applications against a variety of human cancers [34]. However, the adverse effects, missing of the best time of treatments, and low specificity negatively impacted the therapeutic efficacy of these inhibitors [20].

The abnormal degradation of ECM proteins in cartilage, tendon, and bone mediated by MMPs is the cause of arthritis [35]. MMP-1, -3, and -13 are predominately up-regulated in the synovial fluid in patients with rheumatoid arthritis and osteoarthritis [36]. Unlike cancer, however, TIMP levels are not increased and even decreased in some circumstances, resulting in a high MMP/TIMP ratio, which ensures the strong activity of MMPs [37].

MMPs also play a key role in central nervous system diseases due to their functions in the disruption of the blood brain barrier (a physical tight barrier), destruction and remodeling of ECM proteins, and tissue inflammation during brain ischemia [38]. Among the MMPs, MMP-8 and -9 are particularly important in the central nervous tissue injury caused by bacterial infections [37].

In addition to as biomarkers and therapeutic targets, the overexpression of MMPs has also provided an opportunity for the delivery of drug molecules and imaging agents to pathological sites via an enzyme-triggered mechanism. Recently, many efforts have been made regarding the development of nanosystems sensitive to various MMPs. The MMP sensitivity allows these delivery systems to "smartly" control drug release, cellular uptake, and the intracellular fate of drugs. The common mechanisms of MMP-sensitive strategies are summarized in Fig. 3.1. In the following text, these emerging MMP-sensitive nanocarriers will be discussed in detail in terms of their delivery strategies, structures and formulations, physical and chemical properties, and biomedical applications.

3.4 Design of MMP-Sensitive Nanosized Drug Delivery and Imaging Systems

3.4.1 *Substrates for MMPs*

MMPs have a wide range of natural substrates. These natural substrates, in most cases, are proteins or protein derivatives, including ECM proteins

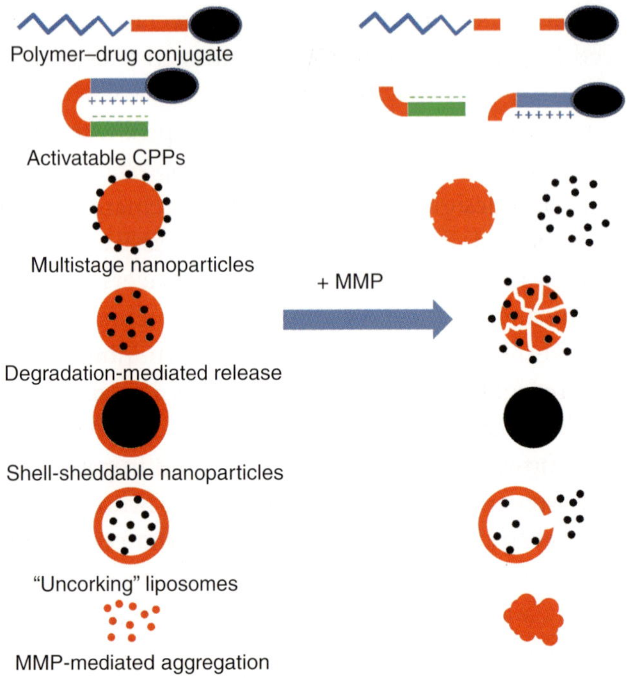

Figure 3.1. The mechanisms of MMP-sensitive strategies used in nanocarriers. Red: MMP-responsive peptide/protein/polymer; Black: Payloads; Blue: CPPs; Green: Polyanion.

(e.g. collagen, gelatin, fibronectin, aggrecan, etc.) and other proteins (e.g. fibrinogen, fibrin, casein, plasminogen, etc.). The cleavage spectra and specificity of individual MMPs vary. For example, most collagens, except collagen-6 and -9, can be efficiently cleaved by MMP-2. Fibronectin can be cleaved by most MMPs, except MMP-8 and -9. A summary table of the natural substrates of MMPs can be found in [39]. The cleavability of MMP substrates is significantly dependent on their amino acid sequences, especially the region around the site of cleavage [40]. It has been suggested that simple collagen-like peptide sequences (PLG~L) could be the substrates of most MMPs [41]. However, it was still not accurate and sufficient for designing an effective synthetic MMP substrate. To further clarify the favorable amino acid sequences labile to various MMPs, an excellent work has been done by Turk *et al.* [40]. In their study, six enzymes from the MMP family, including MMP-1, -2, -3, -7, -9, and MT1-MMP, were thoroughly analyzed for their cleavage (site) specificity by the oriented peptide library mixtures, which were previously used for the determination of the target sequence preferences of protein kinases and protein interaction domains. Eight positions surrounding the cleavage site of the substrates were evaluated for these MMPs by these libraries. Based on k_{cat}/K_m values (k_{cat}, the catalytic constant; K_m, the Michaelis constant), the consensus sequences for each MMP have been identified [40]. However, the advanced structures of the substrates and the properties of physically or covalently attached molecules (e.g. oligo- or poly- saccharides) may cause steric hindrance and block enzyme access, thus also playing an important role in the MMP-induced enzymatic reaction.

Recently, the artificial/synthetic MMP substrates including proteins, peptides and polymers have been designed and used for drug and imaging reagent delivery [14, 42, 43]. Although the consensus sequences for MMPs have been discovered, the sequences reported in MMP-sensitive drug delivery systems are usually different. These sequences are summarized in Table 3.1. These peptide sequences are usually short (e.g., 5–10 amino acids) and linear, probably due to the concerns of the cost and difficulty of peptide synthesis as well as the future engineering processes. They can be easily

Table 3.1. The MMP-sensitive substrates used in the nanoparticles.

Sequence	Specificity	Type of nanocarriers	References
PLGLAG	MMP-2, -9, MT1-MMP	Activatable cell penetrating peptides (ACPPs)	[14, 49, 50]
		ACPPs-modified dendrimers	[51–54]
		ACPPs-doxorubicin conjugate	[55]
		PEG–PCL nanoparticles	[56]
		PEG–PLA nanoparticles	[57]
		Nanostructured lipid carriers	[51]
		Ferritin nanoparticles	[58]
		NIR fluorescence probe	[59]
PVGLIG	MMP-2, -9	Methotrexate-dextran conjugates	[60–62]
		Doxorubicin-PEG conjugate	[63]
		PEG-PCL nanoparticles	[64]
		Methotrexate-dendrimer conjugate	[65]
GPLGIAGQ	MMP-2, -9	Doxorubicin-albumin conjugates	[43, 66]
		Liposomes	[46, 67, 68]
		Paclitaxel-PEG polymeric micelles	[28]
		Paclitaxel-peptide conjugates	[69]
		PEG–pp–PEI–PE polymeric micelles	[70]

(Continued)

Table 3.1. (*Continued*)

Sequence	Specificity	Type of nanocarriers	References
GPLGVRG	MMP-2, -9, -13	Doxorubicin-PEG micelles	[44]
		Silica nanoparticles	[71]
		Fluorescence sensors	[48]
		Au-Fe_3O_4 nanoparticles	[47]
		PLGA nanoparticles	[72]
		Gold nanorods	[73]
		PEGylated quantum dots	[74]
PLGLWA	MMP-1, -2, -9	Viral vectors	[75]
		Activatable cytokine	[76]
		Plasmid encoding an activatable cytokine	[77]
PLGL	MMP-2, -9, -14	Doxorubicin-peptide conjugate	[78]
		Drug-peptide conjugates	[79]
GPQGIAGQR	MMP-9	Liposomes	[80, 81]
PLGVR	MMP-2, -9, -13	Ferritin nanoparticles	[82]
GPLGV	MMP-2, -9, -13	Doxorubicin-PEG micelles	[44]
RIGFLR	MT1-MMP	Imaging probe	[83]
GRIGFLR	MT1-MMP	Cyclized membrane lytic peptide	[84]
GPQGIAG	MMP-1	MMP-cleavable composite peptide	[85]
DGPLGVC	MMP-2, -9	Nanofibrous matrix	[86]
HPVGLLAR	MMP-2	Albumin nanoparticles	[87]
GPVGLIGK	MMP-2	Polymeric gel	[88]
RPLALWRS	MMP-7	PEGylated quantum dots	[89]
GPQGIFGQ	MMP-2, -9	Protein polymers	[90]
VPLSLYSGCG	MMP-7	Polymeric nanoparticles	[91]
GPLGMRGLGK	MMP-2, -13	NIR fluorescence probe	[92]
VLVPMAMMAS	MMP-1	Pro-antibody	[45]
GGGVPLSLYSGGGG	MMP-2	Multifunctional Envelope-type Nano Device	[93]
Gelatin	MMP-2, -9	Gelatin/quantum dot nanoparticles	[94, 95]
		Supramolecular nanogel	[96]
Collagen (type-I and -IV)	MMP-2, -9	Dendrimer/collagen hybrid gel	[97]
Ascorbyl palmitate (Asc-Pal)	MMP-2, -9	Nanofibrous gel	[98]

conjugated or incorporated with various molecules or nanoparticles, including small drugs [28, 44], peptides [14] or proteins [45], hydrophilic polymers [46], or inorganic nanoparticles [47, 48]. Interestingly, the cyclic peptidic moieties embodying MMP cleavable sites have also been synthesized and showed the specificity to certain MMPs, such as MMP-2 and -9. Moreover, these cyclic peptides presented a very good plasma stability, which might benefit *in vivo* drug delivery [42].

3.4.2 *MMP-sensitive drug conjugates*

To impart the MMP sensitivity to the drug or imaging system, the straightforward method is covalent conjugation or physical incorporation of the MMP-sensitive peptide to the drug molecule or imaging agent. To do so, both the target molecule and MMP-sensitive peptide must contain reactive or functional groups. The free terminal amine and carboxyl groups of the peptide are ideal and widely used for the peptide/protein conjugation reaction. To accelerate the reaction and increase the yield, the coupling reagents, *N*,*N*′-dicyclohexylcarbodiimide (DCC) or 1-ethyl-3-(3-dimethylaminopropyl) carbodiimide (EDC) and *N*-hydroxysuccinimide (NHS) [46], and *N*,*N*′-diisopropylcarbodiimide (DIC) or DCC and 4-dimethylaminopyridine (DMAP) [28, 69], are commonly used. However, to minimize the non-specific reaction such as conjugation onto undesired groups, these functional groups should be capped first before running the conjugation and liberated thereafter [69]. Alternatively, the functionalized drug or polymer derivatives containing the maleimide [55, 66] or NHS [46] are synthesized before their conjugation with the peptides/proteins containing cysteine or amine. In addition to chemical methods, for MMP-sensitive peptide or protein conjugates, the biological methods (e.g. peptide/protein biosynthesis) can be another effective way [45, 76].

The direct conjugation of an MMP-sensitive peptide with the poorly water-soluble chemotherapeutic, paclitaxel, increased its solubility and cytotoxicity, as evidenced by the significantly decreased half maximal inhibitory concentrations (IC_{50}) and arrest of G_2/M-phase of the cell cycle in MMP-2 overexpressing cancer cells [69]. In another study, a 4-amino acid peptide was conjugated with doxorubicin to impart the MMP-9 sensitivity to the prodrug [79]. To further increase the MMP selectivity, the doxorubicin–hexapeptide conjugates were capped at the positively charged amino (NH_2) terminus to prevent the non-specific aminopeptidase degradation, increase their solubility, and decrease their acute toxicity. The MMP selective conjugates were preferentially metabolized in HT1080 xenografts, leading to a ten-fold increase in the tumor/heart ratio of doxorubicin [78].

However, insufficient cellular uptake of drug–peptide conjugates [78], an undesirable pharmacokinetic profile, and low *in vivo* tumor growth inhibition [69] implicated that further improvements might be needed. Biologically compatible molecules, including proteins and hydrophilic polymers, have been widely used as the drug carrier to improve the *in vivo* pharmacokinetic characteristics of drug molecules. Albumin has been found to significantly increase the solubility and stability of bound drugs. In addition, these macromolecules typically have hydrodynamic diameters of 2–10 nm, allowing extravasation into tumor tissue, but not into normal tissue. Therefore, they are ideal drug carriers for the development of tumor-targeted macromolecular prodrugs [99]. The octapeptide (GPLGIAGQ) has been successfully used as a labile linker in a human serum albumin (HSA, ~66 kDa)-doxorubicin prodrug, which was sensitive to MMP-2 and MMP-9 [43]. To covalently link doxorubicin to albumin, Mansour *et al.* synthesized a water-soluble maleimide derivative of doxorubicin containing the same MMP-2 specific peptide sequence. This doxorubicin–peptide conjugate could bind rapidly and selectively to the cysteine-34 position of circulating albumin. After forming the albumin–doxorubicin conjugate, its serum stability was significantly increased, as evidenced by a half-life of about 16 h. This conjugate was completely and specifically cleaved by MMP-2 within 1.5 h, and liberated a doxorubicin tetrapeptide and subsequently free doxorubicin. The long circulation and MMP-2 sensitivity resulted in the improved antitumor activity compared to the parent compound doxorubicin in an A375 human melanoma xenograft [66]. To achieve intracellular delivery, a cell penetrating peptide (CPP) was linked to a short stretch of acidic peptide residues via an MMP-sensitive peptide linker (PLGLAG) to form the ACPPs (Fig. 3.2). In this design, the CPP was masked by a polyanion via electrostatic binding before activation, while it would be exposed after the MMP-induced cleavage resulting in the enhanced cell penetration [14]. The ACPPs have shown significant tumor targeting when incorporated in various drug delivery and imaging systems [50, 53–55].

The use of hydrophilic polymers as drug carriers is another efficient way. In a systematic study, Chau *et al.* synthesized a dextran–peptide (PVGLIG)–methotrexate conjugate for tumor-targeted delivery of methotrexate via the mediation of MMP-2 (k_{cat}/K_m, $1.21 \times 10^5\ M^{-1}S^{-1}$) and MMP-9 ($k_{cat}/K_m$, $3.60 \times 10^3\ M^{-1}S^{-1}$) [62]. Both of the MMP-sensitive and non-sensitive conjugates had prolonged blood circulation and higher tissue accumulation compared to free methotrexate, indicating the passive drug targeting of these conjugates via the effect of EPR [60]. The MMP-sensitive conjugates demonstrated the effective tumor growth inhibition by 83% in two separate tumor models with MMP overexpression (HT-1080 and U-87). Although non-sensitive conjugates also

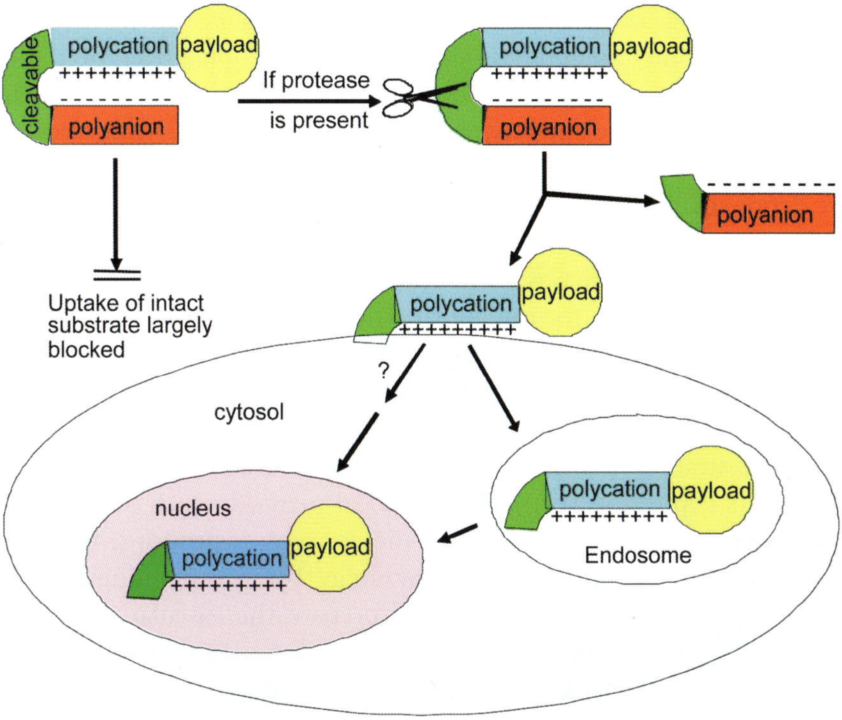

Figure 3.2 Schematic diagram of ACPPs. Cellular uptake induced by a cationic peptide is blocked by a short stretch of acidic residues attached by a cleavable linker. Once the linker is cleaved, the acidic inhibitory domain drifts away, and the cationic CPP is free to carry its cargo into cells. Reproduced with permission of Ref. [14].

showed the tumor targeting via the EPR effect, they were found to be accumulated in the small intestine and bone marrow, causing toxicity. In contrast, free methotrexate didn't show tumor growth inhibition in these models. Furthermore, no tumor inhibition effect was observed in the mice bearing the RT-112 tumor, which had no MMP overexpression [61]. A methoxy–polyethylene–glycol-derivatized poly-L-lysine (PEG-PLL) based near-infrared fluorescence imaging probe containing an MMP-2 cleavable peptide (PLGVR) was prepared to sense the MMP activity in intact tumors. Due to its excellent MMP targeting, it was also used to evaluate the *in vivo* effect of MMP inhibitors in nude mice [100].

PEGylation is a well-known technique that can enhance drug solubility and stability, prolong drug blood circulation time, and prevent drug's non-specific interaction with biological systems, via the so-called "steric stabilization" [101]. To improve drug solubility, the PEG2000 was conjugated to paclitaxel via the MMP-2 cleavable peptide linker (GPLGIAGQ) [28].

As expected, the resultant polymer–drug conjugate had a significantly increased (apparent) solubility. Despite the protection effect of the PEG chain, the conjugate could be efficiently cleaved by human MMP-2 and liberated the toxic paclitaxel moiety, resulting in the MMP-2-triggered tumor cell-specific cytotoxicity. Most importantly, the conjugate could self-assemble into a "core-shell" nano-structure, in which paclitaxel was entrapped in the hydrophobic core and isolated by the hydrophilic PEG chain. Therefore, the PEG–paclitaxel conjugate was not only an MMP-sensitive prodrug, but could also be a building block for the construction of the MMP-2 sensitive paclitaxel-containing nanomedicine [28]. By the similar method, an MMP-sensitive PEG5000–hexapeptide–doxorubicin conjugate was prepared for targeted drug delivery against malignant glioma cells [63].

Another strategy is to incorporate MMP-cleavable sequences into the protein or its expression vector by biological methods, such as protein biosynthesis or vector construction. Like the concept of prodrug, Erster *et al.* proposed a so-called "pro-antibody" strategy by incorporating an MMP-sensitive sequence into an antibody. To prepare the vascular cell adhesion molecule 1 (VCAM-1) pro-antibody, an expression plasmid encoding a scFv, a MMP-1 susceptible peptide, and a binding site-masking peptide, was constructed. The gene product (pro-antibody) contained a VCAM-1 antibody masked by a masking peptide via an MMP-cleavable linker. Activation of this pro-antibody by MMP-1 yielded a 200-fold increase in its *in vitro* binding affinity and restored anti-VCAM-1 binding in tissue sections in an *ex vivo* experiment [45]. In another study, for targeted delivery of therapeutics to the rheumatoid tissue, novel complementary deoxyribonucleic acid (cDNA) sequences encoding human interleukin (IL)-4 linked to one or more synovial endothelium targeting fusion peptides (SyETPs) via an MMP-cleavable sequence were constructed. The generated fusion proteins were able to specifically deliver IL-4 to human synovial tissues and showed the enhanced bioactivity in these tissues, which were transplanted into severe combined immunodeficiency (SCID) mice [76]. Instead of encoding peptide sequences in pDNAs, interestingly, the peptide could be directly incorporated into the viral vectors to impart the MMP sensitivity to them for the MMP-sensitive gene delivery [75].

3.4.3 *MMP-sensitive liposomal nanocarriers*

Given their excellent biocompatibility and biodegradability, low toxicity and immunogenicity, and versatile drug loading capacity, liposomes have been one of the most widely used pharmaceutical carriers in the past few decades.

The successes of liposomes in biomedical applications include several commercial liposomal drugs, such as liposomal doxorubicin (Myocet®) and its PEGylated counterpart (DOXIL®), liposomal amphotericin B (AmBisome®), liposomal morphine (DepoDur®), etc., as well as many biomedical products and liposome-based technologies [5].

Although the "plain" (unmodified) liposomes can load and protect drugs from degradation and control drug release, they are readily taken up by macrophages of the reticuloendothelial system (RES) rather than by their target cells, therefore losing their efficacies [9]. To improve their performance, liposomes are usually engineered to show different surface properties and functions. To prolong their blood circulation, the most commonly used strategy is to incorporate the PEGylated lipid derivatives into liposomes, such as various lengths of dioleoyl-*N*-(monomethoxy polyethylene glycol succinyl)-phosphatidylethanolamine (PEG–PE) and PEG–cholesterol derivatives. However, recent studies have indicated that the stable PEGylation of nanoparticles may not always enhance drug delivery. The PEG "corona" was reported to interfere with the cell internalization process resulting in low cellular uptake of PEGylated liposomes [46, 102]. Additionally, flexible hydrophilic PEG chains have been found to interfere with intracellular events, such as endosomal escape [103], resulting in lowered drug efficacy. Ideally, the protective PEG should be selectively removed at the target site before cell entry and subsequent intracellular processes.

A variety of stimulus-sensitive linkers have been incorporated between the PEG and nanocarriers to respond to local stimuli, including MMPs, and to deshield PEG chains [16]. Terada *et al.* synthesized an MMP-2 cleavable lipopolymer (PEG2000-peptide-DOPE) and incorporated it into the galactosylated liposomes for the purpose of the MMP-2 sensitive drug delivery. They found that 0.5% of this lipopolymer completely masked the galactose-induced cellular uptake of the liposomes, while the MMP-2 pretreatment led to a significantly high cellular uptake in human liver carcinoma (HepG2) cells. In contrast, the cellular uptake of these liposomes in peritoneal macrophages was significantly inhibited. The PEGylation and MMP-2 sensitivity enabled the targeted delivery of N^4-octadecyl-1-β-D-arabinofuranosylcytosine to liver cancer cells by these liposomal nanocarriers [67]. More recently, a functional lipopolymer, maleimide-PEG2000-peptide-DOPE, was synthesized for the further modification of liposomes. Using the novel polymer, a multifunctional liposomal nanocarrier containing the MMP-2 responsive moiety, anti-nucleosome antibody (mAb 2C5), and cell penetrating peptide (TATp) was developed. In the normal physiological conditions, the antibodies were exposed on the liposome surface via the conjugation to the distal end of the long chain PEG3400, while the TAT peptides on the distal end of the short PEG2000 were completely blocked. However, after the MMP-mediated cleavage, the PEG3400 was removed and TATp was exposed for the enhanced cell internalization. This design ensured that all the functions, including long blood circulation time, the EPR effect-induced passive targeting, antibody-mediated active tumor targeting, MMP-2 sensitivity, and CPP-mediated cell internalization, were used in the same nanocarrier and these functions worked in a collaborative manner [46]. The MMP-sensitive liposomes were also used for the stimulus-responsive gene delivery. For tumor-targeted delivery of the E1/E3-deleted recombinant adenovirus vectors coding a luciferase reporter gene, the adenoviruses were encapsulated into the anionic liposomes via the calcium-induced phase change method, followed by post-insertion of MMP-2 sensitive PEG-peptide-cholesterol conjugates into the surface of the liposomes. The resultant MMP-2 sensitive liposomal viruses exhibited higher *in vitro* gene expression than the non-cleavable liposomal viruses in MMP-2 overexpressing cancer cells. Unlike cationic liposomes, the anionic liposomal viruses were very safe even at a high concentration of lipids. Due to the protection of lipid bilayers, the virus-induced innate immune responses were markedly reduced. Furthermore, no significant hepatotoxicity was observed after intravenous injection of the liposomal viruses in C57BL/6N mice [68].

The MMP sensitivity strategy has also been used to control the release of liposomal contents. To switch on/off the drug release from liposomes in response to MMP-9, Sarkar *et al.* synthesized a novel lipopeptide conjugate by linking an MMP-cleavable triple helical peptide to a stearic acid. After insertion into the lipid bilayer of liposomes, the lipopeptide could be specifically cleaved by MMP-9, but not by the trypsin. This was probably because the MMP-9 preferred to cleave triple helical peptides, while the trypsin failed to unwind the triple helical structure to affect the cleavage [80]. After the cleavage, the liposomes underwent destabilization, leading to "uncorking" and release of the encapsulated cargoes [104].

3.4.4 *MMP-sensitive polymeric micelles*

About 40% of newly developed drugs experience low solubility and bioavailability due to their high hydrophobicity. Among many attempts, "core-shell" micellar nanocarriers have shown successes, not only in drug solubilization, but also in drug delivery [6].

Amphiphilic molecules, such as phospholipids, have the potential to (self-) assemble into "core-shell" structures or other structures, such as bilayers, in an aqueous environment. This property has been used to design many different drug delivery systems including micelles, liposomes, emulsions, and other nanoparticles [105, 106]. Since many drugs have high hydrophobicity, the conjugation of a hydrophilic moiety (e.g. PEG) to a hydrophobic drug molecule (e.g. doxorubicin and paclitaxel) should improve the drug solubility, as well as impart amphiphilicity to the formed conjugates, so that the resultant amphiphilic molecules assemble into a "core-shell" structure. Doxorubicin has been conjugated with PEG2000 via an MMP-2/-9 cleavable peptide linker. Due to the amphiphilicity, the doxorubicin conjugates could form a micellar structure, which was ready for loading of extra free doxorubicin using an o/w emulsion method. The nanoparticles with the size of 100–350 nm and total doxorubicin loading of more than 60 wt% showed MMP-2 sensitive cytotoxicity in the Lewis lung carcinoma (LLC) cells. In a LLC xenograft model, the doxorubicin conjugate micelles significantly increased the blood circulation half life of doxorubicin for more than ten folds. After loaded with free doxorubicin, the nanoparticles showed a similar *in vivo* tumor growth inhibition effect compared to free doxorubicin due to the improved PK characteristics, although their *in vitro* activity was much less than that of free doxorubicin [44]. In a recent study, Zhu *et al.* synthesized a novel MMP-2 sensitive PEG–peptide–paclitaxel conjugate. In addition to as a prodrug, the conjugate was successfully inserted into the PEG–lipid micelles as a functional building block. The 20 nm MMP-2 sensitive mixed micelles containing a paclitaxel conjugate and TAT peptide were able to efficiently penetrate the cell membrane in cell monolayers, as well as in three-dimensional (3D) tumor cell spheroids, resulting in a more than two-fold increase in their cytotoxicity in the spheroids compared to other paclitaxel formulations. The MMP-2 sensitivity as well as negligible drug leakage and serum protein adsorption ensured the successful *in vivo* tumor growth inhibition and reduced side effects compared to free or conventional micellar paclitaxel, as evidenced by excellent tumor-specific drug accumulation and cell internalization, high tumor size reduction, significant tumor cell apoptosis/death, and low side toxicity (Fig. 3.3) [28].

The synthesis of MMP-sensitive polymers as the nanocarrier building block is another effective way to impart MMP sensitivity to micelles. An MMP-9 cleavable block copolymer, poly(ethylene glycol)-bl-poly(propylene sulfide) (PEG–PPS), was synthesized and used to prepare polymeric micelles for the loading of iron oxide. The iron oxide micelles were able to shed 10–20 nm of the hydrodynamic PEG "shell" in response to MMP-9, leading to up to a six-fold decrease in nanoparticle internalization by macrophages [107].

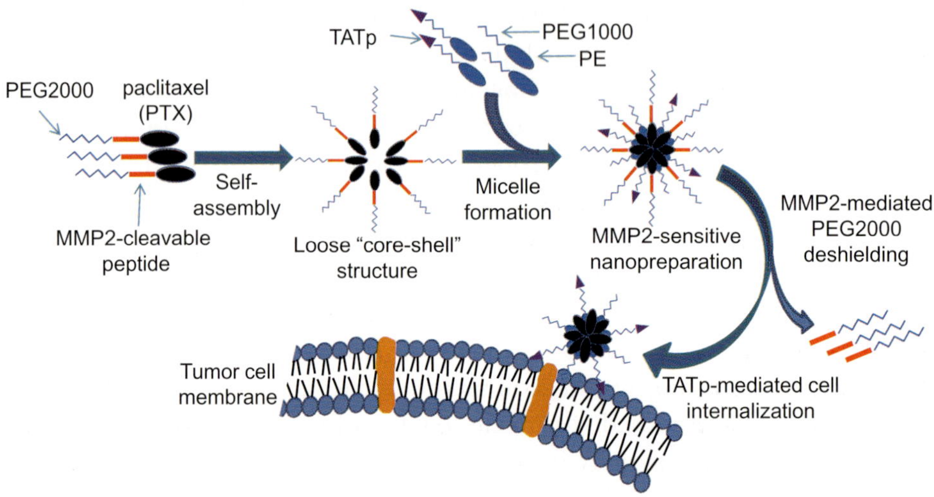

Figure 3.3. Tumor-targeted delivery of paclitaxel by MMP-2 sensitive polymeric micelles. Reproduced with permission of Ref. [28].

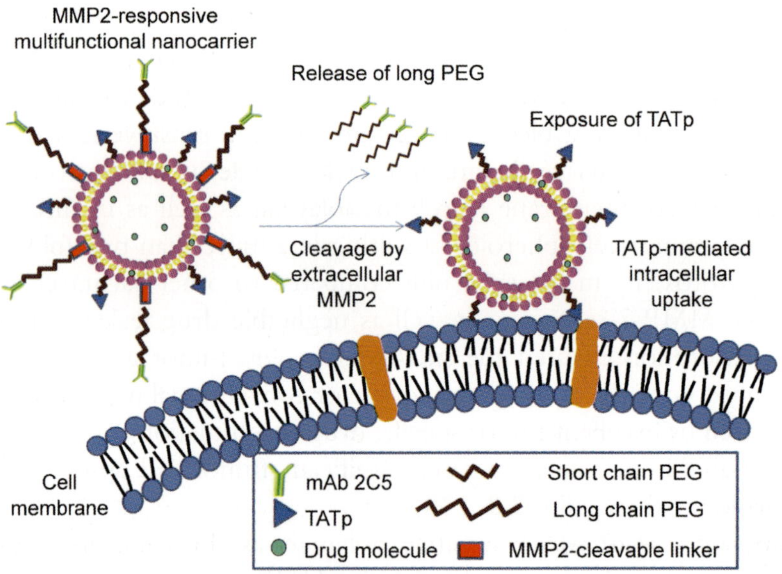

Figure 3.4. MMP-2 sensitive multifunctional liposomal nanocarrier. Reproduced with permission of Ref. [46].

To deliver pDNA, the MMP-responsive polyplex micelles were prepared from the peptide-linked block copolymer, PEG-PLGVRG-PAsp(DET) [poly (aspartamide) with a flanking N-(2-aminoethyl)-2-aminoethyl group], which was synthesized by click reaction and ring-opening polymerization. The outer PEG layer was deshielded after MMP-2-induced cleavage to expose positively charged polyplex for the enhanced transfection and endosomal escape [108].

In another interesting work, Chien *et al.* synthesized the self-assembling peptide-polymer amphiphiles (PPAs) which could self-assemble into a micellar structure. To monitor the MMP-cleavage-mediated micelle/particle aggregation, the otherwise similar block copolymer amphiphiles were labeled with either fluorescein or rhodamine. When mixed together, signals from the fluorescence resonance energy transfer (FRET)-pairs were utilized to detect particle aggregation and, hence, enzymatic activity. This technique can be used to estimate particle stability (e.g. critical aggregation concentration) and intraparticle distances between amphiphiles in the micellar aggregates, especially when light scattering and electron microscopy are interfered by exogenous particulates and detritus [109]. Most importantly, in response to MMP-induced cleavage, the nanosized polymeric micelles underwent morphological change and aggregated to microparticles [110]. Using the same idea, the 20 nm micelles formed by only the Alexa Fluor 647-labeled PPAs were prepared. Upon intratumoral injection into HT-1080 xenograft tumors, the morphology of the micelles was changed from the 20 nm spherical micelles to micrometer-scale aggregates due to MMP-induced cleavage, resulting in the tumor accumulation of the micelles [111].

3.4.5 *MMP-sensitive protein-based nanoparticles*

Unlike the self-assembly of amphiphilic peptide/polymer–drug conjugates into the nanosized structure, proteins themselves can be a particulate system with the size falling into the nano scale, due to their large molecular weight. This property could be used to improve the *in vivo* behavior of the protein-based drug delivery systems, e.g. the albumin-bound paclitaxel, Abraxane®. In addition to the aforementioned MMP-sensitive albumin–drug conjugates [43, 66, 99], albumin has also been modified with MMP-sensitive moieties and is used for loading molecules via physical forces. The bovine serum albumin (BSA) nanoparticles have been coated with positively charged lysine-containing MMP-2 sensitive peptides for the immobilization of peptides or proteins (payloads) on BSA nanoparticles via lysine residue-mediated electrostatic interactions. The release of a model drug (BSA) could be tuned from 5% to 90% over 14 days, depending on the conditions of the peptide coating, especially the specificity and sensitivity of the MMP-2 sensitive peptide sequences [112].

Ferritin, a ubiquitous intracellular protein (~450 kDa) consisting of 24 subunits, has a spherical cavity with an external diameter of 12 nm and an internal diameter of 8 nm. A variety of inorganic materials, including iron oxide or small organic molecules, can be loaded into this protein "cage", making ferritin an attractive nanocarrier for biomedical applications [113]. Matsumura *et al.* designed a ferritin-based nanoparticle as a tumor environment-responsive nanocarrier for both drug delivery and magnetic resonance imaging (MRI). Doxorubicin-loaded ferritin nanoparticles were surface-modified with the MMP-2 cleavable PEG-peptide, ensuring the sharp particle size distribution and MMP-2 responsiveness. The MMP-2-induced cleavage and PEG deshielding initiated and facilitated the agglomeration, resulting in the tumor-targeted drug accumulation and enhancement of T2 relaxivity of the contrast agent (iron, Fe^{3+}) [58]. Besides, Lin *et al.* prepared an MMP-sensitive ferritin-based hybrid nanocarrier by a novel disassembly/reassembly method for near-infrared fluorescence (NIRF) imaging of tumors. The Cy5.5-tagged MMP-cleavable peptide and black hole quencher (BHQ-3) were incorporated into ferritin separately. The subunits (conjugates) of ferritin were disassembled under the acidic condition (pH 2.0), followed by reassembly at normal physiological condition (pH 7.4), resulting in the hybrid ferritin nanoparticle containing both Cy5.5-tagged MMP-cleavable peptide and BHQ-3. The conjugated Cy5.5 would be liberated upon MMP-induced cleavage and showed fluorescence. The *in vivo* data indicated that, using this method, the noise background was lowered and the specificity of the dye (tumor targeting) was significantly increased in a head and neck squamous cell carcinoma (UM-SCC-22B) xenograft tumor model [82].

3.4.6 *Dendrimer-based nanoparticles*

Dendrimers are highly branched, globular synthetic macromolecules with an accurate molecular weight, a uniform size, and well-defined peripheral groups, making them promising for the delivery of various types of molecules [114].

To image the MMP-7 expression in tumor cells, a fluorogenic proteolytic substrate (PB-M7VIS) was built on a generation-4 (G4) Starburst® PAMAM dendrimer (14.2 kDa) covalently coupled with a fluorescein-labeled peptide (Fl) sensitive to MMP-7 and with tetramethylrhodamine (as a fluorescence reference). PB-M7VIS was efficiently and selectively cleaved by MMP-7 as evidenced by a k_{cat}/K_m value of 56- and 13-fold higher than MMP-2 and -3, respectively. In the mouse xenograft models derived from SW480 human colon cancer cells, intravenous administration of PB-M7VIS significantly increased Fl fluorescence in the MMP-7-positive tumor, but not in the control tumor. In contrast, pre-administration of Batimastat (BB-94), an MMP inhibitor, dramatically decreased the Fl fluorescence in the MMP-7-positive tumor by about 60% [115]. Using the same idea, the same group prepared a MMP-7 specific near-infrared polymer-based proteolytic nanobeacon, PB-M7NIR. To increase the *in vivo* performance, a G4 PAMAM dendrimer was first PEGylated and then coupled with a Cy5.5-labeled MMP-7 cleavable peptide and Alexa Fluor® 750 (AF750) (as an internal reference). In the *in vivo* study, the histological examination displayed an approximately 300-fold enhancement in the fluorescence in MMP-7 overexpressing tumors compared to non-expressing tumors after intravenous injection of PB-M7NIR. Furthermore, injection of PB-M7NIR was able to image adenomas as small as 0.01cm^2 in the multiple intestinal neoplasia (APC^{Min}) mice, as evidenced by a six-fold increase in the ratio of signal-to-reference in the adenomas of APC^{Min} mice, compared to the control intestinal tissue or adenomas from MMP-7-null Min mice. The signal was found to be independent of the tumor size [116].

An MMP-sensitive collagen peptide-modified G4 PAMAM dendrimer that attached doxorubicin via a pH-degradable linkage was synthesized as a polymer prodrug for metastasis-associated drug delivery. Conjugation of doxorubicin to the dendrimer significantly decreased its diffusion rate compared to free doxorubicin, ensuring its tumor targeting. With the collagen coat, the prodrug was more sensitive to the highly invasive breast cancer cells (MDA-MB-231) with a high MMP expression compared to the poorly invasive breast cancer cells (MCF-7). In the MDA-MB-231 tumor-bearing mice, subcutaneous injection of doxorubicin-conjugated dendrimer/collagen hybrid gels into the tumor area not only suppressed the tumor growth, but also attenuated the metastatic activity [97]. In another study, a G5 PEGylated polylysine dendrimer was conjugated with methotrexate (MTX) via an MMP-2 and -9 cleavable linker for tumor targeting. It was found that the capping of the α-carboxyl group on MTX–dendrimer conjugates prevented rapid liver accumulation of the dendrimer and prolonged the blood circulation time without attenuating *in vitro* or *in vivo* antitumor activities [117].

3.4.7 *MMP-sensitive nanogels*

Hydrogels have been successfully used as 3D scaffolds and carriers for organizing cells and for the delivery of various biomolecules to mimic the native ECM in tissue engineering [118]. Hydrogels have also been used to prepare the MMP-sensitive drug delivery systems. In addition to coating nanoparticles with collagen or gelatin [97, 119], some amphiphilic polymers sensitive to MMPs have been used to prepare nanogels for loading and on-demand release of drug and gene molecules.

A nanofibrous gel was prepared using an amphiphilic polymer, Asc-Pal, via self-assembly. The nano fiber with a diameter of around 5 nm could efficiently encapsulate molecules into its hydrophobic region and be sturdy enough to withstand shear forces in joints. Differently, unlike the cleavage of peptide linkers, ascorbyl palmitate could be proteolyzed at the ester bond by MMP-2, MMP-9, and esterase. The nanogels were found to be disassembled to gradually release encapsulated agents in response to synovial fluid from arthritic patients, while remaining stable following injection into healthy joints. The stimuli-sensitive and prolonged drug release would be beneficial for localized treatment of proteolytic diseases, like rheumatoid arthritis [98].

Various block copolymers have been used to prepare the gel matrix for entrapping pDNA for sustained gene release. To efficiently load and deliver pDNAs, the linear PEI was conjugated on the surface of the Poly (ε-caprolactone) [PCL] (70,000–100,000 Da)–PEG2000 block copolymer-based nanofibrous gel via an MMP-cleavable peptide linker. The release of the fluorescent dye-labeled DNA was significantly prolonged due to the slow diffusion of DNA and strong electrostatic binding between PEI and DNA. The resultant nano device was found to be more efficient in gene transfection through the wounds in response to the high level of MMPs in diabetic mice compared to normal mice [120]. The same nanogel was also successfully used for dermal delivery of therapeutic siRNA. Treating diabetic animals with dorsal burns with anti-MMP-2 siRNA-incorporated nanofibrous meshes for 7 days dramatically down-regulated MMP-2 level and decreased neo-collagen accumulation at the wound sites, resulting in the significant increase in the wound recovery rates of diabetic ulcers [121].

A versatile method was developed to prepare a supramolecular nanosized hydrogel with a well-controlled particle size via electrostatic interaction between positive surfactant micelles and negative gelatin polypeptides. The core-shell-like structures of the nanogels contained stable separated hydrophobic cores facilitating the encapsulation of hydrophobic guest molecules and a gelatin shell for the loading of hydrophilic molecules. In response to MMPs, the gelatin shell could be degraded, resulting in the destabilization of nanogels and the release of noncovalently encapsulated molecules for tumor targeting [96].

3.4.8 *Other polymeric nanoparticles sensitive to MMPs*

Similar to the concept of ACPPs [14], an MMP-activatable low molecular weight protamine (ALMWP) and the ALMWP-modified poly(ethylene glycol)-poly(lactic acid) (PEG–PLA) nanoparticles were developed for the enhanced permeability and site-specific targeting delivery of anticancer drugs. Unlike positively charged unmasked protamine, the ALMWP or its modified nanoparticles, were negatively charged and showed the MMP-dependent cellular uptake in the MMP overexpressing HT-1080 cells. In rats, after *i.v.* injection, the paclitaxel-loaded ALMWP-nanoparticles significantly increased paclitaxel blood circulation for more than two folds and increased AUC for more than ten folds, compared to free paclitaxel. In a HT-1080 xenograft mouse model, unlike accumulation in the liver and lung of unmodified nanoparticles or LMWP-nanoparticles, ALMWP-nanoparticles were mainly accumulated in the tumor, resulting in high tumor growth inhibition [57]. The same idea was used to prepare a brain-targeted delivery system by conjugation of ALMWP to PEG–PCL nanoparticles. Like ALMWP-modified PEG–PLA nanoparticles, the improved PK parameters were observed for ALMWP-modified PEG–PCL nanoparticles. In an orthotopic glioma model, ALMWP-modified PEG–PCL nanoparticles were specifically accumulated in the brain and efficiently penetrated into glioma regions, leading to the significant prolongation in the survival time of tumor-bearing mice [56].

The polymerized dopamine has been found to have latent reactivity towards nucleophiles, such as amines and thiols, allowing for the immobilization of functional molecules onto nanoparticles with various types of surfaces [122]. Unlike the step-by-step conjugation reaction strategy, Gullotti *et al.* prepared an MMP-activatable poly(lactic-co-glycolic acid) (PLGA) nanoparticle using a simple and versatile dopamine polymerization method. In their studies, TAT peptide and MMP-sensitive PEG-peptide were conjugated onto the surface of PLGA nanoparticles simultaneously during dopamine polymerization. The MMP sensitivity and TAT peptide-mediated enhanced cellular uptake were observed *in vitro*. However, the expected increased cytotoxicity of the nanoparticles was lacking, probably due to the burst drug release from PLGA nanoparticles [72].

Due to the excellent biocompatibility and drug loading capacity of lipid nanoparticles, they have been also modified with MMP-sensitive lipopolymers, e.g. PEG–peptide–PE [46, 51]. A docetaxel-loaded nanostructured lipid carrier (NLC) with a high encapsulation efficiency and sustained drug release was modified with tumor microenvironment-sensitive polypeptides (TMSP), an idea similar to ACPPs [14], and the tumor-targeting ligand, folate, on its surface, for tumor-targeted delivery of chemotherapeutics. The fact of the overexpression of both folate receptor and MMPs ensured the enhanced cellular uptake and cytotoxicity of the constructed nanocarriers in various cancer cell lines. Furthermore, these nanocarriers could efficiently penetrate into the 3D tumor cell spheroids due to the exposed cell-penetrating peptides after MMP-induced cleavage. As a result, the nanoparticles modified with both TMSP and folate significantly enhanced the *in vivo* antitumor efficacy and tumor cell apoptosis, compared to free drug and the nanoparticles modified with only one functional moiety, in a human mouth epidermal carcinoma (KB) xenograft mouse model [51].

3.4.9 *MMP-sensitive inorganic nanoparticles*

Due to unique physicochemical properties, such as material- or size-dependent optical and magnetic properties, as well as stability and ease of functionalization, inorganic nanoparticles have been widely used in the fields of magnetic resonance and fluorescence imaging, drug delivery, and magnetic manipulation [123]. Inorganic nanoparticles, such as quantum dot, gold nanoparticles, and silica nanoparticles have been modified with MMP-sensitive proteins, peptides, or polymers for MMP-responsive delivery of drugs and imaging agents. For example, quantum dots modified with dye-labeled MMP-7 cleavable PEG–peptide conjugates were found to be almost completely cleaved after incubation at physiologically relevant concentrations of MMP-7 for 24 h [89]. Using a similar strategy, Yi *et al.* prepared an MMP-sensitive gold nanoparticle by conjugation of the well-studied activatable NIR fluorescent probes on the surface of gold nanorods. Due to the unique nature of gold nanoparticles, including the localized surface plasmon resonance (LSPR) and ability of conversion of NIR laser into heat, these gold nanorods were activated to show the NIR fluorescence and, most importantly, worked as a hyperthermal therapeutic agent, in response to the up-regulated MMPs, such as MMP-2, -3, -7, -9, and -13. After laser irradiation, the tissue damage around tumors and the accumulation of gold nanorods in the damaged tissue were observed. However, the hyperthermia caused by irradiation was unable to kill tumor cells [73]. The data suggests that an intracellular delivery strategy may be needed, so as to increase the real tumor cell internalization of the gold nanorods, but not only the tumor tissue accumulation [28].

A well-defined MMP-2 sensitive PEGylated quantum dot nanoparticle containing the CPP was prepared. The degree of PEGylation and the number of the CPP on the surface of streptavidin-modified quantum dots were stoichiometrically controlled by varying the feed amount of biotinylated conjugates. Nine PEG chains per quantum dot brought down the cellular uptake to only 20% of that of non-PEGylated CPP-modified quantum dots. Upon MMP-2 pre-treatment, the cellular uptake was restored to a level similar to that of non-PEGylated CPP-modified quantum dots due to the PEG deshielding and CPP exposure [74].

To fully utilize the advantages of gold nanoparticles and iron oxide nanoparticles, a unique flower-shaped gold (Au)–Fe_3O_4 hybrid nanoparticle containing Cy5.5-labeled MMP-13 cleavable polymers anchored on the iron oxide surface and PEG5000 anchored on the gold surface was prepared. The novel flower-like activatable nanoparticles showed significantly higher tumor targeting due to their excellent MMP sensitivity after *i.v.* injection in a head and neck squamous cell carcinoma (SCC-7) xenograft model, compared to co-administration of the MMP inhibitors or the gold-only nanoparticles [47].

Proteins, like collagen and gelatin, can also be used as scaffolds for the preparation of MMP-sensitive inorganic nanoparticles. A multistage nanoparticle delivery system containing a gelatin core and a surface covered with quantum dots was developed to respond to the elevated gelatinase in the tumor mass. The degradation of the gelatin core resulted in the collapse of the 100-nm nanoparticles and the release of 10-nm quantum dots. This unique MMP-sensitive "shrunken" property significantly improved the nanoparticles' diffusion in an *in vitro* collagen diffusion experiment upon MMP-2 pre-treatment. This novel design allowed 100-nm nanoparticles' passive tumor targeting via the EPR effect, as well as the fast and efficient diffusion of the 10-nm quantum dots in tumoral ECM after the nanoparticle extravasation [94]. Alternatively, the quantum dots could be embedded into the hydrogel-like interior of supramolecular gelatin nanoparticles by a controllable desolvation method. Upon surface crosslinking with glutaraldehyde, a stable quantum dot-encapsulated gelatin nanoparticle sensitive to MMPs was prepared [95].

3.4.10 *MMP-sensitive multifunctional nanocarriers*

Although the MMP-sensitive strategy has resulted in the improved efficiency and specificity of drug delivery, recent studies suggested that better and/or special clinical outcomes could even be achieved by the combined use of the MMP sensitivity with other compatible strategies [28, 46, 51, 53, 71, 91, 119]. The MMP-sensitive moieties can "switch on or off" other functions of the nanocarrier in response to local MMPs, resulting in drug release and/or cell internalization of drug-loaded nanocarriers in the right place at the right time. The use of the MMP sensitivity in the nanocarriers such as protein- or hydrophilic polymer-based nanoparticles showed long blood circulation, increased passive tumor targeting, and MMP-dependent drug release [60, 66]. To improve the tumor active targeting, the targeting ligands, including mAbs

[45, 46] and small ligands (e.g. sugar residues [67] and folate [51]), were conjugated onto the surface of the MMP-sensitive nanocarriers for direct binding with their receptors on the cell membrane or on-demand interaction with target cells via the MMP-induced PEG deshielding.

The nanocarrier designed to be sensitive to multi-stimuli is another strategy to improve drug targetability and delivery efficiency. A nanocarrier system engineered with pH-sensitive MMP-2 ACPPs on its surface was successfully developed to sense both the lowered extracellular pH and up-regulated tumoral MMP-2 in the tumor microenvironment for the tumor-targeted DNA/siRNA and drug co-delivery [52, 119]. Another novel pH-responsive polymeric nanoparticle with MMP-7 dependent proximity-activated targeting was developed for MMP-induced cellular uptake and pH-triggered cytosolic delivery of siRNA [91].

More sophisticated multifunctional nanocarriers have been proposed for therapeutic and/or diagnostic purposes. The delivery strategies, including longevity in the bloodstream, stimuli sensitivity, passive and/or active targeting, cellular penetration, intracellular targeting, and the versatile loading capacity of small molecules, large molecules, and imaging reagents, can be incorporated into a single nanocarrier and collaborate in an efficient manner for the best clinical outcomes [11]. For example, a novel multifunctional liposomal nanocarrier, composed of a protective long chain PEG, an antitumor monoclonal antibody (mAb 2C5), an intracellular penetrating moiety (TAT peptide), and an MMP-2 sensitive PEG-lipid conjugate, has been designed to enhance tumor targeting (Fig. 3.4) [46]. In another study, an MMP-2 labile PEG2000–peptide–paclitaxel prodrug used as a micelle building block was incorporated into a TAT peptide–PEG1000 conjugate-engineered polymeric micelle for the tumor-targeted delivery of paclitaxel [28]. The well-designed multifunctional nanocarrier named Multifunctional Envelope-type Nano Device (MEND) has also been engineered with MMP-cleavable PEG-peptide-lipid for *in vivo* cancer gene therapy [93]. Most recently, an MMP-2 sensitive multifunctional block copolymer, PEG-peptide-polyethylenimine-phosphatidylethanolamine (PEG–pp–PEI–PE), was designed. The micellar nanocarriers constructed by this polymer via self-assembly possessed several key features for siRNA and drug delivery. These cooperative functions improved the tumor targeting, tumor cell internalization, and synergistic antitumor activity of co-loaded siRNA and hydrophobic drugs [70]. In these aforementioned nanocarriers, the prolonged circulation half-life, thanks to PEGylation, facilitates the EPR effect-mediated passive tumor targeting and antibody-mediated active tumor targeting. Upon tumor accumulation of the nanocarriers, tumoral up-regulated extracellular MMPs cleave the peptide linker, resulting in the liberation of the protective long

PEG chains and/or bound drugs. The exposure of the previously hidden surface-attached CPPs or PEI mediates the enhanced intracellular delivery of the nanocarriers after PEG deshielding. Furthermore, the fusogenic lipids [e.g. DOPE and cationic polymers (e.g. protamine or PEI)] facilitate the endosomal escape and intracellular targeting. These functions in nanocarriers have been "pre-programmed" to fit well within the cascade pathway of drug delivery based on pathological conditions and expected outcomes.

3.5 Challenges and Future Directions

As therapeutic targets, MMPs have been successfully used for more than 40 years [34], as evidenced by several MMPIs applied clinically [20]. However, the use of MMPs as the stimuli for drug delivery and imaging has drawn significant attention recently [16]. Compared to the conventional drug delivery systems, the MMP-sensitive nanocarriers show a superior ability for control and adjustment of the location and time of drug release and cell internalization. However, the current MMP-sensitive systems are still imperfect. The challenges include: (i) low specificity of the MMP-sensitive substrates, e.g. most substrates are labile to several enzymes (Table 3.1); (ii) slow and insufficient stimulus–response, e.g. long contact time and high enzyme concentration are needed to ensure the complete cleavage of MMP-labile linkers [46, 66, 94]; (iii) the off-target delivery due to drug treatment/harsh condition-stimulated up-regulation of MMPs in non-target tissues and/or cells [124, 125]; (iv) the insufficient information of the *in vivo* patterns and levels of MMPs; and (v) less *in vitro* and *in vivo* correlation. Therefore, to ensure the success of an MMP-sensitive nanocarrier, at least, the *in vivo* distribution, expression level, and activity of desired MMPs, properties of designed MMP-sensitive substrates and their engineered nanocarriers, pathological conditions, and expected clinical outcomes have to be thoroughly considered. From a practical point of view, a reliable and efficient preparation procedure is also important in the construction of a successful nanocarrier/nanomedicine. Besides, nanoparticle engineering without a substantial increase in its complexity and without the influence of the cost and commercial scale-up is preferred [1]. Due to the limitations of individual drug delivery strategies, the combined use of the MMP sensitivity with other strategies is of great interest [16]. As shown in the current studies and as expected in the future, the "smart" MMP-sensitive strategy will make multi-functional/-componential nanocarriers more controllable and efficient for the diagnosis and therapy. These functions of nanocarriers may also provide an option of the flexible combination of various delivery strategies (personalized nanomedicine) for improved therapeutic outcomes and reduced costs.

References

1. Peer, D., Karp, J. M., Hong, S., Farokhzad, O. C., Margalit, R. and Langer, R. (2007). Nanocarriers as an emerging platform for cancer therapy. *Nat. Nanotechnol.*, **2**, 751–760.
2. Shi, J., Votruba, A. R., Farokhzad, O. C. and Langer, R. (2010). Nanotechnology in drug delivery and tissue engineering: From discovery to applications. *Nano Lett.*, **10**, 3223–3230.
3. LoRusso, P. M., Weiss, D., Guardino, E., Girish, S. and Sliwkowski, M. X. (2011). Trastuzumab emtansine: A unique antibody-drug conjugate in development for human epidermal growth factor receptor 2-positive cancer. *Clin. Cancer Res.*, **17**, 6437–6447.
4. Zhu, L. and Mahato, R. I. (2010). Lipid and polymeric carrier-mediated nucleic acid delivery. *Expert Opin. on Drug Del.*, 7, 1209–1226.
5. Torchilin, V. P. (2005). Recent advances with liposomes as pharmaceutical carriers. *Nat. Rev. Drug Discov.*, **4**, 145–160.
6. Torchilin, V. P. (2007). Micellar nanocarriers: Pharmaceutical perspectives. *Pharmaceut. Res.*, **24**, 1–16.
7. Harris, J. M. and Chess, R. B. (2003). Effect of pegylation on pharmaceuticals. *Nat. Rev. Drug Discov.*, **2**, 214–221.
8. Allen, T. M. and Cullis, P. R. (2004). Drug delivery systems: Entering the mainstream. *Science*, **303**, 1818–1822.
9. Zhu, L., Movassaghian, S. and Torchilin, V. P. (2012). Overcoming biological barriers with parenteral nanomedicines: Physiological and mechanistic issues; in Alonso, M. J., Csaba, N. S. (eds.), *Nanostructured Biomaterials for Overcoming Biological Barriers*, London: Royal Society of Chemistry, pp. 435–455.
10. Zolnik, B. S., Gonzalez-Fernandez, A., Sadrieh, N. and Dobrovolskaia, M. A. (2010). Nanoparticles and the immune system. *Endocrinology*, **151**, 458–465.
11. Torchilin, V. P. (2006). Multifunctional nanocarriers. *Adv. Drug Deliver Rev.*, **58**, 1532–1555.
12. Thurber, G. M., Schmidt, M. M. and Wittrup, K. D. (2008). Antibody tumor penetration: Transport opposed by systemic and antigen-mediated clearance. *Adv. Drug Deliver Rev.*, **60**, 1421–1434.
13. Oldham, R. K. and Dillman, R. O. (2008). Monoclonal antibodies in cancer therapy: 25 years of progress. *J. Clin. Oncol.*, **26**, 1774–1777.
14. Jiang, T., Olson, E. S., Nguyen, Q. T., Roy, M., Jennings, P. A. and Tsien, R. Y. (2004). Tumor imaging by means of proteolytic activation of cell-penetrating peptides. *Proceedings of the National Academy of Sciences of the United States of America*, **101**, 17867–17872.
15. Ruoslahti, E., Bhatia, S. N. and Sailor, M. J. (2010). Targeting of drugs and nanoparticles to tumors. *J. Cell Biol.*, **188**, 759–768.
16. Zhu, L. and Torchilin, V. P. (2013). Stimulus-responsive nanopreparations for tumor targeting. *Integr. Biol.*, **5**, 96–107.
17. Nagase, H., Visse, R. and Murphy, G. (2006). Structure and function of matrix metalloproteinases and TIMPs. *Cardiovasc. Res.*, **69**, 562–573.
18. Egeblad, M. and Werb, Z. (2002). New functions for the matrix metalloproteinases in cancer progression. *Nat. Rev. Cancer*, **2**, 161–174.

19. Vu, T. H. and Werb, Z. (2000). Matrix metalloproteinases: Effectors of development and normal physiology. *Genes Dev.*, **14**, 2123–2133.
20. Roy, R., Yang, J. and Moses, M. A. (2009). Matrix metalloproteinases as novel biomarkers and potential therapeutic targets in human cancer. *J. Clin. Oncol.*, **27**, 5287–5297.
21. Thiery, J. P. (2002). Epithelial-mesenchymal transitions in tumour progression. *Nat. Rev. Cancer*, **2**, 442–454.
22. La Rocca, G., Pucci-Minafra, I., Marrazzo, A., Taormina, P. and Minafra, S. (2004). Zymographic detection and clinical correlations of MMP-2 and MMP-9 in breast cancer sera. *Br. J. Cancer*, **90**, 1414–1421.
23. Fernandez, C. A., Yan, L., Louis, G., Yang, J., Kutok, J. L. and Moses, M. A. (2005). The matrix metalloproteinase-9/neutrophil gelatinase-associated lipocalin complex plays a role in breast tumor growth and is present in the urine of breast cancer patients. *Clin. Cancer Res.*, **11**, 5390–5395.
24. Yokoyama, M., Ochi, K., Ichimura, M., Mizushima, T., Shinji, T., Koide, N., Tsurumi, T., Hasuoka, H. and Harada, M. (2002). Matrix metalloproteinase-2 in pancreatic juice for diagnosis of pancreatic cancer. *Pancreas*, **24**, 344–347.
25. Kuhlmann, K. F., van Till, J. W., Boermeester, M. A., de Reuver, P. R., Tzvetanova, I. D., Offerhaus, G. J., Ten Kate, F. J., Busch, O. R., van Gulik, T. M., Gouma, D. J. and Crawford, H. C. (2007). Evaluation of matrix metalloproteinase 7 in plasma and pancreatic juice as a biomarker for pancreatic cancer. *Cancer Epidemiol. Biomarkers Prev.*, **16**, 886–891.
26. Tian, M., Cui, Y. Z., Song, G. H., Zong, M. J., Zhou, X. Y., Chen, Y. and Han, J. X. (2008). Proteomic analysis identifies MMP-9, DJ-1 and A1BG as overexpressed proteins in pancreatic juice from pancreatic ductal adenocarcinoma patients. *BMC Cancer*, **8**, 241.
27. Jumper, C., Cobos, E. and Lox, C. (2004). Determination of the serum matrix metalloproteinase-9 (MMP-9) and tissue inhibitor of matrix metalloproteinase-1 (TIMP-1) in patients with either advanced small-cell lung cancer or non-small-cell lung cancer prior to treatment. *Respir Med*, **98**, 173–177.
28. Zhu, L., Wang, T., Perche, F., Taigind, A. and Torchilin, V. P. (2013). Enhanced anticancer activity of nanopreparation containing an MMP2-sensitive PEG-drug conjugate and cell-penetrating moiety. *Proceedings of the National Academy of Sciences of the United States of America*, **110**, 17047–17052.
29. Liu, D., Nakano, J., Ishikawa, S., Yokomise, H., Ueno, M., Kadota, K., Urushihara, M. and Huang, C. L. (2007). Overexpression of matrix metalloproteinase-7 (MMP-7) correlates with tumor proliferation and a poor prognosis in non-small cell lung cancer. *Lung Cancer*, **58**, 384–391.
30. Morgia, G., Falsaperla, M., Malaponte, G., Madonia, M., Indelicato, M., Travali, S. and Mazzarino, M. C. (2005). Matrix metalloproteinases as diagnostic (MMP-13) and prognostic (MMP-2, MMP-9) markers of prostate cancer. *Urol. Res.*, **33**, 44–50.
31. Riddick, A., Shukla, C. J., Pennington, C. J., Bass, R., Nuttall, R. K., Hogan, A., Sethia, K. K., Ellis, V., Collins, A. T., Maitland, N. J., Ball, R. Y. and Edwards, D. R. (2005). Identification of degradome components associated with prostate cancer progression by expression analysis of human prostatic tissues. *Br. J. Cancer*, **92**, 2171–2180.
32. Friedberg, M. H., Glantz, M. J., Klempner, M. S., Cole, B. F. and Perides, G. (1998). Specific matrix metalloproteinase profiles in the cerebrospinal fluid correlated with the presence of malignant astrocytomas, brain metastases and carcinomatous meningitis. *Cancer*, **82**, 923–930.

33. Jäälinojä, J., Herva, R., Korpela, M., Höyhtyä, M. and Turpeenniemi-Hujanen, T. (2000). Matrix metalloproteinase 2 (MMP-2) immunoreactive protein is associated with poor grade and survival in brain neoplasms. *J. Neuro Oncol.*, **46**, 81–90.
34. Coussens, L. M., Fingleton, B. and Matrisian, L. M. (2002). Matrix metalloproteinase inhibitors and cancer: Trials and tribulations. *Science*, **295**, 2387–2392.
35. Khokha, R., Murthy, A. and Weiss, A. (2013). Metalloproteinases and their natural inhibitors in inflammation and immunity. *Nat. Rev. Immunol.*, **13**, 649–665.
36. Pozgan, U., Caglic, D., Rozman, B., Nagase, H., Turk, V. and Turk, B. (2010). Expression and activity profiling of selected cysteine cathepsins and matrix metalloproteinases in synovial fluids from patients with rheumatoid arthritis and osteoarthritis. *Biol. Chem.*, **391**, 571–579.
37. Malemud, C. J. (2006). Matrix metalloproteinases (MMPs) in health and disease: An overview. *Front. in Biosci.*, **11**, 1696–1701.
38. Yong, V. W., Power, C., Forsyth, P. and Edwards, D. R. (2001). Metalloproteinases in biology and pathology of the nervous system. *Nat. Rev. Neurosci.*, **2**, 502–511.
39. Sternlicht, M. D. and Werb, Z. (2001). How matrix metalloproteinases regulate cell behavior. *Annu. Rev. Cell Dev. Biol.*, **17**, 463–516.
40. Turk, B. E., Huang, L. L., Piro, E. T. and Cantley, L. C. (2001). Determination of protease cleavage site motifs using mixture-based oriented peptide libraries. *Nat. Biotechnol.*, **19**, 661–667.
41. Nagase, H. and Fields, G. B. (1996). Human matrix metalloproteinase specificity studies using collagen sequence-based synthetic peptides. *Pept. Sci.*, **40**, 399–416.
42. Moustoifa el, F., Alouini, M. A., Salaün, A., Berthelot, T., Bartegi, A., Albenque-Rubio, S. and Déléris, G. (2010). Novel cyclopeptides for the design of MMP directed delivery devices: A novel smart delivery paradigm. *Pharm. Res.*, **27**, 1713–1721.
43. Kratz, F., Drevs, J., Bing, G., Stockmar, C., Scheuermann, K., Lazar, P. and Unger, C. (2001). Development and *in vitro* efficacy of novel MMP2 and MMP9 specific doxorubicin albumin conjugates. *Bioorg. Medi. Chem. Lett.*, **11**, 2001–2006.
44. Lee, G. Y., Park, K., Kim, S. Y. and Byun, Y. (2007). MMPs-specific PEGylated peptide-DOX conjugate micelles that can contain free doxorubicin. *Eur. J. Pharmaceut. Biopharmaceut.*, **67**, 646–654.
45. Erster, O., Thomas, J. M., Hamzah, J., Jabaiah, A. M., Getz, J. A., Schoep, T. D., Hall, S. S., Ruoslahti, E. and Daugherty, P. S. (2012). Site-specific targeting of antibody activity *in vivo* mediated by disease-associated proteases. *J. Control. Release*, **161**, 804–812.
46. Zhu, L., Kate, P. and Torchilin, V. P. (2012). Matrix metalloprotease 2-responsive multifunctional liposomal nanocarrier for enhanced tumor targeting. *ACS Nano*, **6**, 3491–3498.
47. Xie, J., Zhang, F., Aronova, M., Zhu, L., Lin, X., Quan, Q., Liu, G., Zhang, G., Choi, K. Y., Kim, K., Sun, X., Lee, S., Sun, S., Leapman, R. and Chen, X. (2011). Manipulating the power of an additional phase: A flower-like Au-Fe_3O_4 optical nanosensor for imaging protease expressions in vivo. *ACS Nano*, **5**, 3043–3051.
48. Feng, D., Zhang, Y., Feng, T., Shi, W., Li, X. and Ma, H. (2011). A graphene oxide-peptide fluorescence sensor tailor-made for simple and sensitive detection of matrix metalloproteinase 2. *Chem. Commun.*, **47**, 10680–10682.

49. Aguilera, T. A., Olson, E. S., Timmers, M. M., Jiang, T. and Tsien, R. Y. (2009). Systemic in vivo distribution of activatable cell penetrating peptides is superior to that of cell penetrating peptides. *Integr. Biol.*, **1**, 371–381.
50. Olson, E. S., Aguilera, T. A, Jiang, T., Ellies, L. G., Nguyen, Q. T., Wong, E. H., Gross, L. A. and Tsien, R. Y. (2009). *In vivo* characterization of activatable cell penetrating peptides for targeting protease activity in cancer. *Integr. Biol.*, **1**, 382–393.
51. Gao, W., Xiang, B., Meng, T. T., Liu, F. and Qi, X. R. (2013). Chemotherapeutic drug delivery to cancer cells using a combination of folate targeting and tumor microenvironment-sensitive polypeptides. *Biomaterials*, **34**, 4137–4149.
52. Huang, S., Shao, K., Liu, Y., Kuang, Y., Li, J., An, S., Guo, Y., Ma, H. and Jiang, C. (2013). Tumor-targeting and microenvironment-responsive smart nanoparticles for combination therapy of antiangiogenesis and apoptosis. *ACS Nano*, 7, 2860–2871.
53. Olson, E. S., Jiang, T., Aguilera, T. A., Nguyen, Q. T., Ellies, L. G., Scadeng, M. and Tsien, R. Y. (2010). Activatable cell penetrating peptides linked to nanoparticles as dual probes for *in vivo* fluorescence and MR imaging of proteases. *Proceedings of the National Academy of Sciences of the United States of America*, **107**, 4311–4316.
54. Nguyen, Q. T., Olson, E. S., Aguilera, T. A., Jiang, T., Scadeng, M., Ellies, L. G. and Tsien, R. Y. (2010). Surgery with molecular fluorescence imaging using activatable cell-penetrating peptides decreases residual cancer and improves survival. *Proceedings of the National Academy of Sciences of the United States of America*, **107**, 4317–4322.
55. Shi, N. Q., Gao, W., Xiang, B. and Qi, X. R. (2012). Enhancing cellular uptake of activable cell-penetrating peptide-doxorubicin conjugate by enzymatic cleavage. *Int. J. Nanomedicine*, 7, 1613–1621.
56. Gu, G., Xia, H., Hu, Q., Liu, Z., Jiang, M., Kang, T., Miao, D., Tu, Y., Pang, Z., Song, Q., Yao, L., Chen, H., Gao, X. and Chen, J. (2013). PEG-co-PCL nanoparticles modified with MMP-2/9 activatable low molecular weight protamine for enhanced targeted glioblastoma therapy. *Biomaterials*, **34**, 196–208.
57. Xia, H., Gu, G., Hu, Q., Liu, Z., Jiang, M., Kang, T., Miao, D., Song, Q., Yao, L., Tu, Y., Chen, H., Gao, X. and Chen, J. (2013). Activatable cell penetrating peptide-conjugated nanoparticles with enhanced permeability for site-specific targeting delivery of anticancer drug. *Bioconjugate Chem.*, **24**, 419–430.
58. Matsumura, S., Aoki, I., Saga, T. and Shiba, K. (2011). A tumor-environment-responsive nanocarrier that evolves its surface properties upon sensing matrix metalloproteinase-2 and initiates agglomeration to enhance T 2 relaxivity for magnetic resonance imaging. *Mol. Pharm.*, **8**, 1970–1974.
59. Myochin, T., Hanaoka, K., Komatsu, T., Terai, T. and Nagano, T. (2012). Design strategy for a near-infrared fluorescence probe for matrix metalloproteinase utilizing highly cell permeable boron dipyrromethene. *J. Am. Chem. Soc.*, **134**, 13730–13737.
60. Chau, Y., Dang, N. M., Tan, F. E. and Langer, R. (2006). Investigation of targeting mechanism of new dextran-peptide-methotrexate conjugates using biodistribution study in matrix-metalloproteinase-overexpressing tumor xenograft model. *J. Pharm. Sci.*, **95**, 542–551.
61. Chau, Y., Padera, R. F., Dang, N. M. and Langer, R. (2006). Antitumor efficacy of a novel polymer-peptide-drug conjugate in human tumor xenograft models. *Int. J. Cancer*, **118**, 1519–1526.

62. Chau, Y., Tan, F. E. and Langer, R. (2004). Synthesis and characterization of dextran-peptide-methotrexate conjugates for tumor targeting via mediation by matrix metalloproteinase II and matrix metalloproteinase IX. *Bioconjugate Chem.*, **15**, 931–941.
63. Lim, S. H., Jeong, Y. I., Moon, K. S., Ryu, H. H., Jin, Y. H., Jin, S. G., Jung, T. Y., Kim, I. Y., Kang, S. S. and Jung, S. (2010). Anticancer activity of PEGylated matrix metalloproteinase cleavable peptide–conjugated adriamycin against malignant glioma cells. *Int. J. Pharmaceutics*, **387**, 209–214.
64. Li, R., Wu, W., Liu, Q., Wu, P., Xie, L., Zhu, Z., Yang, M., Qian, X., Ding, Y., Yu, L., Jiang, X., Guan, W. and Liu, B. (2013). Intelligently targeted drug delivery and enhanced antitumor effect by gelatinase-responsive nanoparticles. *PloS One*, **8**, e69643.
65. Kaminskas, L. M., Kelly, B. D., McLeod, V. M., Sberna, G., Boyd, B. J., Owen, D. J. and Porter, C. J. (2011). Capping methotrexate alpha-carboxyl groups enhances systemic exposure and retains the cytotoxicity of drug conjugated PEGylated polylysine dendrimers. *Molecular Pharm.*, **8**, 338–349.
66. Mansour, A. M., Drevs, J., Esser, N., Hamada, F. M., Badary, O. A., Unger, C., Fichtner, I. and Kratz, F. (2003). A new approach for the treatment of malignant melanoma: Enhanced antitumor efficacy of an albumin-binding doxorubicin prodrug that is cleaved by matrix metalloproteinase 2. *Cancer Res.*, **63**, 4062–4066.
67. Terada, T., Iwai, M., Kawakami, S., Yamashita, F. and Hashida, M. (2006). Novel PEG-matrix metalloproteinase-2 cleavable peptide-lipid containing galactosylated liposomes for hepatocellular carcinoma-selective targeting. *J. Control. Release*, **111**, 333–342.
68. Wan, Y., Han, J., Fan, G., Zhang, Z., Gong, T. and Sun, X. (2013). Enzyme-responsive liposomes modified adenoviral vectors for enhanced tumor cell transduction and reduced immunogenicity. *Biomaterials*, **34**, 3020–3030.
69. Yamada, R., Kostova, M. B., Anchoori, R. K., Xu, S., Neamati, N. and Khan, S. R. (2010). Biological evaluation of paclitaxel-peptide conjugates as a model for MMP2-targeted drug delivery. *Cancer Biol. Ther.*, **9**, 192–203.
70. Zhu, L., Perche, F., Wang, T. and Torchilin, V. P. (2014). Matrix metalloproteinase 2-sensitive multifunctional polymeric micelles for tumor-specific co-delivery of siRNA and hydrophobic drugs. *Biomaterials*, **35**, 4213–4222.
71. Zhang, J., Yuan, Z. F., Wang, Y., Chen, W. H., Luo, G. F., Cheng, S. X., Zhuo, R. X. and Zhang, X. Z. (2013). Multifunctional envelope-type mesoporous silica nanoparticles for tumor-triggered targeting drug delivery. *J. Am. Chem. Soc.*, **135**, 5068–5073.
72. Gullotti, E., Park, J. and Yeo, Y. (2013). Polydopamine-based surface modification for the development of peritumorally activatable nanoparticles. *Pharm. Res.*, **30**, 1956–1967.
73. Yi, D. K., Sun, I. C., Ryu, J. H., Koo, H., Park, C. W., Youn, I. C., Choi, K., Kwon, I. C., Kim, K. and Ahn, C. H. (2010). Matrix metalloproteinase sensitive gold nanorod for simultaneous bioimaging and photothermal therapy of cancer. *Bioconjugate Chem.*, **21**, 2173–2177.
74. Mok, H., Bae, K. H., Ahn, C. H. and Park, T. G. (2008). PEGylated and MMP-2 specifically dePEGylated quantum dots: Comparative evaluation of cellular uptake. *Langmuir*, **25**, 1645–1650.
75. Peng, K. W., Morling, F. J., Cosset, F. L., Murphy, G. and Russell, S. J. (1997). A gene delivery system activatable by disease-associated matrix metalloproteinases. *Hum. Gene Ther.*, **8**, 729–738.

76. Wythe, S. E., DiCara, D., Taher, T. E., Finucane, C. M., Jones, R., Bombardieri, M., Man, Y. K., Nissim, A., Mather, S. J., Chernajovsky, Y. and Pitzalis, C. (2013). Targeted delivery of cytokine therapy to rheumatoid tissue by a synovial targeting peptide. *Ann. Rheum. Dis.*, **72**, 129–135.
77. Adams, G., Vessillier, S., Dreja, H. and Chernajovsky, Y. (2003). Targeting cytokines to inflammation sites. *Nat. Biotechnol.,* , **21**, 1314–1320.
78. Albright, C. F., Graciani, N., Han, W., Yue, E., Stein, R., Lai, Z., Diamond, M., Dowling, R., Grimminger, L., Zhang, S. Y., Behrens, D., Musselman, A., Bruckner, R., Zhang, M., Jiang, X., Hu, D., Higley, A., Dimeo, S., Rafalski, M., Mandlekar, S., Car B., Yeleswaram, S., Stern, A., Copeland, R. A., Combs, A., Seitz, S. P., Trainor, G. L., Taub, R., Huang, P. and Oliff, A. (2005). Matrix metalloproteinase-activated doxorubicin prodrugs inhibit HT1080 xenograft growth better than doxorubicin with less toxicity. *Mol. Cancer Ther.*, **4**, 751–760.
79. Kline, T., Torgov, M. Y., Mendelsohn, B. A., Cerveny, C. G. and Senter, P. D. (2004). Novel antitumor prodrugs designed for activation by matrix metalloproteinases-2 and -9. *Mol. Pharm.,* **1**, 9–22.
80. Elegbede, A. I., Banerjee, J., Hanson, A. J., Tobwala, S., Ganguli, B., Wang, R., Lu, X., Srivastava, D. K. and Mallik, S. (2008). Mechanistic studies of the triggered release of liposomal contents by matrix metalloproteinase-9. *J. Am. Chem. Soc.*, **130**, 10633–10642.
81. Sarkar, N, Banerjee, J., Hanson, A. J., Elegbede, A. I., Rosendahl, T., Krueger, A. B., Banerjee, A. L., Tobwala, S., Wang, R., Lu, X., Mallik, S. and Srivastava, D. K. (2008). Matrix metalloproteinase-assisted triggered release of liposomal contents. *Bioconjugate Chem.*, **19**, 57–64.
82. Lin, X., Xie, J., Zhu, L., Lee, S., Niu, G., Ma, Y., Kim, K. and Chen, X. (2011). Hybrid ferritin nanoparticles as activatable probes for tumor imaging. *Angew. Chem.*, **123**, 1607–1610.
83. Zhu, L., Zhang, F., Ma, Y., Liu, G., Kim, K., Fang, X., Lee, S. and Chen, X. (2011). In vivo optical imaging of membrane-type matrix metalloproteinase (MT-MMP) activity. *Mol. Pharm.*, **8**, 2331–2338.
84. Zhong, J. and Chau, Y. (2008). Antitumor activity of a membrane lytic peptide cyclized with a linker sensitive to membrane type 1-matrix metalloproteinase. *Mol. Cancer Ther.*, 7, 2933–2940.
85. Nam, E. J., Kang, J. H., Sung, S., Sa, K. H., Kim, K. H., Seo, J. S., Kim, J. H., Han, S. W., Kim, I. S. and Kang, Y. M. (2013). A matrix metalloproteinase 1-cleavable composite peptide derived from transforming growth factor beta-inducible gene h3 potently inhibits collagen-induced arthritis. *Arthritis Rheum.*, **65**, 1753–1763.
86. Kim, H. S. and Yoo, H. S. (2010). MMPs-responsive release of DNA from electrospun nanofibrous matrix for local gene therapy: *In vitro* and *in vivo* evaluation. *J. Control. Release*, **145**, 264–271.
87. Singh, H. D., Bushnak, I. and Unsworth, L. D. (2012). Engineered peptides with enzymatically cleavable domains for controlling the release of model protein drug from "soft" nanoparticles. *Acta Biomater.*, **8**, 636–645.
88. Garripelli, V. K., Kim, J. K., Son, S., Kim, W. J., Repka, M. A. and Jo, S. (2011). Matrix metalloproteinase-sensitive thermogelling polymer for bioresponsive local drug delivery. *Acta Biomater.*, 7, 1984–1992.

89. Smith, R., Sewell, S. L. and Giorgio, T. D. (2008). Proximity-activated nanoparticles: *In vitro* performance of specific structural modification by enzymatic cleavage. *Int. J. Nanomedicine*, **3**, 95–103.
90. Gustafson, J. A., Price, R. A., Frandsen, J., Henak, C. R., Cappello, J. and Ghandehari, H. (2013). Synthesis and characterization of a matrix-metalloproteinase responsive silk-elastin-like protein polymer. *Biomacromolecules*, **14**, 618–625.
91. Li, H., Yu, S. S., Miteva, M., Nelson, C. E., Werfel, T., Giorgio, T. D. and Duvall, C. L. (2013). Matrix metalloproteinase responsive, proximity-activated polymeric nanoparticles for siRNA delivery. *Advanced Functional Materials*, **23**, 3040–3052.
92. Yhee, J. Y., Kim, S. A., Koo, H., Son, S., Ryu, J. H., Youn, I. C., Choi, K., Kwon, I. C. and Kim, K. (2012). Optical imaging of cancer-related proteases using near-infrared fluorescence matrix metalloproteinase-sensitive and cathepsin B-sensitive probes. *Theranostics*, **2**, 179–189.
93. Hatakeyama, H., Akita, H., Kogure, K. and Harashima, H. (2006). Development of a novel systemic gene delivery system for cancer therapy with a tumor-specific cleavable PEG-lipid. *Gene Ther.*, **14**, 68–77.
94. Wong, C., Stylianopoulos, T., Cui, J., Martin, J., Chauhan, V. P., Jiang, W., Popovic, Z., Jain, R. K., Bawendi, M. G. and Fukumura, D. (2011). Multistage nanoparticle delivery system for deep penetration into tumor tissue. *Proceedings of the National Academy of Sciences of the United States of America*, **108**, 2426–2431.
95. Xu, J. H., Gao, F. P., Liu, X. F., Zeng, Q., Guo, S. S., Tang, Z. Y., Zhao, X. Z. and Wang, H. (2013). Supramolecular gelatin nanoparticles as matrix metalloproteinase responsive cancer cell imaging probes. *Chem. Commun.*, **49**, 4462–4464.
96. Kim, K., Bae, B., Kang, Y. J., Nam, J. M., Kang, S. and Ryu, J. H. (2013). Natural polypeptide-based supramolecular nanogels for stable noncovalent encapsulation. *Biomacr.*, **14**, 3515–3522.
97. Kojima, C., Suehiro, T., Watanabe, K., Ogawa, M., Fukuhara, A., Nishisaka, E., Harada, A., Kono, K., Inui, T. and Magata, Y. (2013). Doxorubicin-conjugated dendrimer/collagen hybrid gels for metastasis-associated drug delivery systems. *Acta Biomater.*, **9**, 5673–5680.
98. Vemula, P. K., Boilard, E., Syed, A., Campbell, N. R., Muluneh, M., Weitz, D. A., Lee, D. M. and Karp, J. M. (2011). On-demand drug delivery from self-assembled nanofibrous gels: A new approach for treatment of proteolytic disease. *J. Biomed. Mater. Res., A*, **97**, 103–110.
99. Kratz, F. (2008). Albumin as a drug carrier: Design of prodrugs, drug conjugates and nanoparticles. *J. Control. Release*, **132**, 171–183.
100. Bremer, C., Tung, C. H. and Weissleder, R. (2001). *In vivo* molecular target assessment of matrix metalloproteinase inhibition. *Nat. Med.*, **7**, 743–748.
101. Veronese, F. M. and Pasut, G. (2005). PEGylation, successful approach to drug delivery. *Drug Discov. Today*, **10**, 1451–1458.
102. Sawant, R. M., Hurley, J. P., Salmaso, S., Kale, A., Tolcheva, E., Levchenko, T. S. and Torchilin, V. P. (2006). "SMART" drug delivery systems: Double-targeted pH-responsive pharmaceutical nanocarriers. *Bioconjugate Chem.*, **17**, 943–949.
103. Mishra, S., Webster, P. and Davis, M. E. (2004). PEGylation significantly affects cellular uptake and intracellular trafficking of non-viral gene delivery particles. *Eur. J. Cell Biol.*, **83**, 97–111.

104. Sarkar, N. R., Rosendahl, T., Krueger, A. B., Banerjee, A. L., Benton K., Mallik, S. and Srivastava, D. K. (2005). "Uncorking" of liposomes by matrix metalloproteinase-9. *Chem. Commun.*, **8**, 999–1001.
105. Rosler, A., Vandermeulen, G. W. and Klok, H. A. (2001). Advanced drug delivery devices via self-assembly of amphiphilic block copolymers. *Adv. Drug Deliv. Rev.*, **53**, 95–108.
106. Torchilin, V. P. (2007). Targeted pharmaceutical nanocarriers for cancer therapy and imaging. *The AAPS J.*, **9**, E128–E147.
107. Yu, S. S., Lau, C. M., Thomas, S. N., Jerome, W. G., Maron, D. J., Dickerson, J. H., Hubbell, J. A. and Giorgio, T. D. (2012). Size- and charge-dependent non-specific uptake of PEGylated nanoparticles by macrophages. *Int. J. Nanomedicine*, 7, 799–813.
108. Li, J., Ge, Z. and Liu, S. (2013). PEG-sheddable polyplex micelles as smart gene carriers based on MMP-cleavable peptide-linked block copolymers. *Chem. Commun.*, **49**, 6974–6976.
109. Chien, M. P., Thompson, M. P., Lin, E. C. and Gianneschi, N. C. (2012). Fluorogenic enzyme-responsive micellar nanoparticles. *Chem. Sci.*, **3**, 2690–2694.
110. Ku, T. H., Chien, M. P., Thompson, M. P., Sinkovits, R. S., Olson, N. H., Baker, T. S. and Gianneschi, N. C. (2011). Controlling and switching the morphology of micellar nanoparticles with enzymes. *J. Am. Chem. Soc.*, **133**, 8392–8395.
111. Chien, M. P., Carlini, A. S., Hu, D., Barback, C. V., Rush, A. M., Hall, D. J., Orr, G. and Gianneschi, N. C. (2013). Enzyme-directed assembly of nanoparticles in tumors monitored by in vivo whole animal imaging and ex vivo super-resolution fluorescence imaging. *J. Am. Chem. Soc.*, **135**, 18710–18713.
112. Singh, H. D., Bushnak, I. and Unsworth, L. D. (2012). Engineered peptides with enzymatically cleavable domains for controlling the release of model protein drug from "soft" nanoparticles. *Acta Biomater.*, **8**, 636–645.
113. MaHam, A., Tang, Z., Wu, H., Wang, J. and Lin, Y. (2009). Protein-based nanomedicine platforms for drug delivery. *Small*, **5**, 1706–1721.
114. Gillies, E. R. and Frechet, J. M. (2005). Dendrimers and dendritic polymers in drug delivery. *Drug Discov. Today*, **10**, 35–43.
115. McIntyre, J. O., Fingleton, B., Wells, K. S., Piston, D. W., Lynch, C. C., Gautam, S. and Matrisian, L. M. (2004). Development of a novel fluorogenic proteolytic beacon for in vivo detection and imaging of tumour-associated matrix metalloproteinase-7 activity. *Biochem. J.*, **377**, 617–628.
116. Scherer, R. L., VanSaun, M. N., McIntyre, O. and Matrisian, L. M. (2008). Optical imaging of matrix metalloproteinase-7 activity *in vivo* using a proteolytic nanobeacon. *Mol. Imaging*, 7, 118–131.
117. Kaminskas, L. M., Kelly, B. D., McLeod, V. M., Sberna, G., Boyd, B. J., Owen, D. J. and Porter, C. J. (2011). Capping methotrexate α-carboxyl groups enhances systemic exposure and retains the cytotoxicity of drug conjugated PEGylated polylysine dendrimers. *Mol. Pharm.*, **8**, 338–349.
118. Drury, J. L. and Mooney, D. J. (2003). Hydrogels for tissue engineering: Scaffold design variables and applications. *Biomaterials*, **24**, 4337–4351.
119. Huang, S., Shao, K., Kuang, Y., Liu, Y., Li, J., An, S., Guo, Y., Ma, H., He, X. and Jiang, C. (2013). Tumor targeting and microenvironment-responsive nanoparticles for gene delivery. *Biomaterials*, **34**, 5294–5302.

120. Kim, H. S. and Yoo, H. S. (2010). MMPs-responsive release of DNA from electrospun nanofibrous matrix for local gene therapy: *In vitro* and *in vivo* evaluation. *J. Control. Release,* **145**, 264–271.
121. Kim, H. and Yoo, H. (2012). Matrix metalloproteinase-inspired suicidal treatments of diabetic ulcers with siRNA-decorated nanofibrous meshes. *Gene Ther.,* **20**, 378–385.
122. Lee, H., Rho, J. and Messersmith, P. B. (2009). Facile conjugation of biomolecules onto surfaces via mussel adhesive protein inspired coatings. *Adv. Mater.,* **21**, 431–434.
123. Huang, H. C., Barua, S., Sharma, G., Dey, S. K. and Rege, K. (2011). Inorganic nanoparticles for cancer imaging and therapy. *J. Control. Release,* **155**, 344–357.
124. Spallarossa, P., Altieri, P., Garibaldi, S., Ghigliotti, G., Barisione, C., Manca, V., Fabbi, P., Ballestrero, A., Brunelli, C. and Barsotti, A. (2006). Matrix metalloproteinase-2 and -9 are induced differently by doxorubicin in H9c2 cells: The role of MAP kinases and NAD(P)H oxidase. *Cardiovasc. Res.,* **69**, 736–745.
125. Cheng, Y. C., Chen, L. M., Chang, M. H., Chen, W. K., Tsai, F. J., Tsai, C. H., Lai, T. Y., Kuo, W. W., Huang, C. Y. and Liu, C. J. (2009). Lipopolysaccharide upregulates uPA, MMP-2 and MMP-9 via ERK1/2 signaling in H9c2 cardiomyoblast cells. *Mol. Cell. Biochem.,* **325**, 15–23.

Chapter 4

Redox-Sensitive Nanosystems

J. H. Jeong[1], *T. I. Kim*[2], *J. W. Bae*[3] *and K. D. Park*[3]

[1]*School of Pharmacy, Sungkyunkwan University, Suwon, Republic of Korea*
[2]*Department of Biosystems and Biomaterials Science and Engineering, College of Agriculture and Life Sciences, Seoul National University, Seoul, Republic of Korea*
[3]*Department of Molecular Science and Technology, Ajou University, Suwon, Republic of Korea*

4.1 Introduction

For improved therapeutic efficacy of drugs (small molecule drugs, genes, and proteins), tremendous progress has been made in recent decades to overcome several hurdles encountered using polymer-based drug delivery nanosystems. After drug-carrying polymeric nanosystems reach the target tissue, they need to internalize into cells and release encapsulated or complexed drugs. All these steps should take place sequentially and efficiently to maximize the therapeutic efficacy of delivered drugs, while minimizing potential adverse effects. In this regard, programmed or "on-demand" drug release has attracted great interest as one of the robust drug delivery strategies [1, 2]. Once taken up by cells, polymeric nanosystems can liberate free drugs in response to appropriate stimuli (e.g. pH, temperature, and redox potential). These stimuli-responsive properties can be tailored by incorporating unique chemical functionalities (e.g. environment-sensitive moieties) into polymer structures.

Of the stimuli listed above and others, the redox potential has been extensively used to develop redox-sensitive nanosystems for programmed drug release [3, 4]. The redox potential (or more accurately, reduction potential) is used to infer the direction and free energy cost of reactions involving electron transfer. Such redox reactions, also called reduction–oxidation reactions, are a matched set of redox couples, and a redox state of the redox couples is known to tightly regulate the direction of many cellular processes. Among several redox couples, the glutathione disulfide–glutathione (GSSG–2GSH) couple plays a pivotal role in the two distinct redox environments of intra- and extracellular spaces, which encourages us to develop redox-sensitive nanosystems for efficient intracellular drug delivery. In general, most cytosols are maintained as reductive environments by various intracellular redox-couples for stabilization of correct protein structures, redox cycles, etc. while extracellular space is oxidative [5–7]. Glutathione (GSH, γ-glutamyl-cysteinyl-glycine) is the most abundant intracellular reducing molecule (intracellular GSH concentration: 0.5–10 mM; extracellular milieu GSH: 2–20 μM) and the redox couple between GSH and GSSG (30:1–100:1) is regulated by glutathione reductase [8, 9].

Therefore, bioreducible polymer-based drug delivery nanocarriers can be degraded in the cytosol by cleavage of disulfide linkages via thiol–disulfide exchange reactions due to the high level of GSH, releasing drugs from the redox-sensitive nanosystems. They can also maintain the sufficient stability in the extracellular environment, contrary to their hydrolytically degradable counterpart polymers containing ester, amino ester, or phosphoester linkages [10–12]. This intracellular reducing environment by a unique distribution of GSH is of particular importance with respect to gene delivery systems that require stable condensation of genes in extracellular environments and the programmed release of genes from the systems in intracellular environments. In addition, the reducing environment is also beneficial for targeting small molecule drugs and proteins to the nucleus because endosomal escape is a key step in both intracellular drug and gene therapy. Figure 4.1 shows the schematic diagram of the gene delivery mechanism using bioreducible polymers. Bioreducible polymers can also acquire additional advantages, such as decreased cytotoxicity by avoiding the accumulation of cationic polymers with high molecular weights due to the degradation inside cells.

So far, various kinds of bioreducible polymers have been developed for redox-sensitive drug delivery nanosystems [13, 14]. Herein, the initiation and recent advances in the development and application of bioreducible polymer-based drug delivery systems will be reviewed according to their molecular structures, particularly focusing on gene delivery systems. Their drug delivery mechanism, chemical structure–activity relationships, and therapeutic efficacy will also be discussed.

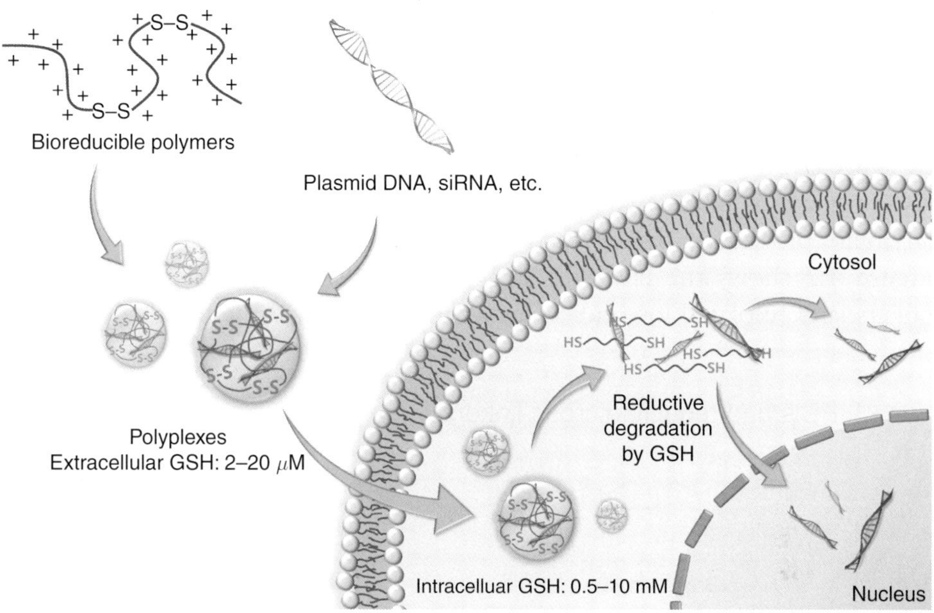

Figure 4.1. Schematic diagram for gene delivery mechanism of bioreducible polymers. DNA = deoxyribonucleic acid; siRNA = short interfering ribonucleic acid, GSH = glutathione.

4.2 Redox-Sensitive Polyethylenimines (PEIs)

PEIs are one of the most frequently used polymers in the field of non-viral gene delivery since the Jean-Paul Behr group utilized these polymers to deliver deoxyribonucleic acid (DNA) and oligonucleotides to cells for the first time [15]. They are composed of repeating ethylenimine units with branched or linear forms and possess high charge density and lots of amines, including primary, secondary, and tertiary amines (1:2:1). This unique structure of PEIs enables the condensation of nucleic acids at low weight ratios and efficient transfection by endosomal escape of polyplexes through a proton sponge effect, which means that internal amines of PEIs can buffer the endosomal pH by an influx of protons, leading to an increase of osmotic pressure in endosomes, the mechanical swelling of PEIs, and finally, endosome escape of polyplexes by endosomal disruption [16]. However, high molecular weight PEIs, such as PEI25k, has shown severe cytotoxicity even at low concentrations, which is caused by intracellular accumulation of the polycations interacting with cellular compartments or molecules. On the other hand, it was reported that low molecular weight PEIs with low cytotoxicity showed poor transfection efficiencies in comparison with PEI25k, probably due to easy dissociation of plasmid DNA (pDNA) from polyplexes [17]. Therefore, lots of high molecular weight PEI derivatives with biodegradable linkages, such as ester bonds, have been developed for the maintenance of high transfection efficiency and decrease of cytotoxicity by cross-linking low molecular weight PEIs [18]. Among them, PEI derivatives with bioreducible linkages have been also synthesized for gene delivery systems in order to apply the selective degradability of disulfide bonds in the reducing environment of the cytoplasm. Figure 4.2 shows the chemical structures of PEIs, several reducible cross-linkers, and thiolated low molecular weight PEIs mentioned in this review.

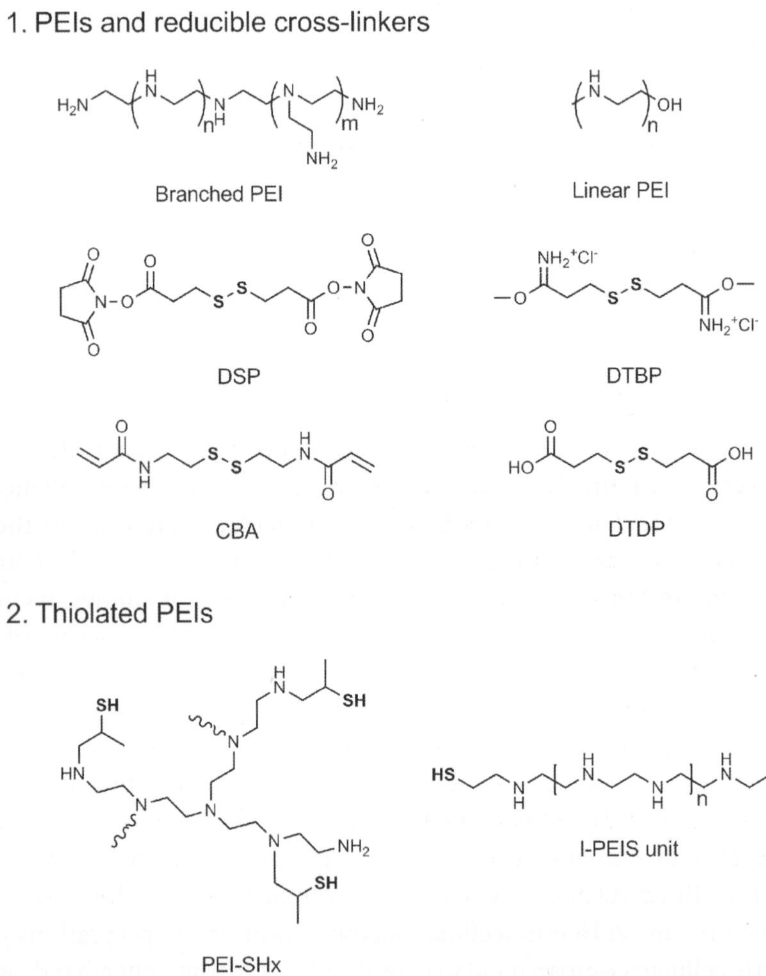

Figure 4.2. Chemical structures of PEIs, several reducible cross-linkers, and thiolated low molecular weight PEIs. DSP = dithiobis(succinimidylpropionate), DTBP = dimethyl 3,3′ dithiobispropionimidate·2HCl, CBA = cystamine bisacrylamide, DTDP = 3,3′-dithiodipropionic acid.

Lee and coworkers synthesized bioreducible PEI derivatives by cross-linking low molecular weight PEIs (800 Da) with amine-reactive and reducible cross-linkers, dithiobis(succinimidylpropionate) (DSP) or dimethyl·3,3′-dithiobispropionimidate·2HCl (DTBP) to improve gene delivery efficiency [19]. They found that these cross-linked polymers could not condense pDNA in the presence of a reducing agent, dithiothreitol (DTT), and the transfection efficiency was proportional to the conjugate size. DTBP conjugates exhibited higher gene expression than DSP conjugates in the almost equivalent condition, presumably due to the maintenance of net polymer charges by DTBP cross-linking through amidine bond formation, unlike the DSP case. Breunig *et al.* synthesized disulfide cross-linked low molecular weight linear PEIs using similar amine reactive cross-linkers (DSP or mixture of *N*, *N′*-bis-(*tert*-butoxycarbonyl)cystine and 4-(4,6-dimethoxy[1.3.5]triazin-2-yl)4-methyl-morpholiniumchlorid hydrate (DMT-MM)) [20]. They confirmed that the intracellular reductive degradation of the polymers produced nontoxic fragments and these cross-linked bioreducible PEIs showed a superior transfection efficiency compared with commercial transfection reagents, such as PolyFect, Lipofectamine, FuGENE6, or JetPEI. Degradable disulfide-containing PEIs (SS-PEIs) were also synthesized for gene delivery systems by a Michael addition of cystamine bisacrylamide (CBA) with a PEI (800 Da) [21]. Although the transfection efficiency of SS-PEIs was comparable to that of PEI25k, the SS-PEIs showed a lower cytotoxicity. Moreover, the level of transfection efficiency decreased by the addition of RGD tripeptide, demonstrating that a surplus of RGD peptide might inhibit the specific endocytosis of RGD in polyplexes and nonspecific endocytosis of SS-PEI polyplexes in Henrietta Lacks (HeLa) cells possessing $\alpha_v\beta_3$ and $\alpha_v\beta_5$ integrins. This reducible polymer was further modified with mannosylated PEG by the Xiao group for tumor necrosis factor-α (TNF-α) short interfering ribonucleic acid (siRNA) delivery targeted to macrophages in which the mannose receptor is highly overexpressed [22]. In addition, they stabilized the polyplex with sodium triphosphate. These reducible polyplex nanoparticles could inhibit the expression of TNF-α in macrophages specifically, showing the potential for macrophage-targeted inflammatory bowel disease (IBD). Park *et al.* also synthesized reducible PEIs by the cross-linking of PEIs (2 kDa) by CBA and conjugated the polymer to hyaluronic acid (HA) via amide bond formation [23]. The resulting (PEI-SS)-g-HA could be internalized into cells by HA receptor-mediated endocytosis and the (PEI-SS)-g-HA/transforming growth factor-β (TGF-β) siRNA complex showed a feasible therapeutic effect on liver cirrhosis with drastic reduction of messenger ribonucleic acid (mRNA) level.

On the other hand, bioreducible PEIs were synthesized by the direct cross-linking of PEIs (800 Da) with 3,3′-dithiodipropionic acid (DTDP) using

EDC/NHS (*N*-ethyl-*N'*-(3-dimethylaminopropyl)carbodiimide/*N*-hydroxysuccinimide) chemistry and were used to deliver human telomerase reverse transcriptase (hTERT) siRNAs for the inhibition of HepG2 tumor growth *in vitro* and *in vivo* [24]. Other groups also used this DTDP as a reducible linker to conjugate Tween 85 and PEI (2 kDa) by EDC chemistry [25]. Tween 85 was conjugated to stabilize polyplexes in the blood circulation and increase cellular uptake via interactions with low-density lipoprotein receptors. Polyplexes formed by the synthesized Tween 85-s-s-PEI 2K (TSP) and p65 (subunit of NF-κB promoting cancer cell proliferation and metastasis) shRNA, could significantly inhibit p65 expression in cancer cells, induce cell apoptosis, and eventually inhibit growth and metastasis of breast cancer cells in a mouse xenograft model.

The Kissel group reported another strategy for bioreducible gene delivery systems based on PEI cross-linking [26]. They formulated PEI (5 kDa)/DNA polyplexes by mixing pre-cross-linked polymers with DNA or using polyplexes cross-linked after complexation. Cross-linking reactions were performed using amine reactive reducible cross-linker DSP. Only the latter method showed enhanced resistance against polyanion exchange and high ionic strength, and displayed reduced interactions with major blood components like albumin and erythrocytes. Moreover, it was revealed in further study that the transfection efficiency was significantly affected by intracellular status, and cross-linked PEI/DNA polyplexes appeared to be stable in the blood circulation, showing accumulation in the liver and the lungs of mice and predominated liver transfection [27]. The same group performed an additional PEGylation step to increase the stability of reversibly cross-linked PEI polyplexes [28]. They prepared polyplexes with the resulting AB-type and ABA-type copolymers (A: PEG20k or PEG30k; B: PEI25k) and cross-linked them with DSP. These PEI–PEG copolymers displayed reduced cytotoxicity and hemolytic activity, and increased transfection efficiency and circulation times in the blood stream. However, the cross-linked polyplexes without PEG showed increased blood levels of plasmid deoxyribonucleic acid (pDNA).

Lee *et al.* employed a similar strategy for siRNA delivery, which is based on the stabilization of polyplexes by surface cross-linking with DSP [29]. They prepared the cross-linked polyplexes (Cr PEI-DA) by using deoxycholic acid-conjugated PEI1.8k (PEI-DA) to improve gene transfection efficiency via endocytosis- and energy-dependent manners due to facially amphipathic DA [30]. Matrix metalloproteinase-2 (MMP-2) siRNA was complexed with Cr PEI-DA to inhibit smooth muscle cell (SMC) invasion in neointimal hyperplasia [31]. It was found that Cr PEI-DA/MMP-2 siRNA polyplexes were efficiently taken up by SMCs and showed effective silencing of MMP-2 gene expression under serum conditions, in comparison with non-reducible PEI-DA polyplexes, indicating the therapeutic potential of MMP-2 siRNA delivery by Cr PEI-DA.

Bioreducible PEI can also be prepared without using cross-linkers. Zhong and coworkers first synthesized thiolated PEI0.8k (PEI-SH_x) via a ring-opening reaction of methylthiirane (thiolation) and then cross-linked PEI-SH_x by mild dimethyl sulfoxide (DMSO) oxidation [32]. These cross-linked PEIs (molecular weight: 7.1–8.4 kDa) showed a lower cytotoxicity and a higher gene transfection efficiency, as compared with PEI25k. In a similar way, Kim and coworkers attempted to cross-link branched PEIs (1.2k) via the ring-opening of propylene sulfide and DMSO oxidation, and further modified the polymers with cNGR-PEG5k-maleimide for tumor targeted gene delivery systems [33]. Cyclic NGR (cNGRCK) is known to recognize the CD13/aminopeptidase N receptor over-expressed in neo-vasculature, showing potential as a tumor-targeting moiety [34]. The transfection efficiency of these cross-linked PEIs was highly enhanced with low cytotoxicity. Free cNGR treatment reduced the cellular uptake and transfection efficiency of BPEI-SS-PEG-cNGR polyplex in CD13-positive cells, showing its tumor-target specificity. After additional treatment of buthionine sulfoximine (BSO) depleting intracellular GSH, the transfection efficiency was reduced synergistically, indicating its GSH-dependent transfection pathway. They also conjugated a neuronal cell specific targeting moiety, RVG peptide (YTIWMPENPRPGTPCDIFTNSRGKfRASNG), to disulfide-cross-linked BPEI (PEI-SS) with a PEG linker for gene delivery to brain, and found that this polymer could deliver pDNA to mice brains through a receptor-mediated pathway [35].

On the other hand, Lee *et al.* synthesized linear poly(ethylenimine sulfide) (l-PEIS) using oxidation of thiolated linear polyamines for gene delivery systems [36]. The transfection efficiency of l-PEIS increased with increase in the amine density and it was comparable to that of ExGen500 without high cytotoxicity due to the biodegradable disulfide bonds. Intracellular degradation of l-PEIS in cells was also confirmed by visualization using the probe–probe dequenching effect of BODIPY-FL fluorescence dye. In a later study, the same group reported the synthesis and characterization of bioreducible branched poly(ethylenimine sulfide) (b-PEIS) by cross-linking l-PEIS using bisepoxide as a cross-linker [37]. It was observed that b-PEIS could be degraded in reducing conditions with dependency on the degree of cross-linking. The b-PEIS (M_w: 13.6–191.2 kDa) showed higher transfection efficiency and lower cytotoxicity than l-PEIS (M_w: 9.5 kDa). Intracellular trafficking observation results indicated that b-PEIS was finally excreted from the cells during transfection process without accumulation in the cell interior.

4.3 Redox-Sensitive Polypeptides

Bioreducible polypeptides have many advantages for the development of gene delivery carriers. As a building block, amino acids are non-toxic and almost non-immunogenic endogenous molecules found to be rich in body. In addition, they have more than 20 kinds of chemical structures with various functional side chains (e.g. cationic residues for the condensation of nucleic acids) that provide the diversity or flexibility for designing gene delivery carriers. Peptides are generally known to be very stable and have already been used as clinical therapeutics. Synthesis of peptides by well-established solid-phase chemistry leads us to produce efficiently sequence-specific and almost monodispersed peptide products. Some motif structures of peptides can also be introduced in this sequence-specific manner for specific interactions with biologically important molecules. Cross-linking reactions can also be simply performed in mild conditions by oxidizing thiol-terminated short peptides containing cysteine terminals or by using commercial cross-linkers. However, the high cost of peptide synthesis and the low synthesis yield are still major issues for the development of polypeptide-based gene delivery carriers. Using a similar strategy with synthesized polymers, the cross-linking of short peptides with reducible linkages has been extensively studied with the aims of developing novel gene delivery systems with high stability and transfection efficiency and low cytotoxicity.

The Rice group first reported the bioreducible gene delivery carriers based on peptide [38]. PEG-peptides (PEG–SS–Cys–Trp–Lys(18), PEG–SS–CWK_{18}, where Cys = cysteine, Trp = tryptophan, Lys = lysine) possessing disulfide linkages displayed a much more enhanced gene transfection efficiency compared to PEG-peptides connected by non-reducible cross-linkers (PEG-VS-CWK_{18}), probably due to the intracellular release of DNA triggered by a disulfide linkage reduction. In a following study, they synthesized low molecular weight polypeptides by substitution of one to four lysine residues of CWK_{18} with cysteine residues [39]. These polypeptides could form stable polyplexes with pDNA after cross-linking by spontaneous oxidization of surficial thiols of cysteines and exhibit significantly enhanced transfection efficiencies compared to the uncross-linked polyplexes. In addition, it was found that a minimal peptide composed of four lysines and two terminal cysteines could also result in similar transgene expression with long peptide chains when polyplexes were formed after spontaneous cross-linking [40]. Furthermore, optimal peptides of Cys–His–$(Lys)_6$–His–Cys (where His = histidine) showed enhanced transfection efficiency in the absence of chloroquine, endosomolytic agent, due to enough buffering capacity, indicating that low molecular weight cross-linking peptides can also buffer the endosomal pH.

In contrast to the above reports, Seymour and coworkers synthesized reducible polycations (RPCs) by oxidation of terminal cysteinyl thiol groups of Cys(Lys)$_{10}$Cys as a DNA delivery vector [41]. The coated and surface-cross-linked RPC polyplexes with *N*-(2-hydroxypropyl)methacrylamide (PHPMA) showed reversible and lateral stabilization, which released free DNA after incubation with polyanions in reducing conditions, contrary to non-reducible poly-L-lysine (PLL) polyplexes. The coated RPC polyplexes exhibited enhanced transgene expression, compared to the non-reducible PLL complexes due to the intracellular degradation. In a following study, Read *et al.* reported that RPCs composed of peptide (Cys-Lys$_{10}$-Cys) showed enhanced gene delivery efficiency compared with PLLs in the presence of cationic lipid (*N*-(1-(2,3-dioleoyloxy)propyl)-*N*,*N*,*N*-trimethylammonium chloride) (DOTAP), demonstrating the potential of lipid-mediated RPC-based transfection [42]. Several HIS RPCs consisting of histidine and polylysine residues were also developed for gene delivery systems by oxidation of terminal cysteinyl thiol groups [43]. Low cytotoxic HIS RPC vector ([Cys–His$_6$–Lys$_3$–His$_6$–Cys]$_n$) with at least 70% histidine content could mediate efficient gene transfer without chloroquine, endosomolytic agent, and also show enhanced delivery of green fluorescent protein (GFP) messenger ribonucleic acid (mRNA) and siRNA directed against the neurotrophin receptor p75 (NTR) in comparison with PEI25k. Moreover, it was found that the molecular weight and cleavage rate of HIS RPC were important parameters for gene transfection. Efficient enhanced green fluorescent protein (EGFP) siRNA delivery by HIS RPC was further performed and showed necessity of polymer reducibility for siRNA activity [44]. In this work, the incorporation of liver-targeting peptide onto HIS RPC significantly improved the delivery of DNA or siRNA into hepatocytes. As a result, HIS RPC/siRNA (for hepatitis B virus surface antigen) polyplexes could decrease both mRNA and protein expression, suggesting the potential for viral hepatitis therapy. Another RPC formed from Cys(Lys)$_{16}$Cys units was combined with an anionic fusogenic peptide for endocytic escaping effect [45]. RPC/DNA peptide particles were stable in the presence of anionic fusogenic peptide. RPC/DNA/fusogenic peptide particles showed serum-tolerant and highly efficient transfection efficiency.

Reducible copolypeptides (rCPP) composed of a histidine-rich peptide (HRP, CKH$_3$KH$_3$KC) and a nuclear localization sequence (NLS, CGAGPK$_3$RKVC) peptide with different relative contents were synthesized by an oxidative copolymerization [46]. The minimal cytotoxicity of rCPPs was thought to be due to lower charge density and higher chain rigidity of the rCPPs than those of PEI25k, as well as reductive intracellular and plasma membrane degradation. Transfection efficiency of rCPPs increased with increasing content of the HRP in rCPP, probably due to the enhanced buffering capacity. No clear effect of the NLS sequence on transfection activity suggested that NLS peptides might be dissociated from rCPP polyplexes by reductive cleavage in cytosol prior to nuclear transportation of DNA. In a following study, rCPP/(siRNA or pri-miRNA) polyplex also showed increased gene knock down with a greater HRP content [47]. Interestingly, it was observed that including NLS could facilitate nuclear localization of delivered siRNA or primary micro ribonucleic acid (pri-miRNA).

Park *et al.* utilized a fusogenic peptide (KALA peptide) with two terminal cysteine residues to cross-link the peptide by mild oxidation [48]. Cytotoxicity of reducible cross-linked KALA (cl-KALA) was low and cl-KALA/siRNA polyplexes exhibited comparable gene silencing efficiency. In addition, cl-KALA/siRNA-PEG conjugate polyplexes showed superior gene down-regulation efficiency in serum conditions due to the shielding effect of PEG on the polyplex surface. In a later study, cl-KALA was complexed with 6PEG-siRNA-Hph1 (siRNA-conjugated six-arm PEG functionalized with cell penetrating peptide, Hph1) and this multifunctional polyelectrolyte complex showed superior physical stability and enhanced gene silencing efficiency in serum [49].

Tat sequence, as a cell-penetrating peptide known to enhance the cellular uptake of various molecules, was incorporated into bioreducible peptides for gene delivery [50]. POLYTAT peptide was synthesized by oxidative cross-linking of Tat-containing peptide, CTATC (CGRKKRRQRRRGC). High molecular weight POLYTAT showed enhanced transfection efficiency compared to control Tat peptide, suggesting that internalization of Tat-based polyplexes might be mediated by endocytosis. In another work, Tat-containing peptide (Tat-10H) with ten histidine residues and terminal cysteine residues was used [51]. Interestingly, Tat-10H displayed 7,000-fold improvement in gene transfection efficiency over Tat peptide. Cross-linked Tat-10H via disulfide bond formation showed more efficient transfection efficiency than Tat-10H because of the enhanced stability of peptide/DNA polyplexes.

Another cell penetrating peptide was used for the synthesis of reducible polypeptide-based gene delivery systems [52]. Reducible poly(oligo-D-arginine) (rPOA) was synthesized by oxidative cross-linking between terminal cysteinyl thiol groups of Cys-(D-R9)-Cys repeating units. rPOA polyplexes showed a higher level of transgene expression than PEI25k. In addition, the mouse lungs did not show any significant morphological or inflammatory changes after

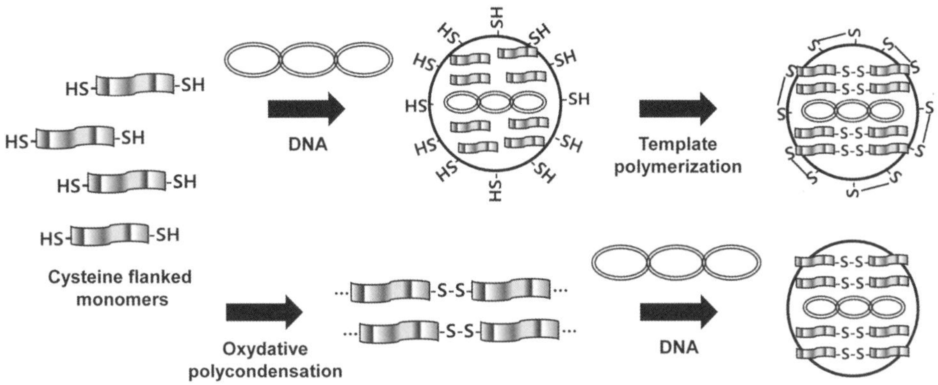

Figure 4.3. Schematic illustration of template polymerization and oxidative polycondensation methods of cysteine flanked peptide monomers.

treatment with the rPOA polyplexes. Evaluation of this rPOA as a carrier of the heme oxygenease-1 (HO-1) gene was also performed [53]. The rPOA/pHO-1 polyplex showed a significantly reduced infarct volume and TNF-α level in an ischemia/reperfusion (I/R)-induced brain stroke rat model, suggesting the potential for protection of the brain from I/R related neuronal injury with anti-inflammatory effects.

Recently, template polymerization and oxidative poly-condensation methods for the formation of bioreducible peptide-based gene delivery carriers were compared [54]. Figure 4.3 shows the scheme for template polymerization and oxidative poly-condensation methods of cysteine-flanked peptide monomers. Cys–His–(Lys)6–His–Cys, Cys–His–(Lys)6–His–His–(Lys)6–His–Cys, and Cys–His–(Arg)6–His–Cys peptides were used for characterization. In this work, polyplexes based on template polymerization showed more resistance to dissociation by polyanions, and polyplexes based on oxidative poly-condensation displayed higher transgene expression efficiency. Among the peptides, arginine-rich peptide showed the most efficient transfection efficiency.

4.4 Redox-Sensitive Poly(amido amine)s

Bioreducible poly(amido amine)s can be synthesized by the addition polymerization of amine donors to a disulfide compound having two acrylamide groups. Although the synthesis of poly(amido amine)s was reported several decades ago [55], their reducible form has recently drawn much attention in the field of gene delivery due to their ability to unload the nucleic acid cargo in intracellular environments. A series of bioreducible poly(amido amine)s containing multiple disulfide bonds (SS-PAAs) was synthesized by Michael-type addition polymerizations [56]. The structure of SS-PAA can be controlled by selecting different amine donors. Branched type poly(amido ethylenimine)s (SS-PAEI) can be synthesized by addition polymerizations between amine donors with more than two amine groups, such as ethylenediamine (EDA), diethylenetriamine (DETA), and triethylenetetramine (TETA), and disulfide bond-containing CBA [57]. The branched structure can be generated since the amine donors contain multiple primary and secondary amines reactive to the acrylamide group of CBA. It is interesting that poly(DETA/CBA) showed the lowest proton buffering capacity, but the highest degree of branching among the three polymers. All of the polymers were able to interact with pDNA and form tight complexes having sizes less than 200 nm. The plasmid pDNA could be readily released from the complex in the presence of reductive forces, such as 2.5 mM dithiothreitol (DTT) due to the reduction of disulfide linkages in the polymer backbone. The SS-PAEIs demonstrated a transfection efficiency up to 20 folds higher than in PEI25k in mammalian cells. It should be noticed that the transfection efficiency of SS-PAEIs was maintained even in the presence of 10% serum, while that of PEI25k significantly reduced. Confocal microscopy results using fluorescently labelled plasmid DNA demonstrated that the plasmid DNA was distributed over the whole cytoplasm, suggesting that the reductive degradation of the polymers could facilitate the unloading of plasmid DNA in the cytoplasm by reduced GSH. However, the polyplexes formed from PEI25k were localized in a specific area in the form of a globular cluster.

The concept of the active cytosolic release was also applied for the delivery of siRNA, since the cytosol is where siRNA takes action [58]. Poly(TETA-CBA) could form stable complexes via electrostatic interactions with siRNA in the physiological condition. The complexes were dissociated to release siRNA under reductive environments containing 2.5 mM DTT. Poly(TETA-CBA)/siRNA formulation containing siRNA targeting vascular endothelial growth factor (VEGF) exhibited significantly higher target gene silencing effects in human prostate cancer cells (PC-3) than linear PEI25k (L-PEI 25k). The increased gene silencing efficiency was offset by the pretreatment of D,L-buthionine sulfoxamine (BSO), an agent causing intracellular GSH depletion. The result shows the critical role of intracellular GSH concentration for efficient gene delivery using the bioreducible polymer. In addition, substantial improvement in the cytosolic distribution of siRNA was observed with the poly(TETA-CBA)/siRNA complexes in a confocal microscopy analysis, while L-PEI 25k/siRNA complexes existed in several areas in the form of globular aggregates. The enhanced gene silencing activity by poly(TETA-CBA) may be due to the rapid cytosolic release of siRNA, which facilitates the access and incorporation of siRNA to RNA-induced silencing complexes (RISC) to induce specific RNAi phenomena. The reductive degradation of poly(TETA-CBA) in the cytoplasm is an interesting property since it can increase the bioavailability of plasmid DNA and siRNA in the cell cytoplasm. In conjunction with plasmid DNA that contains binding motifs for a karyophilic protein, nuclear factor kappa B (NFκB), poly(TETA-CBA) was used for improved intracellular trafficking and nuclear localization of the plasmid DNA [59]. The reductive degradation of the polymer and cytosolic release of plasmid DNA would provide a chance for the DNA to bind to NFκB, since the nuclear transcription factor normally exists in the cytosol and translocates to the nucleus upon stimulation by several signaling factors such as cytokines. It should be noted that poly(TETA-CBA)-based transfection of plasmid DNA resulted in rapid cytosolic release and enhanced nuclear localization upon the stimulation with interleukin-1β (IL-1β), which may lead to enhanced transfection and expression of encoded gene, glucagon-like peptide (GLP-1).

Among the SS-PAEI, poly(amido ethylenediamine) (SS-PAED) was used as a carrier for plasmid DNA encoding vascular endothelial growth factor (VEGF) under the regulation of hypoxia-inducible promoter (RTP801) [60]. The RTP801 promoter is known to up-regulate transcription of a downstream gene under the oxygen depletion condition of cells [61]. The transfection with the SS-PAED/RTP-VEGF complex resulted in a significant increase of VEGF production in primary rat cardiomyoblast cells (H9C2) and rat aortic smooth muscle cells (A7R5) under hypoxic conditions, compared to the cells under normoxic conditions. Local gene delivery in a rat myocardium with infarction at the left ventricle area using SS-PAED/RTP-VEGF complexes demonstrated *ca.* four-fold increase of VEGF production in the infarcted region compared to the complex containing a control plasmid DNA (SS-PAEI/RTP-Luc). The result suggests the possibility of improved viability of cardiomyocytes and neovascularization in the ischemic myocardium. Linear type SS-PAA was also synthesized by the same addition polymerization reaction with various monomers containing primary amine groups, such as 4-amino-1-butanol (ABOL), 5-amino-1-pentanol (APOL), *N*,*N*-dimethyl-1,3-ethylenediamine (DMEA), 2-(2-aminoethoxy) ethanol (AEEOL), 3-methoxypropylamine (MOPA), 3-morpholinopropylamine (MPA), or histamine (HIS)) with CBA containing a disulfide bond. The linear polymers formed stable complexes via electrostatic interactions with plasmid DNA. Although the complexes exhibited a desirable level of stability in the physiological condition, their rapid degradation was observed in the presence of a reducing agent, such as 2.5-mM DTT. High buffering capacity SS-PAAs in the endosomal pH range may contribute to the escape of complexes from the endosomal compartment by hypothetical proton sponge effect [15]. Most SS-PAAs showed low cytotoxicity with desirable levels of transfection efficiency. The highest transfection efficiency could be achieved by using SS-PAAs containing amino alcohol pendent groups (pABOL and pAPOL). A dual bio-responsive gene delivery system composed of a reducible polymer (pABOL) and a plasmid DNA encoding soluble VEGF receptor, sFlt-1, under the control of a cancer-specific survivin-inducible promoter showed cancer-specific expression in the mouse renal carcinoma cells (RENCA), as the system was responsive to reductive intracellular environments and survivin, of which expression is known to be up-regulated in most cancer cells [62].

The synthesis of linear SS-PAA in the form of random and block copolymers composed of two different amine donors with distinct basicity, histamine (HIS) and 3-(dimethylamino)-1-propylamine (DMPA), and reducible CBA was carried out by Engbersen and coworkers [63]. The transfection efficiency of the copolymers, p(CBA-HIS/DMPA)s, increased with the length of HIS block and buffering capacity. The random and block copolymers with a HIS/DMPA composition ratio of 70/30 showed the highest transfection efficiency in COS-7 cells with desirable plasmid DNA condensation ability and buffering capacity, compared to homopolymers. They also synthesized a series of SS-PAA homo- and copolymers having bisacrylamide segments consisting of nonreducible monomers containing secondary and tertiary amine groups such as *N,N′*-methylenebisacrylamide (MBA), 1,4-bis(acryloyl)piperazine (BAP), *N,N′*-hexamethylenebisacrylamide (HMBA) and varying amounts of bioreducible CBA monomers, and investigated the composition effect on transfection efficiency [64]. A single amine donor, 1-(2-aminoethyl) piperazine (AEP), was used for the addition polymerization with the bisacrylamides. Among the copolymers, poly(CBA-AEF) showed significantly higher transfection efficiency in COS-7 cells, compared to homo- or copolymers without reducible disulfide linkages. The result suggests that the addition of appropriate amounts of bisacrylamide monomers containing a disulfide linkage can achieve improvements in transfection efficiency and cytotoxicity profiles. In a separate study, *N,N′*-dimethylcystamine (DMC) as a disulfide-containing bioreducible unit and various bis-acrylamides were used for the synthesis of linear SS-PAA [65]. Owing to its disulfide bond located close to a tertiary amine group that lowers the pK_a value of the basic unit, the DMC-based SS-PAA demonstrated a significantly high buffering capacity within the endosomal pH range, suggesting facilitated endosomal escape of transfection complexes via the proton sponge effect. In addition to its ability to condense plasmid DNA and release DNA in the reductive environment, the DMC-based SS-PAA were capable of transfecting cells (COS-7) with a transfection efficiency up to four folds higher, compared to PEI and pDMAEMA.

Ou *et al.* published a series of linear type reducible poly(disulfide amine) s having pendant primary amines with different lengths of alkyl chains [66]. The polymers were synthesized by addition polymerizations of CBA and *N*-Boc protected diamines, (*tert*-Butyl-*N*-(2-aminoethyl) carbamate (*N*-Boc-1,2-diaminoethane, N-Boc-DAE), *tert*-butyl-*N*-(4-aminobutyl) carbamate (*N*-Boc-1,4-diaminobutane, N-Boc-DAB), *tert*-butyl-*N*-(6-aminohexyl) carbamate (*N*-Boc-1,6-diaminohexane, N-Boc-DAH). The poly(disulfide amine)s could form compact complexes with plasmid DNA and be dissociated to release DNA in reductive conditions, such as in the presence of 5-mM DTT. The polymer also showed a high buffering capacity in the endosomal pH region, which might facilitate the escape of complexes from the endosomal compartment. The transfection efficiency of poly(disulfide amine)s was dependent on the length of the alkyl chain of the *N*-Boc protected diamine monomers. Among the poly(disulfide amine)s, poly(CBA-DAH) having hexaethylene spacers demonstrated the highest transfection efficiency than PEI 25k, while maintaining significantly low cytotoxicity. Poly(CBA-DAH), in conjunction with a prostaglandin E2 (PGE2)-Fas siRNA conjugate, was used for targeted siRNA delivery to cardiomyocytes under a hypoxic condition [67]. The enhanced cellular uptake by rat cardiomyocyte cells (H9C2) was observed in poly(CBA-DAH)/PEG2-siRNA complexes, owing to specific interactions between PEG2 and its cellular receptors, which led to enhanced transfection and increased levels of target gene silencing. The inhibition of Fas gene expression in H9C2 cells could effectively prevent apoptotic cell death in a hypoxic condition, since Fas is known to be one of the most important regulators for ischemia-induced apoptosis in heart myocardium. In addition to the anti-apoptotic activity, the complex did not induce any immune responses involved with the induction of interferon α (IFN-α).

In subsequent research, Nam *et al.* synthesized a conjugate of primary cardiomyocyte-specific peptide (WLSEAGPVVTVRALRGTGSW) and poly(CBA–DAH). Owing to primary amines in poly(CBA-DAH), the polymer could be readily modified to have additional functions. Arginine-grafted poly(CBA-DAH) (ABP) [68–70] and guanidylated bioreducible polymer (GBP) [71] were synthesized by introducing either an arginine or guanidine group to the pendant primary amine group of poly(CBA-DAH). Arginine and guanidine can be found in many cell penetrating peptides (CPPs) and protein transduction domains (PTDs) as main components and are considered to have a critical function in intracellular transport phenomena by CPPs and PDTs [72, 73]. Similar to poly(CBA-DAH), ABP, and GBP were able to form compact nano-complexes with plasmid DNA and siRNA, and the genes could be efficiently released in a reductive environment. They showed significantly improved transfection efficiency in mammalian cells compared to its backbone polymers, poly(CBA-DAH) and PEI25k, while maintaining low cytotoxicity. Especially, GBP was able to mediate high cellular uptake and enhanced localization of plasmid DNA in the nucleus, suggesting the potential role of the guanidine unit for facilitated intracellular trafficking and nuclear localization. The potential use of ABP in gene therapy was accessed in a recent publication, in which ABP was used as a carrier for a VEGF-silencing siRNA for anti-angiogenic cancer therapy. APB was also employed for the *ex vivo* delivery of plasmid DNA encoding VEGF gene to primary myoblasts. This research showed potential application of the polymeric gene delivery system in gene therapy for the ischemic heart disease by VEGF-mediated induction of angiogenesis and myogenesis.

4.5 Redox-Sensitive Polyion Complex (PIC) Micelles

PIC micelles can be differentiated from conventional polymeric micelles by the composition of the core, which consists of a polyelectrolyte complex of two oppositely charged macromolecules [74]. The core structure that is formed by the reduced solubility of two oppositely charged polyelectrolytes via charge canceling can be generally stabilized by the surrounding shells of hydrophilic polymers. Therefore, at least one of the polymers should be a copolymer having a neutral hydrophilic block and a charged block. Owing to their superior colloidal stability and solubility in an aqueous medium, PIC micelles were used as a carrier for plasmid DNA and siRNA [75].

Thiolated poly(ethylene glycol)-block-poly(L-lysine) (PEG-PLL) was synthesized and employed GSH-sensitive PIC micelles by interacting with either antisense oligonucleotides (ODNs) [76] or plasmid DNA [77]. The PIC micelles, having stabilized PIC cores by cross-linking of the thiol groups, showed a significantly enhanced stability against nucleases as compared to PIC micelles without cross-linking. The dissociation of the PIC micelles and release of ODNs and plasmid DNA could be achieved in the presence of reductive forces, such as 2.5-mM DTT. A different type of PIC micelle was also designed to deliver ODNs and siRNA by conjugating PEG to either ODN [78] or siRNA [79] via a disulfide linkage. The PIC micelles based on PEGylated ODN or siRNA have an advantage of selecting various functional cationic polymers, such as PEI with improved endosomal escape function via a proton buffering effect. After intracellular delivery, the PEG segment can be removed and release intact ODN or siRNA by the action of GSH, which could effectively avoid potential interference of PEG. The PIC micelles also demonstrated long circulation in the bloodstream after systemic administration, which may facilitate the accumulation of the micelles in solid tumors via an enhanced permeation and retention (EPR) effect [79]. The systemic injection of PIC micelles containing VEGF-targeting siRNA led to the efficient silencing of VEGF gene expression and significant growth retardation of solid tumors in an animal xenograft model.

Antisense ODN conjugated to HA via a reducible disulfide linkage was also reported [80]. The reducible HA-SS-ODN based PIC micelles could mediate a desirable level of transfection and showed the potential use of the PIC micelle system for targeted delivery of therapeutic ODNs to CD44-positive cancer. Recently, Takae *et al.* synthesized a PEG-sheddable block catiomer, PEG-SS-P[Asp(DET)], and used the catiomer for PIC micelle-based gene delivery [81]. The disulfide linkage allows for the release of PEG after intracellular delivery, and aspartamide with a pendent *N*-(2-aminoethyl)-2-aminoethyl group (Asp(DET)) acts as a proton buffering unit, facilitating escape from the endosome. The PIC micelles with a disulfide linkage (PEG-SS-P[Asp(DET)]) showed higher transfection efficiencies than those without a disulfide linkage, probably due to improved endosomal escape by the cleavage of PEG in the endosome.

4.6 Redox-Sensitive Nanogels

Nanogels are hydrogel-based nanoparticles with a high water content, which are chemically or physically cross-linked. Cross-linking reactions for the preparation of nanogels include ionic cross-linking, self-assembly, crystallization, cross-linking polymerization, radiation cross-linking, and functional group cross-linking [82]. Due to the hydrophilic nature of nanogels and the controllable size for the EPR effect, they are considered suitable for tumor-targeted delivery of hydrophilic drugs, such as proteins, peptides, and genes.

The Park group synthesized a heparin-pluronic (HP) conjugate by conjugating heparin to carboxylated Pluronic via EDC/NHS chemistry [83, 84]. Heparin, a highly negatively charged polysaccharide, has a high binding affinity to various therapeutic proteins including cytokines and growth factors. In addition, amphiphilic Pluronic can encapsulate hydrophobic drugs into the micelle core during self-assembly. These unique features would make the HP conjugate particularly advantageous for simultaneous delivery of multiple drugs. The HP conjugate could form nanogels having 89.1 nm in diameter via self-assembly of Pluronic chains. Since HP nanogels had little cytotoxic activity against HeLa cells due to the high cellular uptake and loading of

RNase A, bioreducible cross-linked HP (DHP) nanogels were prepared by further conjugating cystamine to HP and then treatment with DTT [85], as illustrated in Figure 4.4. The DHP nanogels showed an augmentable release profile of the encapsulated RNase A in response to GSH concentrations, thereby resulting in increased cytotoxicity against NIH3T3 cells. In another study, they functionalized redox-sensitive HP nanogels with cRGDfC peptide that specifically binds to endothelial $\alpha_v\beta_3$ and $\alpha_v\beta_5$ integrins exposed in tumor vessels of several tumor types [86]. The result showed that cRGDfC-HP nanogels could release RNase in response to the GSH treatment and appeared to be more cytotoxic than unmodified HP nanogels. As mentioned above, HP nanogels have the ability to simultaneously encapsulate hydrophilic and hydrophobic drugs due to the negatively charge of heparin and hydrophobic segments of Pluronic. Joung *et al.* recently reported HP nanogels as multi-drug nanocarriers for combination chemotherapy using more than two therapeutic agents with different modes of action [87]. The HP nanogels encapsulating both paclitaxel and deoxyribonuclease showed a dose-dependent synergistic cytotoxicity against HeLa cells, as compared to single drug formulations. Although this system is not redox-sensitive, one can speculate that redox-sensitive HP nanogels carrying both drugs would be more effective to kill cells.

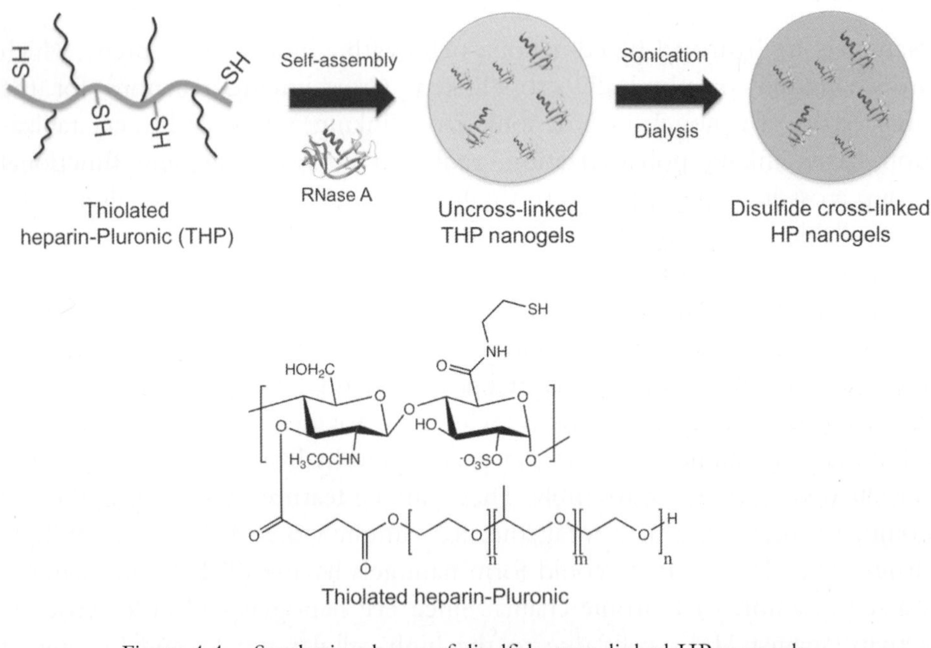

Figure 4.4. Synthetic scheme of disulfide cross-linked HP nanogels.

As heparin is involved in diverse physiological processes through interacting with many proteins, Bae *et al.* developed reducible heparin nanogels for efficient cellular uptake of therapeutic heparin [88]. The heparin nanogels with an average diameter of *ca.* 250 nm were prepared by mixing thiolated heparin and PEG in DMSO, allowing for selective oxidation of D-glucuronic acid residues of heparin and subsequent disulfide cross-linking of cysteamine. The reducible heparin nanogels that disintegrated rapidly and released free heparin under reducing environments significantly inhibited the proliferation of mouse melanoma B16F10 cells by inducing caspase-mediated apoptotic cell death.

Oh *et al.* described the preparation of redox-sensitive nanogels using inverse mini-emulsion atom transfer radical polymerization (ATRP) of oligo(ethylene oxide)monomethyl ether methacrylate (OEOMA) in the presence of disulfide-functionalized dimethacrylate (DMA) as a cross-linker [89]. The OEOMA-DMA nanogels encapsulating rhodamine B isothiocyanate-dextran (RITC-Dx) were degraded into polymeric sols in a reducing environment to release RITC-Dx, and the released carbohydrate drugs could interact with concanavalin A in water, showing the ability to bind to pathogens based on lectins. In another study, they also showed that these nanogels could trigger the release of rhodamine 6G and doxorubicin to kill HeLa cells effectively [90]. Morimoto *et al.* prepared dual stimuli-responsive nanogels based on poly(N-isopropylacrylamide)-g-thiolated pullulan, which was synthesized via reversible addition-fragmentation chain transfer (RAFT) polymerization of N-isopropylacrylamide (NIPAM) from thiol groups of pullulan and aminolysis for disulfide cross-linking [91]. The resultant nanogels could be relaxed either upon cooling to room temperature or upon treatment with a reducing agent (tris(2-carboxyethyl)phosphine, TCEP). Redox/pH dual stimuli-responsive nanogels were also prepared from methacrylic acid and *N,N*-bis(acryloyl) cystamine cross-linker via distillation precipitation polymerization [92]. The doxorubicin-loaded nanogels showed an accelerated drug release (over 91% in 5 h) at pH 5.0 and in reducing environments (10 mM GSH) due to a synergistic effect of reduction and charge exchange of GSH at a low pH. The nanogels could be quickly internalized into human glioma cells and exhibited a comparably anti-tumor efficacy. Recent advances in developing dual or multiple stimuli-responsive nanosystems to improve therapeutic efficacy of drugs have been thoroughly reviewed by the Zhong group [93]. Although this approach has shown great promise for programmed site-specific drug delivery, most systems are proof-of-concept studies showing just how they work *in vitro* and *in vivo*. To achieve clinical impacts, it is thus necessary to overcome current major problems including non-biodegradability and low drug loading capacity.

4.7 Summary and Future Perspective

Redox-sensitive polymeric nanosystems possessing disulfide linkages have been extensively studied for drug delivery applications since they can maintain their structural integrity in extracellular environment and perform the release of active ingredients by triggered cleavage of disulfide linkages in the reductive intracellular environment. In addition, the use of bioreducible polymers in drug delivery can be expected to reduce polymer-related cytotoxicity by avoiding the intracellular accumulation of high molecular weight polymeric materials. As these features of bioreducible polymers are suitable for designing redox-sensitive nanosystems, a number of bioreducible polymers and conjugates that can improve therapeutic efficacy of drugs have been reported. However, it should be noticed that more studies on pharmacokinetic behavior, apparent toxicity, and *in vivo* toxicity in different organs of bioreducible polymers should be carefully carried out for their use in clinical settings. In addition, spacio-temporal reactions involved with the intracellular cleavage of disulfide linkages, such as intracellular compartments of the reactions, and reaction kinetics should also be understood in detail.

Acknowledgements

This work was supported by Basic Science Research Program (2009-0093826) and Priority Research Centers Program (2012- 0006687) through the National Research Foundation of Korea (NRF) funded by the Ministry of Science, ICT & Future Planning, Republic of Korea.

References

1. Ganta, S., Devalapally, H., Shahiwala, A. and Amiji, M. (2008). A review of stimuli-responsive nanocarriers for drug and gene delivery. *J. Control. Release,* **126**, 187–204.
2. Fleige, E., Quadir, M. A. and Haag, R. (2012). Stimuli-responsive polymeric nanocarriers for the controlled transport of active compounds: Concepts and applications. *Adv. Drug Deliv. Rev.,* **64**, 866–884.
3. Meng, F., Hennink, W. E. and Zhong, Z. (2009). Reduction-sensitive polymers and bioconjugates for biomedical applications. *Biomaterials,* **30**, 2180–2198.
4. Cheng, R., Feng, F., Meng, F., Deng, C., Feijen, J. and Zhong, Z. (2011). Glutathione-responsive nano-vehicles as a promising platform for targeted intracellular drug and gene delivery. *J. Control. Release,* **152**, 2–12.
5. Manickam, D. S. and Oupický, D. (2006). Polyplex gene delivery modulated by redox potential gradients. *J. Drug Target.,* **14**, 519–526.
6. Arnér, E. S. J. and Holmgren, A. (2000). Physiological functions of thioredoxin and thioredoxin reductase. *Eur. J. Biochem.,* **267**, 6102–6109.

7. Go, Y. M. and Jones, D. P. (2008). Redox compartmentalization in eukaryotic cells. *Biochim. Biophys. Acta,* **1780**, 1273–1290.
8. Wu, G., Fang, Y., Yang, S., Lupton, J. R. and Turner, N. D. (2004). Glutathione metabolism and its implications for health. *J. Nutr.,* **134**, 489–492.
9. Schafer, F. Q. and Buettner, G. R. (2001). Redox environment of the cell as viewed through the redox state of the glutathione disulfide/glutathione couple. *Free Radic. Biol. Med.,* **30**, 1191–1212.
10. Lim, Y., Kim, C., Kim, K., Kim, S. W. and Park, J. (2000). Development of a safe gene delivery system using biodegradable polymer, poly [α-(4-aminobutyl)-L-glycolic acid]. *J. Am. Chem. Soc.,* **122**, 6524–6525.
11. Lynn, D. M. and Langer, R. (2000). Degradable poly(β-amino esters): Synthesis, characterization, and self-assembly with plasmid DNA. *J. Am. Chem. Soc.,* **122**, 10761–10768.
12. Wang, J., Mao, H. Q. and Leong, K. W. (2001). A novel biodegradable gene carrier based on polyphosphoester. *J. Am. Chem. Soc.,* **123**, 9480–9481.
13. Kim, T. and Kim, S. W. (2011). Bioreducible polymers for gene delivery. *React. Funct. Polym.,* **71**, 344–349.
14. Son, S., Namgung, R., Kim, J., Singha, K. and Kim, W. J. (2012). Bioreducible polymers for gene silencing and delivery. *Acc. Chem. Res.,* **45**, 1100–1112.
15. Boussif, O., Lezoualc'h, F., Zanta, M. A., Mergny, M. D., Scherman, D., Demeneix, B. and Behr, J. P. (1995). A versatile vector for gene and oligonucleotide transfer into cells in culture and in vivo: polyethylenimine. *Proc. Natl. Acad. Sci. USA.,* **92**, 7297–7301.
16. Nel, A. E., Madler, L., Velegol, D., Xia, T., Hoek, E. M. V., Somasundaran, P., Klaessig, F., Castranova, V. and Thompson, M. (2009). Understanding biophysicochemical interactions at the nano-bio interface. *Nat. Mater.,* **8**, 543–557.
17. Godbey, W. T., Wu, K. K. and Mikos, A. G. (1999). Size matters: Molecular weight affects the efficiency of poly(ethylenimine) as a gene delivery vehicle. *J. Biomed. Mater. Res.,* **45**, 268–275.
18. Jere, D., Jiang, H. L., Arote, R., Kim, Y. K., Choi, Y. J., Cho, M. H., Akaike, T. and Cho, C. S. (2009). Degradable polyethylenimines as DNA and small interfering RNA carriers. *Expert Opin. Drug Deliv.,* **6**, 827–834.
19. Gosselin, M. A., Guo, W. and Lee, R. J. (2001). Efficient gene transfer using reversibly cross-linked low molecular weight polyethylenimine. *Bioconjugate Chem.,* **12**, 989–994.
20. Breunig, M., Lungwitz, U., Liebl, R. and Goepferich, A. (2007). Breaking up the correlation between efficacy and toxicity for nonviral gene delivery. *Proc. Natl. Acad. Sci. USA.,* **104**, 14454–14459.
21. Sun, Y. X., Zeng, X., Meng, Q. F., Zhang, X. Z., Cheng, S. X. and Zhuo R. X. (2008). The influence of RGD addition on the gene transfer characteristics of disulfide-containing polyethyleneimine/DNA complexes. *Biomaterials,* **29**, 4356–4365.
22. Xiao. B., Laroui, H., Ayyadurai, S., Viennois, E., Charania, M. A., Zhang, Y. and Merlin D. (2013). Mannosylated bioreducible nanoparticle-mediated macrophage-specific TNF-α RNA interference for IBD therapy. *Biomaterials,* **34**, 7471–7482.
23. Park, K., Hong, S. W., Hur, W., Lee, M. Y., Yang, J. A., Kim, S. W., Yoon, S. K. and Hahn, S. K. (2011). Target specific systemic delivery of TGF-β siRNA/(PEI-SS)-g-HA complex for the treatment of liver cirrhosis. *Biomaterials,* **32**, 4951–4958.
24. Xia, W., Wang, P., Lin, C., Li, Z., Gao, X., Wang, G. and Zhao, X. (2012). Bioreducible polyethylenimine-delivered siRNA targeting human telomerase reverse transcriptase inhibits HepG2 cell growth *in vitro* and *in vivo*. *J. Control. Release,* **157**, 427–436.

25. Xiao, J., Duan, X., Yin, Q., Miao, Z., Yu, H., Chen, C., Zhang Z., Wang, J. and Li, Y. (2013). The inhibition of metastasis and growth of breast cancer by blocking the NF-κB signaling pathway using bioreducible PEI-based/p65 shRNA complex nanoparticles. *Biomaterials,* **34**, 5381–5390.
26. Neu, M., Sitterberg, J., Bakowsky, U. and Kissel, T. (2006). Stabilized nanocarriers for plasmids based upon cross-linked poly(ethylene imine). *Biomacromolecules,* 7, 3428–3438.
27. Neu, M., Germershaus, O., Mao, S., Voigt, K., Behe, M. and Kissel, T. (2007). Crosslinked nanocarriers based upon poly(ethyleneimine) for systemic plasmid delivery: *In vitro* characterization and *in vivo* studies in mice. *J. Control. Release,* **118**, 370–380.
28. Neu, M., Germershaus, O., Behe, M. and Kissel, T. (2007). Bioreversibly crosslinked polyplexes of PEI and high molecular weight PEG show extended circulation times *in vivo. J. Control. Release,* **124**, 69–80.
29. Lee, D., Kim, D., Mok, H., Jeong, J. H., Choi, D. and Kim, S. H. (2012). Bioreducible crosslinked polyelectrolyte complexes for MMP-2 siRNA delivery into human vascular smooth muscle cells. *Pharm. Res.,* **29**, 2213–2224.
30. Chae, S. Y., Kim, H. J., Lee, M. S., Jang, Y. L., Lee, Y., Lee, S. H., Lee, K., Kim, S. H., Kim, H. T., Chi, S., Park, T. G. and Jeong, J. H. (2011). Energy-independent intracellular gene delivery mediated by polymeric biomimetics of cell-penetrating peptides. *Macromol. Biosci.,* **11**, 1169–1174.
31. Kuzuya, M., Kanda, S., Sasaki, T., Tamaya-Mori, N., Cheng, X. W., Itoh, T., Itohara, S. and Iguchi, A. (2003). Deficiency of gelatinase A suppresses smooth muscle cell invasion and development of experimental intimal hyperplasia. *Circulation,* **108**, 1375–1381.
32. Peng, Q., Zhong, Z. and Zhuo, R. (2008). Disulfide cross-linked polyethylenimines (PEI) prepared via thiolation of low molecular weight PEI as highly efficient gene vectors. *Bioconjugate Chem.,* **19**, 499–506.
33. Son, S., Singha, K. and Kim, W. J. (2010). Bioreducible BPEI-SS-PEG-cNGR polymer as a tumor targeted nonviral gene carrier. *Biomaterials,* **31**, 6344–6354.
34. Buehler, A., van Zandvoort, M. A. M. J., Stelt, B. J., Hackeng, T. M., Schrans-Stassen, B. H. G. J., Bennaghmouch, A., Hofstra, L., Cleutjens, J. P. M., Duijvestijn, A., Smeets, M. B., de Kleijn, D. P. V., Post, M. J. and de Muinck, E. D. (2006). cNGR: A novel homing sequence for CD13/APN targeted molecular imaging of murine cardiac angiogenesis in vivo. *Arterioscler. Thromb. Vasc. Biol.,* **26**, 2681–2687.
35. Son, S., Hwang, D. W., Singha, K., Jeong, J. H., Park, T. G., Lee, D. S. and Kim, W. J. (2011). RVG peptide tethered bioreducible polyethyleneimine for gene delivery to brain. *J. Control. Release,* **155**, 18–25.
36. Lee, Y., Mo, H., Koo, H., Park J. Y., Cho, M. Y., Jin, G. and Park, J. S. (2007). Visualization of the degradation of a disulfide polymer, linear poly(ethylenimine sulfide), for gene delivery. *Bioconjugate Chem.,* **18**, 13–18.
37. Koo, H., Jin, G., Kang, H., Lee, Y., Nam, K., Bai, C. Z. and Park, J. S. (2010). Biodegradable branched poly(ethylenimine sulfide) for gene delivery. *Biomaterials,* **31**, 988–997.
38. Kwok, K. Y., Mckenzie, D. L., Evers, D. L. and Rice, K. G. (1999). Formulation of highly soluble poly(ethylene glycol)-peptide DNA condensates. *J. Pharm. Sci.,* **88**, 996–1003.
39. McKenzie, D. L., Kwok, K. Y. and Rice, K. G. (2000). A potent new class of reductively activated peptide gene delivery agents. *J. Biol. Chem.,* **275**, 9970–9977.
40. McKenzie, D. L., Smiley, E., Kwok, K. Y. and Rice, K. G. (2000). Low molecular weight disulfide cross-linking peptides as nonviral gene delivery carriers. *Bioconjugate Chem.,* **11**, 901–909.

41. Oupický, D., Parker, A. L. and Seymour, L. W. (2002). Laterally stabilized complexes of DNA with linear reducible polycations: Strategy for triggered intracellular activation of DNA delivery vectors. *J. Am. Chem. Soc.*, **124**, 8–9.
42. Read, M. L., Bremner, K. H., Oupický, D., Green, N. K., Searle P. F. and Seymour, L. W. (2003). Vectors based on reducible polycations facilitate intracellular release of nucleic acids. *J. Gene Med.*, **5**, 232–245.
43. Read, M. L., Singh, S., Ahmed, Z., Stevenson, M., Briggs, S. S., Oupický, D., Barrett, L. B., Spice, Rachel, Kendall, M., Berry, M., Preece, J. A., Logan, A. and Seymour, L W. (2005). A versatile reducible polycation-based system for efficient delivery of a broad range of nucleic acids. *Nucleic Acids Res.*, **33**, e86.
44. Stevenson, M., Ramos-Perez, V., Singh, S., Soliman, M., Preece, J. A., Briggs, S. S., Read, M. L. and Seymour, L. W. (2008). Delivery of siRNA mediated by histidine-containing reducible polycations. *J. Control. Release*, **130**, 46–56.
45. Parker, A. L., Eckley, L., Singh, S., Preece, J. A., Collins, L. and Fabre, J. W. (2007). (LYS)(16)-based reducible polycations provide stable polyplexes with anionic fusogenic peptides and efficient gene delivery to post mitotic cells. *Biochim. Biophys. Acta*, **1770**, 1331–1337.
46. Manickam, D. S. and Oupický, D. (2006). Multiblock reducible copolypeptides containing histidine-rich and nuclear localization sequences for gene delivery. *Bioconjugate Chem.*, **17**, 1395–1403.
47. Rahbek, U. L., Howard, K. A., Oupický, D., Manickam, D. S., Dong, M., Nielsen, A. F., Hansen, T. B., Besenbacher, F. and Kjems J. (2008). Intracellular siRNA and precursor miRNA trafficking using bioresponsive copolypeptides. *J. Gene Med.*, **10**, 81–93.
48. Mok, H. and Park, T. G. (2008). Self-crosslinked and reducible fusogenic peptides for intracellular delivery of siRNA. *Biopolymers*, **89**, 881–888.
49. Choi, S. W., Lee, S. H., Mok, H. and Park, T. G. (2010). Multifunctional siRNA delivery system: polyelectrolyte complex micelles of six-arm PEG conjugate of siRNA and cell penetrating peptide with crosslinked fusogenic peptide. *Biotechnol. Prog.*, **26**, 57–63.
50. Manickam, D. S., Bisht, H. S., Wan, L., Mao, G. and Oupický, D. (2005). Influence of TAT-peptide polymerization on properties and transfection activity of TAT/DNA polyplexes. *J. Control. Release*, **102**, 293–306.
51. Lo, S. L. and Wang, S. (2008). An endosomolytic Tat peptide produced by incorporation of histidine and cysteine residues as a nonviral vector for DNA transfection. *Biomaterials*, **29**, 2408–2414.
52. Won, Y. W., Kim, H. A., Lee, M. and Kim, Y. H. (2010). Reducible poly(oligo-D-arginine) for enhanced gene expression in mouse lung by intratracheal injection. *Mol. Ther.*, **18**, 734–742.
53. Hyun, H., Won, Y. W., Kim, K. M., Lee, J., Lee, M. and Kim Y. H. (2010). Therapeutic effects of a reducible poly (oligo-D-arginine) carrier with the heme oxygenase-1 gene in the treatment of hypoxic-ischemic brain injury. *Biomaterials*, **31**, 9128–9134.
54. Kiselev, A., Egorova, A., Laukkanen, A., Baranov, V. and Urtti, A. (2013). Characterization of reducible peptide oligomers as carriers for gene delivery. *Int. J. Pharm.* **441**, 736–747.
55. Danusso, F. and Ferruti, P. (1970). Synthesis of tertiary amine polymers. *Polymer*, **11**, 88–113.
56. Lin, C., Zhong, Z., Lok, M. C., Jiang, X., Hennink, W. E., Feijen, J. and Engbersen, J. F. (2007). Novel bioreducible poly(amido amine)s for highly efficient gene delivery. *Bioconjugate Chem.*, **18**, 138–145.

57. Christensen, L. V., Chang, C. W., Kim, W. J. and Kim, S. W. (2006). Reducible poly(amido ethylenimine)s designed for triggered intracellular gene delivery. *Bioconjugate Chem.,* **17**, 1233–1240.
58. Jeong, J. H., Christensen, L. V., Yockman, J. W., Zhong, Z., Engbersen, J. F. J., Kim, W. J., Feijen, J. and Kim, S. W. (2007). Reducible poly(amido ethylenimine) directed to enhance RNA interference. *Biomaterials,* **28**, 1912–1917.
59. Jeong, J. H., Kim, S. H., Christensen, L. V., Feijen, J. and Kim, S. W. (2010). Reducible poly(amido ethylenimine)-based gene delivery system for improved nucleus trafficking of plasmid DNA. *Bioconjugate Chem.,* **21**, 296–301.
60. Christensen, L. V., Chang, C. W., Yockman, J. W., Conners, R., Jackson, H., Zhong, Z., Feijen, J., Bull, D. A. and Kim, S. W. (2007). Reducible poly(amido ethylenediamine) for hypoxia-inducible VEGF delivery. *J. Control. Release,* **118**, 254–261.
61. Shoshani, T., Faerman, A., Mett, I., Zelin, E., Tenne, T., Gorodin, S., Moshel, Y., Elbaz, Shlomo, Budanov, A., Chajut, A., Kalinski, H., Kamer, I., Rozen, A., Mor, O., Keshet, E., Leshkowitz, D., Elinat, Paz., Skaliter, R. and Feinstein, E. (2002). Identification of a novel hypoxia-inducible factor 1-responsive gene, RTP801, involved in apoptosis. *Mol. Cell. Biol.,* **22**, 2283–2293.
62. Namgung, R., Brumbach, J. H., Jeong, J. H., Yockman, J. W., Kim, S. W., Lin, C., Zhong, Z., Feijen, J., Engbersen, J. F. J. and Kim, W. J. (2010). Dual bio-responsive gene delivery via reducible poly(amido amine) and survivin-inducible plasmid DNA. *Biotechnol. Lett.,* **32**, 755–764.
63. Lin, C., Zhong, Z., Lok, M. C., Jiang, X., Hennink, W. E., Feijen, J. and Engbersen, J. F. J. (2007). Random and block copolymers of bioreducible poly(amido amine)s with high- and low-basicity amino groups: study of DNA condensation and buffer capacity on gene transfection. *J. Control. Release,* **123**, 67–75.
64. Lin, C., Zhong, Z., Lok, M. C., Jiang, X., Hennink, W. E., Feijen, J. and Engbersen, J. F. J. (2006). Linear poly(amido amine)s with secondary and tertiary amino groups and variable amounts of disulfide linkages: Synthesis and in vitro gene transfer properties. *J. Control. Release,* **116**, 130–137.
65. Piest, M., Lin, C., Mateos-Timoneda, M. A., Lok, M. C., Hennink, W. E., Feijen, J. and Engbersen, J. F. J. (2008). Novel poly(amido amine)s with bioreducible disulfide linkages in their diamino-units: structure effects and in vitro gene transfer properties. *J. Control. Release,* **132**, E12–E13.
66. Ou, M., Wang, X. L., Xu, R., Chang, C. W., Bull, D. A. and Kim, S. W. (2008). Novel biodegradable poly(disulfide amine)s for gene delivery with high efficiency and low cytotoxicity. *Bioconjugate Chem.,* **19**, 626–633.
67. Kim, S. H., Jeong, J. H., Ou, M., Yockman, J. W., Kim, S. W. and Bull, D. A. (2008). Cardiomyocyte-targeted siRNA delivery by prostaglandin E(2)-Fas siRNA polyplexes formulated with reducible poly(amido amine) for preventing cardiomyocyte apoptosis. *Biomaterials,* **29**, 4439–4446.
68. Kim, S. H., Jeong, J. H., Kim, T. I., Kim, S. W. and Bull, D. A. (2009). VEGF siRNA delivery system using arginine-grafted bioreducible poly(disulfide amine). *Mol. Pharm.,* **6**, 718–726.
69. Kim, T. I., Ou, M., Lee, M. and Kim, S. W. (2009). Arginine-grafted bioreducible poly(disulfide amine) for gene delivery systems. *Biomaterials,* **30**, 658–664.

70. Nam, K., Nam, H. Y., Kim, P. H. and Kim, S. W. (2012). Paclitaxel-conjugated PEG and arginine-grafted bioreducible poly (disulfide amine) micelles for co-delivery of drug and gene. *Biomaterials,* **33**, 8122–8130.
71. Kim, T. I., Lee, M. and Kim, S. W. (2010). A guanidinylated bioreducible polymer with high nuclear localization ability for gene delivery systems. *Biomaterials,* **31**, 1798–1804.
72. Zhao, M. and Weissleder, R. (2004). Intracellular cargo delivery using tat peptide and derivatives. *Med. Res. Rev.,* **24**, 1–12.
73. Zorko, M. and Langel, U. (2005). Cell-penetrating peptides: Mechanism and kinetics of cargo delivery. *Adv. Drug Deliver. Rev.,* **57**, 529–545.
74. Harada, A. and Kataoka, K. (1995). Formation of polyion complex micelles in an aqueous milieu from a pair of oppositely-charged block copolymers with poly(ethylene glycol) segments. *Macromolecules,* **28**, 5294–5299.
75. Kakizawa, Y. and Kataoka, K. (2002). Block copolymer micelles for delivery of gene and related compounds. *Adv. Drug Deliver. Rev.* **54**, 203–222.
76. Kakizawa, Y., Harada, A. and Kataoka, K. (2001). Glutathione-sensitive stabilization of block copolymer micelles composed of antisense DNA and thiolated poly(ethylene glycol)-block-poly(L-lysine): A potential carrier for systemic delivery of antisense DNA. *Biomacromolecules,* **2**, 491–497.
77. Miyata, K., Kakizawa, Y., Nishiyama, N., Harada, A., Yamasaki, Y., Koyama, H. and Kataoka, K. (2004). Block catiomer polyplexes with regulated densities of charge and disulfide cross-linking directed to enhance gene expression. *J. Am. Chem. Soc.,* **126**, 2355–2361.
78. Oishi, M., Hayama, T., Akiyama, Y., Takae, S., Harada, A., Yamasaki, Y., Nagatsugi, F., Sasaki, S., Nagasaki, Y. and Kataoka, K. (2005). Supramolecular assemblies for the cytoplasmic delivery of antisense oligodeoxynucleotide: Polyion complex (PIC) micelles based on poly(ethylene glycol)-SS-oligodeoxynucleotide conjugate. *Biomacromolecules,* **6**, 2449–2454.
79. Kim, S. H., Jeong, J. H., Lee, S. H., Kim, S. W. and Park, T. G. (2006). PEG conjugated VEGF siRNA for anti-angiogenic gene therapy. *J. Control. Release,* **116**, 123–129.
80. Mok, H., Park, J. W. and Park, T. G. (2007). Antisense oligodeoxynucleotide-conjugated hyaluronic acid/protamine nanocomplexes for intracellular gene inhibition. *Bioconjugate Chem.,* **18**, 1483–1489.
81. Takae, S., Miyata, K., Oba, M., Ishii, T., Nishiyama, N., Itaka, K., Yamasaki, Y., Koyama, H. and Kataoka, K. (2008). PEG-detachable polyplex micelles based on disulfide-linked block catiomers as bioresponsive nonviral gene vectors. *J. Am. Chem. Soc.,* **130**, 6001–6009.
82. Yallapu, M. M., Jaggi, M. and Chauhan, S. C. (2011). Design and engineering of nanogels for cancer treatment. *Drug Discov. Today,* **16**, 457–463.
83. Choi, J. H., Jang, J. Y., Joung, Y. K., Kwon, M. H. and Park, K. D. (2010). Intracellular delivery and anti-cancer effect of self-assembled heparin-Pluronic nanogels with RNase A. *J. Control. Release,* **147**, 420–427.
84. Choi, J. H., Joung, Y. K., Bae, J. W., Choi, J. W., Quyen, T. N. and Park, K. D. (2011). Self-assembled nanogel of pluronic-conjugated heparin as a versatile drug nanocarrier. *Macromol. Res.,* **19**, 180–188.
85. Nguyen, D. H., Choi, J. H., Joung, Y. K. and Park, K. D. (2011). Disulfide-crosslinked heparin-Pluronic nanogels as a redox-sensitive nanocarrier for intracellular protein delivery. *J. Bioact. Compat. Polym.,* **26**, 287–300.

86. Nguyen, D. H., Joung, Y. K., Choi, J. H., Moon, H. T. and Park, K. D. (2011). Targeting ligand-functionalized and redox-sensitive heparin-Pluronic nanogels for intracellular protein delivery. *Biomed. Mater.,* **6**, 055004.
87. Joung, Y. K., Jang, J. Y., Choi, J. H., Han, D. K. and Park, K. D. (2013). Heparin-conjugated Pluronic nanogels as multi-drug nanocarriers for combination chemotherapy. *Mol. Pharm.,* **10**, 685–693.
88. Bae, K. H., Mok, H. and Park, T. G. (2008). Synthesis, characterization, and intracellular delivery of reducible heparin nanogels for apoptotic cell death. *Biomaterials,* **29**, 3376–3383.
89. Oh, J. K., Siegwart, D. J. and Matyjaszewski, K. (2007). Synthesis and biodegradation of nanogels as delivery carriers for carbohydrate drugs. *Biomacromolecules,* **8**, 3326–3331.
90. Oh, J. K., Siegwart, D. J., Lee, H. I., Sherwood, G., Peteanu, L., Hollinger, J. O., Kataoka, K. and Matyjaszewski, K. (2007). Biodegradable nanogels prepared by atom transfer radical polymerization as potential drug delivery carriers: Synthesis, biodegradation, in vitro release, and bioconjugation. *J. Am. Chem. Soc.,* **129**, 5939–5945.
91. Morimoto, N., Qiu, X. P., Winnik, F. M. and Akiyoshi, K. (2008). Dual stimuli-responsive nanogels by self-assembly of polysaccharides lightly grafted with thiol-terminated poly(N-isopropylacrylamide) chains. *Macromolecules,* **41**, 5985–5987.
92. Pan, Y., Chen, Y., Wang, D., Wei, C., Guo, J., Lu, D., Chu, C. and Wang, C. (2012). Redox/pH dual stimuli-responsive biodegradable nanohydrogels with varying responses to dithiothreitol and glutathione for controlled drug release. *Biomaterials,* **33**, 6570–6579.
93. Cheng, R., Meng, F., Deng, C., Klok, H. A. and Zhong, Z. (2013). Dual and multi-stimuli responsive polymeric nanoparticles for programmed site-specific drug delivery. *Biomaterials,* **34**, 3647–3657.

Chapter 5

Temperature-Sensitive Pharmaceutical Nanocarriers

Gemma Navarro[1]

[1]*Center for Pharmaceutical Biotechnology and Nanomedicine, Northeastern University, Boston, MA, USA*

5.1 Introduction

Pharmaceutical nanocarriers, such as liposomes, micelles, and polymeric or lipidic nanoparticles, have demonstrated enhanced *in vivo* stability and efficiency of drugs, genes, and diagnostic agents [1]. Surface engineering of nanocarriers can provide unique properties to such systems, such as longevity in the blood, specific recognition of different tissues, cell types, and even cellular organelles, as well as real-time tracking of carriers and payloads by the incorporation of contrast materials. The benefits of nano-scaled delivery systems are clearly supported by numerous pre-clinical and clinical data [2, 3], and several commercialized products (Table 5.1). Thus, many trials that combine these properties with a triggered drug release that can provide a greater spatial and temporal control of the drug therapy have been undertaken. To modulate drug release, nanocarriers can be designed to take advantage of "physiological triggers" such as changes in pH, redox balance, or hypoxia, or increased levels of certain enzymes that are characteristic of particular tissues, organelles, or pathologically altered areas [4–6]. In addition, nanocarriers sensitive to extracorporeal temperature or magnetic stimuli have shown promising results in clinical trials [7]. This chapter focuses on drug delivery systems responsive to temperature changes (Fig. 5.1). Recent advances in the design of such thermo-sensitive nanocarriers, the main classes of temperature-responsive polymers and lipids developed to date for their

Table 5.1. Examples of nanocarrier-based formulations approved for clinical application.

Nanocarrier	Drug	Route	Trade Name	Indication
Liposome	Cytarabine	Intrathecal	DepoCyt®	Malignant lymphomatous meningitis
	Daunorubicin	Intravenous (i.v)	DaunoXome®	Acquired immunodeficiency syndrome (AIDS)-related Kaposi's sarcoma
	Vincristine	i.v	ONCO-TSC®	Aggressive Non-Hodgkin's lymphoma
	Doxorubicin	i.v	Myocet®	Metastatic breast cancer
Polyethylene glycol (PEG) ylated Liposome	Doxorubicin	i.v	Doxil®	Ovarian cancer, AIDS-related Kaposi's sarcoma, Multiple myeloma
	Doxorubicin	Intramuscular (i.m)	Caelyx®	Metastatic breast and ovarian cancer
Albumin-Nanoparticle	PTX	i.v	Abraxane®	Metastatic breast cancer
Polymeric micelles	PTX	i.v	Genexol®	Metastatic breast cancer
Drug–polymer conjugates	PEG-Asparaginase	i.v	Oncaspar®	Acute lymphoblastic leukemia
Iron-NP	Contrast agent	i.v	Feridex®	Magnetic resonance imaging (MRI) imaging of liver tumors
	Contrast agent	i.v	Magnevist®	MRI tumor imaging

construction, their mechanism of triggered drug release, and their clinical potential in combination with the different heat sources will be discussed.

5.2 Pharmaceutical Nanocarriers: The Need for Targeted and Triggered Drug Delivery

The most critical parameters that a drug candidate must meet for its clinical translation are solubility in aqueous media, stability in plasma (and during storage), and an appropriate delivery and therapeutic profile after administration.

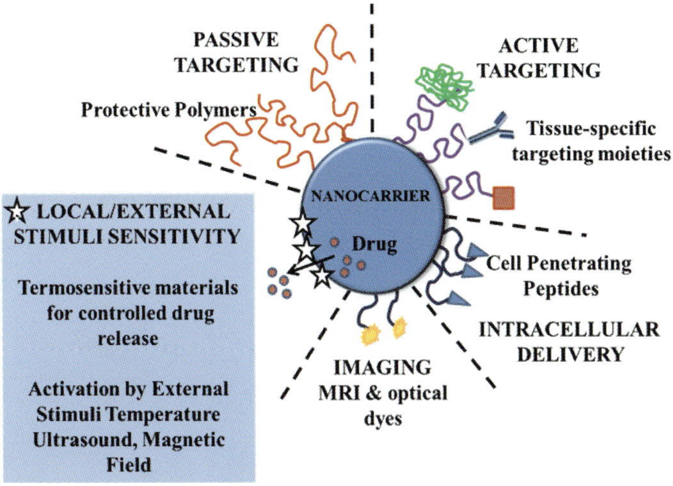

Figure 5.1. Pharmaceutical nanocarriers for enhanced stability, tissue-specificity, real-time tracking, and drug triggered drug release. MRI = magnetic resonance imaging.

Unfortunately, a large percentage of chemical compounds or biomolecules that show potent pharmacological activity are abandoned from further development because of their poor solubility and poor *in vivo* stability [8]. Aggregation of poorly soluble drugs upon i.v injection is associated with serious risks, such as embolism or local toxicities. Low solubility of drugs is also problematic for oral and other administration routes. At the end of the day, any drug, regardless of the route of administration, must be present in its active soluble form to be absorbed at the site of action. Premature degradation and elimination, especially in the case of protein- or nucleic acid-based drugs, are also problematic. For instance, naked short interfering ribonucleic acid (siRNA) has an *in vivo* half-life of <5 min due to a rapid renal clearance, and is degraded within 1 minute after i.v injection [9]. Even for relatively stable and soluble drugs, high distribution volumes (meaning significant generalized uptake of the drug by tissues) and fast clearance rates often results in low levels of drug at the site of action and significant interaction with non-specific organs. With certain drugs (e.g. chemotherapeutics, antibiotics), such pharmacokinetic profiles are associated with severe side effects, limited therapeutic effects, and increased risk of drug resistance development.

The inclusion of drugs into nano-scaled delivery systems has emerged as a way to solve these problems. Pharmaceutical nanocarriers are engineered constructs ranging in size from 10 to 20 nm to a few hundred nanometers. The ability to tailor their composition, size, shape, and surface charge has led to a

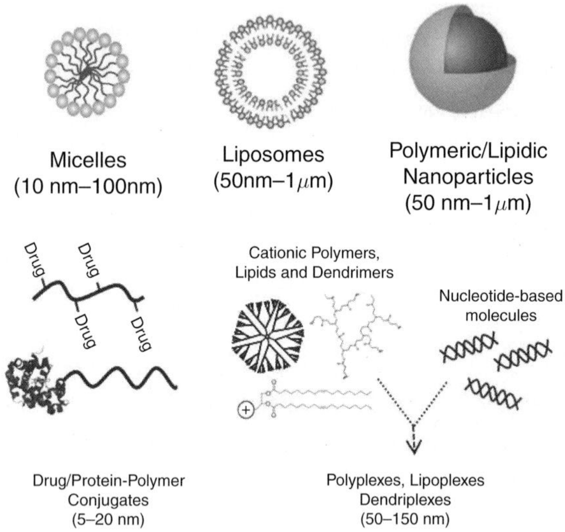

Figure 5.2. Classes of nanoscale delivery systems.

growing number of nano-scaled delivery systems, including liposomes [10], micelles [1], polymeric/lipidic nanoparticles [11], drug–polymer conjugates [12], and cationic polymer/lipid/dendrimer gene delivery systems [13] (Fig. 5.2). They are useful in drug delivery because they confine and protect the drug from metabolism and inactivation in plasma. Precise changes in the composition of nanocarriers can lead to substantial changes in drug biodistribution. By means of their sizing, pharmaceutical nanocarriers can slow drug elimination, avoiding glomerular filtration size cut off (4.5 nm, corresponding to a molecular weight (MW) <40 kDa) [14] and accumulating in certain pathological areas (inflamed or infarcted areas and highly vascularized solid tumors) [15]. To be effective, nanocarriers should balance their stability in the circulation with drug's bioavailability at the action site. They should promote accumulation of intact drug in the area of interest to then provide a controlled release of the drug.

5.3 Passive/Active Targeting of Nanocarriers

Site-specific targeting of nanocarriers can be achieved by passive or active targeting [16] . Since the great majority of pharmaceutical nanocarriers are developed for intravenous administration of anti-cancer agents, passive and active targeting of solid tumors are the most advanced and well-investigated approaches. Passive targeting of nanocarriers to tumors usually relies on the so-called "enhanced permeability and retention (EPR) effect", by which macromolecular drugs and nanomedicines within a certain size range tend to accumulate in solid tumors more than in healthy tissues. The heterogeneity of the blood supply, the disturbed architecture and increased permeability of tumor vessels together with poor lymphatic drainage are the main reasons for the abnormal microenvironment that exists in solid tumors. Tumor overexpression of certain pro-angiogenic molecules, such as vascular endothelial growth factor (VEGF), leads to a rapid formation of a vasculature necessary for the tumor's survival. Typically, the new vasculature is chaotic, dilated, and irregularly distributed, leaving low- or non-perfused regions in the tumor. This latter fact is the main cause of the characteristic hypoxia, low pH, and necrosis in the tumor [17, 18]. The structure of tumor vessels is also abnormal. The endothelial cells lining the vessels have an aberrant morphology, the basement membrane is either unusually thick or absent, and the endothelial pore diameters may be several hundred nanometers or even a few micrometers, depending on the tumor type [19, 20]. Overall, the tumor vasculature is leaky and hypermeable. In addition, proliferating tumor cells can compress blood and lymphatic vessels, especially in the inner regions of a tumor, and cause their collapse. Therefore, the blood flow can be slow and the lymphatic drainage inefficient, promoting a high interstitial fluid pressure in tumor tissues. This phenomenon led to the coining of the term "enhanced permeability and retention (EPR) effect" of macromolecules in solid tumors. Macromolecules can be entrapped in solid tumors and retained there at high concentrations for prolonged periods (more than 100 h), whereas low MW substances are not retained, but are returned to the circulating blood by diffusion [21, 22]. However, the usefulness of nanoparticulate size in the accumulation by the EPR effect may be impaired by their slow diffusion through the tumoral matrix [23]. Passive targeting of nanocarriers was shown to be strongly dependent on the cutoff size of the tumor blood vessel wall, which also varies among tumor types [24]. Clearly, an EPR-based type of targeting requires nanocarriers with a prolonged systemic circulation time to provide for a sufficient level of accumulation at an abnormal site. Since the immune system recognizes most unmodified nanocarriers as foreign particles and rapidly removes them from the circulation, almost all nanocarriers are constructed with a long-circulating property to facilitate the EPR effect.

The most common method for prolonging the circulation of carriers is to "mask" them by modifying their surface with certain water-soluble polymers, such as polyethylene glycol (PEG) also known as poly(oxyethylene) or poly(ethylene oxide) (PEO) [25, 26]. In addition to its Food and Drug Association (FDA) approval, PEGylation provides a very attractive combination of properties as a protective polymer, including excellent solubility in water, high flexibility of the polymeric chain, and a very low toxicity and immunogenicity [27]. A PEG coating protects the nanocarrier and decreases the opsonization of drugs and drug carriers and reduces their clearance by the mononuclear phagocyte system (MPS) by several mechanisms, including shielding of the surface charge, increased surface hydrophobicity [28], enhanced repulsive interaction between coated nanocarriers and blood [29], and the formation of a polymeric layer over the carrier surface, which is impermeable to opsonins even at low PEG concentrations [30]. The commonly used MW range of PEG for the modification of drugs is 1,000 to 20,000 Da, with 2,000 Da as the most frequently used.

Although passive targeting approaches form the basis for the clinical translation of many pharmaceutical nanocarriers, they suffer from several limitations. On the one hand, the accumulation and passive extravasation of nanocarriers depends on the degree of vascularization and permeability of the vessels that can vary with the stage and tumor type [31], which make this approach difficult to control. On the other hand, the PEG chains that protect the carriers from opsonization in the blood (providing a long-circulating property) and facilitate the EPR effect may also prevent their interaction with cancer cells and delay the release of the drug content in the cytoplasm [32, 33]. One way to overcome these limitations is to program the nanocarriers so that they actively bind to specific cells (active targeting) after extravasation. This binding can be achieved by the attachment of targeting molecules, such as antibodies, sugars, or ligand–receptor binding moieties to the surface of the nanocarrier by a variety of conjugation chemistries [10]. A number of targeting ligands have been tested in recent years, including antibodies against growth factor receptors, such as anti-epidermal growth factor receptor (EGFR) and anti-human epidermal growth factor receptor 2 (Her2) [34], antibodies targeting the tumor cell surface [5], and natural ligands to commonly over-expressed receptors in cancer cells, such as luteinizing hormone-releasing hormone, somatostatin, transferrin, and folate receptors [35–37]. Another approach to overcome passive targeting limitations is to include certain polymers or lipids in the composition of nanocarriers that are activated by the presence of physiological triggers at the site of interest (e.g. pH, partial pressure of oxygen (pO_2), redox balance, or enzymatic catalysis), by the application of a rapidly oscillating magnetic field, or by the application of a light or external heat source, resulting in the disruption of the carrier and accelerated drug release.

5.4 Functional Properties of Thermo-sensitive Polymers and Lipids

Temperature-responsive polymers and lipids, which change their conformation and physical properties in response to temperature changes, have been used for the construction of so-called smart materials with a thermo-responsive function, (Fig. 5.3).

5.4.1 *Thermo-sensitive polymers*

Thermo-sensitive polymers undergo coil-to-globule transitions above or below a specific temperature, called the "lower critical solution temperature (LCST)" and "upper critical solution temperature (UCST)", respectively. Polymers that become soluble upon heating have a LCTS, and polymers that become soluble upon heating have a UCST. These processes are entropically- and enthalpically-driven, respectively [38].

Examples of polymers with transition phases at UCST are gelatin or certain polysaccharides, such as agarose or carrageen [39]. They are soluble at

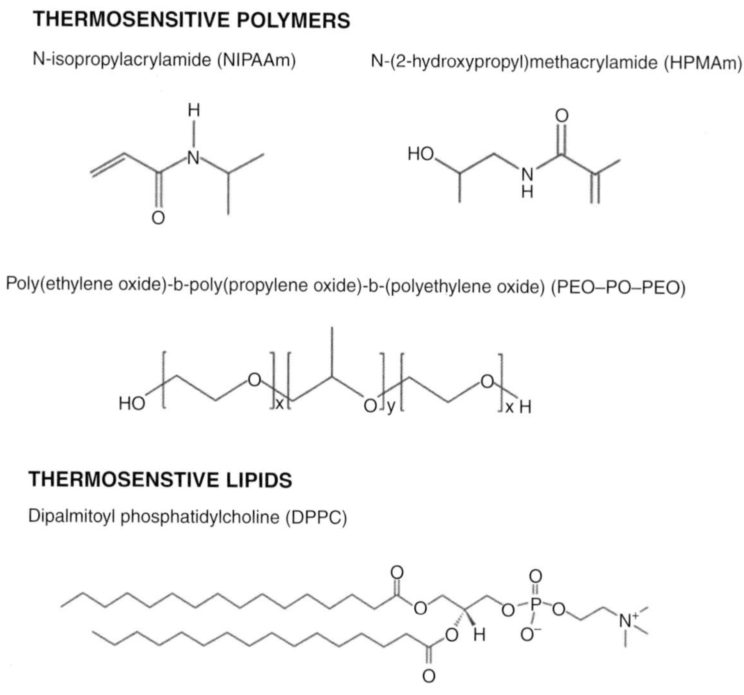

Figure 5.3. Representative chemical structures of temperature-responsive polymers and lipids used for the formulation of thermosensitive nanocarriers.

elevated temperatures, and form gel matrices upon cooling, showing great potential in tissue engineering applications, such as gelatin cryogel cell carriers [40, 41] or for agarose scaffolds for the growing of pancreatic islets and insulinoma cells [42].

The majority of polymers employed in drug delivery are based in LCST transition phases. Their transition arises from the entropic gain from the water molecules associated with the chains of the polymer that are released into the bulk aqueous phase when the temperature becomes close to the LCST (Fig. 5.4A). Below the LCST, the enthalpic contribution of hydrogen bonds between water molecules and polymer chains is higher than that of the entropic gain of the system and is sufficient to solubilize the polymer. As the temperature increases, the efficiency of the hydrogen bounds reduces and phase separation occurs [43]. The polymer solution that was clear below the LCST becomes cloudy when the LCST (also referred to as the "cloud point") is reached. From a thermodynamic perspective, using the Gibbs equation $\Delta G = \Delta H - T\Delta S$ (where G = Gibbs free energy, H = enthalpy, and S = entropy), the reason why phase separation is more favorable above LCST is due to the entropy of the released water molecules. When the polymer is not in solution, the water is less ordered and has higher entropy. This is also called the "hydrophobic effect" [38].

The search for new thermo-sensitive materials is in continuous development. A great variety of smart polymers covering transition temperatures from 0°C to 100°C has been reported. A detailed description of their syntheses, structures, and LCST/UCST has been reported [44]. Thermo-sensitive polymers can be the main component of the nanocarrier or a modifing component to endow thermo-sensitive behavior (i.e. thermo-sensitive polymers modifying liposomes or micelles). For drug delivery purposes, the transition temperature of a given polymer should be sharp (i.e. burst release) and close to physiological temperatures, or in the range of 39–42°C for extracorporeal heat stimulation. A sharp transition temperature is also important to reduce the exposure time of the applied hyperthermia and to avoid damage to healthy adjacent tissues [45].

Thermo-sensitive polymers can be modified by anchoring or by the co-polymerization of hydrophobic or hydrophilic monomers in order to tune their transition temperatures from non-clinically practical ones to temperatures close to 37–42°C. By the addition of hydrophilic monomers, the efficiency of hydrogen bonds between water and the polymer increases and so does the LCST, whereas, the inclusion of hydrophobic groups favors aggregation and decreases the LCST of the system. Other factors that affect the hydrogen bonding ability or the electrostatic interaction of the polymers in solution are polymer concentration and the presence of salts [46].

PNIPAAm is one of the most heavily investigated thermo-sensitive polymers. It has a sharp coil-to-globule transition in water at its LCST of 32°C. To be clinical useful, the LCST of PNIPAAm can be adjusted to temperatures around body temperature by the co-polymerization of hydrophilic monomers, such as acrylic acid [47, 48] and acrylamide [49]. For instance, Hayashi and co-workers demonstrated that the co-polymerization of PNIPAAm with acrylamide monomers (AAm) raised the LCST of pure PNIPAAm from 32°C to 39°C (with 10% AAm), 47.2°C (with 20% AAm), and 53.2°C (with 30%AAm) [49]. The major drawback of PNIPAAm is that it is not biodegradable. To overcome this challenge, several strategies have been proposed. Neradovic and co-workers synthesized a copolymer of NIPAAm and N-(2-hydroxypropyl)methacrylamide (HPMA, chemically similar to NIPAAm but biodegradable) that contained hydrolysable lactate ester side groups. Poly(NIPAAm-co-HPMAm-lactate) with 35 mol% HPMAm-lactate had an LCST below body temperature and converted into poly(NIPAAm-co-HPMAm) with an LCST above 37°C due to an increment of polymer hydrophobicity upon hydrolysis [50, 51]. Soga and co-workers realized that poly(HPMAm-mono/di lactate) without NIPAAm also shows LCST behavior in aqueous solution. Similar to what Neradovi's group observed, the hydrolysis of lactic acid side groups led to an increase in the LCST, but in this case, the degradation products were lactic acid and the pHPMAm [52]. pHPMAm is a well-known non-toxic polymer that has already been tested in clinical trials as pro-drug polymeric delivery systems for anticancer agents [53]. Finally, enzymatically cleavable groups, such as 2-methylene-1,3-dioxepane (by proteinase K) [54] or polyamino acids (by elastase) were introduced to PNIPAAm for improved biodegradability [55].

5.4.2 *Thermo-sensitive lipids*

Thermo-sensitive properties for lipid-based nanocarriers arise from the fact that lipid membranes undergo phase transitions upon heating. Phospholipids, depending on their fatty acid composition, have a specific main phase transition temperature (T_m), or the melting temperature between the gel and the liquid crystalline states. The melting transition is accompanied by enthalpy and volume changes. In the gel phase, alkyl chain tails are fully extended in *trans* orientation. In this state, the alkyl chains are rigid and straight. The head groups of the phospholipids are tightly packed, which results in a lower

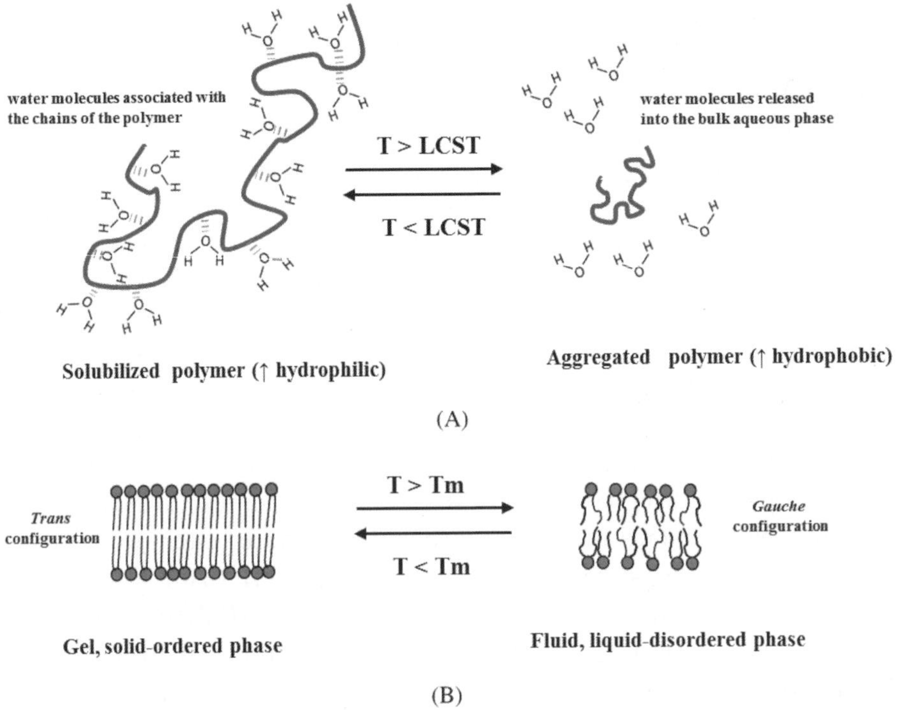

Figure 5.4. Schematic representation of the phase transition of thermosensitive polymers (A) and lipids (B). T = temperature; LCST = lower critical solution temperature; T_m = main phase transition temperature.

area per molecule than in the liquid phase (Fig. 5.4B). As the temperature increases, the mobility of the head groups increases, and the *trans* orientation of the alkyl chains switches to *gauche* configuration. The head groups in the liquid phase are loosely packed with rotational freedom, giving rise to a larger area per molecule at room temperature than that observed in the gel phase. The alkyl chain tails are not rigid and no longer efficiently packed due to the presence of "kinks" caused by the *gauche* orientation. As a result of these packing defects, the membrane becomes fluid and leaky [56, 57]. The T_m depends on the length and the degree of saturation of the alkyl chains in the phospholipid. Saturated phospholipids with long alkyl tails have higher transition temperatures than unsaturated and short ones. Egg or soy phosphatidylcholines (eggPC, 16:0–18:1 or soyPC 18:2) have transition temperatures below room temperature (about −10°C for eggPC). Dimyristoyl phosphatidylcholine (DMPC, 14:0) has a transition temperature close to room temperature (21 to 24°C). Dipalmitoyl phosphatidylcholine (DPPC, 16:0) and distearoyl phosphatidylcholine (DSPC, 18:0) are examples of phospholipids with transition temperatures well above normal body temperature (T_m = 41 and 51°C, respectively).

Controlling the transition temperatures of lipid membranes is crucial for the use of lipid vesicles as drug delivery systems. Modification of the lipid membrane composition allows for the adjustment of the lipid membrane fluidity for a faster or slower diffusion of the drug loaded inside. Liposomal drug carriers are usually prepared by mixing different phospholipids and cholesterol. The presence of cholesterol has a deep influence on the properties of the lipid bilayers. The inclusion of cholesterol to a fluid phase bilayer decreases its permeability. Cholesterol molecules fill in the free space that is formed due to the kinks of the *gauche* lipid conformation and impair their lateral movement. As a consequence, the rigidity of the membrane increases. EggPCs or soyPCs with low transition temperatures and PEG-modified lipids are usually employed as the main component of liposomal bilayers to ensure their stability *in vivo.* As it will be explained in further sections, the goal with thermo-sensitive liposomal therapies is to prepare drug liposomes that 1) are stable at body temperature to avoid premature drug leakage during circulation, 2) accumulate at the site of action, and 3) undergo fast destabilization upon local heating of the area for a rapid release of the drug. Therefore, thermo-sensitive liposomes frequently include DPPC as the thermo-sensitive lipid component because it has a sharp melting temperature at 41°C, ideal for the combination of such liposomes with non-damaging mild hyperthermia (39–42°C).

5.5 Thermo-Sensitive Micelles

Micelles are spherical carriers with amphiphile-based unilamellar structures that separate the inner aqueous compartment from the external aqueous solvent [1]. Micelles are colloidal dispersions that are spontaneously formed by the thermodynamically favored aggregation of amphiphiles at or above the critical micellar concentration (CMC). Micelle amphiphilies are often constructed from lipid or polymeric moieties as hydrophobic blocks capping hydrophilic polymer chains. Common examples are PEG-PLA-or PEG-PE-based micelles [58, 59]. In this way, micelles can solubilize molecules of poorly soluble nonpolar drugs within the micelle core (with polar molecules adsorbed on the micelle surface and substances with intermediate polarity can be distributed along the amphiphilic molecules in intermediate positions). Because of their small sizes, micelles demonstrate significant spontaneous penetration into the tumor interstitium [60, 1]. Thus, they are attractive carriers for tumor targeting, although their stability *in vivo* can be an issue due to their dilution below the CMC after intravenous injection. Biodistribution studies have revealed significant accumulation of the micelles at the tumor site via passive targeting, but it was limited primarily to the tumor periphery [61].

The modification of micelles with thermo-sensitive polymers (i.e. PNIPAAm) can improve the delivery of drugs to the area of interest by promoting their disintegration after the application of either local hyperthermia or hypothermia. Two main approaches are employed to design thermo-sensitive micelles based on PNIPAAm. In the first one, the copolymers of PNIPAAm are conjugated with hydrophobic blocks, such as poly(methylmethacrylate), poly(butylmethacrylate), and poly (D,L-lactide). At temperatures below LCST, the PNIPAAm in the micellar corona is soluble and in an expanded conformation. At temperatures above LCST, the PNIPAAm corona aggregates with the subsequent disruption of the micellar structure and the triggered release of the drug content (Fig. 5.5A). In the second approach, the copolymers of PNIPAAm are conjugated with hydrophilic blocks, such as PEG. At temperatures below LCST, both PNIPAAm and PEG blocks are soluble. Upon heating, the PNIPAAm block becomes more and more hydrophobic and micellization occurs. These micelles can be easily prepared by heating a solution of polymer and drug above 32°C, and they are stable at body temperatures. Upon their arrival at the site of interest, the drug content can be released by local hypothermia [62, 51] (Fig. 5.5B).

As mentioned before, in order to modify the 32°C LCST of PNIPAAm, the PNIPAAm block in micelles is often modified with hydrophobic monomers to achieve an a LCST of >37°C to allow their disruption at 40–43°C by local hyperthermia. Another strategy used to tune the LCST of PNIPAAm micelles is to include biodegradable chains that promote the hydrophobic-to-hydrophilic conversion of the polymer. The micelles are stable at body temperatures and destabilize over time due to the degradation of the chains. This strategy was first reported by the co-polymerization of PNIPAAm with *N*-acryloxy succinimide to yield polymers with increased LCST upon the hydrolysis of acryloxy succinimide into acrylic acid [63]. However, the modification of PNIPAAm's transition temperature has been best developed by using hydrolysable lactate groups. Some illustrative examples of LCST modulation in PNIPAAm micelles and other thermo-sensitive micelles are summarized in Table 5.2.

Another class of thermo-sensitive micelles is based on the tri-block copolymer, poly(ethylene oxide)-b-poly(propylene oxide)-*b*-(polyethylene oxide) (PEO-PPO-PEO). These polymers are also called Poloxamer or Pluronic, and constitute a family of 30 different copolymers with varying MW and PEO/PPO ratios. In aqueous solution, above their critical micelle temperature, they self-assemble into micelles due to the dehydration of PPO blocks

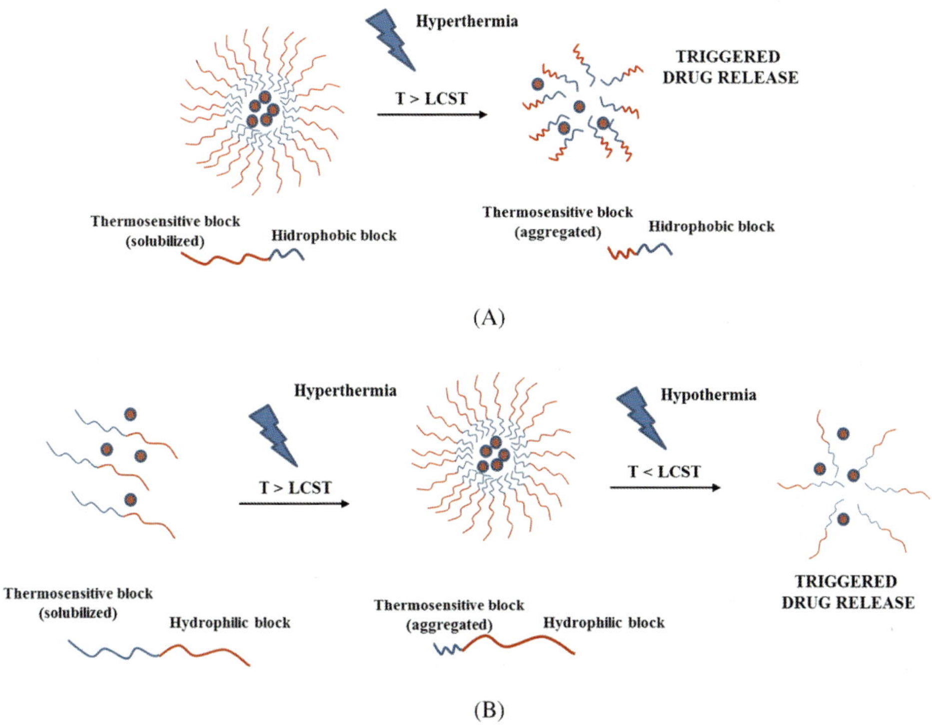

Figure 5.5. Design of thermosensitive micelles for triggered drug release. T = temperature, LCST = lower critical solution temperature.

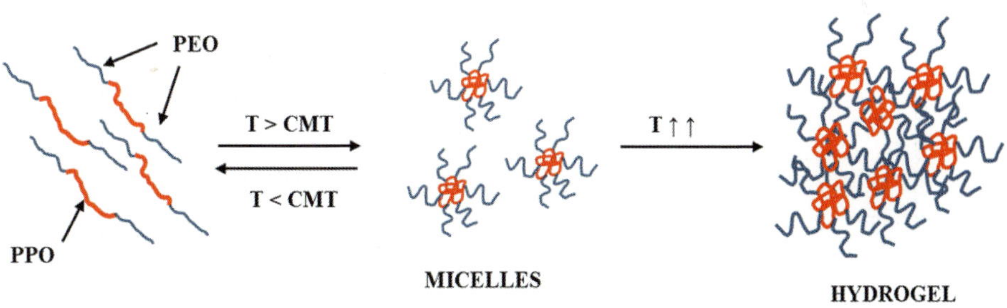

Figure 5.6. Schematic representation of the micellation and gel formation of Poloxamer (PEO-PPO-PEO). T = temperature, CMT = critical micelle temperature.

(Fig. 5.6). The micelle formation mechanisms of different Pluronic mixtures as well as their numerous applications in the pharmaceutical field have been described [68–71]. A micellar nanocarrier composed of Pluronic L61 (MW 2 kDa) and Pluronic F127 (MW 12.6 kDa) for tumor delivery of doxorubicin is currently being evaluated in clinical trials [72, 73].

Table 5.2. Examples of LCST modulation in PNIPAAm micelles and other thermo-sensitive micelles.

Polymer/Micelle Composition	Block Functionality		Property/Outcome/Application	Reference
NIPAAm-co-HEMA-Lac-co-AAc	N-isopropylacrylamide	Thermo-sensitive	— Temperature dependent drug release due to hydrophobic to hydrophilic conversion. — Pure NIPAAm has an LCST of 32°C. — ↓ LCST with ↑% Ac. — ↑ LCST with ↑ hydrolysis lactate group. — LCST before hydrolysis: 23–28°C , after hydrolysis: 43–44 °C for 20% HEMA-Lac copolymer. — Incubation time for LCST change via lactate hydrolysis: 1–10 days (for 10% or 20% HEMA-Lac, respectively) (PBS, pH 7.4).	[64]
	2-hydroxyethyl methacryl lactate	Hydrolysable		
	Acrylic acid	Hydrophobic		
(NIPAAm-co-HPMAm–Lac)–b-PEG	N-isopropylacrylamide	Thermo-sensitive	— Temperature dependent drug release due to hydrophobic to hydrophilic conversion. — Pure NIPAAm has an LCST of 32°C. — ↓ LCST with ↑% HPMA-Lac. — ↑ LCST with ↑ hydrolysis lactate group. — LCST before hydrolysis: 17°C, after hydrolysis 45°C for 35% HPMA-Lac copolymer. — Incubation time for LCST change via lactate hydrolysis: 2 days (PBS, pH 8.5). — The copolymers form micelles (200 nm) at physiological T. The destabilization of the micelles can be triggered by hypothermia or by hydrolysis of the lactate side group.	[50]
	2-hydroxypropyl methacrylamide-lactate	Hydrolysable		
	PEG	Hydrophilic		

(HPMAm–di-Lac)–b-PEG	2-hydroxypropyl methacrylamide-di-lactate	Thermo-sensitive Hydrolysable	— Temperature-dependent drug release due to hydrophobic to hydrophilic conversion. — Pure HPMAm-di-Lac has a LCST 10°C. — Pure HPMAm-mono-Lac has a LCST 65°C. — ↑ LCST with ↑ hydrolysis lactate side group. — LCST before hydrolysis: 10°C , after hydrolysis 65°C. — The copolymers form micelles (50 nm) at physiological T. The destabilization of the micelles is mediated by the hydrolysis of the di-lactate into mono-lactate. — HPMA has been shown to be non-toxic in clinical trials[65].	[52, 66]
	Poly(ethylene glycol)	Hydrophilic		
	PTX	Anti-cancer drug Hydrophobic	— Micelles were loaded by mixing concentrated PTX ethanol solution with the polymer solution and subsequent heating above 50°C for 1 min and cooled to RT. — PTX was loaded in the micelles at a concentration of 2 mg/mL. — PTX-loaded micelles had a mean size around 60 nm. — At pH 8.8 and 37°C, PTX-loaded micelles destabilized within 10 h due to the hydrolysis of the lactic acid side group. — The micelles were stable for about 200 h at physiological conditions. — The PTX-loaded micelles showed a release of the incorporated 70% of PTX during 20 h at 37°C and at pH 7.4. — PTX-loaded micelles were internalized by B16F10 cells and showed a cytotoxicity comparable to that of Taxol® (commercial formulation PTX containing Cremophor EL), but were less toxic than the Cremophor EL vehicle.	[67]

5.6 Thermo-Sensitive Hydrogels

Thermo-sensitive hydrogels are three-dimensional (3D) polymer networks with the ability to collapse or expand in response to temperature changes. They are useful as devices for local and controlled release of drugs [74–77], but also for cell encapsulation and tissue engineering [78]. One main advantage over other gel systems is that they do not require complex procedures, such as chemical cross-linking or photo-polymerization for their gelification. They are formed *in situ* by temperature-induced coil-to-globule transitions (i.e. PNIPAAm) or by micelle packing of block copolymers (i.e. PEO–PPO–PEO) [79, 80] (Fig. 5.6).

Swollen hydrogels keep the drug solution embedded at low temperatures and allow low rate of diffusion of the drug. Upon heating, they display a rapid release of the embedded drugs due to the matrix collapse, followed by a slower release due to the limited diffusion of the drug though the compacted gel (Fig. 5.7).

This pulsatile on–off drug release was first described for PNIPAAm hydrogels loaded with the hydrophobic drug, indomethacin [81–84]. Drug solubility is one of the most important factors to control drug release from thermo-sensitive hydrogels [85]. Verestiuc and co-workers evaluated the efficacy of PNIPAAm-chitosan-based hydrogels as controlled delivery carriers for ophthalmic drug delivery. From the panel of drugs evaluated, water-soluble ionic compounds, such as pilocarpine hydrochloride and atropine sulfate, presented fast-burst release within the first hour, whereas drugs with low water solubility, such as chloramphenicol and norfloxacin, showed prolonged release profiles (over 48 h) [86]. PNIPAAm-based hydrogels can be injected intramuscularly, intramyocardially, or subcutaneously for *in situ* formation of scaffolds for the slow release of drugs and proteins [87–90], but can also be intravenously injected. Fang and co-workers developed PNIPAAm-based hydrogels for the parenteral administration of nalbuphine. This analgesic drug has a short half-life and poor oral bioavailability and, thus, requires frequent injections (e.g. every 3–6 h). The injected hydrogels, with an LCST around 32°C, facilitated the controlled release of nalbuphine and improved the duration of the analgesic action in rats from 1.5 h for the aqueous solution of nalbuphine to 3 h for the drug loaded hydrogels [91]. In another study, PNIPAAm hydrogels were modified with acrylamide monomers to obtain LCST values of 37°C to 42°C. The nanohydrogels had an average size of 50 nm and were loaded with a near infrared dye to track their *in vivo* behavior. The nanohydrogels were intravenously administered in tumor-bearing mice with or without local hyperthermia treatment. The free

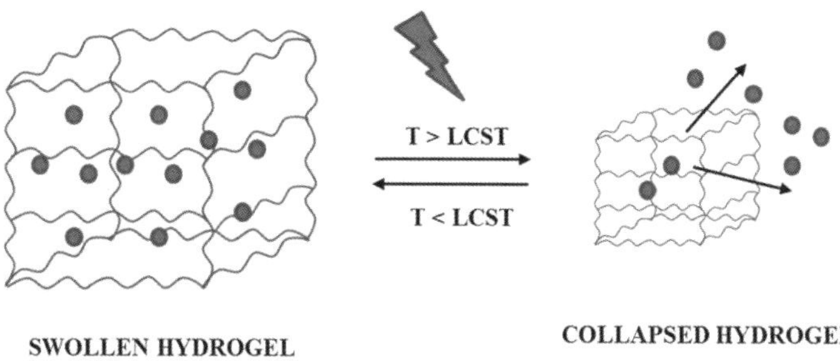

Figure 5.7. Schematic representation of the pulsatile on–off drug release from thermo-sensitive hydrogels. T = temperature, LCST = lower critical solution temperature.

dye had no tumor targeting effect even with the help of hyperthermia treatment. The nanohydrogels displayed a thermally targeted behavior. They accumulated for up to 12 h in heated tumors, whereas no fluorescence signal was observed in non-heated tumors [92]. In a later study [93], the same hydrogels were loaded with docetaxel and injected in tumor-bearing mice. Weekly injection of docetaxel nanohydrogels for a period of 3 weeks showed a tumor inhibition rate of 78% when the treatment was combined with local hyperthermia, 48% in the absence of hyperthermia, and 40% in the case of free docetaxel.

Poloxamer (PEO–PPO–PEO) hydrogels are an alternative to PNIPAAm hydrogels. In particular, Poloxamer 407 (MW 12,600 and PEO/PPO ratio 2:1) is used in the majority of hydrogel formulations. It shows thermo-reversible gelation at concentrations of 20–30wt%. Below 25°C, the polymer solution is a free flow liquid that turns into gel at body temperature. In aqueous solution, P407 molecules self-assemble into micelles at the critical micelle temperature due to the dehydration of PPO blocks. When the temperature rises above a certain point, micelles lose their water, become dehydrated, and form a gel (Fig. 5.6). Some recent examples of Poloxamer-based hydrogels in drug delivery are summarized in Table 5.3. Besides its weak mechanical strength and rapid erosion, the major disadvantages of Poloxamer gels are that they are not biodegradable and that, despite being considered as non-toxic, they can produce hyperlipidemia at doses ≥140 mg/kg [94, 95].

5.7 Thermo-Sensitive Liposomes

Liposomes are spherical carriers with amphiphile-based multilamellar structures that separate the inner aqueous compartment from the external aqueous solvent. Similar to micelles, they can be loaded with a variety of water soluble

Table 5.3. Recent examples of Poloxamer-based hydrogels in drug delivery.

Hydrogel composition	Functionality	Drug/route	Property/outcome/application	Reference
Pluronic (PF127)	Thermo-sensitive	Doxorubicin Subcutaneous	— Sol-to-gel transition at 37°C. — Mean size of micelles 100–200 nm (previous to hydrogel formation). — Sustained release of doxorubicin for up to 28 days, order zero-release. — Tumor growth inhibition (s.c. tumor) significantly higher than free drug, non-croslinked P127/Hyaluronic Acid doxorubicin.	[96]
Hexamethylene diisocyanate	Cross-linking of P127 chains Improve mechanical stability			
Hyaluronic Acid	Promote gel formation			
Pluronic (P407)	Thermo-sensitive	Doxorubicin Intravesical	— Floating hydrogel containing doxorubicin encapsulated in albumin nanoparticles for intravesical delivery of the drug without urinary obstruction. — Sol-to-gel transition at 10°C. Gelation time was 2 min at 37°C. — Sustained release of doxorubicin for up to 10 h. — The inclusion of Doxorubicin nanoparticles in the floating hydrogel could significantly extend the residence time of the drug in bladder after intravesical instillation. Non-floating hydrogel produced urinary obstruction whereas floating hydrogel did not.	[97]
$NaHCO_3$	Production of carbon dioxide (CO_2), in acidic environment			

Pluronic (PF127) **Pluronic (P188)**	Thermo-sensitive	Curcumin Intranasal	— Sol-to-gel transition at 32°C. Gelation time was 20 s. — Sustained release of curcumin for up to 6 h. — The inclusion of curcumin in hydrogels increased *in vivo* mucociliary transport time (10-fold higher than curcumin solution) without any signs of severe toxicity in the nasal mucose of the rats. — The in-situ nasal gel demonstrated an enhanced brain uptake (olfactory bulb, cerebellum, hippocampus) of curcumin as compared to i.v administration.	[98]
Pluronic-Heparine (P407)	Thermo-sensitive	Heparine Intramuscular	Pluronic Heparine copolymer displayed 1 gel-sol-gel phase transition behavior similar as P407.The hydrogel was superior to P407 hydrogel for restoring the functions and epithelial structure of the broken vessel junctions after end-to-end anastomosis in the rabbit common jugular artery.	[99]

drugs (into the inner compartment) or insoluble drugs (into the hydrophobic compartment of the lipid bilayer). Liposomes are constructed as phospholipid vesicles, mimicking the plasma membrane of mammalian cells. They are biologically inert and cause few, if any, toxic or antigenic reactions [10, 100].

Among thermo-sensitive nanocarriers, temperature sensitive liposomes in combination with mild hyperthermia for triggered and targeted drug release to solid tumors have achieved the most advanced levels of development with one formulation currently undergoing phase III clinical trials [101]. Triggering drug release with hyperthermia has advantages over physiological triggers (i.e. pH, pO_2, redox balance, or enzymatic catalysis) because it can be externally modulated and modified for a personalized treatment. Hyperthermia can be used to destroy cells (i.e. tumor ablation >50°C, 10 min), or at lower temperatures (mild hyperthermia, 39 to 42°C, 60 min), to induce physiological changes in the tissues where applied, such as increased blood flow, perfusion, oxygenation and permeability [102]. In fact, hyperthermia has long been used in synergy with cancer chemotherapy and radiotherapy [103, 104].

Hyperthermia combined with a thermostable liposomal formulation has demonstrated positive results due to the increased tumor vasculature permeability, tumor accumulation of liposomes, and hyperthermia-induced chemosensitization [105–107]. Van Bree and co-workers compared the pharmacokinetics and the anti-tumor effect of liposomal daunorubicin versus free drug in combination with hyperthermia in a solid sarcoma rat model. Pharmacokinetic studies revealed that liposomal accumulation in tumor tissue was similar to that of free drug. A significant increase in tumor drug delivery was observed when liposomal daunorubicin was combined with hyperthermia (42°C, 60 min). For free drug, hyperthermia effected some tumor growth delay. The combination of hyperthermia and drug-loaded liposomes led to a complete regression of the tumor [108]. The same beneficial effect of hyperthermia was observed for thermostable PEGylated liposomal doxorubicin in a mouse tumor model [109].

In the case of thermo-sensitive liposomes, hyperthermia can augment the accumulation of the liposomes in the tumor [110] and trigger the burst release of the drug, creating a high local concentration of the drug within the tumor vasculature that promotes the extravasation of the drug into the tumor interstitium [111]. The hyperthermia also increases the cell membrane permeability, which in turn, increases the cellular uptake [102]. Therefore, such liposomes need to be stable at body temperature while in the circulation and release their payload at temperatures >39°C when they reach the tumor. Modification of the lipid membrane composition allows for the adjustment of the lipid membrane fluidity and the control of their transition temperatures.

Earlier studies used DPPC and DSPC (3:1 ratio, T_m of 41 and 51°C, respectively) to prepare liposomes to be able to release the drug in the temperature range of 42.5 to 44.5°C. Similar liposomes (DPPC/DSPC 7:3) loaded with methotrexate improved the stability of the drug in blood (clearance to 10% of injected dose was 1 h versus 3 min post-injection for free drug) and showed a 3.6-fold higher drug accumulation in heated tumors 20 h post-injection, compared to non-heated tumors [112]. Further development of these formulations introduced polyethylenglycol-grafted lipids ($DSPEG_{2000}$) and cholesterol in their lipid composition. Inclusion of PEG-PE had little influence in the drug release but transformed the liposomes into long-circulating ones with greater accumulation and greater drug release in the heated region. Li and co-workers evaluated the effect of PEG content in the surface of DPPC/DSPC liposomes [113]. Liposomes containing 5 mol% DSPE-PEG_{2000} were stable at 37°C. At 42°C, the liposomes released 60% of their content in 1 min and almost 100% after 1 h. Higher concentrations of PEG led to premature drug leakage while lower concentrations promoted long circulation of such liposomes. As mentioned in previous sections, the addition of cholesterol tends to increase the rigidity of the liposomal membranes and interferes with their temperature dependent behavior. The percentage of cholesterol in thermo-sensitive liposomes is usually kept below 25–30% mol or avoided completely [114].

A new generation of thermo-sensitive liposomes was reported by Needham and co-workers by incorporating lysolipids into PEGylated DPPC liposomes [115, 101]. Inclusion of lysolipids (phospholipids that have only one alkyl chain) in particular, mono-stearoyl-phosphatidylcholine (MSPC), increased the fluidity of the lipid membrane and slightly decreased the T_m (<1°C) of the liposomes (DPPC/MSPC/$DSPEG_{2000}$, 90:10:4 mol ratio) . Inclusion of lysolipids also improved the rate and the extent of drug release. It is known that the maximum rate of release occurs at the lower end of the gel-to-liquid phase transition (39°C) when initial melting defects start to form in the liposome membrane. Lysolipids may assist in this process by the stabilization of such pores [116]. Because their single tail and relatively large head group, lysolipids tend to form micelles. As the temperature approaches the T_m, the lateral mobility of the lipids in the membrane increases and lysolipids accumulate in the liquid–gel boundaries that are formed in the lipid bilayer. There, the lysolypids' tendency to form micelles results in the stabilization of the defects in the liposome membrane. The $DSPEG_{2000}$, which tends to form inverted micelles, is also required to enhance the formation of stable pores [117]. In summary, the stabilization of these pores by MSPC and $DSPEG_{2000}$ leads to an increase in the permeability of the membrane and the rapid escape of the liposome contents. The lysolipid liposomes had a triggered drug release at the ideal hyperthermia temperatures of 39 to 42°C. They were stable at body temperatures (20% release in 30 min at 37 to 38°C), but had an extensive rate of drug release (60% release in 20 sec) at 41.3°C. The 20-second burst release at mild hyperthermia temperatures is critical for the efficacy of the liposomes because it matches the transit time of the liposomes through the tumor microvasculature and gives enough time for the triggering of drug release from the liposome during their passage through the heated microvasculature. The median red cell velocity through tumor microvessels is 0.54 mm/sec [118]. The typical transit length through a tumor is many millimeters. Therefore, the transit time will typically be far longer than 20 seconds.

The lysolipid liposomal formulation developed by Needham and Dewhirst is the most studied and advanced thermo-sensitive formulation marketed as ThermoDox ®, as doxorubicin loaded in lysolipid thermally sensitive liposomes (LTSLs). It is currently in Phase III of clinical trials in combination with radiofrequency ablation for primary or metastatic liver disease, and in Phase II of trials for breast cancer. Preclinical studies demonstrated the superiority of ThermoDox over free doxorubicin, as well as its superiority over thermostable and non-lysolipid-containing thermo-sensitive liposomes at reducing tumor growth in multiple mouse xenograft models [119, 120, 115]. Pre-clinical studies in larger animals showed the importance of the tumor vasculature in the efficacy of these liposomes. The lack of vascularity in certain areas of the tumor, especially if the tumor is large, can prevent the homogenous distribution of heat and decrease the drug release and efficacy [121]. This observation also highlights the need for precise and controllable heating methods for hyperthermia modes where the range of temperatures between the on–off switch for triggered release is only a few degrees. Another concern in lysolipid-based formulations is the drug leakage from the liposomes during circulation (~20% at 37°C) due to lipid desorption of the liposomal membrane and the potential associated systemic toxicity. Banno and co-workers demonstrated that 70% of lysolipids are desorbed from the liposomes 1 h post-injection [122]. In addition, the liposomes recovered from plasma 1 h and 4 h post-injection, released 20% and 50% less doxorubicin upon heating, meaning that the transfer of lysolipids into the biological milieu can compromise the liposome's thermo-sensitivity and delivery of the drug. In order to improve the temperature sensitivity and stability of lysolipids and liposomes, alternative formulations have been developed by the inclusion of thermo-responsive lipids in the lipid membrane or by the grafting or encapsulation thermo-responsive polymers (Table 5.4).

Table 5.4. Recent examples of thermo-sensitive liposomes for drug delivery.

Liposome composition/ratio		Lipid functionality	Property/outcome/application	Reference
DPPC/DSPC/ $DPPG_2$ 5:3:2	DPPC	Thermo-sensitive	— Inclusion of $DPPG_2$ (analog to DPPG containing an extra glycerol moiety) to render similar function as $DSPEG_{2000}$. — Higher mol % than $DSPEG_{2000}$ without compromising the stability of the membrane. — Temperature of transition = 40.2°C. — 80% release in 3 min at 42°C and <15% release in 1 h at 39°C.	[123–125]
	DSPC	Thermo-sensitive		
	$DPPG_2$	Long-circulation and stability *in vivo*		
DPPC/Brij 7896:4 (HaT)	DPPC	Thermo-sensitive	— Inclusion of Brij 78 (surfactant) to rend similar function as $DSPEG_{2000}$ and lysolipids. — Temperature of transition = 41°C. — 100% release in <3 min at 42°C and < 20% release in 30 min at 37°C. — Improved doxorubicin release compare to lysolipid liposomes (2-fold and 1.2-fold greater at 40°C and 41°C). — Improved delivery of Dox in heated tumor and improved anti-tumor efficacy compare to lysolipid liposomes. — A second generation of liposomes (HaT-II) was developed for improved stability. The drug is loaded using copper (Cu^{+2}) gradient (instead of pH gradient in HaT).	[126–129]
	Brij 78	Stabilization of pores during T_m		
		Long-circulation and stability in vivo		

(*Continued*)

Table 5.4. (*Continued*)

Liposome composition/ratio	Lipid functionality		Property/outcome/application	Reference
(PINAAm-co-AAm)-modified-liposomes	DOPE/EggPC[a] (6:4) or DPPC[b]	Non- thermo-sensitive (in practical terms)[a] Thermo-sensitive[b]	— Incorporation of NIPAAm copolymers in the liposome membrane endowed temperature sensitivity[a]. Unmodified liposomes released <10% of their content at 40°C versus 45% release for modified liposomes. — Incorporation of NIPAAm copolymers in the liposome membrane led to synergistic effects of the polymer and DPPC[b]. — Liposomes exhibited ↑LCST with ↑%AAm[a,b].	[49, 130]
	(PINAAm-co-AAm) 10 to 30 mol%	Thermo-sensitive polymer		
Liposome encapsulated Poloxamer	DOPC/cholesterol	Non- thermo-sensitive (in practical terms)	— At T < CMC (critical micelle concentration), the liposomes encapsulated Poloxamer did not associate with the lipid bilayer. — At T > CMC, the encapsulated Poloxamer interact with the membrane with subsequent disruption and content release.	[131, 132]
	Pluronic F127	Thermo-sensitive polymer		

5.8 Modalities for Hyperthermia Application

The success of temperature sensitive drug nanocarriers depends on their careful design and particular properties (controlled release rate, stability, circulation time), but also on the ability to homogenously heat the site of interest to a precise temperature, without damaging adjacent tissues.

Hyperthermia modalities can be categorized depending on their invasiveness and their capacity to heat deep-seated areas. Regional hyperthermia is the less invasive modality and is employed in the majority of pre-clinical studies. The source of heat (external) is a heated water bath where the area of interest is immersed. This method provides a poorly localized heat and is limited to superficial areas. As an alternative, preclinical models of hyperthermia also employed external array antennae that emit microwaves [121]. This modality has a greater capacity for tissue penetration (maximum penetration of 3 cm) and is more suitable for large tumor sizes than are water baths. Heating of deep seated areas is achieved by the so-called localized interstitial hyperthermia using microwave antennae, radiofrequency electrodes, or laser fibers [103]. This is more invasive than the previous methods and is limited to well-localized tumors smaller than 5 cm. Localized interstitial hyperthermia is used in the clinic for the ablation of head, neck and prostate tumors, since the insertion of the probe is minimally invasive. In addition, radiofrequency ablation in combination with ThermoDox is currently being evaluated in clinical trials for hepatocellular carcinoma, despite its greater invasiveness, since these tumors are usually non-resectable and radiotherapy and chemotherapy are largely ineffective. More recently, focused ultrasound (FUS) and high intensity focused ultrasound (HIFU) technologies [133] have provided a way to deliver heat into localized deep-seated areas, with minimal damage and recovery times [134]. In addition, MR-imaging to guide FUS or HIFU heating and to map the temperature contours in tumors is currently under investigation [135]. MR-imaging has also been validated as a potential method for real-time tracking of drug release. For instance, lysolipid liposomes were loaded with doxorubicin by utilizing manganese (Mn^{+2}) gradients between the inner and the outer liposomal compartment. The release of Mn^{+2} in heated tumors enhanced MR contrast. The release of Mn^{+2} correlated well with doxorubicin release [136]. It was shown that when hyperthermia was administered before liposome injection, the accumulation of doxorubicin was located in the center of the tumor, whereas when the liposomes were administered after hyperthermia, the drug distribution was more peripheral. Similar effects were reported with thermo-sensitive liposomes loaded with doxorubicin and gadopentetic acid (Gd-DPTA). The authors proposed the comparison of MR images before and after ultrasound treatments to adjust the heating dose and the doxorubicin dose as a potential way to personalize the treatment [127].

5.9 Future Perspectives

The combination of hyperthermia and thermo-sensitive drug delivery technologies is close to becoming a common treatment in the standard care of cancer patients. In addition, it is not difficult to envision the expansion of such therapies to other areas (i.e. cardiovascular, pain, infectious diseases) sharing physiopathological similarities with solid tumors. It is clear from the multiple examples illustrated in this chapter that the design and development of temperature-sensitive carriers and heating modalities have achieved a high level of sophistication and mastery, especially in the area of thermo-sensitive liposomes. However, there remains plenty of room for improvement. In the case of temperature responsive polymer-based carriers (micelles, hydrogels), the major need is to validate the safety and clinical application of the growing number of new polymer chemistries by moving the available *in vitro* data towards late pre-clinical stages. For thermo-sensitive liposomes, new approaches to improve their stability during circulation are needed to avoid premature loss of drugs and associated toxicities. In addition, advances in multifunctional nanocarriers, such as those combining thermo-responsiveness and imaging properties, would dramatically improve the efficacy and accuracy of current traditional systems.

References

1. Torchilin, V. P. (2007). Micellar nanocarriers: Pharmaceutical perspectives. *Pharm. Res.*, **24**, 1–16.
2. Fanciullino, R. and Ciccolini, J. (2009). Liposome-encapsulated anticancer drugs: Still waiting for the magic bullet? *Curr. Med. Chem.*, **16**, 4361–4371.
3. Peer, D., Karp, J. M., Hong, S., Farokhzad, O. C., Margalit, R. and Langer, R. (2007). Nanocarriers as an emerging platform for cancer therapy. *Nat. Nanotechnol*, **2**, 751–760.
4. Jhaveri, A., Deshpande, P. and Torchilin, V. (2014). Stimuli-sensitive nanopreparations for combination cancer therapy. *J. Control Release*, **190**, 352–357.
5. Torchilin, V. (2009). Multifunctional and stimuli-sensitive pharmaceutical nanocarriers. *Eur. J. Pharm. Biopharm.*, **71**, 431–444.
6. Zhu, L. and Torchilin, V. P. (2013). Stimulus-responsive nanopreparations for tumor targeting. *Integr. Biol. (Camb.)*, **5**, 96–107.
7. Mura, S., Nicolas, J. and Couvreur, P. (2013). Stimuli-responsive nanocarriers for drug delivery. *Nat. Mater.*, **12**, 991–1003.
8. Williams, H. D., Trevaskis, N. L., Charman, S. A., Shanker, R. M., Charman, W. N., Pouton, C. W. and Porter, C. J. (2013). Strategies to address low drug solubility in discovery and development. *Pharmacol. Rev.*, **65**, 315–499.

9. Gao, S., Dagnaes-hansen, F., Nielsen, E. J., Wengel, J., Besenbacher, F., Howard, K. A. and Kjems, J. (2009). The effect of chemical modification and nanoparticle formulation on stability and biodistribution of siRNA in mice. *Mol. Ther,* **17,** 1225–1233.
10. Torchilin, V. P. (2005). Recent advances with liposomes as pharmaceutical carriers. *Nat. Rev. Drug Discov.,* **4**, 145–160.
11. Battaglia, L. and Gallarate, M. (2012). Lipid nanoparticles: State of the art, new preparation methods and challenges in drug delivery. *Expert Opin. Drug Deliv.,* **9**, 497–508.
12. Duncan, R., Ringsdorf, H. and Satchi-Fainaro, R. (2006). Polymer therapeutics — polymers as drugs, drug and protein conjugates and gene delivery systems: Past, present and future opportunities. *J. Drug Target.,* **14**, 337–341.
13. Tros De Ilarduya, C., Sun, Y. and Duzgunes, N. (2010). Gene delivery by lipoplexes and polyplexes. *Eur. J. Pharm. Sci.,* **40**, 159–170.
14. Haraldsson, B., Nystrom, J. and Deen, W. M. (2008). Properties of the glomerular barrier and mechanisms of proteinuria. *Physiol. Rev.,* **88**, 451–487.
15. Maeda, H. (2012). Vascular permeability in cancer and infection as related to macromolecular drug delivery, with emphasis on the EPR effect for tumor-selective drug targeting. *Proc. Jpn. Acad. Ser. B. Phys. Biol. Sci.,* **88**, 53–71.
16. Danhier, F., Feron, O. and Preat, V. (2010). To exploit the tumor microenvironment: Passive and active tumor targeting of nanocarriers for anti-cancer drug delivery. *J. Control. Release,* **148**, 135–146.
17. Hockel, M. and Vaupel, P. (2001). Tumor hypoxia: Definitions and current clinical, biologic, and molecular aspects. *J. Natl. Cancer Inst.,* **93,** 266–276.
18. Brown, J. M. (2007). Tumor hypoxia in cancer therapy. *Methods Enzymol.,* **435**, 297–321.
19. Jain, R. K. (2005). Normalization of tumor vasculature: An emerging concept in antiangiogenic therapy. *Science,* **307**, 58–62.
20. Hobbs, S. K., Monsky, W. L., Yuan, F., Roberts, W. G., Griffith, L., Torchilin, V. P. and Jain, R. K. (1998). Regulation of transport pathways in tumor vessels: Role of tumor type and microenvironment. *Proc. Natl. Acad. Sci. USA.,* **95**, 4607–4612.
21. Noguchi, Y., Wu, J., Duncan, R., Strohalm, J., Ulbrich, K., Akaike, T. and Maeda, H. (1998). Early phase tumor accumulation of macromolecules: A great difference in clearance rate between tumor and normal tissues. *Jap. J. Cancer Res.,* **89**, 307–314.
22. Matsumura, Y. and Maeda, H. (1986). A new concept for macromolecular therapeutics in cancer-chemotherapy — mechanism of tumoritropic accumulation of proteins and the antitumor agent smancs. *Cancer Res.,* **46**, 6387–6392.
23. Jain, R. K. and Stylianopoulos, T. (2010). Delivering nanomedicine to solid tumors. *Nat. Rev. Clin. Oncol.,* **7**, 653–664.
24. Yuan, F., Dellian, M., Fukumura, D., Leunig, M., Berk, D. A., Torchilin, V. P. and Jain, R. K. (1995). Vascular-permeability in a human tumor xenograft — molecular-size dependence and cutoff size. *Cancer Res.,* **55**, 3752–3756.
25. Klibanov, A. L., Maruyama, K., Torchilin, V. P. and Huang, L. (1990). Amphipathic polyethyleneglycols effectively prolong the circulation time of liposomes. *FEBS Lett.,* **268**, 235–237.
26. Zalipsky, S. (1995). Functionalized poly(ethylene glycol) for preparation of biologically relevant conjugates. *Bioconj. Chem.,* **6**, 150–165.
27. Yamaoka, T., Tabata, Y. and Ikada, Y. (1994). Distribution and tissue uptake of poly(ethylene glycol) with different molecular-weights after intravenous administration to mice. *J. Pharm. Sci.,* **83**, 601–606.

28. Gabizon, A. and Papahadjopoulos, D. (1992). The role of surface-charge and hydrophilic groups on liposome clearance *In vivo*. *Biochim. Biophys. Acta,* **1103**, 94–100.
29. Needham, D., Mcintosh, T. J. and Lasic, D. D. (1992). Repulsive interactions and mechanical stability of polymer-grafted lipid-membranes. *Biochim. Biophys. Acta,* **1108**, 40–48.
30. Torchilin, V. P., Omelyanenko, V. G., Papisov, M. I., Bogdanov, A. A., Trubetskoy, V. S., Herron, J. N. and Gentry, C. A. (1994). Poly(ethylene glycol) on the liposome surface — on the mechanism of polymer-coated liposome longevity. *Biochimi. Biophys. Acta.,* **1195**, 11–20.
31. Bae, Y. H. (2009). Drug targeting and tumor heterogeneity. *J Control Release,* **133**, 2–3.
32. Hong, R. L., Huang, C. J., Tseng, Y. L., Pang, V. F., Chen, S. T., Liu, J. J. and Chang, F. H. (1999). Direct comparison of liposomal doxorubicin with or without polyethylene glycol coating in C-26 tumor-bearing mice: Is surface coating with polyethylene glycol beneficial? *Clin. Cancer Res.,* **5**, 3645–3652.
33. Pozzi, D., Colapicchioni, V., Caracciolo, G., Piovesana, S., Capriotti, A. L., Palchetti, S., De Grossi, S., Riccioli, A., Amenitsch, H. and Lagana, A. (2014). Effect of polyethyleneglycol (PEG) chain length on the bio-nano-interactions between PEGylated lipid nanoparticles and biological fluids: From nanostructure to uptake in cancer cells. *Nanoscale,* **6**, 2782–2792.
34. Baselga, J. and Arteaga, C. L. (2005). Critical update and emerging trends in epidermal growth factor receptor targeting in cancer. *J. Clin. Oncol.,* **23**, 2445–2459.
35. Schally, A. V. (2008). New approaches to the therapy of various tumors based on peptide analogues. *Horm. Metab. Res.,* **40**, 315–322.
36. Daniels, T. R., Delgado, T., Helguera, G. and Penichet, M. L. (2006). The transferrin receptor part II: Targeted delivery of therapeutic agents into cancer cells. *Clin. Immunol.,* **121**, 159–176.
37. Zhao, X. B., Li, H. and Lee, R. J. (2008). Targeted drug delivery via folate receptors. *Expert Opin. Drug Deliv.,* **5**, 309–319.
38. Ward, M. A. and Georgiu, T. K. (2011). Thernoresposive polymers for biomedical applications. *Polymers,* **3**, 1215–1242.
39. Ruel-Gariepy, E. and Leroux, J. C. (2004). *In situ*-forming hydrogels — review of temperature-sensitive systems. *Eur. J. Pharm. Biopharm.,* **58**, 409–426.
40. Dubruel, P., Unger, R., Vlierberghe, S. V., Cnudde, V., Jacobs, P. J., Schacht, E. and Kirkpatrick, C. J. (2007). Porous gelatin hydrogels: 2. *In vitro* cell interaction study. *Biomacromolecules,* **8**, 338–344.
41. Vlierberghe, S. V., Cnudde, V., Dubruel, P., Masschaele, B., Cosijns, A., Paepe, I. D., Jacobs, P. J., Hoorebeke, L. V., Remon, J. P. and Schacht, E. (2007). Porous gelatin hydrogels: 1. Cryogenic formation and structure analysis. *Biomacromolecules,* **8**, 331–337.
42. Bloch, K., Lozinsky, V. I., Galaev, I. Y., Yavriyanz, K., Vorobeychik, M., AZarov, D., Damshkaln, L. G., Mattiasson, B. and Vardi, P. (2005). Functional activity of insulinoma cells (INS-1E) and pancreatic islets cultured in agarose cryogel sponges. *J. Biomed. Mater. Res. A,* **75**, 802–809.
43. De Las Heras Alarcon, C., Pennadam, S. and Alexander, C. (2005). Stimuli responsive polymers for biomedical applications. *Chem. Soc. Rev.,* **34**, 276–285.
44. Roy, D., Brooks, W. L. and Sumerlin, B. S. (2013). New directions in thermoresponsive polymers. *Chem. Soc. Rev.,* **42**, 7214–7243.
45. Yarmolenko, P. S., Moon, E. J., Landon, C., Manzoor, A., Hochman, D. W., VIglianti, B. L. and Dewhirst, M. W. (2011). Thresholds for thermal damage to normal tissues: an update. *Int. J. Hyperthermia,* **27**, 320–343.

46. Livney, Y. D., Portnaya, I., Faupin, B., Ramon, O., Cohen, Y., Cogan, U. and Mizrahi, S. (2003). Interactions between inorganic salts and polyacrylamide in aqueous solutions and gels. *J. Polym. Sci. Polym. Phys.*, **41**, 508–519.
47. Lue, S. J., Chen, C. H. and Shih, C. M. (2011). Tuning of lower critical solution temperature (LCST) of poly(Nisopropylacrylamide-co-acrylic acid) hydrogels. *J. Macromol. Sci. B*, **50**, 563–579.
48. Ta, T., Convertine, A. J., Reyes, C. R., Stayton, P. S. and Porter, T. M. (2010). Thermosensitive liposomes modified with poly(N-isopropylacrylamide-co-propylacrylic acid) copolymers for triggered release of doxorubicin. *Biomacromolecules*, **11**, 1915–1920.
49. Hayashi, H., Kono, K. and Takagishi, T. (1999). Temperature sensitization of liposomes using copolymers of N-isopropylacrylamide. *Bioconjug. Chem.*, **10**, 412–418.
50. Neradovic, D., Van Nostrum, C. F. and Hennink, W. E. (2001). Thermoresponsive Polymeric Micelles with Controlled Instability Based on Hydrolytically Sensitive N-Isopropylacrylamide Copolymers. *Macromolecules*, **34**, 7589–7591.
51. Neradovic, D., Soga, O., Van Nostrum, C. F. and Hennink, W. E. (2004). The effect of the processing and formulation parameters on the size of nanoparticles based on block copolymers of poly(ethylene glycol) and poly(N-isopropylacrylamide) with and without hydrolytically sensitive groups. *Biomaterials*, **25**, 2409–2418.
52. Soga, O., Van Nostrum, C. F. and Hennink, W. E. (2004). Poly(N-(2-hydroxypropyl) methacrylamide mono/di lactate): A new class of biodegradable polymers with tuneable thermosensitivity. *Biomacromolecules*, **5**, 818–821.
53. Seymour, L. W., Ferry, D. R., Kerr, D. J., Rea, D., Whitlock, M., Poyner, R., Boivin, C., Hesslewood, S., Twelves, C., Blackie, R., Schatzlein, A., Jodrell, D., Bissett, D., Calvert, H., Lind, M., Robbins, A., Burtles, S., Duncan, R. and Cassidy, J. (2009). Phase II studies of polymer-doxorubicin (PK1, FCE28068) in the treatment of breast, lung and colorectal cancer. *Int. J. Oncol.*, **34**, 1629–1636.
54. Sun, L. F., Zhuo, R. X. and Liu, Z. L. (2003). Studies on the synthesis and properties of temperature responsive and biodegradable hydrogels. *Macromol. Biosci.*, **3**, 725–728.
55. Yoshida, T., Aoyagi, T., Kokufuta, E. and Okano, T. (2003). Newly designed hydrogel with both sensitive thermoresponse and biodegradability. *J. Polym. Sci. Polym. Chem.*, **41**, 779–787.
56. Nagle, J. F. (1976). Theory of lipid monolayer and bilayer phase transitions: Effect of headgroup interactions. *J. Membr. Biol.*, **27**, 233–250.
57. John, K., Schreiber, S., Kubelt, J., Herrmann, A. and Muller, P. (2002). Transbilayer movement of phospholipids at the main phase transition of lipid membranes: Implications for rapid flip-flop in biological membranes. *Biophys. J.*, **83**, 3315–3323.
58. DIscher, D. E. and Ahmed, F. (2006). Polymersomes. *Annu. Rev. Biomed. Eng.*, **8**, 323–341.
59. Sawant, R. R. and Torchilin, V. P. (2010). Polymeric micelles: Polyethylene glycol-phosphatidylethanolamine (PEG-PE)-based micelles as an example. *Methods Mol. Biol.*, **624**, 131–149.
60. Tang, N., Du, G. J., Wang, N., Liu, C. C., Hang, H. Y. and Liang, W. (2007). Improving penetration in tumors with nanoassemblies of phospholipids and doxorubicin. *J. Natl. Cancer Inst.*, **99**, 1004–1015.
61. Hoang, B., Lee, H., Reilly, R. M. and Allen, C. (2009). Noninvasive monitoring of the fate of in-111-labeled block copolymer micelles by high resolution and high sensitivity MicroSPECT/CT imaging. *Mol. Pharm.*, **6**, 581–592.

62. Zhu, P. W. (2004). Particle formation and aggregation-collapse behavior of poly(N-isopropylacrylamide) and poly(ethylene glycol) block copolymers in the presence of cross-linking agent. *J. Mater. Sci. Mater. Med.,* **15**, 567–573.
63. Shah, S. S., Wertheim, J., Wang, C. T. and Pitt, C. G. (1997). Polymer-drug conjugates: manipulating drug delivery kinetics using model LCST systems. *J. Control. Release,* **45**, 95–101.
64. Lee, B. H. and Vernon, B. (2005). Copolymers of N-isopropylacrylamide, HEMA-lactate and acrylic acid with time-dependent lower critical solution temperature as a bioresorbable carrier. *Polym. Int.,* **54**, 418–422.
65. Duncan, R. (2003). The dawning era of polymer therapeutics. *Nat. Rev. Drug. Discov.,* **2**, 347–360.
66. Soga, O., Van Nostrum, C. F., Ramzi, A., Visser, T., Soulimani, F., Frederik, P. M., Bomans, P. H. and Hennink, W. E. (2004). Physicochemical characterization of degradable thermosensitive polymeric micelles. *Langmuir,* **20**, 9388–9395.
67. Soga, O., Van Nostrum, C. F., Fens, M., Rijcken, C. J., Schiffelers, R. M., Storm, G. and Hennink, W. E. (2005). Thermosensitive and biodegradable polymeric micelles for paclitaxel delivery. *J. Control. Release,* **103**, 341–353.
68. Verma, P., Nath, S., Singh, P. K., Kumbhakar, M. and Pal, H. (2008). Effects of block size of pluronic polymers on the water structure in the corona region and its effect on the electron transfer reactions. *J. Phys. Chem. B,* **112**, 6363–6372.
69. Nolan, S. L., Phillips, R. J., Cotts, P. M. and Dungan, S. R. (1997). Light Scattering Study on the Effect of Polymer Composition on the Structural Properties of PEO-PPO-PEO Micelles. *J. Colloid Interface Sci.,* **191**, 291–302.
70. Alvarez-Lorenzo, C., Sosnik, A. and Concheiro, A. (2011). PEO-PPO block copolymers for passive micellar targeting and overcoming multidrug resistance in cancer therapy. *Curr. Drug Targets,* **12**, 1112–1130.
71. Almeida, H., Amaral, M. H., Lobao, P. and Sousa Lobo, J. M. (2013). Applications of poloxamers in ophthalmic pharmaceutical formulations: An overview. *Expert Opin. Drug Deliv.,* **10**, 1223–1237.
72. Danson, S., Ferry, D., Alakhov, V., Margison, J., Kerr, D., Jowle, D., Brampton, M., Halbert, G. and Ranson, M. (2004). Phase I dose escalation and pharmacokinetic study of pluronic polymer-bound doxorubicin (SP1049C) in patients with advanced cancer. *Br. J. Cancer,* **90**, 2085–2091.
73. Valle, J. W., Armstrong, A., Newman, C., Alakhov, V., Pietrzynski, G., Brewer, J., CAmpbell, S., Corrie, P., Rowinsky, E. K. and RAnson, M. (2011). A phase 2 study of SP1049C, doxorubicin in P-glycoprotein-targeting pluronics, in patients with advanced adenocarcinoma of the esophagus and gastroesophageal junction. *Invest. New Drugs,* **29**, 1029–1037.
74. Rauck, B. M., Friberg, T. R., Medina Mendez, C. A., Park, D., Shah, V., Bilonick, R. A. and Wang, Y. (2014). Biocompatible reverse thermal gel sustains the release of intravitreal bevacizumab *in vivo*. *Invest. Ophthalmol. Vis. Sci.,* **55**, 469–476.
75. Rahimi, S., Sarraf, E. H., Wong, G. K. and Takahata, K. (2011). Implantable drug delivery device using frequency-controlled wireless hydrogel microvalves. *Biomed. Microdevices,* **13**, 267–277.
76. Chen, Y. Y., Wu, H. C., Sun, J. S., Dong, G. C. and Wang, T. W. (2013). Injectable and thermoresponsive self-assembled nanocomposite hydrogel for long-term anticancer drug delivery. *Langmuir,* **29**, 3721–3729.

77. Payyappilly, S., Dhara, S. and Chattopadhyay, S. (2014). Thermoresponsive biodegradable PEG-PCL-PEG based injectable hydrogel for pulsatile insulin delivery. *J. Biomed. Mater. Res. A,* **102**, 1500–1509.
78. Van Vlierberghe, S., Dubruel, P. and Schacht, E. (2011). Biopolymer-based hydrogels as scaffolds for tissue engineering applications: A review. *Biomacromolecules,* **12**, 1387–1408.
79. Cabana, A., AItkadi, A. and Juhasz, J. (1997). Study of the gelation process of polyethylene oxide(a) polypropylene oxide(b) polyethylene oxide(a) copolymer (Poloxamer 407) aqueous solutions. *J. Colloid Interface Sci.,* **190**, 307–312.
80. Li, H., Yu, G. E., Price, C., Booth, C., Hecht, E. and Hoffmann, H. (1997). Concentrated aqueous micellar solutions of diblock copoly(oxyethylene/oxybutylene) E(41)B(8): A study of phase behavior. *Macromolecules,* **30**, 1347–1354.
81. Yoshida, R., Kaneko, Y., Sakai, K., Okano, T., Sakurai, Y., Bae, Y. H. and Kim, S. W. (1994). Positive thermosensitive pulsatile drug-release using negative thermosensitive hydrogels. *J. Control. Release,* **32**, 97–102.
82. Dinarvand, R. and Demanuele, A. (1995). The Use of Thermoresponsive Hydrogels for on-Off Release of Molecules. *J. Control. Release,* **36**, 221–227.
83. Shin, Y. S., Chang, J. H., Liu, J., Williford, R., Shin, Y. K. and Exarhos, G. J. (2001). Hybrid nanogels for sustainable positive thermosensitive drug release. *J. Control. Release,* **73**, 1–6.
84. Fundueanu, G., Constantin, M., Asmarandei, I., Bucatariu, S., Harabagiu, V., Ascenzi, P. and Simionescu, B. C. (2013). Poly(N-isopropylacrylamide-co-hydroxyethylacrylamide) thermosensitive microspheres: The size of microgels dictates the pulsatile release mechanism. *Eur. J. Pharm. Biopharm.,* **85**, 614–623.
85. Coughlan, D. C., Quilty, F. P. and Corrigan, O. I. (2004). Effect of drug physicochemical properties on swelling/deswelling kinetics and pulsatile drug release from thermoresponsive poly(N-isopropylacrylamide) hydrogels. *J. Controlled Release,* **98**, 97–114.
86. Verestiuc, L., Nastasescu, O., Barbu, E., Sarvaiya, I., Green, K. L. and Tsibouklis, J. (2006). Functionalized chitosan/NIPAM (HEMA) hybrid polymer networks as inserts for ocular drug delivery: Synthesis, *in vitro* assessment, and *in vivo* evaluation. *J. Biomed. Mater. Res. A,* **77A**, 726–735.
87. Wu, D. Q., Qiu, F., Wang, T., Jiang, X. J., Zhang, X. Z. and Zhuo, R. X. (2009). Toward the development of partially biodegradable and injectable thermoresponsive hydrogels for potential biomedical applications. *ACS Appl. Mater. Inter.,* **1**, 319–327.
88. Nelson, D. M., Hashizume, R., Yoshizumi, T., Blakney, A. K., Ma, Z. W. and Wagner, W. R. (2014). Intramyocardial injection of a synthetic hydrogel with delivery of bFGF and IGF1 in a rat model of ischemic cardiomyopathy. *Biomacromolecules,* **15**, 1–11.
89. Garbern, J. C., Minami, E., Stayton, P. S. and Murry, C. E. (2011). Delivery of basic fibroblast growth factor with a pH-responsive, injectable hydrogel to improve angiogenesis in infarcted myocardium. *Biomaterials,* **32**, 2407–2416.
90. Na, K., Park, J. H., Kim, S. W., Sun, B. K., Woo, D. G., Chung, H. M. and Park, K. H. (2006). Delivery of dexamethasone, ascorbate, and growth factor (TGF beta-3) in thermo-reversible hydrogel constructs embedded with rabbit chondrocytes. *Biomaterials,* **27**, 5951–5957.
91. Fang, J. Y., Chen, J. P., Leu, Y. L. and Hu, J. W. (2008). Temperature-sensitive hydrogels composed of chitosan and hyaluronic acid as injectable carriers for drug delivery. *Eur. J. Pharm. Biopharm.,* **68**, 626–636.

92. Zhang, J., Chen, H., Xu, L. and Gu, Y. (2008). The targeted behavior of thermally responsive nanohydrogel evaluated by NIR system in mouse model. *J. Control. Release,* 131, 34-40.
93. Zhang, J., Qian, Z. and Gu, Y. (2009). *In vivo* anti-tumor efficacy of docetaxel-loaded thermally responsive nanohydrogel. *Nanotechnology,* **20**, 325102.
94. Johnston, T. P. and Palmer, W. K. (1993). Mechanism of poloxamer 407-induced hypertriglyceridemia in the rat. *Biochem. Pharmacol.,* **46**, 1037–1042.
95. Blonder, J. M., Baird, L., Fulfs, J. C. and Rosenthal, G. J. (1999). Dose-dependent hyperlipidemia in rabbits following administration of poloxamer 407 gel. *Life Sci.,* **65**, PL261–266.
96. Chen, Y. Y., Wu, H. C., Sun, J. S., Dong, G. C. and Wang, T. W. (2013). Injectable and thermoresponsive self-assembled nanocomposite hydrogel for long-term anticancer drug delivery. *Langmuir,* **29**, 3721–3729.
97. Lin, T., Wu, J., Zhao, X., Lian, H., Yuan, A., Tang, X., Zhao, S., Guo, H. and Hu, Y. (2014). In situ floating hydrogel for intravesical delivery of adriamycin without blocking urinary tract. *J. Pharm. Sci.,* **103**, 927–936.
98. Chen, X., Zhi, F., Jia, X., Zhang, X., Ambardekar, R., Meng, Z., Paradkar, A. R., Hu, Y. and Yang, Y. (2013). Enhanced brain targeting of curcumin by intranasal administration of a thermosensitive poloxamer hydrogel. *J. Pharm. Pharmacol.,* **65**, 807–816.
99. Zhao, Y. Z., Lv, H. F., Lu, C. T., Chen, L. J., Lin, M., Zhang, M., Jiang, X., Shen, X. T., Jin, R. R., Cai, J., Tian, X. Q. and Wong, H. L. (2013). Evaluation of a novel thermosensitive heparin-poloxamer hydrogel for improving vascular anastomosis quality and safety in a rabbit model. *Plos One,* **8**, e73178.
100. Torchilin, V. (2012). Liposomes in Drug Delivery. *In:* Siepmann, J., SIegel, R. A. and Rathbone, M. J. (eds.) *Fundamentals and Applications of Controlled Release Drug Delivery.* Springer.
101. Landon, C. D., Park, J. Y., Needham, D. and Dewhirst, M. W. (2011). Nanoscale Drug Delivery and Hyperthermia: The materials design and preclinical and clinical testing of low temperature-sensitive liposomes used in combination with mild hyperthermia in the treatment of local cancer. *Open Nanomed J,* **3**, 38–64.
102. May, J. P. and Li, S. D. (2013). Hyperthermia-induced drug targeting. *Expert Opin. Drug Deliv.,* **10**, 511–527.
103. Wust, P., Hildebrandt, B., Sreenivasa, G., Rau, B., Gellermann, J., Riess, H., Felix, R. and Schlag, P. M. (2002). Hyperthermia in combined treatment of cancer. *Lancet. Oncol.,* **3**, 487–497.
104. Hildebrandt, B., Wust, P., Ahlers, O., Dieing, A., Sreenivasa, G., Kerner, T., Felix, R. and Riess, H. (2002). The cellular and molecular basis of hyperthermia. *Crit. Rev. Oncol. Hematol.,* **43**, 33–56.
105. Andriyanov, A. V., Koren, E., Barenholz, Y. and Goldberg, S. N. (2014). Therapeutic efficacy of combining pegylated liposomal doxorubicin and radiofrequency (RF) ablation: comparison between slow-drug-releasing, non-thermosensitive and fast-drug-releasing, thermosensitive nano-liposomes. *Plos One,* **9**.
106. Ahmed, M. and Goldberg, S. N. (2004). Combination radiofrequency thermal ablation and adjuvant IV liposomal doxorubicin increases tissue coagulation and intratumoural drug accumulation. *Int. J. Hyperthermia,* **20**, 781–802.

107. Monsky, W. L., Kruskal, J. B., Lukyanov, A. N., Girnun, G. D., Ahmed, M., Gazelle, G. S., Huertas, J. C., Stuart, K. E., Torchilin, V. P. and Goldberg, S. N. (2002). Radiofrequency ablation increases intratumoral liposomal doxorubicin accumulation in a rat breast tumor model. *Radiology,* **224**, 823–829.
108. Van Bree, C., Krooshoop, J. J., Rietbroek, R. C., Kipp, J. B. and Bakker, P. J. (1996). Hyperthermia enhances tumor uptake and antitumor efficacy of thermostable liposomal daunorubicin in a rat solid tumor. *Cancer Res.,* 56, 563–568.
109. Huang, S. K., Stauffer, P. R., Hong, K., Guo, J. W., Phillips, T. L., Huang, A. and Papahadjopoulos, D. (1994). Liposomes and hyperthermia in mice: Increased tumor uptake and therapeutic efficacy of doxorubicin in sterically stabilized liposomes. *Cancer Res.,* **54**, 2186–2191.
110. Kong, G., Braun, R. D. and Dewhirst, M. W. (2000). Hyperthermia enables tumor-specific nanoparticle delivery: Effect of particle size. *Cancer Res.,* **60**, 4440–4445.
111. Manzoor, A. A., Lindner, L. H., Landon, C. D., Park, J. Y., Simnick, A. J., Dreher, M. R., DaS, S., Hanna, G., Park, W., Chilkoti, A., Koning, G. A., Ten Hagen, T. L. M., Needham, D. and Dewhirst, M. W. (2012). Overcoming limitations in nanoparticle drug delivery: triggered, intravascular release to improve drug penetration into tumors. *Cancer Res.,* **72**, 5566–5575.
112. Weinstein, J. N., Magin, R. L., Yatvin, M. B. and Zaharko, D. S. (1979). Liposomes and local hyperthermia: Selective delivery of methotrexate to heated tumors. *Science,* **204**, 188–191.
113. Li, L., Ten Hagen, T. L., Schipper, D., Wijnberg, T. M., Van Rhoon, G. C., Eggermont, A. M., Lindner, L. H. and Koning, G. A. (2010). Triggered content release from optimized stealth thermosensitive liposomes using mild hyperthermia. *J. Control. Release,* **143**, 274–279.
114. Gaber, M. H., Hong, K., Huang, S. K. and Papahadjopoulos, D. (1995). Thermosensitive sterically stabilized liposomes: Formulation and *in vitro* studies on mechanism of doxorubicin release by bovine serum and human plasma. *Pharm. Res.,* **12**, 1407–1416.
115. Needham, D., Anyarambhatla, G., Kong, G. and Dewhirst, M. W. (2000). A new temperature-sensitive liposome for use with mild hyperthermia: Characterization and testing in a human tumor xenograft model. *Cancer Res.,* **60**, 1197–1201.
116. Mills, J. K. and Needham, D. (2005). Lysolipid incorporation in dipalmitoylphosphatidylcholine bilayer membranes enhances the ion permeability and drug release rates at the membrane phase transition. *Biochim. Biophys. Acta,* **1716**, 77–96.
117. Needham, D., Park, J. Y., Wright, A. M. and Tong, J. (2013). Materials characterization of the low temperature sensitive liposome (LTSL): Effects of the lipid composition (lysolipid and DSPE-PEG2000) on the thermal transition and release of doxorubicin. *Faraday Discuss.,* **161**, 515–534; discussion 563–589.
118. Dewhirst, M. W., Tso, C. Y., Oliver, R., Gustafson, C. S., Secomb, T. W. and Gross, J. F. (1989). Morphologic and hemodynamic comparison of tumor and healing normal tissue microvasculature. *Int. J. Radiat. Oncol. Biol. Phys.,* **17**, 91–99.
119. Yarmolenko, P. S., Zhao, Y., Landon, C., Spasojevic, I., Yuan, F., Needham, D., Viglianti, B. L. and Dewhirst, M. W. (2010). Comparative effects of thermosensitive doxorubicin-containing liposomes and hyperthermia in human and murine tumours. *Int. J. Hyperthermia,* **26**, 485–498.

120. Kong, G., Anyarambhatla, G., Petros, W. P., Braun, R. D., Colvin, O. M., Needham, D. and Dewhirst, M. W. (2000). Efficacy of liposomes and hyperthermia in a human tumor xenograft model: importance of triggered drug release. *Cancer Res.,* **60**, 6950–6957.
121. Hauck, M. L., Larue, S. M., Petros, W. P., Poulson, J. M., Yu, D., Spasojevic, I., Pruitt, A. F., Klein, A., Case, B., Thrall, D. E., Needham, D. and Dewhirst, M. W. (2006). Phase I trial of doxorubicin-containing low temperature sensitive liposomes in spontaneous canine tumors. *Clin. Cancer Res.,* **12**, 4004–4010.
122. Banno, B., Ickenstein, L. M., Chiu, G. N., Bally, M. B., Thewalt, J., Brief, E. and Wasan, E. K. (2010). The functional roles of poly(ethylene glycol)-lipid and lysolipid in the drug retention and release from lysolipid-containing thermosensitive liposomes *in vitro* and *in vivo. J. Pharm. Sci.,* **99**, 2295–2308.
123. Hossann, M., Wiggenhorn, M., Schwerdt, A., Wachholz, K., Teichert, N., Eibl, H., Issels, R. D. and Lindner, L. H. (2007). *In vitro* stability and content release properties of phosphatidylglyceroglycerol containing thermosensitive liposomes. *Biochim. Biophys. Acta,* **1768**, 2491–2499.
124. Hossann, M., Wang, T., Wiggenhorn, M., Schmidt, R., Zengerle, A., Winter, G., Eibl, H., Peller, M., Reiser, M., Issels, R. D. and Lindner, L. H. (2010). Size of thermosensitive liposomes influences content release. *J. Control. Release,* **147**, 436–443.
125. Lindner, L. H., Eichhorn, M. E., Eibl, H., Teichert, N., Schmitt-sody, M., Issels, R. D. and Dellian, M. (2004). Novel temperature-sensitive liposomes with prolonged circulation time. *Clin. Cancer Res.,* **10**, 2168–2178.
126. Tagami, T., Foltz, W. D., Ernsting, M. J., Lee, C. M., Tannock, I. F., May, J. P. and Li, S. D. (2011). MRI monitoring of intratumoral drug delivery and prediction of the therapeutic effect with a multifunctional thermosensitive liposome. *Biomaterials,* **32**, 6570–6578.
127. Tagami, T., Ernsting, M. J. and Li, S. D. (2011). Optimization of a novel and improved thermosensitive liposome formulated with DPPC and a Brij surfactant using a robust *in vitro* system. *J. Control. Release,* **154**, 290–297.
128. Tagami, T., Ernsting, M. J. and Li, S. D. (2011). Efficient tumor regression by a single and low dose treatment with a novel and enhanced formulation of thermosensitive liposomal doxorubicin. *J. Control. Release,* **152**, 303–309.
129. Tagami, T., May, J. P., Ernsting, M. J. and Li, S. D. (2012). A thermosensitive liposome prepared with a Cu(2)(+) gradient demonstrates improved pharmacokinetics, drug delivery and antitumor efficacy. *J. Control. Release,* **161**, 142–149.
130. Kono, K., Nakai, R., Morimoto, K. and Takagishi, T. (1999). Thermosensitive polymer-modified liposomes that release contents around physiological temperature. *Biochim. Biophys. Acta,* **1416**, 239–250.
131. Wells, J., Sen, A. and Hui, S. W. (2003). Localized delivery to CT-26 tumors in mice using thermosensitive liposomes. *Int. J. Pharm.,* **261**, 105–114.
132. Chandaroy, P., Sen, A. and Hui, S. W. (2001). Temperature-controlled content release from liposomes encapsulating Pluronic F127. *J. Control. Release,* **76**, 27–37.
133. Kennedy, J. E. (2005). High-intensity focused ultrasound in the treatment of solid tumours. *Nat. Rev. Cancer,* **5**, 321–327.
134. Illing, R. O., Kennedy, J. E., Wu, F., Ter Haar, G. R., Protheroe, A. S., Friend, P. J., Gleeson, F. V., Cranston, D. W., Phillips, R. R. and Middleton, M. R. (2005). The safety and feasibility of extracorporeal high-intensity focused ultrasound (HIFU) for the treatment of liver and kidney tumours in a Western population. *Br. J. Cancer,* **93**, 890–895.

135. Staruch, R., Chopra, R. and Hynynen, K. (2012). Hyperthermia in bone generated with MR imaging-controlled focused ultrasound: Control strategies and drug delivery. *Radiology*, **263**, 117–127.
136. Ponce, A. M., Viglianti, B. L., Yu, D., Yarmolenko, P. S., Michelich, C. R., Woo, J., Bally, M. B. and Dewhirst, M. W. (2007). Magnetic resonance imaging of temperature-sensitive liposome release: Drug dose painting and antitumor effects. *J. Natl. Cancer Inst.*, **99**, 53–63.

Chapter 6

Ultrasound-Controlled Nanosystems

Shao-Ling Huang, Hyunggun Kim, Melvin Klegerman, David D. McPherson

Division of Cardiovascular Medicine, Department of Internal Medicine, The University of Texas Health Science Center at Houston, TX, USA

6.1 Introduction

Ultrasound is a widely used and accepted biomedical imaging methodology. Over the past number of years, ultrasound has been implemented for pharmaceutical delivery. Ultrasound is non-invasive and allows for the controlled release of therapeutics, both spatially and temporally. Ultrasound-controlled nanosystems refer to specialized nanosized active delivery vehicles that are equipped with pharmaceutics inside the nanoparticle structure and that have the ability to respond to ultrasound pressure for the release of their payloads. The primary physical principle of ultrasound-controlled nanosystems lies in the fact that the structural composition of the nanocarriers, usually composed with thermo-sensitive materials or nanobubbles, can promote the delivery and release of payload contents to a specific biologic environment.

In general, an ultrasound-controlled nanosystem should have the following characteristics:

(1) Stable encapsulation: Ultrasound-responsive nanoparticles (NPs) can be loaded with various therapeutics, such as hydrophilic drugs [1], hydrophobic drugs [2], proteins [3], short interfering ribonucleic acids (siRNAs) [4], and plasmid deoxyribonucleic acids (DNAs) [5, 6]. When injected into the human body, delivery vehicles interact with blood cells,

serum proteins, lipid membranes, and several potential hydrophobic components, all of which can cause a leakage of their contents. Stable encapsulation is required for effective drug delivery resulting from local release of the loaded therapeutic at the disease sites.

(2) Active targeting: Ultrasound-responsive NPs containing therapeutic agents can be conjugated to antibodies and targeted to specific disease sites, allowing for high local concentrations with low systemic toxicity.

(3) Response to ultrasound: Ultrasound provides an effective means to control the release of contents from nanoparticles. Therapeutic delivery using loaded nanoparticles demands enhanced local therapeutic concentrations over a period of time [7]. Ultrasound-triggered cavitation induces the destruction of gas bubbles, increases the permeability of cells and tissues, and facilitates therapeutic delivery into cells and tissues. In particular for gene delivery, transfection efficiency can be also enhanced by sonification [6]. Ultrasound-responsive nanoparticles containing both gases and pharmaceutics can not only enhance cavitation effects but also are suitable for image-guided delivery.

(4) Little toxicity: Ultrasound-controlled nanosystems should be non-toxic and biodegradable.

This review will discuss the characteristics of these nanosystems, targeting abilities of the systems, and ultrasound applications with the systems for enhanced therapeutic delivery.

6.2 Structure and Properties of Ultrasound-Controlled Nanosystems

A number of ultrasound-responsive nanoparticles have been developed, with or without gas encapsulation, at an average diameter of 0.8 to 10 μm. With respect to material compositions, ultrasound-responsive nanoparticles can be divided into liposomal nanoparticles [8], polymeric nanoparticles [9], protein-shelled nanoparticles, and others. Ultrasound-responsive nanoparticles can be also categorized into thermo-sensitive nanoparticles and ultrasound-responsive nanoparticles, which can be further divided into hard-shelled nanoparticles and soft-shelled nanoparticles (Table 6.1). Many therapeutic agents and genes can be loaded into these nanocarriers (Fig. 6.1). An agent containing both imaging and therapeutic materials can be utilized as a multifunctional platform for theranostics [10]. Multiple therapeutic modalities, such as a combination of chemotherapy and hyperthermia can be utilized together with

Table 6.1. Ultrasound-responsive nanoparticles.

Nanoparticles without gas		Nanoparticles with gas		
Shell materials	Response to ultrasound		Shell materials	Response to ultrasound
Phospholipid (thermosensitive liposome, gas generating liposome)	Heating cavitation	Soft shell	Phospholipid (lipid shelled bubble, gas-containing liposome)	
			Protein (Albumin)	
Polymer micelle	Heating cavitation	Hard shell	Polymer bubble	Cavitation

ultrasound-controlled nanosystems to facilitate the synergistic effects of the combined therapies [10].

Microbubbles are gas bubbles stabilized by a surfactant such as albumin, a polymer, or a monolayer lipid. Drugs or genes can be associated within the thin-walled protein or lipid shell of the microbubble by hydrophobicity or electrostaticity (Fig. 6.1A). However, the amount of pharmaceutical payload per microbubble is limited due to their inherently hydrophobic molecules. The hydrophobic reservoir of the bubbles can be increased by thickening the bubble shell, such as by adding soybean oil to the bubble composition (Fig. 6.1B). Other types of agents have been developed, including polymer-shelled microbubbles with hydrophobic agents on the polymer shell (Fig. 6.1C), microbubbles with liposomes or nanoparticles attached on the surface via a biotin–avidin–biotin bridging system (Fig. 6.1D) [11], or by forming a multilayered structure of a lipid/polymer microbubble sequentially coated with DNAs and polymer layers (Fig. 6.1E). Liposomes composed with bilayer phospholipids have been used as surfactants for gas stabilization, as gas resides between the bilayer or within the aqueous compartment as a monolayer-covered air bubble or by forming a micelle-like structure or a gas-containing liposome structure. Hydrophobic drugs can be encapsulated within the lipid bilayer and hydrophilic agents can be encapsulated within the aqueous core of liposomes. (Fig. 6.1F).

6.2.1 *Polymer-shelled nanoparticles*

Ultrasound-mediated micelles for therapeutic delivery are based on two physical characteristics of ultrasound: thermal and non-thermal (mechanical).

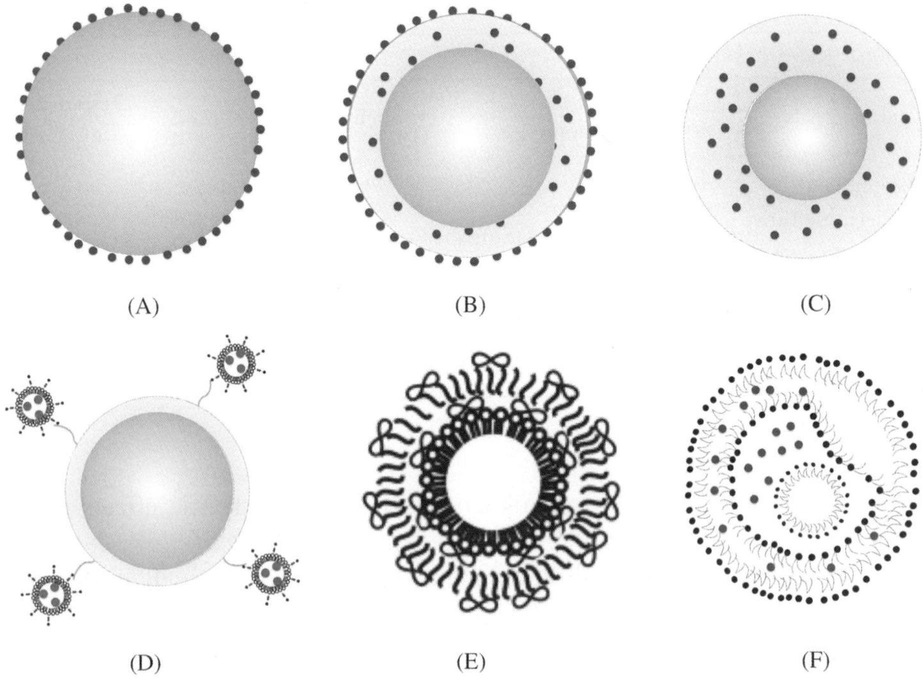

Figure 6.1. Schematics of therapeutic loading strategies of loading into ultrasound-responsive nanosystems. (A) Noncovalent binding of drugs or genes onto the surface of a thin-walled protein-shelled microbubble. (B) A micelle with encapsulated therapeutics in the oil layer coated with protein, lipid, or polymer shell. (C) A thick-walled polymer microbubble with therapeutics in the lipophilic shell. (D) Attachment of liposomes or nanoparticles to the surface of a microbubble via a biotin–avidin–biotin bridging system. (E) Multilayered structure of a lipid microbubble sequentially coated with DNAs and poly-L-lysine layers. (F) A liposome with therapeutics in the aqueous core or in the lipid bilayer.

Ultrasonic thermal effects are useful in combination with thermo-responsive micelles. The mechanical effects of ultrasound are useful for ultrasound-triggered drug release from bubble micelles (Table 6.1).

6.2.1.1 *Polymer micelles*

Copolymer micelles are formed by the self-assembly of amphiphilic block copolymer molecules in an aqueous milieu when the copolymer concentration exceeds a certain threshold value called the critical micelle concentration (CMC). Hydrophobic blocks form the micelle cores, while hydrophilic blocks (usually polyethylene glycol, PEG) form the micelle corona (or shells). Lipophilic drugs are solubilized in the micelle cores.

A primary shortcoming of polymer micelles as a systemic therapeutic carrier is the characteristic of decreased concentration below the CMC, leading to self-dissociation in the circulatory system. This problem has resulted in several disappointing clinical trial outcomes [12, 13]. To overcome these shortcomings, stable polymer micelles have been developed using hydrophobic or hydrogen bonds in the micelle cores or by cross-linking the core-forming blocks [24, 30] in combination with external triggering (e.g. light [14] and ultrasound [9]) to induce therapeutic release at the desired target site.

Pluronic type polymers are block copolymers composed of ethylene oxide and propylene oxide. Pluronic copolymer P-105 micelles can encapsulate and release hydrophobic drugs [15] when exposed to continuous wave ultrasound [16] or pulsed ultrasound [17]. Adding PEG2000-DSPE to pluronic P-105 micelles results in effective tumor-selective drug delivery *in vivo* due to the increased stability of micelles in circulation [18]. Ultrasound application can then trigger local therapeutic release from the polymer micelles facilitating crossing the tumor cell membranes and enhancing intracellular drug uptake by the tumor cells [17].

6.2.1.2 *Polymer-shelled microbubbles*

Polymer microspheres are thick polymer-shelled gas bubbles, also called hard-shelled bubbles. As therapeutic carriers, hydrophobic or hydrophilic drugs are usually incorporated inside the thick cross-linked polymer matrix, and plasmid DNAs are encapsulated within the cationic polymer (Polylysine, protamin sulfate, or polyethylenimine (PEI)) microbubbles [19].

The thickness of the polymer microbubbles ranges between 50 mm and 200 nm. Due to such hard shell compositions, a high acoustic pressure is required for the release and update of the entrapped drug. The drug is released from these polymer microbubbles by fragmentation and pushed into the cells via a jetting effect. Kooiman *et al.* found that phospholipid-stabilized microbubbles oscillated when insonified at 1.4 MPa of 2.25 MHz ultrasound, polymer-shelled microcapsules remained largely intact at a pressure up to 1.5 MPa, and under a larger pressure than this value, the shells were fragmented and the contained gas was released [20].

6.2.2 *Liposomes*

6.2.2.1 *Temperature-sensitive liposomes*

Temperature-sensitive liposomes release the encapsulated payload at the melting phase transition temperature (T_m) of the lipid bilayer (Fig. 6.2). In general,

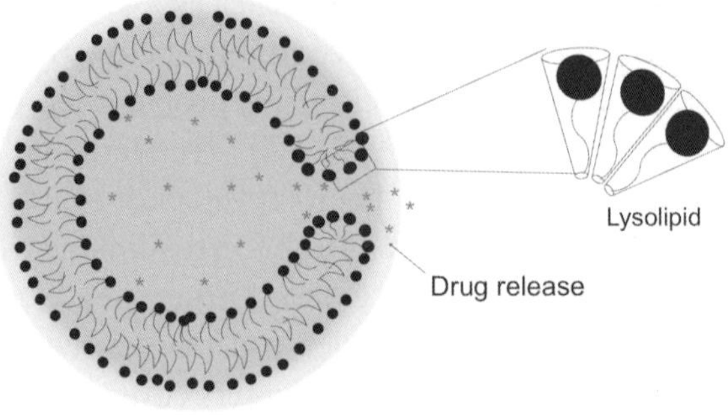

Figure 6.2. Schematic of a temperature-sensitive liposome for drug encapsulation and temperature-triggered release.

every lipid has a gel phase and a liquid phase. The transition temperature refers to the temperature at which the lipid transitions from the gel phase to the liquid–crystalline phase. At the phase transition temperature, the membrane permeability of the lipid bilayer increases by several orders of magnitude. The phase transition temperature of the phospholipid is related to its acyl chain length and saturation. Lipids commonly used in temperature-sensitive liposomal formulations include dipalmitoylphosphatidylcholine (DPPC, T_m = 41.5°C), dipalmitoylphosphatidylglycerol (T_m = 41.5°C), and distearoylphosphatidylcholine (DSPC, T_m = 54.9°C). As each lipid has a different T_m, these temperature-sensitive liposomes can be made sensitive to various degrees of heating by adjusting the lipid compositions [21].

The first generation of temperature-sensitive liposomes were composed of pure DPPC or DSPC [22]. However, this composition had relatively slow drug-release kinetics and unstable drug release at T_m. Cholesterol is a liposomal component commonly used to increase liposomal stability while broadening the range of T_m. Hence, cholesterol is not recommended for temperature-sensitive liposomal formulations as the broad transition temperature range is not desirable. Instead, lysolipids monostearoylphosphatidylcholine and/or myristoylpalmitoylphosphatidylcholine are often used to stabilize the formation while eliminating transient pores in the liposomes. Needham *et al.* found that lysolipid-based temperature-sensitive liposomes provided fast and effective drug release within a few seconds in response to mild hyperthermia (40–45°C) [23]. During the gel-to-liquid phase transition in response to mild hyperthermia, pores form along the highly compressible liquid–solid boundaries on the liposomal membranes. These pores are hydrophobic and the membranes have a tendency to reseal. Due to the conical molecular shape (large head group, single chain) and the tendency to form spherical micelles in aqueous solution, lysolipids line up on the pores, while the head group points toward the pores. The structure of the stable lysolipid hydrophilic head group-lined aqueous pores facilitates the release of the entrapped hydrophilic drugs from the liposomes (Fig. 6.2).

6.2.2.2 *Liposome-shelled nanobubbles*

Liposomes have a lipid bilayer structure which clearly separates the hydrophilic and hydrophobic regions. Hydrophilic portions of the lipid bilayer are directed towards the internal and external aqueous phases, whereas hydrophobic portions of both lipid layers are directed towards each other, forming the internal core of the membrane. These characteristics provide great potential for therapeutic deliveries, as the liposome has the ability to encapsulate and localize both water-soluble and water-insoluble substances, either together or separately. Water-soluble materials are entrapped in the aqueous core, while water-insoluble and oil-soluble hydrophobic drugs or other substances reside inside the bilayer [7] (Fig. 6.1F).

Liposomes have long been acknowledged as excellent therapeutic delivery vehicles for reliability, safety, high therapeutic-loading capacity, and superior targeting capabilities [24]. Phospholipids are the key components to these liposomal vesicle formulations. When phospholipids are used as a surfactant for gas stabilization, they form micelle-like structures or gas-containing liposomes. Due to the similar energetic characteristics, gases encapsulated within liposomes behave like hydrophobic drugs. Here, gas resides between the bilayers or within the aqueous compartment as a monolayer-covered air bubble (Fig. 6.1E).

Liposomes are currently the most popular system of ultrasound-responsive nanoparticles. Their lipid surfaces can be easily modified by adding a variety of lipid formulations for specific drug or gene delivery. For targeted delivery, antibodies can be conjugated to the surface of the lipid membrane. Hydrophobic drugs can be encapsulated inside the lipid membrane, whereas hydrophilic drugs can be entrapped in the aqueous interior of liposomes. Modifications of particle surface with positively charged lipids can induce electrostatic interactions, allowing for the incorporation of plasmid DNA for gene delivery [7, 25].

When gas is entrapped in therapeutic-loaded liposomes, these liposomes (ELIPS) not only retain the comparable therapeutic loading capacity of the conventional liposomes, but are more sensitive to ultrasound-triggered release. For example, the freeze-lyophilization method to create echogenic liposomes in the presence of mannitol allows for the encapsulation of air from the ambient environment [26], resulting in acoustically active liposomes containing two compartments. The smaller compartment contains air and is bounded by a monolayer, and the larger compartment contains the aqueous phase. The average diameter of such liposomes is approximately 800 nm. These procedures to create echogenic (i.e. acoustically active) liposomes have an encapsulation efficiency of 15% for calcein, a fluorescent marker for the aqueous compartment where a hydrophilic drug would be contained [27]. When these liposomes are exposed to continuous wave ultrasound, a third of the aqueous contents are released. This results in a more clinically beneficial therapeutic release profile with an initial robust and rapid release followed by a slow, but constant release.

Modifications of the lipid composition (or the composition of the lipid–gene complex) can provide a variety of therapeutic release profiles more sensitive or less sensitive to ultrasound bioeffects [7]. Lin *et al.* demonstrated that ultrasound-triggered leakage of liposomes was enhanced by the incorporation of a covalently-linked polyethylene glycol moiety (PEGylated lipids) into the bilayer [28]. Pong *et al.* found great differences in the leakage from liposomes, which was dependent on the amount of unsaturated lipid [29]. Cholesterol content is another important factor in determining the sensitivity of the liposomal membrane to ultrasound. Ultrasound-triggered leakage occurs either preferentially within cholesterol-rich regions, termed “lipid rafts” of the bilayer, or in the perimeter of the lipid rafts [29]. PEG provides a steric brush that minimizes the interactions between the lipid shell and blood components, increasing the circulating time of the liposomes.

6.2.3 *Protein-shelled microbubbles*

Protein microbubbles are formed by disulfide cross-linking of proteins during sonication. Albumin is the most popular protein for this purpose. The formed protein shell has a thickness of approximately 15 nm. Drug molecules and nucleic acids can be loaded fully or partially within the shell during covalent cross-linking of proteins during bubble formulation [30, 31]. However, protein-shelled microbubbles have poor efficiency in therapeutic loading, and low protection for degradation of therapeutics from exposure to the circulation. Lentacker *et al.* utilized a cationic polymer and the layer-by-layer assembly technique to bind plasmid DNAs to the microbubble surface to overcome these issues, and demonstrated an improved loading capacity of the microbubbles compared to conventional methods [32].

6.3 Active Targeting of Ultrasound-Responsive Nanoparticles

Ultrasound-responsive nanoparticles containing therapeutic agents can be conjugated to antibodies and targeted to specific disease sites, allowing for high local concentrations with low systemic toxicity.

6.3.1 *Targeting strategies*

The kind of chemical modifications that can be made for the conjugation of targeting ligands depends on the nature of the nanoparticulate surface exposed to the physiologic environment. NPs provide a great variety of functional groups that can be coupled to a bifunctional linker for conjugation to a targeting ligand. For instance, protein-, peptide-, peptidoglycan-, and lipoprotein-containing matrices have exposed primary amino and carboxy groups that can be easily linked to a number of different moieties. Peptide bonds can be formed between exposed amino groups (amino terminus or side chain) or carboxy groups (carboxy terminus or side chain) on the protein/peptide and the complementary group on the ligand by use of a water-soluble carbodiimide compound EDC [33–35].

As carbodiimide forms peptide bonds directly between two molecules, the conjugate lacks any spacer function. This can be especially problematic for molecular targeting of nanoparticles, where the targeting ligand can be sterically hindered by the particle. Such a problem can be avoided by a two-step EDC coupling of a spacer peptide between the NP and the ligand. Another drawback of carbodiimide coupling procedures is extensive cross-linking of protein, resulting in aggregation and precipitation, which can be minimized by a two-step procedure utilizing *N*-hydroxysuccinimide (NHS) [34].

Amino functions can be cross-linked by the use of glutaraldehyde, involving highly pH-dependent complex chemistries and often requiring the introduction of cyclized linkers [36]. Glutaraldehyde is also well known to cause extensive protein cross-linking, denaturation, aggregation, and precipitation, so care should be taken to optimize protein-specific multi-step approaches to conjugation using this agent. Organic isothiocyanates can also be easily coupled to amino groups, and fluorophores, such as fluorescein and Rhodamine B, have long been manufactured as isothiocyanate derivatives for protein labeling. Recently, however, a relatively simple method for isothiocyanate derivatization of primary amines, using cyanuric chloride, has been reported [37], making a complete laboratory-scale amino–amino conjugation through a stable thiourea linkage possible.

The most commonly used and versatile approach to nanoparticle targeting involves thiol conjugation techniques. A great many agents are marketed for both the conversion of other functional groups to thiols and the linkage of thiols to other functional groups. Although disulfide linkages are often utilized, they tend to be unstable *in vivo* because of the possibility of oxidation and disulfide exchange with compounds, such as glutathione and cystine [38]. The preferred conjugation, where possible, is a thioether linkage by a Michael addition of a sulfhydryl function across the maleimide double bond, which tends to be very stable *in vivo*. This was recently verified by a research team at Genentech that developed the antibody–drug conjugate Trastuzumab-DM1 [39].

A popular reagent for the conversion of amino functions to thiols is 3-(2-pyridyldithiolpropionic acid)-N-hydroxysuccinimide ester (SPDP), which alkylates primary amines with a methylpyridyl disulfide. Reduction with agents such as dithiothreitol leaves a methyl sulfhydryl with the release of 2-thiopyridone, which strongly absorbs ultraviolet light at 343 nm, enabling convenient determination of thiolation efficiency [40]. Alternatively, N-succinimidyl (-4-[*p*-maleimidophenyl]) butyrate (SMPB) can be used to amidate primary amines with the 1,2-dipalmitoyl-sn-glycero-3-phosphoethanolamine-N-[4-(p-maleimidophenyl) butyrate] (MPB) moiety. Diacyl phosphatidylethanolamines, such as 1,2-dipalmitoyl-3-phosphatidylethanolamine (DPPE), are now marketed as their MPB derivatives for conjugation of thiols to lipid-based formulations, such as liposomes [41–46].

Immunoglobulin G (IgG) antibodies, which are often used to target nanoparticles, can be reduced to monomers possessing free sulfhydryl groups. These IgG monomers, fully retaining their antigen-binding capability, can then be conjugated to maleimido groups on nanoparticles by this approach. Antigen-binding Fab' antibody fragments also possess free sulfhydryls and can likewise be conjugated to lipid-based nanoparticles [40]. Other maleimido cross-linkers are available for the coupling of thiols to other functional groups. For instance, 4-(maleimidomethyl)cyclohexane-1-carboxyl-hydrazide (MMCCH or SMCC hydrazide) can be used to link thiols to carbonyl functions [47].

Carbohydrates, such as glycan-derived nanoparticles, can be linked to proteins, peptides, and other amines via their hydroxyl functions using cyanogen bromide (CNBr). CNBr activation of the carbohydrate involves the formation of cyanate esters and cyclic imidocarbonates with conjugation through the formation of isourea derivatives or substituted imidocarbonates, respectively. The latter involves displacement of the amine [48]. CNBr is extremely toxic, such that the entire conjugation procedure must be carried out in a fume hood, including the weighing of the compound.

A safer and more convenient method for carbohydrate coupling to amines involves the use of triazines, such as cyanuric chloride, involving serial substitution of up to three chloride groups [49]. In practice, the substitution of the first two chlorides makes the third relatively nonreactive, such that the agent is essentially a bifunctional linker. The reactivity of cyanuric chloride, which has become a preferred approach to targeting PEGylated lipid-based nanoparticles, is limited to alkaline conditions (up to pH 8.8) and is non-reactive at physiological pH, making the conjugates safe and stable *in vivo* [50]. Conjugate stability has been confirmed in cell culture [51].

Another popular method for coupling ligands to carbohydrates through their hydroxyl functions employs bisoxiranes, such as 1,4-butanediol diglycidyl ether [52]. Different bisoxiranes can be used to provide variable linker lengths. The method is not good for coupling proteins or alkali-labile carbohydrates, since basic conditions (pH >9 or 11–13, respectively) are required for the conjugation of amines and hydroxyl groups. Where alkali-stable amino ligands can be used, the method has the advantage of being able to couple secondary amines. Protein thiols can readily be used, however, since sulfhydryl groups react well with epoxy-activated carbohydrates at neutral pH. This conjugation method provides extremely stable linkages [53].

As previously mentioned, hydrazide linkers react with carbonyl functions. This includes aldehydes, so that both reducing and keto sugars can be conjugated to functionalized ligands by this method. Since polylactides do not possess free hydroxyl groups (except at the end), this would be a preferred method for targeting such nanoparticles. All the carbohydrate conjugation methods can be used to functionalize polymerized silica, which forms the matrix of silica nanoparticles. In the form of mesoporous multifunctional silica nanoparticles, this material has become the basis of a burgeoning nanotechnology for targeted biomedical applications, especially in the field of cancer diagnosis and therapy [54].

Gas-filled, lipid-stabilized microbubbles have proven difficult to target with protein ligands, since proteins tend to denature when they are in close proximity to gases [55]. This is because the non-polar environment provided by the juxtaposed gas does not favor the thermodynamically most stable configuration of the folded protein [56, 57]. This obstacle has been successfully overcome in many cases by coupling the targeting ligands to preformed microbubbles using biotin-avidin linkage techniques [58–60]. Both the ligand and the nanoparticle may be biotinylated and then linked with a streptavidin bridge, or one component may be biotinylated and the other coupled to streptavidin by conjugation techniques, such as those summarized above.

Biotin-avidin targeting approaches have been used for a broad spectrum of applications, taking advantage of the very high affinity of the noncovalent association interaction, in the femtomolar range [61]. A popular approach is the radiological practice of "pretargeting" tumors with avidin-coupled antibodies specific for tumor antigens, followed by the administration of biotinylated radionuclides for targeted cancer radiotherapy. The use of avidin in clinical formulations, however, poses potential problems because of the protein's allergenicity (especially egg white avidin) and immunogenicity, but a number of clinical trials of biotin–avidin constructs have been done and are in progress without serious difficulties [62, 63].

6.3.2 *Targeted nanoparticle evaluation*

Putatively targeted nanoparticles should first be evaluated for retention of essential functions. Diagnostic functions could include retention of the contrast agent with consequent imaging enhancement, while therapeutic functions may involve drug encapsulation and release characteristics. Next, conjugation efficiency should be assessed and expressed as mass (grams or moles) of ligand per NP mass or number of ligand molecules per NP. An *in vitro* measure of targeting efficiency, preferably including ligand–target binding affinity, should assess the number of NPs binding to the target. The latter should be in a form that is consistent with the ultimate *in vivo* application. For instance, if the target molecules are expressed in blood vessels, they should be immobilized on a surface for evaluation, either statically in an enzyme-linked immunosorbent assay (ELISA) format or dynamically in a flow simulator. Targeting efficiency can then be expressed as NP number per unit area, number of ligands per NP binding to target or some product of affinity and ligand number per NP [46, 64].

Non-specific ionic interactions can exhibit affinities up to $K_D = 20$ nM [46], where K_D is the dissociation constant. Therefore, ligand–target binding affinities below this threshold tend to lack clinical utility and should be in the low nanomolar range. Ligands having relatively low affinities for their targets can be used if they are conjugated to nanoparticles in sufficient quantities, to provide optimal aggregate avidities. Overall targeting avidities (monomolecular affinity * number of ligand molecules per liposome) for liposomes of $>1 * 10^{12}$ M^{-1} have been associated with optimal targeting in a physiologic flow simulator for vascular applications [46, 65]. The most accurate measure of targeting efficiency is the specific avidity of NP ligands binding to immobilized target molecules [64].

Initial *in vitro* assessments of both conjugation and targeting efficiency should be quantitative, to lend themselves readily to optimization of production procedures and then serve as essential quality control measures. Colorimetric procedures, immunoassays, flow cytometry, and radiotracer methods have all been used for these purposes [45, 46, 64, 66]. Such methods often require alteration or optimization for reliable measurements of ligand number or concentration when coupled to nanoparticulate formulations [45].

Next, an *in vitro* or *ex vivo* demonstration of NP targeting should be performed more within the context of the ultimate application, which can be relatively semi-qualitative in nature. For instance, ultrasound imaging of a fibrin-targeted contrast agent can be done after incubation of the agent with a clot to demonstrate functional utility [67], fluorescently labeled targeted NPs can be incubated with appropriate tissue sections and relevant binding demonstrated by fluorescent microscopy [68], and fluorescently labeled aptamer-targeted NPs can be screened for cancer cell uptake by flow cytometry [69].

Finally, the formulation should be tested in a context approximating its ultimate intended clinical application in an appropriate animal model. For therapeutic applications, relative efficacy compared to an untargeted control formulation should be assessed. Such a control can consist of the NP formulation without any conjugated construct, conjugated to a nonspecific form of the ligand, or conjugated to an irrelevantly targeted ligand. For both diagnostic and therapeutic applications, many approaches to assessing targeting efficiency *in vivo* can be employed. For instance, ultrasound imaging of atheroma-targeted ELIPS has been performed in a Yucatan miniswine atherosclerosis model [70–73], targeted NP loaded with a contrast agent has been used to detect infarcted muscle by computed tomography in a mouse model [74], tumor uptake of radiolabeled CD105-targeted graphene oxide NP in a mouse breast cancer model was found to be greater than that of the same NP after antibody blocking or of untargeted NP [75], and lipid NP delivery of siRNAs specific for PTEN protein into the brains of rats resulted in lower levels of the protein, measured by western blot analysis, compared to untreated tissue and to NP alone [76].

6.4 Therapeutic Ultrasound

Ultrasound provides an effective means to control the release of contents from nanoparticles. For example, a single high amplitude pulse allows bolus release, a series of low amplitude pulses generate sustained release, and a combination of high and low amplitude pulses produce two different types of release profiles simultaneously. Therapeutic delivery using loaded nanoparticles demands enhanced local therapeutic concentration over defined time periods [7]. Ultrasound-triggered cavitation induces the destruction of gas bubbles, increases the permeability of cells and tissues, and facilitates therapeutic delivery into the cells and tissues. In particular for gene delivery, transfection efficiency can be also enhanced by ultrasound application [6].

6.4.1 *Biophysics of therapeutic ultrasound*

Sound is best described as pressure waves induced by rapid motion of molecules in a medium. Compression is when the molecules get closer to each other. Rarefaction indicates when they move away from each other. Compression corresponds to regions of high pressure, whereas rarefactions are related to regions of low pressure. Rapid and cyclic mechanical motions of molecules due to compression and rarefaction is called mechanical sound. Sound is categorized into infrasound, audible sound, and ultrasound according to their frequencies. Infrasound has frequencies less than 20 Hz. Audible sound has a frequency range between 20 Hz and 20 kHz, which is audible to the human ear. Ultrasound is sound with frequencies greater than 20 kHz. Ultrasound for use in medical therapeutics ranges between 20 kHz and 40 MHz [77].

When mechanical sound waves are applied to human tissue, two primary bioeffects occur: mechanical bioeffects and/or thermal bioeffects. In clinical ultrasound medical devices, the mechanical index (MI) and thermal index are used to indicate mechanical and thermal bioeffects occur. The MI is calculated by the maximum value of peak negative pressure and the center frequency of the ultrasound field. To decrease ultrasound induced bioeffects, an acoustic amplitude is considered low (no bioeffects) when the MI is lower than 0.3. Minor bioeffects (e.g. hemorrhage) can occur when the MI is between 0.3–0.7 in neonatal lung or intestine. An MI greater than 0.7 may have considerable bioeffects [78]. The Food and Drug Association (FDA) guidelines state that an approved MI upper limit is 1.9 in the absence of microbubbles [79]. As microbubbles can increase ultrasound-induced bioeffects, it is necessary to have low MI values in the presence of these agents.

6.4.2 *Cavitation*

Ultrasound cavitation is the formation of gas cavities by ultrasound application in liquid. There are two types of cavitation: stable (or non-inertial) cavitation and transient (or inertial) cavitation. Stable cavitation causes cavitation bubbles to expand and contract upon acoustic rarefaction and compression, without breaking the bubble. Transient cavitation is defined as a rapid growth of cavitation bubbles over the acoustic pressure cycles, eventually leading to fragmentation of the bubbles. In the presence of nanobubbles or microbubbles, thresholds for cavitation bubble fragmentation are found at lower energy levels of ultrasound [80]. For the delivery of therapeutic-loaded nanoparticle vehicles that contain air or gas inside, cavitation can generate partial disruption of the vehicles followed by the release of therapeutics.

6.4.3 *Acoustic radiation force*

Acoustic radiation forces refer to primary forces induced by acoustic waves that direct particles away from the acoustic source, as well as secondary forces that trigger attraction between ultrasound-responsive particles [60]. Primary radiation forces are utilized to thrust untargeted ultrasound-responsive particles from the luminal center of the blood vessel towards the vessel wall [81]. These marinating effects can markedly increase the binding efficiency of P-selectin-conjugated lipid microbubbles to the endothelium by up to 80 times [16].

6.4.4 *Thermal bioeffects*

Under certain conditions, ultrasound can induce temperature changes in tissue. When an ultrasound wave travels in a medium, the medium tends to resist the vibrational motion of the wave particles, and a portion of the mechanical energy of the acoustic wave is converted into heat [82]. High-intensity focused ultrasound (HIFU) is utilized for site-directed therapeutic delivery for chemotherapy when used with temperature-sensitive liposomes or genes containing a heat-inducible promoter [83].

The thermal index (TI) is an index to describe the potential bioeffects induced by tissue temperature increases when exposed to ultrasound. TI refers to a ratio of the power used by the system to the power required to increase the temperature in the tissue by one degree (°C). Thermal effects should be carefully evaluated for all ultrasound-responsive nanoparticles [84].

6.4.5 *Acoustic parameters for enhanced therapeutic delivery*

6.4.5.1 *Peak rarefactional pressure and center frequency*

The definition of MI clearly describes the dependence of peak rarefactional pressure amplitude and center frequency, as it determines various bioeffects [85]. The ratio of peak rarefactional pressure amplitude to center frequency (PRP/f) can be used as a threshold index for bioeffects [85, 86]. It is important to carefully determine appropriate dosages and ultrasound parameters prior to performing any *in vivo* studies when nanobubbles or microbubbles are utilized for cavitation.

6.4.5.2 *Pulse duration and duty cycle*

Nanobubbles and microbubbles can fragment and reaggregate repeatedly under certain acoustic pulses, or can remain as discrete fragments following these pulses. Pulse duration and duty cycle are important acoustic parameters in determining the sequences and efficacies of therapeutic delivery as well as potential biologic effects. For imaging, commercial ultrasound imaging unites using short pulses (in the order of microseconds) and a pulse repetition frequency (PRF) in the kHz range. Longer pulses have a larger effect when identical duty cycles are applied. For example, a PRP threshold of 6.3 MPa was found for blood–brain barrier (BBB) permeability with a 10-μs pulse, while the PRP threshold decreased to 1.1 MPa when longer acoustic pulses were applied [87]. The threshold for BBB disruption decreases from 0.7 to 0.4 MPa as pulse length increases from 0.1 to 10 ms [88].

6.5 Therapeutic Applications of Ultrasound-Controlled Nanosystems

6.5.1 *Drug and gene delivery to the brain*

It is a major challenge to develop efficient therapeutic delivery systems to the central nervous system (CNS) due to the inability of therapeutic agents to adequately cross the BBB. Due to the specificity and intrinsic restrictive characteristics of the BBB, only lipophilic drugs with molecular weights smaller than 500 Daltons can cross the BBB and demonstrate pharmacologic effects. The majority of candidate neurologic therapeutics do not satisfy these requirements [89].

Transcranial ultrasound can help therapeutics permeate the BBB [90]. A potential application with such an ultrasound effect is the use of ultrasound-responsive carriers for BBB permeability enhancement. Cavitation can disrupt the BBB, and acoustic radiation forces can push drugs across the BBB. Ultrasound-responsive nanoparticles can lower the pressure threshold for BBB disruption, allowing for ultrasound effects at two orders of magnitude lower than without these carriers [91]. Inertial cavitation is not the only physical basis to stimulate ultrasound-induced BBB disruption when these carriers (nanobubbles) are present [92]. While nanobubbles or microbubbles traverse the capillaries, they can stretch the vascular walls during ultrasound induced bubble oscillation, resulting in the opening of the BBB. Radiation forces and acoustic streaming are also important for ultrasound local vascular effects [93]. Enhancement of BBB permeability using ultrasound-responsive nanoparticles can last for up to 4 h, and permit the passage of molecules with weights ranging from 139 to 40,000 Da [94] and nanoparticles with diameters ranging from 50 nm [95] to several μm [96].

Although ultrasound sonoporation is a promising method for facilitating therapeutic delivery across the BBB, several concerns have been raised in terms of potential damage to brain tissue. These include ultrasound-induced edema, hemorrhage, ischemia, and apoptosis [92, 97–99]. These side effects are directly associated with ultrasound frequency, pulse length, duty cycle, duration, and most importantly, the peak negative pressure [100]. When the burst length is shorter than 10 ms, there is less brain damage from the ultrasound [101]. Tung *et al.* demonstrated BBB permeability without cellular damage at acoustic pressures from 0.30 to 0.45 MPa when a 1.5-MHz focused ultrasound pulse (20% duty cycle, PRF 10 Hz, duration 1 min) was applied [102].

Another delivery methodology is the use of focused ultrasound to facilitate the transport of chemotherapeutic agents. This requires selective opening of the tumor area while preserving an intact peri-tumor BBB. BBB disruption using HIFU allows for effective therapeutic delivery into the brain tumor with decreases in tumor size at less than half the concentration required without HIFU application [103].

All species, including humans, have different sizes of skull bone thickness. Thus, ultrasound travels at different speeds across the skull. As humans have thicker skulls, the efficiency of ultrasound transmission through the human skull is important for ultrasound applications to the brain [104]. Swines have similar skull thickness as humans and are often used as surrogates to translate into humans. Xie *et al.* used high frequency high-intensity focused transtemporal ultrasound in combination with microbubbles to disrupt the BBB in a swine model [105]. They found that the peak negative pressure of the ultrasound wave was reduced nearly ten-fold with the use of microbubbles. Microbubbles can decrease ultrasound parameters for therapeutic delivery to the brain across the skull.

6.5.2 *Drug and gene delivery to the kidneys*

The kidneys receive approximately 20% of the cardiac output from the abdominal aorta. However, only 5% of lipid microbubble shell components can be deposited in the kidneys in normal physiologic conditions [106]. Targeted therapeutic delivery is thus required to achieve sufficient therapeutic effect.

Up-regulation of P-selectin occurs immediately following renal damage. Several studies have demonstrated the effect of P-selectin modulation in acute treatment of renal disease such as ischemia-reperfusion injury [107]. Inflammation is commonly involved in chronic kidney diseases [59, 108]. Microbubbles targeted to the vascular cell adhesion molecule-1 (VCAM-1) allow for the imaging of renal inflammation following ischemia-reperfusion injury [109]. Deelman *et al.* demonstrated the specific accumulation of microbubbles in chronic type 1 diabetic kidneys, utilizing neutralized anti-transforming growth factor (TGF)-β antibodies targeted to overexpression of TGF-β in the diabetic kidneys [110].

Targeted drug delivery to the kidneys in combination with ultrasound has great potential for enhanced endothelial permeability and facilitated binding of targeted microbubbles by acoustic radiation forces.

6.5.3 *Drug and gene delivery to the lungs and the pleural space*

Ultrasound-facilitated therapeutic delivery into the lungs is challenging. The lung tissue contains air that reflects and scatters ultrasound. In particular, the use of transcutaneous ultrasound for the enhancement of therapeutic delivery efficiency to the lung tissue can cause serious tissue damage and must be very carefully considered. Mechanical ultrasound can result in hemorrhage in the lung tissue due to cavitation effects [111]. Pitt *et al.* demonstrated a novel methodology to avoid these issues by using an aerosol with an ultrasonic nebulizer to deliver cationic DNA complexes for transfection into the epithelial cells [25]. This methodology is likely to be applicable for therapeutic delivery to tumor cells in the lungs that are located close to epithelial lining cells.

The value of ultrasound for the diagnosis of pleural effusion is well documented. Pleural loculation is a serious complication of infectious or non-infectious pleural diseases and remains an important clinical problem without effective management [112]. Patients with pleural loculation have impaired pleural drainage, which predisposes them to poor outcomes, including prolonged hospitalization, respiratory compromise due to lung restriction, persistent sepsis, and death [113]. Treatment options are currently unsatisfactory [114]. Single chain urokinase (scuPA) can improve outcomes in patients with pleural loculation based on its increased bioavailability and resistance to inhibition by plasminogen activator inhibitor-1 (PAI-1) [115]. Encapsulation of scuPA into carriers (liposomes) and application of transthoracic ultrasound can promote therapeutic efficacy at lower, otherwise ineffective doses of intrapleural scuPA. Liposomal carrier delivery combined with ultrasound-facilitated release of tissue plasminogen activator (tPA) also showed improved outcomes with reduced intrapleural dosing.

6.5.4 *Gene therapy applications*

Gene therapies are generally accomplished by introducing foreign plasmid DNAs and synthetic oligonucleotides using genetic delivery vectors. Viral vectors are efficient, but have undesired biolocalization issues. Non-viral vectors are relatively safe, but often provide insufficient gene expression. To overcome these barriers, ultrasound-responsive nanosystems have been investigated.

In cancer, nanoparticle delivery augments gene expression in a site-specific manner for a variety of tumors, including brain cancer [116], prostate cancer [117], hepatocellular carcinoma [118], pancreatic cancer [119], and kidney cancer [120]. Ultrasound-responsive nanoparticles protect genes from degradation in the circulation and improve therapeutic delivery to target sites. Ultrasound enhances extravasation of the ultrasound-responsive nanoparticles and gene complexes into extravascular tissue and into tumor cells.

In earlier studies, the plasmid DNA was incorporated onto the surface of a protein-shelled microbubble contrast agent [121]. Although *in vitro* and *ex vivo* studies demonstrated efficient delivery outcomes, a lack of gene uptake was found beyond the blood vessels and luminal tissue *in vivo* [122]. Similarly, systemic injection of naked plasmid DNAs with localized ultrasound application is inefficient, presumably as the naked DNAs are degraded by serum nucleases resulting in a low DNA concentration near the sonoporated cells.

Non-viral vectors, such as liposomes and polymers, play an important role in extending the circulation times of DNA and facilitating intracellular delivery. DNA complexes containing cationic lipids enhance uptakes, both *in vitro* and *in vivo* [123, 124]. *In vitro* studies demonstrate increased transfection up to 18-fold and 90-fold with ultrasound alone and DNA complexation with PEI. The combination of ultrasound and PEI synergistically increases transfection as much as 200-fold, with reporter gene expression in 34% of the cells. Kinetic studies show rapid effects by ultrasound alone and beneficial effects by PEI after 4 h of incubation with the DNA plasmids following sonication [125]. This indicates that liposomal transfection agents can be optimized to maximize synergistic ultrasound effects for enhanced transitional membrane permeability.

Other novel approaches use non-viral vectors with microbubble shells to improve DNA loading capacity and *in vivo* gene transfection efficiency. Vandenbroucke *et al.* demonstrated a novel liposomal microbubble hybrid for enhanced siRNA loading and improved siRNA knockdown [126]. They found that ultrasound produced a 60% increase in expression of green fluorescence protein plasmid DNA in rat eyes with the use of liposomes (polyethylene glycol and DSPC) containing perfluoropropane gas. Ultrasound-responsive liposomes provide higher transfection efficiency than that of standard lipofection agents [127].

Transfection efficiency is influenced by ultrasound parameters, the presence of a contrast agent, and the local concentration of plasmid DNA, as well as by the transfection agent [128]. This exciting field has great potential for improved gene transfection into tumor tissue.

6.5.5 *Drug and gene delivery in atherosclerosis and myocardium*

6.5.5.1 *Atherosclerosis*

Although much research has been done and reported on the use of targeted nanoparticulate ultrasound contrast agents, especially microbubbles [129–132] and liposomes [70–72, 133–135], for the imaging of atheroma, relatively little has actually been accomplished in demonstrating the feasibility of such formulations for drug and gene delivery to the same targets.

Microbubbles have been used for ultrasound-enhanced delivery of rapamycin to inhibit smooth muscle cell proliferation *in vitro* [136] and of reporter genes (coding for green fluorescent protein or luciferase) to inflamed endothelium *in vitro*, *ex vivo*, and *in vivo* [137–139]. In latter studies, microbubbles were targeted to CD9, VCAM-1, ICAM-1, and P-selectin. Unger *et al.* reviewed ultrasound-enhanced microbubble drug delivery [140].

The cationic amphiphilic lipid SAINT-C18 has been incorporated into both liposomes [141] and lipoplexes [142] for the delivery of anti-inflammatory siRNAs *in vivo* and *in vitro*, respectively. The liposomes were targeted to VCAM-1 and carried siRNA to knock down vascular endothelial (VE)-cadherin messenger ribonucleic acid (mRNA) in the renal microvasculature of tumor necrosis factor (TNF) α-challenged mice or siRNA specific for NFκB p65 in lipopolysaccharide-treated mice. The lipoplexes, containing siRNA specific for VE-cadherin, were targeted to both VCAM-1 and E-selectin expressed by inflammation-activated primary endothelial cells in culture.

Intrinsically ELIPs have been used for ultrasound-enhanced, atheroma-targeted delivery of drugs, genes, bioactive gases, and stem cells *in vitro* and *in vivo* [7, 143–145]. Drugs thus far tested in this formulation are papavarine for stroke-related vasospasm [146] and tPA as a thrombolytic agent [67, 147–150]. Nucleic acid constructs delivered include plasmids [151] and decoy oligonucleotides [152].

Ultrasound-facilitated release of nitric oxide (NO) from NO-loaded ELIPs has been shown to ameliorate atheroma progression in rabbits [153] and to enhance targeted ELIP uptake by atheroma in a miniswine atherosclerosis model [70, 154]. $CD34^+$ stem cells have been delivered to ICAM-1-expressing porcine endothelium *ex vivo* [155], and through TNFα-activated endothelial cell monolayers in transwell plates [156] by bifunctionally targeted ELIP.

6.5.5.2 *Myocardium*

Most drug and gene delivery to the myocardium has involved microbubbles. Vancraeynest *et al.* [157] investigated the feasibility and safety of delivering colloid nanoparticles to rat hearts *in vivo* in the presence of microbubbles subjected to ultrasound-targeted microbubble destruction (UTMD) by a 1.3 MHz ultrasound at peak pressures of 0.6–1.8 MPa for 1–9 min. They confirmed myocardial delivery, but found microvessel rupture and reversible contractile dysfunction at all parameters tested. Shohet *et al.* [31] were able to deliver β-galactosidase complementary deoxyribonucleic acid (cDNA) in an adenovirus vector to rat myocardium by UTMD using albumin-coated perfluoropropane-filled microbubbles.

Angiogenesis in the rat myocardium was demonstrated after delivery of human $VEGF_{165}$ cDNA by UTMD using lipid-stabilized microbubbles [158]. Hypercellular foci were observed associated with human $VEGF_{165}$ expression and endothelial cell markers. Capillary density increased by 33% and arteriolar density increased by 86% at 10 days. Acidic fibroblast growth factor (aFGF) protein was delivered to the rat myocardium in a heparin-modified MB construct subjected to external ultrasound. Enhanced angiogenesis was observed by M-mode echocardiography [159].

6.5.6 *Bioactive gas delivery*

There are a number of bioactive gases for biomedical use. These include xenon, hydrogen sulfide (H_2S), NO, and carbon monoxide. In cancer, recent research has demonstrated that H_2S inhibit various stages of cancer development [160]. Other important biologic effects of H_2S include increased immunocompetence, inhibition of procarcinogen activation by oxidases, inhibition of Nf-κB nuclear translocation, epigenetic silencing of proto-oncogenes, protection and repair of DNAs, and abolishment of chronic inflammation [161].

Gaseous therapeutics can readily cross biologic barriers and rapidly reach an effective local concentration at target tissue. Obstacles for using bioactive gases for therapeutic treatment are the lack of suitable administration and delivery methodologies and the adverse effects associated with toxic amounts of the administered gas.

Ultrasound contrast agents have been used for ultrasonic imaging for years as air interfaces produce a high contrast between the circulation and the surrounding tissue. When air in these contrast agents is replaced with a bioactive gas, these carriers become gaseous therapeutic carriers or gas donors [134, 153]. These carriers have great potential to protect and deliver bioactive gases in the circulation for an enhanced local effect. The development of ultrasound-responsive liposomes for bioactive gas delivery is currently in its infancy, but several studies have demonstrated bioactive gas release from ultrasound-responsive liposomes [134, 135]. There are three potential mechanisms for gas release from ultrasound-responsive liposomes: static gas leakage, acoustically driven diffusion, and rapid fragmentation. Smith *et al.* demonstrated acoustically driven diffusion of gas out of ultrasound-responsive liposomes using Doppler ultrasound [96]. This indicates that low power ultrasound can produce little damage to the surrounding tissue without inertial cavitation. Targeting of these carriers can further increase local concentration of the bioactive gas near the diseased tissue. Bioactive gas carriers can increase therapeutic permeability into various types of tissue, such as those found in tumors, myocardium infarctions, and stroke injuries [162, 163].

6.6 Conclusions

Over the past several decades, therapeutic ultrasound has greatly improved our ability to deliver and activate drugs and genes to enhance therapeutic efficacy. Focused and unfocused ultrasound, with and without acoustic carriers, have improved therapeutic delivery to target tissues. Ultrasound can also cause biologic tissue damage if inappropriate ultrasound parameters are used, especially in the presence of ultrasound contrast agents. Therefore, it is important to perform thorough studies for each type of target tissue to determine the appropriate ultrasound parameters and suitable therapeutic delivery agents for optimized ultrasound-based therapeutic delivery. Ultrasound-controlled nanosystems have great potential to become a standard armamentarium in disease treatment.

References

1. Escoffre, J. M., Piron, J., Novell, A. and Bouakaz, A. (2011). Doxorubicin delivery into tumor cells with ultrasound and microbubbles. *Mol. Pharm.*, **8**, 799–806.
2. Cochran, M. C., Eisenbrey, J., Ouma, R. O., Soulen, M. and Wheatley, M. A. (2001). Doxorubicin and paclitaxel loaded microbubbles for ultrasound triggered drug delivery. *Int. J. Pharm.*, **414**, 161–170.
3. Heath, C. H., Sorace, A., Knowles, J., Rosenthal, E. and Hoyt, K. (2012). Microbubble therapy enhances anti-tumor properties of cisplatin and cetuximab *in vitro* and *in vivo*. *Otolaryngol. Head Neck Surg.*, **146**, 938–945.
4. Chauhan, V. P., Stylianopoulos, T., Boucher, Y. and Jain, R. K. (2011). Delivery of molecular and nanoscale medicine to tumors: Transport barriers and strategies. *Annu. Rev. Chem. Biomol. Eng.*, **2**, 281–298.

5. Liao, Z. K., Tsai, K. C., Wang, H. T., Tseng, S. H., Deng, W. P., Chen, W. S. and Hwang, L.-H. (2012). Sonoporation-mediated anti-angiogenic gene transfer into muscle effectively regresses distant orthotopic tumors. *Cancer Gene Ther.*, **19**, 171–180.
6. Liu, Y., Bowen, N. J., Matyunina, L., McDonald, J. and Prausnitz, M. R. (2012). Gene transfection enhanced by ultrasound exposure combined with drug treatment guided by gene chip analysis. *Int. J. Hyperthermia*, **28**, 349–361.
7. Huang, S. L. (2010). Ultrasound-responsive liposomes. *Methods Mol. Biol.*, **605**, 113–128.
8. Ibsen, S., Benchimol, M., Simberg, D., Schutt, C., Steiner, J. and Esener, S. (2011). A novel nested liposome drug delivery vehicle capable of ultrasound triggered release of its payload. *J. Control Release*, **155**, 358–366.
9. Rapoport, N., Nam, K. H., Gupta, R., Gao, Z., Mohan, P., Payne, A., Todd, N., Liu X, Kim, T., Shea, J., Scaife, C., Parker, D. L., Jeong, E. K. and Kennedy, A. M. (2011). Ultrasound-mediated tumor imaging and nanotherapy using drug loaded, block copolymer stabilized perfluorocarbon nanoemulsions. *J. Control Release*, **153**, 4–15.
10. Fernandez-Fernandez, A., Manchanda, R. and McGoron, A. J. (2011). Theranostic applications of nanomaterials in cancer: Drug delivery, image-guided therapy, and multifunctional platforms. *Appl. Biochem. Biotechnol.*, **165**, 1628–1651.
11. Bohmer, M. R., Chlon, C. H., Raju, B. I., Chin, C. T., Shevchenko, T. and Klibanov, A. L. (2010). Focused ultrasound and microbubbles for enhanced extravasation. *J. Control Release*, **148**, 18–24.
12. Danson, S., Ferry, D., Alakhov, V., Margison, J., Kerr, D., Jowle, D. l., Brampton, M., Halbert, G. and Ranson, M. (2004). Phase I dose escalation and pharmacokinetic study of pluronic polymer-bound doxorubicin (SP1049C) in patients with advanced cancer. *Br. J. Cancer*, **90**, 2085–2091.
13. Matsumura, Y., Hamaguchi, T., Ura, T., Muro, K., Yamada, Y., Shimada, Y., Shirao, K., Okusaka, T., Ueno, H., Ikeda, M. and Watanabe, N. (2004). Phase I clinical trial and pharmacokinetic evaluation of NK911, a micelle-encapsulated doxorubicin. *Br. J Cancer*, **91**, 1775–1781.
14. Jiang, J., Tong, X. and Zhao, Y. (2005). A new design for light-breakable polymer micelles. *J. Am. Chem. Soc.*, **127**, 8290–8291.
15. Muthu, M. S., Rajesh, C. V., Mishra, A. and Singh, S. (2009). Stimulus-responsive targeted nanomicelles for effective cancer therapy. *Nanomedicine (Lond).*, **4**, 657–667.
16. Husseini, G. A., Myrup, G. D., Pitt, W. G., Christensen, D. A. and Rapoport, N. Y. (2000). Factors affecting acoustically triggered release of drugs from polymeric micelles. *J. Control Release*, **69**, 43–52.
17. Marin, A., Muniruzzaman, M. and Rapoport, N. (2001). Acoustic activation of drug delivery from polymeric micelles: Effect of pulsed ultrasound. *J. Control Release*, **71**, 239–249.
18. Rapoport, N. (2004). Combined cancer therapy by micellar-encapsulated drug and ultrasound. *Int. J. Pharm.*, **277**, 155–162.
19. Seemann, S., Hauff, P., Schultze-Mosgau, M., Lehmann, C. and Reszka, R. (2002). Pharmaceutical evaluation of gas-filled microparticles as gene delivery system. *Pharm. Res.*, **19**, 250–257.
20. Kooiman, K., Vos, R., Versluis, M. and de Jong, N. (2014). Acoustic behavior of microbubbles and implications for drug delivery. *Adv. Drug Deliv. Rev.*, **72**, 28–48.
21. Grull, H. and Langereis, S. (2012). Hyperthermia-triggered drug delivery from temperature-sensitive liposomes using MRI-guided high intensity focused ultrasound. *J. Control Release*, **161**, 317–327.

22. Mabrey, S. and Sturtevant, J. M. (1976). Investigation of phase transitions of lipids and lipid mixtures by sensitivity differential scanning calorimetry. *Proc. Natl. Acad. Sci. U. S. A.*, **73**, 3862–3866.
23. Needham, D., Park, J. Y., Wright, A. M. and Tong, J. (2013). Materials characterization of the low temperature sensitive liposome (LTSL, effects of the lipid composition (lysolipid and DSPE-PEG2000) on the thermal transition and release of doxorubicin. *Faraday Discuss.*, **161**, 515–534, discussion 63–89.
24. Luk, B. T., Fang, R. H. and Zhang, L. (2012). Lipid- and polymer-based nanostructures for cancer theranostics. *Theranostics*, **2**, 1117–1126.
25. Pitt, W. G., Husseini, G. A. and Staples, B. J. (2004). Ultrasonic drug delivery — a general review. *Expert Opin. Drug Deliv.*, **1**, 37–56.
26. Alkan-Onyuksel, H., Demos, S. M., Lanza, G. M., Vonesh, M. J., Klegerman, M. E., Kane, B. J., Kuszak, J. and McPherson D. D. (1996). Development of inherently echogenic liposomes as an ultrasonic contrast agent. *J. Pharm. Sci.*, **85**, 486–490.
27. Huang, S. L., McPherson, D. D. and Macdonald, R. C. (2008). A method to co-encapsulate gas and drugs in liposomes for ultrasound-controlled drug delivery. *Ultrasound Med. Biol.*, **34**, 1272–1280.
28. Lin, H. Y. and Thomas, J. L. (2004). Factors affecting responsivity of unilamellar liposomes to 20 kHz ultrasound. *Langmuir.*, **20**, 6100–6106.
29. Pong, M., Umchid, S., Guarino, A. J., Lewin, P. A., Litniewski, J., Nowicki, A. and Wrenn, S. P. (2006). *In vitro* ultrasound-mediated leakage from phospholipid vesicles. *Ultrasonics.*, **45**, 133–145.
30. Chen, S., Kroll, M. H., Shohet, R. V., Frenkel, P., Mayer, S. A. and Grayburn, P. A. (2002). Bioeffects of myocardial contrast microbubble destruction by echocardiography. *Echocardiography*, **19**, 495–500.
31. Shohet, R. V., Chen S, Zhou Y. T, Wang Z., Meidell R. S., Unger R. H. and Grayburn, P. A. (2000). Echocardiographic destruction of albumin microbubbles directs gene delivery to the myocardium. *Circulation*, **101**, 2554–2556.
32. Lentacker, I., De Geest, B. G., Vandenbroucke, R. E., Peeters, L., Demeester, J., De Smedt, S. C. and Sanders, N. N. (2006). Ultrasound-responsive polymer-coated microbubbles that bind and protect DNA. *Langmuir*, **22**, 7273–7278.
33. DeSilva, N. S., Ofek, I. and Crouch, E. C. (2003). Interactions of surfactant protein D with fatty acids. *Am. J. Respir. Cell Mol. Biol.*, **29**, 757–770.
34. Grabarek, Z. and Gergely, J. (1990). Zero-length crosslinking procedure with the use of active esters. *Anal. Biochem.*, **185**, 131–135.
35. Taniuchi. M., Clark, H. B. and Johnson, E. M., Jr. (1986). Induction of nerve growth factor receptor in Schwann cells after axotomy. *Proc. Natl. Acad. Sci. USA.*, **83**, 4094–4098.
36. Migneault, I., Dartiguenave, C., Bertrand, M. J. and Waldron, K. C. (2004). Glutaraldehyde: Behavior in aqueous solution, reaction with proteins, and application to enzyme crosslinking. *Biotechniques*, **37**, 790–796, 8–802.
37. Sun, N., Li, B., Shao, J., Mo, W., Hu, B., Shen, Z. and Hu, X. (2012). A general and facile one-pot process of isothiocyanates from amines under aqueous conditions. Beilstein *J. Org. Chem.*, **8**, 61–70.
38. Ducry, L. and Stump, B. (2010). Antibody-drug conjugates: Linking cytotoxic payloads to monoclonal antibodies. *Bioconjug. Chem.*, **21**, 5–13.

39. Lewis Phillips, G. D., Li, G., Dugger, D. L., Crocker, L. M., Parsons, K. L., Mai, E., Blättler, W. A., Lambert, J. M., Chari, R. V., Lutz, R. J., Wong, W. L., Jacobson, F. S., Koeppen, H., Schwall, R. H., Kenkare-Mitra, S. R., Spencer, S. D. and Sliwkowski, M. X. (2008). Targeting HER2-positive breast cancer with trastuzumab-DM1, an antibody-cytotoxic drug conjugate. *Cancer Res.*, **68**, 9280–9290.
40. Martin, F. J., Heath, T. D. and New, R. R. C. (1990). 'Covalent attachment of proteins to liposomes' in New R. R. C. (ed.), *Liposomes: A Practical Approach*, IRL Press, New York, p. 301.
41. Betageri, G. V., Black, C. D., Szebeni, J., Wahl, L. M. and Weinstein, J. N. (1993a). Fc-receptor-mediated targeting of antibody-bearing liposomes containing dideoxycytidine triphosphate to human monocyte/macrophages. *J. Pharm. Pharmacol.*, **45**, 48–53.
42. Betageri, G. V. and Burrell, L. S. (1993b). Stability of antibody-bearing liposomes containing dideoxyinosine triphosphate. *Int. J. Pharmaceutics*, **98**, 149–155.
43. Demos, S. M., Onyuksel, H., Gilbert, J., Roth, S. I., Kane, B., Jungblut, P., Pinto, J. V., McPherson, D. D. and Klegerman, M. E. (1997). *In vitro* targeting of antibody-conjugated echogenic liposomes for site-specific ultrasonic image enhancement. *J. Pharm. Sci.*, **86**, 167–171.
44. Heath, T. D., Montgomery, J. A., Piper, J. R. and Papahadjopoulos, D. (1983). Antibody-targeted liposomes: Increase in specific toxicity of methotrexate-gamma-aspartate. *Proc. Natl. Acad. Sci. USA.*, **80**, 1377–1381.
45. Klegerman, M. E., Hamilton, A. J., Huang, S. L., Tiukinhoy, S. D., Khan, A. A., MacDonald, R. C. and McPherson, D. D. (2002). Quantitative immunoblot assay for assessment of liposomal antibody conjugation efficiency. *Anal. Biochem.*, 300, 46–52.
46. Klegerman, M. E., Huang, S., Parikh, D., Martinez, J., Demos, S. M., Onyuksel, H. A. and McPherson, D. D. (2007). Lipid contribution to the affinity of antigen association with specific antibodies conjugated to liposomes. *Biochim. Biophys. Acta.*, **1768**, 1703–1716.
47. Griffiths, G. L., Mattes, M. J., Stein, R., Govindan, S. V., Horak, I. D., Hansen, H. J. and Goldenberg, D. M. (2003). Cure of SCID mice bearing human B-lymphoma xenografts by an anti-CD74 antibody-anthracycline drug conjugate. *Clin. Cancer Res.*, **9**, 6567–6571.
48. Kohn, J. and Wilchek, M. (1982). A new approach (cyano-transfer) for cyanogen bromide activation of Sepharose at neutral pH, which yields activated resins, free of interfering nitrogen derivatives. *Biochem. Biophys. Res. Commun.*, **107**, 878–884.
49. Finlay, T. H., Troll, V., Levy, M., Johnson, A. J. and Hodgins, L. T. (1978). New methods for the preparation of biospecific adsorbents and immobilized enzymes utilizing trichloro-s-triazine. *Anal. Biochem.*, **87**, 77–90.
50. Bendas, G., Krause, A., Bakowsky, U., Vogel, J. and Rothe, U. (1999). Targetability of novel immunoliposomes prepared by a new antibody conjugation technique. *Int. J. Pharm.*, **181**, 79–93.
51. Hood, E., Gonzalez, M., Plaas, A., Strom, J. and VanAuker, M. (2007). Immunotargeting of nonionic surfactant vesicles to inflammation. *Int. J. Pharm.*, **339**, 222–230.
52. Sundberg, L. and Porath, J. (1974). Preparation of adsorbents for biospecific affinity chromatography. Attachment of group-containing ligands to insoluble polymers by means of bifuctional oxiranes. *J. Chromatogr.*, **90**, 87–98.
53. Hermanson, G. T., Mallia, A. K. and Smith, P. K. (1992). *Immobilized Affinity Ligand Techniques*. Academic Press, San Diego, p. 454.

54. Argyo, C., Weiss, V., Brauchle, C. and Bein, T. (2014). Multifunctional mesoporous silica nanoparticles as a universal platform for drug delivery. *Chem. Mater.*, **26**, 435–451.
55. Klibanov, A. L. (2005). Ligand-carrying gas-filled microbubbles: Ultrasound contrast agents for targeted molecular imaging. *Bioconjug. Chem.*, **16**, 9–17.
56. Clarkson, J. R., Cui, Z. F. and Darton, R. C. (1999). Protein denaturation in foam. *J. Colloid Interface Sci.*, **215**, 323–332.
57. Klegerman, M. E., Wassler, M., Huang, S., Zou, Y., Kim, H., Shelat, H. S., Holland, C. K, Geng, Y. J. and McPherson, D. D. (2010). Liposomal modular complexes for simultaneous targeted delivery of bioactive gases and therapeutics. *Journal of Controlled Release*, **142**, 326–331.
58. Lanza, G. M. and Wickline, S. A. (2001). Targeted ultrasonic contrast agents for molecular imaging and therapy. *Prog. Cardiovasc. Dis.*, **44**, 13–31.
59. Lindner, J. R., Song, J., Christiansen, J., Klibanov, A. L., Xu, F. and Ley, K. (2001). Ultrasound assessment of inflammation and renal tissue injury with microbubbles targeted to P-selectin. *Circulation*, **104**, 2107–2112.
60. Rychak, J. J., Klibanov, A. L., Ley, K. F. and Hossack, J. A. (2007). Enhanced targeting of ultrasound contrast agents using acoustic radiation force. *Ultrasound Med. Biol.*, **33**, 1132–1139.
61. Livnah, O., Bayer, E. A., Wilchek, M. and Sussman, J. L. (1993). Three-dimensional structures of avidin and the avidin-biotin complex. *Proc. Natl. Acad. Sci. U. S. A.*, **90**, 5076–5080.
62. Grana, C. M., Chinol, M., De Cicco, C., Bartolomei, M., Cremonesi, M., Bodei, L. and Paganelli, G. (2012). Eleven-year experience with the avidin-biotin pretargeting system in glioblastoma: Toxicity, efficacy and survival. *Open Nuclear Med. J.*, **4**, 14–20.
63. Moro, M., Pelagi, M., Fulci, G., Paganelli, G., Dellabona, P., Casorati, G., Siccardi, A. G. and Corti, A. (1997). Tumor cell targeting with antibody-avidin complexes and biotinylated tumor necrosis factor alpha. *Cancer Res.*, **57**, 1922–1928.
64. Klegerman, M. E., Zou, Y., Golunski, E., Peng, T., Huang, S. L. and McPherson, D. D. (2014). Use of thermodynamic coupling between antibody-antigen binding and phospholipid acyl chain phase transition energetics to predict immunoliposome targeting affinity. *J. Liposome Res.*, **24**, 216–222.
65. Demos, S. M., Dagar, S., Klegerman, M., Nagaraj, A., McPherson, D. D. and Onyuksel, H. (1998). In vitro targeting of acoustically reflective immunoliposomes to fibrin under various flow conditions. *J. Drug Target.*, **5**, 507–518.
66. Huang, X., Schwind, S., Yu, B., Santhanam, R., Wang, H., Hoellerbauer, P., Mims, A., Klisovic, R., Walker, A. R., Chan, K. K., Blum, W., Perrotti, D., Byrd, J. C., Bloomfield, C. D., Caligiuri, M. A., Lee, R. J., Garzon, R., Muthusamy, N., Lee, L. J. and Marcucci, G. (2013). Targeted delivery of microRNA-29b by transferrin-conjugated anionic lipopolyplex nanoparticles: A novel therapeutic strategy in acute myeloid leukemia. *Clin. Cancer Res.*, **19**, 2355–2367.
67. Tiukinhoy-Laing, S. D., Buchanan, K., Parikh, D., Huang, S., MacDonald, R. C., McPherson, D. D. and Klegerman, M. E. (2007). Fibrin targeting of tissue plasminogen activator-loaded echogenic liposomes. *J. Drug Target.*, **15**, 109–114.
68. Hitchcock, K. E., Caudell, D. N., Sutton, J. T., Klegerman, M. E., Vela, D., Pyne-Geithman, G. J., Abruzzo, T., Cyr, P. E., Geng, Y. J., McPherson, D. D. and

Holland, C. K. (2010). Ultrasound-enhanced delivery of targeted echogenic liposomes in a novel ex vivo mouse aorta model. *J. Control Release.*, **144**, 288–295.

69. Valencia, P. M., Pridgen, E. M., Rhee, M., Langer, R., Farokhzad, O. C. and Karnik, R. (2013). Microfluidic platform for combinatorial synthesis and optimization of targeted nanoparticles for cancer therapy. *ACS Nano.*, 7, 10671–10680.
70. Kim, H., Kee, P. H., Rim, Y., Moody, M. R., Klegerman, M. E., Vela, D., Huang, S. L., McPherson, D. D. and Laing, S. T. (2013). Nitric oxide improves molecular imaging of inflammatory atheroma using targeted echogenic immunoliposomes. *Atherosclerosis.*, **231**, 252–260.
71. Demos, S. M., Alkan-Onyuksel, H., Kane, B. J., Ramani, K., Nagaraj, A., Greene, R., Klegerman, M. E. and McPherson, D. D. (1999). In vivo targeting of acoustically reflective liposomes for intravascular and transvascular ultrasonic enhancement. *J. Am. Coll. Cardiol.*, **33**, 867–875.
72. Hamilton, A. J., Huang, S. L., Warnick, D., Rabbat, M., Kane, B., Nagaraj, A., Klegerman, M. E. and McPherson, D. D. (2004). Intravascular ultrasound molecular imaging of atheroma components *in vivo*. *J. Am. Coll. Cardiol.*, **43**, 453–460.
73. Kim, H., Moody, M. R., Laing, S. T., Kee, P. H., Huang, S. L., Klegerman, M. E. and McPherson, D. D. (2010). *In vivo* volumetric intravascular ultrasound visualization of early/inflammatory arterial atheroma using targeted echogenic immunoliposomes. *Invest. Radiol.*, **45**, 685–691.
74. Danila, D., Johnson, E. and Kee, P. (2013). CT imaging of myocardial scars with collagen-targeting gold nanoparticles. *Nanomedicine.*, **9**, 1067–1076.
75. Hong, H., Yang, K., Zhang, Y., Engle, J. W., Feng, L., Yang, Y., Nayak, T. R., Goel, S., Bean, J., Theuer, C. P., Barnhart, T. E., Liu, Z. and Cai, W. (2012). *In vivo* targeting and imaging of tumor vasculature with radiolabeled, antibody-conjugated nanographene. *ACS Nano.*, **6**, 2361–2370.
76. Rungta, R. L., Choi, H. B., Lin, P. J., Ko, R. W., Ashby, D., Nair, J., Manoharan, M., Cullis, P. R. and Macvicar, B. A. (2013). Lipid nanoparticle delivery of siRNA to silence neuronal gene expression in the brain. *Mol. Ther. Nucleic Acids*, **2**, e136.
77. Polat, B. E., Hart, D., Langer, R. and Blankschtein, D. (2011). Ultrasound-mediated transdermal drug delivery: Mechanisms, scope, and emerging trends. *J. Control Release*, **152**, 330–348.
78. Postema, M. and Gilja, O. H. (2011). Contrast-enhanced and targeted ultrasound. *World J. Gastroenterol.*, **17**, 28–41.
79. Qin, S., Caskey, C. F. and Ferrara, K. W. (2009). Ultrasound contrast microbubbles in imaging and therapy: Physical principles and engineering. *Phys. Med. Biol.*, **54**, R27–R57.
80. Hernot, S. and Klibanov, A. L. (2008). Microbubbles in ultrasound-triggered drug and gene delivery. *Adv. Drug Deliv. Rev.*, **60**, 1153–1166.
81. Dayton, P., Klibanov, A., Brandenburger, G. and Ferrara, K. (1999). Acoustic radiation force in vivo: A mechanism to assist targeting of microbubbles. *Ultrasound Med. Biol.*, **25**, 1195–1201.
82. Rome, C., Deckers, R. and Moonen, C. T. (2008). The use of ultrasound in transfection and transgene expression. *Handb. Exp. Pharmacol.*, **185**, 225–243.
83. Kennedy, J. E., Ter Haar, G. R. and Cranston, D. (2003). High intensity focused ultrasound: Surgery of the future? *Br. J. Radiol.*, **76**, 590–599.

84. Bigelow, T. A., Church, C. C., Sandstrom, K., Abbott, J. G., Ziskin, M. C., Edmonds, P. D., Herman, B., Thomenius, K. E. and Teo, T. J. (2011). The thermal index: Its strengths, weaknesses, and proposed improvements. *J. Ultrasound Med.*, **30**, 714–734.
85. Miller, D. L., Dou, C. and Wiggins, R. C. (2008). Frequency dependence of kidney injury induced by contrast-aided diagnostic ultrasound in rats. *Ultrasound Med. Biol.*, **34**, 1678–1687.
86. Stieger, S. M., Caskey, C. F., Adamson, R. H., Qin, S., Curry, F. R., Wisner, E. R. and Ferrara, K. W. (2007). Enhancement of vascular permeability with low-frequency contrast-enhanced ultrasound in the chorioallantoic membrane model. *Radiology*, **243**, 112–121.
87. Hynynen, K., McDannold, N., Vykhodtseva, N. and Jolesz, F. A. (2003). Non-invasive opening of BBB by focused ultrasound. *Acta. Neurochir. Suppl.*, **86**, 555–558.
88. McDannold, N., Vykhodtseva, N. and Hynynen, K. (2008). Effects of acoustic parameters and ultrasound contrast agent dose on focused-ultrasound induced blood-brain barrier disruption. *Ultrasound Med. Biol.*, **34**, 930–937.
89. Patel, T., Zhou, J., Piepmeier, J. M. and Saltzman, W. M. (2012). Polymeric nanoparticles for drug delivery to the central nervous system. *Adv. Drug Deliv. Rev.*, **64**, 701–705.
90. Park, E. J., Zhang, Y. Z., Vykhodtseva, N. and McDannold, N. (2012). Ultrasound-mediated blood-brain/blood-tumor barrier disruption improves outcomes with trastuzumab in a breast cancer brain metastasis model. *J. Control Release*, **163**, 277–284.
91. Yang, F. Y., Lin, G. L., Horng, S. C., Chang, T. K., Wu, S. Y., Wong, T. T. and Wang, H. E. (2011). Pulsed high-intensity focused ultrasound enhances the relative permeability of the blood-tumor barrier in a glioma-bearing rat model. *IEEE Trans. Ultrason. Ferroelectr. Freq. Control*, **58**, 964–970.
92. McDannold, N. J., Vykhodtseva, N. I. and Hynynen, K. (2006). Microbubble contrast agent with focused ultrasound to create brain lesions at low power levels: MR imaging and histologic study in rabbits. *Radiology*, **241**, 95–106.
93. Sheikov, N., McDannold, N., Vykhodtseva, N., Jolesz, F. and Hynynen, K. (2004). Cellular mechanisms of the blood-brain barrier opening induced by ultrasound in presence of microbubbles. *Ultrasound Med. Biol.*, **30**, 979–989.
94. Etame, A. B., Diaz, R. J., Smith, C. A., Mainprize, T. G., Hynynen, K. and Rutka, J. T. (2012). Focused ultrasound disruption of the blood–brain barrier: A new frontier for therapeutic delivery in molecular neurooncology. *Neurosurg. Focus*, **32**, E3.
95. Etame, A. B., Diaz, R. J., O'Reilly, M. A., Smith, C. A., Mainprize, T. G., Hynynen, K., et al. (2012). Enhanced delivery of gold nanoparticles with therapeutic potential into the brain using MRI-guided focused ultrasound. *Nanomedicine*, **8**, 1133–1142.
96. Yang, F. Y., Wong, T. T., Teng, M. C., Liu, R. S., Lu, M., Liang, H. F. and Wei, M. C. (2012). Focused ultrasound and interleukin-4 receptor-targeted liposomal doxorubicin for enhanced targeted drug delivery and antitumor effect in glioblastoma multiforme. *J. Control Release*, **160**, 652–658.
97. Liu, H. L., Wai, Y. Y., Chen, W. S., Chen, J. C., Hsu, P. H., Wu, X. Y., Huang, W. C., Yen, T. C. and Wang, J. J. (2008). Hemorrhage detection during focused-ultrasound induced blood-brain-barrier opening by using susceptibility-weighted magnetic resonance imaging. *Ultrasound Med. Biol.*, **34**, 598–606.
98. Yoshino, S., Fukushima, T., Hayashi, S., Nonaka, M., Ogawa, K., Sasaki, K. and Umemura, S. (2009). Effects of focused ultrasound sonodynamic treatment on the rat blood-brain barrier. *Anticancer Res.*, **29**, 889–895.

99. Vykhodtseva, N., McDannold, N. and Hynynen, K.. (2006). Induction of apoptosis in vivo in the rabbit brain with focused ultrasound and Optison. *Ultrasound Med. Biol.*, **32**, 1923–1929.
100. Weng, J. C., Wu, S. K., Lin, W. L. and Tseng, W. Y. (2011). Detecting blood-brain barrier disruption within minimal hemorrhage following transcranial focused ultrasound: A correlation study with contrast-enhanced MRI. *Magn. Reson. Med.*, **65**, 802–811.
101. McDannold, N., Vykhodtseva, N. and Hynynen, K. (2007). Use of ultrasound pulses combined with Definity for targeted blood-brain barrier disruption: A feasibility study. *Ultrasound Med. Biol.*, **33**, 584–590.
102. Tung, Y. S., Vlachos, F., Choi, J. J., Deffieux, T., Selert, K. and Konofagou, E. E. (2010). In vivo transcranial cavitation threshold detection during ultrasound-induced blood-brain barrier opening in mice. *Phys. Med. Biol.*, **55**, 6141–6155.
103. Barqawi, A. B. and Crawford, E. D. (2008). Emerging role of HIFU as a noninvasive ablative method to treat localized prostate cancer. *Oncology (Williston Park)*, **22**, 123–129, discussion 9, 33, 37 passim.
104. Ammi, A. Y., Mast, T. D., Huang, I. H., Abruzzo, T. A., Coussios, C. C., Shaw, G. J. and Holland, C. K. (2008). Characterization of ultrasound propagation through *ex-vivo* human temporal bone. *Ultrasound Med. Biol.*, **34**, 1578–1589.
105. Xie, F., Boska, M. D., Lof, J., Uberti, M. G., Tsutsui, J. M. and Porter, T. R. (2008). Effects of transcranial ultrasound and intravenous microbubbles on blood brain barrier permeability in a large animal model. *Ultrasound Med. Biol.*, **34**, 2028–2034.
106. Tartis, M. S., Kruse, D. E., Zheng, H., Zhang, H., Kheirolomoom, A., Marik, J. and Ferrara, K. W. (2008). Dynamic microPET imaging of ultrasound contrast agents and lipid delivery. *J. Control Release*, **131**, 160–166.
107. Fuller, T. F., Sattler, B., Binder, L., Vetterlein, F., Ringe, B. and Lorf, T. (2001). Reduction of severe ischemia/reperfusion injury in rat kidney grafts by a soluble P-selectin glycoprotein ligand. *Transplantation*, **72**, 216–222.
108. Lindner, J. R., Song, J., Xu, F., Klibanov, A. L., Singbartl, K., Ley, K. and Kaul, S. (2000). Noninvasive ultrasound imaging of inflammation using microbubbles targeted to activated leukocytes. *Circulation*, **102**, 2745–2750.
109. Zhang, Y., Ye, C., Wang, G., Gao, Y., Tan, K., Zhuo, Z., Liu, Z., Xia, H., Yang, D. and Li, P. (2013). Kidney-targeted transplantation of mesenchymal stem cells by ultrasound-targeted microbubble destruction promotes kidney repair in diabetic nephropathy rats. *Biomed. Res. Int.*, **2013**, 526367.
110. Deelman, L. E., Decleves, A. E., Rychak, J. J. and Sharma, K. (2010). Targeted renal therapies through microbubbles and ultrasound. *Adv. Drug Deliv. Rev.*, **62**, 1369–1377.
111. Marum, S. and Price, S. (2011). The use of echocardiography in the critically ill, the role of FADE (Fast Assessment Diagnostic Echocardiography) training. *Curr. Cardiol. Rev.*, 7, 197–200.
112. Tucker, T. and Idell, S. (2013). Plasminogen-plasmin system in the pathogenesis and treatment of lung and pleural injury. *Semin. Thromb. Hemost.*, **39**, 373–381.
113. Light, R. W. (2006). Parapneumonic effusions and empyema. *Proc. Am. Thorac. Soc.*, **3**, 75–80.
114. Davies, H. E., Rosenstengel, A. and Lee, Y. C. (2011). The diminishing role of surgery in pleural disease. *Curr. Opin. Pulm. Med.*, **17**, 247–254.
115. Komissarov, A. A., Mazar, A. P., Koenig, K., Kurdowska, A. K. and Idell, S. (2009). Regulation of intrapleural fibrinolysis by urokinase-alpha-macroglobulin complexes in

tetracycline-induced pleural injury in rabbits. *Am. J. Physiol. Lung Cell Mol. Physiol.*, **297**, L568–L577.

116. O'Reilly, M. A. and Hynynen, K. (2012). Ultrasound enhanced drug delivery to the brain and central nervous system. *Int. J. Hyperthermia*, **28**, 386–396.
117. Palermo, G., Pinto, F., Totaro, A., Miglioranza, E., Calarco, A., Sacco, E., Daddessi, A., Vittori, M., Racioppi, M., Dagostino, D., Gulino, G., Giustacchini, M. and Bassi, P. (2013). High-intensity focused ultrasound in prostate cancer: Today's outcomes and tomorrow's perspectives. *Scand. J. Urol.*, **47**, 179–187.
118. Cochran, M. C., Eisenbrey, J. R., Soulen, M. C., Schultz, S. M., Ouma, R. O., White, S. B., Furth, E. E. and Wheatley, M. A. (2011). Disposition of ultrasound sensitive polymeric drug carrier in a rat hepatocellular carcinoma model. *Acad. Radiol.*, **18**, 8.
119. Boujaoude, J. (2007). Role of endoscopic ultrasound in diagnosis and therapy of pancreatic adenocarcinoma. *World J. Gastroenterol.*, **13**, 3662–3666.
120. Al-Bataineh, O., Jenne, J. and Huber, P. (2012). Clinical and future applications of high intensity focused ultrasound in cancer. *Cancer Treat. Rev.*, **38**, 346–353.
121. Frenkel, V. (2008). Ultrasound mediated delivery of drugs and genes to solid tumors. *Adv. Drug Deliv. Rev.*, **60**, 1193–1208.
122. Sirsi, S. R. and Borden, M. A. (2012). Advances in ultrasound mediated gene therapy using microbubble contrast agents. *Theranostics.*, **2**, 1208–1222.
123. Anwer, K., Kao, G., Proctor, B., Anscombe, I., Florack, V., Earls, R., Wilson, E., McCreery, T., Unger, E., Rolland, A. and Sullivan, S. M. (2000). Ultrasound enhancement of cationic lipid-mediated gene transfer to primary tumors following systemic administration. *Gene Ther.*, **7**, 1833–1839.
124. Greenleaf, W. J., Bolander, M. E., Sarkar, G., Goldring, M. B. and Greenleaf, J. F. (1998). Artificial cavitation nuclei significantly enhance acoustically induced cell transfection. *Ultrasound Med. Biol.*, **24**, 587–595.
125. Deshpande, M. C. and Prausnitz, M. R. (2007). Synergistic effect of ultrasound and PEI on DNA transfection in vitro. *J. Control Release*, **118**, 126–135.
126. Vandenbroucke, R. E., Lentacker, I., Demeester, J., De Smedt, S. C. and Sanders, N. N. (2008). Ultrasound assisted siRNA delivery using PEG-siPlex loaded microbubbles. *J. Control Release*, **126**, 265–273.
127. Suzuki, R., Takizawa, T., Negishi, Y., Hagisawa, K., Tanaka, K., Sawamura, K., Utoguchi, N., Nishioka, T. and Maruyama, K. (2007). Gene delivery by combination of novel liposomal bubbles with perfluoropropane and ultrasound. *J. Control Release*, **117**, 130–136.
128. Newman, C. M. and Bettinger, T. (2007). Gene therapy progress and prospects: Ultrasound for gene transfer. *Gene Ther.*, **14**, 465–475.
129. Khanicheh, E., Mitterhuber, M., Xu, L., Haeuselmann, S. P., Kuster, G. M. and Kaufmann, B. A. (2013). Noninvasive ultrasound molecular imaging of the effect of statins on endothelial inflammatory phenotype in early atherosclerosis. *PLoS One*, **8**, e58761.
130. McCarty, O. J., Conley, R. B., Shentu, W., Tormoen, G. W., Zha, D., Xie, A., Qi, Y., Zhao, Y., Carr, C., Belcik, T., Keene, D. R., de Groot, P. G. and Lindner, J. R. (2010). Molecular imaging of activated von Willebrand factor to detect high-risk atherosclerotic phenotype. *JACC Cardiovasc. Imaging*, **3**, 947–955.
131. Weller, G. E., Villanueva, F. S., Tom, E. M. and Wagner, W. R. (2005). Targeted ultrasound contrast agents: In vitro assessment of endothelial dysfunction and multi-targeting to ICAM-1 and sialyl Lewisx. *Biotechnol. Bioeng.*, **92**, 780–788.

132. Yang, H., Xiong, X., Zhang, L., Wu, C. and Liu, Y. (2011). Adhesion of bio-functionalized ultrasound microbubbles to endothelial cells by targeting to vascular cell adhesion molecule-1 under shear flow. *Int. J. Nanomedicine*, **6**, 2043–2051.
133. Huang, S., Hamilton, A. J., Tiukinhoy, S. D., Nagaraj, A., Kane, B. J., Klegerman, M., McPherson, D. D. and MacDonald, R. C. (2002). Liposomes as ultrasound imaging contrast agents and as ultrasound-sensitive drug delivery agents. *Cell Mol. Biol. Lett.*, 7, 233–235.
134. Britton, G. L., Kim, H., Kee, P. H., Aronowski, J., Holland, C. K., McPherson, D. D. and Huang, S. L. (2010). In vivo therapeutic gas delivery for neuroprotection with echogenic liposomes. *Circulation*, **122**, 1578–1587.
135. Smith, D. A., Porter, T. M., Martinez, J., Huang, S., MacDonald, R. C., McPherson, D. D. and Holland, C. K. (2007). Destruction thresholds of echogenic liposomes with clinical diagnostic ultrasound. *Ultrasound Med. Biol.*, **33**, 797–809.
136. Phillips, L. C., Klibanov, A. L., Wamhoff, B. R. and Hossack, J. A. (2011). Localized ultrasound enhances delivery of rapamycin from microbubbles to prevent smooth muscle proliferation. *J. Control Release*, **154**, 42–49.
137. Barreiro, O., Aguilar, R. J., Tejera, E., Megias, D., de Torres-Alba, F., Evangelista, A. and Sánchez-Madrid, F. (2009). Specific targeting of human inflamed endothelium and *in situ* vascular tissue transfection by the use of ultrasound contrast agents. *JACC Cardiovasc. Imaging*, **2**, 997–1005.
138. Phillips, L. C., Klibanov, A. L., Wamhoff, B. R. and Hossack, J. A. (2012). Intravascular ultrasound detection and delivery of molecularly targeted microbubbles for gene delivery. *IEEE Trans. Ultrason. Ferroelectr. Freq. Control*, **59**, 1596–1601.
139. Xie, A., Belcik, T., Qi, Y., Morgan, T. K., Champaneri, S. A., Taylor, S., Davidson, B. P., Zhao, Y., Klibanov, A. L., Kuliszewski, M. A., Leong-Poi, H., Ammi, A. and Lindner, J. R. (2012). Ultrasound-mediated vascular gene transfection by cavitation of endothelial-targeted cationic microbubbles. *JACC Cardiovasc. Imaging*, **5**, 1253–1262.
140. Unger, E., Porter, T., Lindner, J. and Grayburn, P. (2014). Cardiovascular drug delivery with ultrasound and microbubbles. *Adv. Drug Deliv. Rev.*, **72**, 110–126.
141. Kowalski, P. S., Zwiers, P. J., Morselt, H. W., Kuldo, J. M., Leus, N. G., Ruiters, M. H., Molema, G. and Kamps, J. A. (2014). Anti-VCAM-1 SAINT-O-Somes enable endothelial-specific delivery of siRNA and downregulation of inflammatory genes in activated endothelium in vivo. *J. Control Release*, **176**, 64–75.
142. Leus, N. G., Talman, E. G., Ramana, P., Kowalski, P. S., Woudenberg-Vrenken, T. E., Ruiters, M. H., Molema, G. and Kamps, J. A. (2014). Effective siRNA delivery to inflamed primary vascular endothelial cells by anti-E-selectin and anti-VCAM-1 PEGylated SAINT-based lipoplexes. *Int. J. Pharm.*, **459**, 40–50.
143. Huang, S. L. (2008). Liposomes in ultrasonic drug and gene delivery. *Adv. Drug Deliv. Rev.*, **60**, 1167–1176.
144. Huang, S. L. and MacDonald, R. C. (2004). Acoustically active liposomes for drug encapsulation and ultrasound-triggered release. *Biochim. Biophys. Acta.*, **1665**, 134–141.
145. Tiukinhoy, S. D., Khan, A. A., Huang, S., Klegerman, M. E., MacDonald, R. C. and McPherson, D. D. (2004). Novel echogenic drug-immunoliposomes for drug delivery. *Invest. Radiol.*, **39**, 104–110.
146. Kee, P. H., Abruzzo, T. A., Smith, D. A., Kopechek, J. A., Wang, B., Huang, S. L., MacDonald, R. C., Holland, C. K. and McPherson, D. D. (2008). Synthesis, acoustic

stability, and pharmacologic activities of papaverine-loaded echogenic liposomes for ultrasound controlled drug delivery. *J. Liposome Res.*, **18**, 263–277.

147. Laing, S. T., Moody, M., Smulevitz, B., Kim, H., Kee, P., Huang, S., *et al.* (2011). Ultrasound-enhanced thrombolytic effect of tissue plasminogen activator-loaded echogenic liposomes in an *in vivo* rabbit aorta thrombus model — brief report. *Arterioscler. Thromb. Vasc. Biol.*, **31**, 1357–1359.
148. Laing, S. T., Moody, M. R., Kim, H., Smulevitz, B., Huang, S. L., Holland, C. K., McPherson, D. D. and Klegerman, M. E. (2012). Thrombolytic efficacy of tissue plasminogen activator-loaded echogenic liposomes in a rabbit thrombus model. *Thromb. Res.*, **130**, 629–635.
149. Smith, D. A., Vaidya, S. S., Kopechek, J. A., Huang, S. L., Klegerman, M. E., McPherson, D. D. and Holland, C. K. (2010). Ultrasound-triggered release of recombinant tissue-type plasminogen activator from echogenic liposomes. *Ultrasound Med. Biol.*, **36**, 145–157.
150. Tiukinhoy-Laing, S. D., Huang, S., Klegerman, M., Holland, C. K. and McPherson, D. D. (2007). Ultrasound-facilitated thrombolysis using tissue-plasminogen activator-loaded echogenic liposomes. *Thromb. Res.*, **119**, 777–784.
151. Tiukinhoy, S. D., Mahowald, M. E., Shively, V. P., Nagaraj, A., Kane, B. J., Klegerman, M. E. and McPherson, D. D. (2000). Development of echogenic, plasmid-incorporated, tissue-targeted cationic liposomes that can be used for directed gene delivery. *Invest. Radiol.*, **35**, 732–738.
152. Buchanan, K. D., Huang, S. L., Kim, H., McPherson, D. D. and MacDonald, R. C. (2010). Encapsulation of NF-kappaB decoy oligonucleotides within echogenic liposomes and ultrasound-triggered release. *J. Control Release*, **141**, 193–198.
153. Huang, S. L., Kee, P. H., Kim, H., Moody, M. R., Chrzanowski, S. M., Macdonald, R. C., et al. (2009). Nitric oxide-loaded echogenic liposomes for nitric oxide delivery and inhibition of intimal hyperplasia. *J. Am. Coll. Cardiol.*, **54**, 652–659.
154. Kee, P. H., Kim, H., Huang, S., Laing, S. T., Moody, M. R., Vela, D., Klegerman, M. E. and McPherson, D. D. (2014). Nitric oxide pretreatment enhances atheroma component highlighting *in vivo* with intercellular adhesion molecule-1-targeted echogenic liposomes. *Ultrasound Med. Biol.*, **40**, 1167–1176.
155. Herbst, S. M., Klegerman, M. E., Kim, H., Qi, J., Shelat, H., Wassler, M., Moody, M. R., Yang, C. M., Ge, X., Zou, Y., Kopechek, J. A., Clubb, F. J., Kraemer, D. C., Huang, S., Holland, C. K., McPherson, D. D. and Geng, Y. J. (2010). Delivery of stem cells to porcine arterial wall with echogenic liposomes conjugated to antibodies against CD34 and intercellular adhesion molecule-1. *Mol. Pharm.*, 7, 3–11.
156. Klegerman, M. E., Wassler, M., Huang, S. L., Zou, Y., Kim, H., Shelat, H. S., Holland, C. K., Geng, Y. J. and McPherson, D. D. (2010). Liposomal modular complexes for simultaneous targeted delivery of bioactive gases and therapeutics. *J. Control Release*, **142**, 326–331.
157. Vancraeynest, D., Havaux, X., Pouleur, A. C., Pasquet, A., Gerber, B., Beauloye, C., Rafter, P., Bertrand, L. and Vanoverschelde, J. L. (2006). Myocardial delivery of colloid nanoparticles using ultrasound-targeted microbubble destruction. *Eur. Heart J.*, **27**, 237–245.
158. Korpanty, G., Chen, S., Shohet, R. V., Ding, J., Yang, B., Frenkel, P. A. and Grayburn, P. A. (2005). Targeting of VEGF-mediated angiogenesis to rat myocardium using ultrasonic destruction of microbubbles. *Gene Ther.*, **12**, 1305–1312.

159. Zhao, Y. Z., Lu, C. T., Li, X. K., Tang, Q. Q., Tian, X. Q., Zhao, Y. P., Zhang, Y., Tian, J. L., Yang, W., Ge, S., Nair, C. K. and Shen, X. (2012). Improving the cardio protective effect of aFGF in ischemic myocardium with ultrasound-mediated cavitation of heparin modified microbubbles: Preliminary experiment. *J. Drug Target.*, **20**, 623–631.
160. Kashfi, K. (2014). Anti-cancer activity of new designer hydrogen sulfide-donating hybrids. *Antioxid. Redox Signal.*, **20**, 831–846.
161. Predmore, B. L., Lefer, D. J. and Gojon, G. (2012). Hydrogen sulfide in biochemistry and medicine. *Antioxid. Redox Signal.*, **17**, 119–140.
162. Kim, H., Britton, G. L., Peng, T., Holland, C. K., McPherson, D. D. and Huang, S. L. (2014). Nitric oxide-loaded echogenic liposomes for treatment of vasospasm following subarachnoid hemorrhage. *Int. J. Nanomedicine*, **9**, 155–165.
163. Peng, T., Britton, G. L., Kim, H., Cattano, D., Aronowski, J., Grotta, J., McPherson, D. D. and Huang, S. L. (2013). Therapeutic time window and dose dependence of xenon delivered via echogenic liposomes for neuroprotection in stroke. *CNS Neurosci. Ther.*, **19**, 773–784.

Chapter 7

Plasmonic Nanobubble-Controlled on Demand Drug Delivery and Release with High Target Cell Specificity

Ekaterina Y. Lukianova-Hleb[1] *and Dmitri O. Lapotko*[1,2]

[1]*Department of Biochemistry and Cell Biology, Rice University, Houston, TX, USA*
[2]*Department of Physics and Astronomy, Rice University, Houston, TX, USA*

7.1 Introduction

The drug and radiation resistance and aggressiveness of some cancers (head and neck, brain, prostate, lung, and others) result in their high lethality, especially when doses of current therapies are limited by their high non-specific toxicity. Surgery often cannot fully resect such tumors if they are intertwined with important normal tissues, while incomplete tumor resection results in dangerous microscopic residual disease [1–4]. The inability to safely resect or treat aggressive cancers ultimately reduces patients' survival rates and quality of life [5–7]. Therefore, it is highly desirable to develop a novel approach that (i) selectively amplifies the efficacy of standard therapies only in cancer cells, (ii) preserves the functionality of co-localized normal tissues, and (iii) reduces non-specific toxicity and treatment time.

Our approach employs two strategic innovations:

1. Using the mechanical, rather than chemical, radiation, or thermal, impact for cancer cell destruction and drug delivery in order to overcome a wide spectrum of drug resistance mechanisms (Fig. 7.1);

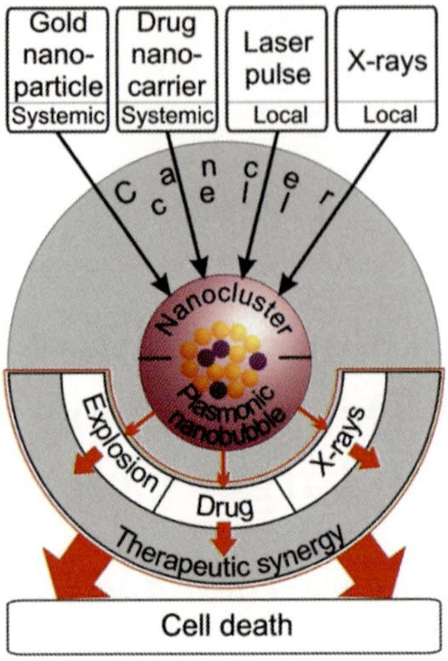

Figure 7.1. Principle of the quadrapeutics: four components — colloidal gold, encapsulated drugs, low-energy short laser pulses, and X-rays — are administered in a simple three-step protocol. First, cancer cells self-assemble antibody-functionalized gold nanoparticles (GNPs) and drug-loaded nanocarriers into intracellular nanoclusters via receptor-mediated endocytosis. Second, the on-demand threshold activation of these nanoclusters with an external laser pulse and X-rays triggers co-localized intracellular transient events: mechanical explosion of vapor plasmonic nanobubbles (PNBs), instantaneous ejection of drugs into the cytoplasm, and local amplification of X-rays. The intracellular synergy of these three threshold-activated, mechanical and physical events selectively amplifies the therapeutic effect of standard chemotherapy and chemo-radiation only in cancer cells [50].

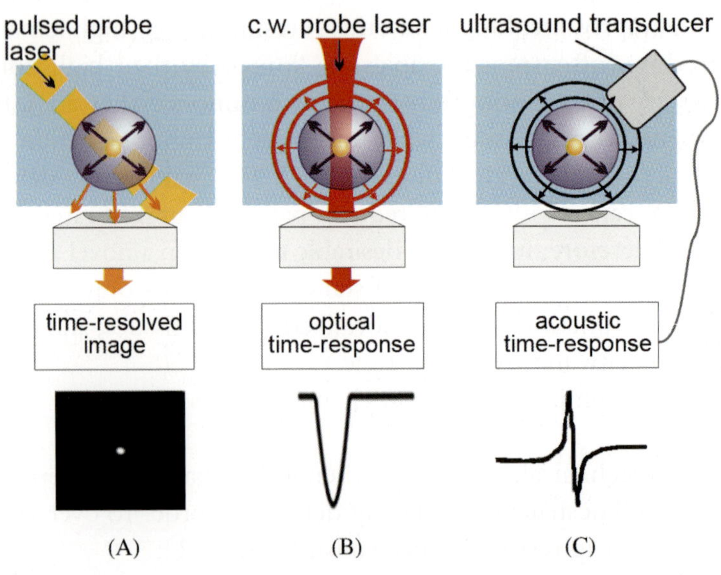

Figure 7.2. Methods for detecting vapor PNBs in liquid [69]: (A) optical time-resolved scattering imaging is obtained with a short probe laser pulse and a slow image detector, (B) optical scattering (extinction) time–response is obtained with a continuous focused probe laser pulse and a fast photodetector, (C) acoustical time–response is obtained with an ultrasound detector.

2. Selectively amplifying chemo- and radiotherapies inside target cancer cells by employing intracellular synergy of four standard components (Fig. 7.1): encapsulated drugs, colloidal GNPs [8–10], near-infrared short laser pulses, and radiations.

These two principles support novel therapeutic mechanisms (Fig. 7.1). Firstly, the cells self-assemble systemically administered antibody-functionalized GNPs and drug-loaded nanocarriers into intracellular nanoclusters via receptor-mediated endocytosis. Secondly, the on-demand threshold activation of these nanoclusters with locally administered laser pulse and radiation triggers co-localized intracellular therapeutic events: the mechanical impact of expanding and collapsing transient vapor nanobubbles (PNBs) generated around GNP clusters in target cells, instantaneous ejection of drugs into the cytoplasm from PNB-disrupted nanocarriers, and local amplification of the radiation. The intracellular synergy of these threshold-activated, mechanical, and physical transient events selectively amplifies the therapeutic effect of standard chemo-radiation only in cancer cells. We termed this technology quadrapeutics since it employs four components (Fig. 7.1).

This chapter considers the physical properties of a key quadrapeutics agent, PNB, and the methods of PNB generation and detection (Section 7.2), target cell specificity of PNBs vs. that of nanoparticles (NPs) and nanocarriers (Section 7.3), the mechanism of PNB-induced intracellular delivery and release of molecular payload (Section 7.4), the therapeutic mechanisms of PNBs at the cell level (Section 7.5), the mechanism of intracellular self-amplification of the therapeutic strength with cancer aggressiveness (Section 7.6), generation and detection of PNBs *in vivo* (Section 7.7), *in vivo* evaluation of the PNB therapeutic and diagnostic potential (Section 7.8), and a discussion of the medical applications of PNBs (Section 7.9).

7.2 Generation and Detection of PNBs

Transient thermal evaporation and the generation of vapor bubbles are one of the basic processes accompanying non-stationary high-temperature heat- and mass-transfer in liquids. The science and the methods for the detection of inertial vapor bubbles of various origins were well developed for macro- and micro-sized bubbles and mainly employ their ability to emit pressure and to scatter the incident light [11–36]. Recent developments in nanoscience reduced the spatial and temporal scale of vapor bubbles to nanometers and nanoseconds [17–20, 33, 37–43]. Unlike their larger analogs, PNBs require much higher sensitivity and resolution of the detection methods for their imaging, quantification, and identification, among other phenomena such as transient heating and the generation of stress waves. Here we analyze several experimental techniques for the imaging and quantitative analysis of transient PNBs as single events. Due to the multiple biomedical applications of PNBs and related phenomena [43–60], it should be noted that we consider the transient events, but not the materials (particles) that are often also called nanobubbles [61–63]. We also do not consider the cavitation of pre-existing bubbles that is well studied elsewhere [64].

While vapor nanobubbles may have various sources of energy (the heating of liquid above the boiling threshold, local rarefaction, and plasma discharge), we employed an experimental model of a single PNB in water. Such a model provides maximal precision, control, and reproducibility in PNB generation through the localized transient photothermal heating of liquid above the evaporation point. This was achieved through the optical excitation of individual 60-nm gold nanospheres (NSPs) in water with single short laser pulses (70 ps, 532 nm) at specific fluences above the PNB generation threshold. We used the plasmonic conversion of optical energy into heat to control the maximal diameter and lifetime of the PNBs through the fluence of a single laser pulse, as described in detail previously [18–20, 43]. This experimental model includes an internal metal nanoparticle that acts as the source of the bubble energy during PNB generation and prevents the development of extreme temperatures and sonoluminescence at the collapse stage [18, 65], unlike "classical" bubbles that are generated in homogeneous media [66, 67]. Another important difference in the model employed is the absence of an external acoustic field that is often used alone or in combination with optical energy to generate a vapor bubble [67, 68]. The above conditions and the nanoscale size resulted in a single cycle of the vapor nanobubble that did not oscillate.

First, we consider imaging methods. Optical scattering phenomena allow both PNB imaging and the monitoring of its dynamics [17–20, 24–28, 69]. Direct imaging of slower and larger macro- and micro-bubbles employed different cameras and light sources [23, 70–76]. There is the possibility of using a continuous light source and of capturing a short transient PNB with high-speed image detectors [77–81]. However, that would require a nanosecond gating speed, sub-micron spatial resolution and the corresponding high optical sensitivity. Dynamic transmission electron microscopy (TEM) and time-resolved diffraction of X-rays [41, 81] can even better provide direct imaging of the dynamics of PNB. Unfortunately, the cost and technical complexity of such solutions turn them into unique tools with very limited availability. Alternatively, the imaging of a transient PNB can also be realized with much more affordable equipment using slow cameras and pulsed light sources (Fig. 7.2A) providing the pulsed illumination with a shorter duration than a PNB lifetime within a nano- and pico-second range. This scheme allows the imaging of a single transient PNB with a lifetime down to 10 ns (Fig. 7.3A).

In this method, the pixel image amplitude is used as a PNB metric, but it is not very practical: (i) any deviation in the alignment of the probe beam would require the recalibration of the system and (ii) this metric will not characterize the maximal diameter of an PNB when one fixed time delay is applied. The advantage of this method is in the relatively precise identification of a PNB and its location. It also shows the number of PNBs if multiple events occur (Fig. 7.3B).

The maximal diameter of a PNB is the most important parameter that also characterizes the amount of energy received by the bubble [11–16, 18]. It can be measured with another optical scattering technique that employs the monitoring of the intensity of a continuous probe beam with a broadband high-speed photo-detector. The beam is focused at the PNB and its scattering by the PNB changes its axial intensity, which is measured as a time–response with the photo-detector. In one method, the beam is pointed at the low noise detector through the aperture and produces a high base level (Fig. 7.2B). A PNB scatters the light and, thus, increasingly reduces the axial intensity of the probe beam during its expansion stage and brings it back to the base level during its collapse stage. This results in the bubble-specific shape of the time–response (Fig. 7.3A-I). Technically, the detector measures the optical extinction of the probe beam, but since a PNB does not absorb any optical energy, the extinction becomes equal to its scattering component. The most precise way to measure the maximal diameter of a single PNB is to measure the duration of the PNB-specific signal (Fig. 7.3A-I), which was shown to correlate to its maximal diameter [11–16, 18].

The simultaneous generation of several PNBs of different maximal sizes influences not only the amplitude, but also the shape of the PNB time–response (Fig. 7.3B-I). Smaller PNBs form a sharp front and fewer large PNBs form a long tail of lower amplitude. While the generation of PNBs under some settings may typically involve such a multi-PNB pattern, it is still possible to estimate the lifetime of the largest PNB(s) that represent the maximal

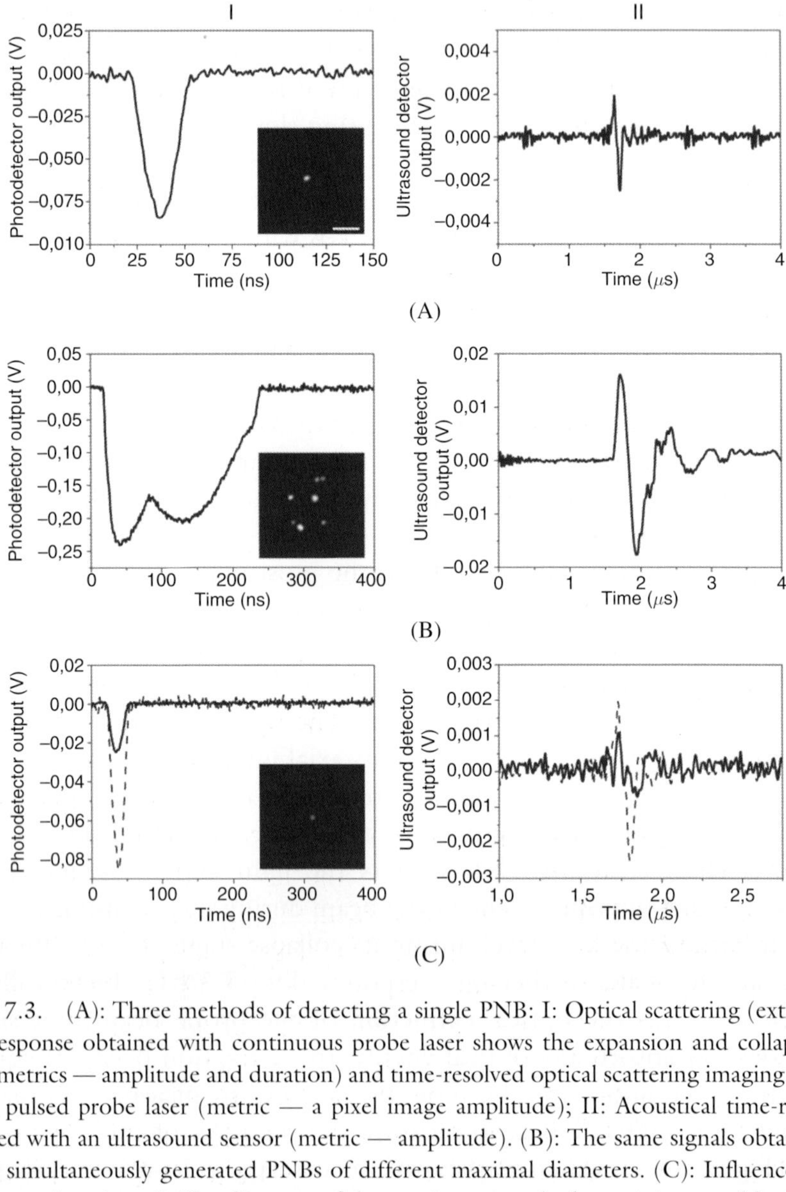

Figure 7.3. (A): Three methods of detecting a single PNB: I: Optical scattering (extinction) time–response obtained with continuous probe laser shows the expansion and collapse of a PNB (metrics — amplitude and duration) and time-resolved optical scattering imaging (insert) with a pulsed probe laser (metric — a pixel image amplitude); II: Acoustical time-response obtained with an ultrasound sensor (metric — amplitude). (B): The same signals obtained for several simultaneously generated PNBs of different maximal diameters. (C): Influence of the experimental settings: I: The diameter of the continuous probe beam is increased by 10–15% and the angle of incidence of the pulsed probe beam increased from 45° to 135°; II: The resonance frequency of the acoustic sensor is decreased from 10 to 5 MHz. Corresponding signals as shown in (A) are shown with dashed lines. Scale bar is 15 μm [69].

deposition of the thermal energy by measuring the duration of the time–response at the reduced level, instead of at half of maximum.

The third method of PNB detection employs the pressure waves generated during the bubble expansion and collapse [33, 36] (Fig. 7.2C). In the case of a single inertial bubble, the pressure and rarefaction waves produce a typical signal (Fig. 7.3A-II). This is definitely the simplest and the least expensive method of the three, because it does not require expensive optical sources, optics, and sensors. In addition, it detects PNBs in optically opaque media, so it is probably the most universal of the methods. The most obvious metric of the acoustic signal is its maximal (peak-to-peak) amplitude. However, the correct identification and measurement of a PNB with a standard acoustic detector can be challenging in a single pulse mode. Most detectors have a resonant frequency of 10 MHz and lower, where their sensitivity is the maximal. For small PNBs with lifetimes from 5 to 100 ns, the corresponding pressure rise times and associated frequencies seem to be higher than the resonant frequencies of most acoustic sensors. Therefore, the sensor signal does not show the real pressure, but rather represents a response function of the sensor. In this case, the amplitude of the output signal of the acoustic sensor significantly depends upon its resonant frequency and may cause a significant error (Fig. 7.3C-II). For a specific acoustic sensor, the amplitude of its output also depends upon the number of simultaneously generated PNBs (Fig. 7.3B-II). Therefore, a reliable measurement of the PNB diameter through the amplitude of acoustic response assumes that a single PNB is generated.

7.2.1 *Mechanism of PNB generation*

A PNB results from several transient nano-scale processes (Fig. 7.4). Plasmonic photon–phonon–heat conversion, nanoparticle surface–liquid heat transfer, evaporation, and vapor bubble expansion and collapse.

PNB generation is determined by two basic parameters: laser excitation duration — from picosecond to continuous, and the state of the nanoparticle system — from isolated GNPs to their suspensions and to large clusters of aggregated GNPs. We studied the basic and most available type of plasmonic GNPs — gold colloids [82] — under resonant and off-resonant single pulse excitation. Individual PNBs were directly detected and quantified to provide reliable and easy-to-interpret data, which could be extrapolated to various applications. To study the influence of the excitation pulse duration on PNB generation, we employed isolated GNPs and single laser pulses at 532 nm, which is close to the wavelength of plasmon resonance in 60-nm gold NSPs. Individual GNPs and PNBs in water were imaged via methods described above (Fig. 7.5A–B).

Under the fixed laser fluence applied (66 mJ/cm^2), PNBs were observed for all three picosecond pulse durations from 20 ps to 400 ps, but were not observed for the nanosecond pulses (Fig. 7.5D). The maximum PNB lifetime

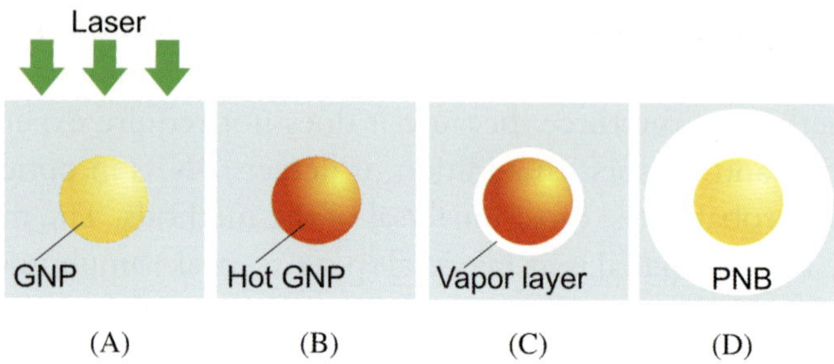

Figure 7.4. A chain of processes involved into the photothermal generation of PNBs: (A) exposure of GNP by short laser pulse, (B) heating of GNP, (C) formation of the initial vapor layer and (D) the expansion of vapor bubble. GNP = gold nanoparticle; PNB = plasmonic nanobubble.

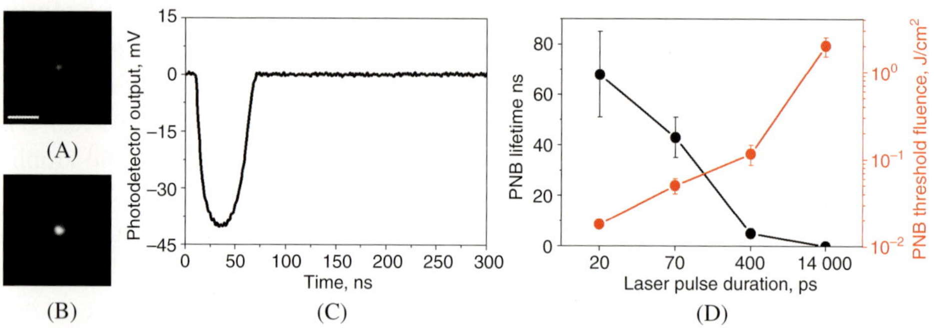

Figure 7.5. Generation of PNBs around isolated GNPs in water [83]. Optical scattering time-resolved images of an individual gold 60-nm sphere in water (A) and of PNB (B) generated around the same sphere in single 20-ps pulse excitation at 532 nm. Scale bar: 2 μm. (C): Optical scattering time-response of the same PNB as shown at (B) was obtained with the (continuous wave) c.w. probe laser at 633 nm. The lifetime is measured as the duration at the level of 0.5 of the maximum amplitude of the PNB-specific signal. (D): Dependences of the PNB lifetime (*black curve*) under specific excitation wavelength of 532 nm and fluence of 66 mJ/cm^2 and dependence of the PNB generation threshold fluence (*red curve*) at the excitation wavelength of 532 nm upon the excitation duration. PNB = plasmonic nanobubble.

(and hence the maximum energy efficacy of PNB generation) was observed for the shortest pulse of 20 ps. Apparently, the fluence applied, 66 mJ/cm^2, was above the PNB generation threshold for 20 and 70 ps pulses, close to the threshold for the 400 ps pulse and below the threshold for the 14 ns pulse. We next measured the PNB generation threshold fluence as a function of the laser pulse duration (red curve in Fig. 7.5D). The minimum threshold, which is less than 18 mJ/cm^2, was observed for the shortest, 20-ps pulse. The 20-fold increase in the pulse duration from 20 ps to 400 ps increases the threshold fluence 6.4-fold to 115 mJ/cm^2, while for the nanosecond pulse, the PNB generation threshold fluence increases by almost two orders of magnitude. Thus, the laser pulse duration radically influences both the energy efficacy and the threshold fluence of PNB generation. To summarize, short picosecond pulses provide the best energy efficacy of photothermal PNB generation.

The isolated GNPs studied above are not typical for real-world photothermal applications, where GNPs are used in the form of suspensions or/and aggregated clusters. For example, active bio-targeting of GNPs results in their intracellular clustering [43–47, 50, 51, 84]. Many industrial applications employ suspensions of GNPs [85, 86, 87–89]. We therefore studied the PNB lifetimes (a metric of their maximal size and the energy efficacy of PNB generation) as a function of the cluster size for isolated GNP clusters (Fig. 7.6A–C) and the concentration for GNP suspension (Fig. 7.6D) under a specific level of the laser fluence, 66 mJ/cm^2, above PNB generation threshold fluence. The PNB lifetime was measured around individual clusters and for the individual PNB in response to a single laser pulse (Fig. 7.6B). The PNB lifetime increased with the GNP cluster size almost linearly and exceeded that of a single GNP by approximately five-fold (black curve in Fig. 7.6C). This significant increase in PNB generation efficacy under identical laser fluence can be explained by: (1) the enhancement of plasmonic properties and the increased optical absorbance of the cluster [90], and (2) the coalescence of the initial vapors around GNPs into the joint vapor blanket around the whole cluster, which increases the bubble radius and hence reduces the surface tension pressure [18]. The clustering of GNPs results in an almost four-fold decrease in the PNB threshold fluence compared to that for a single GNP: from 18 mJ/cm^2 to 5 mJ/cm^2 (red curve in Fig. 7.6C). This effect provides a unique opportunity to selectively generate PNBs only around the largest GNP clusters at the minimum laser fluence, and without generating PNBs around isolated GNPs or their small clusters. In biomedical applications, this cluster size effect allows a dramatic increase in the target cell specificity of PNBs compared to that of GNPs [43–45, 50, 83, 91].

In addition to the resonant excitation of GNPs considered above, we studied PNB generation at off-resonant near-infrared wavelengths in the range of 700–800 nm, which are far from the plasmon resonance of gold colloids and correspond to poor optical absorbance: about 1% of that at the resonance wavelength [92]. The off-resonant excitation of isolated GNPs with a 20-ps pulse revealed an ultra-narrow, just 3 nm wide peak at 782 nm (Fig. 7.7A). In comparison, the spectral width of the optical absorption spectrum of the GNPs employed in our experiments is about 100 nm, with the peak wavelength close to 532 nm. Under identical laser fluence, the PNB lifetime at

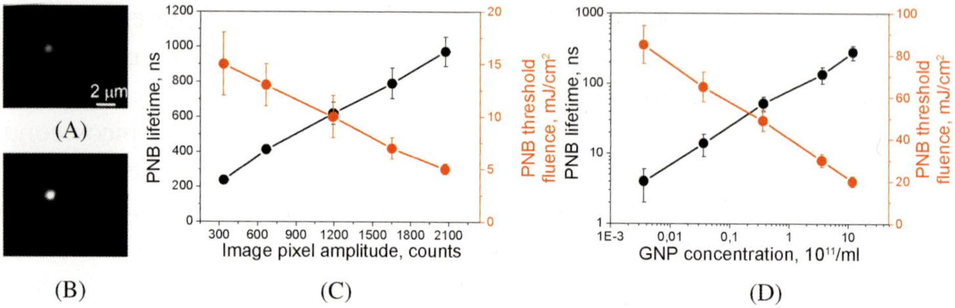

Figure 7.6. Optical scattering time-resolved images of a GNP cluster in water (A) and of PNB (B) generated around the same cluster in single pulse excitation at 532 nm. (C): The lifetime of individual PNBs (*black curve*) generated around isolated GNP clusters in water as a function of the GNP cluster size (quantified via the pixel image amplitude of the optical absorbance of a GNP cluster) under specific duration (20 ps) and fluence (66 mJ/cm^2) of a single laser pulse and the PNB generation threshold fluence (*red curve*) as a function of the GNP cluster size under specific duration (20 ps) of a single laser pulse. (D): The lifetime of PNBs (*black curve*) generated in water suspension of GNPs at laser fluence of 66 mJ/cm^{-2} and the PNB generation threshold fluence (*red curve*) as a function of the GNP concentration. All measurements are performed for 20 ps excitation pulse of 532 nm wavelength [83]. GNP = gold nanoparticle, PNB = plasmonic nanobubble.

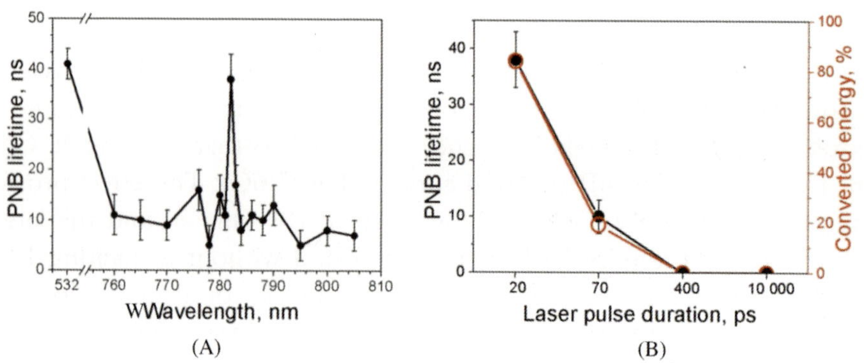

Figure 7.7. Off-resonant generation of PNBs around isolated GNPs in water obtained for 60-nm gold spheres in water irradiated by the 20-ps laser pulse at fluence 66 mJ/cm^2. (A): PNB lifetime spectra for individual gold 60-nm spheres in water. (B): Experimentally measured PNB lifetime (*black curve*) and theoretically calculated fraction of the energy of the laser pulse, which coincides with the estimated time interval from the onset of the vapor formation conditions near the GNP surface to the beginning of active expansion of the PNB — "off-resonant PNB energy" (*red curve*) [83]. PNB = plasmonic nanobubble.

782 nm was comparable to that at 532 nm, thus implying that the levels of optical absorbance of GNPs at 532 nm and 782 nm are similar. Interestingly, we observed earlier a similar effect for longer pulses [92]. Therefore, as with resonant excitation, we studied the influence of the duration of the near-infrared laser pulse on PNB generation (Fig. 7.7B). At a fixed laser fluence, 66 mJ/cm^2, the PNB lifetime rapidly dropped with the pulse length and no PNBs were observed for 400 ps and 14 ns pulses. Compared to resonant excitation (Fig. 7.5D), we observed a more radical dependence upon the laser pulse duration: it was impossible to induce PNBs with pulses longer than 70 ps under laser fluences up to 300 mJ/cm^2.

The above represents a practical guide to PNB generation rather than a complete analysis of all PNB-related mechanisms. The results obtained for 60-nm gold colloids in water are fairly representative for the range of GNP sizes between 10–200 nm and for other liquids. We show how to optimize PNB generation in conditions typical for the majority of GNP aplications. We did not study the influence of GNP size and shapes because these factors were extensively analyzed by us previously [18, 20]. Generally, the solid spheres of a smaller size or larger than 200 nm would require more fluence to generate identical PNBs under the resonant excitation [18]. The pulsed excitation of GNPs of other shapes, such as rods and shells, in PNB mode was reported previously [93–95]. The demonstrated results suggest several practical conclusions:

- The maximum energy efficacy of the generation of PNBs is achieved with picosecond laser pulses.
- Among all nanoparticle systems studied, the clusters of aggregated GNPs provide the maximal energy efficacy of PNB generation compared to isolated GNPs or their suspensions.
- For short laser pulses, the generation of PNBs occurs at the ambient bulk temperature.
- Short picosecond off-resonant optical excitation of gold colloids with solid spherical NPs results in efficient generation of PNBs at a specific off-resonant near-infrared wavelength and in a very narrow spectral interval.

7.3 Target Cell Specificity of PNBs versus NPs

Nanomedicine promises unique abilities to support diagnostic, therapeutic, and theranostic functions at nano-scale, providing molecule- and cell-level resolution, specificity, and selectivity. These functions are usually mediated through NPs that have to be delivered to specific molecular and cellular targets associated with a certain pathology or diagnosis. However, this strategic advantage of nanomedicine is compromised by the principal limitation in NP targeting. This is that no current method can deliver NPs only to target cells and molecules because some amount of NPs always accumulates non-specifically in non-target cells, thus reducing the specificity and selectivity of nanomedicine. The limited specificity of NP targeting, in turn, requires higher loads of NPs in order to achieve the desired diagnostic or therapeutic effect [96–102]. The high loads of NPs required for therapy delivery further induces non-specific accumulation and cause toxicity issues because most NPs are inorganic and of non-biological origin. Increased NP loads and low targeting specificity result in macro- rather than nano-resolution for NP medicines. Among NPs, those based on gold have been exploited most often. GNPs have low toxicity [103, 104] and have been used in a relatively wide spectrum of biomedical applications: optical [101, 105–107] and photoacoustic [108, 109] diagnostics, drug delivery [106, 110–113], the direct destruction of target cells through photothermal effects [15, 33, 38–40, 96–101, 114–119], or in combination with chemotherapy [120, 121]. In addition, GNPs were applied in combination with other NPs, such as drug carriers and diagnostic labels [122–128].

The specificity of GNP targeting to specific (target) cells was improved by chemically attaching target-specific vectors to the GNPs, thus coupling GNPs to specific target receptors at the cellular membrane [44, 51, 84, 106, 129–135]. This active targeting is more effective compared to passive targeting with "bare" non-functionalized GNPs. However, many cellular receptors are widely expressed, albeit at very different levels, on target and non-target cells (bio-heterogeneity). As a result, a considerable number of actively targeted GNPs will still get to non-target cells through various non-specific mechanisms [40, 96–101]. Therefore, the targeting of GNPs to cells so far cannot provide sufficient specificity, which slows the translation of nanomedicine to clinics.

The high biomedical specificity of GNP-based effects could be achieved by using PNBs instead of GNPs to efficiently discriminate between GNPs in target and non-target cells. We studied the six cell models representing four molecular targets (Table 7.1) [91]. Each pair of cells represented target cells with high levels of expression of specific molecular target and non-target cells with low levels of the expression of such molecular targets.

Since the largest NP clusters can be selectively formed in target cells through the receptor-mediated endocytosis of NPs [44, 51, 84, 106, 129–134], we further studied the GNP cluster-PNB mechanism in living cells in order to compare the cellular specificity of GNPs and PNBs under identical conditions of GNP targeting and optical excitation. Several different molecular targets were investigated *in vitro* in cell systems that included cells with a high

Table 7.1. Cell models and conditions of their treatment with GNP and laser pulses. NSP = gold spheres, NS = gold shells, EGFR = epidermal growth factor receptor.

				Incubation conditions		Laser pulse parameters		
Molecular target	GNP	Vector	Cells: Target/ non-target	GNP concentratio, GNP/ml	Incubation time	Duration, ps	Wavelength, nm	Fluence, mJ/cm^2
EGFR	50 nm NS	Panitumumab	HN31/NOM9	$2.4*10^{10}$	24 h	70	820	30
EGFR	60 nm NSP	Panitumumab	HN31/NOM9	$2.4*10^{10}$	24 h	70	532	60
EGFR	60 nm NSP	C225	HN31/NOM9	$2.4*10^{10}$	24 h	70	532	60
—	60 nm NSP	none	HN31/NOM9	$2.4*10^{10}$	24 h	70	532	60
EGFR	110 nm NS	C225	C42B/HS5	$1.2*10^{11}$	30 min	500	787	38
MUC1	60 nm NSP	214D4	HES/HS5	$2.4*10^{10}$	1 h	500	532	40
PSMA	60 nm NSP	Anti-PSMA	C42B/HS5	$1.2*10^{11}$	30 min	500	532	60
CD3	60 nm NSP	OKT3	CD3 + T-cells/ CD3-BMC	$1.2*10^{11}$	30 min	70	532	37
CD3	60 nm NSP	OKT3	J32/ JRT3-T3.5	$1.2*10^{11}$	45 min	500	532	63
EGFR	60 nm NSP	C225	A549/ Fibro-blasts	$6*10^{10}$	30 min	500	532	30

level of molecular target expression (target cells) and a low level of the expression of the same molecular target (non-target cells).

We studied cell models representing lung (A549), head and neck (HN31), prostate (C4-2B), epithelial (HES, a WISH/HeLa derivative), and blood (Jurkat J32) cancers, and also human T-cells [91]. Both solid gold nano spheres (NSPs) and gold nanoshells (NSs) were conjugated to target-specific antibodies (Table 7.1) and were administered under concentrations and incubation times that were experimentally optimized for each cell model for maximal uptake of GNPs. After removing unbound GNPs, the accumulation of GNPs

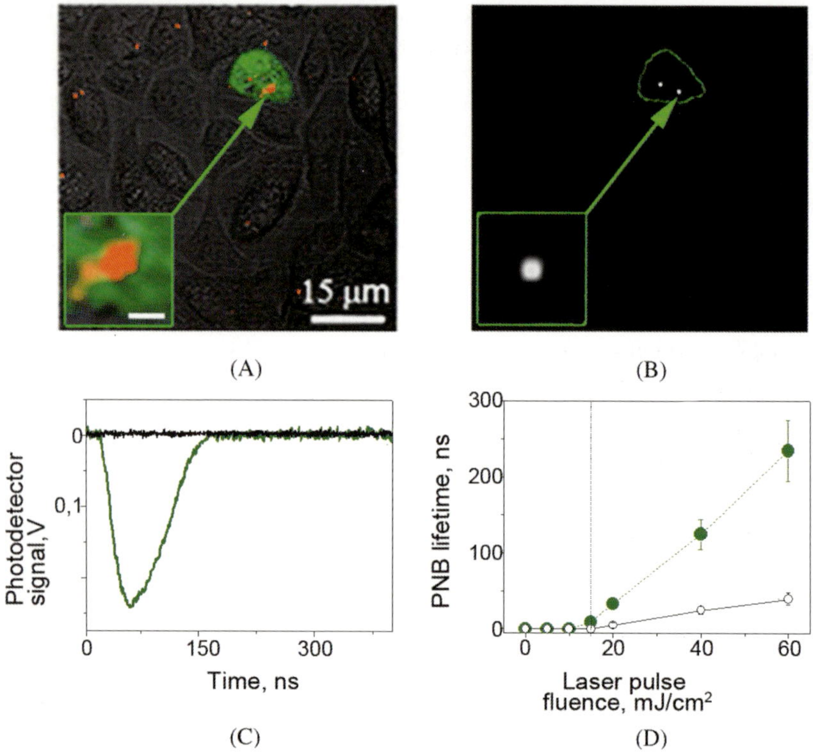

Figure 7.8. Images and signals of GNPs and PNBs in co-culture of target (HN31, labeled with Green Fluorescent Protein for identification) and non-target (NOM9) cells identically treated with 60 nm gold NSP-C225 conjugates (specific to EGFR that is over-expressed in target cells). (A): Overlay of bright field, fluorescent and scattering images shows target cells (*green*) and GNPs (*red*) that can be found in both types of cells (the arrows show GNP clusters in non-target cells). (B): Time-resolved scattering image of the same field shows PNB images (*bright white spots*) only in target cells. (C): Optical scattering time-responses of individual target (*green*) and non-target (*black*) cells show the PNB-specific signal only for target cell and the definition of the PNB lifetime of PNBs; time is measured from the moment of the exposure to the excitation laser pulse. (D): Dependence of the maximal size of PNB (measured as its lifetime through its optical scattering time-response) upon the fluence of excitation (pump) laser pulse (70 ps, 820 nm) for target (*green*) [43] and non-target (*hollow*) cells.

in individual cells was imaged and measured through GNP-specific optical scattering (Fig. 7.8A).

Similar to previous experiments, we measured cell-averaged levels of scattering image amplitudes in target and non-target cells (Fig. 7.9, row a). In all six cases, we observed a higher level of GNP signals in target cells, but all non-target cells also showed a significant level of GNP uptake and formation of GNP clusters (Figs. 7.8A and 7.9A), so that the ratio of the GNP signal for target versus non-target cells was below 10. However, higher pixel image amplitudes in target cells indicated the formation of the largest GNP clusters in target cells.

Next, target and non-target cells were identically treated with single laser pulses within the range of pulse fluences for PNB generation around GNP clusters. For each cell model, we experimentally determined the level of excitation pulse fluence that provided the generation of PNBs mainly in target cells and did not induce PNBs in non-target cells (Fig. 7.8C,D). The optical scattering images and time responses of individual cells were processed to compare the corresponding metrics for GNP accumulation (Fig. 7.9, row a) and PNB generation (Fig. 7.9, row b,c) in target and non-target cells. Compared to GNP signals, the PNB signals showed a much higher discrimination between target and non-target cells in all six cases. The cellular specificities of GNPs and

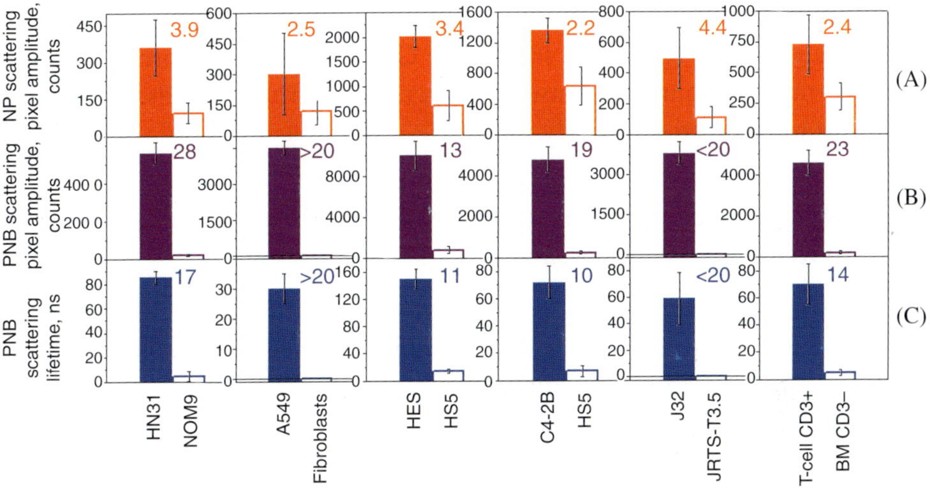

Figure 7.9. Cell population-averaged levels of optical scattering signals obtained for individual target (*solid bar*) and non-target (*hollow bar*) cells in six cell models represented by target/non-target cells/molecular targets: Squamous cell carcinoma, HN31/NOM9/EGFR (treated with 50 nm NS-Panitumumab conjugates); Lung cancer, A549/Fibroblast/EGFR (treated with 60 nm NSP-C225 conjugates); Epithelial cancer, HES/HS5/MUC1 (treated with 60 nm NSP-214D4 conjugates); Prostate cancer, C2-4B/HS5/PSMA (treated with 60 nm NSP-anti-PSMA conjugates); Leukemia, J32/JRT3-T3.5/CD3 and human T-cells, T-cell/BM/CD3 (treated with 60 nm NSP-OKT3 conjugates) for: row (A) (*red*): GNP amplitude of scattering image of GNPs (a metric for the uptake of GNPs by cells); row (B) (*purple*): time-resolved scattering image amplitudes of PNBs; row (C) (*blue*): PNB lifetimes. The ratio of the signals for target/non-target cell is shown for each parameter and cell model and indicates the cellular specificity of GNPs (row (A)) and PNBs (rows (B, C)) [91]. PNB = plasmonic nanobubble, NP = nanoparticle.

PNBs were quantitatively shown through the ratios of the target cell signals to the corresponding signals in non-target cells (shown as colored numbers in each frame of Fig. 7.9). Compared to GNPs, the PNBs improved cellular specificity in some models by more than one order of magnitude. While the non-target cells showed significant uptake of GNPs and even their potential aggregation into small clusters, no PNBs, or very small ones, were observed in non-target cells under identical treatment conditions (Figs. 7.8 and 7.9).

The difference in cellular specificity of GNPs and PNBs can be clearly seen in experiments with a co-culture of target (labeled with green fluorescent protein for identification) and non-target cells (Fig. 7.8). At a specific fluence of the excitation laser pulse (25 mJ/cm^2 at 778 nm), only target cells yielded PNBs, while even adjacent non-target cells with GNPs did not. Such a difference between GNP and PNB signals was observed for all six cell models, adherent (HN31, HES, A549) and suspension (C4-2B, T-cells, Jurkat) cells, and for all molecular targets, receptors (EGFR, CD3, PSMA) and glycoproteins (MUC1). These results indicate the universal nature of the high cellular specificity of PNBs compared to that of GNPs [91]. Therefore, PNB provided better discrimination between target and non-target cells, even when such cells were heterogeneously mixed.

7.4 PNB-Induced Ejection (Release) of an Encapsulated Payload

The high target cell specificity and *in situ* dynamic tunability of PNBs were applied to support an entirely novel mechanism of the intracellular ejection of liposomal cargo. Co-localized with liposomes, PNBs disrupt them almost instantaneously and, thus, eject their cargo into the cytoplasm (Fig. 7.10B) [50, 136]. We studied this in the head and neck squamous cell carcinoma (HNSCC) HN31 cell line. Cancer cells were concurrently targeted by using the clinically-validated target, epithelial growth factor receptor (EGFR), with a C225 antibody [137] conjugated to 60-nm solid NSPs and by liposomal green fluorescent dye conjugated with 2C5 antibody [138]. A high level of EGFR expression resulted in the endocytotic formation of intracellular GNP-liposome clusters that revealed optical scattering of GNPs (Fig. 7.10C, left), but showed no fluorescence due to its quenching in intact liposomes (Fig. 7.10C, center and right). Next, we exposed the cells to a single picosecond laser pulse of the low fluence of 40 mJ/cm^2 at 532 nm. We simultaneously detected PNBs optically with the laser pulse (Fig. 7.10D, center). The PNB location in cells coincided with that of the GNP cluster (Fig. 7.10D, left), and its optical time–response showed PNB expansion and collapse within 50–60 ns (Fig. 7.10D, right). The duration

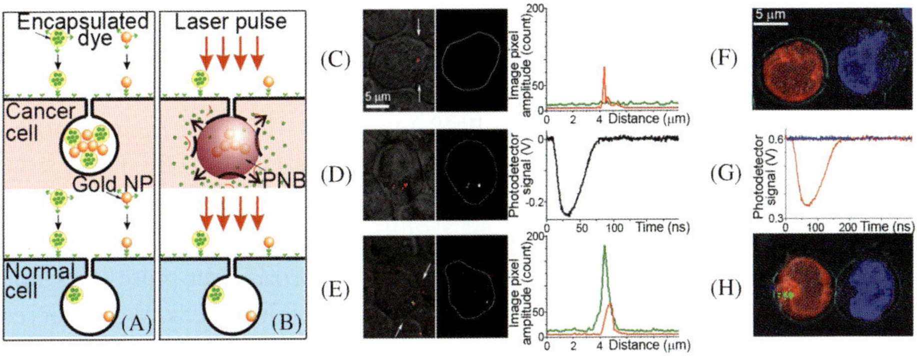

Figure 7.10. PNB-enhanced intracellular on-demand ejection (release) of encapsulated Calcein Green dye. (A) Systemic administration of low doses of GNPs and drug-loaded nanocarriers results in the large mixed intracellular clusters of GNPs and drug nanocarriers, which are self-assembled by cancer cells (top), but not by normal cells (bottom). (B) Local administration of a single laser pulse results in the cancer cell-specific generation of a PNB that delivers the localized intracellular mechanical impact and ejection of the encapsulated drug (green dots) into cytoplasm (in normal cells non-specific uptake of fewer GNPs is insufficient to generate PNBs and no drug release is triggered). Confocal microscopy images of living cells treated with green fluorescent dye-loaded liposomes conjugated with 2C5 antibody (detected as green in fluorescent mode) and GNP-C225 conjugates (detected as red in scattering mode). (C) before the laser pulse (green fluorescence is quenched and dimmed in intact liposomes). (D) during the laser pulse: clusters of GNPs (left) following exposure to a single picosecond laser pulse (532 nm, 40 mJ/cm^2) generate co-localized PNBs (center, optical scattering time-resolved image) that are quantified through their optical scattering time-response (right). (E) immediately after a single laser pulse green fluorescence coincides with GNP cluster and PNB. (F) Cellular specificity of PNB-enhanced intracellular dye release. Merged bright field and confocal tri-color fluorescent image of co-culture of EGFR-positive (*red*) and EGFR-negative (*blue*) cells treated with conjugates of GNPs and liposomal green dye (dimmed due to the fluorescence quenching in intact liposomes). (G) Optical scattering time-responses show the selective generation of PNBs only in red EGFR-positive cancer cells following simultaneous exposure of all cells to a single broad laser pulse (532 nm, 40 mJ/cm^2, *red trace* for EGFR-positive cell and *blue trace* for EGFR-negative cell). (H) Cells at 5–10 min after exposure to a single laser pulse show the localized release of green dye only in EGFR-positive cell due to the selective generation of small PNB that disrupted the dye nanocarriers and endosome and locally ejected the dye (*green*). Non-specific uptake of GNPs in EGFR-negative cell did not generate PNBs and thus did not trigger the dye release [50]. NP = nanoparticle, PNB = plasmonic nanobubble.

of the response, the PNB lifetime, characterized its maximal diameter [18, 69]. Immediately after the PNB generation, re-imaged cells yielded a bright liposome-specific green fluorescence in sub-micrometer zones co-localized with the GNP clusters and PNBs (Fig. 7.10E). No released dye was observed in PNB-negative cells. The small maximal size of the PNB (200–400 nm for a

PNB with a 60-ns lifetime [18]) and its mechanical, non-thermal, nature [18] coupled with the high speed of the PNB expansion (30 ns) and dye ejection resulted in a high intracellular dye concentration and localization.

The cancer cell specificity of the PNB-induced payload release mechanism was studied in a mixed suspension of EGFR-positive HN31 cells (Fig. 7.10F, red) and EGFR-negative suspension-stable J32 cells (blue, Fig. 7.10F). The cells were mixed and treated identically with the conjugates of GNPs and liposomal green dye, as in the previous experiment. A broad laser pulse (532 nm, 40 mJ/cm^2) simultaneously irradiated hundreds of mixed cells. We observed PNBs with a lifetime of 55–60 ns in HN31 (red trace), but none in J32 (blue trace) cells (Fig. 7.10G). The threshold nature of PNBs prevented their generation in EGFR-negative cells because these cells did not form sufficiently large GNP clusters and, therefore, had a PNB generation threshold fluence [43, 50, 91, 136] above the level of the applied laser fluence. The second tricolor imaging, obtained immediately after laser treatment, revealed localized intracellular bright green fluorescence in HN31 cells, but none in J32 cells (Fig. 7.10H). Image cytometry of 150–180 HN31 and J32 cells showed (i) that the onset of local green fluorescence coincided with PNB generation, (ii) a high intracellular localization of the released payload, and (iii) high cancer cell specificity of the PNB-induced release (ejection) of the liposomal payload [50].

7.5 Therapeutic Responses to Drugs, GNPs, Laser Pulses, and X-rays *In Vitro*

The long-term therapeutic effects to drugs, GNPs, laser pulses, and X-rays were analyzed *in vitro* for cancer (HN31) and normal (NOM9) cells [50]. We used liposomal doxorubicin (Doxil) [139] and micellar paclitaxel [140] as drug nanocarriers. Standard Doxil liposomes and the paclitaxel-loaded micelles were functionalized with antibody and incubated separately or simultaneously with GNP-255 conjugates. The used nanocarriers alone revealed low cancer cell specificity and the high drug resistance of HN31 cells (Fig. 7.11A).

PNBs alone revealed the cancer-specific therapeutic effect only at the high laser fluence associated with large, cell-disrupting [44] PNBs with a lifetime of more than 200 ns, while smaller PNBs were non-lethal (Fig. 7.11B). In contrast, the co-localization of small PNBs with drug nanocarriers in cancer cells significantly improved the therapeutic efficacy (Fig. 7.11D, mode D+PNB), especially in the short term. Thus, PNB-induced intracellular release of soluble doxil and insoluble paclitaxel amplified both the efficacy and specificity of chemotherapy. At the same time, the absence of PNBs in normal cells coupled with entry drug doses reduced to 2–3% of their standard dose effectively spared normal cells under identical treatment (Fig. 7.11E). However, this therapeutic

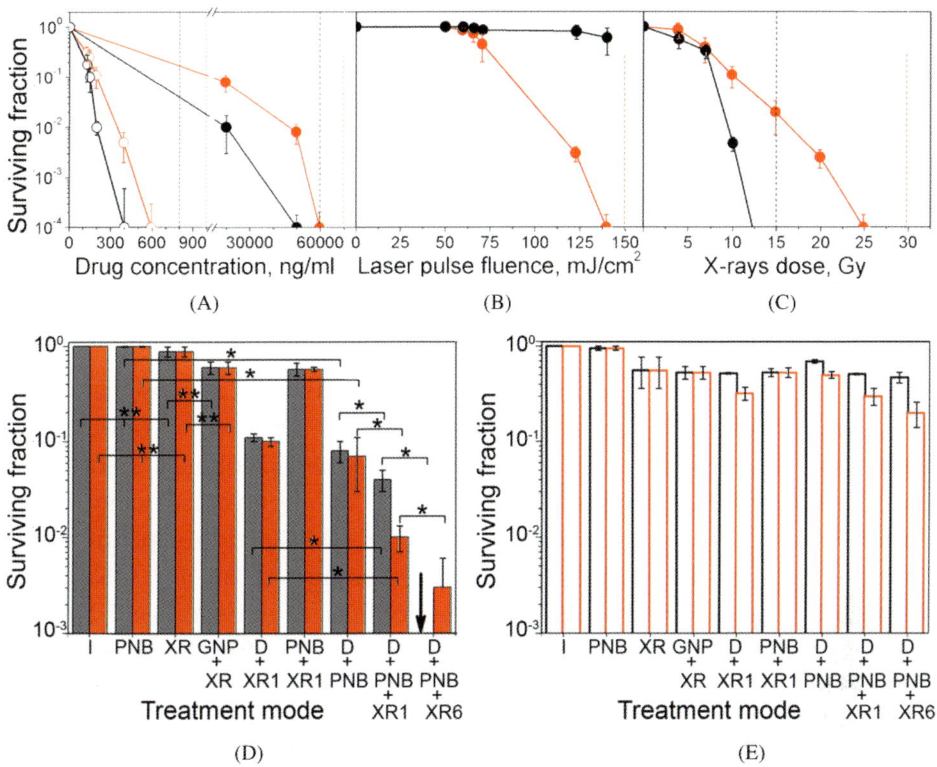

Figure 7.11. Surviving fraction of cancer HN31 cells (*red circles*) and normal NOM9 cells (*black circles*) in response to single therapeutic agents as function of: (A) drug concentration (Doxil: *solid*; Micellar paclitaxel: *hollow*), (B) PNB via the laser pulse fluence, (C) X-ray dose. (D) Surviving fraction of HN31 cells as a function of the treatment mode (*I* — intact cells; *PNB* — GNP-C225 conjugates and single laser pulse (780 nm, 45 mJ cm^{-2}); *XR* — X-rays, 4 Gy; *GNP+XR* — GNP-C225 and X-rays; *D+XR* – drug (*grey* — Doxil, 2 μg mL^{-1}; *red* — Paclitaxel, 33 ng mL^{-1}) and X-rays; *PNB+XR1* — GNP-C225, single laser pulse and X-rays 1 h after laser exposure; *D+PNB* — GNP-C225, drug and laser pulse; *D+PNB+XR1* — GNP-C225, drug, laser pulse and X-rays 1 h after laser exposure; *D+PNB+XR6* – GNP-C225, drug, laser pulse and X-rays 6 h after laser exposure), black arrow indicates a zero surviving fraction. (E) Surviving fraction of normal NOM9 cells as a function of the treatment mode identical to those in (d) [50]. Data are expressed as mean ± S.D. for the three independent experiments, $*p < 0.05$, as determined by two-tailed student's *t*-test. $**p < 0.05$, not significant.

amplification requires effective co-localization of the drug nanocarriers and PNBs. Without such co-localization, the mechanical impact of PNBs apparently does not reach the otherwise randomly internalized drug nanocarriers, resulting in a relatively poor therapeutic effect.

X-rays alone without drugs and GNPs were efficient only at high and unsafe doses due to the resistance of HN31 (Fig. 7.11C). GNPs improved the therapeutic efficacy of X-rays due to the emission of local secondary electrons [141, 142], but only incrementally (Fig. 7.11D, the mode GNP+XR). The combination of X-rays and therapeutic nanocarriers resulted in a predictable therapeutic effect (Fig. 7.11D, the mode D+XR1) due to the well-known synergy of drugs and X-rays [143, 144]. Neither the tested therapeutic components alone, nor their standard combinations described above, prevented the growth of resistant and aggressive cancer cells.

Next, we evaluated the therapeutic efficacy of quadrapeutics. Briefly, quadrapeutics is administered in three simple steps, each of which further amplifies the therapeutic efficacy and specificity [50]: nanoparticle cluster formation, threshold activation of these nanoclusters with a single laser pulse for mechanical cell disruption and drug release, and activation of these nanoclusters with X-rays for the local amplification of the X-ray dose by nanoclusters. These three steps employ four therapeutic components: colloidal gold, drug-loaded nanocarriers, laser pulses, and X-rays. Step 1 involved nanoparticle cluster formation via systemic administration of GNPs and therapeutic nanocarriers employed EGFR-specific antibody-conjugated 60-nm GNPs and therapeutic drug-loaded nanocarriers. Intracellular aggregation of GNP and drug carriers into a mixed cluster was achieved by activating the universal cell defense, receptor-mediated endocytosis. Cells self-assemble nanoclusters by (i) accumulating nanoparticles at the membrane as a result of the interaction of the nanoparticle-conjugated antibody with EGFR, and (ii) through receptor-mediated endocytosis of membrane-bound nanoparticles [43, 44, 51, 91]. As a result (Fig. 7.12C,D), 5–50 GNPs are aggregated into a cluster in endosomes or lysosomes, and the largest clusters are formed by the most aggressive cancer cells due to the maximal GNP accumulation at the membranes. Due to unavoidable non-specific uptake, some GNPs and drug carriers end up in normal cells and tissues (Figs. 7.12E and 7.13B). Formation of this "Trojan horse" is a key step in quadrapeutics: it does not trigger any therapeutic effect and can remain safe and stable until it is externally activated with a laser pulse and X-rays. Step 2 involves the local generation and detection of PNBs with a short near-infrared laser pulse, mechanical destruction of target cells, and intracellular drug release (Fig. 7.10). Such a rapid and localized release creates a high concentration of the ejected free drug in the cytoplasm of a cancer cell. Step 3 involves the amplification of X-rays that employed GNP-C225 conjugates administered systemically and a single X-ray dose administered locally after the formation of GNP clusters. GNP clusters amplify external X-rays by emitting local secondary electrons upon absorption of X-rays [141, 142]. The X-ray amplification is limited to several micrometers and is maximal for large GNP clusters, which are located in the most aggressive cancer cells. Likewise, the X-ray is minimally amplified for the single, small GNPs taken up by normal cells. Thus, the amplification is limited to the host cancer cell and it cannot damage the nearby normal cells. The administration of all four components (gold colloid, incapsulated drug, near-infrared short laser pulses, and X-rays) as quadrapeutics resulted in a radical amplification of the therapeutic efficacy (Fig. 7.11D, the mode D+PNB+XR1) and specificity (Fig. 7.11E, the mode D+PNB+XR1). An even stronger effect, the full suppression of cancer cell proliferation, was achieved when we applied X-rays 6 h after PNB treatment (Fig. 7.11D, the mode D-PNB-XR6). Identically treated normal cells demonstrated high long-term viability (Fig. 7.11E).

7.6 The Mechanism of Intracellular Self-Amplification of the Therapeutic Strength with Cancer Aggressiveness

The intracellular self-amplification of the therapeutic strength with the cancer aggressiveness was studied by analyzing therapeutic responses, GNP clustering, and PNBs in normal epithelial (NOM9), slow-growing indolent cancer (HN30), and fast-growing aggressive cancer (HN31) cells (both HNSCC [145]) that were all identically treated with quadrapeutics and chemoradiation. Compared to indolent cells, aggressive cells were more resistant to standard chemoradiation (Fig. 7.12A). In contrast, they could not resist quadrapeutics, which completely suppressed their growth.

The therapeutic amplification with cancer aggressiveness was quantified by dividing the surviving fractions of chemoradiation-treated cells to those of quadrapeutics-treated cells and analyzed as the function of the growth rate of intact xenograft tumors induced in mice with the corresponding cancer cells [145] (Fig. 7.12B). The therapeutic amplification coefficient rapidly increased above 100 for aggressive HN31 cells, thus revealing the effect of self-regulation of therapeutic efficacy with cancer aggressiveness. This amplification coefficient increased with the size of the intracellular GNP nanoclusters and PNBs (Fig. 7.12B): normal cells non-specifically internalized single GNPs (Fig. 7.12C), which did not generate PNBs and, therefore, normal cells survived the treatment (Fig. 7.12A); indolent HN30 cells formed medium-sized nanoclusters (Fig. 7.12D) that generated very small PNBs, resulting in a moderate therapeutic amplification; aggressive HN31 cells formed the largest nanoclusters (Fig. 7.12E) and generated maximal PNBs, resulting in the maximal therapeutic amplification.

These *in vitro* experiments reveal quadrapeutics' unique combination of on-demand cancer cell-specific therapeutic amplification of the effect of very low entry doses of drugs, 2–3%, and X-rays, 6–7% of their standard doses, in response to a laser pulse and X-rays, with a reduction in non-specific toxicity [50].

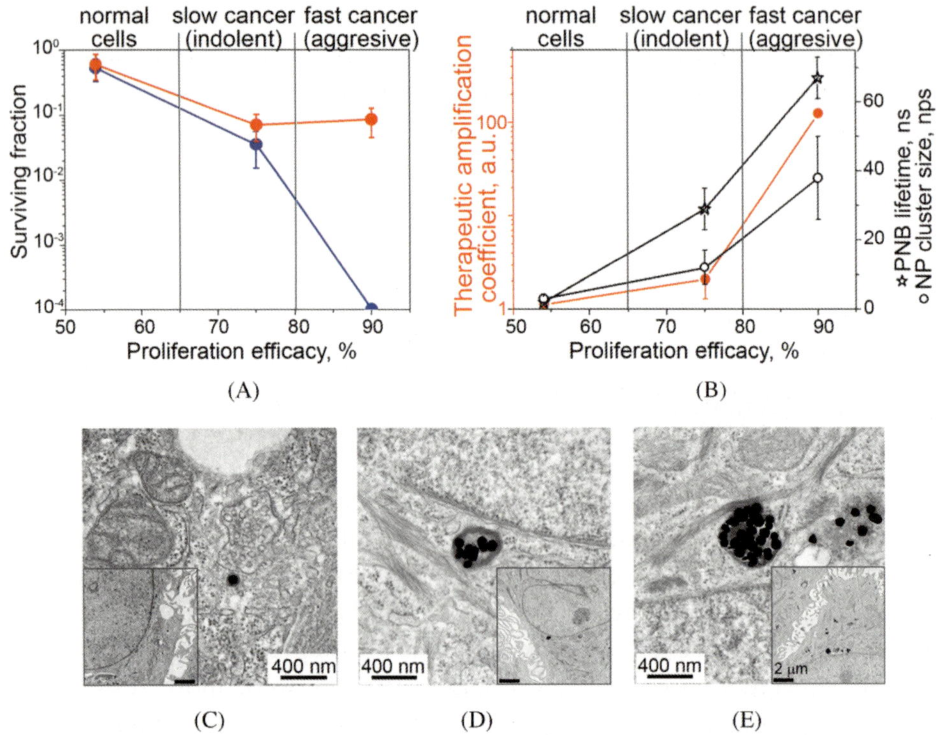

Figure 7.12. (A) Surviving fraction for normal (NOM9), slow growing indolent cancer (HN30) and fast growing aggressive cancer (HN31) cells after standard chemoradiation (*blue*) and quadrapeutics (*red*) treatments measured as function of the cancer aggressiveness (characterized via the growth rate of mouse-xenografted tumors for the corresponding cancer cells), all therapeutic doses as in Fig. 7.11D. (B) Therapeutic amplification coefficient (ratio of the surviving fraction of chemoradiation-treated cells to that of quadrapeutics-treated cells, *red*), population-averaged nanocluster size (measured via the number of nanoparticles in a cluster, *black circle*) and maximal size of PNB (measured via the PNB lifetime, *black star*) in normal (NOM9), slow growing indolent cancer (HN30) and fast growing aggressive cancer (HN31) cells measured as function of the cancer aggressiveness (characterized via the growth rate of mouse-xenografted tumors for the corresponding cancer cells), all therapeutic doses as in in Fig. 7.11D. (C–E) TEM images of GNPs and their clusters in cells after 24 h incubation with GNP-C225 for normal NOM9 cells (C), indolent cancer HN30 cells (D) and aggressive cancer HN31 cells (E), insets show the same cells under low magnification. Sample volume $n = 6$ (for (A) and (B)). Data: mean ± s.d. for the three independent experiments [50]. PNB = plasmonic nanobubble, NP = nanoparticle.

7.7 Generation and Detection of PNBs *In Vivo*

In order to translate the achieved result *in vivo* and to address the impact of tissue on the propagation of light and GNPs, we first studied the delivery of GNPs, laser radiation, and the generation and detection of PNBs in a xenograft mouse model of HNSCC that used the same cancer cell line, HN31.

The formation of GNP clusters in the tumor was imaged and quantified by using TEM and the PNB generation in the tumor slices. GNP-C225 conjugates of solid 60-nm gold spheres were administered systemically (intravenous injection in a tail vein of 200 μL at $4.5*10^{10}$ GNP/mL (0.8 μg/g)). In 24 hours after systemic administering, TEM analysis of the tumor and adjacent normal muscle tissue showed large clusters of GNPs (an average size of 297 ± 135 nm) only in the tumor (Fig. 7.13A), while non-specific uptake resulted in occasional single GNPs in normal tissue (an average size close to that of a single NP, 64 ± 15 nm, Fig. 7.13B). Thus, despite a non-specific uptake of single gold NPs by normal tissues, the sources of PNBs, larger NP clusters, were observed only in tumors and not in normal tissues [45, 50].

Due to the limited optical transparence of real tissue in animals, the PNBs were detected acoustically (Fig. 7.13C). Briefly, the expansion and collapse of a PNB produces pressure pulses that can be remotely detected in opaque tissue in animals with an ultrasound detector as PNB-specific acoustic time–responses (Fig. 7.13D, green). The amplitude of the acoustic response of a PNB was found to be almost linearly proportional to the optically measured lifetime of the same PNB [45]. Thus, we employed the acoustic signal amplitude as an *in vivo* metric of PNB [45, 50, 69].

Next, in order to optimize the laser wavelength for the near-infrared excitation, the acoustic amplitudes of PNBs were obtained as the spectra of the excitation laser wavelength in tumor and normal tissue in animals that received intravenous GNP-C225 conjugates and for an intact tumor in animals that did not receive any GNPs (Fig. 7.13E). Spectra of the tumor in the GNP-treated animal showed a 3-nm wide peak at 780 nm. Other near-infrared wavelengths returned much lower signals comparable to the background level. The normal tissue of the GNP-treated animal showed very small acoustic signals that were also comparable to the background level, while the intact tumor returned no signals under identical optical excitation (Fig. 7.13E). We concluded that the detected acoustic signals of the GNP-treated tumor were associated with GNP clusters that were selectively formed in the tumor cells as we found previously (Fig. 7.13A). We therefore employed the near-infrared wavelength of 780 nm for the experiments *in vivo*.

The specificity and sensitivity of the tumor detection were studied through the amplitude of the acoustic signal of PNBs that was measured as function of tissue type and of the laser excitation fluence. The amplitude of the PNB signal increased with the fluence of the excitation laser pulse (Fig. 7.13F). It should be noted that just a single laser pulse was required to detect a tumor-specific acoustic signal. The duration of this diagnostic process was limited by the speed of the ultrasound detector and was below 1 millisecond.

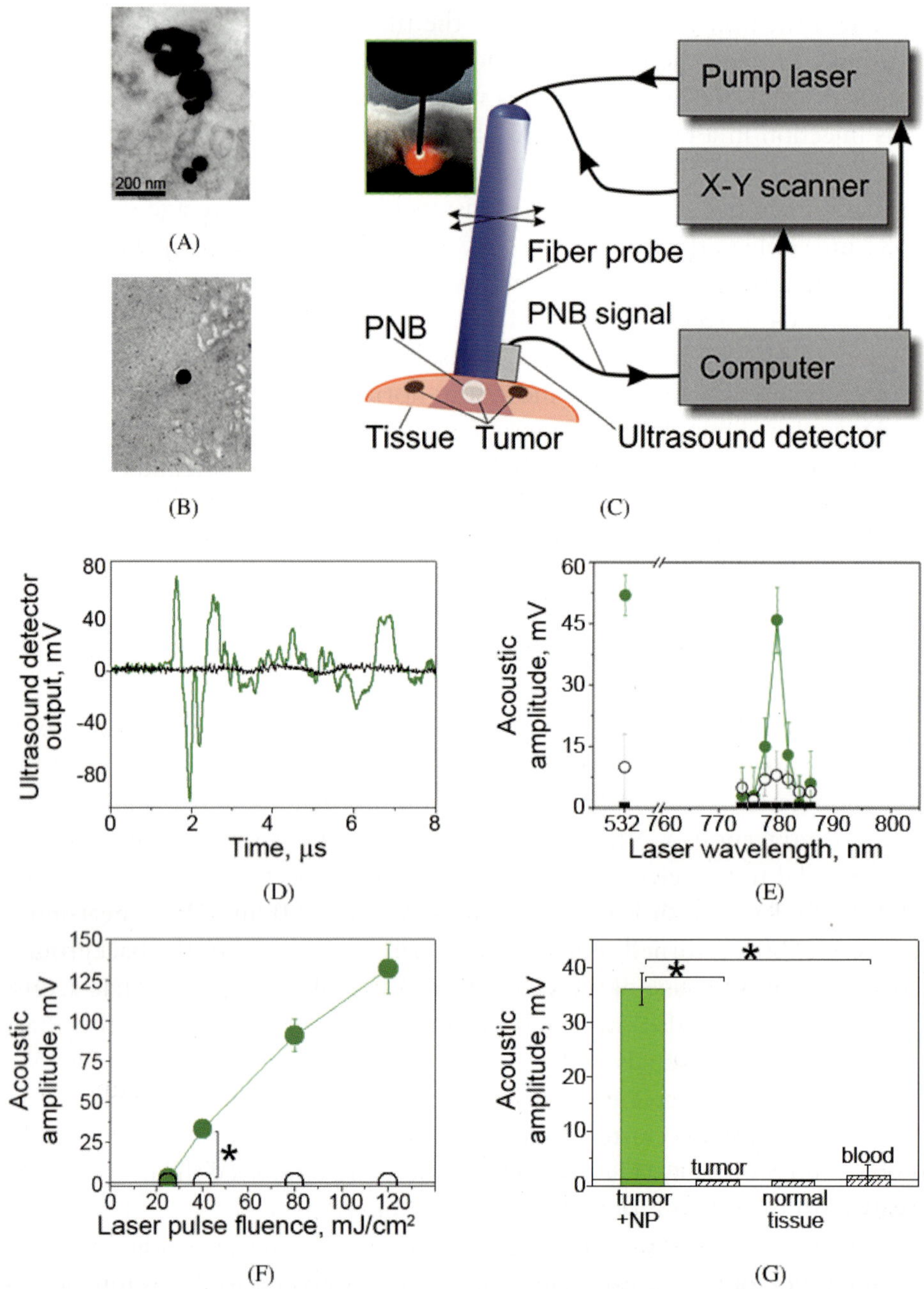

Figure 7.13. PNBs in HNSCC-bearing mouse. TEM image of a tumor (A) and adjacent muscle tissue. (B) in 24 h after systemic injection of NSP-C225 conjugates to mouse. (C) Functional diagram of the experimental setup, optical fiber delivers short laser pulse through the surface of the tissue, PNBs are detected with ultrasound detector, which signals, the scan of the fiber and laser pulse generation are controlled by computer program.

Figure 7.13. (*Continue on facing page*) (D): Acoustic responses to single laser pulses (70 ps, 780 nm, 40 mJ/cm^2) obtained from a tumor (*green*) and adjacent muscle (*black*) in mouse systemically treated with 60-nm solid gold sphere conjugates, GNP-C225. (E): Spectrum of PNB amplitudes of acoustic responses obtained from a squamous cell carcinoma tumor (*solid green*) and normal tissue (*hollow black circles*) of a mouse systemically treated with C225-conjugated 60-nm gold spheres, tumor in intact animal (*solid black squares*) and (40 mJ/cm^2). (F): Dependence of the amplitudes of acoustic responses upon laser pulse fluence (energy per square unit) for the same animals obtained from the tumor (*green*) and adjacent muscle (*black*) (* $p < 0.001$). (G): Acoustic amplitudes obtained during the scans of the tumor, muscle and blood of NP-treated animals and of a tumor of an intact animal (* $p < 0.001$) [45]. PNB = plasmonic nanobubble.

During the laser scans of a tumor and normal adjacent tissue in the GNP-treated animals, the detectable acoustic signal of the PNBs was observed only for the tumor and not for normal muscle tissue and blood (Fig. 7.13G). Tumors in intact animals (that did not receive GNPs) did not yield any detectable acoustic signal under the identical fluence of laser pulse (Fig. 7.13E). Therefore, the optical fluence of 40 mJ/cm^2 was below the bubble generation threshold in blood and other normal tissues for the employed near-infrared wavelength (780 nm) and duration (20–70 ps) of laser pulse. The results demonstrated high sensitivity, specificity, safety, and speed of tumor detection in a single near-infrared laser pulse procedure.

7.8 The Evaluation of PNB Therapy and Diagnostics *In Vivo*

The PNBs and the PNB-based theranostic technology, quadrapeutics, were evaluated for an aggressive and resistant cancer, using HNSCC as a model, with clinically-validated GNPs, encapsulated Doxil, picosecond laser pulses, and X-rays *in vivo*. The basic mechanisms of quadrapeutics and an evaluation of its translational potential for one of the major challenges of HNSCC, microscopic residual disease (MRD) [1], were studied.

We evaluated quadrapeutics against primary HNSCC tumors in two mouse models. In the first model, we injected a low number of HN31 cells s.c. pretreated by GNP-C225, Doxil, PNBs, and X-rays into the mouse flank and monitored the tumor growth (Fig. 7.14). We compared four groups of animals with intact cells, cells treated with 55-ns PNBs alone, cells treated with Doxil (2 μg/mL) and X-rays (4 Gy), and cells treated with quadrapeutics (Doxil-C225, 2 μg/mL, GNP-C225, 2.4*10^{10} GNP/ml, laser pulse, 780 nm, 45 mJ/cm^2 at 24 h after GNP and drug administration, and X-rays, 4 Gy, 6 h after laser treatment). After 15 days, tumors in the animals injected with intact cells reached the maximal size allowed (Fig. 7.14A, left flank). Similarly,

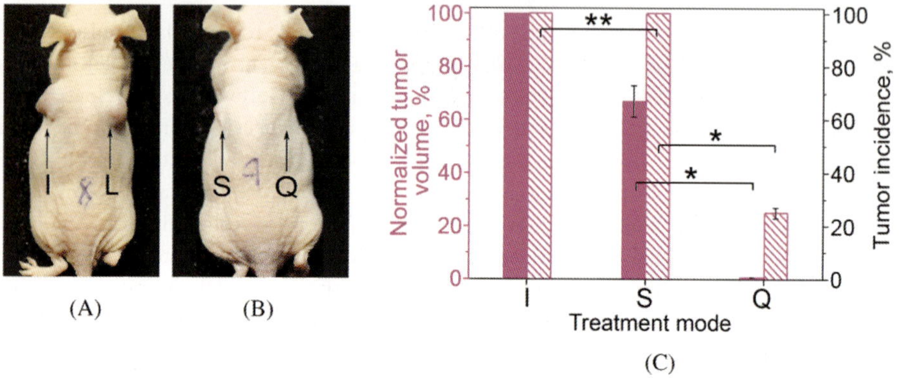

Figure 7.14. Quadrapeutics treatment of HNSCC in mouse models: *in vitro* pre-treated HN31 cells injected into mice: (A, B) images of the animals 15 days after the cell injection: intact HN31 cells (*I*), chemoradiation therapy-treated cells (*S*), PNB-treated cells (*L*) and quadrapeutics-treated (*Q*) cells (Doxil, 2 μg/mL, GNP-C225, 2.4 10^{10} particles/mL, laser pulse, 780 nm, 45 mJ/cm^2, X-rays, 4 Gy), (C) tumor incidence rate (*grey*) and volumes (*magenta*) in three treatment groups measured 15 days after the cell injection. Data: mean ± s.e.m. (*I*: $n = 5$, *S*: $n = 6$, *Q*: $n = 4$) [50]. * $p < 0.05$, ** $p > 0.05$.

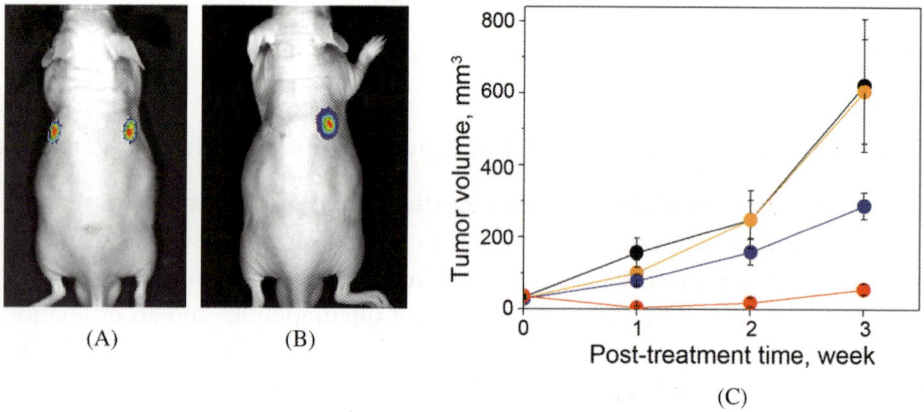

Figure 7.15. Quadrapeutics treatment of HNSCC in mouse models: primary tumor mouse model: bioluminescent images of the animal before (A, and one week after (B) single-time treatments with quadrapeutics (left flank, Doxil-C225, 1 mg/kg and GNP-C225, 0.8 mg/kg, both i.v. injected, laser pulse, 780 nm, 45 mJ/cm^2, local and X-rays, 4 Gy) and chemoradiation (right flank, the same doses of Doxil and X-rays). (C) Time course of the primary tumor volumes after the single-time *in vivo* administration of the following treatments: quadrapeutics (Doxil-C225, 1 mg/kg, GNP-C225, 0.8 mg/kg, laser pulse, 780 nm, 45 mJ/cm^2, X-rays, 4 Gy) — *red* ($n = 11$); chemoradiation (identical to the above drug and X-ray dose) — *blue* ($n = 11$); PNB alone (identical to the above GNP and laser doses) — *orange*: ($n = 4$); untreated animals — *black* ($n = 6$). Data: mean ± S.E.M [50].

large tumors were observed after treatment with PNBs alone (Fig. 7.14A, right flank). Standard chemoradiation did not prevent tumor growth (Fig. 7.14B, left flank) either. In contrast, there were no or very small tumors in the quadrapeutics group (Fig. 7.14B, right flank) that yielded less than 5% of the tumor volume observed in the chemoradiation-treated group (Fig. 7.14C).

In the second model, we induced primary tumors with Luciferase-encoded HN31 cells. The therapeutic response to a single treatment *in vivo* was compared in the quadrapeutics group (GNP-C225, 0.8 μg/g, Doxil-C225, 1mg/kg systemically, laser pulse, 780 nm, 45 mJ/cm^2 in 24 h, and X-rays (4 Gy) in 6 h), the chemoradiation group (same doses of drug and X-rays), the PNB group (same doses of GNPs and laser pulses), and the control group (no treatment). After a single treatment, we monitored all animals weekly for tumor volumes and bioluminescence (Fig. 7.15A,B). The quadrapeutics treatment rapidly suppressed tumor growth after the first week (Fig. 7.15B,C) to 3% of that in the control group and to 6% of that in the chemoradiation group [50]. Small PNBs alone or chemoradiation alone did not prevent tumor growth and demonstrated relatively low therapeutic efficacy (Fig. 7.15C). In contrast, quadrapeutics radically accelerated and improved the effect of chemoradiation by 17-fold after single administration (Fig. 7.15C) [50].

Next, we evaluated the translational potential of quadrapeutics for an intra-operative diagnosis and adjuvant treatment of MRD, a highly aggressive and lethal disease which represents one of the major challenges of HNSCC treatment. A primary HNSCC tumor was grown from HN31 cells. We administered GNP and Doxil conjugates systemically as described above, 24 hours prior to the tumor resection. Immediately after resecting the tumor, we scanned the surgical margins in 20–30 s with a broad near-infrared pulsed laser (780 nm, 45 mJ/cm^2 , Fig. 7.16a). We simultaneously detected PNBs acoustically during the laser scan by using an ultrasound sensor (Figs. 7.13d and 7.16a). We compared the amplitudes of PNB acoustical responses for the primary tumor, surgical margins, and adjacent muscle tissues of GNP-treated and intact animals (Fig. 7.16b). The PNBs detected in surgical margins indicated the presence of MRD (Fig. 7.16b,c), as was later confirmed by observing the local tumor recurrence. At the same time, a standard photoacoustic imaging (Vevo LAZR, Visual Sonics) of the same animals failed to detect MRD in the same animals (Fig. 7.16c). This experiment demonstrated the high cancer sensitivity, specificity, and speed of the PNB-based intra-operative diagnosis of MRD in a biopsy-free and real-time manner [50].

The intra-operative adjuvant treatment of MRD was analyzed for three groups: surgery, surgery with chemoradiation (Doxil, 1 mg/kg, 24 h before surgery and X-rays, 4 Gy, 6 h after surgery), and surgery with quadrapeutics (Doxil, 1 mg/kg concomitantly with GNPs (0.8 μg/g), 24 h before surgery,

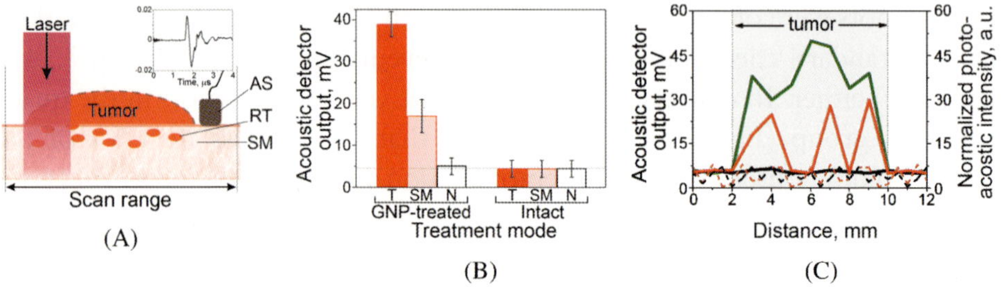

Figure 7.16. Evaluation of quadrapeutics for intra-operative real-time diagnosis of HNSCC MRD. (A) Experimental model of intra-operative diagnosis and treatment of MRD: primary tumor (*tumor*), residual micro-tumor (*RT*) in surgical margins (*SM*), laser beam scan range and PNB detection with acoustic sensor (*AS*) and acoustic response of the PNB (inset). (B) Intra-operative diagnosis of MRD: amplitude photoacousties of the PNB acoustic responses obtained during the laser scans in GNP-treated and untreated animals for: primary tumors before their resection (*T, red*), surgical margins immediately after the tumor resection (*SM, pink*) and adjacent normal tissue (*N, white*); horizontal grey line shows the background level of acoustic signal. Data: mean ± s.e.m (n = 3). (C) Scans of GNP-treated animal: PNB signal amplitudes for primary tumor (*solid green*), surgical margins after tumor resection (*solid red*) and primary tumor in intact animal that was not treated with GNPs (*solid black*), standard photoacoustic small imaging system signals for the same animals (*dashed lines*) [50].

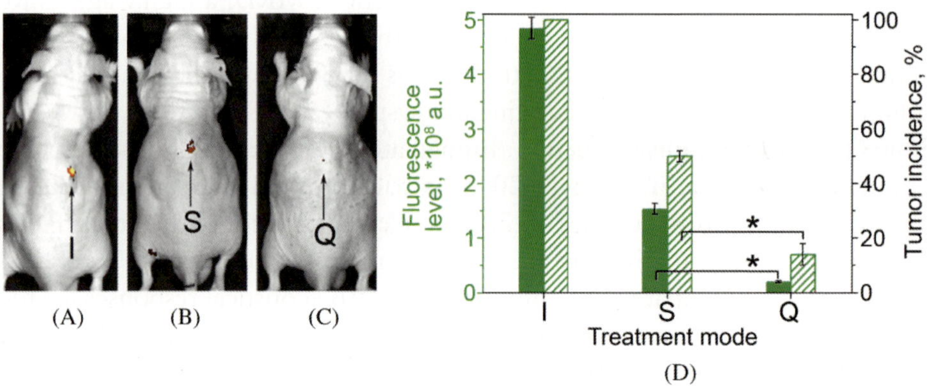

Figure 7.17. Evaluation of quadrapeutics for intra-operative treatment of HNSCC microscopic residual: (A–C) Intra-operative treatment of MRD: fluorescent images of green fluorescent protein-encoded tumors obtained 28 days after (A) surgery alone (*I*), (B) surgery and adjuvant chemoradiation (*S*), (C) surgery and adjuvant quadrapeutics (all therapeutic doses identical to those shown in Figure 7.15). (D) Metrics of recurrent tumors obtained in 28 days after the intra-operative treatment of MRD: *solid green* — level of the tumor fluorescence; *shaded grey* — incidence rate of a recurrence tumor (I: $n = 5$; S: $n = 6$, Q: $n = 7$). Data: mean ± s.e.m. $*P < 0.05$ [50].

laser scan during surgery (780 nm, 45 mJ/cm^2) and X-rays, 4 Gy, 6 h after surgery). We confirmed the presence of MRD intra-operatively via PNB acoustic signals immediately after the tumor resection as described above (Fig. 7.16C). Monitoring of the local recurrence of HNSCC after treatment revealed that surgery alone resulted in tumor recurrence in 2–4 weeks in 100% of the animals (Fig. 7.17A,D) and surgery with adjuvant chemoradiation also failed to prevent tumor recurrence (Fig. 7.17B,D). In contrast, surgery with adjuvant quadrapeutics efficiently suppressed tumor recurrence (Fig. 7.17C,D). Thus, quadrapeutics provided efficient intra-operative detection and treatment of MRD in a single theranostic (diagnosis and treatment) procedure.

7.9 Future Medical Applications of PNBs

The coherence of the *in vitro* (Figs. 7.11 and 7.12) and *in vivo* (Figs. 7.14–7.17) therapeutic data represents a solid proof of principle for the quadrapeutics. The quadrapeutics paradigm lies in the cancer cell-specific intracellular amplification of standard macro-therapies. Its high therapeutic efficacy, specificity, safety, and speed are based on the formation and on-demand threshold activation of intracellular nanoclusters in cancer cells, which remain safe and stable until activated with a laser pulse and X-rays. Nanocluster size increases with cancer aggressiveness (Fig. 7.12C–E), and it is this size which, in turn, determines the threshold activation energy and the therapeutic efficacy of the following non-stationary therapeutic events:

(1) Intracellular mechanical (non-thermal) impact of the laser pulse-induced PNB expansion and collapse around laser pulse-heated GNPs. A PNB is not a particle, but a transient non-stationary event lasting nanoseconds, that mechanically disrupts a host cell, co-localized nanocarriers, and endosomes, to locally and instantaneously eject the drug into the cytoplasm, thus creating its high intracellular concentration (Fig. 7.10H). A near-infrared laser pulse of low, safe energy has the sufficient depth of tissue penetration [146] and can be delivered even deeper with standard optical catheters and endoscopes [147, 148]. The largest nanoclusters, which form in aggressive cancer cells (Fig. 7.12C) [50], require minimal laser energy to generate the large, lethal PNBs [18, 19, 92]. Smaller, non-specific clusters or even single GNPs in normal cells (Figs. 7.12E and 7.13B) require a much higher level of laser energy (fluence), thereby preventing PNBs from being generated with an identical laser pulse (Figs. 7.8D and 7.13F) [43, 45, 50, 91]. The slow diffusive release of the drug from non-specifically internalized nanocarriers in normal cells does not induce notable toxicity due to the radical reduction in drug doses to 2–3% of standard doses [50] (Fig. 7.11E).

(2) Intracellular amplification of X-rays through the local emission of secondary electrons [141, 142] by GNP nanoclusters upon absorption of external X-rays is maximal around the largest nanoclusters. GNPs alone enhance radiotherapy [141, 142, 149], but at the cost of high GNP doses. At lower GNP doses, the therapeutic gain is low (Fig. 7.11d, mode GNP+XR). GNPs can also carry a drug and then amplify X-rays [150], but in the absence of the intracellular co-localization of the released drug and amplified X-rays, no amplification of their therapeutic effect occurs.

The synergy of the transient events described above (mechanical impact, drug ejection, and X-ray amplification) radically amplifies their therapeutic effect due to their intracellular co-localization. This amplification can be explained through well-established macro-effects: the mutual enhancement of chemo- and radiotherapies [144, 151–153] and the mechanical radio- and chemo-sensitization of tumors [63, 154–158]. Thus, an intracellular nanocluster serves as a kind of Trojan horse, whose activation threshold energy decreases with size and, thus, makes it possible to selectively trigger the aforementioned non-stationary therapeutic events only in cancer cells, thereby achieving an efficacy, specificity, and speed impossible with macro-therapies which, in contrast, employ materials with permanent properties [123, 159, 160]. At the same time, GNPs or laser pulses alone are absolutely safe *in vitro* (Fig. 7.11e) and *in vivo* (Fig. 7.15c).

The therapeutic functions of quadrapeutics are complemented by the real-time diagnostic effect of PNBs. Emission of a pressure pulse by PNBs enables a rapid and highly sensitive diagnosis (Fig. 7.16). Current intra-operative diagnostic methods require biopsy and are slow and often inaccurate [1, 161]. While the PNB method is technically close to photo-acoustics [162], the latter, unlike PNBs, has low sensitivity (Fig. 7.16c) due to the weak acoustic emission of GNPs compared to that of PNBs. The acoustic detection of PNBs is also advantageous over optical intra-operative diagnostic methods [163], which are less sensitive and specific.

To summarize, PNBs and the PNB-based technologies, such as quadrapeutics, convert established macro-therapies into on-demand intracellular modalities. Cancer cell-specific gold nanoclusters and PNBs radically amplify the therapeutic efficacy and specificity of drugs and radiation up in highly resistant and aggressive tumors. This amplification effect is self-regulated by the tumor aggressiveness via the PNB dependence upon the size of GNP clusters in cancer cells. PNB-supported diagnosis and treatment are united into a single theranostic procedure, whose speed, sensitivity, and efficacy allow for the real-time intra-operative detection and treatment of MRD, reduction of the efficient entry of therapeutic doses of drugs and radiation to less than 10% of their clinical doses, and minimization of the non-specific toxicity of drugs and radiation. Compared to nanoparticle medicine, PNBs introduce a new dimension of the non-stationary nano-event medicine with the possibility to control the diagnostic, therapeutic, and theranostic procedures at level of the cell.

References

1. Meier, J. D., Oliver, D. A. and Varvares, M. A. (2005). Surgical margin determination in head and neck oncology: Current clinical practice. The results of an International American Head and Neck Society Member Survey. *Head Neck.*, **27**, 952–958.
2. Looser, K. G., Shah, J. P. and Strong, E. W. (1978). The significance of "positive" margins in surgically resected epidermoid carcinomas. *Head Neck Surg.*, **1**, 107–111.
3. Vikram, B., Strong, E. W., Shah, J. P. and Spiro, R. (1984). Failure in the neck following multimodality treatment for advanced head and neck cancer. *Head Neck Surg.*, **6**, 724–729.
4. Yuen, P. W., Lam, K. Y., Chan, A. C., Wei, W. I. and Lam, L. K. (1998). Clinicopathological analysis of local spread of carcinoma of the tongue. *Am. J. Surg.*, **175**, 242–244.
5. Langendijk, J. A., Doornaert, P., Verdonck-de Leeuw, I. M., Leemans, C. R., Aaronson, N. K. and Slotman B. J. (2008). Impact of late treatment-related toxicity on quality of life among patients with head and neck cancer treated with radiotherapy. *J. Clin. Oncol.*, **26**, 3770–3887.
6. Jemal, A., Siegel, R., Xu, J. and Ward, E. (2010). Cancer statistics, 2010. *CA Cancer J. Clin.*, **60**, 277–300.
7. Jemal, A., Bray, F., Center, M. M., Ferlay, J., Ward, E. and Forman, D. (2011). Global cancer statistics. *CA Cancer J. Clin.*, **61**, 69–90.
8. Merchant, B. (1998). Gold, the noble metal and the paradoxes of its toxicology. *Biologicals*, **26**, 49–59.
9. Root, S. W., Andrews, G. A., Kniseley, R. M. and Tyor, M. P. (1954). The distribution and radiation effects of intravenously administered colloidal Au198 in man. *Cancer*, 7, 856–866.
10. Kean, W. F. and Kean, I. R. L. (2008). Clinical pharmacology of gold. *Inflammopharmacology*. **16**, 112–125.
11. Raleigh, L. (1917). On the pressure developed in a liquid during the collapse of a spherical cavity. *Philos. Mag.*, **34**, 94–98.
12. Plesset, M. S. (1949). The dynamics of cavitation bubbles. *J. Appl. Mech.*, **16**, 277–282.
13. Askaryan, G. A., Prokhorov, A. M., Chanturiya, G. F. and Shapiro, G. P. (1963). The effects of a laser beam in a liquid. *Sov. Phys. JEPT*, **17**, 1463–1465.
14. Barnes, P. A. and Rieckhoff, K. E. (1968). Laser induced underwater sparks. *Appl. Phys. Lett.*, **13**, 282–284.
15. Neumann, J. and Brinkmann, R. **(2007)**. Nucleation dynamics around single microabsorbers in water heated by nanosecond laser irradiation. *J. Appl. Phys.*, **101,** 114701.
16. Roider, J., El Hifnawi, E. S. and Birngruber, R. (1998). Bubble formation as primary interaction mechanism in retinal laser exposure with 200-ns laser pulses. *Lasers Surg. Med.*, **22**, 240–248.

17. Lapotko, D., Lukianova, E., Shnip, S., Zheltov, G., Potapnev, M., Savitsky, V., Klimovich, O. and Oraevsky, A. (2005). Photothermal microscopy and laser ablation of leukemia cells targeted with gold nanoparticles. *Proc. SPIE,* **5697**, 82–89.
18. Lukianova-Hleb, E. Y., Hu, Y., Latterini, L., Tarpani, L., Lee, S., Drezek, R., Hafner, J. H. and Lapotko, D. O. (2010). Plasmonic nanobubbles as transient vapor nanobubbles generated around plasmonic nanoparticles. *ACS Nano* **4**, 2109–2123.
19. Hleb, E. Y. and Lapotko, D. O. (2009). Influence of transient environmental photothermal effects on optical scattering by gold nanoparticles. *Nano Lett.*, **9**, 2160–2166.
20. Lapotko, D. O. (2009). Optical excitation and detection of vapor bubbles around plasmonic nanoparticles. *Opt. Express,* **17**, 2538–2556.
21. Kling, C. L. and Hammitt, F. G. (1972). A photographic study of spark-induced cavitation bubble collapse. *Transactions of the ASME D: J. Basic Engineering,* **94**, 825–833.
22. Lauterborn, W. (1972). High-speed photography of laser-induced breakdown in liquids. *Appl. Phys. Lett.,* **21**, 27–29.
23. Ohl, C. D., Philipp, A. and Lauterborn, W. (1995). Cavitation bubble collapse studied at 20-million frames-per-second. *Annalen der Physik,* **4**, 26–34.
24. Yavas, O., Leiderer, P., Park, H. K., Grigoropoulos, C. P., Poon, C. C., Leung, W. P., Do, N. and Tam, A. C. (1993). Optical reflectance and scattering studies of nucleation and growth of bubbles at a liquid-solid interface induced by pulsed laser heating. *Phys. Rev. Lett.,* **70**, 1830–1833.
25. Vogel, A. and Lauterborn, W. (1988). Acoustic transient generation by laser-produced cavitation bubbles. *J. Acoust. Soc. Am.,* **84**, 719–731.
26. Marston, P. L. (1979). Critical angle scattering by a bubble: Physicaloptics approximation and observations. *J. Opt. Soc. Am.,* **69**, 1205–1211.
27. Stroud, J. S. and Marston, P. L. (1993). Optical detection of transient bubble oscillations associated with the underwater noise of rain. *J. Acoust. Soc. Am.,* **94**, 2788–2792.
28. Asaki, T. J., Thiessen, D. B. and Marston, P. L. (1995). Effect of an insoluble surfactant on capillary oscillations of bubbles in water: Observation of a maximum in the damping. *Phys. Rev. Lett.,* **75**, 2686–2689.
29. Fujimoto, J. G., Lin, W. Z., Ippen, E. P., Puliafito, C. A. and Steinert, R. F. (1985). time-resolved studies of Nd:YAG laser-induced breakdown. *Invest. Ophthalmol. Vis. Sci.,* **26**, 1771–1777.
30. Gaitan, D. F., Crum, L. A., Church, C. C. and Roy, R. A. (1992). Sonoluminescence and bubble dynamics for a single stable, cavitation bubble. *J. Acoust. Soc. Am.,* **91**, 3166–3183.
31. Barber, B. P. and Putterman, S. J. (1992). Light scattering measurements of the repetitive supersonic implosion of a sonoluminescing bubble. *Phys. Rev. Lett.,* **69**, 3839–3842.
32. Plech, A., Kotaidis, V., Lorenc, M. and Boneberg, J. (2006). Femtosecond laser near-field ablation from gold nanoparticles. *Nat. Phys.,* **2**, 44–47.
33. Lin, C. P. and Kelly, M. W. (1998). Cavitation and acoustic emission around laser-heated microparticle. *Appl. Phys. Lett.,* **72**, 2800–2802.
34. Neumann, J. and Brinkmann, R. (2008). Self-limited growth of laser-induced vapor bubbles around single microabsorbers. *Appl. Phys. Lett.,* **93**, 033901.
35. Brinkmann, R., Hüttmann, G., Rögener, J., Roider, J., Birngruber, R. and Lin, C. P. (2000). Origin of retinal pigment epithelium cell damage by pulsed laser irradiance in the nanosecond to microsecond time regimen. *Lasers Surg. Med.,* **27**, 451–464.
36. Chen, H. and Diebold, G. (1995). Chemical generation of acoustic waves: A giant photoacoustic effect. *Science.,* **270**, 963–966.

37. Lapotko, D. and Kuchinsky, G. (1995). Photothermal microscopy for cell imaging and diagnostics. *Proc. SPIE,* **2390**, 89–100.
38. Vogel, A., Linz, N., Freidank, S. and Paltauf, G. (2008). Femtosecond-laser-induced nanocavitation in water: Implications for optical breakdown threshold and cell surgery, *Phys. Rev. Lett.,* **100**, 038102.
39. Vogel, A., Noack, J., Hüttmann, G. and Paltauf, G. (2005). Mechanisms of femtosecond laser surgery of cells and tissue. *Appl. Phys. B,* **81**, 1015–1047.
40. Pitsillides, C. M., Joe, E. K., Wie, X., Anderson, R. R. and Lin, C. P. (2003). Selective cell targeting with light-absorbing microparticles and nanoparticles. *Biophys. J.,* **84**, 4023–4032.
41. Kotaidis, V., Dahmen, C., von Plessen, G., Springer, F. and Plech, A. (2006). Excitation of nanoscale vapor bubbles at the surface of gold nanoparticles in water. *J. Chem. Phys.* **124,** 184702.
42. Kotaidis, V. and Plech, A. (2005). Cavitation dynamics on the nanoscale. *Appl. Phys. Lett.,* **87,** 213102.
43. Lukianova-Hleb, E. Y., Ren, X., Zasadzinski, J. A., Wu, X. and Lapotko, D. O. (2012). Plasmonic nanobubbles enhance efficacy and selectivity of chemotherapy against drug-resistant cancer cells. *Adv. Mater.,* **24**, 3831–3837.
44. Wagner, D. S., Delk, N. A., Lukianova-Hleb, E. Y., Hafner, J. H., Farach-Carson, M. C. and Lapotko, D. O. (2010). The *in vivo* performance of plasmonic nanobubbles as cell theranostic agents in zebrafish hosting prostate cancer xenografts. *Biomaterials,* **31**, 7567–7574.
45. Lukianova-Hleb, E. Y., Ren, X., Townley, D., Wu, X., Kupferman, M. and Lapotko, D. (2012). Plasmonic nanobubbles rapidly detect and destroy drug-resistant tumors, *Theranostics,* **2**, 976–987.
46. Lukianova-Hleb, E., Wagner, D., Brenner, M. and Lapotko, D. (2012). Cell-specific transmembrane injection of molecular cargo with gold nanoparticle-generated transient plasmonic nanobubbles. *Biomaterials,* **33**, 5441–5450.
47. Lapotko, D., Lukianova, E. and Oraevsky, A. (2006). Selective laser nano-thermolysis of human leukemia cells with microbubbles generated around clusters of gold nanoparticles. *Lasers Surg. Med.,* **38**, 631–642.
48. Lapotko, D., Shnip, A. and Lukianova, E. (2003). Photothermal detection of laser-induced damage in single intact cells. *Lasers Surg. Med.,* **33**, 320–326.
49. Lukianova-Hleb, E. Y. and Lapotko, D.O. (2014). Nano-theranostics with plasmonic nanobubbles. *IEEE J. Sel. Top. Quant. Electron.,* **20**, 7300412.
50. Lukianova-Hleb, E. Y., Ren, X., Sawant, R. R., Wu, X., Torchilin, V. P. and Lapotko, D. O. (2014). On-demand intracellular amplification of chemoradiation with cancer-specific plasmonic nanobubbles. *Nat. Medicine,* **20**, 778–784.
51. Lapotko, D., Lukianova-Hleb, E. and Oraevsky, A. (2007). Clusterization of nanoparticles during their interaction with living cells. *Nanomedicine,* **2**, 241–253.
52. Braun, G. B., Pallaoro, A., Wu, G., Missirlis, D., Zasadzinski, J. A., Tirrell, M. and Reich, N. O. (2009). Laser-Activated gene silencing via gold nanoshell-siRNA conjugates. *ACS Nano,* **3**, 2007–2015.
53. McLaughlan, J. R., Roy, R. A., Ju, H. and Murray, T. W. (2010). Ultrasonic enhancement of photoacoustic emissions by nanoparticle-targeted cavitation. *Opt. Lett.,* **35**, 2127–2129.
54. Wu, T. H., Kalim, S., Callahan, C., Teitell, M. A. and Chiou, P. Y. (2010). Image patterned molecular delivery into live cells using gold particle coated substrates. *Opt. Express,* **18**, 938–946.

55. Ibrahimkutty, S., Kim, J., Cammarata, M., Ewald, F., Choi, J., Ihee, H. and Plech, A. (2011). Ultrafast structural dynamics of the photocleavage of protein hybrid nanoparticles. *ACS Nano,* **5**, 3788–3794.
56. Rudnitzki, F., Bever, M., Rahmanzadeh, R., Brieger, K., Endl, E., Groll, J. and Hüttmann, G. (2012). Bleaching of plasmon-resonance absorption of gold nanorods decreases efficiency of cell destruction. *J. Biomed. Opt.,* **17**, 058003.
57. Peeters, S., Kitz, M., Preisser, S., Wetterwald, A., Rothen-Rutishauser, B., Thalmann, G. N., Brandenberger, C., Bailey, A. and Frenz, M. (2012). Mechanisms of nanoparticle-mediated photomechanical cell damage. *Biomed. Opt. Express,* **3**, 435–446.
58. Arita, Y., Ploschner, M., Antkowiak, M., Gunn-Moore, F. and Dholakia, K. (2013). Laser-induced breakdown of an optically trapped gold nanoparticle for single cell transfection. *Opt. Lett.,* **38**, 3402–3405.
59. Wang, S., Chen, K. J., Wu, T. H., Wang, H., Lin, W. Y., Ohashi, M., Chiou, P. Y. and Tseng, H. R. (2010). Photothermal effects of supramolecularly assembled gold nanoparticles for the targeted treatment of cancer cells. *Angew. Chem. Int. Ed. Engl.,* **49**, 3777–3781.
60. Chen, C. L., Kuo, L. R., Chang, C. L., Hwu, Y. K., Huang, C. K., Lee, S. Y., Chen, K., Lin, S. J., Huang, J. D. and Chen, Y. Y. (2010). *In situ* real-time investigation of cancer cell photothermolysis mediated by excited gold nanorod surface plasmons. *Biomaterials,* **31**, 4104–4112.
61. Jain, K. K. (2010). Advances in the field of nanooncology. *BMC Med.,* 8, 83.
62. O'Neill, B. E. and Rapoport, N. (2011). Phase-shift, stimuli-responsive drug carriers for targeted delivery. *Ther. Deliv.,* **2**, 1165–1187.
63. Gao, Z., Kennedy, A. M., Christensen, D. A. and Rapoport, N. Y. (2008). Drug-loaded nano/microbubbles for combining ultrasonography and targeted chemotherapy. *Ultrasonics.,* **48**, 260–270.
64. Ashokkumar, M. (2011). The characterization of acoustic cavitation bubbles — an overview. *Ultrason. Sonochem.,* **18**, 864–872.
65. Lapotko, D. and Lukianova, K. (2005). Laser-induced micro-bubbles in cells. *Int. J. Heat Mass Transfer.,* **48**, 227–234.
66. Lauterborn, W., Kurz, T., Geisler, R., Schanz, D. and Lindau, O. (2007). Acoustic cavitation, bubble dynamics and sonoluminescence. *Ultrason. Sonochem.,* **14**, 484–491.
67. Kurz, T., Kröninger, D., Geisler, R. and Lauterborn, W. (2006). Optic cavitation in an ultrasonic field. *Phys. Rev. E Stat. Nonlin. Soft. Matter. Ohys.,* **74**, 066307.
68. Krasovitski, B., Kislev, H. and Kimmel, E. (2007). Modeling photothermal and acoustical induced microbubble generation and growth. *Ultrasonics,* **47**, 90–101.
69. Lukianova-Hleb, E. Y. and Lapotko, D. O. (2012). Experimental techniques for imaging and measuring transient vapor nanobubbles. *Appl. Phys. Lett.,* **101**, 264102.
70. Wong, Z. Z., Kripfgans, O. D., Qamar, A., Fowlkes, J. B. and Bull, J. L. (2011). Bubble evolution in acoustic droplet vaporization at physiological temperature *via* ultra-high speed imaging. Soft Matter, 7, 4009–4016.
71. Marmottant, P., Versluis, M., Jong, de N., Hilgenfeldt, S. and Lohse, D. (2006). High-speed imaging of an ultrasound-driven bubble in contact with a wall: "*Narcissus*" effect and resolved acoustic streaming. *Exp. Fluids.,* **41**, 147–153.
72. Thoroddsen, S. T., Etoh, T. G. and Takehara, K. (2008). High-speed imaging of drops and bubbles. *Annu. Rev. Fluid Mech.,* **40**, 257–285.

73. Stover, R. L., Tobias, C. W. and Denn, M. M. (1997). Bubble coalescence dynamics. *AIChE J.*, **43**, 2385–2392.
74. Keim, N. C., Møller, P., Zhang, W. W. and Nagel, S. R. (2006). Breakup of air bubbles inwater: memory adn breakdown of cylindrical symmetry. *Phys. Rev. Lett.*, **97**, 144503.
75. Lindau, O. and Lauterborn, W. (2003). Cinematographic observation of the collapse and rebound of laser-produced cavitation bubble near a wall. *J. Fluid Mech.*, **479**, 327–348.
76. Lauterborn, W. and Hentschel, W. (1985). Cavitation bubble dynamics studied by high speed photography and holography. Part 1. *Ultrasonics,* **23**, 260–267.
77. Ohtake, H., Hayashida, T., Kitamura, K., Arai, T., Yonai, J., Tanioka, K., Maruyama, H., Eto, T., Poggemann, D., Ruckelshausen, A., van Kuijk, H. and Bosiers, J. T. (2006). Development of a 300 000-pixel ultrahigh-speed high-sensitivity CCD. *Proc SPIE.*, **6119**, 61190E.
78. Kosonocky, W. F., Yang, G., Ye, C., Kabra, R. K., Xie, L., Lawrence, J. L., Mastrocolla, V., Shallcross, F. V. and Patel, V. (1996). 60x360-element very high burst-frame rate image sensor. *Dig. Tech. Pap. IEEE Int. Solid State Circuits Conf.*, **39**, 182–183.
79. Lauterborn, W., Judt, A. and Schmitz, E. (1993). High-speed off-axis holographic cinematography with a copper-vapor-pumped dye laser. *Optics Lett.*, **18**, 4–6.
80. Shiraga, H., Fujioka, S., Jaanimagi, P. A., Stoeckl, C., Stephens, R. B., Nagatomo, H., Tanaks, K. A., Kodama, R. and Azechi, H. (2004). Multi-imaging X-ray streak camera for ultrahigh-speed two-dimensional x-ray imaging of imploded core plasmas. *Rev. Sci. Instrum.*, **75**, 3921–3925.
81. King, W. E., Campbell, G. H., Frank, A., Reed, B., Schmerge, J. F., Siwick, B. J., Stuart, B. C. and Weber, P. M. (2005). Ultrafast electron microscopy in materials science, biology, and chemistry. *J. Appl. Phys.*, **97**, 111101.
82. Faraday, M. (1857). The Bakerian Lecture: Experimental Relations of Gold (and Other Metals) to Light. *Phil. Trans. R. Soc. Lond.*, **147**, 145–181.
83. Lukianova-Hleb, E. Y., Volkov, A. and Lapotko, D. O. (2014). Laser pulse duration is critical for the generation of plasmonic nanobubbles. *Langmuir*, **30**, 7425–7434.
84. Chithrani, B. D., Ghazani, A. A. and Chan, W. C. (2006). Determining the size and shape dependence of gold nanoparticle uptake into mammalian cells. *Nano Lett.*, **6**, 662–668.
85. François, L., Mostafavi, M., Belloni, J. and Delaire, J. A. (2001). Optical limitation induced by gold clusters: Mechanism and efficiency. *Phys. Chem. Chem. Phys.*, **3**, 4965–4971.
86. Neumann, O., Urban, A., Day, J., Lal, S., Nordlander, P. and Halas, N. J. (2013). Solar vapor generation enabled by nanoparticles. *ACS Nano,* **7**, 42–49.
87. Adleman, J. R., Boyd, D. A., Goodwin, D. G. and Psaltis, D. (2009). Heterogenous catalysis mediated by plasmon heating. *Nano Lett.*, **9**, 4417–4423.
88. Yuan, F., Sankin, G. and Zhong, P. (2011). Dynamics of tandem bubble interaction in a microfluidic channel. *J. Acoust. Soc. Am.*, **130**, 3339–3346.
89. Zhao, C., Liu, Y., Zhao, Y., Fang, N. and Huang, T. J. (2013). A reconfigurable plasmofluidic lens. *Nat. Commun.*, **4**, 2305.
90. Akchurin, G., Khlebtsov, B., Akchurin, G., Tuchin, V., Zharov, V. and Khlebtsov, N. (2008). Gold nanoshell photomodification under a single-nanosecond laser pulse accompanied by color-shifting and bubble formation phenomena. *Nanotechnology,* **19**, 015701.
91. Lukianova-Hleb, E. Y., Ren, X., Constantinou, P. E., Danysh, B. P., Shenefelt, D. L., Carson, D. D., Farach-Carson, M. C., Kulchitsky, V. A., Wu, X., Wagner, D. S. and Lapotko, D. O. (2012). Improved cellular specificity of plasmonic nanobubbles versus nanoparticles in heterogeneous cell systems. *PLoS One,* **7**, e34537.

92. Lukianova-Hleb, E. Y., Volkov, A. N., Wu, X. and Lapotko, D. O. (2013). Transient enhancement and spectral narrowing of the photothermal effect of plasmonic nanoparticles under pulsed excitation. *Adv. Mater.,* **25**, 772–776.
93. Lukianova-Hleb, E. Y., Sassaroli, E., Jones, A. and Lapotko, D. O. (2012). Transient photothermal spectra of plasmonic nanobubbles. *Langmuir,* **28**, 4858–4866.
94. Hleb, E. Y. and Lapotko, D. O. (2008). Photothermal properties of gold nanoparticles under exposure to high optical energies. *Nanotechnology,* **19**, 355702.
95. Lukianova-Hleb, E. Y., Anderson, L. J. E., Lee, S., Hafner, J. H. and Lapotko, D. O. (2010). Hot plasmonic interactions: A new look at the photothermal efficacy of gold nanoparticles. *Phys. Chem. Chem. Phys.,* **12**, 12237–12244.
96. Cai, W., Gao, T., Hong, H. and Sun, J. (2008). Applications of gold nanoparticles in cancer nanotechnology. *Nanotechnol. Sci. Appl.,* **1**, 17–32.
97. Boisselier, E. and Astruc, D. (2009). Gold nanoparticles in nanomedicine: Preparations, imaging, diagnostics, therapies and toxicity. *Chem. Soc. Rev.,* **38**, 1759–1782.
98. El-Sayed, I. H., Huang, X. and El-Sayed, M. A. (2005). Surface plasmon resonance scattering and absorption of anti-EGFR antibody conjugated gold nanoparticles in cancer diagnostics: applications in oral cancer. *Nano Lett.,* **5**, 829–834.
99. Loo, C., Lowery, A., Halas, N., West, J. and Drezek, R. (2005). Immunotargeted nanoshells for integrated cancer imaging and therapy. *Nano Lett.,* **5**, 709–711.
100. West, J. L. and Halas, N. J. (2003). Engineered nanomaterials for biophotonics applications: improving sensing, imaging, and therapeutics. *Annu. Rev. Biomed. Eng.,* **5**, 285–292.
101. Sokolov, K., Follen, M., Aaron, J., Pavlova, I., Malpica, A., Lotan, R. and Richards-Kortum, R. (2003). Real-time vital optical imaging of precancer using anti-epidermal growth factor receptor antibodies conjugated to gold nanoparticles. *Cancer Res.,* **63**, 1999–2004.
102. Vigderman, L., Manna, P. and Zubarev, E. R. (2012). Quantitative replacement of cetyl trimethylammonium bromide by cationic thiol ligands on the surface of gold nanorods and their extremely large uptake by cancer cells. *Angew. Chem. Int. Ed. Engl.,* **51**, 636–641.
103. Lewinski, N., Colvin, V. and Drezek, R. (2008). Cytotoxicity of nanoparticles. *Small,* **4**, 26–49.
104. Oberdoester, G., Oberdoester, E. and Oberdoester, J. (2005). Nanotoxicology: An emerging discipline evolving from studies of ultrafine particles. *Environ. Health Perspect.,* **118**, 823–839.
105. Kumar, S., Harrison, N., Richards-Kortum, R. and Sokolov, K. (2007). Plasmonic nanosensors for imaging intracellular biomarkers in live cells. *Nano Lett.,* 7, 1338–1343.
106. Jain, P. K., Huang, X., El-Sayed, I. H. and El-Sayed, M. A. (2008). Noble metals on the nanoscale: optical and photothermal properties and some applications in imaging, sensing, biology, and medicine. *Acc. Chem. Res.,* **41**, 1578–1586.
107. Huang, X., El-Sayed, I. H., Qian, W. and El-Sayed, M. A. (2006). Cancer cell imaging and photothermal therapy in the near-infrared region by using gold nanorods. *J. Am. Chem. Soc.,* **128**, 2115–2120.
108. Mallidi, S., Larson, T., Tam, J., Joshi, P. P., Karpiouk, A., Sokolov, K. and Emelianov, S. (2009). Multiwavelength photoacoustic imaging and plasmon resonance coupling of gold nanoparticles for selective detection of cancer. *Nano Lett.,* **9**, 2825–2831.
109. Xu, X., Liu, H. and Wang, L. V. (2011). Time-reversed ultrasonically encoded optical focusing into scattering media. *Nat. Photonics,* **5**, 154–157.

110. Wang, F., Wang, Y. C., Dou, S., Xiong, M. H., Sun, T. M. and Wang, J. (2011), Doxorubicin-tethered responsive gold nanoparticles facilitate intracellular drug delivery for overcoming multidrug resistance in cancer cells. *ACS Nano,* **5**, 3679–3692.
111. Kim, B., Han, G., Toley, B. J., Kim, C. K., Rotello, V. M. and Forbes, N. S. (2010). Tuning payload delivery in tumour cylindroids using gold nanoparticles. *Nat. Nanotechnol.*, **5**, 465–472.
112. Paasonen, L., Sipila, T., Subrizi, A., Laurinmaki, P., Butcher, S. J., Rappolt, M., Yaghmur, A., Urtti, A. and Yliperttula, M. (2010). Gold-embedded photosensitive liposomes for drug delivery: triggering mechanism and intracellular release. *J. Control. Release,* **147**, 136–143.
113. Pissuwan, D., Niidome, T. and Cortie, M. B. (2011). The forthcoming applications of gold nanoparticles in drug and gene delivery systems. *J. Control. Release,* **149**, 65–71.
114. Yguerabide, J. and Yguerabide, E. E. (1998). Light-scattering submicroscopic particles as highly fluorescent analogs and their use as tracer labels in clinical and biological applications. *Anal. Biochem.*, **262**, 157–176.
115. Tong, L. and Cheng, J. X. (2009). Gold nanorod-mediated photothermolysis induces apoptosis of macrophages via damage of mitochondria. *Nanomedicine (Lond.),* **4**, 265–276.
116. Hutson, M. S. and Ma, X. Y. (2007). Plasma and cavitation dynamics during pulsed laser microsurgery in vivo. *Phys. Rev. Lett.,* **99**, 158104.
117. Yao, C. P., Rahmanzadeh, R., Endl, E., Zhang, Z. X., Gerdes, J. and Huttmann, G. (2005). Elevation of plasma membrane permeability by laser irradiation of selectively bound nanoparticles. *J. Biomed. Opt.*, **10**, 064012.
118. Stevenson, D., Agate, B., Tsampoula, X., Fischer, P., Brown, C. T. A., Sibbett, W., Riches, A., Gunn-Moore, F. and Dholakia, K. (2006). Femtosecond optical transfection of cells: Viability and efficiency. *Opt. Express,* **14**, 7125–7133.
119. Prentice, P., Cuschierp, A., Dholakia, K., Prausnitz, M. and Campbell, P. (2005). Membrane disruption by optically controlled microbubble cavitation. *Nat. Phys.*, **1**, 107–110.
120. Hauck, T. S., Jennings, T. L., Yatsenko, T., Kumaradas, J. C. and Chan, W. C. W. (2008). Enhancing the toxicity of cancer chemotherapeutics with gold nanorod hyperthermia. *Adv. Mater.*, **20**, 3832–3838.
121. Lee, S. M., Kim, H. J., Kim, S. Y., Kwon, M. K., Kim, S., Cho, A., Yun, M., Shin, J. S. and Yoo, K. H. (2014). Drug-loaded gold plasmonic nanoparticles for treatment of multidrug resistance in cancer. *Biomaterials,* **35**, 2272–2282.
122. Cho, K., Wang, X., Nie, S., Chen, Z. G. and Shin, D. M. (2008). Therapeutic nanoparticles for drug delivery in cancer. *Clin. Cancer Res.*, **14**, 1310–1316.
123. Peer, D., Karp, J. M., Hong, S., Farokhzad, O. C., Margalit, R. and Langer, R. (2007). Nanocarriers as an emerging platform for cancer therapy. *Nat. Nanotechnol.*, **2**, 751–760.
124. Meng, H., Liong, M., Xia, T., Li, Z., Ji, Z., Zink, J. I. and Nel, A. E. (2010). Engineered design of mesoporous silica nanoparticles to deliver doxorubicin and P-glycoprotein siRNA to overcome drug resistance in a cancer cell line. *ACS Nano,* **4**, 4539–4550.
125. Liang, X. J., Meng, H., Wang, Y., He, H., Meng, J., Lu, J., Wang, P. C., Zhao, Y., Gao, X., Sun, B., Chen, C., Xing, G., Shen, D., Gottesman, M. M., Wu, Y., Yin, J. J. and Jia, L. (2010). Metallofullerene nanoparticles circumvent tumor resistance to cisplatin by reactivating endocytosis. *Proc. Natl. Acad. Sci. U S A,* **107**, 7449–7454.
126. Qin, G., Li, Z., Xia, R., Li, F., O'Neill, B. E., Goodwin, J. T., Khant, H. A., Chiu, W. and Li, K. C. (2011). Partially polymerized liposomes: Stable against leakage yet capable of instantaneous release for remote controlled drug delivery. *Nanotechnology,* **22**, 155605.

127. Sajja, H. K., East, M. P., Mao, H., Wang, Y. A., Nie, S. and Yang, L. (2009). Development of multifunctional nanoparticles for targeted drug delivery and noninvasive imaging of therapeutic effect. *Curr. Drug Discov. Technol.,* **6**, 43–51.
128. Paasonen, L., Laaksonen, T., Johans, C., Yliperttula, M., Kontturi, K. and Urtti, A. (2007). Gold nanoparticles enable selective light-induced contents release from liposomes. *J. Control. Release,* **122**, 86–93.
129. Hleb, E., Hafner, J. H., Myers, J. N., Hanna, E. Y., Rostro, B. C., Zhdanok, S. A. and Lapotko, D. O. (2008). LANTCET: Elimination of solid tumor cells with photothermal bubbles generated around clusters of gold nanoparticles. *Nanomedicine (Lond.),* **3**, 647–667.
130. Chithrani, B. D. and Chan, W. C. (2007). Elucidating the mechanism of cellular uptake and removal of protein-coated gold nanoparticles of different sizes and shapes. *Nano Lett.*, **7**, 1542–1550.
131. Salmaso, S., Caliceti, P., Amendola, V., Meneghetti, M., Magnusson, J. P., Pasparakis, G. and Alexander, C. (2009). Cell up-take control of gold nanoparticles functionalized with a thermoresponsive polymer. *J. Mater. Chem.,* **19**, 1608–1615.
132. Mandal, D., Maran, A., Yaszemski, M. J., Bolander, M. E. and Sarkar, G. (2009). Cellular uptake of gold nanoparticles directly cross-linked with carrier peptides by osteosarcoma cells. *J. Mater. Sci. Mater. Med.,* **20**, 347–350.
133. Haigler, H. T., McKanna, J. A. and Cohen, S. (1979). Direct visualization of the binding and internalization of a ferritin conjugate of epidermal growth factor in human carcinoma cells A-431. *J. Cell Biol.,* **81**, 382–395.
134. Shukla, R., Bansal, V., Chaudhary, M., Basu, A., Bhonde, R. R. and Sastry, M. (2005). Biocompatibility of gold nanoparticles and their endocytotic fate inside the cellular compartment: a microscopic overview. *Langmuir,* **21**, 10644–10654.
135. Yang, L., Cao, Z., Sajja, H. K., Mao, H., Wang, L., Geng, H., Xu, H., Jiang, T., Wood, W. C., Nie, S. and Wang, Y. A. (2008). Development of Receptor Targeted Magnetic Iron Oxide Nanoparticles for Efficient Drug Delivery and Tumor Imaging, *J. Biomed. Nanotechnology,* **4**, 439–449.
136. Lukianova-Hleb, E. Y., Belyanin, A., Kashinath, S., Wu, X. and Lapotko, D. (2012). Plasmonic nanobubble-enhanced endosomal escape processes for selective and guided intracellular delivery of chemotherapy to drug-resistant cancer cells. *Biomaterials,* **33**, 1821–1826.
137. Sharafinski, M. E., Ferris, R. L., Ferrone, S. and Grandis, J. R. (2010). Epidermal growth factor receptor targeted therapy of squamous cell carcinoma of the head and neck. *Head Neck.,* **32**, 1412–1421.
138. Iakoubov, L., Rokhlin, O. and Torchilin, V. (1995). Anti-nuclear autoantibodies of the aged reactive against the surface of tumor but not normal cells. *Immunol. Lett.,* **47**, 147–149.
139. Centocor Ortho Biotech Products (2011). Doxil: Product Information Book. Centocor Ortho Biotech Products, L.P. USA.
140. Torchilin, V. P., Lukyanov, A. N., Gao, Z. and Papahadjopoulos-Sternberg, B. (2003). Immunomicelles: Targeted pharmaceutical carriers for poorly soluble drugs. *Proc. Natl. Acad. Sci. USA,* **100**, 6039–6044.
141. Hainfeld, J. F., Dilmanian, F. A., Zhong, Z., Slatkin, D. N., Kalef-Ezra, J. A. and Smilowitz, H. M. (2010). Gold nanoparticles enhance the radiation therapy of a murine squamous cell carcinoma. *Phys. Med. Biol.,* **55**, 3045–3059.

142. Hainfeld, J. F., Slatkin, D. N. and Smilowitz, H. M. (2004). The use of gold nanoparticles to enhance radiotherapy in mice. *Phys. Med. Biol.*, **49**, N309–N315.
143. Calais, G., Alfonsi, M., Bardet, E., Sire, C., Germain, T., Bergerot, P., Rhein, B., Tortochaux, J., Oudinot, P. and Bertrand, P. (2000). Stage III and IV cancers of the oropharynx: Results of a randomized study of Gortec comparing radiotherapy alone with concomitant chemotherapy. *Bull. Cancer,* **87**, 48–53.
144. Sano, D., Matsumoto, F., Valdecanas, D. R., Zhao, M., Molkentine, D. P., Takahashi, Y., Hanna, E. Y., Papadimitrakopoulou, V., Heymach, J., Milas, L. and Myers, J. N. (2011). Vandetanib restores head and neck squamous cell carcinoma cells' sensitivity to cisplatin and radiation in vivo and in vitro. *Clin. Cancer Res.,* **17**, 1815–1827.
145. Sano, D., Xie, T. X., Ow, T. J., Zhao, M., Pickering, C. R., Zhou, G., Sandulache, V. C., Wheeler, D. A., Gibbs, R. A., Caulin, C. and Myers, J. N. (2011). Disruptive TP53 mutation is associated with aggressive disease characteristics in an orthotopic murine model of oral tongue cancer. *Clin. Cancer Res.*, **17**, 6658–6670.
146. Welch, A. J. and van Gemert, M. J. C. (2011). *Optical-Thermal Response of Laser-Irradiated Tissue*, 2nd edn., Springer, Berlin.
147. Ambrosch, P. (2007). The role of laser microsurgery in the treatment of laryngeal cancer. *Curr. Opin. Otolaryngol. Head Neck Surg.,* **15**, 82–88.
148. Karni, R. J., Rich, J. T., Sinha, P. and Haughey, B. H. (2011). Transoral laser microsurgery: A new approach for unknown primaries of the head and neck. *Laryngoscope*, **121**, 1194–1201.
149. Rahman, W. N., Bishara, N., Ackerly, T., He, C. F., Jackson, P., Wong, C., Davidson, R. and Geso, M. (2009). Enhancement of radiation effects by gold nanoparticles for superficial radiation therapy. *Nanomedicine,* **5**, 136–142.
150. Jiao, P., Zhou, H., Otto, M., Mu, Q., Li, L., Su, G., Zhang, Y., Butch, E. R., Snyder, S. E., Jiang, G. and Yan, B. (2011). Leading neuroblastoma cells to die by multiple premeditated attacks from a multifunctionalized nanoconstruct. *J. Am. Chem. Soc.,* **133**, 13918–13921.
151. Argiris, A., Karamouzis, M. V., Raben, D. and Ferris, R. L. (2008). Head and neck cancer. *The LANCET,* **371**, 1695–1709.
152. Jegannathen, A., Mais, K., Sykes, A., Lee, L., Yap, B., Birzgalis, A., Homer, J., Ryder, W. D. and Slevin, N. (2011). Synchronous chemoradiotherapy in patients with locally advanced squamous cell carcinoma of the head and neck using capecitabine: A single-centre, open-label, single-group phase II study. *Clin. Oncol. R. Coll. Radiol.,* **23**, 149–158.
153. Chen, Y., Pandya, K., Keng, P. C., Johnstone, D., Li, J., Lee, Y. J., Smudzin, T. and Okunieff, P. (2003). Phase I/II clinical study of pulsed paclitaxel radiosensitization for thoracic malignancy: A therapeutic approach on the basis of preclinical research of human cancer cell lines. *Clin. Cancer Res.,* **9**, 969–975.
154. Paliwal, S. and Mitragotri, S. (2006). Ultrasound-induced cavitation: Applications in drug and gene delivery. *Exper. Opin. Drug. Deliv.,* **3**, 713–726.
155. Czarnota, G. J., Karshafian, R., Burns, P. N., Wong, S., Al Mahrouki, A., Lee, J. W., Caissie, A., Tran, W., Kim, C., Furukawa, M., Wong, E. and Giles, A. (2012). Tumor radiation response enhancement by acoustical stimulation of the vasculature. *Proc. Natl. Acad. Sci. U S A.*, **109**, E2033–E2041.
156. Rapoport, N., Nam, K. H., Gupta, R., Gao, Z., Mohan, P., Payne, A., Todd, N., Liu, X., Kim, T., Shea, J., Scaife, C., Parker, D. L., Jeong, E. K. and Kennedy, A. M. (2011). Ultrasound-mediated tumor imaging and nanotherapy using drug loaded, block copolymer stabilized perfluorocarbon nanoemulsions. *J. Control. Release,* **153**, 4–15.

157. Rapoport. N. Y. (2007). Physical stimuli-responsive polymeric micelles for anti-cancer drug delivery. *Progress in Polymer Science,* **32**, 962–990.
158. Kamaev, P. and Rapoport, N. (2006). Effect of anticancer drug on the cell sensitivity to ultrasound *in vitro* and *in vivo. Am. Inst. Phys. Conf. Proc.,* **829**, 543–547.
159. Allen, T. M. and Cullis, P. R. (2004). Drug delivery systems: Entering the mainstream. *Science,* **303**, 1818–1822.
160. Schroeder, A., Heller, D. A., Winslow, M. M., Dahlman, J. E., Pratt, G. W., Langer, R., Jacks, T. and Anderson, D. G. (2011). Treating metastatic cancer with nanotechnology. *Nat. Rev. Cancer,* **12**, 39–50.
161. de Carvalho, A. C., Kowalski, L. P., Campos, A. H., Soares, F. A., Carvalho, A. L. and Vettore, A. L. (2012). Clinical significance of molecular alterations in histologically negative surgical margins of head and neck cancer patients. *Oral Oncol.,* **48**, 240–248.
162. Zhang, H. F., Maslov, K., Stoica, G. and Wang, L. V. (2006). Functional photoacoustic microscopy for high-resolution and noninvasive *in vivo* imaging. *Nat. Biotechnol.*, **24**, 848–851.
163. Qian, X., Peng, X. H., Ansari, D. O., Yin-Goen, Q., Chen, G. Z., Shin, D. M., Yang, L., Young, A. N., Wang, M. D. and Nie, S. (2008). *In vivo* tumor targeting and spectroscopic detection with surface-enhanced Raman nanoparticle tags. *Nat. Biotechnol.,* **26**, 83–90.

Chapter 8

Light-Activated Nanopreparations

L. A. Wells[1] *and H. Sheardown*[2]

[1]*Department of Chemical Engineering, Queens University, Kingston, ON, USA*
[2]*Department of Chemical Engineering, McMaster University, Hamilton, ON, USA*

8.1 Introduction

The ability to target tissues has the potential to result in more effective therapies, overcoming the issues of non-specific cell and biodistribution, where the drug may be readily metabolized or excreted. Cancer therapies in particular are critically limited by the need for site-specific drug delivery and activation, although other treatments would potentially be used for enhanced control [1]. The use of light as an external stimulus has the potential to be used to control properties of materials in a host of applications, particularly in the delivery of drugs. There has recently been a significant amount of interest in this area due to the relatively non-invasive nature and the potential for spatiotemporal control, with drug delivery that can be turned off and on as necessary [2–4]. Additionally, the ability to control wavelength, intensity, duration, and site of irradiation means that the quantity of drug release, its location and the duration of release can be controlled [1]. Wavelengths corresponding to ultraviolet (UV), visible, near-infrared (NIR), and infrared (IR) have been exploited to generate on-demand payload delivery [2]. These wavelengths are both biocompatible and capable of inducing structural changes in the polymer. An advantage of light responsive materials is that changes to the structure can either be physical (increased solubility and/or conformation change) or chemical (bond cleavage), depending on the nature of the light-sensitive group incorporated [5].

Nanomedicine has significant potential to enhance treatment in a wide variety of applications. Using devices that are within the same size range as the therapeutic being delivered (1–1000 nm versus less than 10 nm) has advantages including higher cell targeting, reduced settling velocity, and potential for functionalization [6]. By combining these two platforms, the potential for targeted, on-demand drug delivery will lead to new and innovative treatments. Potential applications of these materials include ocular and dermal delivery, as both of these tissues are easily light accessible, although other areas, particular for cancer treatment, have been targeted. This chapter will provide an overview of the structure and chemistry used to create light-sensitive nanodelivery vehicles and the intended applications.

8.2 Light-Modulated Self Assembly

Light-responsive vehicles can be developed through the covalent incorporation of light-responsive groups that can be induced to release the payload upon illumination [7]. Light can be also be used to modulate the assembly of delivery vehicles, with the advantages of control, rapidity, and ease of focus [8]. There has been increased interest in the application of photo-dimerization, photo-isomerization, and photo cleavage in supramolecular chemistry [9]. A wide variety of functional chromophores, including coumarin, spiropyran, spirooxazine, and others, have been used. Among the most widely studied polymers in this class contain azobenzene, which, when exposed to light of a particular wavelength, is capable of switching from the *cis* to *trans* form of the isomer.

Azobenzene, for example, has been used to develop materials, which transitions from a micelle to a hollow sphere when exposed to visible light, while the reverse transition occurs upon exposure to UV radiation [10]. These micelles could also be formed by light exposure and subsequently cross-linked to stable delivery vehicles. Photo-induced shape transformation of azo containing materials has also been used [11, 12]. Spherical nanoparticles, when exposed to an argon (Ar) laser with a wavelength of 488 nm, could be induced to adopt an ellipsoidal shape, with the ratio of the length to the radius being dependent on irradiation time.

Coumarin has also been widely used as a photo-responsive group. Hydrophilic block copolymers containing coumarin were assembled into micelles and subsequently cross-linked with UV light (310 nm) [13]. The resulting materials showed interesting intra-particular lower critical solution temperature (LCST) and intra-particular upper critical solution temperature (UCST) behavior.

The incorporation of thymine groups in particular or other small amphiphiles in general [196–198] can be used to induce a chemical change in the material. Specifically, a cycloaddition between the C5 and C6 double bonds of adjacent thymine molecules upon exposure to UV light (200–300 nm) has been used to generate stable microparticles.

8.3 Photo-Responsive Molecules

Molecules that undergo isomerization or dimerization with specific UV and visible light can be exploited to introduce reversible qualities in biomaterials through their covalent incorporation into polymer systems. There is extensive literature exploring biomaterials that incorporate molecules that undergo either photo-isomerization or photo-dimerization.

Azobenzene [14] and spirobenzopyran [15] undergo photo-isomerization with UV and visible light, which can introduce photo-initiated structural and solubility alterations when bound within polymers. Polymer-bound photo-isomerizing groups can modify rheological and cross-linking properties of polymer gels. For example, azobenzene bound to polymer networks can create photo-induced hydrophobic domains that bridge polymer networks to cause cross-linking [16]. When bound to two chains, the length decrease associated with azobenzene *trans* to *cis* isomerization (9.0 to 5.5 angstroms) [17] can alter mesh sizes and, therefore, the cross-linking density of polymer networks [16, 18]. While photo-isomerizing molecules clearly introduce reversible changes into bulk polymer systems, they have shown minimal influence on changing diffusion and delivery rates of molecules [16, 19]. In addition, the ability for visible light to act as stimuli is undesirable in some applications; unintentional changes may occur in ophthalmic applications, for example, since visible light is constantly exposed to the eye. Therefore, it is more likely that molecules that undergo reversible photo-dimerization with other wavelengths of light will find significant application.

Molecules that have the ability to dimerize with a second molecule could be used to form reversible cross-linking systems to create photo-responsive systems with minimum side-product formation, fast curing, and high spatial control [20]. Anthracene [20], cinnamylidene acetate [21, 22], nitrocinnamate [23], and coumarin [24] are popular dimerizing molecules in the investigation of photo-sensitive polymerization and the cross-linking of both synthetic and natural hydrogels. As illustrated in Fig. 8.1, nitrocinnamate, cinnamylidene acetate, and coumarin undergo isomerization followed by reversible [$2\pi+2\pi$] photo-addition to form cyclobutane rings [24–26], while

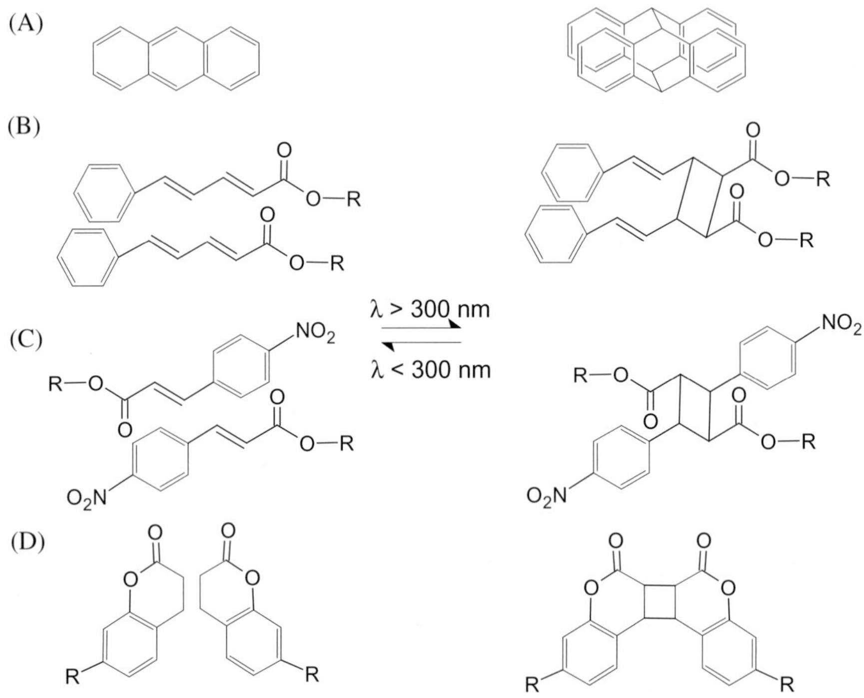

Figure 8.1. Dimerization and dissociation of anthracene (A), cinnamylidene acetate (B), nitrocinnamate (C) and coumarin (D) λ = wavelength.

anthracene undergoes reversible [4π+4π] photo-cycloaddition with dimerization at wavelengths over 300 nm and dissociation at wavelengths less than 300 nm [27, 28]. Table 8.1 briefly summarizes some examples of polymer systems modified with photo-dimerizing molecules.

Incorporation of photo-dimerizing groups with polymers can potentially create materials that can support cell growth and protein attachment and potentially alter the diffusion of molecules. Due to its high compatibility, polyethylene glycol (PEG)-based hydrogels with photo-dimerizing groups have been investigated; for example, cinnamylidene acetate terminated four arm PEG can form gels with UV light exposure [29], coumarin-bound organophosphorus hydrolase could be photo-immobilized [30], and protein flux could be lowered through these materials with UV light treatments [20]. Incomplete solution–gel transitions were noted due to a small amount of cross-linking from dimerization that also occurs at low UV wavelengths of 254 nm [22]. Anthracene terminated 8 arm PEG that undergoes photo-cross-linking has also been synthesized, demonstrating light-responsive reversible surface roughness [21]. Nitrocinnamate capped 8-armed PEG polymers have been used to deliver fibroblast growth

Table 8.1. Overview of investigated photo-reversible-polymer hydrogel systems. PEG = polyethylene glycol, PMAA = polymethacrylic acid, PMMA = polymethyl methacrylate, MA = methacrylate, PS = polystyrene.

Photo-reversible group	Polymer	Results
Cinnamylidene acetate	PEG 4-armed	• Photo-switching behaviour • Physical properties controlled by wavelength exposure • Long exposure times required
	MA copolymer	• Photo-sensitive • Effects swelling and gelation
Nitrocinnamate	PEG 8-armed	• Photo-scissable in the absence of photo-initiators or catalysts • Rapid reversible crosslinking • FGF-2 drug release over short term
	Gelatin	• Control swelling with UV light exposure • Low cytotoxicity
Anthracene	PMAA	• For reversible molecular imprinting
	PEG 8-armed	• Photo-reversible up to 20% • No sol-gel-sol physical transition
	Dextran	• Hydrophilic with hydrophobic microdomains
	PMMA	• Examined critical number of cross-links and photo-scission — Not a water based system • Indicated more investigations required
	MA copolymer	• Small spacer between backbone • Examined side reactions
	PS	• Complete photo-reversibility • Used as chain builder, not cross-linker
	PS/PMMA	• End capped polymer • Used to form diblock copolymers
Coumarin	PEG	• For photo-reversible chain extension

factor 2 at rates that are dependent on UV light exposure [31]. Release of growth factors from these materials showed UV-induced decreased release with 50–70% delivery at 120 h (5 days) [31].

8.4 Vehicles for Light-Activated Drug Delivery

8.4.1 *Liposomes*

Varying in size from nanometers to micrometers, liposomes are lipid assemblies comprised of aqueous cores that are surrounded by phospholipid bilayers, with the drug entrapped in either the core or the bilayer. They can be prepared as unilamellar vesicles or multilamellar vesicles with various significant variations in surface charge, membrane fluidity, and lipid composition, making them relatively amenable to a variety of drug delivery challenges. There is also a great deal of flexibility in the nature of the drug that can be released from these vehicles; both hydrophobic and hydrophilic drugs can be accommodated by the amphiphilic lipids, while factors such as charge of the lipid layer can be used to control loading and drug release. Liposomal systems can help protect vulnerable compounds from degradation or can be used to solubilize poorly soluble molecules [32].

Recently, Leung, and Romanowski reported the development of a series of gold-coated ligand liposomes. These materials were capable of spectral tenability, meaning that the wavelength of light used to initiate release could be carefully controlled. Using NIR light increased the depth of penetration, making this system more versatile. By focusing the light on a small cell-sized region, it was possible to control the spatial release of the drug of interest [33, 34].

While photo-dynamic therapy has a great deal of potential in the treatment of disease, including tumors, the drugs are typically hydrophobic and aggregate in their monomeric state. Since the monomeric form is typically necessary for efficacy, an appropriate carrier is often necessary to ensure that the active drugs reach the site of interest. Therefore, various carriers and drug delivery systems have been developed in order to protect the drugs and ensure its activity at the site of action [35]. One of the more successful examples of a photo-sensitizer encapsulated in a non-PEGylated liposomal formulation is the QLT Inc. drug Visudyne, currently approved for the treatment of wet age related macular degeneration (AMD) and useful in the treatment of other back-of-the-eye diseases. The incorporation of PEG groups into this formulation and into other formulations used in the treatment of various diseases, presumably by protecting the active pharmaceutical ingredient from metabolic and other biological processes, has been found to significantly enhance the residence time [36, 37].

Active targeting has been widely used with photo-dynamic formulations, as the cytotoxic singlet oxygen generated in the reaction has a relatively short effective distance, meaning that accumulation via the enhanced permeability and retention (EPR) effect may not be sufficient to generate a robust reaction, particularly with aggressive tumor cells. Modification of the liposomal membrane with various functional groups to target cells of interest has therefore been widely used to enhance the efficacy of the treatment. Targeting ligands including antibodies, antibody fragments, aptamers, peptides, glycoproteins, and small molecules have been conjugated to the lipid surface to ensure maximum efficacy [38]. Coupled with the liposomal release of the photo-active agent, mechanisms of triggering the release, including the incorporation of pH, temperature, or enzyme sensitivity, have been used [38–42]. The incorporation of photo-sensitivity into the liposome structure can further enhance the release, as well as the spatial localization of the molecule of interest.

One of the more novel advances in the field of liposomal materials for delivery of photo-sensitive agents is the development of a novel porphysome [35, 43–47]. With the properties of conventional liposomes, these materials also have multimodal imaging and therapeutic properties. In contrast to conventional vehicles where the free photo-sensitizer is encapsulated inside the liposomal vesicle, in these materials, the pyropheophorbide was conjugated directly to the phospholipid. The pyro-lipid conjugates retain the properties to allow for self assembly into liposomal vesicles, but upon photo-activation, the release of heat can be used to generate particles that are both photo-thermal sensitizers and photo-acoustic contrast agents, allowing for the potential for visualization and therapy.

8.4.2 *Micelles*

Micelles are similar to liposomes in that they are self assemblies of amphiphilic molecules formed at concentrations greater than the critical micelle concentration. The amphiphiles assemble in aqueous media to form hydrophobic cores surrounded by hydrophilic shells. Different structures, including unimolecular, standard, and reverse micelles, can form depending on the solvent and the amphiphilic molecule used to form the micelle [48]. Typically, PEG is included in polymeric micelles to minimize aggregation and prolong the circulation time. By selecting the proper molecular weight of the PEG shell, such that it is greater than the hydrophobic core, control over the size of the micelles formed, as well as the critical micelle concentration, can be achieved [6].

There are fewer applications of light in controlling the release from or the assembly of micelles. In one application, the reduced non-specific uptake of micelles in normal tissues and their preferential uptake in solid tumors due to the EPR effect were combined with photo-chemical internalization (PCI), which led to selective photo-chemical rupture of endocytosed molecules. This combination minimizes drug degradation in the lysosome [1].

Matteini *et al.* [49] combined thermo-responsiveness and light activation in a novel sponge-like nanocomposite. The sponges were fabricated from chitosan, and incorporated gold nanorods to include light responsiveness. The drug was contained within amphiphilic micelles comprised of a poly caprolactone — poly ethylene glycol — poly caprolactone (PCL–PEG–PCL) polymer. Laser activation of scaffolds loaded with the drug doxorubicin led to the death of Henrietta Lacks (HeLa) cells.

Similar to micelles, but with the advantage of having the capability to encapsulate both hydrophilic and hydrophobic drugs, polymersomes represent an emerging class of self assembling drug delivery vehicles. Using synthetic amphiphilic copolymers, which have a similar structure to lipids, but which have the advantage of flexibility in terms of its potential for modification and functionalization, these materials have the opportunity for enhanced encapsulation and targeted drug delivery [50]. The synthetic nature of the polymers used in the preparation of polymersomes means that the incorporation of functional groups allows for external stimulation to be used. As with other nanopreparations, the use of NIR has shown the most promise for translation to the clinic. To overcome the problems with the relatively slow NIR-triggered photochemical reactions, combination materials involving lanthanide upconverting nanoparticles (UCNPs) have been used [51]. When irradiated by NIR light, the UCNPs absorb the light and convert it to high energy photons in the UV and visible spectra. Alternatively, the tethering of coumarin derivatives along the chain leads to materials that can be activated by two-photon NIR radiation [52]. Similarly, a combination of thermally and photo-activated systems has been developed in which the photon energy absorbed by a metal-containing nanoparticle is converted to heat, which can lead to disruption of the particle and release of the contents [50].

8.4.3 *Nanoparticles*

Gold nanoparticles have the potential to be a promising carrier for cancer therapies in particular because like other nanoparticles, they easily accumulate in the immune system [53], they can be easily functionalized to incorporate drugs or other functional groups, and they can be activated by the application of light. Light can then penetrate the nanoparticles, heating them up, leading to the destruction of nearby cancer cells. Similar work has demonstrated the functionality of gold nanorods and gold nanocages in this application [54, 55].

Guardado-Alvarez *et al.* have demonstrated the use of light to open mesoporous silica nanoparticle carriers containing disulfide bonds [56]. The systems are described as "snap top" with the release being mediated either by UV light or NIR two-photon excitation. In another study, the incorporation of light responsive functional groups was shown to lead to photo-induced cross-linking and decross-linking of silicone materials [57]. In other work, photo-induced cross-linking and decross-linking of alginate and hyaluronic acid-based microparticles facilitated by the incorporation of a photo-active anthracene or a photo-active coumarin group into the polymers led to control over the release of both large and small molecules [58–62]. By tethering the anthracene to a multifunctional PEG-based cross-linker, the systems were shown to be relatively generic and suitable for crosslinking various polymers [63].

8.5 Challenges

Among the first clinical applications of light responsive therapies is photo-dynamic therapy (PDT), in which photo-sensitizers are activated via light to generate highly reactive oxygen singlet molecules, which have the potential to damage biological membranes, potentially useful in the treatment of cancer [64] and in drug delivery to the back of the eye. Photo-dynamic therapy with verteporfin is an approved treatment for a variety of choroidal vascular disorders that now rarely includes wet-AMD since the emergence of anti-vascular endothelial growth factor (VEGF) agents onto the market. In addition to ophthalmology, systems in gynaecology, dermatology, otolaryngology, gastroenterology, and physiotherapy are also amenable to light and laser responsive devices and are medical areas that are already well-versed in the application of lasers. However, the tissue damage induced by UV radiation is potentially significant. Therefore, the discovery that near-infrared light with wavelengths between 700 and 1,000 nm can penetrate several centimeters into tissue without causing damage [7] has opened the door for other potential treatments using this technology. While results with systems activated in this manner are promising, most published studies to date using NIR radiation use irradiation times on the order of hours, complicating the practical application of these technologies [7]. Therefore, as with most stimuli responsive materials, light responsive systems are limited to some extent by the rapidity of the response [65]. While it is possible to increase the response time by making the materials thinner or smaller, this often renders the system too fragile to be useful. Furthermore, in biomedical applications, the ultimate end use and the need for biocompatibility must be carefully considered when designing the vehicle. Currently, there is a significant lack of *in vivo* evaluation of these materials.

8.6 Conclusions

The potential for photo-activated materials to play a role in the treatment of a host of diseases is significant, although significant limitations do exist. Combining photo-responsiveness with nanomaterials will lead to the generation of materials that are spatially and temporally specific. The use of near infrared light is likely to be the most common method of application in the future. Combining photo-sensitivity with another responsive methodology, such as thermal or pH, has the potential to overcome some of the limitations associated with materials that only respond to light.

References

1. Cabral, H., Nakanishi, M., Kumagai, M., Jang, W. D., Nishiyama, N. and Kataoka, K. A. (2009). Photo-activated targeting chemotherapy using glutathione sensitive camptothecin-loaded polymeric micelles. *Pharm. Res.*, **26**, 82–92.
2. Joglekar, M. and Trewyn, B. G. (2013). Polymer-based stimuli-responsive nanosystems for biomedical applications. *Biotechnol. J.*, **8**, 931–945.
3. Mura, S., Nicolas, J. and Couvreur, P. (2013). Stimuli-responsive nanocarriers for drug delivery. *Nat. Mater.*, **12**, 991–1003.
4. Cheng, R., Meng, F. H., Deng, C., Klok, H. A. and Zhong, Z. Y. (2013). Dual and multi-stimuli responsive polymeric nanoparticles for programmed site-specific drug delivery. *Biomaterials*, **34**, 3647–3657.
5. Kelley, E. G., Albert, J. N. L., Sullivan, M. O. and Epps, T. H. (2013). Stimuli-responsive copolymer solution and surface assemblies for biomedical applications. *Chem. Soc. Rev.*, **42**, 7057–7071.
6. Kompella, U. B., Amrite, A. C., Ravi, R. P. and Durazo, S. A. (2013). Nanomedicines for back of the eye drug delivery, gene delivery, and imaging. *Prog. Retin. Eye Res.*, **36**, 172–198.
7. Lehner, R., Wang, X. Y., Wolf, M. and Hunziker, P. (2012). Designing switchable nanosystems for medical application. *J. Control. Release*, **161**, 307–316.
8. Zhang, J. X., Li, X. D. and Li, X. H. (2012). Stimuli-triggered structural engineering of synthetic and biological polymeric assemblies. *Prog. Polym. Sci.*, **37**, 1130–1176.
9. Yagai, S. and Kitamura, A. (2008). Recent advances in photoresponsive supramolecular self-assemblies. *Chem. Soc. Rev.*, **37**, 1520–1529.
10. Liu, X. K. and Jiang, M. (2006). Optical switching of self-assembly: Micellization and micelle-hollow-sphere transition of hydrogen-bonded polymers. *Angew. Chem. Int. Ed.*, **45**, 3846–3850.
11. Wang, D. R. and Wang, X. G. (2013). Amphiphilic azo polymers: Molecular engineering, self-assembly and photoresponsive properties. *Prog. Polym. Sci.*, **38**, 271–301.
12. Che, P. C., He, Y. N. and Wang, X. G. (2006). Synthesis of a main-chain photo-responsive azo polymer through azo-coupling reaction. *Acta Polym. Sin.*, 977–981.
13. He, J., Yan, B., Tremblay, L. and Zhao, Y. (2011). Both Core- and Shell-Cross-Linked Nanogels: Photoinduced Size Change, Intraparticle LCST, and Interparticle UCST Thermal Behaviors. *Langmuir*, **27**, 436–444.
14. Kumar, G. S. and Neckers, D. C. (1989). Photochemistry of azobenzene-containing polymers. *Chem. Rev.*, **89**, 1915–1925.
15. Minkin, V. I. (2004). Photo-, thermo-, solvato-, and electrochromic spiroheterocyclic compounds. *Chem. Rev.*, **104**, 2751–2776.

16. Tomer, R. and Florence, A. T. (1993). Photo-responsive hydrogels for potential responsive release applications. *Int. J. Pharm.*, **99**, R5–R8.
17. El Halabieh, R. H., Mermut, O. and Barrett, C. J. (2004). Using light to control physical properties of polymers and surfaces with azobenzene chromophores. *Pure Appl. Chem.*, **76**, 1445–1465.
18. Suzuki, T., Shinkai, S. and Sada, K. (2006). Supramolecular crosslinked linear poly(trimethylene iminium trifluorosulfonimide) polymer gels sensitive to light and thermal stimuli. *Adv. Mater.*, **18**, 1043–+.
19. Patnaik, S., Sharma, A. K., Garg, B. S., Gandhi, R. P. and Gupta, K. C. (2007). Photoregulation of drug release in azo-dextran nanogels. *Int. J. Pharm.*, **342**, 184–193.
20. Zheng, Y. J., Mieie, M., Mello, S. V., Mabrouki, M., Andreopoulos, F. M., Konka, V., *et al.* (2002). PEG-based hydrogel synthesis via the photodimerization of anthracene groups. *Macromolecules*, **35**, 5228–5234.
21. Andreopoulos, F. M., Beckman, E. J. and Russell, A. J. (1998). Light-induced tailoring of PEG-hydrogel properties. *Biomaterials*, **19**, 1343–1352.
22. Andreopoulos, F. M., Beckman, E. J. and Russell, A. J. (2000). Photoswitchable PEG-CA hydrogels and factors that affect their photosensitivity. *J. Polym. Sci. Polym. Chem.*, **38**, 1466–1476.
23. Zheng, Y. J., Andreopoulos, F. M., Micic, M., Huo, Q., Pham, S. M. and Leblanc, R. M. (2001). A novel photoscissile poly(ethylene glycol)-based hydrogel. *Adv. Funct. Mater.*, **11**, 37–40.
24. Trenor, S. R., Long, T. E. and Love, B. J. (2004). Photoreversible chain extension of poly(ethylene glycol). *Macromol. Chem. Phys.*, **205**, 715–723.
25. Ziffer, H., Bax, A., Highet, R. J. and Green, B. (1988). Investigation by two-dimensional nmr of the structure and stereochemistry of a methyl para-nitrocinnamate photodimer. *J. Org. Chem.*, **53**, 895–896.
26. Trenor, S. R., Shultz, A. R., Love, B. J. and Long, T. E. (2004). Coumarins in polymers: From light harvesting to photo-cross-linkable tissue scaffolds. *Chem. Rev.*, **104**, 3059–3077.
27. Bouas-Laurent, H., Castellan, A., Desvergne, J. P. and Lapouyade, R. (2000). Photodimerization of anthracenes in fluid solution: Structural aspects. *Chem. Soc. Rev.*, **29**, 43–55.
28. Bouas-Laurent, H., Castellan, A., Desvergne, J. P. and Lapouyade, R. (2001). Photodimerization of anthracenes in fluid solutions: (Part 2) Mechanistic aspects of the photocycloaddition and of the photochemical and thermal cleavage. *Chem. Soc. Rev.*, **30**, 248–263.
29. Andreopoulos, F. M., Deible, C. R., Stauffer, M. T., Weber, S. G., Wagner, W. R., Beckman, E. J., *et al.* (1996). Photoscissable hydrogel synthesis via rapid photopolymerization of novel PEG-based polymers in the absence of photoinitiators. *J. Am. Chem. Soc.*, **118**, 6235–6240.
30. Andreopoulos, F. M., Roberts, M. J., Bentley, M. D., Harris, J. M., Beckman, E. J. and Russell, A. J. (1999). Photoimmobilization of organophosphorus hydrolase within a PEG-based hydrogel. *Biotechnol. Bioeng.*, **65**, 579–588.
31. Andreopoulos, F. M. and Persaud, I. (2006). Delivery of basic fibroblast growth factor (bFGF) from photoresponsive hydrogel scaffolds. *Biomaterials*, **27**, 2468–2476.

32. Blenke, E. O., Mastrobattista, E. and Schiffelers, R. M. (2013). Strategies for triggered drug release from tumor targeted liposomes. *Expert Opin. Drug Deliv.*, **10**, 1399–1410.
33. Leung, S.J. and Romanowski, M. (2012). NIR-activated content release from plasmon resonant liposomes for probing single-cell responses. *Acs Nano*, **6**, 9383–9391.
34. Leung, S. J. and Romanowski, M. (2012). Light-activated content release from liposomes. *Theranostics*, **2**, 1020–1036.
35. Jin, C. S. and Zheng, G. (2011). Liposomal nanostructures for photosensitizer delivery. *Lasers Surg. Med.*, **43**, 734–748.
36. Immordino, M. L., Dosio, F. and Cattel, L. (2006). Stealth liposomes: Review of the basic science, rationale, and clinical applications, existing and potential. *Int. J. Nanomedicine*, **1**, 297–315.
37. Woodle, M. C. (1998). Controlling liposome blood clearance by surface-grafted polymers. *Adv. Drug Deliv. Rev.*, **32**, 139–152.
38. Kraft, J. C., Freeling, J. P., Wang, Z. Y. and Ho, R. J. Y. (2014). Emerging research and clinical development trends of liposome and lipid nanoparticle drug delivery systems. *J. Pharm. Sci.*, **103**, 29–52.
39. Dou, Y. N., Zheng, J. Z., Foltz, W. D., Weersink, R., Chaudary, N., Jaffray, D. A., *et al.* (2014). Heat-activated thermosensitive liposomal cisplatin (HTLC) results in effective growth delay of cervical carcinoma in mice. *J. Control. Release*, **178**, 69–78.
40. Li, L., ten Hagen, T. L. M., Haeri, A., Soullie, T., Scholten, C., Seynhaeve, A. L. B., et al. (2014). A novel two-step mild hyperthermia for advanced liposomal chemotherapy. *J. Control. Release*, **174**, 202–208.
41. Offerman, S. C., Verma, A. V. K., Telfer, B. A., Berk, D. A., Clarke, D. J. and Aojula, H. S. (2014). Ability of co-administered peptide liposome nanoparticles to exploit tumour acidity for drug delivery. *Rsc Advances*, **4**, 10779–10790.
42. You, J., Zhang, P. Z., Hu, F. Q., Du, Y. Z., Yuan, H., Zhu, J., *et al.* (2014). Near-infrared light-sensitive liposomes for the enhanced photothermal tumor treatment by the combination with chemotherapy. *Pharm. Res.*, **31**, 554–565.
43. Carter, K. A., Shao, S., Hoopes, M. I., Luo, D., Ahsan, B., Grigoryants, V. M., *et al.* (2014). Porphyrin-phospholipid liposomes permeabilized by near-infrared light. *Nat. Commun.*, **5**.
44. Ng, K. K., Lovell, J. F., Vedadi, A., Hajian, T. and Zheng, G. (2013). Self-assembled porphyrin nanodiscs with structure-dependent activation for phototherapy and photodiagnostic applications. *ACS Nano*, **7**, 3484–3490.
45. Hsu, C. Y., Nieh, M. P. and Lai, P. S. (2012). Facile self-assembly of porphyrin-embedded polymeric vesicles for theranostic applications. *Chem. Commun.*, **48**, 9343–9345.
46. Lovell, J. F., Jin, C. S., Huynh, E., Jin, H. L., Kim, C., Rubinstein, J. L., *et al.* (2011). Porphysome nanovesicles generated by porphyrin bilayers for use as multimodal biophotonic contrast agents. *Nat. mater.*, **10**, 324–332.
47. Lovell, J. F., Jin, C. S., Huynh, E., MacDonald, T. D., Cao, W. G. and Zheng, G. (2012). Enzymatic regioselection for the synthesis and biodegradation of porphysome nanovesicles. *Angew. Chem. Int. Ed.*, **51**, 2429–2433.
48. Trivedi, R. and Kompella, U. B. (2010). Nanomicellar formulations for sustained drug delivery: Strategies and underlying principles. *Nanomedicine*, **5**, 485–505.
49. Matteini, P., Martina, M. R., Giambastiani, G., Tatini, F., Cascella, R., Ratto, F., *et al.* (2013). Light-responsive nanocomposite sponges for on demand chemical release with high spatial and dosage control. *J. Mater. Chem. B*, **1**, 1096–1100.

50. Feng, A. C. and Yuan, J. Y. (2014). Smart nanocontainers: Progress on novel stimuli-responsive polymer vesicles. *Macromol. Rapid Commun.*, **35**, 767–779.
51. Gohy, J. F. and Zhao, Y. (2013). Photo-responsive block copolymer micelles: Design and behavior. *Chem. Soc. Rev.*, **42**, 7117–7129.
52. Babin, J., Pelletier, M., Lepage, M., Allard, J. F., Morris, D. and Zhao, Y. A (2009). New two-photon-sensitive block copolymer nanocarrier. *Angew. Chem. Int. Ed.*, **48**, 3329–3332.
53. Almeida, J. P. M., Figueroa, E. R. and Drezek, R. A. (2014). Gold nanoparticle mediated cancer immunotherapy. *Nanomed. Nanotech. Biol. Med.*, **10**, 503–514.
54. Parak, W. J., Carregal-Romero, S., Hartmann, R. and Nazarenus, M. (2013). Light-activated opening of gold-nanoparticle modified capsules. *Abstr. Pap. Am. Chem. Soc.*, **245**.
55. Wang, H., Zhao, Y. L. and Nie, G. J. (2013). Multifunctional nanoparticle systems for combined chemo- and photothermal cancer therapy. *Front. Mater. Sci.*, 7, 118–128.
56. Guardado-Alvarez, T. M., Devi, L. S., Vabre, J. M., Pecorelli, T. A., Schwartz, B. J., Durand, J. O., *et al.* (2014). Photo-redox activated drug delivery systems operating under two photon excitation in the near-IR. *Nanoscale*, **6**, 4652–4658.
57. Fawcett, A. S. and Brook, M. A. (2014). Thermoplastic silicone elastomers through self-association of pendant coumarin groups. *Macromolecules*, **47**, 1656–1663.
58. Wells, L. A., Furukawa, S. and Sheardown, H. (2011). Photoresponsive PEG-anthracene grafted hyaluronan as a controlled-delivery biomaterial. *Biomacromolecules*, **12**, 923–932.
59. Wells, L. A., Lasowski, F., Fitzpatrick, S. D. and Sheardown, H. (2010). Responding to change, thermo- and photo-responsive polymers as unique biomaterials. *Crit. Rev. Biomed. Eng.*, **38**, 487–509.
60. Wells, L. A. and Sheardown, H. (2007). Extended release of high pI proteins from alginate microspheres via a novel encapsulation technique. *Eur. J. Pharm. Biopharm.*, **65**, 329–335.
61. Wells, L. A. and Sheardown, H. (2011). Photosensitive controlled release with polyethylene glycol-anthracene modified alginate. *Eur. J. Pharm. Biopharm.*, **79**, 304–313.
62. Kang, M. K. and Kim, J. C. (2014). UV-triggerable alginate beads decorated with basic proteinoid-coumarin conjugates. *J. Biomater. Sci. Polym. Ed.*, **25**, 843–854.
63. Wells, L. A., Brook, M. A. and Sheardown, H. (2011). Generic, anthracene-based hydrogel crosslinkers for photo-controllable drug delivery. *Macromol. Biosci.*, **11**, 988–998.
64. Christie, J. G. and Kompella, U. B. (2008). Ophthalmic light sensitive nanocarrier systems. *Drug Discov. Today*, **13**, 124–134.
65. Qiu, Y. and Park, K. (2012). Environment-sensitive hydrogels for drug delivery. *Adv. Drug Deliv. Rev.*, **64**, 49–60.

Chapter 9

Magnetic Field-Responsive Nanocarriers

Aditi Jhaveri

Center for Pharmaceutical Biotechnology and Nanomedicine,
Department of Pharmaceutical Sciences,
Northeastern University, Boston, MA, USA

9.1 Introduction

Drug delivery practices have undergone rapid advances in the past few decades, largely due to parallel advances in the fields of nanotechnology and nanomedicine, as well as to the tremendous progress in our understanding of fundamental cancer biology [1, 2]. These advances are evident in the development of a diverse array of sophisticated nanocarriers constructed from materials including lipids, polymers, metals, and mesoporous materials, as well as biomolecules like proteins, carbohydrates, and nucleic acids [3–12]. The design of nanocarriers has evolved from simple carriers, which passively accumulate through the leaky vasculature in diseased areas (tumors, infarcts, inflammation sites), to nanocarriers, which are actively targeted and take advantage of over-expressed or differentially expressed antigens in diseased tissues and organs. In addition, "smart" or stimuli-responsive carriers go a step further, by sensing pathophysiological changes in their surrounding environment that trigger the release of therapeutic payloads, thus adding the extra dimensions of spatial and temporal control to drug release. These smart nanocarriers may respond to intrinsic cues like altered pH, redox status, enzymatic activity, or temperature of the cell environment, and can be constructed from materials that enable them to respond to an external stimulus applied to diseased tissues and tumors. The external stimulus can be provided by light (ultraviolet (UV), visible, or near-infrared (NIR)), ultrasound, heat, or an external magnetic field. The stimuli-triggered nanocarriers (both internal and external) provide controlled drug release, enhance the cellular uptake of payloads, control their intracellular fates, and also lead to physical action on the surrounding pathological tissues [1, 13].

Magnetically-responsive or magnetic field responsive nanocarriers (MFRNs) constitute a very promising class of the external stimuli-responsive carriers. MFRNs typically carry a therapeutic payload along with a magnetically active component in a pharmaceutically stable formulation. MFRNs may also be designed to carry one or more therapeutic moieties simultaneously, targeting ligands, peptides for intracellular delivery, fluorescent labeling dyes , radionuclides, or other environmentally-responsive components to form multifunctional MFRNs that bring together a host of useful features for diagnosis, therapy, as well as for imaging across multiple platforms [14, 15]. For the purpose of this chapter, we focus our discussion mainly on various applications of MFRNs, which include drug targeting, medical hyperthermia, imaging, and sophisticated theranostic and multifunctional nanocarrier platforms for drug delivery. The detailed synthesis, preparation, and conjugation strategies for magnetic particles have not been discussed here, but can be found in excellent reviews elsewhere [16–23]. This chapter is organized into two main parts, with the first part serving as a general introduction to magnetic nanoparticles (MNPs), their properties, as well as an introduction to some terms frequently encountered in the field of magnetism. The rest of the chapter is focused on the application of MNPs that are incorporated within nanocarriers to form MFRNs, and covers their use specifically in the fields of drug delivery, imaging, and diagnosis, using select references from the recent literature. For clarity, we have used the term MNPs to describe the nanoparticles that impart the magnetic property (e.g. iron oxide) and MFRN is used for the vehicle/nanocarrier that incorporates MNPs.

9.2 Magnetic Materials and Nanoparticles: A Brief Overview

The physical and chemical properties of a material at the nano-scale range differ significantly from those seen in its bulk form. Such nano-scale materials are said to manifest so-called "quantum size effects"[19]. At the nano-scale range, a material displays greatly increased surface-to-volume ratios, altered surface chemistry, as well as increased chemical reactivity from that in the bulk phase [15, 24]. The increased fraction of surface atoms that results at the nano level is considered to be one of the main reasons for the markedly different physico-chemical properties exhibited compared to the bulk material. A decrease in the particle size leads to an increase in the fraction of surface energy in its chemical potential [19]. Magnetic properties also change significantly between bulk materials and nanomaterials. These properties include magnetization per atom, magnetic anisotropy, Curie (T_C) or Neel (T_N) temperatures (i.e. temperatures of spontaneous parallel or anti-parallel orientations of spins between nanoparticle and the corresponding microscopic phases), remanence (residual magnetism), and the coercivity field (H_C) (the field required to bring the magnetization to zero) [16, 19].

We discuss some common terminology and concepts related to MNPs in the sections that follow.

9.2.1 *Properties and types of magnetic materials*

The magnetic properties of a material exist in the orbital and spin motions of its electrons, whose spin and angular momentum are associated with a magnetic moment [25]. Finite-size effects and surface effects are the dominant determinants of various special properties of the MNPs [16, 26]. A number of factors determine the magnetic properties of the MNPs. The important ones include the chemical composition, the type and degree of defectiveness in the crystal lattice, particle size and shape, morphology, and the interaction of the particle with the surrounding matrix and neighboring particles. Varying the nanoparticle (NP) size, shape, composition, and structure can enable control of the magnetic characteristics of the material to a certain extent. However, it is difficult to maintain control of these factors during the synthesis of NPs, which are nearly equal in size and composition, and may ultimately result in NPs of the same type having remarkably different properties [19]. Magnetism comprises the atomic or sub-atomic response of a material to an applied magnetic field, in which the spin and charge of electrons create a dipole moment and magnetic field. The combination of multiple dipoles can create a magnetic field of a measurable intensity [27].

Magnetic materials are classified based on their magnetic susceptibility (χ), which is defined by the ratio of the induced magnetization (M) to the applied magnetic field (H) [28]. The five main types of magnetic materials based on the orientation of magnetic moments are diamagnetic, ferromagnetic, paramagnetic, anti-ferromagnetic, and ferrimagnetic materials [29, 30]. Most elements, including gold, silver, copper, and many other materials, fall into the diamagnetic group. They usually lack any unpaired electrons, which results in a zero net magnetic moment, and a negative magnetic susceptibility ($\chi < 0$). In the presence of an external magnetic field, the atomic loops created by the orbital motion of electrons tend to oppose the applied field, but only weakly. Usually, any form of magnetism that a material possesses overpowers diamagnetism due to its weak nature [30]. In ferromagnetic materials, an atom has a net magnetic moment due to its unpaired electrons. Such materials are composed of domains, each of which contains large numbers of atoms with parallel magnetic moments that produce a net magnetic moment of the domain that points in some direction. The random distribution of these magnetic moments produces a zero net magnetic moment of the material. In the presence of a magnetic field, the magnetic moments align along the direction of the applied field and form a large net magnetic moment, which exists even after the applied field is removed [29]. In paramagnetic materials, an atom has a net magnetic moment due to unpaired electrons. However, there are no magnetic domains. In the presence of a magnetic field the magnetic moments of the atoms align along the direction of the field giving rise to a weak net magnetic moment that is not retained upon removal of the field. Antiferromagnetic materials are compounds of two different atoms that occupy different lattice positions, such that their magnetic moments are equal in magnitude and opposite in direction, yielding a zero net magnetic moment. Ferrimagnetic materials are similar to antiferromagnetic compounds, except that the magnitudes of their magnetic moments are different, and hence do not cancel each other out, resulting in a net spontaneous magnetic moment. In a magnetic field, antiferromagnetic and ferrimagnetic materials behave like ferromagnetic materials [29].

As the size of ferromagnetic materials, such as MNPs, is reduced below a critical value, they become a single domain and maintain one large magnetic moment [28]. When the size of single-domain particles decrease further, below a critical diameter, the coercivity becomes zero and these particles become superparamagnetic. The phenomenon of superparamagnetism is caused by thermal effects. In such particles, at sufficiently high temperatures (i.e. blocking temperature T_B), the thermal energy is strong enough to spontaneously demagnetize a previously saturated assembly. Therefore, these particles have zero coercivity and show no hysteresis [30]. The superparamagnetic NPs become magnetic in the presence of the external magnetic field, but lack any residual magnetism after removal of the field. This property allows them to maintain colloidal stability and avoid aggregation, making them useful for biomedical applications [28]. Some examples belonging to each of the different types of magnetic materials are enlisted in Table 9.1.

9.2.2 *Commonly employed magnetic materials*

By far, the most commonly employed magnetic components for preparation of MFRNs for biomedical applications are the iron oxide nanoparticles, magnetite (Fe_3O_4) or its oxidized and more stable form, maghemite (Fe_2O_3).

Table 9.1. Representative examples belonging to each class of magnetic materials.

Type of magnetic material	Representative examples
Diamagnetic	Majority of the materials, gold, silver, copper
Ferromagnetic	Iron, nickel, cobalt, rare earth elements
Paramagnetic	Gadolinium, magnesium, lithium, tantalum
Anti-ferromagnetic	Manganese (II) oxide (MnO), cobalt (II) oxide (CoO), nickel (II) oxide (NiO), copper (II) chloride ($CuCl_2$), troilite, iron (II) sulfide (FeS)
Ferrimagnetic	Magnetite and maghemite

They are preferred to other metal oxide nanoparticles due to their excellent biocompatibility and stability [20, 28, 31, 32]. The magnetic susceptibility of elemental iron (218 emu/g), as well as magnetite (90 emu/g), is higher compared to other metals like nickel (Ni) and cobalt (Co). Since iron is also widely distributed in the body, it is devoid of any toxicity. Iron oxide nanoparticles are superparamagnetic, and are better known as superparamagnetic iron oxide nanoparticles (SPIONs). SPIONs are hydrophobic and are, thus, prone to aggregation unless coated with some inert materials. Some of the commonly employed coating materials for SPIONs used to avoid aggregation and improve their biocompatibility include dextran, chitosan, polyethylene glycol (PEG), polyethyleneimine (PEI), silica, poly(vinyl alcohol) (PVA), oleic acid, citric acid, pullulan, and alginate, among others [33–42]. SPIONs may be in the form of magnetic colloids or ferrofluids, which are colloidal suspensions of single-domain MNPs of about 10 nm, dispersed in a polar or non-polar liquid [43]. SPIONs are discussed in detail in the later sections of this chapter, since they are the most widely used magnetic components for drug delivery and imaging applications.

Elemental Ni has a much lower magnetic susceptibility than Fe (55 emu/g) [27]. Ni is mostly employed in the form of Ni oxides [44, 45], or as hybrid composites with arsenic [46], iron (Fe) [47, 48], or Co [49] for various applications ranging from the generation of reactive oxygen species (ROS) and mediating apoptosis, to the generation of hyperthermia in an external magnetic field. However, the use of Ni nanoparticles is limited due to reports of various toxicities. Acute toxicity to the liver and spleen, lung inflammation, and cardiotoxicity were observed in rats upon intravenous (i.v.) administration of Ni NPs [50]. Genotoxicity and inflammation in the lungs, oxidative stress, as well as carcinogenicity have also been reported for Ni NPs [51, 52]. Ni exposure also affects a number of signaling pathways, including hypoxia inducible factor (HIF)-1α [53], leads to increased expression of matrix metalloproteinases (MMP-2 and MMP-9) [54] and secretion of inflammatory cytokines and chemokines [55, 56], and is also responsible for causing oxidative stress [57, 58].

MNPs based on Co also show a high magnetic susceptibility (160 emu/g and 68 emu/g, respectively, for elemental Co and Co ferrite) and have been used for magnetically targeted drug and gene delivery, hyperthermia, and imaging applications [27, 59–63]. The toxicity of Co nanoparticles, however, is better characterized compared to that of Ni nanoparticles. Co nanoparticles have been reported to cause reproductive toxicity [64], carcinogenicity, increased secretion of MMPs and oxidative DNA damage [65], genotoxicity, ROS-dependent and independent toxicity [66], and secretion of cytokines like tumor necrosis factor (TNF)-α and interferon (IFN)-γ [67].

Several other MNPs have been used for various purposes including hyperthermia applications, magnetic targeting, or as magnetic resonance imaging (MRI) contrast agents. These include gadolinium (Gd), manganese (Mn), zinc (Zn), and magnesium (Mg), and their oxides, alloys of Fe with Co, Ni, platinum (Pt), and Zn, ferrimagnetic spinels, Mn–Zn–Gd ferrite particles, copper–nickel, lanthanum complexes, and ferrimagnetic $SrFe_{12}O_{19}/\gamma Fe_2O_3$, among others [19, 20, 68–72]. Gd is a promising coating agent for MNPs due to its popularity as a positive contrast agent for MRI [73]. It has a high magnetic moment, considering that it is a paramagnetic material. However, this magnetic moment is still well below that of superparamagnetic materials, like iron oxide [22].

MNPs are synthesized using many different methods including co-precipitation, microemulsion, chemical vapor deposition, solution–gel, laser pyrolysis synthesis, thermal decomposition, sonochemical, microwave assisted, and solvothermal methods [23, 30]. Details about the synthesis of MNPs can be found in other reviews as suggested earlier.

9.3 Magnetic Drug Targeting (MDT) Using Nanocarriers: General Concepts

Nanocarriers overcome a number of limitations faced by conventional drugs including poor specificity for the target site, impaired diffusion through biological barriers, unfavorable pharmacokinetics, poor biodistribution, sub-optimal pharmacological activity, as well as the toxicity resulting from off-target effects of drugs [6, 23, 74]. A number of nanocarriers including liposomes, polymeric micelles, dendrimers, nanogels, solid lipid nanoparticles, metallic nanoparticles, and nanocapsules are under investigation to aid the delivery of therapeutics [2, 6, 13, 75, 76]. Active targeting strategies that make use of stimuli-responsive nanocarriers are particularly interesting due to their ability to selectively deliver the drug to a specific target (cells, organs, or tissues). MFRNs constitute a very promising class of stimuli-responsive nanocarriers. They are constructed using magnetic materials of which iron oxides are the most popular. Other magnetic materials mentioned in Section 9.2.2 are also employed in the construction of MFRNs.

For MDT, the MFRNs are usually loaded with a therapeutic agent, as well as a magnetically active component in a pharmaceutically stable formulation. Once injected intravenously (i.v.), the MFRNs are subjected to an external magnetic field applied at the pathological site (tumor), so as to concentrate them there and eventually release the encapsulated therapeutic payload in a site-specific manner [23, 77] (Fig. 9.1). Once the MFRNs concentrate at the target, the active payload may be released via enzymatic activity or changes in the pH, temperature, or osmolality. MDT allows a drug to be concentrated

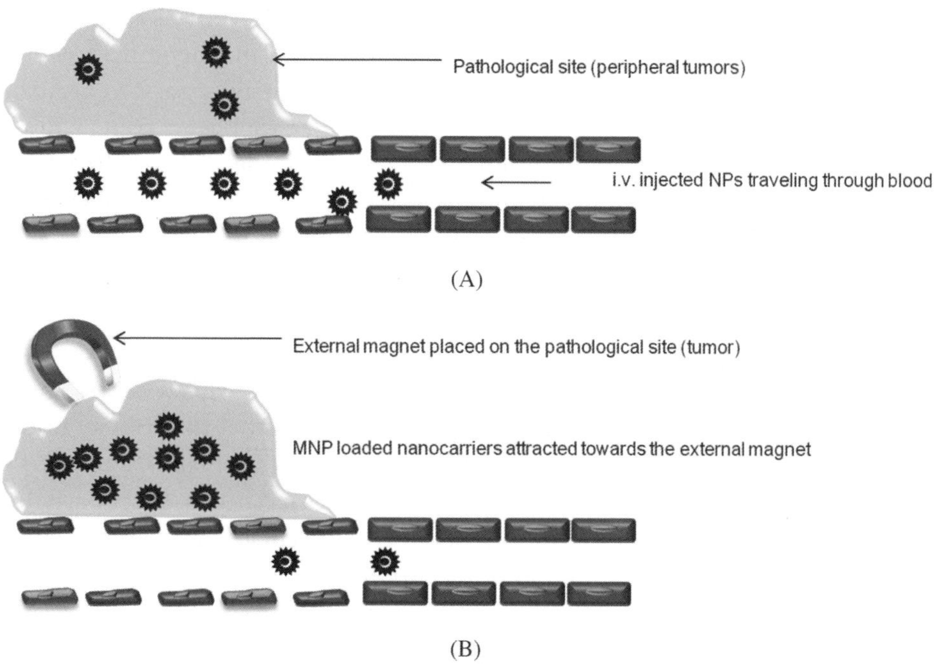

Figure 9.1. Magnetic drug targeting. (A) Nanocarriers in the blood after i.v. administration. In the absence of an external magnet, a small fraction of the administered NPs accumulate within the tumor. (B) MFRNs loaded with magnetic NPs accumulate in the tumor under the influence of an external magnet, resulting in a much higher fraction of the accumulated NPs in the tumor. i.v. = intravenous, MNP = magnetic nanoparticle.

in the desired site, thus minimizing accumulation in non-target or healthy tissues. The overall drug uptake by the mononuclear phagocytic system (MPS) can be significantly reduced by better selectivity and a longer duration of exposure of the drug at the target. MDT also uses much smaller doses of the drug, which results in increased drug efficacy and reduced toxicity [78–80].

Magnetic drug targeting is influenced by certain physiological parameters of the patient, as well as technical requirements of the ferrofluid/MNP drug combination. The properties of the MNP/drug carrier include particle size and surface characteristics, concentration of the MNPs/ferrofluids, volume of the fluid, reversibility and strength of the drug/ferrofluid binding, and finally the route and duration of infusion. The magnetic field properties that influence MDT include the strength of the magnetic field and duration of its application, while the physiological characteristics of the patient that affect MDT include the size, weight, body surface area, blood volume, cardiac output, systemic vascular resistance, circulation time, tumor volume and location, and tumor vascular content and blood flow [81].

Although the benefits of MDT are evident, it comes with its own limitations. Its main limitation relates to the strength of the magnetic field which needs to be applied to establish the magnetic gradient necessary to control the residence time of MFRNs in the desired area, as well as to trigger drug desorption [82]. The magnetic gradient decreases as the distance from the target increases. Permanent neodymium–iron–boron magnets in combination with MNPs increase the effective magnetic field to a depth of 10–15 cm [83]. The geometry of the magnetic field is an important consideration, as MNPs accumulate throughout the cross-section from the applied source to the depth marking the field limit, and not only at the target site [82]. Another limitation to MDT arises from the size of MNPs. They need to be small enough for superparamagnetism, which in turn, is essential to avoid agglomeration upon removal of the magnetic field. However, the small size of MNPs produces a reduced magnetic response, which makes it hard to direct particles to the target due to the drag of blood flow. Larger particles on the other hand can withstand the flow dynamics of blood circulation much more easily. Thus, MDT is likely to be more effective in regions of slower blood flow, especially if the target site is close to the magnet source [80]. Also, MNP accumulation in the target region may lead to the possibility of embolization of the blood vessels in the target region. With respect to the biomedical applications of MDT, the consideration of various physiological parameters, as well as larger distances between the target site and magnet, may complicate the extrapolation of MDT data from lab animals to humans and constitutes an additional limitation [80, 82]. The toxicity of MNPs, both direct toxicity as well as that from its degradation products, is always a concern, and more so, when studies on the fate of MNPs, as well as their characterization, are inadequate. Lastly, MDT is mainly applicable to well defined surface tumors amenable to application of an external magnet. Using MDT for deep-seated tumors, or for smaller tumors in the early stages of their growth is a challenging prospect, which seems to require sophisticated nanocarrier platforms [82].

9.4 Magnetically Responsive Nanocarriers for Specific Applications

In this section, selected examples from the recent literature, which use MFRNs for various applications, are described. This includes passively targeted, as well as ligand modified and actively targeted MFRNs for drug and gene delivery, MFRNs used for the purpose of hyperthermia, and MRI and MFRNs developed as theranostics. Finally, we also briefly discuss additional applications of MFRNs, including those for biomagnetic separations, cell labeling, and tracking.

9.4.1 *Passively targeted magnetic nanocarriers*

Liposomes have been widely investigated as nanocarriers and are the most clinically advanced of all existing nanocarriers. They have also been employed widely for the purpose of magnetic drug targeting [84–88]. Dandamudi *et al.* showed that the tumor vascular uptake of magnetic cationic liposomes (MCLs) loaded with magnetite (MAG-C) and the anti-cancer drug vinblastine increased significantly in the presence of an external magnet in a murine melanoma model. In addition, a significant anti-tumor effect was also observed with reduced tumor nodules in metastatic sites in the presence of an external magnet compared to the no-magnet treatment group [89]. However, magnetic liposomes do suffer from some drawbacks, such as instability related to the lipid bilayers and the fact that their morphology can vary with the concentrations of MNPs and lipids. They may also result in short circulation times and inclusion leakage after i.v administration due to their propensity to aggregate and form large vesicles in suspension [90, 91]. To overcome this drawback, hybrid liposomal cerasome, which is a partially ceramic- or silica-coated liposome, has become of much interest [92, 93]. Like liposomes, cerasomes can also accommodate both hydrophobic and hydrophilic agents. Cao *et al.* reported, for the first time, stabilized magnetic cerasomes for the delivery of anti-cancer drugs using an external magnetic field [94]. The authors constructed

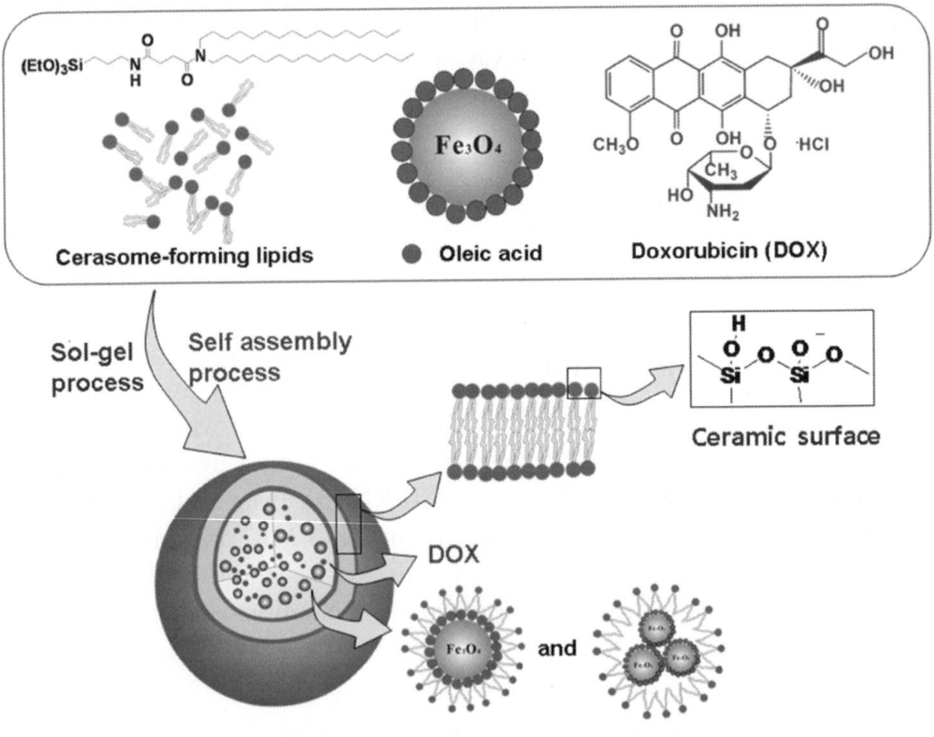

Figure 9.2. Schematic illustration of DOX-loaded magnetic cerasomes. (Reprinted with permission from Cao, Z. *et al.*, Stabilized magnetic cerasomes for drug delivery. *Langmuir*, 2013, 29(48), 14976–14983. Copyright © 2013, American Chemical Society).

doxorubicin (Dox) (DOX)-loaded magnetic cerasomes (DLMCs) by loading both oleic acid-modified hydrophobic Fe_3O_4 NPs and DOX in the aqueous interior of liposomes via a facile one-step construction using the sol–gel process and self-assembly techniques (Fig. 9.2).

The hydrophobic Fe_3O_4 NPs were likely trapped in the cerasome core through the formation of an intermediate Fe_3O_4–micelle complex. The maximum loading of Fe_3O_4 in DLMCs was found at a Fe_3O_4 /lipid molar ratio of 4:1. The DLMCs showed superior superparamagnetism and a strong response to the magnetic fields. They also had a high encapsulation efficacy (43.4 ± 4.7%) and a high drug loading content (3.2 ± 1.3%). The storage stability of DLMCs and their sustained release behavior was significantly better than drug-loaded liposomes. *In vitro* studies in Henrietta Lacks (HeLa) cells showed that application of an external magnetic field caused a rapid and efficient uptake of DLMCs by the cells, resulting in a higher capacity to kill tumor cells than non-magnetic drug-loaded cerasomes [94].

Magnetofection or magnetic transfection is the process by which nucleic acid (deoxyribonucleic acid, DNA; antisense oligodeoxynucleotides, AODN; and short interfering ribonucleic acid, siRNA) delivery can be enhanced in the presence of a magnetic field [23, 95, 96]. Magnetic vectors containing magnetic NPs (typically, iron oxide) are prepared, and under the action of a magnetic field, are concentrated *in vivo,* in the target tissue against the hydrodynamic forces or are allowed to magnetically sediment on target cells in cell culture conditions. Magnetofection helps to overcome the diffusion barrier to nucleic acid delivery and can enhance the gene delivery efficiency by several hundred folds [95–98]. A novel gene delivery system based on hybrid polymer-magnetic micelles was recently reported by Haladjova *et al.* [99]. The hybrid micelles were formed by co-dissolution of iron oxide modified with oleic acid (hydrophobic surface) and amphiphilic polystyrene-*b*-poly(quaternized 2-vinylpyridine) (PS-*b*-P2QVP) block copolymer in an organic solvent. The micelles consisted of a cationic hydrophilic corona of PQVP and a hydrophobic PS core, in which the magnetic NPs were randomly distributed. Complexes of these micelles were then formed with linear DNA (salmon sperm, 2000 bp) and plasmid DNA (pEGFP-N1, 4370 bp). The resulting magnetopolyplexes had phosphate:amine (P:N) ratios in the range of 0.5–20. The hybrid magnetic/cationic copolymer micelles had a narrow size distribution with the zeta potential values varying from negative to positive depending on the P:N ratio. They had low toxicity and were able to introduce pEGFP-N1 DNA into cells. The transfection efficiency was lower than commercially available reagents in the absence of a magnetic field, yet the internalization was higher than PEI indicating that although the magnetopolyplexes were internalized by cells, they were unable to escape the endo/lysosomes. Upon application of a magnetic field, however, the transgene expression efficiency increased drastically (up to 3.5 times), which was superior to commercially available transfection agents, like Lipofectamine and dendritic PEI. This was a clear indication that magnetic forces helped the magnetopolyplexes in escaping from the endolysosomal pathway to a large extent, and allowed for spontaneous transfection and a substantial benefit over non-magnetic systems [99].

Similarly, magnetic lipoplexes (ternary complexes of cationic lipids with plasmid DNA associated with SPIONs) were reported by Wang *et al.* for the delivery of short hairpin RNA (shRNA) to suppress Type-1 insulin-like growth factor receptor (IGF-1R) over-expression in lung cancer *in vitro* and *in vivo* [100]. The authors compared liposomal magnetofection of plasmids expressing green fluorescence protein (GFP) and shRNAs targeting IGF-1R (pGFPshIGF-1R) to lipofection by lipofectamine 2000. Transfection efficiencies were studied in A549 cells and in A549 tumor-bearing mice. Liposomal magnetofection resulted in a three-fold improvement in transgene expression over lipofection *in vitro*, and showed better delivery efficiency *in vivo* over the lipofectamine group [100]. Other carriers like mPEG–PEI–SPION nanoparticles have been investigated for the co-delivery of siRNA against PIN-2 interacting protein (PinX1-siRNA) with DOX in glioma cells [101]. Nanocarriers composed of iron oxide NPs and cationic lipid-like materials known as lipidoids have also been investigated for enhanced DNA and siRNA delivery by magnetofection [102].

9.4.2 *Actively targeted magnetic nanocarriers*

The efficiency of MFRNs can be significantly increased by increasing their local accumulation in diseased tissues. Active targeting of MFRNs involves their surface modification by conjugation of targeting ligands that can identify and possess an enhanced affinity for unique molecular signatures found on diseased cells (e.g. tumor cells) and tissues [28]. Such an active targeting strategy can trigger receptor-mediated endocytosis (RME), resulting in higher intracellular drug concentrations and as a consequence increased cytotoxicity [23]. Some of the widely investigated targeting ligands for MNPs include monoclonal antibodies (mAbs) [103–105], small molecules such as folate [106, 107], proteins (e.g. transferrin) [108, 109], peptides [110, 111], and aptamers [112, 113].

Varshosaz *et al.* reported on the development of micelles self-assembled from folate-conjugated dextran/retinoic acid copolymers (FA/DEX-RA) and encapsulated with oleic acid-coated Fe_3O_4 MNPs and DOX to achieve dual tumor-targeting [114]. A schematic for the synthesis of the dual targeted micelles is depicted in (Fig. 9.3).

The dual targeted micelles showed effective folate-targeting in folate over-expressing MCF-7 cells. The greatest growth inhibitory effect was seen in cells treated with FA/DEX-RA MNPs in the presence of a magnetic field (0.42 T) at a DOX concentration of 1μM, which was about two and four times greater than the effect observed with FA/DEX-RA MNPs without local magnetic field and free DOX at 1μM, respectively. However, in the folate receptor-negative MDA-MB-468 breast cancer cells, there was no difference between magnetically targeted micelles without DOX and free DOX, due to the ineffective internalization in the folate receptor-negative cells. Thus, the micelles showed successful dual targeting, which may lead to a reduction of both the dose of DOX and the side effects. A combination of external magnetic field and targeting with folate resulted in improved cytotoxic effects of the various formulations tested [114].

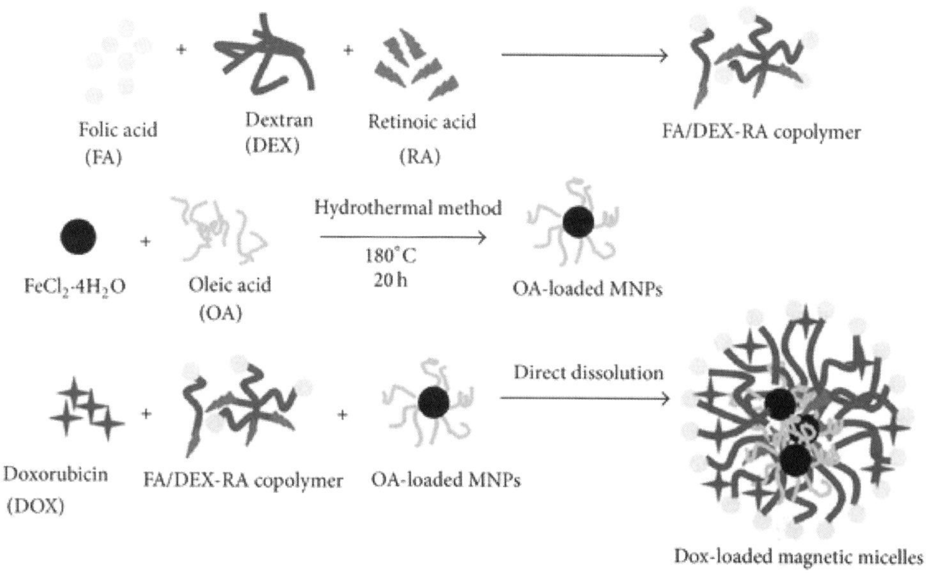

Figure 9.3. Formation of DOX-loaded FA/DEX-RA magnetic micelles. MNP = magnetic nanoparticle. (Reprinted from Varshosaz, J. *et al.*, Use of Magnetic Folate-Dextran-Retinoic Acid Micelles for Dual Targeting of Doxorubicin in Breast Cancer. *BioMed Res. Int.*, 2013, 16).

In another example, a novel dual-targeted drug delivery system was reported for ovarian cancer [115]. An inner core constructed from Fe_3O_4 NPs was surface-functionalized with amine groups to link both a single-chain antibody (scFv) targeted to Endoglin, over-expressed in ovarian cancer, and β-cyclodextrin (β-CD). Docetaxel (TXT) was encapsulated in the grafted β-CD. The particles were confirmed to be superparamagnetic following magnetization studies. TXT exhibited a high encapsulation efficacy and drug loading capacity (98.7% and 10.7%, respectively) and showed a sustained concentration-driven release *in vitro*. The scFv, grafted on the NPs and targeted to Endoglin, retained its ability to bind Endoglin as observed from modified enzyme-linked immunosorbent assay (ELISA) studies. The magnetic targeting property was confirmed using an *in vitro* flow chamber. These studies showed the potential of this system as a dual-targeted system for ovarian cancer using an external magnetic field and binding of scFv to Endoglin in ovarian cancer tissue. Cytotoxicity studies also showed that TXT released from this system successfully inhibited the growth of SKOV-3 ovarian cancer cells with a sustained inhibition for up to 72 h. The NPs that were not loaded with TXT showed no growth inhibition, confirming the non-toxic nature of the carrier. Moreover, the system showed no toxicity to normal human umbilical vein endothelial cells (HUVECs) [115].

Guo *et al.* recently reported a novel drug delivery system comprised of c(RGDyK)-modified Fe_3O_4 NPs with a high DOX load (R-DMP) to combine the benefits of MDT, integrin $\alpha_v\beta_3$ targeting, and high drug-loading properties [111]. The targeting peptide c(RGDyK) and DOX were coupled chemically to the dual function MNPs (DMP) using poly-L-glutamic acid (PLGA). The peptide-modified nanoparticles showed a pH-sensitive DOX release. Compared to the non-peptide modified group, the peptide-modified nanoparticles showed enhanced uptake by integrin expressing tumor cells and had a stronger cytotoxicity [111]. The authors had previously reported a $Fe_3O_4 \cong SiO_2$-based delivery system targeting the transferrin receptor (TfR), which combined magnetic targeting, TfR targeting, and high DOX loading [109]. They immobilized Tf onto $Fe_3O_4 \cong SiO_2$ NPs with high DOX loading (TfDMP), by chemically coupling Tf and DOX with dual-functional magnetic nanoparticles (DMPs) using a multi-armed cross-linker, PLGA. The TfDMP exhibited enhanced uptake by the TfR-expressing cells, as well as a stronger cytotoxic effect than the non-targeted counterparts [109].

9.4.3 *MFRNs for hyperthermia*

Hyperthermia is the application of heat to tissues at temperatures slightly above those found in normal physiological conditions [116]. Cancer cells exhibit a greater sensitivity to the effects of heat compared to normal cells because of their compact and disorganized vascular architecture and a subsequent difficulty in dissipating heat [117]. This has generated interest in treating cancer using artificially induced hyperthermia, which heats malignant cells, but spares the surrounding normal tissues, and has led to some encouraging results so far [118, 119]. A major limitation in the application of hyperthermia is the lack of selectivity in confining heating only to the affected zone (tumor) without damaging the surrounding healthy tissue [120, 121]. To that end, iron oxide NPs can be selectively localized to tumors making them ideal candidates for local heat generation. This procedure, known as magnetic fluid hyperthermia (MFH), involves injecting SPIONs into the body in the form of ferrofluids, followed by the application of an alternating magnetic field (AMF) of sufficient strength and frequency to allow the particles to heat. In the presence of the AMF, the magnetic moment inside the particles oscillates and magnetic energy is liberated in the form of heat through hysteresis losses or Nèel relaxation and is transferred to the tissue environment [80, 83, 122, 123]. By adjusting the strength and duration of exposure of the magnetic field, the degree of heat produced can be manipulated [120, 123, 124]. When a temperature of 41–43°C is reached, the tumor cells are irreversibly damaged, whereas the damage to normal cells is reversible [83]. Magnetic hyperthermia (MHT), thus, presents an attractive option to promote only the intended target tissue heating. Local heating, besides having a direct effect on cells, can cause an increase in perfusion of systemically administered drugs into the core of the tumor and can also lead to increased permeability [125].

Hayashi *et al.* recently reported the development of smart NPs (Fe_3O_4/DOX/PPy-PEG-FA, where PPy = polypyrrole), that generated heat in response to an alternating current magnetic field (ACMF) and sequentially released DOX [126]. The smart NPs were composed of a polymer with a glass transition temperature of (T_g) of 44°C, which contained clustered Fe_3O_4 NPs and DOX.

In response to the ACMF, the clustered Fe_3O_4 NPs produced heat. As their temperature increased above 44°C, it softened the polymer phase, causing the release of DOX (Figs. 9.4a and 9.4b).

To increase the NP retention in the tumor, the NPs were modified with folic acid (FA) by amidation of Fe_3O_4/DOX/PPyCOOH NPs with amine- and FA-heterobifunctionalized PEG (FA-PEG-NH_2) (Fig. 9.5a). The *in vivo* therapeutic efficacy of the combination of MHT and chemotherapy using smart NPs was studied in mice bearing subcutaneous xenograft tumors derived from the RPMI8226 multiple myeloma cell line.

While MHT or chemotherapy inhibited tumor growth somewhat, the combination of MHT and chemotherapy using smart NPs completely abrogated the tumor without recurrence of malignancy (Fig. 9.5b). The combination destroyed the cancer cells in the entire tumor due to a synergistic effect. The Fe_3O_4/DOX/PPy-PEG-FA NPs, which were administered intratumorally, exhibited no significant toxicity [126].

A number of therapeutics are under investigation to target various heat shock protein (HSP)-related pathways. Bortezomib targets the nuclear factor (NF)-κB pathway that is in part regulated by HSP70 and HSP90 [127], and geldanamycin (GM) inhibits HSP90 and the mammalian target of rapamycin/hypoxia inducible factor (mTOR/HIF) pathways [128].

The major problem with hyperthermia for induction apoptosis is the thermoresistance or the acquisition of resistance to thermal stress by the heat-treated cells, which then have an increased survival rate [129]. This necessitates the use of high temperature hyperthermia to achieve therapeutic efficacy, which in turn, initiates cell necrosis that can induce inflammatory disease and cancer metastasis [130]. To counter the challenge of activation of apoptosis at a low temperature without inducing thermoresistance, Yoo *et al.*, reported

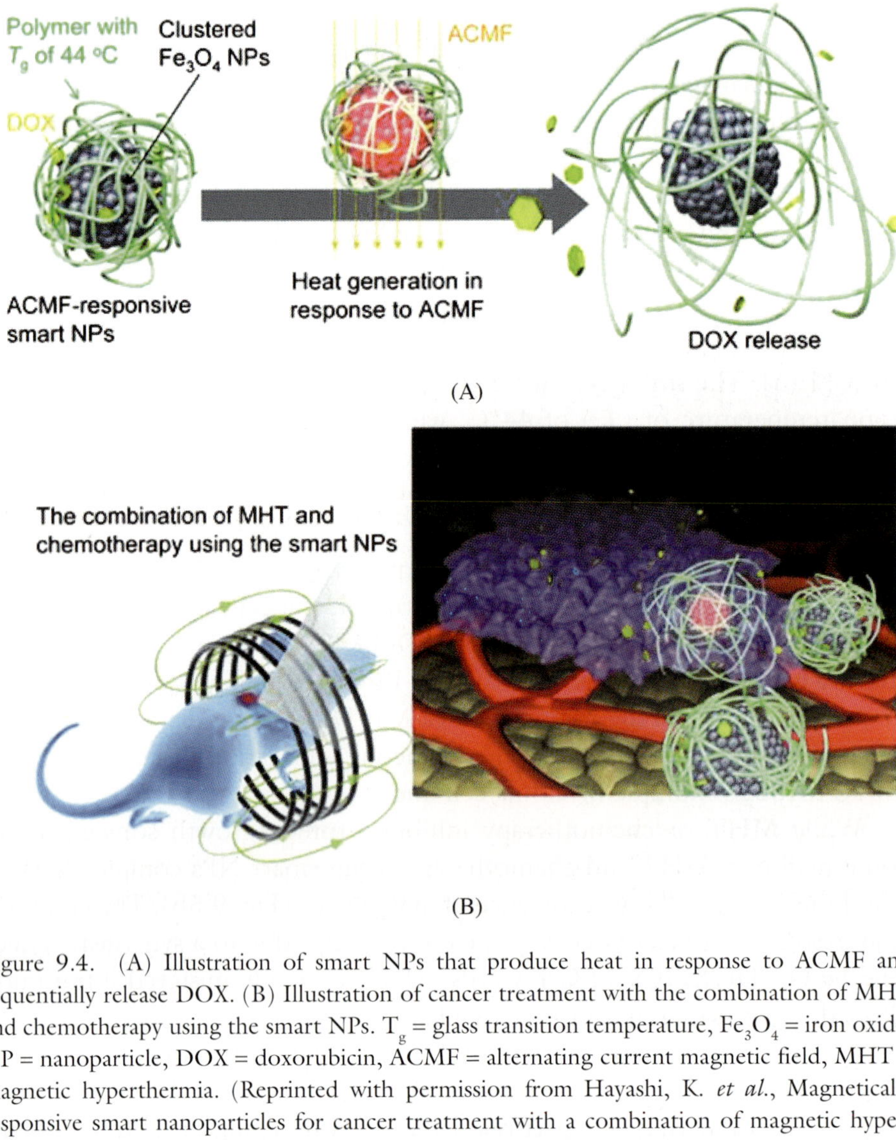

Figure 9.4. (A) Illustration of smart NPs that produce heat in response to ACMF and sequentially release DOX. (B) Illustration of cancer treatment with the combination of MHT and chemotherapy using the smart NPs. T_g = glass transition temperature, Fe_3O_4 = iron oxide, NP = nanoparticle, DOX = doxorubicin, ACMF = alternating current magnetic field, MHT = magnetic hyperthermia. (Reprinted with permission from Hayashi, K. *et al.*, Magnetically responsive smart nanoparticles for cancer treatment with a combination of magnetic hyperthermia and remote-control drug release. *Theranostics*, 2014, **4**(8), 834–844).

the development of a novel resistance-free apoptosis-inducing magnetic nanoparticle (RAIN) that promotes thermoresistance-free apoptosis [131]. RAIN had two functional subunits: (i) for Hsp inhibition and (ii) for heat generation from MNP, such that these two functions were designed to be triggered only when an AMF was applied. The Hsp inhibition strategy was employed to minimize stress-mediated chaperonic functional responses, which inhibit apoptosis. GM, a benzoquinone ansamycin was used to inhibit Hsp90. RAIN

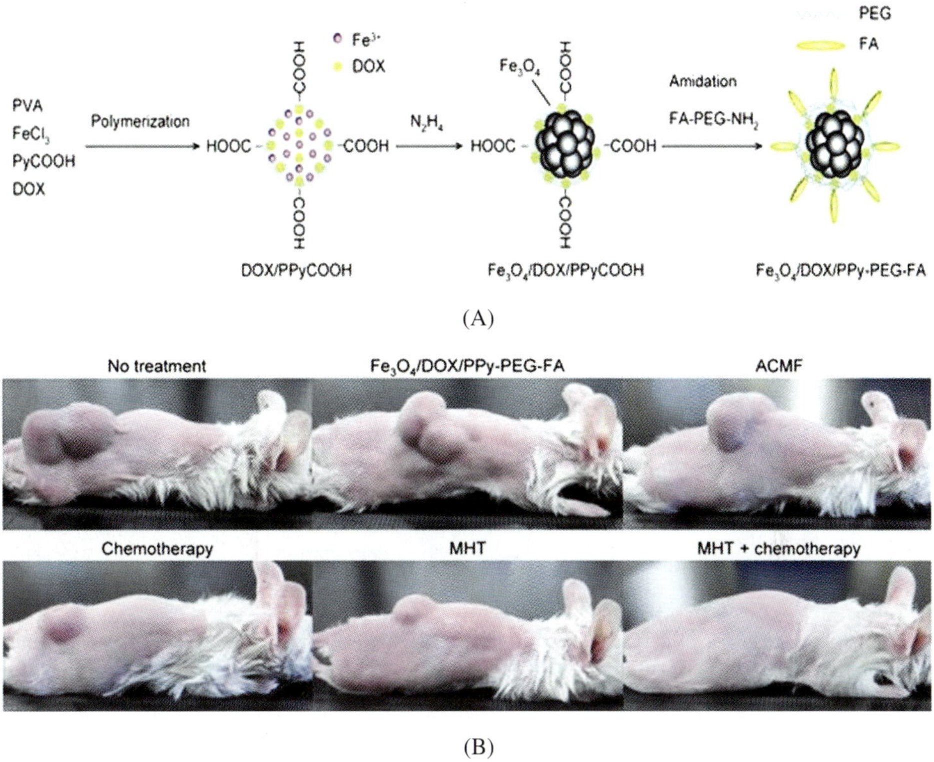

Figure 9.5. (A) Schematic showing the synthesis of Fe_3O_4/DOX/PPy-PEG-FA NPs (B) Photographs of non-treated mice, mice treated with chemotherapy, mice exposed to ACMF, mice injected with Fe_3O_4/DOX/PPy-PEG-FA NPs intratumorally, mice treated with MHT, and mice treated with the combination of MHT and chemotherapy 45 days after treatment. Fe^{3+} = iron (III), DOX = doxorubicin, PEG = polyethylene glycol, FA = folic acid, ACMF = alternating current magnetic field, MHT = magnetic hyperthermia. (Reprinted with permission from Hayashi, K. *et al.*, Magnetically responsive smart nanoparticles for cancer treatment with a combination of magnetic hyperthermia and remote-control drug release. *Theranostics*, 2014, 4(8), 834–844).

successfully promoted exclusive apoptosis and obstructed cell survival by inhibiting HSP90 both *in vitro* and *in vivo* in MDA-MB-231 xenograft bearing nude mice under low temperature hyperthermia conditions (~43°C). To synthesize RAIN, an amine-functionalized MNP, 15nm $Zn_{0.4}Fe_{2.6}O_{4}$, was connected to GM via a thermolabile azo linker. The linker is cleaved with the loss of nitrogen gas upon exposure to an AMF releasing GM (Fig. 9.6).

The RAINs were injected into tumor-bearing mice and subjected to an AMF (37.4 kAm^{-1}, 500 kHz) to generate hyperthermia. RAIN hyperthermia resulted in complete tumor elimination by the eighth day. In the MNP hyperthermia group, there was a slight tumor reduction followed by the re-growth of tumors [131].

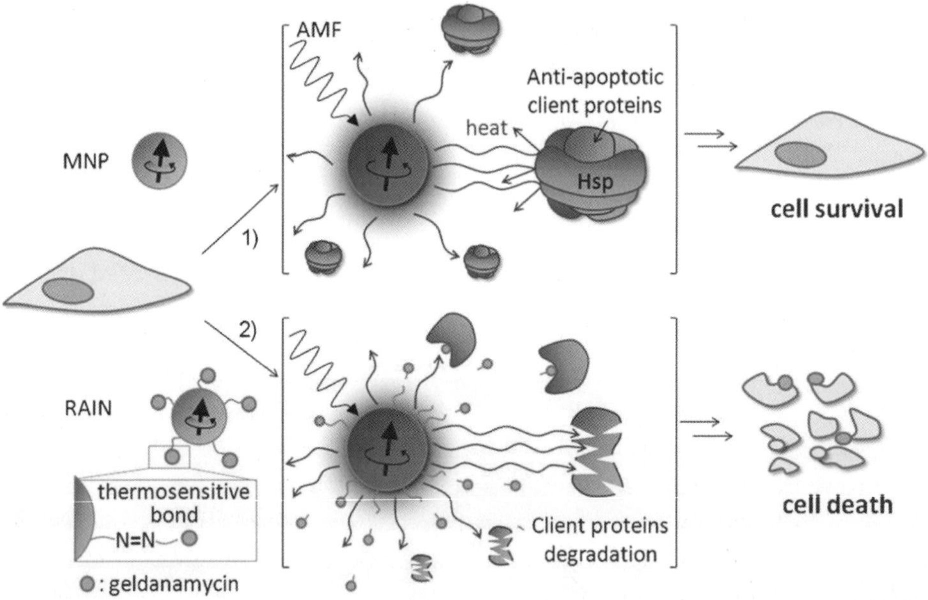

Figure 9.6. RAIN for effective apoptotic hyperthermia (1) Heat-treated cancer cells launch a protective mechanism for cell survival by overexpression of Hsp that acts as a molecular chaperone for the protection of client proteins. As a consequence, a conventional apoptotic hyperthermia only shows a marginal efficacy. (2) The RAIN simultaneously generates heat and releases an Hsp inhibitor to block the protective function of Hsp upon application of an AMF. The RAIN is effective for apoptotic hyperthermia. MNP = magnetic nanoparticle, AMF = alternating magnetic field, Hsp = heat shock protein, RAIN = resistance-free apoptosis-inducing magnetic nanoparticle. (Reprinted from Yoo, D. *et al.*, Magnetically triggered dual functional nanoparticles for resistance-free apoptotic hyperthermia. *Angew. Chem. Int. Ed. Engl.*, 2013, 52(49), 13047–13051. Copyright © 2013 WILEY-VCH Verlag GmbH & Co. KGaA, Weinheim).

To maximize the therapeutic potential of MHT, instead of focusing on just one of the HSPs, Yin *et al.* simultaneously targeted multiple key downstream effectors of HSPs that promote cell survival and inhibit apoptosis following treatment [132]. They reported another novel application of highly magnetic, zinc-doped iron oxide NPs ($ZnFe_2O_4$) for the dual purpose of delivering a microRNA (miRNA), let-7a, (a known tumor suppressor that targets multiple downstream pathways modulated by HSPs) and inducing magnetic hyperthermia to enhance the treatment of cancer cells. Glioblastoma multiforme (GBM) was used as the model system. The combined MNP-based let-7a delivery and magnetic hyperthermia had an additive effect, resulting in significantly more apoptosis in brain cancer cells than either treatment alone [132]. Many other MFRNs have also been reported for the purpose of magnetic hyperthermia [87, 133–135].

9.4.4 *Magnetic field-responsive nanocarriers for imaging*

MRI is a popular imaging modality in medicine that can provide high resolution anatomical images, and it is widely employed to image soft tissue and cartilage pathologies to obtain information with high spatial and temporal resolution [23, 136]. MRI uses high intensity magnetic fields to align the nuclear magnetization of the body's hydrogen atoms, to which a radio frequency (RF) pulse is applied, which changes the alignment of these nuclei. Upon removal of the pulse, the nuclei "relax" back to their original state, which can be measured by either the longitudinal (T1) or transverse (T2) relaxation [17]. Each of these can be used to generate an MR image. Variation in relaxation rates corresponds to image contrast that allows for the discrimination between tissue types [17]. A number of magnetic materials can be employed as MRI contrast materials including paramagnetic Gd-, Europium (Eu)-, neodymium (Nd)- and Mn-containing materials and superparamagnetic iron oxide in the form of maghemite or magnetite [25]. The interaction of these contrast agents with protons of water molecules creates contrast by reducing T1 or T2 relaxation times. SPIONs are generally more effective at reducing the T2 relaxation times [137]. The use of Gd-contrast materials is associated with toxicity concerns and may induce severe adverse effects. SPIONs are preferred due to their excellent contrast properties and better biocompatibility. They also have excellent sensitivity in T2* and T2-weighted images. These relaxation effects cause a signal reduction in T2-weighted images, providing a negative (dark) contrast due to the induction of strong magnetic susceptibility on the water protons around the particles [138].

Polymeric micelles encapsulating SPIONs, as well as Gd/Eu or Gd chelates, have been reported for improving the MRI T2 contrast and/or to facilitate precise tumor targeting [139–141]. PEG-*b*-poly(ε-caprolactone) nanomicelles loaded with DOX and superparamagnetic iron oxide (SPIO) were targeted with Cetuximab, a human-murine chimeric anti-epithelial growth factor receptor (EGFR) immunoglobulin G (IgG) monoclonal antibody (mAb) [142]. A 1.5 T MRI scanner was used to evaluate the SPIO transporting ability of these micelles into cells. The MRI T2 signal of A431 cells decreased significantly when treated with SPIO-loaded immunomicelles. The cetuximab immunomicelles delivered DOX and SPIO to EGFR-over-expressing cells and could serve as a MRI-visible targeted delivery agent for improved tumor imaging and therapy.

In another study, SPIONs were loaded into magnetic cationic liposomes (MCLs) that were guided to the tumor site in metastatic melanoma (B16-F10) tumor-bearing mice using an external magnet [143]. Their accumulation was monitored non-invasively using MRI, as well as with biodistribution analysis [143]. Quantitative MR analysis and biodistribution studies using ^{111}In-labeled MCLs showed up to a two-fold greater accumulation of the MCLs in the tumor mass by magnetic targeting and also resulted in significant decreases in the T2* relaxation time and tumor signal intensity [143].

Imaging of atherosclerotic plaques in cardiovascular disease using MR molecular imaging probes that target cellular and molecular events can provide important information about the molecular mediators of atherosclerosis [144]. Nguyen *et al.* hypothesized that G8 poly(amido amine) (G8-PAMAM) dendrimers targeted to oxidation-specific epitopes (OSEs) (key factors in atherosclerotic plaque progression and destabilization) would deliver high Mn payloads and enable *in vivo* detection of macrophage-rich plaques [145]. G8 dendrimers were modified to allow their conjugation with MnDTPA (758 Mn ion) and an antibody MDA2 targeted to malondialdehyde (MDA)-lysine epitopes. The *in vivo* efficacy of targeted and non-targeted dendrimers was evaluated in apolipoprotein E knockout ($apoE^{-/-}$) mice over a 96-h time period using a clinical 3 T MR system, following a bolus administration of 0.05 mmol Mn/kg dose. Within 72 h of administration of targeted dendrimers, a significant enhancement (>60%) of atherosclerotic lesions was reported. Mn (>8 μg Mn/g) was detected within atherosclerotic lesions using spectroscopic methods. Also, the signal attenuation (<18%) and Mn deposition (<1 μg Mn/g) observed in the arterial wall after injecting untargeted dendrimers was limited. The study, thus, demonstrated the usefulness of Mn-labeled dendrimers, in allowing a high Mn payload to be targeted to OSEs for sensitive *in vivo* detection and monitoring of macrophage-rich atherosclerotic lesions [145].

Some newer classes of nanoplatforms such as the dendrimersomes (which resemble liposomes to some extent) have also been investigated and have shown promise for biomedical imaging [146]. These are formed by the self-assembly of Janus dendrimers and can be loaded with amphiphilic MRI chelates. Filippi *et al.* reported on the ability of dendrimersomes to load hydrophilic and/or amphiphilic Gd-complexes as MRI probes. Gadoteridol (GdHPDO3A), a clinically approved agent, was selected as the hydrophilic probe. Two approaches were pursued for the incorporation of amphiphilic agents into the membrane vesicles: i) synthesis of a novel Janus dendrimer (JD1) covalently conjugated to a Gd-chelate (JD1-$GdDOTAMAC_6$) and ii) incorporation into the bilayer of the lipophilic Gd-DOTAMA$(C_{18})_2$ complex (Fig. 9.7). The authors concluded from their preliminary findings that these vesicles assembled from Janus dendrimers could be considered potential vectors of MRI agents, drugs, and multi-modal imaging probes [146].

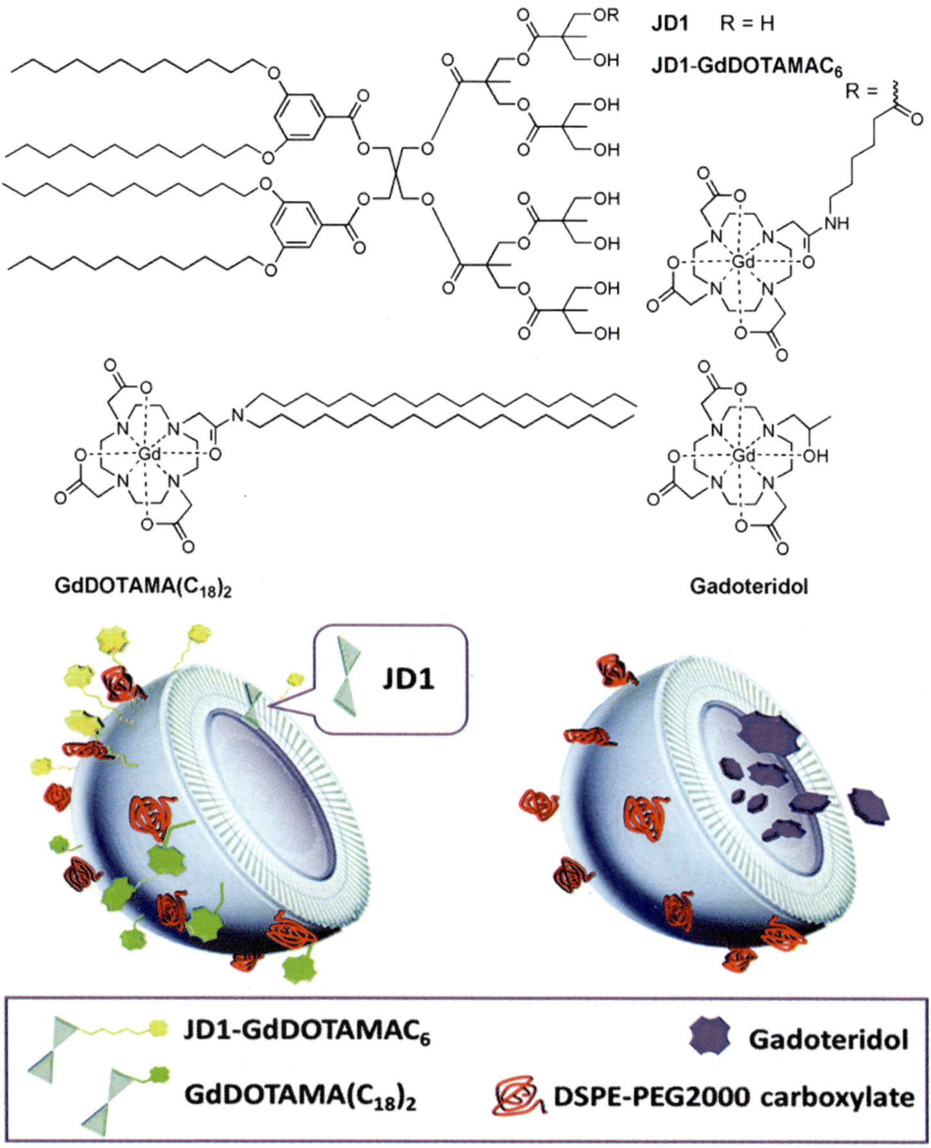

Figure 9.7. (A) Janus dendrimer JD1 and its GdDOTAMAC$_6$ conjugate structures (top), Gd-DOTAMA(C$_{18}$)$_2$ and Gadoteridol (bottom) (B) Dendrimersomes made of JD1 dendrimers incorporating JD1–GdDOTAMAC$_6$ or Gd-DOTAMA(C$_{18}$)$_2$ into the membrane bilayer (left), and encapsulating Gadoteridol in the inner core. (Reprinted from Filippi, M. *et al.*, Dendrimersomes: a new vesicular nano-platform for MR-molecular imaging applications. *Chem. Commun. (Camb.)*, 2014, **50**(26), 3453–3456 with permission of The Royal Society of Chemistry).

MFRNs also find a use in imaging lymph nodes, inflammation in central nervous system (CNS) disorders, and macrophage infiltration in transplanted organs and in knee joints [23]. MFRNs have been explored for multi-modal imaging with magneto-optical or magneto-radioactive probes [25]. A number of MFRNs with near-infrared (NIR), as well as fluorescence or luminescence imaging capability, in addition to MR imaging, have been developed [147, 148]. Commonly used NIR fluorochromes include fluorescent cyanine dyes like Cy5.5, fluorescent semiconductor quantum dots, lanthanides, and gold NPs [25].

9.4.5 *Magnetic field-responsive nanocarriers as theranostics*

Theranostics are essentially defined as a combination of therapeutics and diagnostics and have received increasing interest of late [149]. MFRNs have a number of favorable properties that underscore their usefulness as theranostic agents. MFRNs can be functionalized and simultaneously be guided by an external magnetic field, enabling them to become part of systems that combine drug or gene delivery with selective tumor imaging [31]. SPIONs can bind a myriad of substances including antibodies, dyes, drugs, and nucleic acids. This also allows them to undertake a number of roles including targeting, diagnosis, prevention, and treatment of diseases, including cancer [77]. A number of sophisticated theranostic platforms have been developed with MFRNs to date [150–155]. We discuss some examples that validate the suitability of MFRNs to act as integrative systems.

It is now increasingly evident that the tumor stroma is an integral part of cancer initiation, growth, and progression [156, 157]. Moreover, stromal components also present an interesting opportunity to target cancer [158]. To that end, Mikhaylov *et al.* reported the development of a new delivery platform for targeting both the tumor and its microenvironment, which was based on ferrimagnetic nanoclusters encapsulated in PEGylated liposomes [86]. Ferrimagnetic iron oxide nanoparticles, (magnetite, Fe_3O_4; FMIO) were loaded into liposomes (Ferri-liposomes, ~95 nm diameter) and were used to deliver different types of cargoes *in vivo* under the influence of an external magnet. The nanoparticles exhibited excellent T2 MRI contrast properties, which enabled the ferri-liposomes to provide a non-invasive, real-time MRI capability, unlimited depth penetration, and significantly improved sensitivity. Initially, in genetically engineered mice bearing MMTV-PyMT human breast tumors, a single-dose treatment of doxorubicin (DOX) targeted by ferri-liposomes resulted in a 90% tumor volume reduction two weeks after administration, compared to only a 60% decrease seen with standard DOX administration. For further investigations, the authors selected a poorly bioavailable small molecule, JPM-565, a broad spectrum cysteine cathepsin inhibitor that is very potent in treating pancreatic islet cell cancer. It was delivered by the ferri-liposomes to a mammary tumor and its microenvironment in a mouse model. JPM-565 was targeted by the ferri-liposomes to the peri-tumoral region of mouse breast cancer and resulted in significant tumor growth reduction versus systemic delivery of the same drug. A favorable biodistribution of ferri-liposomes to the tumor and its microenvironment, as well as their excellent MR imaging properties, provides for simultaneous delivery and detection of therapeutic agents *in vivo* [86].

Li *et al.* recently reported an intelligent SPIO-based theranostic NP with an inner SPIO core and a disulfide-containing polyethylenimine (SSPEI) outer layer (SSPEI-SPIO) NP for redox-triggered gene delivery in response to the highly reducing intracellular environment [159]. The SSPEI-SPIO NPs bound genes (GFP) and formed nano-complexes, from which the gene release could be facilitated in the presence of dithiothreitol (5–20 mM) with high transfection efficiencies against different cancer cells. The SSPEI-SPIO NPs also successfully delivered siRNA against human telomerase reverse transcriptase (hTERT) genes in HepG2 cells, leading to apoptosis and growth inhibition. In addition, the NPs could be used as T2-negative contrast agents for MRI of nude mouse tumor xenografts. Even at high concentrations of 100 μg/mL, these NPs had low toxicity. Thus, these SSPEI-SPIO NPs exhibit great promise for tumor gene therapy and imaging, as well as their ability to be utilized as theranostic nanocarriers [159].

Ding *et al.* developed a micelle-based theranostic platform for MRI and drug delivery [160]. The authors designed a cell-internalizable and intracellularly-degradable micellar system, assembled from multiblock polyurethanes. The side chain had cell-penetrating gemini quaternary ammonium pendant groups, and redox-responsive disulfide linkages were present throughout the backbone (Fig. 9.8).

The nanocarrier exhibited a typical "cleavable core-internalizable shell-protective corona" architecture, and had a small size, positive surface charge, high loading capacity, and reduction-triggered destabilization. The micelles rapidly entered the tumor cells and released their cargoes in response to the elevated glutathione levels in the cell, resulting in enhanced drug efficacy *in vitro*. When loaded with oleic acid-modified SPIONs, the magnetic micelles showed an excellent MRI contrast enhancement. The T2 relaxivity was affected by the morphology of SPIO-clustering inside the micelle core. The micellar carrier had excellent cytocompatibility and non-toxic degradation products, and it exhibited the attributes of a promising theranostic platform for efficient intracellular drug delivery and real-time monitoring [160].

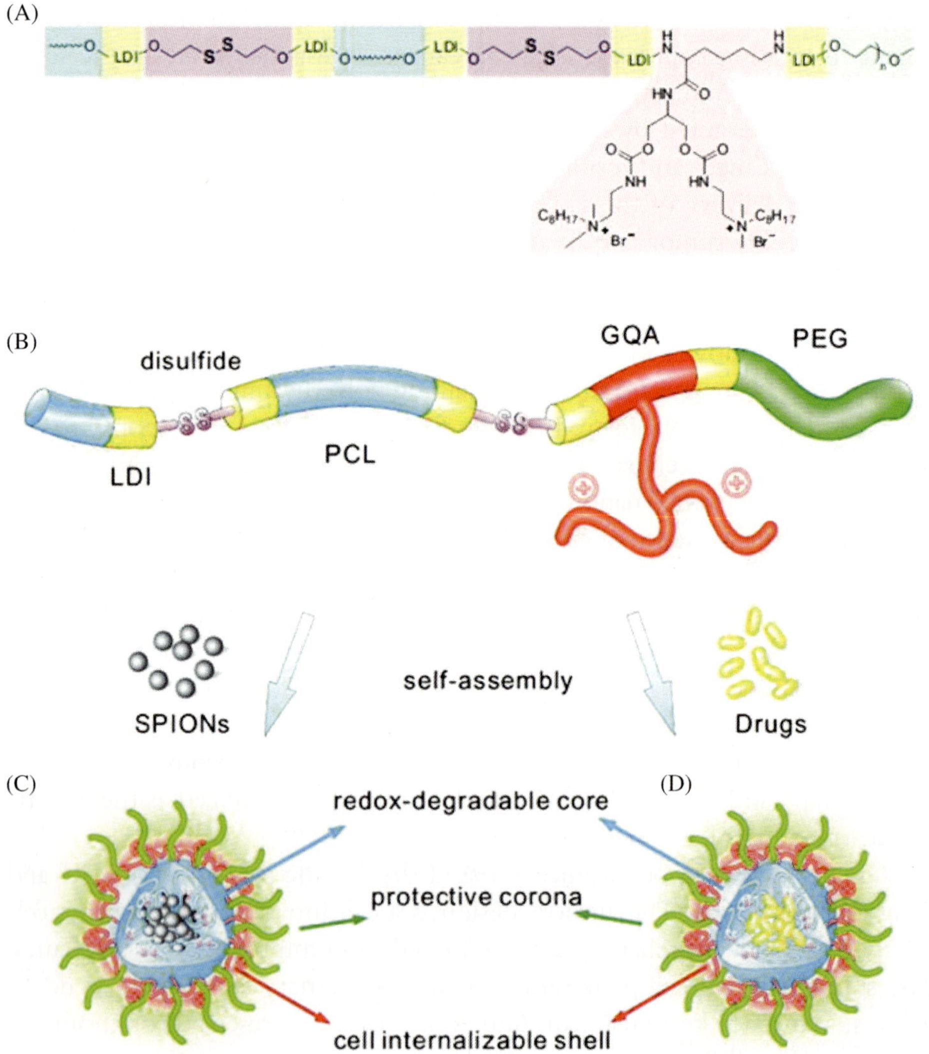

Figure 9.8. (A) A typical molecular structure of cell internalizable and redox-degradable cationic polyurethanes. Color code: blue, PCL; yellow, l-lysine ethyl ester diisocyanate (LDI); purple, bis(2-hydroxyethyl) disulfide (DHDS); red, gemini quaternary ammonium (GQA); green, methoxyl-poly(ethylene glycol) (MPEG). The wavy line represents the PCL segment. (B–D) Schematic illustration of cationic multiblock polyurethanes (B) and their micelles loading SPIO nanoparticles (C) and antitumor drugs (D). SPION = superparamagnetic iron oxide nanoparticle. (Reprinted from Ding, M. *et al.*, Cell Internalizable and Intracellularly Degradable Cationic Polyurethane Micelles as a Potential Platform for Efficient Imaging and Drug Delivery. *Biomacromolecules*, 2014. Copyright © 2013, American Chemical Society).

9.4.6 *Other biomedical applications of MNPs*

Over and above the applications mentioned above, the sheer versatility of MFRNs has led to the development of a myriad of biomedical applications. We discuss these additional applications with regard to SPIONs.

Iron oxide NPs (IONPs) are employed for cell labeling and imaging applications for cellular MRI, a rapidly growing field where cells are tracked and visualized in organisms [161, 162]. Specific cells including transplanted stem cells, pancreatic islet cells, as well as dendritic cells, have been imaged by labeling with SPIONs. Apoptotic cells can be detected by conjugating IONPs or Gd-DTPA with annexin V, which recognizes the phosphatidyl serine receptors on their cell surface [163, 164].

Other applications for ferro/paramagnetic substances are included in the areas of cell separation or cell sensing. Magnetic separation is a quick and gentle method for rapid isolation of target molecules, cell organelles, and cells from complex mixtures of crude samples, such as blood, bone marrow, tissue homogenates, and other biological materials. Magnetic separation is widely employed in immunomagnetic separations (IMSs) [165, 166]. Using positive isolation, the cells of interest are magnetically labeled by specific particles and isolated for analysis and other downstream applications. At times, negative isolation (depletion) may be needed where unwanted cells are removed before analyzing the population of interest. IMS is frequently encountered in the isolation of stem cells, cancer cells, cancer stem cells, or T-lymphocytes [166, 167]. A number of commercial microbeads are available (e.g. Dynabeads®, Life technologies, or MACS® microbeads, Miltenyi Biotec) with various cell surface markers, as well as in a variety of sizes. Bacterial cells can also be separated from cell culture broths using MNPs guided by an external magnet. Nucleic acids and proteins can be separated magnetically, which enables direct isolation of target biomolecules from difficult-to-handle biological samples [166].

The quick magnetic recovery from various mixtures has led to an increased use of magnetic particles as a solid support for proteins. Enzymes and proteins can be immobilized on the surface of MNPs by direct conjugation using simple chemistries [23]. A number of enzymes, including trypsin, horseradish peroxidase, glucose oxidase, papain, triglycerol lipase, pectinase and many others, have been immobilized on magnetic surfaces [123].

The large surface area and magnetic susceptibility of IONPs has also facilitated bioanalysis and immunoassay applications. By utilizing sensitive magnetometers, differences in magnetic fields generated by the magnetically labeled targets can be detected directly [168]. Magnetic engineering of radioimmunoparticles has also aided the delivery to tumor and therapeutic efficacy of radioimmunoconjugates [23].

9.5 Multifunctional Magnetic Nanocarriers

Magnetic nanocarriers have evolved over the years in terms of both their design, as well as the breadth of their applications. It is, thus, a logical step to envision magnetic nanocarriers that allow for the integration of various functionalities in one platform. Magnetic nanocarriers may be modified to incorporate antibodies or ligands to target unique molecular signatures on diseased sites or tumors simultaneously, with fluorescent dyes or radionuclides to construct magneto-optical or magneto-radioactive probes for imaging and/or therapy, with cell penetrating peptides for efficient intracellular delivery, with hydrophilic coatings to allow for long circulation *in vivo*, with small molecule drugs, as well as biomolecules (DNA, siRNA, aptamers) for therapy, and they may also have additional stimuli responsive components (e.g. pH-, redox-, enzyme-, thermo-, ultrasound-, or light-sensitive) within the same carrier.

Multifunctionality may be imparted to individual MNPs (e.g. SPIONs) or to MFRNs that incorporate SPIONs. One of the key requirements in creating a multifunctional platform is to make sure that the integration of various functionalities in the nanocarrier is seamless and that each one functions in a coordinated manner with the other included components [1, 15].

Baldi *et al.* recently reported the chemical synthesis and characterization of a poly (D,L-lactide-*co*-glycolide)-*b*-(poly(ethylene glycol)-based nanocarrier containing iron oxide NPs and modified with human epidermal growth factor receptor on the outer shell (Fe_3O_4-1-PNP-hEGFR) [169] . In addition, the nanocarrier was radiolabeled with ^{99m}Tc. It was investigated for its diagnostic ability *in vivo* and for its therapeutic hyperthermic effects in A431 epidermoid carcinoma cells. The biodistribution and therapeutic hyperthermic effects of this nanosystem were evaluated non-invasively *in vivo* in A431 tumor-bearing mice. Radiolabeling of the nanocarrier with ^{99m}Tc allowed for *in vivo* biodistribution in mice and detection of antibody–receptor interactions. The entrapped MNPs allowed a proof-of-concept evaluation of the hyperthermic efficacy on tumor size reduction by the nanocarrier. A substantial decrease in tumor size and increased survival correlated with a corresponding increase in both nanoparticle concentration and the local temperature achieved, confirming the utility of this multifunctional theranostic system for hyperthermic treatment of epidermoid carcinoma [169].

Another very interesting example of a multifunctional system was reported by Chiang *et al.* [170]. The authors reported pH-sensitive double emulsion nanocapsules (DENCs), conjugated with Trastuzumab for the delivery of the hydrophilic DOX and the hydrophobic drug, paclitaxel (PTX). To enhance the efficacy of this delivery system, magnetic targeting via an external magnet was utilized. The DENCs were formulated in a two-step double emulsion process with amphiphilic poly(vinyl alcohol) (PVA) and hydrophobic SPIO, as seen in Fig. 9.9.

Poly(methacrylic acid) (PMMA), a pH-responsive polymer was functionalized with a thiol group (PMA_{SH}) and was added to DENCs to modify the shell and enable pH-triggered release. Trastuzumab was conjugated to DENCs

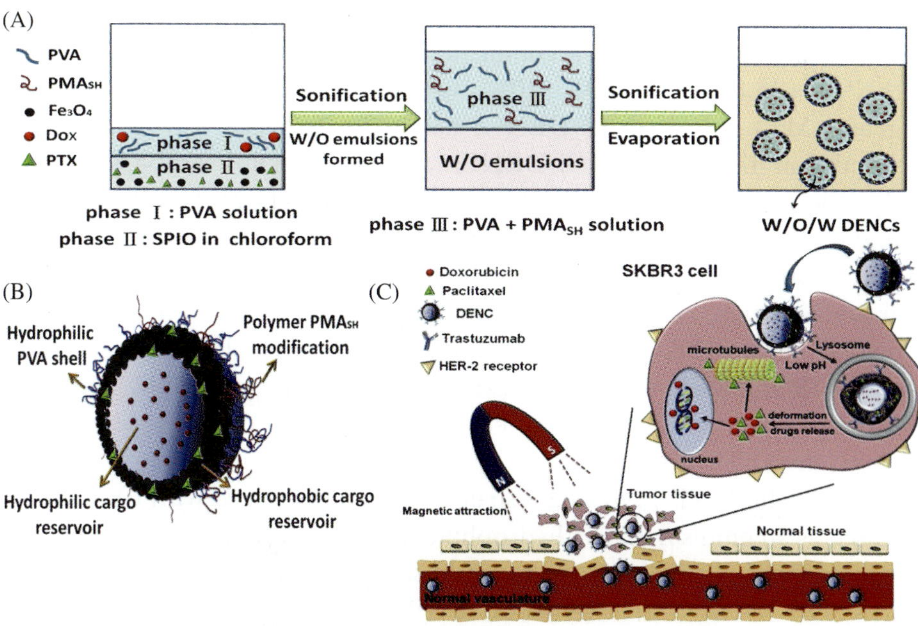

Figure 9.9. (A) Schematic of the synthesis of DENCs using the two-step emulsion process. The hydrophobic phase and the hydrophilic phase were stabilized using PVA polymer with a molecular weight of 16000 g/mol. (B) DENCs carried hydrophilic cargo in the inner core and hydrophobic cargo in the outer shell; PMA_{SH} was modified out on its shell site, which entangled with the hydrophilic polymer PVA. (C) Uptake of DENCs by SKBR3 cells. PVA = poly(vinyl alcohol), PMA_{SH} = PMMA with thiol group, Fe_3O_4 = iron oxide, Dox = doxorubicin, PTX = paclitaxel, SPIO = superparamagnetic iron oxide, W/O = without, DENC = double emulsion nanocapsule, HER-2 = human epidermal growth factor receptor 2, SKBR3 = human breast adenocarcinoma cell line. (Modified from Chiang *et al.* (2014) with permission from *Nanomedicine: Nanotechnology, Biology and Medicine*, 10(1), 99–107. Copyright © 2014 Elsevier).

via the thiol groups in the side chain. DOX and PTX were loaded simultaneously at a neutral pH, so that their release could be triggered in acidic environments. The hydrophilic PMMA exists like a "random coil" at neutral pH, but shrinks to a "globule-like structure" under low pH conditions. The PMA_{SH} was entangled in the shell under acidic conditions to cause deformation of DENCs and trigger drug release. PTX and DOX were loaded in the shell and the core respectively, showing a higher drug release at pH 4.0 compared to pH 7.0, with PTX exhibiting a more dramatic release. The uptake of DENCs by HER-2 positive breast tumor cells was enhanced by the conjugation of Trastuzumab. Near-complete and significant cytotoxicity was achieved by dual drug loaded DENCs (T-PTX-DOX-DENCs) due to the combination of pH-sensitivity and targeted drug delivery. Furthermore, the effect of magnetic targeting was evaluated in nude mice bearing SKBR3 tumors by application of a 2000 G magnetic field for 6 h, immediately following the administration of formulations. An increase in accumulation from the initial dose was seen for both PTX (25.8%) and DOX (20%) with the magnetic field. Magnetically targeted (MT) T-PTX-DOX-DENCs had a 2.4-fold and 1.87-fold higher accumulation than PTX-DOX-DENCs and T-PTX-DOX-DENCs, respectively. The combination of MT therapy with T-PTX-DOX-DENCs demonstrated the best therapeutic efficacy compared with PTX or DOX alone [170].

Yet another multifunctional system was reported by Shen *et al.* to deliver hydrophilic drugs, DOX and verapamil (VER) [171]. MNPs modified with citric acid were initially combined with chitosan through electrostatic interactions. DOX and VER were then adsorbed to the chitosan shell by hydrogen bonds, and finally using a double emulsion (W/O/W) solvent evaporation method, the DOX/VER-loaded MNPs were entrapped in PLGA NPs under sonication. The MNP-PLGA nanostructures were stabilized with PVA, which was used as an emulsifier. The hydrophilic drugs scattered in the PLGA polymer matrix and also adhered to the chitosan shell on the MNPs. To obtain the cRGDfK-DOX/VER-MNP-PLGA NPs, a c(RGDfK) ligand was conjugated by its lysine residues to the end carboxyl groups of PLGA via an amide bond (Figure 10). The MNP-PLGA were reported to have entrapped both drugs with high efficiency (DOX ~75% and VER ~53%), achieved targeted drug delivery via c(RGDfK) with an external magnet, and exhibited NIR-triggered drug release and MFH due to the heating of MNPs.

Compared to free drugs or untargeted NPs, the targeted DOX/VER-MNP-PLGA showed a significant cytotoxic activity *in vitro* due to synergistic effects of DOX/VER and a greater accumulation in tumors due to c(RGDfK). They also demonstrated a stronger anti-tumor activity than the free drug groups and had a significantly higher tumor-inhibitory rate (75.6%) compared to that of free DOX (18.5%) or free DOX/VER (32.7%). Finally, an increased

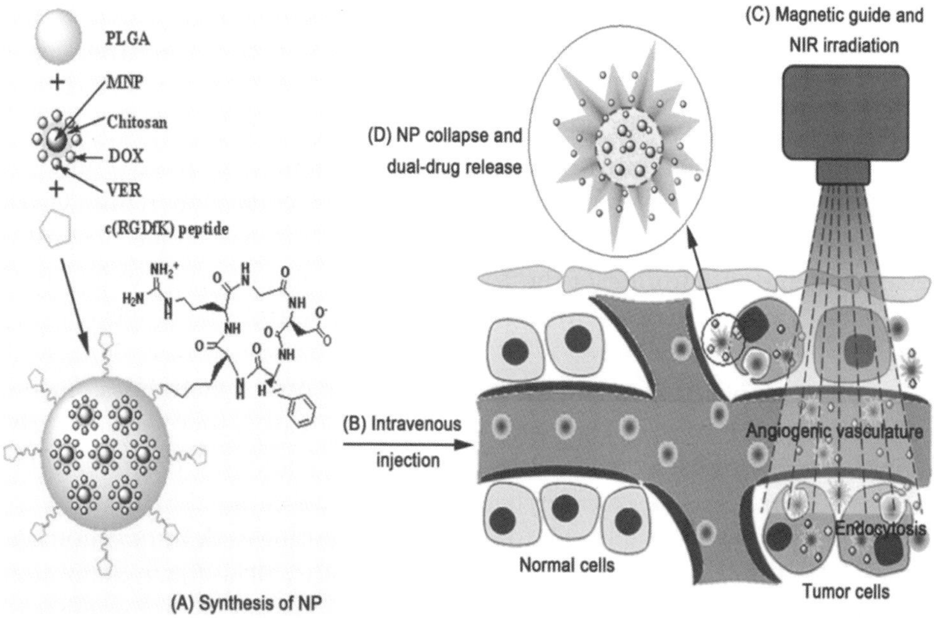

Figure 9.10. cRGD-functionalized polymeric MNPs as a dual drug delivery system PLGA = poly-L-glutamic acid, MNP = magnetic nanoparticle, DOX = doxorubicin, VER = verapamil, NP = nanoparticle, NIR = near-infrared. (Reprinted with permission from Shen *et al.* (2013). *Pharmacological Research*, 70(1), 102–115. Copyright © 2013 Elsevier).

tumor growth inhibition effect was seen, since exposure to the NIR laser beam caused MNP-generated hyperthermia (43°C) that was detrimental to tumor cells *per se* and also resulted in increased drug release due to a lower critical solution temperature (LCST) effect exhibited by the PLGA [171].

9.6 Conclusion and Future Perspectives

The past few decades have seen tremendous developments in the design and variety of nanocarriers. Magnetically-responsive nanocarriers, in particular, have had significant success as MRI contrast agents, as well as drug carriers. Although MRI-based applications of MFRNs have proven clinical utility, other applications including antibody or ligand-targeted MFRNs, MFRNs for magnetic hyperthermia, and multifunctional MFRN platforms are still largely confined to pre-clinical investigations. This is likely to change when the physico-chemical parameters of MNPs are further investigated to reduce or eliminate safety concerns and improve their biocompatibility, especially when utilized for *in vivo* applications. Another interesting point is that in spite of the availability of numerous materials with magnetic properties, only a few are really useful for biomedical applications. To increase the list of materials that we can use for magnetic applications, cost-effective and biocompatible magnetic particles must be developed [166].

Of the available MNPs, SPIONs have the widest utility and are the most versatile. The ease of modifying them with various functionalities to construct multifunctional platforms has enabled researchers to develop a number of novel applications of these MNPs, including those that combine diagnosis, treatment, and monitoring of therapy in a single carrier for theranostic purposes. Such multifunctional MNPs can be useful across a wide spectrum of diseases, in addition to cancer.

Another application for MNP is the generation of hyperthermia in the presence of AMF to selectively heat the diseased tissue. However, this technique is largely restricted to easily-accessible tumor nodules, but not to disseminated tumors. Generation of hyperthermia is also limited by the number of MNPs that can be localized in the tissue, which can result in non-homogenous heat diffusion and areas of high or no heating. The functionalization of MNPs with intracellular targeting ligands will enable a greater number of MNPs to accumulate in the tumor and aid the hyperthermia process. Similarly monitoring the spatial distribution of heat in tissues may also allow for the development of ways for homogeneous heat diffusion throughout the tissues of interest. The application of MNPs in magnetic drug targeting using external magnets also suffers due to a lack of magnets that can target deep-seated tumors in the body. This may be countered by developing magnets with greater strength and perhaps, with internal magnets using SPIONs.

The toxicity of MNPs depends on their physicochemical characteristics including size, charge, surface characteristics, composition, route of administration, and dose. Elaborate long-term toxicity studies need to be undertaken for MNPs if they are to be approved clinically for applications other than MRI. Magnetically-responsive nanocarriers are definitely poised to go a long way as far as theranostic platforms are concerned. Advances in nanocarrier technologies, imaging techniques, as well as the emergence of novel therapeutics, will only aid their usefulness. However, we must not lose sight of the imminent biocompatibility and safety concerns that need to be addressed if we wish to envision a future for these magnetic field-responsive nanocarriers.

References

1. Jhaveri, A., Deshpande, P. and Torchilin, V. (2014). Stimuli-sensitive nanopreparations for combination cancer therapy. *J. Control. Release*, **190**, 352–370.
2. Ferrari, M. (2005). Cancer nanotechnology: Opportunities and challenges. *Nat. Rev. Cancer*, **5**, 161–171.

3. Chou, L. Y., Zagorovsky, K. and Chan, W. C. (2014). DNA assembly of nanoparticle superstructures for controlled biological delivery and elimination. *Nat. Nanotechnol.*, **9**, 148–155.
4. Lohcharoenkal, W., Chen, Y. C. and Rojanasakul, Y. (2014). Protein nanoparticles as drug delivery carriers for cancer therapy. *BioMed. Res. Int.*, **2014**, 12.
5. Marradi, M., Chiodo, F., Garcia, I. and Penades, S. (2013). Glyconanoparticles as multifunctional and multimodal carbohydrate systems. *Chem. Soc. Rev.*, **42**, 4728–4745.
6. Torchilin, V. P. (2006). Multifunctional nanocarriers. *Adv. Drug Deliv. Rev.*, **58**, 1532–1555.
7. Duncan, R. (2006). Polymer conjugates as anticancer nanomedicines. *Nat. Rev. Cancer*, **6**, 688–701.
8. Duncan, R. (2003). The dawning era of polymer therapeutics. *Nat. Rev. Drug Discov.*, **2**, 347–360.
9. Loomba, L. and Scarabelli, T. (2013). Metallic nanoparticles and their medicinal potential. Part II: Aluminosilicates, nanobiomagnets, quantum dots and cochleates. *Ther. Deliver.*, **4**, 1179–1196.
10. Torchilin, V. P. (2005). Recent advances with liposomes as pharmaceutical carriers. *Nat. Rev. Drug Discov.*, **4**, 145–160.
11. Mai, W. X. and Meng, H. (2013). Mesoporous silica nanoparticles: A multifunctional nano therapeutic system. *Integr. Biol. (Camb.)*, **5**, 19–28.
12. Lim, S. B., Banerjee, A. and Onyuksel, H. (2012). Improvement of drug safety by the use of lipid-based nanocarriers. *J. Control. Release*, **163**, 34–45.
13. Torchilin, V. (2009). Multifunctional and stimuli-sensitive pharmaceutical nanocarriers. *Eur. J. Pharm. Biopharm.*, **71**, 431–444.
14. McCarthy, J. R. and Weissleder, R. (2008). Multifunctional magnetic nanoparticles for targeted imaging and therapy. *Adv. Drug Deliv. Rev.*, **60**, 1241–1251.
15. Jhaveri, A. M. and Torchilin, V. P. (2014). Multifunctional polymeric micelles for delivery of drugs and siRNA. *Front. Pharmacol.*, **5**, 77.
16. Lu, A. H., Salabas, E. L. and Schuth, F. (2007). Magnetic nanoparticles: Synthesis, protection, functionalization, and application. *Angew. Chem. Int. Ed. Engl.*, **46**, 1222–1244.
17. Veiseh, O., Gunn, J. W. and Zhang, M. (2010). Design and fabrication of magnetic nanoparticles for targeted drug delivery and imaging. *Adv. Drug Deliv. Rev.*, **62**, 284–304.
18. Tartaj, P., Morales, M. d. P., Veintemillas-Verdaguer, S., González-Carreño, T. and Serna, C. J. (2003). The preparation of magnetic nanoparticles for applications in biomedicine. *J. Phy. D: Appl. Phys.*, **36**, R182.
19. Gubin, S. P., Koksharov, Y. A., Khomutov, G. and Yurkov, G. Y. (2005). Magnetic nanoparticles: Preparation, structure and properties. *Russ. Chem. Rev.*, **74**, 489.
20. Laurent, S., Forge, D., Port, M., Roch, A., Robic, C., Vander Elst, L. and Muller, R. N. (2008). Magnetic iron oxide nanoparticles: Synthesis, stabilization, vectorization, physicochemical characterizations, and biological applications. *Chem. Rev.*, **108**, 2064–2110.
21. Gao, J., Gu, H. and Xu, B. (2009). Multifunctional magnetic nanoparticles: Design, synthesis, and biomedical applications. *Acc. Chem. Res.*, **42**, 1097–1107.
22. Huber, D. L. (2005). Synthesis, properties, and applications of iron nanoparticles. *Small*, **1**, 482–501.
23. Reddy, L. H., Arias, J. L., Nicolas, J. and Couvreur, P. (2012). Magnetic nanoparticles: Design and characterization, toxicity and biocompatibility, pharmaceutical and biomedical applications. *Chem. Rev.*, **112**, 5818–5878.

24. Elsaesser, A. and Howard, C. V. (2012). Toxicology of nanoparticles. *Adv. Drug Deliv. Rev.*, **64**, 129–137.
25. Colombo, M., Carregal-Romero, S., Casula, M. F., Gutierrez, L., Morales, M. P., Bohm, I. B., Heverhagen, J. T., Prosperi, D. and Parak, W. J. (2012). Biological applications of magnetic nanoparticles. *Chem. Soc. Rev.*, **41**, 4306–4334.
26. Xavier, B. and Amílcar, L. (2002). Finite-size effects in fine particles: Magnetic and transport properties. *J. Phys. D: Appl. Phys.*, **35**, R15.
27. Klostergaard, J. and Seeney, C.E. (2012). Magnetic nanovectors for drug delivery. *Nanomedicine*, **8**, S37–S50.
28. Sun, C., Lee, J. S. and Zhang, M. (2008). Magnetic nanoparticles in MR imaging and drug delivery. *Adv. Drug Deliv. Rev.*, **60**, 1252–1265.
29. Issa, B., Obaidat, I. M., Albiss, B. A. and Haik, Y. (2013). Magnetic nanoparticles: Surface effects and properties related to biomedicine applications. *Int. J. Mol. Sci.*, **14**, 21266–21305.
30. Akbarzadeh, A., Samiei, M. and Davaran, S. (2012). Magnetic nanoparticles: Preparation, physical properties, and applications in biomedicine. *Nanoscale Res. Lett.*, 7, 144.
31. Shubayev, V. I., Pisanic, 2nd, T. R. and Jin, S. (2009). Magnetic nanoparticles for theragnostics. *Adv. Drug Deliv. Rev.*, **61**, 467–477.
32. Gupta, A. K. and Gupta, M. (2005). Synthesis and surface engineering of iron oxide nanoparticles for biomedical applications. *Biomaterials*, **26**, 3995–4021.
33. Mojica Pisciotti, M. L., Lima, E., Jr., Vasquez Mansilla, M., Tognoli, V. E., Troiani, H. E., Pasa, A. A., Creczynski-Pasa, T. B., Silva, A. H., Gurman, P., Colombo, L., Goya, G. F., Lamagna, A. and Zysler, R. D. (2014). *In vitro* and *in vivo* experiments with iron oxide nanoparticles functionalized with DEXTRAN or polyethylene glycol for medical applications: Magnetic targeting. *J. Biomed. Mater. Res. B Appl. Biomater.*, **102**, 860–868.
34. Kievit, F. M., Stephen, Z. R., Veiseh, O., Arami, H., Wang, T., Lai, V. P., Park, J. O., Ellenbogen, R. G., Disis, M. L. and Zhang, M. (2012). Targeting of primary breast cancers and metastases in a transgenic mouse model using rationally designed multifunctional SPIONs. *ACS Nano.*, **6**, 2591–2601.
35. Lee, H. J., Nguyen, Y. T., Muthiah, M., Vu-Quang, H., Namgung, R., Kim, W. J., Yu, M. K., Jon, S., Lee, I. K., Jeong, Y. Y. and Park, I. K. (2012). MR traceable delivery of p53 tumor suppressor gene by PEI-functionalized superparamagnetic iron oxide nanoparticles. *J. Biomed. Nanotechnol.*, **8**, 361–371.
36. Xue, S., Wang, Y., Wang, M., Zhang, L., Du, X., Gu, H. and Zhang, C. (2014). Iodinated oil-loaded, fluorescent mesoporous silica-coated iron oxide nanoparticles for magnetic resonance imaging/computed tomography/fluorescence trimodal imaging. *Int. J. Nanomedicine*, **9**, 2527–2538.
37. Schulze, F., Dienelt, A., Geissler, S., Zaslansky, P., Schoon, J., Henzler, K., Guttmann, P., Gramoun, A., Crowe, L. A., Maurizi, L., Vallee, J. P., Hofmann, H., Duda, G. N. and Ode, A. (2014). Amino-polyvinyl alcohol coated superparamagnetic iron oxide nanoparticles are suitable for monitoring of human mesenchymal stromal cells *in vivo*. *Small*, **10**(21), 4340–4351.
38. Chen, J., Shi, M., Liu, P., Ko, A., Zhong, W., Liao, W. and Xing, M. M. (2014). Reducible polyamidoamine-magnetic iron oxide self-assembled nanoparticles for doxorubicin delivery. *Biomaterials*, **35**, 1240–1248.

39. Andreas, K., Georgieva, R., Ladwig, M., Mueller, S., Notter, M., Sittinger, M. and Ringe, J. (2012). Highly efficient magnetic stem cell labeling with citrate-coated superparamagnetic iron oxide nanoparticles for MRI tracking. *Biomaterials*, **33**, 4515–4525.
40. Dias, A. M., Hussain, A., Marcos, A. S., and Roque, A. C. (2011). A biotechnological perspective on the application of iron oxide magnetic colloids modified with polysaccharides. *Biotechnol. Adv.*, **29**, 142–155.
41. Jo, J., Aoki, I. and Tabata, Y. (2010). Design of iron oxide nanoparticles with different sizes and surface charges for simple and efficient labeling of mesenchymal stem cells. *J. Control. Release*, **142**, 465–473.
42. Bar-Shir, A., Avram, L., Yariv-Shoushan, S., Anaby, D., Cohen, S., Segev-Amzaleg, N., Frenkel, D., Sadan, O., Offen, D. and Cohen, Y. (2014). Alginate-coated magnetic nanoparticles for noninvasive MRI of extracellular calcium. *NMR Biomed.*, **27**, 774–783.
43. Scherer, C. and Figueiredo Neto, A.M. (2005). Ferrofluids: Properties and applications. *Braz. J. Phys.*, **35**, 718–727.
44. Siddiqui, M. A., Ahamed, M., Ahmad, J., Majeed Khan, M. A., Musarrat, J., Al-Khedhairy, A. A. and Alrokayan, S. A. (2012). Nickel oxide nanoparticles induce cytotoxicity, oxidative stress and apoptosis in cultured human cells that is abrogated by the dietary antioxidant curcumin. *Food Chem. Toxicol.*, **50**, 641–647.
45. Rodriguez-Llamazares, S., Merchan, J., Olmedo, I., Marambio, H. P., Munoz, J. P., Jara, P., Sturm, J. C., Chornik, B., Pena, O., Yutronic, N. and Kogan, M. J. (2008). Ni/Ni oxides nanoparticles with potential biomedical applications obtained by displacement of a nickel-organometallic complex. *J. Nanosci. Nanotechnol.*, **8**, 3820–3827.
46. Lee, S. M., Lee, O. S., O'Halloran, T. V., Schatz, G. C. and Nguyen, S. T. (2011). Triggered release of pharmacophores from [Ni(HAsO(3))]-loaded polymer-caged nanobin enhances pro-apoptotic activity: A combined experimental and theoretical study. *ACS Nano.*, **5**, 3961–3969.
47. Al-Qubaisi, M. S., Rasedee, A., Flaifel, M. H., Ahmad, S. H., Hussein-Al-Ali, S., Hussein, M. Z., Zainal, Z., Alhassan, F. H., Taufiq-Yap, Y. H., Eid, E. E., Arbab, I. A., Al-Asbahi, B. A., Webster, T. J. and El Zowalaty, M. E. (2013). Induction of apoptosis in cancer cells by NiZn ferrite nanoparticles through mitochondrial cytochrome C release. *Int. J. Nanomedicine*, **8**, 4115–4129.
48. Al-Qubaisi, M. S., Rasedee, A., Flaifel, M. H., Ahmad, S. H., Hussein-Al-Ali, S., Hussein, M. Z., Eid, E. E., Zainal, Z., Saeed, M., Ilowefah, M., Fakurazi, S., Mohd Isa, N. and El Zowalaty, M. E. (2013). Cytotoxicity of nickel zinc ferrite nanoparticles on cancer cells of epithelial origin. *Int. J. Nanomedicine*, **8**, 2497–2508.
49. Kale, S. N., Jadhav, A. D., Verma, S., Koppikar, S. J., Kaul-Ghanekar, R., Dhole, S. D. and Ogale, S. B. (2012). Characterization of biocompatible $NiCo_2O_4$ nanoparticles for applications in hyperthermia and drug delivery. *Nanomedicine*, **8**, 452–459.
50. Magaye, R. R., Yue, X., Zou, B., Sh,i H., Yu, H., Liu, K., Lin, X., Xu, J., Yang, C., Wu, A. and Zhao, J. (2014). Acute toxicity of nickel nanoparticles in rats after intravenous injection. *Int. J. Nanomedicine*, **9**, 1393–1402.
51. Capasso, L., Camatini, M. and Gualtieri, M. (2014). Nickel oxide nanoparticles induce inflammation and genotoxic effect in lung epithelial cells. *Toxicol. Lett.*, **226**, 28–34.
52. Magaye, R. and Zhao, J. (2012). Recent progress in studies of metallic nickel and nickel-based nanoparticles' genotoxicity and carcinogenicity. *Environ. Toxicol. Pharmacol.*, **34**, 644–650.

53. Pietruska, J. R., Liu, X., Smith, A., McNeil, K., Weston, P., Zhitkovich, A., Hurt, R., Kane, A. B. (2011). Bioavailability, intracellular mobilization of nickel, and HIF-1alpha activation in human lung epithelial cells exposed to metallic nickel and nickel oxide nanoparticles. *Toxicol. Sci.*, **124**, 138–148.
54. Wan, R., Mo, Y., Chien, S., Li, Y., Tollerud, D. J. and Zhang, Q. (2011). The role of hypoxia inducible factor-1alpha in the increased MMP-2 and MMP-9 production by human monocytes exposed to nickel nanoparticles. *Nanotoxicology*, **5**, 568–582.
55. Morimoto, Y., Ogami, A., Todoroki, M., Yamamoto, M., Murakami, M., Hirohashi, M., Oyabu, T., Myojo, T., Nishi, K., Kadoya, C., Yamasaki, S., Nagatomo, H., Fujita, K., Endoh, S., Uchida, K., Yamamoto, K., Kobayashi, N., Nakanishi, J. and Tanaka, I. (2010). Expression of inflammation-related cytokines following intratracheal instillation of nickel oxide nanoparticles. *Nanotoxicology*, **4**, 161–176.
56. Glista-Baker, E. E., Taylor, A. J., Sayers, B. C., Thompson, E. A. and Bonner, J. C. (2012). Nickel nanoparticles enhance platelet-derived growth factor-induced chemokine expression by mesothelial cells via prolonged mitogen-activated protein kinase activation. *Am. J. Respir. Cell Mol. Biol.*, **47**, 552–561.
57. Horie, M., Fukui, H., Nishio, K., Endoh, S., Kato, H., Fujita, K., Miyauchi, A., Nakamura, A., Shichiri, M., Ishida, N., Kinugasa, S., Morimoto, Y., Niki, E,, Yoshida, Y. and Iwahashi, H. (2011). Evaluation of acute oxidative stress induced by NiO nanoparticles *in vivo* and *in vitro*. *J. Occup. Health*, **53**, 64–74.
58. Ahamed, M., Akhtar, M. J., Siddiqui, M. A., Ahmad, J., Musarrat, J., Al-Khedhairy, A. A., AlSalhi, M. S. and Alrokayan, S. A. (2011). Oxidative stress mediated apoptosis induced by nickel ferrite nanoparticles in cultured A549 cells. *Toxicology*, **283**, 101–108.
59. Fantechi, E., Innocenti, C., Zanardelli, M., Fittipaldi, M., Falvo, E., Carbo, M., Shullani, V., Di Cesare Mannelli, L., Ghelardini, C., Ferretti, A. M., Ponti, A., Sangregorio, C. and Ceci, P. (2014). A smart platform for hyperthermia application in cancer treatment: Cobalt-doped ferrite nanoparticles mineralized in human ferritin cages. *ACS Nano*, **8**, 4705–4719.
60. Scarberry, K. E., Dickerson, E. B., McDonald, J. F. and Zhang, Z. J. (2008). Magnetic nanoparticle-peptide conjugates for *in vitro* and *in vivo* targeting and extraction of cancer cells. *J. Am. Chem. Soc.,* **130**, 10258–10262.
61. Hwang do, W., Ko, H. Y., Lee, J. H., Kang, H., Ryu, S. H., Song, I. C., Lee, D. S. and Kim, S. (2010). A nucleolin-targeted multimodal nanoparticle imaging probe for tracking cancer cells using an aptamer. *J. Nucl. Med.*, **51**, 98–105.
62. Wu, H., Liu, G., Wang, X., Zhang, J., Chen, Y., Shi, J., Yang, H., Hu, H. and Yang, S. (2011). Solvothermal synthesis of cobalt ferrite nanoparticles loaded on multiwalled carbon nanotubes for magnetic resonance imaging and drug delivery. *Acta Biomater.*, **7**, 3496–3504.
63. Sukoyan, M. A., Khrapov, E. A., Voronina, E. N., Boyarskikh, U. A., Gubanov, A. I., Itin, V. I., Magaeva, A. A., Nayden, E. P., Terekhova, O. G. and Filipenko, M. L. (2013). Magnetofection of human somatic cells with magnetite and cobalt ferrospinel nanoparticles. *Bull. Exp. Biol. Med.*, **154**, 673–676.
64. Wang, Z., Chen, Z., Zuo, Q., Song, F., Wu, D., Cheng, W. and Fan, W. (2013). Reproductive toxicity in adult male rats following intra-articular injection of cobalt-chromium nanoparticles. *J. Orthop. Sci.*, **18**, 1020–1026.

65. Annangi, B., Bach, J., Vales, G., Rubio, L., Marcos, R. and Hernandez, A. (2014). Long-term exposures to low doses of cobalt nanoparticles induce cell transformation enhanced by oxidative damage. *Nanotoxicology*,
66. Raghunathan, V. K., Devey, M., Hawkins, S., Hails, L., Davis, S. A., Mann, S., Chang, I. T., Ingham, E., Malhas, A., Vaux, D. J., Lane, J. D. and Case, C. P. (2013). Influence of particle size and reactive oxygen species on cobalt chrome nanoparticle-mediated genotoxicity. *Biomaterials*, **34**, 3559–3570.
67. Petrarca, C., Perrone, A., Verna, N., Verginelli, F., Ponti, J., Sabbioni, E., Di Giampaolo, L., Dadorante, V., Schiavone, C., Boscolo, P., Mariani Costantini, R. and Di Gioacchino, M. (2006). Cobalt nano-particles modulate cytokine *in vitro* release by human mononuclear cells mimicking autoimmune disease. *Int. J. Immunopathol. Pharmacol.*, **19**, 11–14.
68. Nunes, A. D., Ramalho, L. S., Souza, A. P., Mendes, E. P., Colugnati, D. B., Zufelato, N., Sousa, M. H., Bakuzis, A. F. and Castro, C. H. (2014). Manganese ferrite-based nanoparticles induce *ex vivo*, but not *in vivo*, cardiovascular effects. *Int. J. Nanomedicine*, **9**, 3299–3312.
69. Burdinski, D., Pikkemaat, J. A., Lub, J., de Peinder, P., Nieto Garrido, L. and Weyhermuller, T. (2009). Lanthanide complexes of triethylenetetramine tetra-, penta-, and hexaacetamide ligands as paramagnetic chemical exchange-dependent saturation transfer contrast agents for magnetic resonance imaging: Nona- versus decadentate coordination. *Inorg. Chem.*, **48**, 6692–6712.
70. Sahoo, B., Devi, K. S., Dutta, S., Maiti, T. K., Pramanik, P. and Dhara, D. (2014). Biocompatible mesoporous silica-coated superparamagnetic manganese ferrite nanoparticles for targeted drug delivery and MR imaging applications. *J. Colloid Interface Sci.*, **431**, 31–41.
71. Li, Y., Duong, H. T., Laurent, S., MacMillan, A., Whan, R. M., Elst, L. V., Muller, R. N., Hu, J., Lowe, A., Boyer, C. and Davis, T. P. (2014). Nanoparticles based on star polymers as theranostic vectors: Endosomal-triggered drug release combined with MRI sensitivity. *Adv. Healthc. Mater.*, in press.
72. Kumar, C. S. and Mohammad, F. (2011). Magnetic nanomaterials for hyperthermia-based therapy and controlled drug delivery. *Adv. Drug Deliv. Rev.*, **63**, 789–808.
73. Mi, P., Kokuryo, D., Cabral, H., Kumagai, M., Nomoto, T., Aoki, I., Terada, Y., Kishimura, A., Nishiyama, N. and Kataoka, K. (2014). Hydrothermally synthesized PEGylated calcium phosphate nanoparticles incorporating Gd-DTPA for contrast enhanced MRI diagnosis of solid tumors. *J. Control. Release*, **174**, 63–71.
74. Allen, T. M. and Cullis, P. R. (2004). Drug delivery systems: Entering the mainstream. *Science*, **303**, 1818–1822.
75. Fleige, E., Quadir, M.A. and Haag, R. (2012). Stimuli-responsive polymeric nanocarriers for the controlled transport of active compounds: Concepts and applications. *Adv. Drug Deliv. Rev.*, **64**, 866–884.
76. Wang, A. Z., Langer, R. and Farokhzad, O. C. (2012). Nanoparticle delivery of cancer drugs. *Annu. Rev. Med.*, **63**, 185–198.
77. Santhosh, P. B. and Ulrih, N. P. (2013). Multifunctional superparamagnetic iron oxide nanoparticles: Promising tools in cancer theranostics. *Cancer Lett.*, **336**, 8–17.
78. Šafařík, I. and Šafaříková, M. (2002). 'Magnetic nanoparticles and biosciences', in Hofmann, H., Rahman, Z. and Schubert, U. (eds.), *Nanostructured Materials*, Springer, pp. 1–23.

79. Vatta, L. L., Sanderson, R. D. and Koch, K. R. (2006). Magnetic nanoparticles: Properties and potential applications. *Pure Appl. Chem.*, **78**, 1793.
80. Pankhurst, Q. A., Connolly, J., Jones, S. K. and Dobson, J. (2003). Applications of magnetic nanoparticles in biomedicine. *J. Phys. D: Appl. Phys.*, **36**, R167.
81. Lübbe, A. S., Bergemann, C., Brock, J. and McClure, D. G. (1999). Physiological aspects in magnetic drug-targeting. *J. Magn. Magn. Mater.*, **194**, 149–155.
82. Arruebo, M., Fernández-Pacheco, R., Ibarra, M. R. and Santamaría, J. (2007). Magnetic nanoparticles for drug delivery. *Nano. Today*, **2**, 22–32.
83. Neuberger, T., Schöpf, B., Hofmann, H., Hofmann, M. and von Rechenberg, B. (2005). Superparamagnetic nanoparticles for biomedical applications: Possibilities and limitations of a new drug delivery system. *J. Magn. Magn. Mater.*, **293**, 483–496.
84. Dandamudi, S. and Campbell, R. B. (2007). Development and characterization of magnetic cationic liposomes for targeting tumor microvasculature. *Biochim. Biophys. Acta.*, **1768**, 427–438.
85. Martina, M. S., Wilhelm, C. and Lesieur, S. (2008). The effect of magnetic targeting on the uptake of magnetic-fluid-loaded liposomes by human prostatic adenocarcinoma cells. *Biomaterials*, **29**, 4137–4145.
86. Mikhaylov, G., Mikac, U., Magaeva, A. A., Itin, V. I., Naiden, E. P., Psakhye, I., Babes, L., Reinheckel, T., Peters, C., Zeiser, R., Bogyo, M., Turk, V., Psakhye, S. G., Turk, B. and Vasiljeva, O. (2011). Ferri-liposomes as an MRI-visible drug-delivery system for targeting tumours and their microenvironment. *Nat. Nanotechnol.*, **6**, 594–602.
87. Chen, Y., Bose, A. and Bothun, G. D. (2010). Controlled release from bilayer-decorated magnetoliposomes via electromagnetic heating. *ACS Nano.*, **4**, 3215–3221.
88. Dandamudi, S. and Campbell, R. B. (2007). The drug loading, cytotoxicty and tumor vascular targeting characteristics of magnetite in magnetic drug targeting. *Biomaterials*, **28**, 4673–4683.
89. Dandamudi, S., Patil, V., Fowle, W., Khaw, B. A. and Campbell, R. B. (2009). External magnet improves antitumor effect of vinblastine and the suppression of metastasis. *Cancer Sci.*, **100**, 1537–1543.
90. Wang, B., Zhang, L., Bae, S. C. and Granick, S. (2008). Nanoparticle-induced surface reconstruction of phospholipid membranes. *Proc. Natl. Acad. Sci. U. S. A.*, **105**, 18171–18175.
91. Plassat, V., Martina, M. S., Barratt, G., Menager, C. and Lesieur, S. (2007). Sterically stabilized superparamagnetic liposomes for MR imaging and cancer therapy: Pharmacokinetics and biodistribution. *Int. J. Pharm.*, **344**, 118–127.
92. Cao, Z., Ma, Y., Yue, X., Li, S., Dai, Z. and Kikuchi, J. (2010). Stabilized liposomal nanohybrid cerasomes for drug delivery applications. *Chem. Commun. (Camb.)*, **46**, 5265–5267.
93. Jin, Y., Yue, X., Zhang, Q., Wu, X., Cao, Z. and Dai, Z. (2012). Cerasomal doxorubicin with long-term storage stability and controllable sustained release. *Acta. Biomater.*, **8**, 3372–3380.
94. Cao, Z., Yue, X., Li, X. and Dai, Z. (2013). Stabilized magnetic cerasomes for drug delivery. *Langmuir*, **29**, 14976–14983.
95. Plank, C., Anton, M., Rudolph, C., Rosenecker, J. and Krotz, F. (2003). Enhancing and targeting nucleic acid delivery by magnetic force. *Expert Opin. Biol. Ther.*, **3**, 745–758.

96. Scherer, F., Anton, M., Schillinger, U., Henke, J., Bergemann, C., Kruger, A., Gansbacher, B. and Plank, C. (2002). Magnetofection: Enhancing and targeting gene delivery by magnetic force *in vitro* and *in vivo*. *Gene Ther.*, **9**, 102–109.
97. Mykhaylyk, O., Vlaskou, D., Tresilwised, N., Pithayanukul, P., Möller,W. and Plank, C. (2007). Magnetic nanoparticle formulations for DNA and siRNA delivery. *J. Magn. Magn. Mater.*, **311**, 275–281.
98. Leung, K. C., Chak, C. P., Lee, S. F., Lai, J. M., Zhu, X. M., Wang, Y. X., Sham, K. W., Wong, C. H. and Cheng, C. H. (2013). Increased efficacies in magnetofection and gene delivery to hepatocellular carcinoma cells with ternary organic-inorganic hybrid nanocomposites. *Chem. Asian J.*, **8**, 1760–1764.
99. Haladjova, E., Rangelov, S., Tsvetanov, Ch. B., Posheva, V., Peycheva, E., Maximova, V., Momekova, D., Mountrichas, G., Pispas, S. and Bakandritsos, A. (2014). Enhanced gene expression promoted by hybrid magnetic/cationic block copolymer micelles. *Langmuir*, **30**, 8193–8200.
100. Wang, C., Ding, C., Kong, M., Dong, A., Qian, J., Jiang, D. and Shen, Z. (2011). Tumor-targeting magnetic lipoplex delivery of short hairpin RNA suppresses IGF-1R overexpression of lung adenocarcinoma A549 cells *in vitro* and *in vivo*. *Biochem. Biophys. Res. Commun.*, **410**, 537–542.
101. Long, L., Wang, W., Cai, X. D., Cheng, D., Shuai, X. and Peng, Y. (2014). PinX1-siRNA/mPEG-PEI-SPION combined with doxorubicin enhances the inhibition of glioma growth. *Exp. Ther. Med.*, **7**, 1170–1176.
102. Jiang, S., Eltoukhy, A. A., Love, K. T., Langer, R. and Anderson, D. G. (2013). Lipidoid-coated iron oxide nanoparticles for efficient DNA and siRNA delivery. *Nano. Lett.*, **13**, 1059–1064.
103. Jang, M., Yoon, Y. I., Kwon, Y. S., Yoon, T. J., Lee, H. J., Hwang, S. I., Yun, B. L. and Kim, S. M. (2014). Trastuzumab-conjugated liposome-coated fluorescent magnetic nanoparticles to target breast cancer. *Korean J. Radiol.*, **15**, 411–422.
104. Yang, C., Rait, A., Pirollo, K. F., Dagata, J. A., Farkas, N. and Chang, E. H. (2008). Nanoimmunoliposome delivery of superparamagnetic iron oxide markedly enhances targeting and uptake in human cancer cells *in vitro* and *in vivo*. *Nanomedicine*, **4**, 318–329.
105. Cheng, K., Peng, S., Xu, C. and Sun, S. (2009). Porous hollow Fe(3)O(4) nanoparticles for targeted delivery and controlled release of cisplatin. *J. Am. Chem. Soc.*, **131**, 10637–10644.
106. Yu, P., Xia, X. M., Wu, M., Cui, C., Zhang, Y., Liu, L., Wu, B., Wang, C. X., Zhang, L. J., Zhou, X., Zhuo, R. X. and Huang, S. W. (2014). Folic acid-conjugated iron oxide porous nanorods loaded with doxorubicin for targeted drug delivery. *Colloids Surf. B. Biointerfaces*, **120**, 142–151.
107. Krais, A., Wortmann, L., Hermanns, L., Feliu, N., Vahter, M., Stucky, S., Mathur, S. and Fadeel, B. (2014). Targeted uptake of folic acid-functionalized iron oxide nanoparticles by ovarian cancer cells in the presence but not in the absence of serum. *Nanomedicine*, **10**(7), 1421–1431.
108. Sanchez, C., El Hajj Diab, D., Connord, V., Clerc, P., Meunier, E., Pipy, B., Payre, B., Tan, R. P., Gougeon, M., Carrey, J., Gigoux, V. and Fourmy, D. (2014). Targeting a G-protein-coupled receptor overexpressed in endocrine tumors by magnetic nanoparticles to induce cell death. *ACS Nano.*, **8**, 1350–1363.

109. Ding, W. and Guo, L. (2013). Immobilized transferrin $Fe_3O_4 \cong SiO_2$ nanoparticle with high doxorubicin loading for dual-targeted tumor drug delivery. *Int. J. Nanomedicine*, **8**, 4631–4639.
110. Wang, F., Chen, Z. and Zhu, L. (2014). cRGD-conjugated magnetic-fluorescent liposomes for targeted dual-modality imaging of bone metastasis from prostate cancer. *J. Liposome Res.*, **24**, 1–12.
111. Guo, L., Ding, W. and Zheng, L. M. (2014). The C(RgdyK)-conjugated Fe_3O_4 nanoparticles with high drug load for dual-targeting integrin alpha(v)beta3-expressing cancer cells. *J. Nanosci. Nanotechnol.*, **14**, 4858–4864.
112. Li, Z., Liu, Z., Yin, M., Yang, X., Yuan, Q., Ren, J. and Qu, X. (2012). Aptamer-capped multifunctional mesoporous strontium hydroxyapatite nanovehicle for cancer-cell-responsive drug delivery and imaging. *Biomacromolecules*, **13**, 4257–4263.
113. Gao, H., Qian, J., Cao, S., Yang, Z., Pang, Z., Pan, S., Fan, L., Xi, Z., Jiang, X. and Zhang, Q. (2012). Precise glioma targeting of and penetration by aptamer and peptide dual-functioned nanoparticles. *Biomaterials*, **33**, 5115–5123.
114. Varshosaz, J., Sadeghi-aliabadi, H., Ghasemi, S. and Behdadfar, B. (2013). Use of magnetic folate-dextran-retinoic acid micelles for dual targeting of doxorubicin in breast cancer. *BioMed Res. Int.*, **2013**, 16.
115. Huang, X., Yi, C., Fan, Y., Zhang, Y., Zhao, L., Liang, Z. and Pan, J. (2014). Magnetic Fe_3O_4 nanoparticles grafted with single-chain antibody (scFv) and docetaxel loaded beta-cyclodextrin potential for ovarian cancer dual-targeting therapy. *Mater. Sci. Eng. C Mater. Biol. Appl.*, **42**, 325–332.
116. May, J. P. and Li, S. D. (2013). Hyperthermia-induced drug targeting. *Expert Opin. Drug Deliv.*, **10**, 511–527.
117. Moyer, H. R. and Delman, K. A. (2008). The role of hyperthermia in optimizing tumor response to regional therapy. *Int. J. Hyperthermia*, **24**, 251–261.
118. Fotopoulou, C., Cho, C. H., Kraetschell, R., Gellermann, J., Wust, P., Lichtenegger, W. and Sehouli, J. (2010). Regional abdominal hyperthermia combined with systemic chemotherapy for the treatment of patients with ovarian cancer relapse: Results of a pilot study. *Int. J. Hyperthermia*, **26**, 118–126.
119. Wust, P., Hildebrandt, B., Sreenivasa, G., Rau, B., Gellermann, J., Riess, H., Felix, R. and Schlag, P. M. (2002). Hyperthermia in combined treatment of cancer. *Lancet Oncol.*, **3**, 487–497.
120. Campbell, R. B. (2007). Battling tumors with magnetic nanotherapeutics and hyperthermia: Turning up the heat. *Nanomedicine (Lond.)*, **2**, 649–652.
121. van Vlerken, L. E. and Amiji, M. M. (2006). Multi-functional polymeric nanoparticles for tumour-targeted drug delivery. *Expert Opin. Drug Deliv.*, **3**, 205–216.
122. Shinkai, M., Suzuki, M., Iijima, S. and Kobayashi, T. (1995). Antibody-conjugated magnetoliposomes for targeting cancer cells and their application in hyperthermia. *Biotechnol. Appl. Biochem.*, **21**, 125–137.
123. Laurent, S., Dutz, S., Hafeli, U. O. and Mahmoudi, M. (2011). Magnetic fluid hyperthermia: Focus on superparamagnetic iron oxide nanoparticles. *Adv. Colloid Interface Sci.*, **166**, 8–23.
124. Jordan, A., Scholz, R., Wust, P., Fähling, H. and Roland, F. (1999). Magnetic fluid hyperthermia (MFH): Cancer treatment with AC magnetic field induced excitation of biocompatible superparamagnetic nanoparticles. *J. Magn. Magn. Mater.*, **201**, 413–419.

125. Kong, G. and Dewhirst, M. W. (1999). Hyperthermia and liposomes. *Int. J. Hyperther.*, **15**, 345–370.
126. Hayashi, K., Nakamura, M., Miki, H., Ozaki, S., Abe, M., Matsumoto, T., Sakamoto, W., Yogo, T. and Ishimura, K. (2014). Magnetically responsive smart nanoparticles for cancer treatment with a combination of magnetic hyperthermia and remote-control drug release. *Theranostics*, **4**, 834–844.
127. Mimnaugh, E. G., Xu, W., Vos, M., Yuan, X., Isaacs, J. S., Bisht, K. S., Gius, D. and Neckers, L. (2004). Simultaneous inhibition of hsp90 and the proteasome promotes protein ubiquitination, causes endoplasmic reticulum-derived cytosolic vacuolization, and enhances antitumor activity. *Mol. Cancer Ther.*, **3**, 551–566.
128. Mimnaugh, E. G., Xu, W., Vos, M., Yuan, X. and Neckers, L. (2006). Endoplasmic reticulum vacuolization and valosin-containing protein relocalization result from simultaneous hsp90 inhibition by geldanamycin and proteasome inhibition by velcade. *Mol. Cancer Res.*, **4**, 667–681.
129. Calderwood, S. K. (2013). 'Hyperthermia, the tumor microenvironment and immunity', in Keisari, Y. (ed.), *Tumor Ablation*, Springer, Netherlands, pp. 29–37.
130. Koishi, M., Yokota, S., Mae, T., Nishimura, Y., Kanamori, S., Horii, N., Shibuya, K., Sasai, K. and Hiraoka, M. (2001). The effects of KNK437, a novel inhibitor of heat shock protein synthesis, on the acquisition of thermotolerance in a murine transplantable tumor *in vivo*. *Clin. Cancer Res.*, 7, 215–219.
131. Yoo, D., Jeong, H., Noh, S. H., Lee, J. H. and Cheon, J. (2013). Magnetically triggered dual functional nanoparticles for resistance-free apoptotic hyperthermia. *Angew. Chem. Int. Ed. Engl.*, **52**, 13047–13051.
132. Yin, P. T., Shah, B. P. and Lee, K. B. (2014). Combined magnetic nanoparticle-based microRNA and hyperthermia therapy to enhance apoptosis in brain cancer cells. *Small*, **10**(20), 4106–4112.
133. Pradhan, P., Giri, J., Rieken, F., Koch, C., Mykhaylyk, O., Doblinger, M., Banerjee, R., Bahadur, D. and Plank, C. (2010). Targeted temperature sensitive magnetic liposomes for thermo-chemotherapy. *J. Control. Release*, **142**, 108–121.
134. Yoshida, M., Watanabe, Y., Sato, M., Maehara, T., Aono, H., Naohara, T., Hirazawa, H., Horiuchi, A, Yukumi, S., Sato, K., Nakagawa, H., Yamamoto, Y., Sugishita, H. and Kawachi, K. (2010). Feasibility of chemohyperthermia with docetaxel-embedded magnetoliposomes as minimally invasive local treatment for cancer. *Int. J. Cancer*, **126**, 1955–1965.
135. Bealle, G., Di Corato, R., Kolosnjaj-Tabi, J., Dupuis, V., Clement, O., Gazeau, F., Wilhelm, C. and Menager, C. (2012). Ultra magnetic liposomes for MR imaging, targeting, and hyperthermia. *Langmuir*, **28**, 11834–11842.
136. Harisinghani, M. G., Barentsz, J., Hahn, P. F., Deserno, W. M., Tabatabaei, S., van de Kaa, C. H., de la Rosette, J. and Weissleder, R. (2003). Noninvasive detection of clinically occult lymph-node metastases in prostate cancer. *N. Engl. J. Med.*, **348**, 2491–2499.
137. Frullano, L. and Meade, T. J. (2007). Multimodal MRI contrast agents. *J. Biol. Inorg. Chem.*, **12**, 939–949.
138. Qiao, R., Yang, C. and Gao, M. (2009). Superparamagnetic iron oxide nanoparticles: From preparations to *in vivo* MRI applications. *J. Mater. Chem.*, **19**, 6274–6293.
139. Chen, Y., Li, M., Hong, Y., Lam, J. W., Zheng, Q. and Tang, B. Z. (2014). Dual-modal MRI contrast agent with aggregation-induced emission characteristic for liver specific imaging with long circulation lifetime. *ACS Appl. Mater. Inter.*, **6**, 10783–10791.

140. Debroye, E., Eliseeva, S. V., Laurent, S., Vander Elst, L., Muller, R. N. and Parac-Vogt, T. N. (2014). Micellar self-assemblies of gadolinium(III)/europium(III) amphiphilic complexes as model contrast agents for bimodal imaging. *Dalton Trans*, **43**, 3589–3600.
141. Sawant, R., Gultepe, E., Nagesha, D., Papahadjopoulos-Sternberg, B., Sridhar, S. and Torchilin, V. (2009). Nanosized cancer cell-targeted polymeric immunomicelles loaded with superparamagnetic iron oxide nanoparticles. *J. Nanopart. Res.*, **11**, 1777–1785.
142. Liao, C., Sun, Q., Liang, B., Shen, J. and Shuai, X. (2011). Targeting EGFR-overexpressing tumor cells using Cetuximab-immunomicelles loaded with doxorubicin and superparamagnetic iron oxide. *Eur. J. Radiol.*, **80**, 699–705.
143. Gultepe, E., Reynoso, F. J., Jhaveri, A., Kulkarni, P., Nagesha, D., Ferris, C., Harisinghani, M., Campbell, R. B. and Sridhar, S. (2010). Monitoring of magnetic targeting to tumor vasculature through MRI and biodistribution. *Nanomedicine (Lond.)*, **5**, 1173–1182.
144. Quillard, T. and Libby, P. (2012). Molecular imaging of atherosclerosis for improving diagnostic and therapeutic development. *Circ. Res.*, **111**, 231–244.
145. Nguyen, T. H., Bryant, H., Shapsa, A., Street, H., Mani, V., Fayad, Z. A., Frank, J. A., Tsimikas, S. and Briley-Saebo, K. C. (2014). Manganese G8 dendrimers targeted to oxidation-specific epitopes: *In vivo* MR imaging of atherosclerosis. *J. Magn. Reson. Imaging*, in press.
146. Filippi, M., Martinelli, J., Mulas, G., Ferraretto, M., Teirlinck, E., Botta, M., Tei, L. and Terreno, E. (2014). Dendrimersomes: A new vesicular nano-platform for MR-molecular imaging applications. *Chem. Commun. (Camb.)*, **50**, 3453–3456.
147. Sitbon, G., Bouccara, S., Tasso, M., Francois, A., Bezdetnaya, L., Marchal, F., Beaumont, M. and Pons, T. (2014). Multimodal Mn-doped I-III-VI quantum dots for near infrared fluorescence and magnetic resonance imaging: From synthesis to *in vivo* application. *Nanoscale*, **6**, 9264–9272.
148. Abdukayum, A., Yang, C. X., Zhao, Q., Chen, J. T., Dong, L. X. and Yan, X. P. (2014). Gadolinium complexes functionalized persistent luminescent nanoparticles as a multimodal probe for near-infrared luminescence and magnetic resonance imaging *in vivo*. *Anal. Chem.*, **86**, 4096–4101.
149. Ozdemir, V., Williams-Jones, B., Glatt, S. J., Tsuang, M. T., Lohr, J. B. and Reist, C. (2006). Shifting emphasis from pharmacogenomics to theragnostics. *Nat. Biotechnol.*, **24**, 942–946.
150. Urries, I., Munoz, C., Gomez, L., Marquina, C., Sebastian, V., Arruebo, M. and Santamaria, J. (2014). Magneto-plasmonic nanoparticles as theranostic platforms for magnetic resonance imaging, drug delivery and NIR hyperthermia applications. *Nanoscale*, **6**, 9230–9240.
151. Jing, L., Liang, X., Li, X., Lin, L., Yang, Y., Yue, X. and Dai, Z. (2014). Mn-porphyrin conjugated au nanoshells encapsulating doxorubicin for potential magnetic resonance imaging and light triggered synergistic therapy of cancer. *Theranostics*, **4**, 858–871.
152. Sancey, L., Lux, F., Kotb, S., Roux, S., Dufort, S., Bianchi, A., Cremillieux, Y., Fries, P., Coll, J. L., Rodriguez-Lafrasse, C., Janier, M., Dutreix, M., Barberi-Heyob, M., Boschetti, F., Denat, F., Louis, C., Porcel, E., Lacombe, S., Le Duc, G., Deutsch, E., Perfettini, J. L., Detappe, A., Verry, C., Berbeco, R., Butterworth, K., McMahon, S., Prise, K., Perriat, P. and Tillement, O. (2014). The use of theranostic gadolinium-based nanoprobes to improve radiotherapy efficacy. *Br. J. Radiol.*, 20140134.

153. Wang, Y. X., Zhu, X. M., Liang, Q., Cheng, C. H., Wang, W. and Leung, K. C. (2014). *In vivo* chemoembolization and magnetic resonance imaging of liver tumors by using iron oxide nanoshell/doxorubicin/poly(vinyl alcohol) hybrid composites. *Angew. Chem. Int. Ed. Engl.*, **53**, 4812–4815.
154. Li, W. P., Liao, P. Y., Su, C. H. and Yeh C. S. (2014). Formation of oligonucleotide-gated silica shell-coated Fe_3O_4-au core-shell nanotrisoctahedra for magnetically targeted and near-infrared light-responsive theranostic platform. *J. Am. Chem. Soc.*, **136**, 10062–10075.
155. Ganta, S., Singh, A., Rawal, Y., Cacaccio, J., Patel, N. R., Kulkarni, P., Ferris, C. F., Amiji, M. M. and Coleman, T. P. (2014). Formulation development of a novel targeted theranostic nanoemulsion of docetaxel to overcome multidrug resistance in ovarian cancer. *Drug Deliv.*, 1–13.
156. Bhowmick, N. A., Neilson, E. G. and Moses, H. L. (2004). Stromal fibroblasts in cancer initiation and progression. *Nature*, **432**, 332–337.
157. Goubran, H. A., Kotb, R. R., Stakiw, J., Emara, M E. and Burnouf, T. (2014). Regulation of tumor growth and metastasis: The role of tumor microenvironment. *Cancer Growth Metastasis*, **7**, 9–18.
158. Pietras, K. and Ostman, A. (2010). Hallmarks of cancer: Interactions with the tumor stroma. *Exp. Cell Res.*, **316**, 1324–1331.
159. Li, D., Tang, X., Pulli, B., Lin, C., Zhao, P., Cheng, J., Lv, Z., Yuan, X., Luo, Q., Cai, H. and Ye, M. (2014). Theranostic nanoparticles based on bioreducible polyethylenimine-coated iron oxide for reduction-responsive gene delivery and magnetic resonance imaging. *Int. J. Nanomedicine*, **9**, 3347–3361.
160. Ding, M., Weng, X., He, X., Li, J., Tan, H. and Fu, Q. (2014). Cell internalizable and intracellularly degradable cationic polyurethane micelles as a potential platform for efficient imaging and drug delivery. *Biomacromolecules*, **15**, 2896–2906.
161. Bulte, J. W. and Kraitchman, D. L. (2004). Monitoring cell therapy using iron oxide MR contrast agents. *Curr. Pharm. Biotechnol.*, **5**, 567–584.
162. Bulte, J. W., Arbab, A. S., Douglas, T. and Frank, J. A. (2004). Preparation of magnetically labeled cells for cell tracking by magnetic resonance imaging. *Methods Enzymol.*, **386**, 275–299.
163. Schellenberger, E. A., Bogdanov, A., Jr., Hogemann, D., Tait, J., Weissleder, R. and Josephson, L. (2002). Annexin V-CLIO: A nanoparticle for detecting apoptosis by MRI. *Mol. Imaging*, **1**, 102–107.
164. van Tilborg, G. A., Mulder, W. J., Deckers, N., Storm, G., Reutelingsperger, C. P., Strijkers, G. J. and Nicolay, K. (2006). Annexin A5-functionalized bimodal lipid-based contrast agents for the detection of apoptosis. *Bioconjug. Chem.*, **17**, 741–749.
165. Safarik, I. and Safarikova, M. (2004). Magnetic techniques for the isolation and purification of proteins and peptides. *Biomagn. Res. Technol.*, **2**, 7.
166. Šafařík, I., Horská, K. and Šafaříková, M. (2011). 'Magnetic nanoparticles for biomedicine', in Prokop, A. (ed.), *Intracellular Delivery*, Springer, Netherlands, pp. 363–372.
167. Gamarra, L. F., Pavon, L. F., Marti, L. C., Pontuschka, W. M., Mamani, J. B., Carneiro, S. M., Camargo-Mathias, M. I., Moreira-Filho, C. A. and Amaro, E, Jr. (2008). *In vitro* study of CD133 human stem cells labeled with superparamagnetic iron oxide nanoparticles. *Nanomedicine*, **4**, 330–339.
168. Jain, K. K. (2003). Nanodiagnostics: Application of nanotechnology in molecular diagnostics. *Expert Rev. Mol. Diagn.*, **3**, 153–161.

169. Baldi, G., Ravagli, C., Mazzantini, F., Loudos, G., Adan, J., Masa, M., Psimadas, D., Fragogeorgi, E. A., Locatelli, E., Innocenti, C., Sangregorio, C. and Comes Franchini, M. (2014). *In vivo* anticancer evaluation of the hyperthermic efficacy of anti-human epidermal growth factor receptor-targeted PEG-based nanocarrier containing magnetic nanoparticles. *Int. J. Nanomedicine*, **9**, 3037–3056.
170. Chiang, C. S., Hu, S.-H., Liao, B.-J., Chang, Y.-C. and Chen, S.-Y. (2014). Enhancement of cancer therapy efficacy by trastuzumab-conjugated and pH-sensitive nanocapsules with the simultaneous encapsulation of hydrophilic and hydrophobic compounds. *Nanomed. Nanotechnol. Biol. Med.*, **10**, 99–107.
171. Shen, J. M., Gao, F.-Y., Yin, T., Zhang, H.-X., Ma, M., Yang, Y.-J. and Yue, F. (2013). cRGD-functionalized polymeric magnetic nanoparticles as a dual-drug delivery system for safe targeted cancer therapy. *Pharmacol. Res.*, **70**, 102–115.

Chapter 10

Smart Lipid-Based Drug Delivery Systems

Paloma Benito-Gallo[1], *Pavel Gershkovich*[1], *Maria Marlow*[1], *Vanessa Zann*[2] *and Kishor M. Wasan*[3,4]

[1]*School of Pharmacy, University of Nottingham, University Park, Nottingham, UK*
[2]*Quotient Clinical Ltd., Mere Way, Ruddington Fields, Nottingham, UK*
[3]*Faculty of Pharmaceutical Sciences, University of British Columbia, Vancouver, BC, Canada*
[4]*College of Pharmacy and Nutrition, University of Saskatchewan Saskatoon, SK, Canada*

10.1 Introduction

Since the advent of high throughput techniques and development of combinatorial chemistry in the early 1990s, the number of potential drug candidates has significantly increased [1]. Physical properties of new chemical entities have changed towards higher molecular weight, higher melting point, increased H-bonding capacity, and increased lipophilicity, leading to poorer solubility in aqueous media [2]. Two good examples to illustrate this trend are the list elaborated by Takagi *et al.* in 2006 [3] of the top 200 marketed oral product drugs, of which 40% were poorly water-soluble, and the fact that up to 70% of the new active molecular entities in development exhibit poor aqueous solubility [4].

Lipophilicity is defined as the ability of a compound to solubilize in lipids and non-polar solvents. Therefore, it reflects the *in vivo* partition of drugs from aqueous phases to cell membranes and protein binding sites, which are hydrophobic in nature. Lipophilicity is considered one of the most important physicochemical properties in drug discovery and design [5]. It is known to play a key role in solubility and permeability across biological membranes, to contribute to potency, selectivity and promiscuity (binding to unwanted targets), and to affect pharmacokinetic and pharmacodynamic profiles of drugs [6, 7].

Lipophilic drugs and many biologicals (such as peptides, proteins, nucleic acids or vaccines) are poorly water-soluble compounds. Because low solubility can be associated with poor absorption and hence poor bioavailability, it is clear that the greatest challenge for pharmaceutical scientists nowadays is finding novel formulations capable of improving the intraluminal solubility of poorly soluble drugs. In order to address this issue, there are several conventional formulation approaches, such as particle size reduction [8, 9], solid dispersions [10, 11], salt formation [12], cyclodextrins complexation [13, 14], and prodrug design [15, 16]. However, it is lipid-based drug delivery systems (LBDDSs) technology that has emerged as the most promising formulation strategy for efficient delivery of poorly water-soluble compounds [17–20].

Lipids are a group of naturally occurring hydrophobic and amphipathic small molecules, that include fatty acids (FAs), mono-, di- and triglycerides (TGs), phospholipids, waxes, eicosanoids, sphingolipids, sterols, terpenes, prenols and fat-soluble vitamins (Fig. 10.1). Their main biological functions are to store energy, conform cells membrane, and act as signalling molecules [21]. In terms of drug delivery, as there is such a wide spectrum of lipids with different characteristics, this offers great biocompatibility and versatility of application. Furthermore, LBDDSs are already an accepted and successful commercially viable strategy in formulation of sparingly soluble compounds (Table 10.1). Current improvements in LBDDS technology are focused on the development of "smart" lipidic carriers. The use of intelligently-devised lipid formulations offers an interesting opportunity for drug delivery when the carrier system becomes an active participant in the enhancement of its pharmacological activity rather than a passive vehicle.

LBDDSs have attracted considerable academic attention particularly in the field of oral delivery and, as a result, there are a large number of reviews and original research articles detailing their use for improving oral bioavailability [22–27]. The success of lipidic formulations is based on their numerous advantages including, but not limited to: bypassing the dissolution step (drugs are presented to the gastrointestinal fluids in oil) [28]; promoting drug solubilization in the gastrointestinal tract by providing lipidic components that increase the inherent solubilization capacity of the intestinal fluids [29]; and increasing drug permeability through inhibition of efflux transporters such as P-glycoprotein [30, 31], or through change in the membrane fluidity of enterocytes [32].

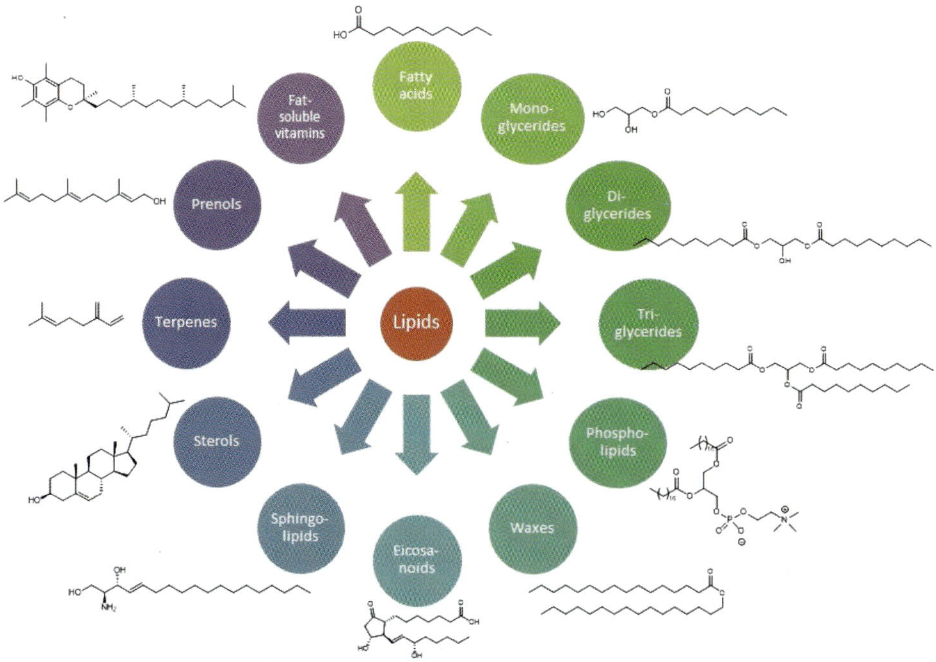

Figure 10.1. Natural occurring lipids and examples. Examples include capric acid (fatty acid), monocaprilin (monoglyceride), dicaprilin (diglyceride), tricaprilin (triglyceride), 1,2-distearoyl-*sn*-glycero-3-phosphocholine (phospholipid), cetyl palmiate (wax), prostaglandin E_1 (eicosanoid), sphingosine (sphingolipid), cholesterol (sterol), myrcene (terpene), farnesol (prenol) and vitamin E (fat-soluble vitamin).

Although not so exhaustively covered, LBDDSs for non-oral routes of administration (systemic, transdermal, ocular and intramuscular) are also present in literature [33, 34]. These reviews usually focus on a certain type of carriers (liposomes [35], solid lipid nanoparticles (SLNs) [36], etc.) and fail to describe the entire range of different lipidic formulations. As a consequence, the full potential of lipidic formulations as drug carriers is usually underestimated.

Despite the existence of the aforementioned reviews, none of them has described the development and application of smart LBDDSs as a whole. Therefore, the aim of this chapter is to fill that gap in the literature by describing smart lipidic formulations for systemic, oral, transdermal, ocular and intramuscular drug delivery.

Two different types of smart lipidic carriers for delivery of drugs and biologicals are considered in the following sections. There are smart-structured lipidic formulations which are target-specific and display stimuli-responsive properties when a triggering agent appears/disappears (e.g. pH-sensitive liposomes, magnetically-sensitive lipid nanoparticles, etc.). On the other hand,

Table 10.1. Selected commercial lipid-based drug delivery systems. (Abbreviations are explained in Table 10.3.)

Active pharmaceutical ingredient	Type of LBDDS	Route of administration	Trade name	Indication	Lipidic components	Reference
Verteporfin	Liposome	Intravenous	Visudyne®	Age-related macular degeneration, pathologic myopia, or ocular histoplasmosis	Ascorbyl palmitate, DMPC, egg PG	[37]
Doxorubicin	Liposome	Intravenous	Doxil®	Ovarian cancer, AIDS-related Kaposi's sarcoma, multiple myeloma	HSPC, $mPEG_{2000}$–DSPE, cholesterol	[38]
Amphotericin B	Liposome	Intravenous	AmBisome®	Fungal and protozoal infections	HSPC, DSPG, cholesterol	[39]
Octafluoropropane	Liposome	Intravenous	Definity®	Ultrasound contrast agent	DPPA, DPPC, $mPEG_{5000}$–DPPE	[40]
Vincristine sulfate	Liposome	Intravenous	Marqibo®	Philadelphia chromosome-negative acute lymphoblastic leukemia	Sphingomyelin, cholesterol	[41]
Daunorubicin	Liposome	Intravenous	DaunoXome®	AIDS-related Kaposi's sarcoma	DSPC, cholesterol	[42]
Cytarabine	Liposome	Spinal	DepoCyt®	Intrathecal treatment of lymphomatous meningitis	Cholesterol, triolein, DOPC, DPPG	[43]

Morphine	Liposome	Epidural	DepoDur®	Treatment of pain following major surgery	DOPC, cholesterol, DPPG, tricaprylin, triolein	[44]
Inactivated hepatitis A virus	Liposome (virosome)	Intramuscular	Epaxal®	Hepatitis A vaccine	DOPC and DOPE	[45]
Inactivated hemaglutinine of influenza virus	Liposome (virosome)	Intramuscular	Inflexal V®	Influenza vaccine	DOPC and DOPE	[45]
Estradiol	Micellar emulsion	Transdermal	Estrasorb®	Moderate to severe vasomotor symptoms of menopause	Soybean oil	[46]
N/A	Emulsion	Intravenous	Intralipid®	Parenteral nutrition, anaesthetic overdose antidote	Soybean oil, α-tocopherol, phytosterols,	[47]
Cyclosporin A	SMEDDS	Oral	Neoral Sandimmune®	Immunosuppressant, prophylaxis of organ transplant rejection	DL-α-tocopherol, corn oil-mono-di-TGs	[48]
Dronabinol	Emulsion	Oral	Marinol®	Appetite stimulation in AIDS-related anorexia	Sesame oil	[49]

there are formulations which might be apparently simple-structured but with such an unusual behavior that are capable of significantly modifying the pharmacokinetic and tissue distribution profile of the incorporated drug (e.g. long-chain triglyceride-based formulations that promote recruitment of lymphatic transport, thus decreasing first pass metabolism).

10.2 Smart Lipid-Based Carriers for Systemic Drug Delivery

The systemic route of administration is the fastest way to deliver therapeutic agents to the body. Furthermore, parenteral delivery of hydrophobic drugs and biological compounds (such as proteins) represents an approach to circumvent intestinal enzymatic degradation and precipitation in the intestinal milieu, respectively.

Currently, research efforts are focused on developing an ideal pharmaceutical biodegradable systemic drug delivery system with small particle size (to prevent embolization of blood vessels and capillary blockage following injection), high loading capacity, long circulation times, and the ability to accumulate in the targeted tissue/organ [50]. Liposomes, micelles, solid lipid nanoparticles, lipid nanocapsules (LCNs), lipid-coated systems, emulsions, and lipid-based liquid crystals are examples of smart LBDDS for the safe and efficient delivery of genes, peptides, proteins and poorly soluble drugs.

10.2.1 *Smart liposomes for systemic delivery*

Liposomes are biocompatible spherical, self-closed structures formed by one or several concentric phospholipid bilayers, ranging from 100 to 5000 nm. Phospholipids are amphiphilic molecules with a hydrophilic head (negatively charged phosphate group or cholesterol) and a hydrophobic tail (two long fatty acid chains) (Tables 10.2 and 10.3). Upon hydration, phospholipids are capable of holding an aqueous phase inside and between the lipid bilayers. Due to this amphipathic character, liposomes allow the delivery of both hydrophilic drugs (entrapped within the aqueous compartment) and hydrophobic counterparts (incorporated into the bilayer membrane) [35]. Importantly, pharmaceuticals delivered by means of liposomal carriers are protected from degradation, so their therapeutic efficacy is increased [51, 52].

10.2.1.1 *First generation of liposomes: Plain liposomes*

Despite their promising properties as drug carriers (Fig. 10.2 A), it was soon realized that practical application of non-modified liposomes was limited due to removal from the circulation by the mononuclear phagocyte system (MPS).

Table 10.2. Characteristic FAs found in phospholipids.

Common name	Lipid nomenclature	Chemical structure
Myristic acid	C14:0	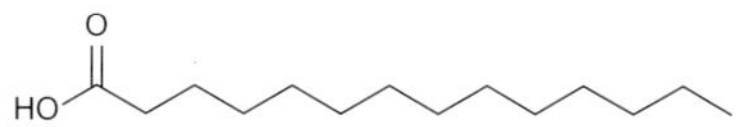
Palmitic acid	C16:0	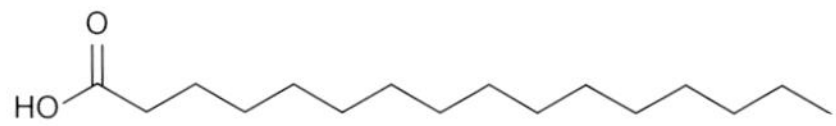
Stearic acid	C18:0	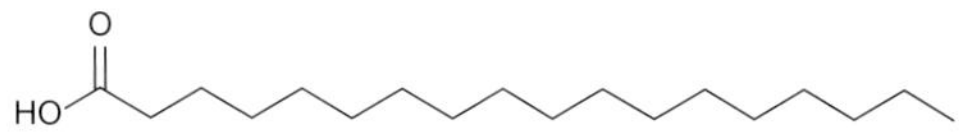
Oleic acid	C18:1 cis-9	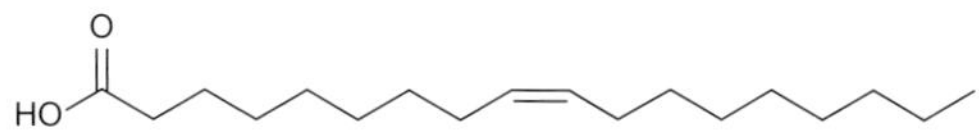

Following intravenous administration, opsonins — blood proteins — adsorb to the liposomal surface, priming their recognition and phagocytosis by mononuclear phagocytes in the liver and the spleen [35, 51, 53]. Although opsonisation is seen as a drawback in most cases, liposome accumulation in the liver might be beneficial for the treatment of MPS localized tumors [54]. To illustrate this fact, Gabizon *et al.* [55, 56] administered doxorubicin-loaded liposomes by intravenous infusion to cancer patients, and found that doxorubicin's antitumors efficacy was maintained, whereas cardiac and gastro-intestinal toxicities were substantially reduced due to phagocytosis by the MPS, thus increasing the therapeutic index of the drug. Opsonization also presents an opportunity for treatment of intracellular infections such as candidiasis, leishmaniasis or listeria since macrophages are directly involved in these parasitic and microbial diseases [57]. For example, AmBisome® is a liposomal formulation of amphotericin B (AmB) used in the treatment against the protozoan parasite *Leishmania* [58]. Liposomes containing AmB are made of hydrogenated soybean oil L-α-phosphatidylcholine (HSPC), cholesterol and 1,2-distearoyl-*sn*-glycero-3-phosphoglycerol (DSPG) [39]. Ambisome® owes its success to the fact that both parasites and liposomes are taken up by phagocytic cells — particularly the Kupffer cells of the liver. Captured liposomes are then capable of fusing their membranes together with the phagolysosome of parasite-hosting macrophages. As a result, encapsulated AmB can gain access into the parasitophorous vacuole and exert its therapeutic effect [59].

10.2.1.2 *Long-circulating liposomes: "Stealth" liposomes*

In order to avoid MPS uptake and increase circulation time, liposomes coated with polyethylene glycol (PEG) were introduced (Fig. 10.2 B). PEG attachment to lipids on the liposome surface leads to the formation of a flexible hydrated

Table 10.3. List of common phospholipids found in LBDDSs.

Type	Chemical structure, examples, and abbreviations				
Phosphatidyl-glycerol (PG)	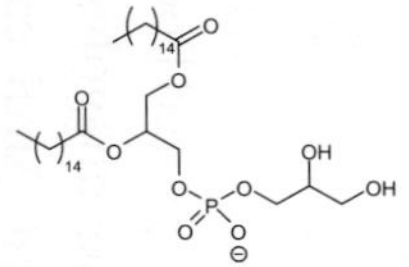 1,2-Dipalmitoyl-*sn*-glycero-3-phosphatidylglycerol (DPPG)				
Phosphatidyl-choline (PC)	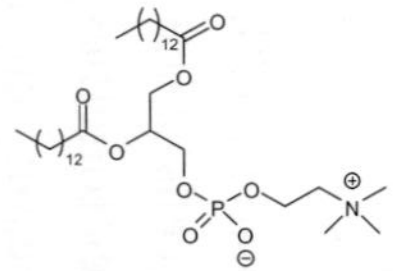 1,2-Dimyristoyl-*sn*-glycero-3-phosphocholine (DMPC)	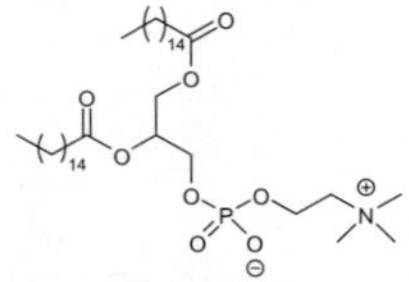 1,2-Dipalmitoyl-*sn*-3-phosphatidylcholine (DPPC)	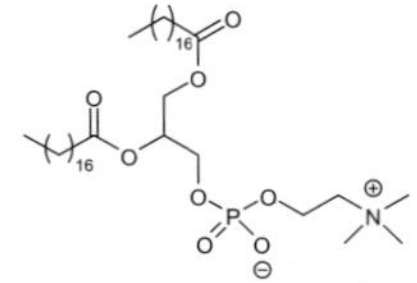 1,2-Distearoyl-*sn*-glycero-3-phosphocholine (DSPC)	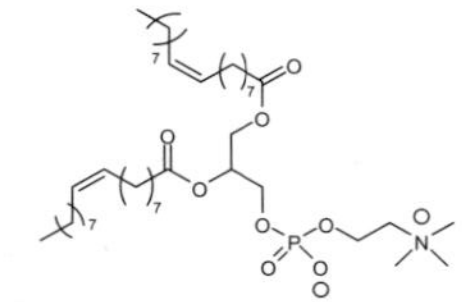 1,2-Dioleoyl-*sn*-glycero-3-phosphocholine (DOPC)	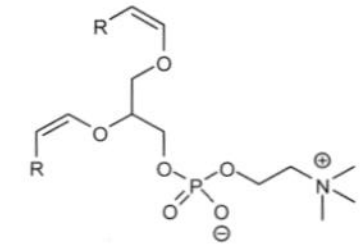 1,2-di-*O*-(*Z*-1′-enyl)-*sn*-glycero-3-phosphocholine; (Diplasmenylcholine, DPPlsC)
Phosphatidyl-ethanolamine (PE)			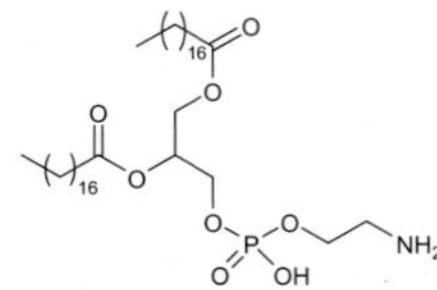 1,2-Distearoyl-*sn*-glycero-3-phosphatidylethanolamine (DSPE)	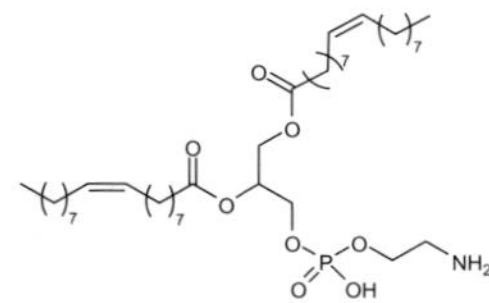 1,2-Dioleoyl-*sn*-3-phosphatidylethanolamine (DOPE)	
Glycerol-phosphate (PA)		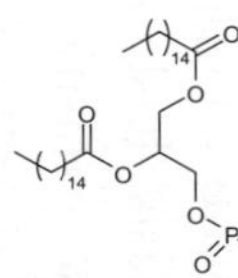 1,2-Dipalmitoyl-*sn*-glycero-3-phosphate (DPPA)	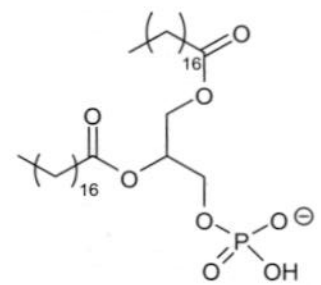 1,2-Distearoyl-*sn*-glycero-3 phosphate (DSPA)		

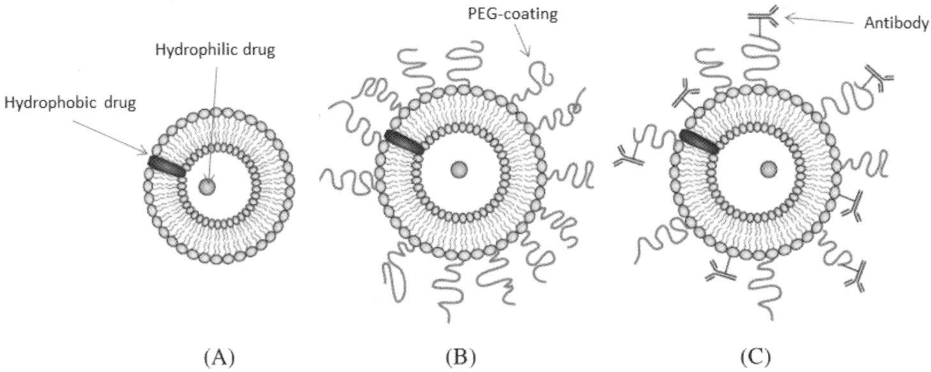

Figure 10.2. Different types of liposomes. A: Plain liposome incorporating a hydrophilic drug inside the aqueous core, and a hydrophobic drug within the phospholipid bilayer. B: Stealth liposome sterically stabilized by PEG-coating. C: Stealth liposome functionalized with antibodies for active targeting. Modified from Ref. [35].

steric barrier which avoids recognition by the immune system, leading to an increase in the circulation time [51, 60–62]. These so-called stealth* liposomes have shown circulating half-lives of up to 45 hours, as opposed to a few hours, or even less, for plain liposomes [54]. As a result of prolonged blood-circulating times, stealth liposomes increase their efficiency through "passive targeting", i.e. selective passive accumulation in tumors and inflamed tissue with leaky vasculature *via* the enhanced permeability and retention effect (EPR) [63, 64]. Clinical applications of PEGylated liposomes include chemotherapeutical agents such as Doxil® (doxorubicin), S-CKD602 (semi-synthetic analog of camptothecin), and Lipoplatin™ (cisplatin) [62]. Although initially thought to be biologically inert, it was shown later that the efficacy and safety of PEGylated liposomes could be severely compromised after repeated administrations due to hypersensitivity, loss of accurate targeting and rapid clearance from the systemic circulation [65].

10.2.1.3 *Long-circulating functionalized liposomes for active targeting*

The next step in liposome evolution was the incorporation of targeting ligands on the surface of stealth liposomes, so encapsulated drugs were delivered to the right cells based on molecular recognition processes (Fig. 10.2 C).

* Stealth® is a registered trademark of Liposome Technology Inc., Menlo Park, CA, USA.

Various targeting moieties have been bound to liposomes through coupling with the water-exposed end of PEG chains in order to increase the therapeutic activity of their just PEGylated counterparts [66–68]. Most research efforts in this area have focused on cancer treatment. Because tumor cells proliferate rapidly, they over-express certain receptors for enhanced absorption of nutrients, like folic acid, sugars and vitamins [69]. By targeting folate receptors, liposomal formulations can penetrate the cell membrane through endocytosis, bypassing efflux transporters and successfully delivering the drug inside tumour cells. Folate-mediated liposomes loaded with daunorubicin [70, 71] and doxorubicin [72] have been reported to increase cytotoxicity against tumour cells. In similar fashion, stealth liposomes coated with transferrin have been developed to target transferrin receptors which are also over-expressed in cancer cells [73]. Another successful example are liposomes coupled with galactosylated lipids targeting asialoglycoprotein receptors, which have been proven to be an effective and promising strategy for inhibiting liver metastases [74]. In all these cases the number of ligand molecules per liposome were optimized to achieve recognition by cell receptors while avoiding detection by the MPS [75].

10.2.1.4 *pH-responsive liposomes*

Healthy tissue has a pH of approximately 7.4, while inflamed, infected or metastasized tissues have a lower pH around 6.5 due to hypoxia. At the intracellular level, endosomes and lysosomes have even lower pH (5.5 and 4.5 respectively) than the cytosol [76]. Based on this pH difference, liposomes that destabilize upon pH change and subsequently release incorporated drug have been studied [77]. The most common pH-sensitive liposomes were of dioleoylphosphatidylethanolamine (DOPE) and weakly acidic amphiphiles such as oleic acid or cholesteryl hemisuccinate (CHEMS). However, these DOPE/CHEMS formulations were rapidly recognized by the MPS. More recently, research has been focused on the development of PEGylated stabilized liposomes with higher pH-sensitivity [78, 79] and on the design of liposomes with detachable PEG which show increased drug cellular uptake [60]. In order to achieve better pH-responsiveness, Rui *et al.* [80] synthesized sterically stable liposomes made of 1,2-di-*O*-(*Z*-1′-hexadecenyl)-*sn*-glycero-3-phosphocholine (DPPlsC), a naturally occurring pH-sensitive derived phospholipid. DPPlsC was hydrolyzed under endosomal acidic conditions to give degradation products glycerophosphocholine, hexadecanal and hexadecanoic acid (Fig. 10.3). This hydrolysis led to significantly increased membrane permeability, and thus, an enhancement in the cytotoxic activity of the incorporated drug.

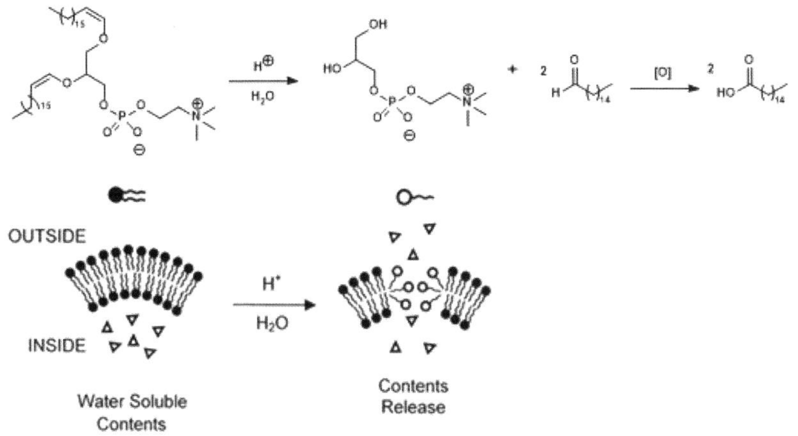

Figure 10.3. Acid-promoted hydrolysis of diplasmenyl phosphocholine (DPPlsC) that results in liposome destabilization and intracellular drug delivery from liposomal carriers via endosomal disruption. Modified from Ref. [81].

10.2.1.5 *Temperature-responsive liposomes: Thermoliposomes*

Yatvin *et al.* [82] in 1978 introduced the idea of using thermo-sensitive carriers triggered by the distinct local hyperthermia of pathological areas. Temperature-sensitive liposomes are stable at normal body temperature but at higher temperatures suffer phase transition (gel to liquid crystalline, and lamellar to hexagonal [69]), become leaky and release their drug content. A typical component of this kind of liposome is dipalmitoylphosphatidylcholine (DPPC) whose phase transition from gel to liquid occurs at 41°C [83]. The other type of thermoliposomes liberates the drug when attached thermosensitive polymers (such as poly[2-(2-ethoxy)ethoxyethyl vinyl ether] (EOEOVE)) precipitate at mild hyperthermia temperatures (40°C for EOEOVE) damaging the liposomal membrane [84].

A different approach which has been found to have more practical applications is inducing liposomal destabilization *via* external localized heating [85], such as radiofrequency ablation [83]. As an example, Kono *et al.* administered temperature-sensitive liposomes (Fig. 10.4) to mice for tumor-specific chemotherapy with doxorubicin which exhibited significant drug release at 45°C (temperature was reached using a radiofrequency oscillator) but very limited at 37°C [86], showing at the same time prolonged circulation properties.

10.2.1.6 *Redox-responsive liposomes*

Another stimuli mechanism for smart drug delivery is the redox potential difference that exists between both sides of the cell membrane, due to higher

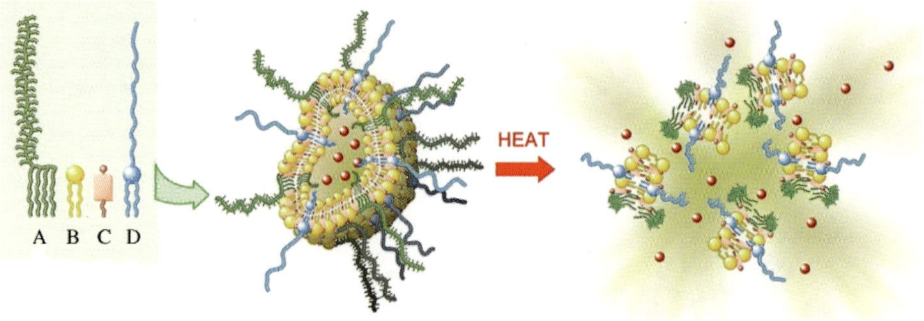

Figure 10.4. Illustration of heat-triggered release of doxorubicin (red spheres) from thermoliposomes composed of (A) themosensitive poly(EOEOVE), (B) egg yolk phosphatidylcholine, (C) cholesterol, and (D) N-[methoxy (PEG) 5000]-distearoyl phosphatidylethanolamine (PEG_{5000}–PE). Modified from Ref. [86].

concentration of reductases in the intracellular environment [87]. Redox-sensitive liposomes rely on reduction and cleavage of disulphide linkages to disrupt the bilipid membrane and subsequently leak the drug content. Common molecules included in redox-stimulated liposomes that bear a disulphide linkage are dithiopropionate (DTP) [88], dithio-s-hexanol (DTH) [89] and dithiobenzyl (DTB) [90] (Table 10.4). Reductive thiolysis may be used as well to shed the PEG coating within the cell and enhance leakage of the incorporated drug [89, 90]. In this regard, Kirpotin *et al.* [88] designed a cleavable PEG-DTP-DSPE (1,2-distearoyl-*sn*-glycero-3-phosphoethanolamine) lipid attached to DOPE/CHEMS liposomes that disassembled under *in vitro* reductive conditions.

In relation to gene delivery, although still in research phase, thiocholesterol-based cationic lipids have been incorporated into liposomes to deliver DNA inside the intracellular space [91].

10.2.1.7 *Magnetically-responsive liposomes: Magnetoliposomes*

Externally applied magnetic fields can be utilized to trigger liposomes that incorporate magnetic materials either in the aqueous core (Fe_2O_3 or Fe_3O_4 nanoparticles) or within the phospholipid bilayer (magnetized polymers) [60, 92]. Initially, magnetoliposomes where synthesized to facilitate targeting and enhanced therapeutical potency by concentrating these carriers in the desired area *via* a magnet [93]. The next step in magnetoliposome evolution was making use of magnetic hyperthermia [94, 95]. By applying an alternating magnetic field, magnetic particles transfer heat only to the surrounding localized area, inducing cell death [96, 97]. This technique is also known as magnetic thermal

Table 10.4. Examples of disulphide linkages included in redox-responsive liposomes.

Name and abbreviation	Chemical structure
1,2-Distearoyl-*sn*-glycero-3-phosphate dithio-*s*-hexanol polyethylene glycol (DSPA–DTH–PEG)	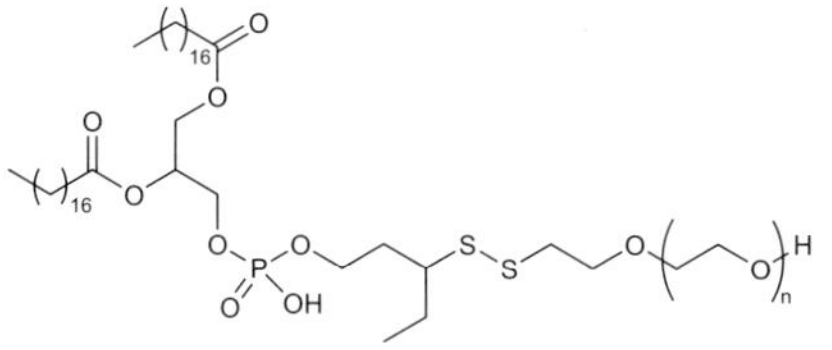
1,2-Distearoyl-*sn*-glycero-3-phosphoethanolamine dithiopropionate polyethylene glycol (DSPE–DTP–PEG)	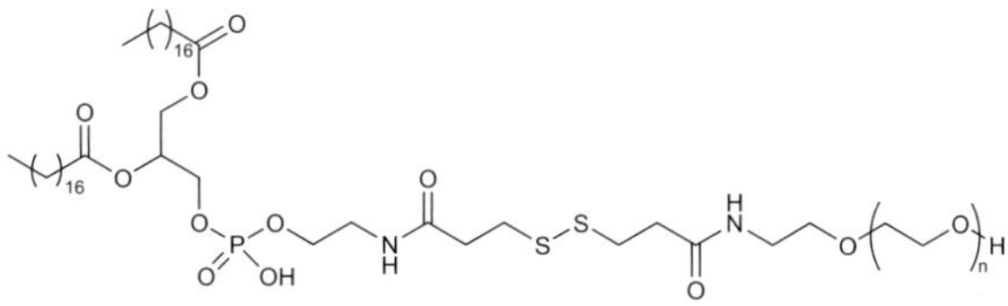
1,2-Distearoyl-*sn*-glycero-3-phosphoethanolamine dithiobenzyl polyethylene glicol (DSPE–DTB–PEG)	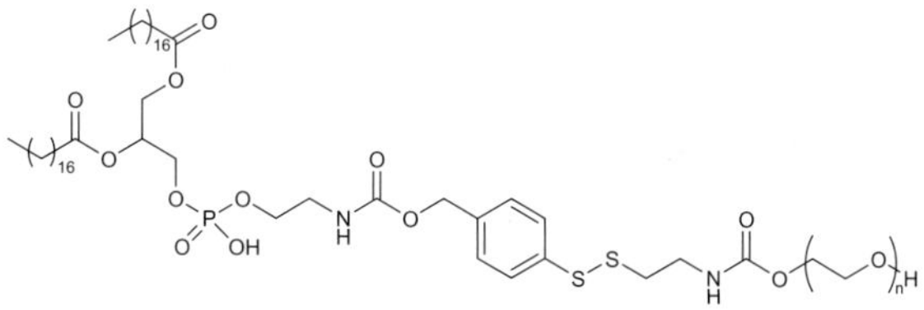
Dioleoylphosphatidylethanolamine dithiopropionate polyethylene glycol (DOPE–DTP–PEG)	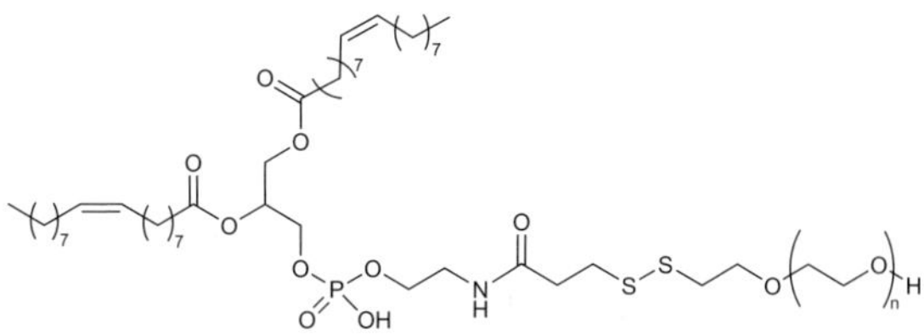

ablation. Another approach of magnetic hyperthermia relies on thermosensitive magnetoliposomes, in which the heat of the magnetic particles is used to induce a phase-transition of the liposomal bilayer, triggering liposome disruption and subsequent drug release [92, 98]. To illustrate this, methotrexate was encapsulated with Fe_2O_3 nanoparticles inside thermo-magnetoliposomes made of DPPC and cholesterol, and its release was studied *in vitro* and *in vivo*. Results showed that thermo-magnetoliposomes significantly increased the accumulation of methotrexate in the skeletal muscular tissue when exposed to an external constant magnetic field that increased the temperature up to 41°C, compared to the absence of the magnetic field and 37°C [92].

10.2.1.8 *Ultrasound-responsive liposomes: Ultrasonic liposomes*

Focused ultrasounds are capable of controlling drug release from liposomes by inducing either thermal or non-thermal effects [99]. Under intense ultrasonically induced stress, the temperature rises and liposomes might undergo a phase-transition and become leaky. For example, Dromi *et al.* [100] combined heat-sensitive liposomes with pulsed-high intensity focused ultrasound exposures and demonstrated enhanced delivery of doxorubicin into tumor cells.

For non-thermal effects, low-frequency ultrasound (LFUS) is suggested to be capable of creating pore-like defects in the liposome membrane through which the drug is rapidly released [101]. Schroeder and co-workers [101] loaded HSPC/cholesterol/PEG-DSPE liposomes with three different drugs — cisplatin, doxorubicin and methylprednisolone hemisuccinate — and demonstrated their controlled release under short exposure to LFUS. Furthermore, the pore-like defects in the membrane resealed once LFUS irradiation had stopped, proving focused ultrasounds could be used as a switch for liposome disruption and thus, drug release.

10.2.1.9 *Light-responsive liposomes*

Photo-triggerable liposomes rely on light to disrupt the phospholipid bilayer and liberate drug content [83]. The mechanisms involved in creating liposomal light-responsive drug delivery systems include photopolymerization, photochemical triggering and photoisomerization [102] (Fig. 10.5). Liposomal light-stimulation based on photopolymerization of membrane lipids relies on attaching polymerizable chemical groups (styril, sorbyl, diacetylenyl, dienoyl) to the hydrophobic domains of liposomes (Table 10.5). Application of light leads to photo-crosslinking of these groups, making liposomes assemble into larger clusters and creating temporal pores through which the drug can diffuse [103, 104]. The effect lasts until surrounding free lipids reassemble the membrane [105]. As an example, when 1,2-bis[10-(2′,4′-hexadienoyloxy) decanoyl]-*sn*-glycero-3-phosphatidylcholine (Bis-Sorb PC) is included in the liposomal composition, and UV light is applied for two minutes, the bilipid membrane sees its permeability increased 200-fold compared with the absence of light [105].

Photochemical triggering or photodegradation can be done by photooxidation or photodeprotection. Photooxidation depends on a liposomal phase-transition to disrupt the bilayer, when singlet oxygen formed upon radiation of a certain sensitizer, it cleaves lipid chains from the membrane [102]. Interestingly, previously mentioned semi-synthetic plasmalogens (DPPlsC) are not only pH-sensitive but also degrade under photooxidative conditions [106]. If lipid hydrolysis is not singlet oxygen-mediated but is done through direct irradiation, the phenomenon is called photodeprotection [105].

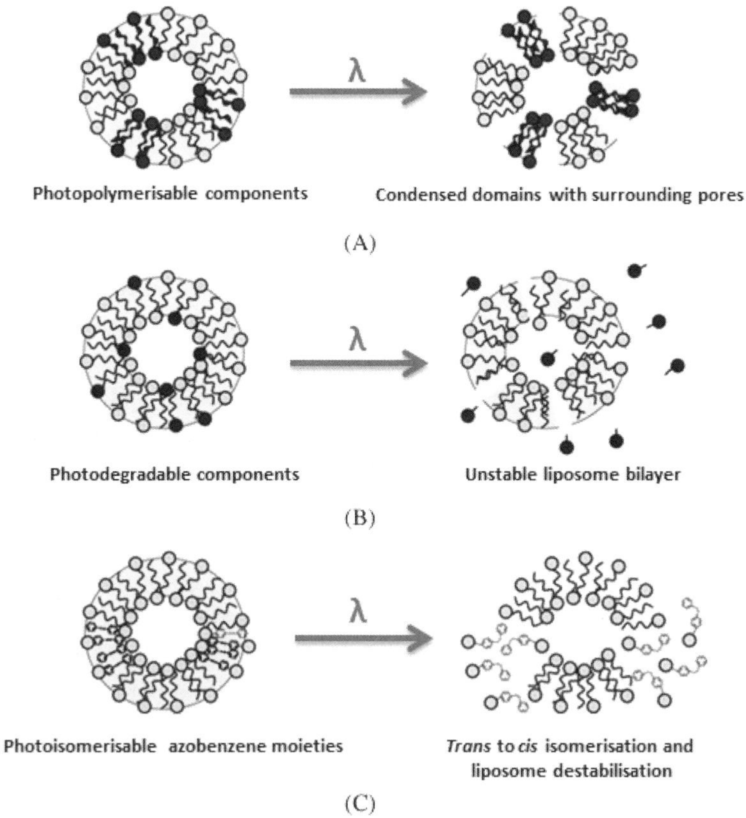

Figure 10.5. Schematic representation of the different mechanisms involved in light-sensitive liposomes: (A) photopolymerization, (B) photochemical triggering (photodegradation), and (C) photoisomerization. Modified from Ref. [105].

In the third type of mechanism, light induces transitional conformational changes (*trans-to-cis* photoisomerization) in phospholipids — particularly 1,2-bis[4-(n-butyl)phenylazo-4′-phenylbutyroyl]phosphatidylcholine (Bis-Azo PC) and analogues — causing temporal leaking of encapsulated drugs [107, 108]. When liposomes containing small amounts of Bis-Azo PC are irradiated with UV-light, the isomerization to the less stable conformation *cis* form occurs, and the release of the entrapped compound is increased.

10.2.1.10. *Cationic liposomes: Lipid-DNA complexes for gene delivery*

Some diseases are caused by cells missing or having abnormal genes. Therefore, in theory, by delivering genes into these cells, the absent or mutated gene can

Table 10.5. List of light-sensitive phospholipids used in light-responsive liposomes.

Name and abbreviation	Chemical structure
1,2-bis(10,12-tricosadiynoyl)-sn-glycero-3-phosphocholine ($DC_{8,9}PC$)	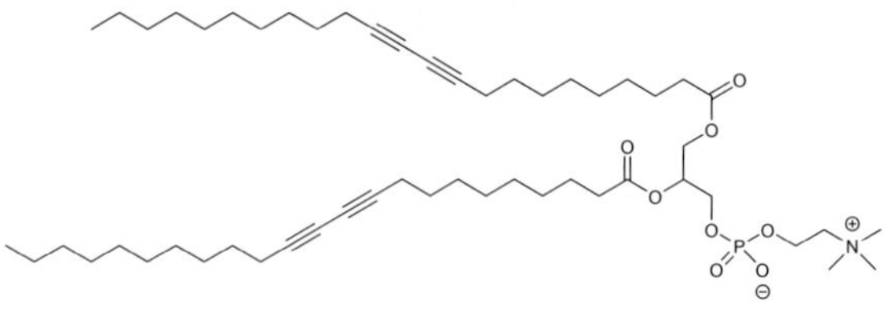
1,2-Distearoyl-sn-glycero-3-phosphoethanolamine styrene (DPPE-Styrene)	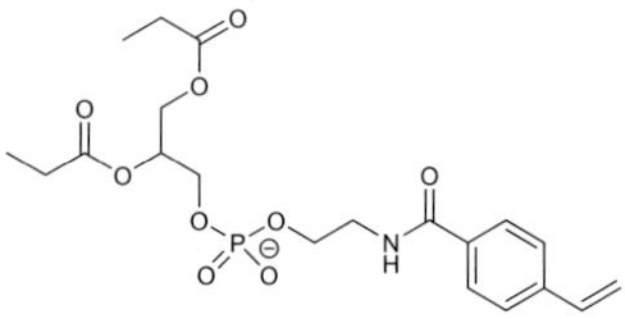
Dioleoylphosphatidylethanolamine styrene (DOPE-Styrene)	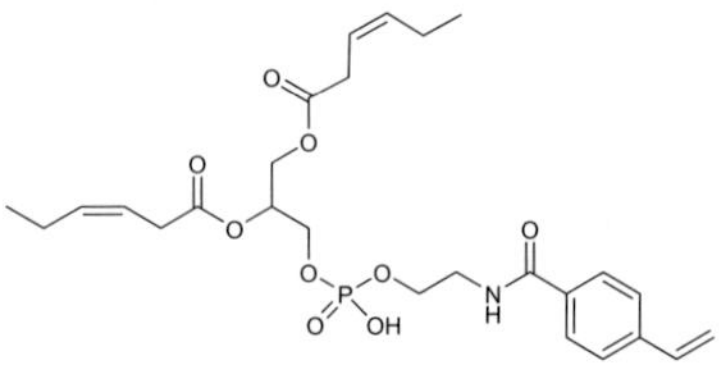
1,2-Bis(4-(n-butyl)phenylazo-4′-phenylbutyroyl) phosphatidylcholine ("Bis-Azo" PC)	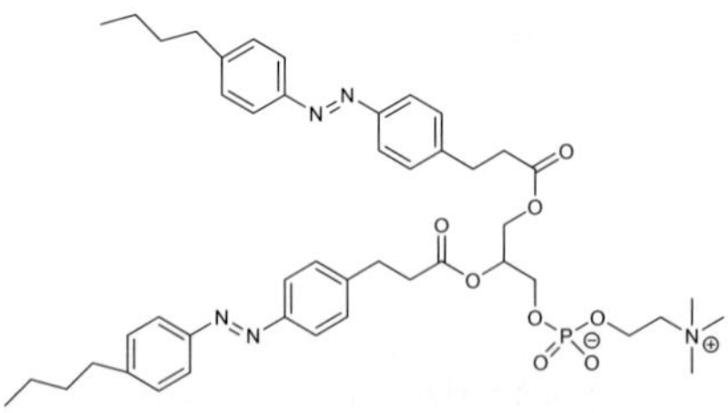
1,2-bis [10-(2′,4′-hexadienoyloxy) decanoyl]-sn-glycero-3-phosphatidylcholine (Bis-Sorb PC, bis-sorbyl PC)	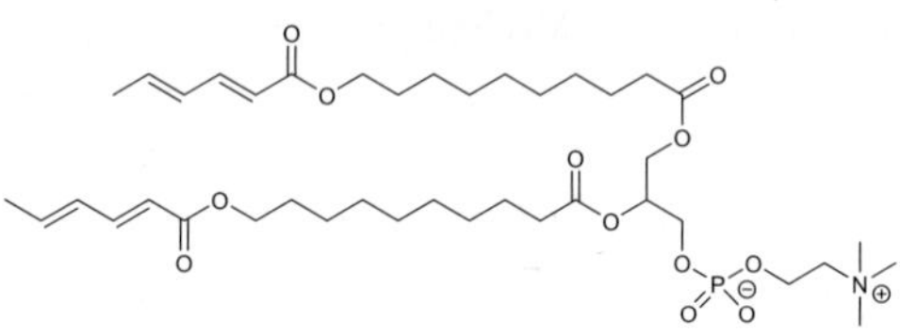

be restored thereby eradicating the symptoms or the disease itself. In addition, genes can be also "switched off", i.e., undesirable gene expression can be stopped and thus the protein that it encodes [39]. The therapeutic application of gene delivery is highly dependent on the development of an efficient vehicle that can carry the nucleic acids to target cells. Nowadays, efforts are devoted to the improvement of non-viral vectors to overcome the disadvantages of viruses used for gene therapy [109]. In this regard, liposomes are an attractive alternative because nucleic acids of any size can be loaded, they are safe (don't provoke an immune response), biodegradable, versatile, cheap and easy to manufacture in large scale [110]. Liposomes made of cationic lipids were seldom used, but in 1987 Felgner and co-workers [111, 112] described the first efficient lipofection (cationic liposome-mediated transfection) (Fig. 10.6) which gave rise to the synthesis of hundreds of new cationic lipids. Cationic liposomes are more correctly termed lipoplexes (LPs) or cationic lipid-DNA complexes, because they do not always include phospholipids in their structure. LPs are made of a cationic lipid (Table 10.6) and usually a helper lipid (cholesterol, DOPE), which is added to facilitate the fusion between the liposome membrane and the endosome, and to decrease its cytotoxicity [113, 114]. Nucleic acids packed inside the LP are sandwiched between lipid bilayers, generating ordered multillamellar domains that result in successful gene transfer [115].

The delivery of small interfering RNA (siRNA) and the delivery of DNA differ in some aspects [116]. Firstly, the destination of siRNA molecules is the cytoplasm, while DNA must generally also cross the nuclear membrane. Secondly, RNA molecules are more prone to degradation due to the action of ribonucleases. Finally, the molecular weight of DNA molecules is often hundreds of times greater than that of siRNA strands. Owing to these differences, some research groups are committed to the design of cationic lipids for

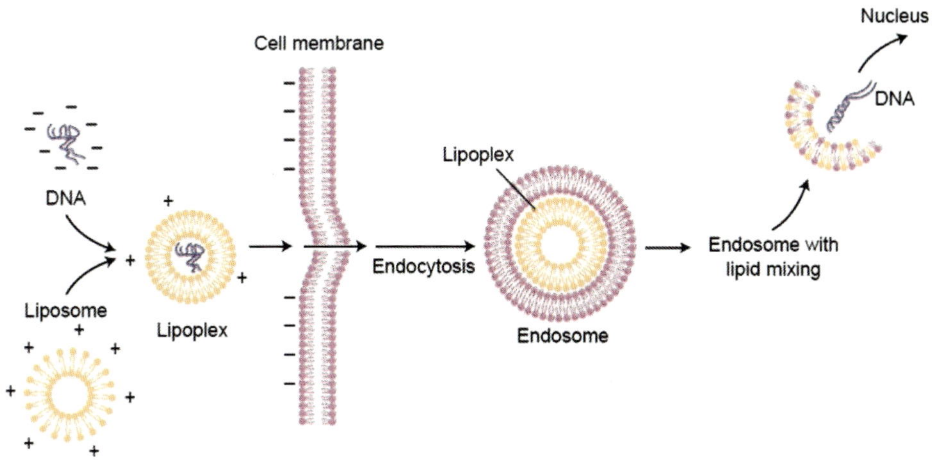

Figure 10.6. Schematic representation of lipoplex-mediated transfection and endocytosis. Lipofection makes use of cationic lipids to form aggregates with negatively charged oligonucleotides. Due to their fusogenecity, lipoplexes merge easily with the cell membrane and gain access to the intracellular space. Subsequent endosome destabilization releases nucleic acids from the lipid complex so they could interact with the appropriate cellular targets. The destination of siRNA is the cytoplasm, while DNA must enter the cell nucleus. Modified from Ref. [120].

Table 10.6. List of cationic lipids included in lipoplexes.

Name and abbreviation	Chemical structure
Dimethyldioctadecylammonium bromide (DDAB)	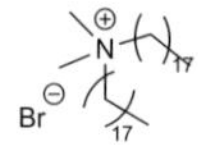
N-[1-(2,3-dimyristyloxy)propyl]-N,N-dimethyl-N-(2-hydroxyethyl) ammonium (DMRIE)	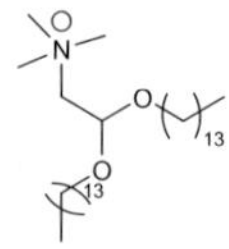
Dioleyldimethylammonium chloride or bromide; also Dioleyldimethylammonium bromide or chloride (DODAC or DODAB)	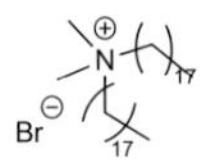
Dioctadecyl amido glycyl spermine (DOGS)	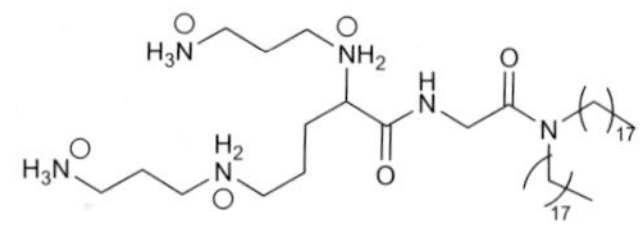
2,3-dioctadecyloxy-N [2(sperminecarboxamido) ethyl]-N,N-dimethyl-1-propanaminium; also 2,3-Dioleyloxy-N-[2(sperminecarboxamido) ethyl]-N,N-dimethyl-1-propanaminiumtrifluoroacetate (DOSPA)	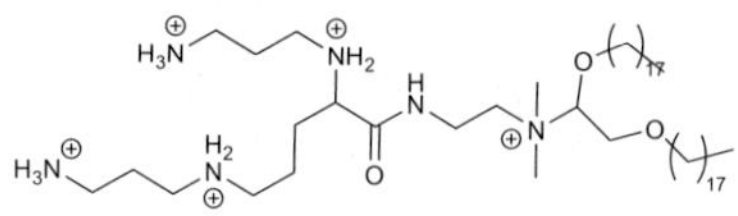
1,3-di-oleoyloxy-2-(6-carboxy-spermyl) propylamid (DOSPER)	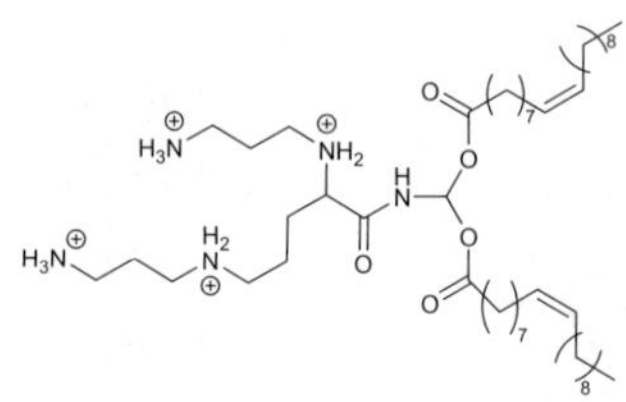
1,2-dioleoyloxy-3-(trimethylammonio) propane (DOTAP)	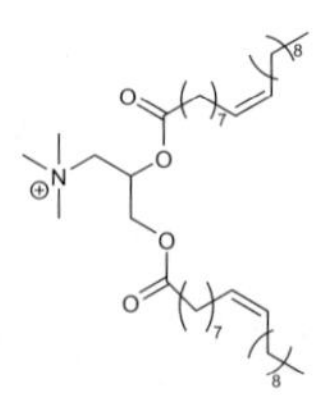
1-[2-(9 (Z) — octadecenoyloxy)-ethyl]-2-(8(Z)-heptadecenyl)-3-(2-hydroxyethyl) imidazolinium chloride (DOTIM)	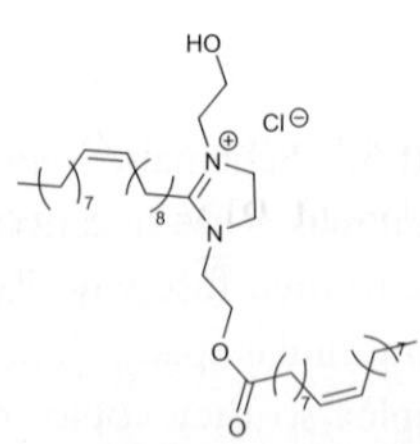

(Continued)

Table 10.6. (*Continued*)

Name and abbreviation	Chemical structure
N-[1-(2,3-dioleyloxy)propyl]- N,N,N-trimethylammonium chloride (DOTMA)	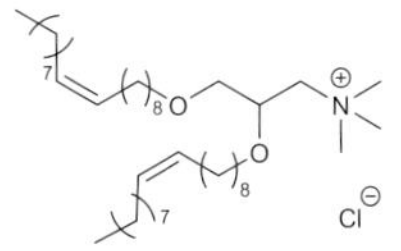
Dipalmytoylphosphatidylethanolamyl spermine (DPPES)	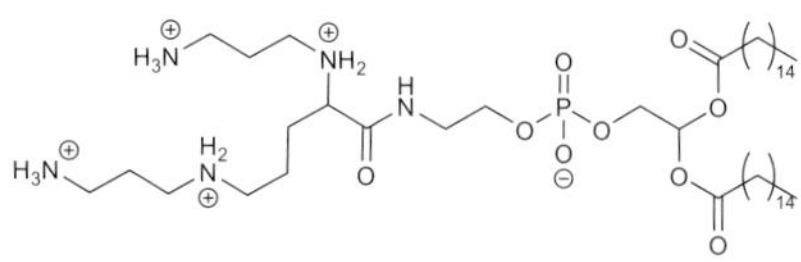

siRNA delivery [117, 118]. However, common cationic lipids such as DOTMA (N-[1-(2,3-dioleyloxy)propyl]-N,N,N-trimethylammonium chloride) and DOTAP (1,2-dioleoyloxy-3-(trimethylammonio) propane) have been shown to be useful for the delivery of both DNA and siRNA molecules [116, 119].

Cationic liposomes must fulfil certain requirements before reaching the cell nucleus, such as protection of DNA from degradation, adhesion onto the cellular membrane, cell internalization, fusion with the endosome membrane, release of the polynucleotide from the endosome, and access of the polynucleotide into the nucleus [121, 122]. Even if the LP is able to overcome all these barriers, delivery of DNA could be poorly correlated to subsequent gene expression [109]. A combination of these hurdles is the reason why first generation LPs produced such low levels of transfection compared to viral vectors. Nevertheless, researchers soon conceived smart modifications to classic LPs that greatly enhanced lipofection efficiencies [123]. LPs may be PEGylated to shield their positive charge, avoid aggregation and increase circulation time. In addition, PEG-LPs can be improved by incorporating targeting ligands on the surface to selectively deliver their payload to target cells. As an example, Yu *et al.* [124] developed immunolipoplexes in which transferrin receptor antibodies were conjugated to the distal end of PEG chains attached to a DOTAP/DOPE-DNA complexes. The complex's antibody ligands enhanced targeting while PEG-coating resulted in steric stabilization, thereby improving the potency of *in vivo* gene delivery to tumor cells. Another direction of research in this area is on the development of cationic lipid-cationic polymer-DNA complexes (LPDs), a novel formulation that results in the formation of a condensed polycation-DNA core surrounded by a lipidic shell, resembling the structure of a virus or sperm [125]. The addition of polycations (such as protamine [125–127], polyethylenimine [128] or poly(L-lysine) [129]) to cationic liposomes dramatically reduced the particle size of the complex and rendered the DNA contained resistant to nuclease activity. The latest modification to lipoplexes is based on replacing the polycation core in LPDs for calcium phosphate [130]. This cutting-edge formulation, named liposome/calcium/phosphate (LCP) nanoparticle, relies on the fast dissolution rate of calcium phosphate in the acidic environment of the endosomes to release the encapsulated polynucleotides. After intravenous administration to mice, siRNA-loaded LCPs showed a similar down-regulation effect than that achieved by LPDs but with decreased immunotoxicity, proving their potential application in clinical trials.

Finally, the nuclear barrier to DNA plasmid entry can be overcome through incorporation of nuclear localization signal peptides inside the liposome [131].

Despite all the aforementioned improvements, in terms of transfection efficiency lipid-based gene carriers cannot compete with viral vectors yet. Therefore, further work needs to be done to allow optimization of cationic lipid based formulations.

10.2.2 *Smart lipid-based micelles for systemic delivery*

Micelles are spherical nanoassemblies ranging from 5 to 100 nm in size, formed spontaneously under certain conditions of temperature (CMT, critical micellization temperature) and concentration (CMC, critical micellar concentration) when amphiphilic molecules are dispersed in aqueous media [132]. In water, hydrophilic fragments of amphiphiles form the micelle's corona, whereas hydrophobic portions form the core, which may be utilized as a loading space for poorly water-soluble drugs. The use of lipids (such as phosphatidylethanolamine (PE) or palmityl and stearyl residues) as hydrophobic blocks attached to hydrophilic polymers (PEG, polyvinylpyrrolidone) provide additional thermodynamic stability compared with pure polymeric micelles due to the relatively short but very hydrophobic diacyl phospholipid region, which creates very strong interactions among the polymeric chains in the micelle's core [133].

PEG-DSPE micelles, the most widely studied micellar systems, have CMCs in a 10^{-5} M range [134] which is around one hundred times lower than those of regular surfactants [135], meaning they maintain their integrity upon strong dilution (e.g. upon intravenous administration). As was discussed with stealth liposomes, the PEG corona provides protection to micelles and determines their longevity *in vivo* by allowing them to avoid blood opsonizing agents. In addition to this, owing to their small size, micelles make better use of the EPR effect compared to carriers with larger size, like liposomes. Thereby, long-circulating micelles are able to spontaneously penetrate into poorly permeable tumors with higher efficiency than other larger stealth nanocarriers [136, 137]. Accordingly, Lukyanov *et al.* [137] intravenously administered PEG_{750}-DSPE and PEG_{2000}–DSPE micelles to tumor-bearing mice and successfully demonstrated their spontaneous accumulation in the tumors *via* the EPR mechanism. With the objective to further increase the drug loading capacity of PEG-DSPE micelles, Krishnadas and co-workers [138] augmented the hydrophobic content by adding egg-phosphatidylcholine (ePC) to the composition. These mixed micelles (molar ratio of 90:10 PEG_{2000}–DSPE:ePC) increased the solubilization potential of paclitaxel by one and a half times when compared with simple micelles and showed *in vitro* cytotoxicity against tumor cells comparable with that of free drug. PEG-DSPE micelles may also be functionalized with antibodies to further enhance the therapeutic index of the incorporated drug. To illustrate this, PEG-DSPE immunomicelles based on cancer-specific antinucleosome 2C5 loaded with paclitaxel were successfully tested *in vivo* and *in vitro* against different cell lines [139]. Because PEG-DSPE micelles have been found to carry a net negative charge [140] which may hamper their internalization by cells, Torchilin's group [141] hypothesized the addition of positively charge lipids would enhance the intracellular uptake. Accordingly, they prepared paclitaxel-loaded mixed polymeric micelles consisting of PEG_{2000}–DSPE, solid triglycerides and lipofectin® (equimolar mixture of DOTMA and DOPE). Results demonstrated that cationic mixed micelles, contrary to plain micelles, could escape from endosomes and access the cytoplasm of cancer cells, hence boosting the chemotherapeutic efficiency of micellar paclitaxel. Finally, PEG-DSPE micelles have been proven to be useful for targeted photodynamic therapy (PDT). In PDT, an administered photosensitizer is activated with light of an appropriate wavelength and intensity, so it transfers its excess of energy to nearby oxygen molecules, leading to the production of reactive molecules such as singlet oxygen, which is cytotoxic [142]. Since photosensitizers are usually poorly water-soluble, and lack of tumor-targeting causes undesirable photodamage of adjacent healthy tissue, it was hypothesized that micelles loaded with a photosensitizer and a magnetic material for remote targeting would significantly enhance the therapeutic efficiency [143]. In this regard, the co-encapsulation of the photosensitizer 2-[1-hexyloxyethyl]-2 devinyl pyropheophorbide-a (HPPH) with magnetite (Fe_3O_4) nanoparticles within PEG_{2000}–DSPE micelles, made magnetically guiding delivery of drugs to target cells possible.

10.2.3 *Smart SLNs for systemic delivery*

SLNs are spheres in the submicron size range (50–1000 nm) composed of a solid lipid core (made of tricaprin, tripalmitin, partial glycerides or waxes) stabilized by surfactants (poloxamers, Tween 80, sodium dodecyl sulphate) embedded in their surface [144]. When SLNs are surrounded by a phospholipid bilayer they are known as lipospheres [145]. Their appeal lies mostly in their efficient sustained drug release owing to reduced drug mobility within the solid core [146]. Other advantages include physiological tolerability and possibility of being manufactured on industrial large scale by high pressure homogenization [36]. SLNs present certain disadvantages however, which are mainly related to the manufacturing process, such as drug degradation due to high pressures and rapid temperature changes, lipid crystallization or presence of further colloidal structures [147].

Numerous drugs (camptothecin [148], actarit [149], etc.) have been solubilized or dispersed in classic SLNs for parenteral administration. However, following intravenous administration, these SLNs are rapidly removed from the circulation by agents of the RES reducing drug bioavailability like liposomes and any other colloidal carrier. To increase circulation time, stealth SLNs have been manufactured by coupling PEG chains to their surface. Smart PEGylated SLNs have led to higher and prolonged plasma levels of doxorubicin [150, 151], paclitaxel [152] and daidzein [153], among others.

Another type of smart nanocarriers with a promising future are the thermo- and magnetically-responsive SLNs developed by Hsu and Su [154]. This technology depends on γ-Fe_2O_3 particles loaded within the solid lipid core which, upon application of an external magnetic field, heats up. The rise in temperature melts the lipid matrix leading to sustained drug release.

10.2.4 *Smart LNCs for systemic delivery*

LNCs are a new platform for nanomedicine created by Heurtault and co-workers in 2000 [155], as an alternative to liposomes and polymeric nanoparticles. They are colloidal particles smaller than 50 nm, consisting of a liquid or semiliquid core surrounded by a solid shell. The core is made of triglycerides (tricaprin, tricaprylin) or fatty acid esters (ethyl palmiate, ethyl oleate, ethyl myristate), protected by a lipophilic surfactant coating (such as lecithin) anchored in the oily phase, and a non-ionic hydrophilic surfactant (such as Kolliphor® HS 15) orientated towards the aqueous phase (Fig. 10.7) [156]. LNCs may carry either lipophilic drugs solubilized in the oily phase, or hydrophilic drugs bound to the surface of the capsule, with much higher encapsulation rates than that of liposomes [156]. Representative examples of incorporated drugs include indinavir [157], tamoxifen [158], paclitaxel [159] and ibuprofen [160]. An organic solvent-free technique based on the phase

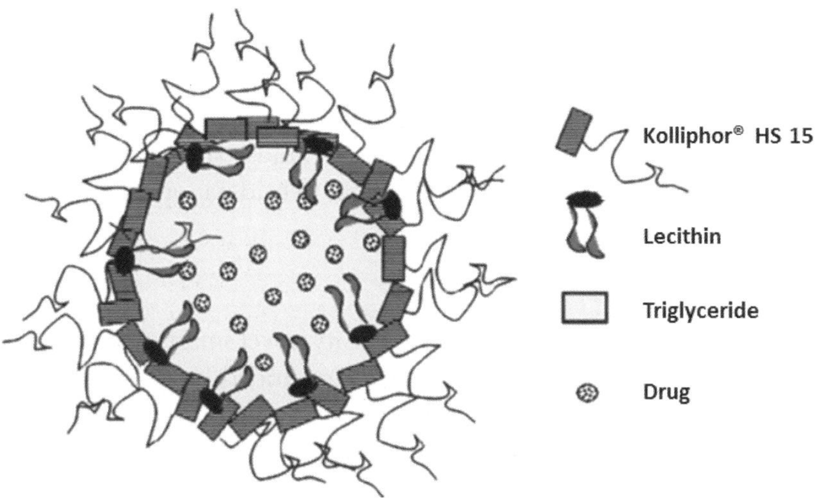

Figure 10.7. Schematic representation of a lipid nanocapsule (LNC). The incorporated drug is solubilized in a core of liquid/semiliquid triglycerides, protected by a shell of lipophilic (lecithin) and hydrophilic (Kolliphor® HS 15) surfactants. Modified from Ref. [156].

inversion of an emulsion, generated by temperature changes and a fast cooling-dilution process has been developed for the manufacture of LNCs.

The presence of hydrophilic surfactants — a mixture of PEG 660 and PEG 660 hydroxystearate — make LNCs smart nanodevices *per se*. Firstly, these PEG-derived surfactants are capable of reverse multidrug resistance by inhibiting P-gp-related drug efflux [161], so the exposure of the incorporated drug is enhanced. Secondly, since PEG represents a barrier to blood proteins adsorption, LNCs are partially invisible to the MPS. Also, if the surfactant Kolliphor® HS 15 is replaced in the formulation by DUB SPEG 30S® (PEG_{1500} stearate), a longer PEG derivative, the blood circulation time substantially increases [162]. In addition to this, PEG chains can also be grafted onto the nanocapsule surface to enhance their stealth properties [163, 164]. LNCs can be modified for active targeting by coupling site-specific ligands to the distal end of PEG chains. Béduneau *et al.* [165] designed immune-LNCs to allow drug delivery to the central nervous system through specific binding between OX26 monoclonal antibodies and transferrin receptors. LNCs have also been used as gene carriers. DOTAP/DOPE-DNA lipoplexes have been loaded into LNCs and their *in vitro* transfection have been tested [166, 167]. Successful DNA delivery was achieved due to the small size of the carriers and the protection from opsonization and degradation provided by the solid outer shell.

10.2.5 *Smart lipid-coated systems for systemic delivery*

Lipids are not only used to form the core of smart pharmaceutical carriers, but they are also used for coating of the surface of other particles to increase their bioavailability and stability. Examples of lipid-coated drug delivery systems are gas microbubbles and lipid–polymer hybrid nanoparticles (LPHNs).

10.2.5.1 *Lipid-coated gas microbubbles*

Lipid-coated gas microbubbles are low-density novel smart carriers formed by a core of gas (normally perfluorocarbon) surrounded by a lipid layer, which is in turn stabilized by a layer of phospholipids [168]. Because lipid-coated microbubbles are efficient reflectors of ultrasound energy, they can be used as therapeutic cavitation nuclei with ultrasound. Cavitation destroys the microbubbles and can be used to release particles trapped within the system (or coated onto the surface), and increase transvascular passage of therapeutic agents and cellular uptake [169]. Potential applications of parenteral lipid-coated microbubbles include sonothrombolysis, based on acceleration of clot lysis [170], and drug delivery to the brain by reversible opening of the blood brain barrier [171]. In addition to this, if lipid-soluble drugs are incorporated in the oil phase and ultrasound energy is focused on the target tissue, the preferential release of the drug in that specific tissue may be achieved [172]. These ultrasound-responsive carriers can be even more useful for drug delivery if they are PEGylated to avoid recognition by the MPS, and incorporate ligands onto the surface for active targeting [173, 174]. In addition, encouraging results have been obtained using microbubbles containing DNA (Fig. 10.8) bound to cationic lipids (DPPC) for gene delivery to the heart and other tissues in dogs and mice [175, 176].

10.2.5.2 *LPHNs*

LPHNs were developed to combine the characteristics of liposomes and polymeric nanoparticles while overcoming their limitations, i.e. lack of integrity and biocompatibility, respectively. As such, LPHNs are made of a stable polymeric core that controls the sustained release of the incorporated drug, an inner lipid layer surrounding the nucleus which confers biocompatibility, and an outer PEGylated lipid layer that provides stealth coating to the system [177]. Although LPHNs are a young technology, there are already publications describing successful targeted delivery to cancer cells either *in vitro* [178] or following intravenous administration *in vivo* [179]. Efficient DNA transfection

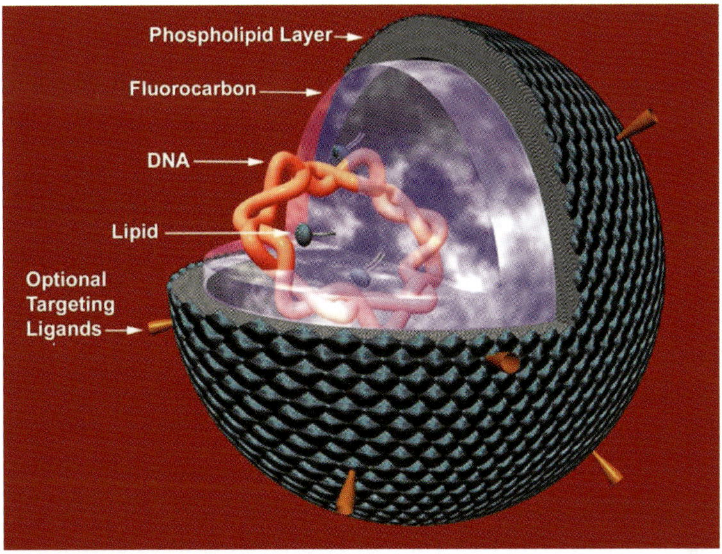

Figure 10.8. Illustration of a lipid-coated fluorocarbon microbubble for gene delivery. Genetic material is entrapped in the centre of the fluorocarbon material and condensed into small electron dense structures within fluorocarbon cores. The outer surface is stabilized by amphiphathic lipid. Ligands may be incorporated onto the head groups of the lipids for active targeting. Reproduced from Ref. [176].

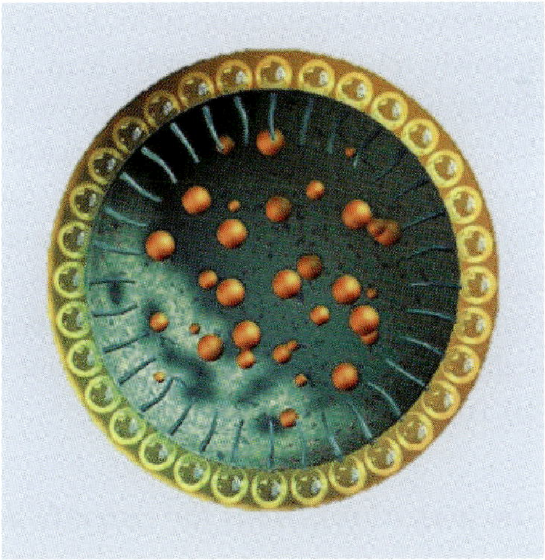

Figure 10.9. Illustration representing a tocol-based emulsion particle in an aqueous environment. Drug molecules (orange spheres) are dissolved in the oily solution (green). Oil droplets are stabilized by surfactants (yellow), which orientate their hydrophobic tails towards the oil phase and their polar heads towards the surrounding aqueous environment (blue). Reproduced from Ref. [193].

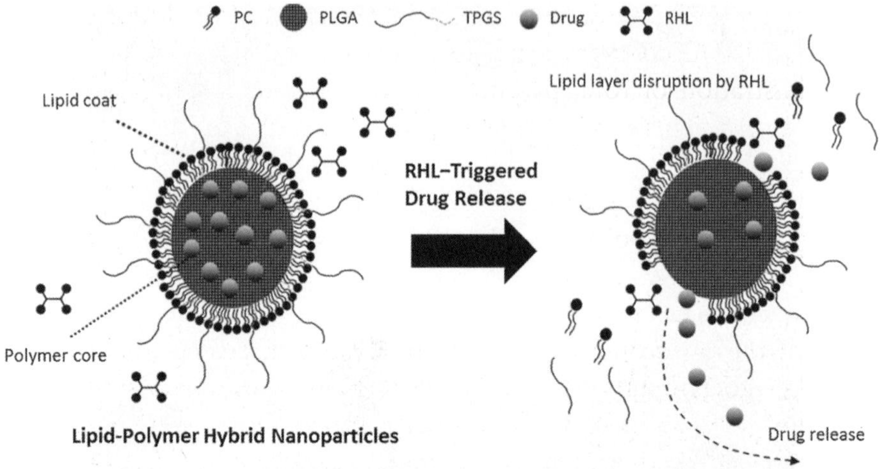

Figure 10.10. Schematic representation of rhamnolipid-estimulated drug release from lipid-polymer hybrid nanoparticles (LPHP). Polymer: PLGA (poly(lactic-co-glycolic acid); lipid: PC (phosphatidylcholine); PEG coating: TPGS (D-α-tocopheryl polyethylene glycol succinate); Trigger: RHL (rhamnolipid). Reproduced from Ref. [182].

results have been obtained as well [180]. LPHNs have also been developed for releasing their content following application of a certain stimulus. For example, Deok Kong *et al.* [181] included iron nanoparticles inside the polymeric nucleus, and upon external application of localized radio-frequency, the core partially melted slowly releasing its drug payload. Although not developed for systemic delivery but for lung delivery, Cheow and Hadinoto [182] have interestingly described LPHNs with trigger-release characteristics in response to encountering rhamnolipids, which are common surfactants present in biofilm colonies of major respiratory pathogens. The triggered release of encapsulated antibiotics is accomplished by the dissolution of the phospholipid constituents of the lipid shell upon interaction with rhamnolipids, which allows the antibiotic to diffuse out of the nanoparticle compartment (Fig. 10.10).

10.2.6 *Smart oil-in-water emulsions for systemic delivery*

An emulsion is a heterogeneous mixture of two or more immiscible liquids, with a third component (emulsifier) used to stabilize the dispersed droplets. Emulsions fall into two general categories: (i) water-in-oil systems in which droplets of water are dispersed in a continuous oil phase, and (ii) oil-in-water emulsions consisting of droplets of oil within a continuous aqueous environment [183]. Only the latter ones can be used for intravenous administration [33]. Originally, lipid emulsions were developed for use in the administration of total parenteral nutrition, but it was soon realized that they could also be used as carriers for hydrophobic drugs due to their high loading capacity, biocompatibility, biodegradability, and facility to handle and manufacture [184]. Components of parenteral fat emulsions are triglycerides (soybean, sesame seed, cotton seed and safflower oils), which form the internal oil phase, surfactants (phospholipids and derivatives) and co-surfactants (spans, tweens, pluronics), which increase the stability of the oil droplets [185]. As previously discussed for other lipid systems, phagocytic uptake by the MPS represents an obstacle for the use of emulsions as drug carriers. Additionally, they would acquire apolipoproteins which endow them with different particulate characteristics, hence altering their *in vivo* disposition profiles [186]. Several authors have modified the colloidal surface of the oil droplets with the aim of prolonging their blood circulation time. These methods are based on the incorporation of PEG [187–190] or sphingomyelin [189, 191] to create a hydrophilic barrier that repels blood opsonising agents. However, these emulsions still lacked specificity toward target tissues. To overcome this issue, for example Lundberg *et al.* [192] conjugated the anti-B-cell lymphoma monoclonal antibody LL2 to long circulating emulsions and tested them *in vitro*, demonstrating active targeting as a promising method to enhance the therapeutic index of incorporated drugs. Another smart approach in the development of lipid emulsions is based on the addition of α-tocopherol (vitamin E) and derivatives in the oil phase. Vitamin E-based emulsions are capable of increasing the stability and solubility of tocophilic drugs (cyclosporin A, paclitaxel, etc.) [193], and even reduce its toxic side effects [190, 194] (Fig. 10.9).

An interesting application of intravenous lipid emulsions is their use as an empty vehicle (i.e. without a drug) as an antidote in cases of local anaesthetic and other lipophilic drug toxicicosis. Although the exact mechanism remains unknown, one of the most accepted theories proposes a "lipid sink condition" into which sequestered lipophilic drug is redistributed from an effective compartment [195]. Temporary compartmentalization of the drug results in higher blood plasma concentration of the drug and less drug available in the toxicity compartments, thus reducing its toxic effects. This therapy is considered relatively safe and is commonly used in veterinary medicine, but in humans it is only utilized in life threatening conditions and when conventional therapies have failed. In contrast to most of the animal publications, which are in the form of experimental studies [196, 197], human published results are found mainly in case reports. For example, in 2006 a case was published in which a woman suffered asystolic cardiac arrest caused by ropivacaine intoxication. After ten minutes of unsuccessful cardiopulmonary resuscitation, the patient was administered 200 mL (100 mL as a bolus and 100 mL as an infusion) of Intralipid 20%, which led to electrical activity and cardiac output restoration [198]. Similarly, a girl developed cardiovascular collapse because of intentional oral overdose of bupropion and lamotrigine. Following unsuccessful conventional resuscitation, she was given the same dose of emulsion vehicle as above, which resulted in rapid cardiovascular improvement [199]. Both patients recovered almost completely.

10.3 Smart Lipid-Based Carriers for Oral Drug Delivery

Parenteral therapy might lead to local toxicity as a result of high drug concentration at the site of injection or because of the nature of the solubilizing agents. In addition to this, intravenous formulations are required to be sterile, and usually hospitalization of the patient is needed, which results in a dramatic increase in the therapy costs. Owing to these drawbacks, there is a need for safer and more patient-convenient routes of administration.

Oral drug delivery is the most acceptable route of administration due to patient compliance and ease of administration. Moreover, oral formulations are less expensive to manufacture, since they do not need to be produced under sterile conditions [200]. However, pre-systemic enzymatic degradation and poor penetration across the intestinal mucosa represent major challenges for effective oral delivery of therapeutic agents. In order to overcome these barriers, researchers have focused their attention on LBDDS.

Digestion and absorption of lipids occurs mostly at the upper part of the gastrointestinal tract [26]. Enzymatic hydrolysis of triglycerides (TGs) — the major components of dietary lipids — begins in the stomach, where they stay for up to two hours, and around 30% of the total TGs may be digested there [201]. The action of lingual and gastric lipases together with peristaltic and segmental contractions promotes lipid emulsification, facilitating subsequent intestinal lipolysis. When chyme (stomach-leaving contents) enters the duodenum, cholecystokinin is secreted from epithelial cells stimulating the release of pancreatic lipase and bile into the duodenum from the pancreas and gallbladder, respectively. Bile salts (mainly sodium glycholate and sodium taurocholate), phospholipids (essentially PC) and cholesterol are the major solutes of bile. Bile components decrease the lipid-water interfacial tension facilitating the formation of smaller oil droplets and stabilizing the oil-in-water emulsion. Additionally, bile contains bicarbonate that neutralises the acidic chyme providing an optimum pH for the action of pancreatic lipase. Pancreatic co-lipase then binds to the interface of droplets acting as an anchor site for pancreatic lipase. This enzyme hydrolyses TGs at the *sn*-1 and *sn*-3 positions generating one 2-monoglyceride (2-MG) and two FAs per each TG. Lysophospholipids (phospholipid digestion products formed by action of phospholipase A_2), cholesterol and bile salts then form a range of colloidal phases (vesicles and mixed micelles) capable of solubilizing FAs and 2-MG. If a poorly-water soluble active compound is loaded in a lipidic formulation or co-administered with a high-fat meal, it gets solubilized within these colloidal phases as well, and thereby its precipitation in the aqueous intestinal milieu is avoided [24, 202].

The exact mechanism by which lipid products are absorbed into the enterocytes remains unknown. Mixed micelles are not thought to be absorbed intact. It is assumed that due to the lower pH area next to the membrane of the enterocyte, micelles are destabilized so FAs, 2-MGs and lipophilic drugs can cross the apical membrane [203]. It has been proposed that FAs, 2-MGs and lipophilic drugs can be absorbed by passive diffusion or by carrier-mediated transport. Alternatively, they can be directly transferred through collision of the carrier against the brush border; or colloidal vehicles may undergo vesicular-mediated uptake and this may be initiated by binding to a transport protein on the apical membrane [24].

Once in the enterocyte, digestion products can either diffuse across the cell and get access to the portal vein, or be re-esterified to triglycerides and incorporated into lipoproteins [204].

In vitro assessment of LBDDS is challenging. In conventional absorption models, such as the Caco-2 cell line [205, 206] or the Ussing chamber technique [207], digestion of LBDDS is not taken into consideration. Therefore, the membrane of the enterocytes in these *in vitro* studies is not exposed to the same substances as membranes of the epithelial cells within a human intestine. *In vitro* lipolysis is the only dissolution technique capable of mimicking the lipid digestion process, and is consequently the most appropriate method to trace the solubilization state of drugs delivered by means of LBDDS. This model has been used by different research groups [208–215], and although experimental conditions and parameters vary among them, the concept and fundamental principles remain the same.

The general protocol for *in vitro* lipolysis is based on the dispersion of the tested LBDDS in the experimental medium consisting of fasted-state simulated intestinal fluids. The addition of the digestive enzyme (pancreatic lipase/co-lipase) to the medium initiates the lipolysis process which is allowed to proceed until the triglyceride breakdown is completed. As a consequence of the triglycerides being hydrolysed, fatty acids are released, inducing a drop in pH. In order to keep pH at a constant value throughout the experiment (to mimic *in vivo* conditions), a pH-stat titrator is used. The instrument continuously measures and controls this transient drop in pH by equimolar titration. Once the process is finished, the resulting solution is ultracentrifuged and separated into three distinct layers:

1. An upper undigested lipid phase,
2. A middle aqueous phase, containing colloidal structures within which poorly-water soluble drug molecules are solubilized, and
3. A lower sediment phase, comprising fatty acids' calcium soaps.

It is assumed that drug molecules solubilized in the aqueous micellar phase are most readily available for absorption. After density-gradient separation, each phase is subjected to analysis for drug content. Finally, the percentage of drug dose solubilized in the aqueous phase of the *in vitro* experiment is then compared with the *in vivo* pharmacokinetic data obtained following oral administration of the LBDDS to an animal.

The key to success of any *in vitro* prediction model is the evaluation of its *in vivo* relevance. If *in vivo* predictive, the *in vitro* lipolysis model may rationalise and shorten the drug delivery development stage for LBDDS by saving resources, reducing dramatically the need for *in vivo* experiments to evaluate the intestinal absorption of poorly-water soluble drugs, and lead to improved product quality.

Liposomes, lipid-based lyotropic liquid crystals (LLCs), self-microemulsifying drug delivery systems (SMEDDSs) and long chain triglyceride emulsions are smart LBDDS whose performances can be evaluated and ranked, in terms of enhanced solubilization of lipophilic drugs, with the aforementioned lipolysis model. The success of these smart carriers is based, among other mechanisms, on avoidance of macrophage removal from the circulation, stimulation of the intestinal lymphatic process, and P-glycoprotein inhibition (if the drug in question is a P-gp substrate).

10.3.1 *Smart liposomes for oral delivery*

Although proteins and peptides have great therapeutic potency, their oral bioavailability is very low due to enzymatic degradation and poor absorption in the gastrointestinal tract [216]. In order to address this problem, liposomes are thought to be the best solution [217–219], but their destabilization through interaction with bile salts usually prevents liposomes from reaching the circulation intact. Despite these challenges, many researchers have focused their attention on oral delivery of insulin utilizing liposomes [216, 220–224]. Current treatment of diabetes is the subcutaneous injection of insulin, but this route of administration does not lead to immediate arrival of the hormone to the liver, and also causes some patient stress, for example pain on injection. An oral route of administration of insulin is therefore considered more suitable. Iwanaga *et al.* [216, 225] conceived a smart way of circumventing insulin-loaded liposome digestion by coating the phospholipid membrane with PEG 2000 or the sugar chain of mucin. Their results proved that while mucin-coated liposomes adhered preferentially to the stomach surface, PEG-coated liposomes moved slowly along the gastrointestinal tract, owing to the strong interaction with the intestinal mucous layer, leading in both cases to enhanced and sustained hypoglycaemic effects of insulin. In the same line of thinking, Li *et al.* [217] loaded recombinant human epidermal growth factor in PEG-coated liposomes with successful results. These initial findings demonstrated the potential use of smart-devised liposomes for oral delivery of peptides and proteins.

Chen and Langer [226] conceived a different approach to increase oral delivery efficiency of liposomes. They demonstrated in mice that oral administration of magnetic-responsive liposomes and subsequent application of external magnetic field to retain the liposomes within the intestine, increased the delivery performance of these liposomes. However, the practical application of this idea is yet to be seen.

10.3.2 *Smart lipid-based liquid crystals for oral delivery*

Amphiphilic lipids when dispersed in water spontaneously self-assemble in thermodynamically stable structures, which may acquire different geometric shapes. This has already been observed with phospholipids, which re-organize into spherical bilayers to form liposomes. Nonetheless, there are other lipidic amphiphiles, like monoglycerides, that can create a variety of LLCs, depending on pH, temperature or ionic strength [227]. At high concentration of amphiphiles, the morphology of these structures range from a simple monodimensional lamellar phase (L_α), to a bidimensional reversed hexagonal phase (H_2) and more complicated tridimensional reversed bicontinuous cubic ones (v_2) [228] (Fig. 10.11). While H_2 consists of reversed cylindrical micelles organized in a hexagonal lattice, v_2 is based on a single continuous curved

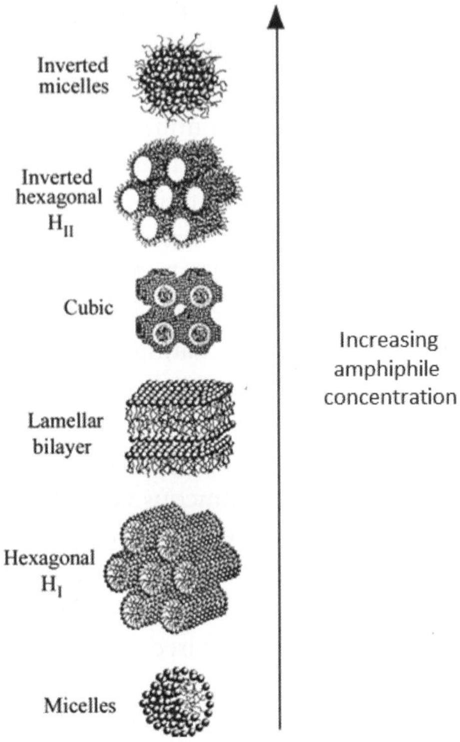

Figure 10.11. Schematic illustration of lipid aggregates formed by self-assembly processes in water. The morphology of the lipid structure changes with increasing amphiphile concentration. Lamellar (L_α), reversed bicontinuous cubic (v_2), and reversed hexagonal (H_2) phases are the most common morphologies used as nanopharmaceutical carriers. Modified from Ref. [240].

lipid bilayer forming a cubic network, which separates two continuous but non-intersecting water channels. Since both phases contain discrete aqueous and lipidic domains, they allow the incorporation of drugs with very different polarities. The best studied phase, as a drug delivery system is v_2 because of its stiffness, high viscosity, and its ability to co-exist in equilibrium with excess water, which allows slow and sustained release of incorporated drugs [229]. The successful sustained *in vitro* release from v_2 has been demonstrated with several compounds, including melatonin, pindolol, propranolol and pyrimethamine [230], dexamethasone [231], hemoglobin [232], and insulin [233] among others. Although different classes of amphiphiles have been studied (monoacylglycerols, glycolipids, phosphatidylethanolamine amphiphiles, urea amphiphiles), almost all publications about lipidic LLCs as drug carriers have used glycerol monooleate (GMO).

Some lipidic LLCs have the property of swelling and changing phases upon change in pH [234, 235] or temperature [236, 237]. Furthermore, those changes can be modulated by the addition of small amounts of a third compound [238, 239]. This characteristic of lipid-based LLCs is what makes them promising stimuli-responsive nanocarriers.

In terms of oral delivery, Kumar M. *et al.* [239] studied the potential of GMO as a gastro-retentive carrier in healthy volunteers. Swelling of the GMO system due to water uptake reduces its density. As a result, the GMO cubic phase floats on top of the gastric contents for a prolonged period of time, protecting the incorporated drug from gastric degradation. Results showed excipients, depending on the polarity of drug, could tune the drug-elution profiles and floatability of the LLC. Negrini and Mezzenga [227] developed the idea of a pH-stimulated lipidic LLC for oral administration for gastric-sensitive drugs consisting of monolinolein and linoleic acid as excipients (97:3 wt%). The system is conceived to change from v_2 to H_2 when the pH conditions go from acidic (pH 2) to neutral (pH 7), to mimic the stomach and small intestine environments, respectively (Fig. 10.12). This structural modification is accompanied by a four-fold increment in the drug-release rate, thereby reducing or preventing the release in the stomach and boosting the release in the intestine. The pH responsiveness is provided by the linoleic acid ($pK_a \sim 5$), which deprotonates at neutral pH conditions causing structural changes from hexagonal to cubic on the LLC. Despite this interesting concept, the experiments where not performed under biorelevant conditions, i.e. in the presence of enzymes, bile salts and phospholipids. However, while it is known that pepsin and pancreatic enzymes do not destroy LLC matrixes and therefore do not modify drug release, bile salts and other endogenous surface active agents are able to micellize monoglycerides, thus affecting the integrity of the matrix and introducing unwanted variability to the performance of orally administered lipidic LLCs [234].

10.3.3 *Smart long-chain triglyceride or fatty acid-based delivery systems for oral delivery*

Although the vast majority of orally administered drugs are absorbed directly into the portal vein, there are a number of highly lipophilic compounds that can access the systemic circulation through the lymphatics [241].

As has been previously mentioned earlier in this section, after enterocyte uptake some digestion products — specifically long-chain fatty acids and monoacylglycerols derived from long fatty acids — are reesterified and incorporated into chylomicrons (largest lipoproteins found in the body) [204]. Chylomicrons mainly consist of a triglyceride core surrounded by apoproteins and phospholipids [242]. After formation, chylomicrons are expelled from

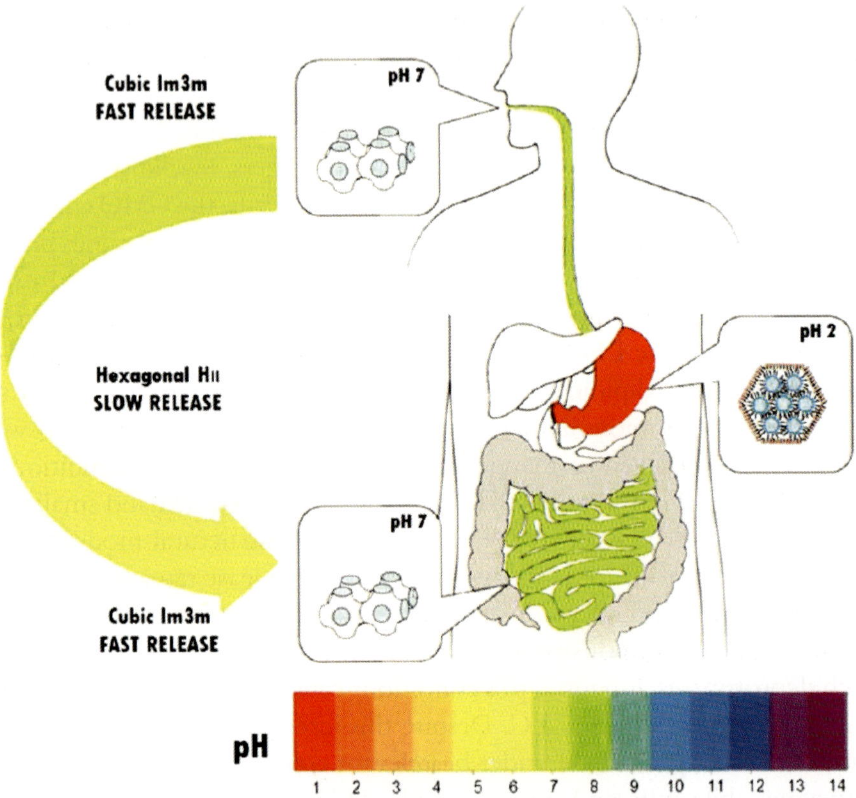

Figure 10.12. Illustration of the proposed pH-responsive lipidic liquid crystal drug delivery system across the gastrointestinal tract. At pH 2, in the stomach, drug carriers are presented as reversed hexagonal phases (H_2), slowing down diffusion and preventing premature release of the drug. By contrast, at pH 7, in the intestine, they change to reversed bicontinuous phases (v_2) triggering the release of the loaded drug. Reprinted with permission from Ref. [227].

enterocytes into the intracellular space. Due to their large size, chylomicrons are unable to enter the blood capillaries and therefore enter the more permeable lymphatic capillaries near the small intestine (lacteals) instead.

When lipophilic drugs are incorporated into lipidic formulations or co-administered with high-fat meals, they are absorbed along with fatty acids and triglycerides. Taking into account the differences in flow between blood and lymph (500:1, v/v), and that only 1% of lymph is made of chylomicrons, a drug is required a partition coefficient of at least 50000 in favor of lymph lipid rather than blood (log $D_{7.4} > 4.7$) to be transported *via* the lymphatic pathway [24]. Another requirement specified by Charman and Stella [243] is that drug solubility in triglycerides needs to be higher than 50 mg/mL for solubilization within the chylomicron core. However, it has been shown that these two physicochemical properties are not sufficient alone to quantitatively predict association with chylomicrons [244], which seems to be a critical step in estimating the degree of intestinal lymphatic transport of lipophilic molecules [245]. The combination of log $D_{7.4}$, degree of ionization, polar surface area, number of hydrogen acceptors and donors, density, molar volume and freely rotatable bonds describes the process of uptake of drugs by chylomicrons more accurately than any single physicochemical property [244].

Intestinal lymphatic transport provides two clear advantages over portal blood transport. First, the mesenteric lymph drains directly into the systemic circulation bypassing the liver. Therefore, lymphatic transport of drugs with significant first pass metabolism results in increased bioavailability due to reduced first pass metabolism [241]. The second advantage is the possibility of effectively targeting the lymphatic system with drugs transported in chylomicrons. Lymphatic targeting can be potentially beneficial in the treatment of autoimmune diseases [24], HIV [246] and metastatic processes [247].

There are some ways to promote drug lymphatic transport, including co-administration with high-fat meals, and use of lipidic prodrugs that increase drug lipophilicity [248]. Lately, lipid-based formulations are growing in popularity because not only do they enhance overall intestinal absorption, but they are also capable of stimulating lipoprotein formation and therefore intestinal lymphatic lipid flux [24]. In terms of carbon chain length of lipids constituting the formulation, Caliph *et al.* [249] demonstrated that long-chain lipids (C_{18}) gave greater enhancement of intestinal lymphatic transport of halofantrine than medium- (C_{8-10}) and short-chain (C_4) lipids. Sheehe and co-workers [250] suggested that unsaturated fatty acids create larger lipoproteins when compared to their saturated counterparts. In regard to lipid class, Charman and Stella [243] showed the rate of DDT lymphatic transport was faster with fatty acids compared to triglycerides. Examples of compounds formulated in long-chain lipids, in which lymphatic transport has been shown to improve oral bioavailability in animal models, include lipophilic cannabinoids in peanut oil [251], halofantrine in soybean oil and glycerol monolinoleate [252], probucol in peanut oil [253], and DDT in oleic acid [254], among others.

10.3.4 *Smart SMEDDSs for oral delivery*

SMEDDS are homogenous mixtures of lipids and surfactants, with optional solvents and co-solvents. Following dilution and mild motility in the gastrointestinal tract they are capable of forming fine oil-in-water emulsions with droplets ranging in size from 100 to 300 nm, although they can be as small as 50 nm [255]. SMEDDSs are suggested to increase the oral absorption and thus bioavailability of hydrophobic drugs through a number of synergistic and synchronized mechanisms including: (i) due to their small size, SMEDDSs can reach more inter-villous space, therefore increasing the surface available for absorption, (ii) increase in gastrointestinal solubilization, (iii) interference of enterocyte membrane fluidity, (iv) opening of epithelial tight junctions, (v) reduction in intra-enterocyte metabolism, (vi) decrease in P-gp efflux activity (applicabale if drug is a P-gp substrate), and (vii) stimulation of intestinal lymphatic transport [32, 252, 255–257].

The most acknowledged example of SMEDDS is Sandimmune Neoral®, a lipidic formulation of cyclosporin A — a hydrophobic polypeptide used as an immunosuppressant in transplantation surgery. As the original Sandimmune® formulation led to considerable inter- and intra-patient variability, it was considered necessary to reformulate it. The optimized Neoral® formulation, which is made of dl-α-tocopherol, corn oil derivatives and polyoxyl 40 hydrogenated castor oil, results in a more predictable and more extensive drug absorption [258]. Other successful commercial examples include the SMEDDS of the anti-HIV drugs ritonavir and saquinavir, Norvir® and Fortovase®, respectively.

One of the representative examples of formulations under development are the SMEDDS (namely iCo-009, iCo-010, and iCo-011) developed by K.M. Wasan's group for the oral administration of amphotericin B (AmB) [32, 213, 257, 259–262]. AmB is one the most effective antibiotic agents in the treatment of systemic fungal infections, or parasitic diseases such as visceral leishmaniasis (for which it has a 97% cure rate). This infection is almost always lethal in humans and is the second-largest parasitic killer in the world after malaria. Because AmB has limited water solubility and endothelial permeability (it belongs to class IV in the biopharmaceutical classification system), it has been administered parenterally for over 50 years. Intravenous therapeutic formulations of AmB include Fungizone® (micellar dispersion with sodium deoxycholate), AmBisome® (liposome-based), and Abelcet® (AmB complexed with two phospholipids in a 1:1 drug-to-lipid molar ratio). However, parenteral administration of AmB leads to side-effects such as acute nephrotoxicity, requires hospitalization, and in the case of lipidic formulations, is considered prohibitively expensive [263]. Therefore, drawbacks of parenteral AmB formulations outweigh their benefits. In light of these facts, the development of a cheaper, safer and more effective oral treatment of AmB is highly needed. With this in mind, Wasan's group came up with the iCo-formulations, SMEDDS of AmB which overcame the problem of low oral bioavailability, achieved higher therapeutic concentration in infected tissues, and decreased kidney toxicity in animal studies. The advantage of encapsulating

AmB in iCo-formulations was that even after a single oral dose, high concentrations of the drug were detected in visceral tissue, which would build-up upon multiple oral administrations to reach high therapeutic levels. Briefly, iCo-009 was made of glycerol monooleate and DSPE-PEG$_{2000}$, while iCo-010 and iCo-011 incorporated GMo and gelucire 44/14® (lauroyl polyoxyl-32 glycerides NF) at ratios of 50/50 (v/v) and 60/40 (v/v), respectively, in addition to 5% (v/v) d-α-tocopherol polyethylene glycol succinate. Results suggested that iCo-010 was the best from all SMEDDS due to higher tissue exposure and prolonged mean residence time. Oral administration of iCo-010 is therefore a promising alternative to current treatments in the battle against leishmanial parasites and other systemic fungal infections.

10.4 Smart Lipid-Based Carriers for Dermal and Transdermal Delivery

Transdermal drug delivery (TDD) depends on the controlled release of drugs through the skin to affect specified targets for the purpose of, for example, blood pressure control, pain management and others [264]. TDD offers many advantages over other routes of administration. This includes bypassing gastrointestinal incompatibility and hepatic first-pass metabolic effects, reduction of side effects due to the optimization of the blood concentration-time profile, minimization of inter- and intra-patient variability, reversibility of drug delivery by removal of the drug source, and patient compliance [264, 265].

The skin is a multilayered biomembrane with particular absorption characteristics. The stratum corneum (SC) is the outermost layer of the skin, and despite being relatively thin (10–15 μm on most areas of the body), it provides a highly efficient barrier for water and protection against microbial and chemical assault, making drug penetration extremely difficult [266]. If a compound is capable of passing through the SC, it still must cross the remaining layers of the epidermis and the dermis to reach the systemic circulation in order to exert therapeutic effect [267].

10.4.1 *Smart liposomes for transdermal delivery*

Classic liposomes have been found to be unsuitable for TDD because they are not able to penetrate deep enough into the skin [267, 268]. To overcome this problem, Cevc and collaborators [269–272] synthesized a highly flexible and elastic modification to plain liposomes designed to mimic the characteristics of cell vesicles. This new smart-devised nanocarrier, called Transfersome®† consists, like liposomes, of a phospholipid bilayer; however it also incorporates a surfactant (called edge activator) that behaves as a softening compound conferring the Transfersome® its characteristic ultra-deformability. The surfactant accumulates at the sites of high stress and lowers the bilipid interfacial tension, thus increasing its deformability. Transfersomes® have been proven to penetrate intact through the skin and successfully deliver its drug content (compounds of different sizes and polarities, macromolecules, proteins and vaccines) achieving comparable efficacies to subcutaneous administration [33].

Ethosomes [273–275] represent another kind of smart vesicles able to squeeze and cross the skin barrier reaching systemic circulation. They are also made of water and phospholipids, but on this occasion ethanol is also incorporated. Although its mechanism of action remains unknown, it is thought to be based on a synergistic effect among the ethosome components and the skin lipids [268]. Several studies in human volunteers have demonstrated the enhanced efficiency of ethosomes when delivering drugs (such as acyclovir [276] or ammonium glycyrrhizinate [277]) across the skin.

10.4.2 *Smart lipid-based liquid crystals for dermal delivery*

Ten years ago it was discovered that the structure of the SC resembles inverted cubic (v_2) lipid/water phases [278]. The observation that the skin itself acquires a cubic architecture suggest that dermal application of v_2 LLCs could probably result in a strong interaction between stratum corneum-LLCs [279].

As previously described in section 10.3.2, v_2 LLCs are highly viscous and rigid, attributes that make them suitable as drug nanocarriers for sustained release, but also limits their application since direct intravenous or subcutaneous dosing of cubic phases is almost impossible [229]. However, the less viscous L_α (lamellar) phase can be used as a precursor system for mucosal [280], vaginal [281], periodontal [282], and dermal [238] delivery since it readily transforms into the v_2 system upon exposure to excess water, such as interstitial fluid (Fig. 10.11). As an example, Fong and co-workers [238] injected glucose subcutaneously to rats incorporated in L_α precursors of v_2/H_2 systems in order to be able to study the temperature-triggered switch between these two phases. In the study, the content of vitamin E added to the lipid phytantriol (3,7,11,15-tetramethyl-1,2,3-hexadecanetriol) was optimized, in order to lower the phase transition and bring it closer to physiological range. Then, the optimized formulation was administered to rats and the temperature-triggered

† Transfersome® is a trademark registered by IDEA AG.

transition between nanostructures (V_2, 30°C ⇄ H_2, 40°C) was induced by application of a small heat pack or an ice pack. The difference in release behavior (V_2 freed its cargo faster than H_2) was evaluated providing an initial proof of concept for the use of lipidic LLCs as on-demand temperature-responsive nanocarriers for hydrophilic drugs.

10.4.3 *Smart solid lipid nanoparticles for dermal delivery*

SLNs have been introduced as carriers for active cosmetic ingredients and topically applied pharmaceutical compounds [36]. In this regard, the smartest application is their use as sunscreen carriers. Müller and co-workers [283, 284] discovered that highly crystalline SLNs efficiently scattered light acting as particulate UV blockers on their own. Following this observation, it was decided to further increase sunlight protection by SNLs *via* incorporation of a molecular sunscreen into the lipid matrix. *In vivo* and *in vitro* studies demonstrated the existence of a great synergistic effect between the sunscreen and the nanoparticle, which might be used to further reduce the amount of molecular sunscreen in the formulation, hence reducing side effects.

Nanostructed lipid carriers (NLCs) are considered the smarter new generation of SLNs [285], created to overcome the problems of relatively low drug loading and drug expulsion during storage of SLNs [286]. The lipid core of NLCs is made of a blend of two spatially very different lipids — a solid lipid and an oil — still solid at body temperature but with lower melting point. This way, the lipid core acquires a certain nanostructure and does not crystallise, resulting in an increased drug payload and reduced drug expulsion [283]. NLCs have been successfully evaluated *in vivo* and *in vitro* for percutaneous delivery of indomethacin [285] and retinol [284].

10.5 Smart Lipid-Based Carriers for Ocular Delivery

With regard to treatment of local ophthalmic diseases, ocular drug carriers are the most desirable technology due to immediate absorption into the eye, avoidance of hepatic first-pass metabolism, lower systemic exposure, and arguably patient compliance and ease of administration [33]. However, corneal and conjunctival epithelia together with tears constitute an almost impenetrable barrier to debris, microorganisms and drugs [287], making ocular drug delivery highly challenging. The eye can be divided in four sites in terms of drug delivery: (i) the conjunctiva, (ii) the cornea (constituted of both hydrophobic and hydrophilic tissues that restrict the access of drugs with polarities of any kind), (iii) the anterior and posterior chambers, and (iv) the vitreous cavity [33].

Despite eye drops being the most common dosage form, their bioavailability is rather low (less than 5%) due to poor pre-corneal retention and penetration [288]. Therefore, there is a clear need to develop new and smarter drug delivery systems aimed at overcoming the ocular barrier.

10.5.1 *Smart liposomes for ocular delivery*

Several publications have demonstrated liposomes' capability to enhance bioavailability of different agents as ocular carriers such as cyclosporin A [289], amphotericin B [290], or insulin [291]. Later on, when positively charged liposomes were developed, it was found that they were preferentially captured at the negatively charged corneal surface compared with neutral or negatively charged liposomes [292], leading to enhanced drug release [293].

Apart from size and charge, another smart way to further improve liposomal ocular delivery is prolonging residence time by coating the liposomes with bioadhesive polymers [294]. As such, chitosan-coated liposomes [295] and complexes of liposomes and chitosan nanoparticles [287] have shown promising results in terms of enhanced pre-corneal retention and trans-corneal penetration. Good retention properties were also found for hyaluronic acid modified liposomes [296]. Collasomes [297, 298], liposomes coupled to collagen matrices, are also based on the principle of enhancing bioavailability through bioadhesiveness.

Other interesting work is by Norley *et al.* [299, 300] who successfully delivered antiviral drugs using immunoliposomes. By coating the phospholipid bilayers with monoclonal antibodies, herpes simplex virus-infected tissue cells were targeted, getting site-specific, sustained release of acyclovir.

Finally, smart liposomes have also been used for ocular gene delivery [301]. For example, Liu *et al.* [302] designed PEGylated liposome-protamine-hyaluronic acid nanosystems encapsulating siRNA targeted against vascular endothelial growth factor receptor 1 (VEGFR1), and reported enhanced VEGFR1 knockdown and accelerate delivery to human retinal pigment epithelium cells over free siRNA *in vitro*.

10.5.2 *Smart solid lipid nanoparticles for ocular delivery*

Recent publications have demonstrated the potential of SLNs and NLCs to increase the ocular bioavailability of hydrophobic drugs such as ibuprofen [303], cyclosporin A [304], and tobramycin [305]. To enhance the therapeutic potency of incorporated drugs, researchers have implemented smart modifications to SLNs and NLCs. These changes depend on coating the particles either with polymers or cationic lipids to prolong pre-corneal residence time, thus delivering higher drug payload into the eye. As an example, Shen *et al.* [306] surface-covered NLCs with thiolated polyethylene glycol monostearate (Cys-PEG-SA) conjugates and suggested that this technology could be a more effective therapeutic strategy for the treatment of intraocular diseases with cyclosporin A. Luo *et al.* [307], relying on chitosan mucoadhesive properties, studied the capability of chitosan-coated NLCs to enhance flurbiprofen ophthalmic delivery.

In terms of gene therapy, cationically modified SLNs are included in the group of non-viral systems for DNA delivery, and they are considered by some authors as a promising alternative to liposomes [308]. However, the ability of these particles to form stable complexes with plasmid DNA has been questioned [309].

10.6 Smart Lipid-Based Carriers for Intramuscular Delivery of Vaccines: Virosomes

Most vaccines are administered intramuscularly or subcutaneously. Compared with subcutaneous administration, the intramuscular route optimises the immunogenicity of the vaccine and minimises adverse reactions at the injection site [310].

Vaccines can be classified in the three groups: live, killed and subunit vaccines. Live attenuated (weakened) vaccines are not very popular because of the inherent risk of introducing live viruses into healthy subjects. Inactivated or killed vaccines are safer and used worldwide, but tend to induce a less robust immune response. Finally, subunit vaccines consist of synthetic or purified specific antigens in the form of proteins, peptides, carbohydrates or nucleic acids. These subunit antigens are capable of evoking an immune response, without other virus components that might cause side effects. Nevertheless, subunit antigens need an adjuvant or delivery system to be protected from extracellular degradation and to target the pertinent immune cells (antigen-presenting cells, APCs, and naïve B-lymphocytes) [311–313].

The antigens responsible for stimulating protective immunity in membrane-enveloped viruses, like influenza, are the spike glycoproteins on their surface. These glycoproteins act as targeting ligands to immune cells and mediate fusion of the viral membrane with that of endosomes. Ever since the importance of spike glycoproteins in virus functionality was discovered, there has been an increasing interest in reconstituting viral spike proteins in liposomal formulations [314]. Liposomes that incorporate viral envelope proteins in their bilayered membranes are called virosomes. The first synthesized virosomes [315], and the most advanced ones, are those based on the influenza virus. Immunopotentiating reconstituted influenza virosomes (IRIVs) are liposomes that carry the protein hemmagglutinin (HA). These nanocarriers are smart for two reasons. First, APCs are targeted through interaction of HA with sialic acid residues of the immune cell membrane. Second, following receptor-mediated endocytosis, HA undergoes a pH-triggered conformation change at the slightly acidic endosomal environment. This modification leads to the fusion of the membranes of the virosome and endosome, releasing the contents of the virosome into the cytosol [316, 317]. Inflexal V® is a commercially available influenza vaccine made of DOPC (1,2-dioleoyl-*sn*-glycero-3-phosphocholine), DOPE, and HA subunits [45].

IRIVs are not only used as influenza vaccines, but they can also act as delivery systems for other viruses, drugs and nucleic acids. As such, the marketed hepatitis A vaccine Epaxal® is based on the same virosome as Inflexal V®, incorporating the inactivated virus coupled to the membrane surface [318]. Waelti and co-workers [319] studied *in vivo* the ability of IRIVs as doxorubicin carrier systems into cancer cells. The formulation relied on PE-PEG attached to Fab fragments for targeting and on HA for effective cytoplasmatic delivery of the drug load. Finally, IRIVs containing DNA and cationic lipids such as DODAC have been demonstrated to successfully promote uptake and intracellular delivery of virosome-plasmid DNA [320].

10.7 Summary

The ever-increasing pressure for discovering novel entities with pharmaceutical applications has led to a significant rise in the number of poorly water-soluble drug candidates. LBDDS have emerged as the most suitable strategy for the delivery of lipophilic compounds due to their biocompatibility, solubilization capacity, and ability to interact with cell membranes. The further improvement of LBDDS has led to the development of smart formulations in which the lipidic excipient assumes an active role in the enhancement of the therapeutic activity of incorporated drugs. In this chapter, two types of intelligent-devised LBDDS have been described. One group is constituted by those formulations which are smartly structured and act as "magic bullets" by delivering the drug where and when it is needed, triggered by certain stimuli. Liposomes, micelles, solid lipid nanoparticles, LNCs, lipid-coated gas microbubbles, LPHNs, and lipid-based liquid crystals are included in this group. The other class is represented by lipidic formulations (such as emulsions, fatty acid-based drug delivery systems, and self-microemulsifying drug delivery systems) which may be simply structured but dramatically modify the pharmacokinetic and tissue distribution profiles of the delivered drug.

Although there is still much work to be done, the encouraging results obtained *in vitro*, *in vivo* and in clinical trials indicate that more marketed smart lipidic formulations are not far away.

Acknowledgements

The authors would like to thank Jonathan CM Wong for editing the English language of the text.

References

1. Lipinski, C. A., Lombardo, F., Dominy, B. W. and Feeney, P. J. (2001). Experimental and computational approaches to estimate solubility and permeability in drug discovery and development settings. *Adv. Drug Deliv. Rev.*, **46**, 3–26.
2. Lipinski, C. A. (2000). Drug-like properties and the causes of poor solubility and poor permeability. *J. Pharm. Toxicol. Meth.*, **44**, 235–249.
3. Takagi, T., Ramachandran, C., Bermejo, M., Yamashita, S., Yu, L. X., Amidon, G. L. (2006). A Provisional Biopharmaceutical Classification of the Top 200 Oral Drug Products in the United States, Great Britain, Spain, and Japan. *Mol. Pharm.*, **3**, 631–643.
4. Hauss, D. J. (2007). *Oral Lipid-Based Formulations: Enhancing the Bioavailability of Poorly Water-Soluble Drugs*, Bristol-Myers Squibb Company, Princeton, New Jersey, USA, pp. vii–viii.
5. Testa, B., Crivori, P., Reist, M. and Carrupt, P.-A. (2000). The influence of lipophilicity on the pharmacokinetic behavior of drugs: Concepts and examples. *Perspect. Drug Discov.*, **19**, 179–211.
6. Leeson, P. (2012). Drug discovery: Chemical beauty contest. *Nature*, **481**, 455–456.
7. Arnott, J. A. and Planey, S. L. (2012). The influence of lipophilicity in drug discovery and design. *Expert Opin. Drug Discov.*, **7**, 863–875.
8. Liversidge, G. G. and Cundy, K. C. (1995). Particle size reduction for improvement of oral bioavailability of hydrophobic drugs: I. Absolute oral bioavailability of nanocrystalline danazol in beagle dogs. *Int. J. Pharm.*, **125**, 91–97.
9. Merisko-Liversidge, E. M. and Liversidge, G. G. (2008). Drug nanoparticles: Formulating poorly water-soluble compounds. *Toxicol. Pathol.*, **36**, 43–48.
10. Leuner, C. and Dressman, J. (2000). Improving drug solubility for oral delivery using solid dispersions. *Eur J. Pharm. Biopharm.*, **50**, 47–60.
11. Vasconcelos, T., Sarmento, B. and Costa, P. (2007). Solid dispersions as strategy to improve oral bioavailability of poor water soluble drugs. *Drug Discovery Today*, **12**, 1068–1075.

12. Serajuddin, A. T. M. (2007). Salt formation to improve drug solubility. *Adv. Drug Deliv. Rev.*, **59**, 603–616.
13. Davis, M. E. and Brewster, M. E. (2004). Cyclodextrin-based pharmaceutics: Past, present and future. *Nat. Rev. Drug Discov.*, **3**, 1023–1035.
14. Tommasini, S., Raneri, D., Ficarra, R., Calabrò, M. L., Stancanelli, R. and Ficarra, P. (2004). Improvement in solubility and dissolution rate of flavonoids by complexation with β-cyclodextrin. *J. Pharm. Biomed. Anal.*, **35**, 379–387.
15. Stella, V. J. and Nti-Addae, K.W. (2007). Prodrug strategies to overcome poor water solubility. *Adv. Drug Del. Rev.*, **59**, 677–694.
16. Jana, S., Mandlekar, S. and Marathe, P. (2010). Prodrug design to improve pharmacokinetic and drug delivery properties: Challenges to the discovery scientists. *Curr. Med. Chem.*, **17**, 3874–3908.
17. Pouton, C. W. (2000). Lipid formulations for oral administration of drugs: Non-emulsifying, self-emulsifying and 'self-microemulsifying' drug delivery systems. *Eur.J. Pharm. Sci.*, **11**, S93–S98.
18. Hauss, D. J. (2007). Oral lipid-based formulations. *Adv. Drug Deliv. Rev.*, **59**, 667–676.
19. Kuentz, M. (2012). Lipid-based formulations for oral delivery of lipophilic drugs. *Drug Discov. Today Technol.*, **9**, e97–e104.
20. Mu, H., Holm, R. and Müllertz, A. Lipid-based formulations for oral administration of poorly water-soluble drugs. *Int. J. Pharm.*, **45**, 215–224.
21. Fahy, E., Subramaniam, S., Murphy, R. C., Nishijima, M., Raetz, C. R. H., Shimizu, T., Spener, F., van Meer, G., Wakelam, M. J. O. and Dennis, E. A. (2009). Update of the LIPID MAPS comprehensive classification system for lipids. *J. Lipid Res.*, **50**, S9–S14.
22. Pouton, C. W. (2006). Formulation of poorly water-soluble drugs for oral administration: Physicochemical and physiological issues and the lipid formulation classification system. *Eur. J. Pharm. Sci.*, **29**, 278–287.
23. Pouton, C. W. and Porter, C. J. H. (2008). Formulation of lipid-based delivery systems for oral administration: Materials, methods and strategies. *Adv. Drug Deliv. Rev.*, **60**, 625–637.
24. Porter, C. J. H., Trevaskis, N. L. and Charman, W. N. (2007). Lipids and lipid-based formulations: Optimizing the oral delivery of lipophilic drugs. *Nat. Rev. Drug Discov.*, **6**, 231–248.
25. Porter, C. J. H., Pouton, C. W., Cuine, J. F. and Charman, W. N. (2008). Enhancing intestinal drug solubilisation using lipid-based delivery systems. *Adv. Drug Deliv. Rev.*, **60**, 673–691.
26. Dahan, A. and Hoffman, A. (2008). Rationalizing the selection of oral lipid based drug delivery systems by an *in vitro* dynamic lipolysis model for improved oral bioavailability of poorly water soluble drugs. *J. Control. Release*, **129**, 1–10.
27. O'Driscoll, C. M. and Griffin, B. T. (2008). Biopharmaceutical challenges associated with drugs with low aqueous solubility — The potential impact of lipid-based formulations. *Adv. Drug Deliv. Rev.*, **60**, 617–624.
28. Wassvik, C. M., Holmén, A. G., Bergström, C. A. S., Zamora, I. and Artursson, P. (2006). Contribution of solid-state properties to the aqueous solubility of drugs. *European J. Pharm. Sci.*, **29**, 294–305.

29. Welling, P. (1977). Influence of food and diet on gastrointestinal drug absorption: A review. *J. Pharm. Biopharm.*, **5**, 291–334.
30. Williams, H., Sassene, P., Kleberg, K., Bakala-N'Goma, J.-C., Calderone, M., Jannin, V., Igonin, A., Partheil, A., Marchaud, D., Jule, E., Vertommen, J., Maio, M., Blundell, R., Benameur, H., Carrière, F., Müllertz, A., Porter, C.J.H. and Pouton, C. W. (2012). Toward the establishment of standardized *in vitro* tests for lipid-based formulations, Part 1: Method parameterization and comparison of *in vitro* digestion profiles across a range of representative formulations. *J. Pharm. Sci.*, **101**, 3360–3380.
31. Wasan, K. M. (2001). Formulation and physiological and biopharmaceutical issues in the development of oral lipid-based drug delivery systems. *Drug Development and Industrial Pharmacy*, **27**, 267–276.
32. Ibrahim, F., Gershkovich, P., Sivak, O., Wasan, E. K. and Wasan K. M. (2013). Pharmacokinetics and tissue distribution of amphotericin B following oral administration of three lipid-based formulations to rats. *Drug Dev. Ind. Pharm.*, **39**, 1277–1283.
33. Gershkovich, P., Wasan, K. M. and Barta, C. A. A. (2008). Review of the application of lipid-based systems in systemic, dermal/transdermal, and ocular drug delivery. *Crit. Rev. Ther. Drug Carrier Syst.* **25**, 545–584.
34. Torchilin, V. P. (2007). 'Lipid-Based Parenteral Drug Delivery Systems: Biological Implications', Wasan, K. M. (ed.), *Role of Lipid Excipients in Modifying Oral and Parenteral Drug Delivery: Basic Principles and Biological Examples*, Hoboken: John Wiley, pp. 48–87.
35. Torchilin, V. P. (2005). Recent advances with liposomes as pharmaceutical carriers. *Nat. Rev. Drug Discov.*, **4**, 145–160.
36. Müller, R. H., Mäder, K. and Gohla, S. (2000). Solid lipid nanoparticles (SLN) for controlled drug delivery — a review of the state of the art. *Eur. J. Pharm.Biopharm.*, **50**, 161–177.
37. Valeant Pharmaceuticals North America LLC (2013). Visudyne® Verteporfin for injection, available at: http://www.visudyne.com/ (Accessed February 2014).
38. Janssen Products LP (2013). Doxil® (Doxorubicin HCl Liposome injection), available at: http://www.doxil.com/ (Accessed February 2014).
39. Lasic, D. D. (1997). Recent developments in medical applications of liposomes: Sterically stabilized liposomes in cancer therapy and gene delivery *in vivo*. *J. Control. Release*, **48**, 203–222.
40. Lantheus Medical Imaging Inc. (2014). Definity® Perflutren Lipid Microsphere (injectable suspension), available at: http://www.definityimaging.com/. (Accessed February 2014.)
41. Spectrum Pharmaceuticals Inc. (2014). Marqibo® Vincristine sulfate (liposome injection), available at: http://www.marqibo.com/ (Accessed February 2014).
42. Galen Limited (2014). DaunoXome® (liposomal daunorubicin), available at: http://daunoxome.com/. (Accessed February 2014.)
43. Sigma-Tau Pharmaceuticals Inc. (2014). DepoCyt® (cytarabine liposome injection), available at: http://www.depocyt.com/ (Accessed October 2014).
44. Pacira Pharmaceuticals (2014). DepoDur® (morphine sulfate extended-release liposome injection), available at: http://www.pacira.com/products/depodur.php (Accessed February 2014).
45. Chang, H.-I. and Yeh, M.-K. (2012). Clinical development of liposome-based drugs: Formulation, characterization, and therapeutic efficacy. *Int J Nanomed.*, 7, 49–60.

46. RxList Inc. (2014). Estrasorb® (estradiol topical emulsion), available at: http://www.rxlist.com/estrasorb-drug.htm (Accessed February 2014).
47. Meisel, J. A., Le, H. D., de Meijer, V. E., Nose, V., Gura, K. M., Mulkern, R. V., Sharif, M. R. A. and Puder, M. (2011). Comparison of 5 intravenous lipid emulsions and their effects on hepatic steatosis in a murine model. *J. Pediatr Surg.*, **46**, 666–673.
48. RxList Inc. (2014). Neoral®, available at: http://www.rxlist.com/neoral-drug.htm (Accessed Februrary 2014).
49. U.S. Food and Drug Administration (2014). Marinol® (Dronabinol capsules), available at: http://www.accessdata.fda.gov/drugsatfda_docs/label/2006/018651s025s026lbl.pdf (Accessed Februrary 2014).
50. Torchilin, V. P. (2004). Targeted polymeric micelles for delivery of poorly soluble drugs. *Cell. Mol Life Sci. CMLS*, **61**, 2549–2559.
51. Perche, F. and Torchilin, V. P. (2013). Recent trends in multifunctional liposomal nanocarriers for enhanced tumor targeting. *J. Drug Deliv.*, **2013**, 32.
52. Allen, T. M. and Cullis, P. R. (2013). Liposomal drug delivery systems: From concept to clinical applications. *Adv. Drug Deliv. Rev.*, **65**, 36–48.
53. Patel, H. M. (1992). Serum opsonins and liposomes: Their interaction and opsonophagocytosis. *Crit. Rev. Ther. Drug Carrier Syst.*, **9**, 39–90.
54. Couvreur, P. and Vauthier, C. (2006). Nanotechnology: Intelligent design to treat complex disease. *Pharm. Res.*, **23**, 1417–1450.
55. Gabizon, A., Peretz, T., Sulkes, A., Amselem, S., Ben-Yosef, R., Ben-Baruch, N., Catane, R., Biran, S. and Barenholz, Y. (1989). Systemic administration of doxorubicin-containing liposomes in cancer patients: a phase I study. *Eur. J. Cancer Clin. Oncol.*, **25**(12), 1795–1803.
56. Gabizon, A., Amselem, S., Goren, D., Cohen, R., Druckmann, S., Fromer, I., Chisin, R., Peretz, T., Sulkes, A. and Barenholz, Y. (1990). Preclinical and clinical experience with a doxorubicin-liposome preparation. *J. Liposome Res.*, **1**, 491–502.
57. Davis, S. S. (1997). Biomédical applications of nanotechnology — implications for drug targeting and gene therapy. *Trends Biotechnol.*, **15**, 217–224.
58. Mukherjee, A. K., Gupta, G., Bhattacharjee, S., Guha, S. K., Majumder, S., Adhikari, A., Bhattachrya, P., Majumdar, S. B. and Majumdar, S. (2010). Amphotericin b regulates the host immune response in visceral leishmaniasis: reciprocal regulation of protein kinase C isoforms. *J. Infect.*, **61**, 173–184.
59. New, R. R. C., Chance, M. L. and Heath, S. (1981). Antileishmanial activity of amphotericin and other antifungal agents entrapped in liposomes. *J. Antimicrob. Chemother.*, **8**, 371–381.
60. Sawant, R. R. and Torchilin, V. P. (2010). Liposomes as 'smart' pharmaceutical nanocarriers. *Soft Matter*, **6**, 4026–4044.
61. Jeon, S. I. and Andrade, J. D. (1991). Protein — surface interactions in the presence of polyethylene oxide: II. Effect of protein size. *J. Colloid Interface Sci.*, **142**, 159–166.
62. Immordino, M. L., Dosio, F. and Cattel, L. (2006). Stealth liposomes: Review of the basic science, rationale, and clinical applications, existing and potential. *Int. J. Nanomed.*, **1**, 297–315.

63. Palmer, T. N., Caride, V. J., Caldecourt, M. A. Twickler, J. and Abdullah, V. (1984). The mechanism of liposome accumulation in infarction. *Biochim. Biophys. Acta (BBA) — General Subjects*, **797**, 363–368.
64. Maeda, H., Wu, J., Sawa, T., Matsumura, Y. and Hori, K. (2000). Tumor vascular permeability and the EPR effect in macromolecular therapeutics: A review. *J. Control. Release*, **65**, 271–284.
65. Judge, A., McClintock, K., Phelps, J. R. and MacLachlan, I. (2006). Hypersensitivity and loss of disease site targeting caused by antibody responses to PEGylated liposomes. *Mol. Ther.*, **13**, 328–337.
66. Torchilin, V. P. (2010). Antinuclear antibodies with nucleosome-restricted specificity for targeted delivery of chemotherapeutic agents. *Ther. Deliv.*, **1**, 257–272.
67. Gabizon, A., Horowitz, A. T., Goren, D., Tzemach, D., Mandelbaum-Shavit, F., Qazen, M. M. and Zalipsky, S. (1999). Targeting folate receptor with folate linked to extremities of poly(ethylene glycol)-grafted liposomes: *In vitro* studies. *Bioconjug. Chem.*, **10**, 289–298.
68. Hatakeyama, H., Akita, H., Ishida, E., Hashimoto, K., Kobayashi, H., Aoki, T., Yasuda, J., Obata, K., Kikuchi, H., Ishida, T., Kiwada, H. and Harashima, H. (2007). Tumor targeting of doxorubicin by anti-MT1-MMP antibody-modified PEG liposomes. *Int. J. Pharm.*, **342**, 194–200.
69. Ganta, S., Devalapally, H., Shahiwala, A. and Amiji, M. (2008). A review of stimuli-esponsive nanocarriers for drug and gene delivery. *J. Control. Release*, **126**, 187–204.
70. Ni, S., Stephenson, S. M. and Lee, R. J. (2002). Folate receptor targeted delivery of liposomal daunorubicin into tumor cells. *Anticancer Res.*, **22**, 2131–2135.
71. Pan, X. Q. and Lee, R. J. (2005). *In vivo* antitumor activity of folate receptor-targeted liposomal daunorubicin in a murine leukemia model. *Anticancer Res.*, **25**, 343–346.
72. Pan, X. Q., Wang, H. and Lee, R. J. (2003). Antitumor activity of folate receptor-targeted liposomal doxorubicin in a KB oral carcinoma murine xenograft model. *Pharm. Res.*, **20**(3), 417–422.
73. Ishida, O., Maruyama, K., Tanahashi, H., Iwatsuru, M., Sasaki, K., Eriguchi, M. and Yanagie, H. (2001). Liposomes bearing polyethyleneglycol-coupled transferrin with intracellular targeting property to the solid tumors *in vivo*. *Pharm. Res.*, **18**, 1042–1048.
74. I. Matsuda, H. Konno, T. Tanaka, S. Nakamura (2001). Antimetastatic effect of hepatotropic liposomal adriamycin on human metastatic liver tumors. *Surg. Today*, **31**, 414–420.
75. Andresen, T. L., Jensen, S. S. and Jørgensen, K. (2005). Advanced strategies in liposomal cancer therapy: Problems and prospects of active and tumor specific drug release. *Prog. Lipid Res.*, **44**, 68–97.
76. Gerweck, L. E. and Seetharaman, K. (1996). Cellular pH gradient in tumor versus normal tissue: Potential exploitation for the treatment of cancer. *Cancer Res.*, **56**, 1194–1198.
77. Yatvin, M. B., Kreutz, W., Horwitz, B. A. and Shinitzky, M. (1980). pH-sensitive liposomes: Possible clinical implications. *Science*, **210**, 1253–1255.
78. Júnior, Á. D. C., Mota, L. G., Nunan, E. A., Wainstein, A. J. A., Wainstein, A. P. D. L., Leal, A. S., Cardoso, V. N. and De Oliveira, M. C. (2007). Tissue distribution evaluation

of stealth pH-sensitive liposomal cisplatin versus free cisplatin in Ehrlich tumor-bearing mice. *Life Sci.*, **80**, 659–664.

79. Leite, E. A., dos Santos Giuberti, C., Wainstein, A. J. A., Wainstein, A. P. D. L., Coelho, L. G. V., Lana, Â. M. Q., Savassi-Rocha, P. R. and de Oliveira, M. C. (2009). Acute toxicity of long-circulating and pH-sensitive liposomes containing cisplatin in mice after intraperitoneal administration. *Life Sci.*, **84**, 641–649.
80. Rui, Y., Wang, S., Low, P. S. and Thompson, D. H. (1998). Diplasmenylcholine–folate liposomes: An efficient vehicle for intracellular drug delivery†. *J. Am. Chem. Soc.*, **120**, 11213–11218.
81. Shin, J., Shum, P., Thompson, D. H. (2003). Acid-triggered release via dePEGylation of DOPE liposomes containing acid-labile vinyl ether PEG–lipids, *J. Control. Release*, **91**, 187–200.
82. Yatvin, M. B., Weinstein, M. J., Dennis, W. H. and Blumenthal, R. (1978). Design of liposomes for enhanced local release of drugs by hyperthermia. *Science*, **202**, 1290–1293.
83. Puri, A. (2013). Phototriggerable liposomes: Current research and future perspectives. *Pharmaceutics*, **6**, 1–25.
84. Kono, K., Nakai, R., Morimoto, K. and Takagishi, T. (1999). Thermosensitive polymer-modified liposomes that release contents around physiological temperature. *Biochim. Biophys Acta (BBA) — Biomembranes*, **1416**, 239–250.
85. Drummond, D. C., Noble, C. O., Hayes, M. E., Park, J. W. and Kirpotin, D. B. (2008). Pharmacokinetics and *in vivo* drug release rates in liposomal nanocarrier development. *J. Pharm. Sci.*, **97**, 4696–4740.
86. Kono, K., Ozawa, T., Yoshida, T., Ozaki, F., Ishizaka, Y., Maruyama, K., Kojima, C., Harada, A. and Aoshima S. (2010). Highly temperature-sensitive liposomes based on a thermosensitive block copolymer for tumor-specific chemotherapy. *Biomaterials*, **31**, 7096–7105.
87. Saito, G. Swanson, J. A. and Lee, K. D. (2003). Drug delivery strategy utilizing conjugation via reversible disulfide linkages: Role and site of cellular reducing activities. *Adv. Drug Deliv. Rev.*, **55**, 199–215.
88. Kirpotin, D., Hong, K., Mullah, N., Papahadjopoulos, D. and Zalipsky, S. (1996). Liposomes with detachable polymer coating: Destabilization and fusion of dioleoylphosphatidylethanolamine vesicles triggered by cleavage of surface-grafted poly(ethylene glycol). *FEBS Letters*, **388**, 115–118.
89. Zhang, J. X., Zalipsky, S., Mullah, N., Pechar, M. and Allen, T. M. (2004). Pharmaco attributes of dioleoylphosphatidylethanolamine/cholesterylhemisuccinate liposomes containing different types of cleavable lipopolymers. *Pharmacol. Res.*, **49**, 185–198.
90. Zalipsky, S., Qazen, M., Walker, J. A., Mullah, N., Quinn, Y. P. and Huang, S. K. (1999). New detachable poly(ethylene glycol) conjugates: cysteine-cleavable lipopolymers regenerating natural phospholipid, diacyl phosphatidylethanolamine. *Bioconjug. Chem.*, **10**, 703–707.
91. Huang, Z., Li, W., MacKay, J. A. and Szoka, F. C. J. (2005). Thiocholesterol-based lipids for ordered assembly of bioresponsive gene carriers. *Mol. Ther.*, **11**, 409–417.
92. Zhu, L., Huo, Z., Wang, L., Tong, X., Xiao, Y. and Ni, K. (2009). Targeted delivery of methotrexate to skeletal muscular tissue by thermosensitive magnetoliposomes. *Int. J. Pharm.*, **370**, 136–143.
93. Zhang, J. Q., Zhang, Z. R., Yang, H., Tan, Q. Y., Qin, S. R. and Qiu, X. L. (2005). Lyophilized Paclitaxel Magnetoliposomes as a Potential Drug Delivery System for

Breast Carcinoma via Parenteral Administration: *In Vitro* and *in Vivo* Studies. *Pharm. Res*, **22**, 573–583.

94. Viroonchatapan, E., Sato, H., Ueno, M., Adachi, I., Tazawa, K. and Horikoshi, I. (1996). Magnetic targeting of thermosensittve magnetoliposomes to mouse livers in an *in situ* on-line perfusion system. *Life Sci.*, **58**, 2251–2261.
95. Viroonchatapan, E., Sato, H., Ueno, M., Adachi, I., Murata, J., Saiki, I., Tazawa, K. and Horikoshi, I. (1998). Microdialysis assessment of 5-fluorouracil release from thermosensitive magnetoliposomes induced by an electromagnetic field in tumor-bearing mice. *J. Drug Target.*, **5**, 379–390.
96. Latorre, M. and Rinaldi, C. (2009). Applications of magnetic nanoparticles in medicine: Magnetic fluid hyperthermia. *P. R. Health Sci. J.*, **28**, 227–238.
97. Hilger, I., Hiergeist, R., Hergt, R., Winnefeld, K., Schubert, H. and Kaiser, W. A. (2002). Thermal ablation of tumors using magnetic nanoparticles: An *in vivo* feasibility study. *Invest. Radiol.,* **37**, 580–586.
98. Masuko, Y., Tazawa, K., Sato, H., Viroonchatapan, E., Takemori, S., Shimizu, T., Ohkami, H., Nagae, H., Fujimaki, M., Horikoshi, I. and Weinstein, J. N. (1997). Antitumor activity of selective hyperthermia in tumor-bearing rats using thermosensitive magnetoliposomes as a new hyperthermic material. *Drug Deliv.*, **4**, 37–42.
99. Schroeder, A., Kost, J. and Barenholz, Y. (2009). Ultrasound, liposomes, and drug delivery: Principles for using ultrasound to control the release of drugs from liposomes. *Chem. Phys. Lipids*, **162**, 1–16.
100. Dromi, S., Frenkel, V., Luk, A., Traughber, B., Angstadt, M., Bur, M., Poff, J., Xie, J., Libutti, S. K., Li, K. C. P. and Wood, B. J. (2007). Pulsed-High Intensity Focused Ultrasound and Low Temperature–Sensitive Liposomes for Enhanced Targeted Drug Delivery and Antitumor Effect. *Clin. Cancer Res.*, **13**, 2722–2727.
101. Schroeder, A., Avnir, Y., Weisman, S., Najajreh, Y., Gabizon, A., Talmon, Y., Kost, J. and Barenholz, Y. (2007). Controlling Liposomal Drug Release with Low Frequency Ultrasound: Mechanism and Feasibility. *Langmuir*, **23**, 4019–4025.
102. Shum, P., Kim, J.-M. and Thompson, D. H. (2001). Phototriggering of liposomal drug delivery systems. *Adv. Drug Deliv. Rev.*, **53**, 273–284.
103. Lamparski, H., Liman, U., Barry, J. A., Frankel, D. A., Ramaswami, V., Brown, M. F. and O'Brien, D. F. (1992). Photoinduced destabilization of liposomes. *Biochemistry*, **31**, 685–694.
104. Bondurant, B., Mueller, A. and O'Brien, D. F. (2001). Photoinitiated destabilization of sterically stabilized liposomes. *Biochim. Biophys. Acta (BBA) — Biomembranes*, **1511**, 113–122.
105. Alvarez-Lorenzo, C., Bromberg, L. and Concheiro, A. (2009). Light-sensitive Intelligent Drug Delivery Systems. *Photochem. Photobiol.*, **85**, 848–860.
106. Anderson, V. C. and Thompson, D. H. (1992). Triggered release of hydrophilic agents from plasmologen liposomes using visible light or acid. *Biochim Biophys. Acta (BBA) — Biomembranes*, **1109**, 33–42.
107. Bisby, R. H., Mead, C. and Morgan, C. G. (2000). Wavelength-Programmed Solute Release from Photosensitive Liposomes. *Biochem. Biophys. Res. Commun.*, **276**, 169–173.
108. Bisby, R. H., Mead, C. and Morgan, C. G. (2000). Active Uptake of Drugs into Photosensitive Liposomes and Rapid Release on UV Photolysis. *Photochem. Photobiol.*, **72**, 57–61.

109. Templeton, N. S. (2002). Cationic liposome-mediated gene delivery *in vivo*. *Biosci. Rep.*, **22**, 283–295.
110. Templeton, N. S., Lasic, D. D., Frederik, P. M., Strey, H. H., Roberts, D. D. and Pavlakis, G. N. (1997). Improved DNA: Liposome complexes for increased systemic delivery and gene expression. *Nat. Biotech.*, **15**, 647–652.
111. Felgner, P. L., Gadek, T. R., Holm, M., Roman, R., Chan, H. W., Wenz, M., Northrop, J. P., Ringold, G. M. and Danielsen, M. (1987). Lipofection: A highly efficient, lipid-mediated DNA-transfection procedure. *Proceedings of the National Academy of Sciences*, **84**, 7413–7417.
112. Felgner, P. L. and Ringold, G. M. (1989). Cationic liposome-mediated transfection. *Nature*, **337**, 387–388.
113. Audouy, S. L., de Leij, L. M. H., Hoekstra, D. and Molema, G. (2002). *In Vivo* Characteristics of Cationic Liposomes as Delivery Vectors for Gene Therapy. *Pharm. Res.*, **19**, 1599–1605.
114. Midoux, P., Pichon, C., Yaouanc, J.-J. and Jaffrès, P.-A. (2009). Chemical vectors for gene delivery: a current review on polymers, peptides and lipids containing histidine or imidazole as nucleic acids carriers. *Br. J. Pharcol.*, **157**, 166–178.
115. Zhdanov, R. I., Podobed, O. V. and Vlassov, V. V. (2002). Cationic lipid–DNA complexes — lipoplexes — for gene transfer and therapy. *Bioelectrochemistry*, **58**, 53–64.
116. Schroeder, A., Levins, C. G., Cortez, C., Langer, R. and. Anderson, D. G (2010). Lipid-based nanotherapeutics for siRNA delivery. *J. Inter. Med.*, **267**, 9–21.
117. Semple, S. C., Akinc, A., Chen, J., Sandhu, A. P., Mui, B. L., Cho, C., Sah, D. W. Y., Stebbing, D., Crosley, E. J., Yaworski, E., Hafez, I. M., Dorkin, J. R., Qin, J., Lam, K., Rajeev, K. G., Wong, K. F., Jeffs, L. B., Nechev, L., Eisenhardt, M. L., Jayaraman, M., Kazem, M., Maier, M. A., Srinivasulu, M., Weinstein, M. J., Chen, Q., Alvarez, R., Barros, S. A., De, S., Klimuk, S. K., Borland, T., Kosovrasti, V., Cantley, W. L., Tam, Y. K., Manoharan, M., Ciufolini, M. A., Tracy, M. A., de Fougerolles, A., MacLachlan, I., Cullis, P. R., Madden, T. D. and Hope, M. J. (2010). Rational design of cationic lipids for siRNA delivery. *Nat. Biotech.*, **28**, 172–176.
118. Akinc, A., Zumbuehl, A., Goldberg, M., Leshchiner, E. S., Busini, V., Hossain, N., Bacallado, S. A., Nguyen, D. N., Fuller, J., Alvarez, R., Borodovsky, A., Borland, T., Constien, R., de Fougerolles, A., Dorkin, J. R., Narayanannair Jayaprakash, K., Jayaraman, M., John, M., Koteliansky, V., Manoharan, M., Nechev, L., Qin, J., Racie, T., Raitcheva, D., Rajeev, K. G., Sah, D. W. Y., Soutschek, J., Toudjarska, I., Vornlocher, H.-P., Zimmermann, T. S., Langer, R. and Anderson, D. G. (2008). A combinatorial library of lipid-like materials for delivery of RNAi therapeutic. *Nat. Biotech.*, **26**, 561–569.
119. Zhang, S., Zhao, B., Jiang, H., Wang, B. and Ma, B. (2007). Cationic lipids and polymers mediated vectors for delivery of siRNA. *J. Control. Rel.*, **123**, 1–10.
120. Parker, A. L., Newman, C., Briggs, S., Seymour, L. and Sheridan, P. J. (2003). Nonviral gene delivery: Techniques and implications for molecular medicine. *Expert Rev. Mol. Med.*, **5**, 1–15.
121. Hui, S. W., Langner, M., Zhao, Y. L., Ross, P., Hurley, E. and Chan, K. (1996). The role of helper lipids in cationic liposome-mediated gene transfer. *Biophys J.*, **71**, 590–599.
122. Roth, C. M. and Sundaram, S. (2004). Engineering synthetic vectors for improved dna delivery: Insights from Intracellular Pathways. *Ann. Rev. Biomed. Eng.*, **6**, 397–426.

123. Ramesh, R., Saeki, T., Templeton, N., Ji, L., Stephens, L. C., Ito, I., Wilson, D. R., Wu, Z., Branch, C. D., Minna, J. D. and Roth, J. A. (2001). Successful treatment of primary and disseminated human lung cancers by systemic delivery of tumor suppressor genes using an improved liposome vector. *Mol. Ther.*, **3**, 337–350.
124. Yu, W., Pirollo, K., Rait, A., Yu, B., Xiang, L., Huang, W., Zhou, Q., Ertem, G. and Chang, E. (2004). A sterically stabilized immunolipoplex for systemic administration of a therapeutic gene. *Gene Ther.*, **11**, 1434–1440.
125. Li, S.-D. and Huang, L. (1997). *In vivo* gene transfer via intravenous administration of cationic lipid-protamine-DNA (LPD) complexes. *Gene Ther.*, **4**, 891–900.
126. Li, S.-D., Rizzo, M.A., Bhattacharya, S. and Huang, L. (1998). Characterization of cationic lipid-protamine-DNA (LPD) complexes for intravenous gene delivery. *Gene Ther.*, **5**, 930–937.
127. Li, S.-D. and Huang, L. (2006). Surface-modified lpd nanoparticles for tumor targeting. *Ann. N. Y. Acad. Sci.*, **1082**, 1–8.
128. Matsuura, M., Yamazaki, Y., Sugiyama, M., Kondo, M., Ori, H., Nango, M. and Oku, N. (2003) Polycation liposome-mediated gene transfer *in vivo. Biochim. Biophys. Acta (BBA) — Biomembranes*, **1612**, 136–143.
129. Gao, X. and Huang, L. (1996). Potentiation of cationic liposome-mediated gene delivery by polycations†. *Biochemistry*, **35**, 1027–1036.
130. Li, J., Chen, Y.-C., Tseng, Y.-C., Mozumdar, S. and Huang, L. (2010). Biodegradable calcium phosphate nanoparticle with lipid coating for systemic siRNA delivery. *J. Control. Release*, **142**, 416–421.
131. Li, S.-D. and Huang, L. (2006). Gene therapy progress and prospects: Non-viral gene therapy by systemic delivery. *Gene Ther.*, **13**, 1313–1319.
132. Torchilin, V. P. (2007). Micellar nanocarriers: pharmaceutical perspectives. *Pharm. Res.*, **24**, 1–16.
133. Lukyanov, A. N. and Torchilin, V. P. (2004). Micelles from lipid derivatives of water-soluble polymers as delivery systems for poorly soluble drugs. *Adv. Drug Deliv. Rev.*, **56**, 1273–1289.
134. Durazo, S. A. and Kompella, U. B. (2011). 'Nanotechnology and Nanoparticles', in Kompella, U. B. and Edelhauser, H. F. (eds.), *Drug Product Development for the Back of the Eye*, Springer, New York, pp. 271–291.
135. Rosen, M. J. (1989). 'Micelle formation by surfactants', in Rosen. M. J. (ed.), *Surfactants and Interfacial Phenomena*, Wiley Interscience, New York, pp. 105–177.
136. Weissig, V., Whiteman, K. and Torchilin, V. (1998). Accumulation of Protein-Loaded Long-Circulating Micelles and Liposomes in Subcutaneous Lewis Lung Carcinoma in Mice. *Pharm. Res.*, **15**, 1552–1556.
137. Lukyanov, A., Gao, Z., Mazzola, L. and Torchilin, V. (2002). Polyethylene glycol-diacyllipid micelles demonstrate increased accumulation in subcutaneous tumors in Mice. *Pharm. Res.*, **19**, 1424–1429.
138. Krishnadas, A., Rubinstein, I. and Önyüksel, H. (2003). Sterically stabilized phospholipid mixed micelles: *in vitro* evaluation as a novel carrier for water-insoluble drugs. *Pharm. Res.*, **20**, 297–302.
139. Torchilin, V. P., Lukyanov, A. N., Gao, Z. and Papahadjopoulos-Sternberg, B. (2003). Immunomicelles: Targeted pharmaceutical carriers for poorly soluble drugs. *Proceedings of the National Academy of Sciences*, **100**, 6039–6044.

140. Lukyanov, A. N., Hartner, W. C. and Torchilin, V. P. (2004). Increased accumulation of PEG–PE micelles in the area of experimental myocardial infarction in rabbits. *J. Control. Rel.*, **94**, 187–193.
141. Wang, J., Mongayt, D. and Torchilin, V. P. (2005). Polymeric micelles for delivery of poorly soluble drugs: Preparation and anticancer activity *in vitro* of paclitaxel incorporated into mixed micelles based on poly(ethylene glycol)-lipid conjugate and positively charged lipids. *J. Drug Target.*, **13**, 73–80.
142. Allison, R. R., Downie, G. H., Cuenca, R., Hu, X.-H., Childs, C. J. H. and Sibata, C. H. (2004). Photosensitizers in clinical PDT. *Photodiagnosis Photodyn Ther.*, **1**, 27–42.
143. Cinteza, L. O., Ohulchanskyy, T. Y., Sahoo, Y., Bergey, E. J., Pandey, R. K. and Prasad, P. N. (2006). Diacyllipid micelle-based nanocarrier for magnetically guided delivery of drugs in photodynamic therapy. *Mol. Pharm.*, **3**, 415–423.
144. Buse, J. and El-Aneed, A. (2010). Properties, engineering and applications of lipid-based nanoparticle drug-delivery systems: Current research and advances. *Nanomedicine*, **5**, 1237–1260.
145. Bekerman, T., Golenser, J. and Domb, A. (2004). Cyclosporin nanoparticulate lipospheres for oral administration. *J. Pharm. Sci.*, **93**, 1264–1270.
146. Hu, L., Tang, X. and Cui, F. (2014). Solid lipid nanoparticles (SLNs) to improve oral bioavailability of poorly soluble drugs. *J. Pharm. Pharmacol.*, **56**, 1527–1535.
147. Elgart, A., Cherniakov, I., Aldouby, Y., Domb, A. J. and Hoffman, A. (2012). Lipospheres and pro-nano lipospheres for delivery of poorly water soluble compounds. *Chem. Phys. Lipids*, **165**, 438–453.
148. Yang, S. C., Lu, L. F., Cai, Y., Zhu, J. B., Liang, B. W. and Yang, C. Z. (1999). Body distribution in mice of intravenously injected camptothecin solid lipid nanoparticles and targeting effect on brain. *J. Control. Release*, **59**, 299–307.
149. Ye, J., Wang, Q., Zhou, X. and Zhang, N. (2008). Injectable actarit-loaded solid lipid nanoparticles as passive targeting therapeutic agents for rheumatoid arthritis. *Int. J. Pharm.*, **352**, 273–279.
150. Fundarò, A., Cavalli, R., Bargoni, A., Vighetto, D., Zara, G. P. and Gasco, M. R. (2000). Non-stealth and stealth solid lipid nanoparticles (SLN) carrying doxorubicin: Pharmacokinetics and tissue distribution after i.v. administration to rats. *Pharmacol. Res.*, **42**, 337–343.
151. Zara, G. P., Cavalli, R., Bargoni, A., Fundarò, A., Vighetto, D. and Gasco, M. R. (2002). Intravenous administration to rabbits of non-stealth and stealth doxorubicin-loaded solid lipid nanoparticles at increasing concentrations of stealth agent: pharmacokinetics and distribution of doxorubicin in brain and other tissues, *J. Drug Target.*, **10**, 327–335.
152. Chen, D.-B., Yang, T.-Z., Lu, W.-L. and Zhang, (Q. 2001). *In vitro* and *in vivo* study of two types of long-circulating solid lipid nanoparticles containing paclitaxel. *Chem. Pharm. Bull.*, **49**, 1444–1447.
153. Gao, Y., Gu, W., Chen, L., Xu, Z. and Li, Y. (2008). The role of daidzein-loaded sterically stabilized solid lipid nanoparticles in therapy for cardio-cerebrovascular diseases. *Biomaterials*, **29**, 4129–4136.
154. Hsu, M.-H. and Su, Y.-C. (2008). Iron-oxide embedded solid lipid nanoparticles for magnetically controlled heating and drug delivery. *Biomed. Microdevices*, **10**, 785–793.

155. Heurtault, B., Saulnier, P., Benoit, J. P., Proust, J. E., Pech, B. and Richard, J. (2011). Lipid nanocapsules, preparation process and use as medicine, in: Google Patents, US 20090238865 A1, 2011.
156. Huynh, N. T., Passirani, C., Saulnier, P. and Benoit, J. P. (2009). Lipid nanocapsules: A new platform for nanomedicine. *Int. J. Pharm.*, **379**, 201–209.
157. Pereira de Oliveira, M., Garcion, E., Venisse, N., Benoît, J.-P., Couet, W. and Olivier, J.-C. (2005). Tissue Distribution of Indinavir Administered as Solid Lipid Nanocapsule Formulation in mdr1a (+/+) and mdr1a (−/−) CF-1 Mice. *Pharm. Res.*, **22**, 1898–1905.
158. Allard, E., Passirani, C., Garcion, E., Pigeon, P., Vessières, A., Jaouen, G. and Benoit, J.-P. (2008). Lipid nanocapsules loaded with an organometallic tamoxifen derivative as a novel drug-carrier system for experimental malignant gliomas. *J. Control. Release*, **130**, 146–153.
159. Lacoeuille, F., Hindre, F., Moal, F., Roux, J., Passirani, C., Couturier, O., Cales, P., Le Jeune, J. J., Lamprecht, A. and Benoit, J. P. (2007). *In vivo* evaluation of lipid nanocapsules as a promising colloidal carrier for paclitaxel. *Inter. J. Pharm.,* **344**, 143–149.
160. Lamprecht, A., Saumet, J.-L., Roux, J. and Benoit, J.-P. (2004). Lipid nanocarriers as drug delivery system for ibuprofen in pain treatment, *Inter. J. Pharm.* , **278**, 407–414.
161. Coon, J. S., Knudson, W., Clodfelter, K., Lu, B. and Weinstein, R. S. (1991). Solutol HS 15, nontoxic polyoxyethylene esters of 12-hydroxystearic acid, reverses multidrug resistance. *Cancer Res.*, **51**, 897–902.
162. Béduneau, A., Saulnier, P., Anton, N., Hindré, F., Passirani, C., Rajerison, H., Noiret, N. and Benoit, J.-P. (2006). Pegylated nanocapsules produced by an organic solvent-free method: evaluation of their stealth properties. *Pharm. Res.*, **23**, 2190–2199.
163. Khalid, M., Simard, P., Hoarau, D., Dragomir, A. and Leroux, J.-C. (2006). Long circulating poly(ethylene glycol)-decorated lipid nanocapsules deliver docetaxel to solid tumors. *Pharm. Res.*, **23**, 752–758.
164. Hoarau, D., Delmas, P., David, S.x.E., Roux, E. and Leroux, J.-C. (2004). Novel long-circulating lipid nanocapsules. *Pharm. Res.*, **21**, 1783–1789.
165. Béduneau, A., Saulnier, P., Hindré, F., Clavreul, A., Leroux, J.-C. and Benoit, J.-P. (2007). Design of targeted lipid nanocapsules by conjugation of whole antibodies and antibody Fab' fragments. *Biomaterials*, **28**, 4978–4990.
166. Vonarbourg, A., Passirani, C., Desigaux, L., Allard, E., Saulnier, P., Lambert, O., Benoit, J.-P. and Pitard, B. (2009). The encapsulation of DNA molecules within biomimetic lipid nanocapsules. *Biomaterials*, **30**, 3197–3204.
167. Morille, M., Passirani, C., Dufort, S., Bastiat, G., Pitard, B., Coll, J.-L. and Benoit, J.-P. (2011). Tumor transfection after systemic injection of DNA lipid nanocapsules. *Biomaterials*, **32**, 2327–2333.
168. Unger, E. C., Porter, T., Culp, W., Labell, R., Matsunaga, T. and Zutshi, R. (2004). Therapeutic applications of lipid-coated microbubbles. *Adv. Drug Deliv. Rev.*, **56**, 1291–1314.
169. Unger, E. C., Hersh, E. , Vannan, M., Matsunaga, T. O. and McCreery, T. (2001). Local drug and gene delivery through microbubbles, Progress in cardiovascular diseases. *Prog. Cardiovas. Dis.*, **44**, 45–54.
170. Porter, T. R. and Xie, F. (2001). Ultrasound, microbubbles, and thrombolysis. *Prog. Cardiovas. Dis.*, **44**, 101–110.

171. Hynynen, K., McDannold, N., Vykhodtseva, N. and Jolesz, F. A. (2001). Noninvasive MR imaging–guided focal opening of the blood-brain barrier in rabbits. *Radiology*, **220**, 640–646.
172. Unger, E. C., Mccreery, T. P., Sweitzer, R. H., Caldwell, V. E. and Wu, Y. (1998). Acoustically active lipospheres containing paclitaxel: a new therapeutic ultrasound contrast agent. *Invest. Radiol.*, **33**, 886–892.
173. Schumann, P. A., Christiansen, J. P., Quigley, R. M., McCreery, T. P., Sweitzer, R. H., Unger, E. C., Lindner, J. R. and Matsunaga, T. O. (2002). Targeted-microbubble binding selectively to GPIIb IIIa receptors of platelet thrombi. *Invest. Radiol.*, **37**, 587–593.
174. Lindner, J. R., Song, J., Xu, F., Klibanov, A. L., Singbartl, K., Ley, K. and Kaul, S. (2000). Noninvasive ultrasound imaging of inflammation using microbubbles targeted to activated leukocytes. *Circulation*, **102**, 2745–2750.
175. Vannan, M., McCreery, T., Li, P., Han, Z., Unger, E., Kuersten, B., Nabel, E. and Rajagopalan, S. (2002). Ultrasound-mediated transfection of canine myocardium by intravenous administration of cationic microbubble-linked plasmid DNA. *J. Am. Soc. Echocardiogr.*, **15**, 214–218.
176. Unger, E. C., Hersh, E., Vannan, M. and McCreery, T. (2001). Gene Delivery Using Ultrasound Contrast Agents. *Echocardiography*, **18**, 355–361.
177. Hadinoto, K., Sundaresan, A. and Cheow, W. S. (2013). Lipid–polymer hybrid nanoparticles as a new generation therapeutic delivery platform: A review. *Eur. J. Pharm. Biopharm.*, **85**, 427–443.
178. Hu, C.-M. J., Kaushal, S., Cao, H. S. T., Aryal, S., Sartor, M., Esener, S., Bouvet, M. and Zhang, L. (2010). Half-Antibody Functionalized Lipid–Polymer Hybrid Nanoparticles for Targeted Drug Delivery to Carcinoembryonic Antigen Presenting Pancreatic Cancer Cells, *Mol. Pharm.*, 7, 914–920.
179. Chan, J. M., Rhee, J.-W., Drum, C. L., Bronson, R. T., Golomb, G., Langer, R. and Farokhzad, O. C. (2011). *In vivo* prevention of arterial restenosis with paclitaxel-encapsulated targeted lipid–polymeric nanoparticles. *Proceedings of the National Academy of Sciences*, **108**, 19347–19352.
180. Li, J., He, Y.-Z., Li, W., Shen, Y.-Z., Li, Y.-R. and Wang, Y.-F. (2010). A novel polymer-lipid hybrid nanoparticle for efficient nonviral gene delivery. *Acta Pharmacol. Sin.*, **31**, 509–514.
181. Deok Kong, S., Sartor, M., Jack Hu, C.-M., Zhang, W., Zhang, L. and Jin, S. (2013). Magnetic field activated lipid–polymer hybrid nanoparticles for stimuli-responsive drug release. *Acta Biomater.*, **9**, 5447–5452.
182. Cheow, W. S. and Hadinoto, K. (2012). Lipid-polymer hybrid nanoparticles with rhamnolipid-triggered release capabilities as anti-biofilm drug delivery vehicles. *Particuology*, **10**, 327–333.
183. Collins-Gold, L. C., Lyons, R. T. and Bartholow, L. C. (1990). Parenteral emulsions for drug delivery. *Adv. Drug Deliv. Rev.*, **5**, 189–208.
184. Lundberg, B. (1994). Preparation of drug — carrier emulsions stabilized with phosphatidylcholine — surfactant mixtures. *J. Pharm. Sci.*, **83**, 72–75.
185. Hansrani, P. K., Davis, S. S. and Groves, M. J. (1983). The preparation and properties of sterile intravenous emulsions. *J. Parenter. Sci. Technol.*, **37**, 145–150.

186. Nishikawa, M., Takakura, Y. and Hashida, M. (1998). 'Biofated of fat emulsions', in Benita, S. (ed.), *Submicron Emulsions in Drug Targeting and Delivery*, Harwood Academic Publisher, Amsterdam, pp. 99–118.
187. Liu, F. and Liu, D. (1995). Long-circulating emulsions (oil-in-water) as carriers for lipophilic drugs. *Pharm. Res.*, **12**, 1060–1064.
188. Lundberg, B. B. (1997). A submicron lipid emulsion coated with amphipathic polyethylene glycol for parenteral administration of paclitaxel (taxol). *J. Pharm. Pharmacology*, **49**, 16–21.
189. Rossi, J., Giasson, S., Khalid, M. N., Delmas, P., Allen, C. and Leroux, J.-C. (2007). Long-circulating poly(ethylene glycol)-coated emulsions to target solid tumors. *Eur.J. Pharm. Biopharm.*, **67**, 329–338.
190. Constantinides, P., Lambert, K., Tustian, A., Schneider, B., Lalji, S., Ma, W., Wentzel, B., Kessler, D., Worah, D. and Quay, S. (2000). Formulation development and antitumor activity of a filter-sterilizable emulsion of paclitaxel. *Pharm. Res.*, **17**, 175–182.
191. Takino, T., Konishi, K., Takakura, Y. and Hashida, M. (1994). Long circulating emulsion carrier systems for highly lipophilic drugs. *Biol. Pharm. Bull.*, **17**, 121–125.
192. Lundberg, B. B., Griffiths, G. and Hansen, H. J. (1999). Conjugation of an anti-B-cell lymphoma monoclonal antibody, LL2, to long-circulating drug-carrier lipid emulsions. *J. Pharm. Pharm.*, **51**, 1099–1105.
193. Constantinides, P. P., Tustian, A. and Kessler, D. R. (2004). Tocol emulsions for drug solubilization and parenteral delivery. *Adv. Drug Deliv. Rev.*, **56**, 1243–1255.
194. Li, J., Nie, S., Yang, X., Wang, C., Cui, S. and Pan, W. (2008). Optimization of tocol emulsions for the intravenous delivery of clarithromycin. *Inter. J. Pharm.*, **356**, 282–290.
195. Fernandez, A. L., Lee, J. A., Rahilly, L., Hovda, L., Brutlag, A. G. and Engebretsen, K. (2011). The use of intravenous lipid emulsion as an antidote in veterinary toxicology. *J. Vet. Emerg. Crit. Care*, **21**, 309–320.
196. Bania, T. C., Chu, J., Perez, E., Su, M. and Hahn, I.-H. (2007). Hemodynamic effects of intravenous fat emulsion in an animal model of severe verapamil toxicity resuscitated with atropine, calcium, and saline. *Acad. Emerg. Med.*, **14**, 105–111.
197. Weinberg, G. L., VadeBoncouer, T., Ramaraju, G. A., Garcia-Amaro, M. F. and Cwik, M. J. (1998). Pretreatment or resuscitation with a lipid infusion shifts the dose-response to bupivacaine-induced asystole in rats. *Anesthesiology*, **88**, 1071–1075.
198. Litz, R. J., Popp, M., Stehr, S. N. and Koch, T. (2006). Successful resuscitation of a patient with ropivacaine-induced asystole after axillary plexus block using lipid infusion*. *Anaesthesia*, **61**, 800–801.
199. Sirianni, A. J., Osterhoudt, K. C., Calello, D. P., Muller, A. A., Waterhouse, M. R., Goodkin, M. B., Weinberg, G. L. and Henretig, F. M. (2008). Use of lipid emulsion in the resuscitation of a patient with prolonged cardiovascular collapse after overdose of bupropion and lamotrigine. *Ann. Emerg. Med.*, **51**, 412–415.e1.
200. Fasano, A. (1998). Innovative strategies for the oral delivery of drugs and peptides. *Trends Biotechnol.*, **16**, 152–157.
201. Carey, M. C., Small, D. M. and Bliss, C. M. (1983). Lipid digestion and absorption. *Annu. Rev. Physiol.*, **45**, 651–677.

202. Kaukonen, A., Boyd, B., Porter, C. H. and Charman, W. (2004). Drug solubilization behavior during *in vitro* digestion of simple triglyceride lipid solution formulations. *Pharm. Res.*, **21**, 245–253.
203. Shiau, Y. F., Kelemen, R. J. and Reed, M. A. (1990). Acidic mucin layer facilitates micelle dissociation and fatty acid diffusion. *Am. J. Physiol — Gastrointest. Liver Physiol.*, **259**, G671–G675.
204. Mansbach, C. M. and Siddiqi, S. A. (2010). The biogenesis of chylomicrons. *Ann. Rev. Physiol.*, **72**, 315–333.
205. Artursson, P., Palm, K. and Luthman, K. (2001). Caco-2 monolayers in experimental and theoretical predictions of drug transport. *Adv. Drug Deliv. Rev.*, **46**, 27–43.
206. Pade, V. and Stavchansky, S. (1998). Link between drug absorption solubility and permeability measurements in Caco-2 cells. *J. Pharm. Sci.*, **87**, 1604–1607.
207. Lennernäs, H. (2007). Animal data: The contributions of the Ussing Chamber and perfusion systems to predicting human oral drug delivery *in vivo*. *Adv. Drug Deliv. Rev.*, **59**, 1103–1120.
208. Zangenberg, N. H., Müllertz, A., Kristensen, H. G. and Hovgaard, L. (2001). A dynamic *in vitro* lipolysis model: I. Controlling the rate of lipolysis by continuous addition of calcium. *Eur. J. Pharm. Sci.*, **14**, 115–122.
209. Zangenberg, N. H., Müllertz, A., Gjelstrup Kristensen, H. and Hovgaard, L. (2001). A dynamic *in vitro* lipolysis model: II: Evaluation of the model. *Eur. J. Pharm. Sci.*, **14**, 237–244.
210. Larsen, A. T., Sassene, P. and Müllertz, A. (2011). *In vitro* lipolysis models as a tool for the characterization of oral lipid and surfactant based drug delivery systems. *Inter. J. Pharm.*, **417**, 245–255.
211. Dahan, A. and Hoffman, A. (2007). The effect of different lipid based formulations on the oral absorption of lipophilic drugs: The ability of *in vitro* lipolysis and consecutive *ex vivo* intestinal permeability data to predict *in vivo* bioavailability in rats. *Eur. J. Pharm. Biopharm.*, **67**, 96–105.
212. Christensen, J. Ø., Schultz, K., Mollgaard, B., Kristensen, H. G. and Mullertz, A. (2004). Solubilisation of poorly water-soluble drugs during *in vitro* lipolysis of medium- and long-chain triacylglycerols. *Eur. J. Pharm. Sci.*, **23**, 287–296.
213. Ibrahim, F., Gershkovich, P., Sivak, O., Wasan, E. K. and Wasan, K. M. (2012). Assessment of novel oral lipid-based formulations of amphotericin B using an *in vitro* lipolysis model. *Eur. J. Pharm. Sci.*, **46**, 323–328.
214. Porter, C. J. H., Kaukonen, A. M., Taillardat-Bertschinger, A., Boyd, B. J., O'Connor, J. M., Edwards, G. A. and Charman, W. N. (2004). Use of *in vitro* lipid digestion data to explain the *in vivo* performance of triglyceride-based oral lipid formulations of poorly water-soluble drugs: Studies with halofantrine. *J. Pharm. Sci.*, **93**, 1110–1121.
215. Dahan, A. and Hoffman, A. (2006). Use of a dynamic *in vitro* lipolysis model to rationalize oral formulation development for poor water soluble drugs: Correlation with *in vivo* data and the relationship to intra-enterocyte processes in rats. *Pharm. Res.*, **23**, 2165–2174.
216. Iwanaga, K., Ono, S., Narioka, K., Kakemi, M., Morimoto, K., Yamashita, S., Namba, Y. and Oku, N. (1999). Application of surface-coated liposomes for oral delivery of peptide: Effects of coating the liposome's surface on the GI transit of insulin. *J. Pharm. Sci.*, **88**, 248–252.

217. Li, H., Song, J.-H., Park, J.-S. and Han, K. (2003). Polyethylene glycol-coated liposomes for oral delivery of recombinant human epidermal growth factor. *Inter.J. Pharm.*, **258**, 11–19.
218. Rogers, J. A. and Anderson, K. E. (1998). The potential of liposomes in oral drug delivery. *Crit. Rev. Ther. Drug Carrier Syst.*, **15**, 421–480.
219. Woodley, J. F. (1985). Liposomes for oral administration of drugs. *Crit. Rev. Ther. Drug Carrier Syst.*, **2**, 1–18.
220. Kisel, M. A., Kulik, L. N., Tsybovsky, I. S., Vlasov, A. P., Vorob'yov, M. S., Kholodova, E. A. and Zabarovskaya, Z. V. (2001). Liposomes with phosphatidylethanol as a carrier for oral delivery of insulin: Studies in the rat. *Int. J. Pharm.*, **216**, 105–114.
221. Cho, Y. and Flynn, M. (1989). Oral delivery of insulin. *Lancet*, **334**, 1518–1519.
222. Choudhari, K. B., Labhasetwar, V. and Dorle, A. K. (1994). Liposomes as a carrier for oral administration of insulin: Effect of formulation factors. *J. Microencapsul.*, **11**, 319–325.
223. Dapergolas, G. and Gregoriadis, G. (1976). Hypoglycæmic effect of liposome-entrapped insulin administered intragastrically into rats. *Lancet*, **308**, 824–827.
224. Patel, H. M. and Ryman, B. E. (1976). Oral administration of insulin by encapsulation within liposomes. *FEBS Letters*, **62**, 60–63.
225. Iwanaga, K., Ono, S., Narioka, K., Morimoto, K., Kakemi, M., Yamashita, S., Nango, M., Oku, N. (1997). Oral delivery of insulin by using surface coating liposmes: Improvement of stability of insulin in GI tract. *Int. J. Pharm.*, **157**, 73–80.
226. Chen, H. and Langer, R. (1997). Magnetically-responsive polymerized liposomes as potential oral delivery vehicles. *Pharm. Res.*, **14**, 537–540.
227. Negrini, R. and Mezzenga, R. (2011). pH-responsive lyotropic liquid crystals for controlled drug delivery. *Langmuir*, **27**, 5296–5303.
228. Kaasgaard, T. and Drummond, C. J. (2006). Ordered 2D and 3D nanostructured amphiphile self-assembly materials stable in excess solvent. *Phys. Chem. Chem. Phys.*, **8**, 4957–4975.
229. Shah, J. C., Sadhale, Y. and Chilukuri, D. M. (2001). Cubic phase gels as drug delivery systems. *Adv. Drug Deliv. Rev.*, **47**, 229–250.
230. Burrows, R., Collett, J. H. and Attwood, D. (1994). The release of drugs from monoglyceride-water liquid crystalline phases. *Int. J. Pharm.*, **111**, 283–293.
231. Gan, L., Han, S., Shen, J., Zhu, J., Zhu, C., Zhang, X. and Gan, Y. (2010). Self-assembled liquid crystalline nanoparticles as a novel ophthalmic delivery system for dexamethasone: Improving preocular retention and ocular bioavailability. *Int. J. Pharm.*, **396**, 179–187.
232. Lesile, S. B., Puvvada, S., Ratna, B. R. and Rudolph, A. S. (1996). Encapsulation of hemoglobin in a bicontinuous cubic phase lipid. *Biochim. Biophys. Acta.*, **1285**, 246–254.
233. Sadhale, Y. and Shah, J. C. (1999). Stabilization of insulin against agitation-induced aggregation by the GMO cubic phase gel. *Int. J. Pharm.*, **191**, 51–64.
234. Chang, C. M. and Bodmeier, R. (1997). Effect of dissolution media and additives on the drug release from cubic phase delivery systems. *J. Control. Release*, **46**, 215–222.
235. Kwon, T. K. and Kim, J. C. (2011). Monoolein cubic phase containing acidic proteinoid: pH-dependent release. *Drug Dev. Ind. Pharm.*, **37**, 56–61.
236. Angelov, B., Angelova, A., Garamus, V. M., Lebas, G., Lesieur, S., Ollivon, M., Funari, S. S., Willumeit, R. and Couvreur, P. (2007). Small-angle neutron and X-ray scattering from

amphiphilic stimuli-responsive diamond-type bicontinuous cubic phase. *J. Am. Chem. Soc.*, **129**, 13474–13479.

237. Fong, W.-K., Hanley, T. L., Thierry, B., Kirby, N. and Boyd, B. J. (2010). Plasmonic nanorods provide reversible control over nanostructure of self-assembled drug delivery materials. *Langmuir*, **26**, 6136–6139.
238. Fong, W. K., Hanley, T. and Boyd, B. J. (2009). Stimuli responsive liquid crystals provide 'on-demand' drug delivery *in vitro* and *in vivo*. *J. Control. Release*, **135**, 218–226.
239. Kumar K., Shah, M. H., Ketkar, A., Mahadik, K. R. and Paradkar, A. (2004). Effect of drug solubility and different excipients on floating behaviour and release from glyceryl monooleate matrices. *Int. J. Pharm.*, **272**, 151–160.
240. Mouritsen, O.G. (2011) Lipids, curvature, and nano-medicine, *Eur. J. Lipid Sci. Technol.*, **113**, 1174–1187.
241. Charman, W. N. and Porter, C. J. H. (1996). Lipophilic prodrugs designed for intestinal lymphatic transport. *Adv. Drug Deliv. Rev.*, **19**, 149–169.
242. Wasan, K. M. and Cassidy, S. M. (1998). Role of plasma lipoproteins in modifying the biological activity of hydrophobic drugs. *J. Pharm. Sci.*, **87**, 411–424.
243. Charman, W. N. A. and Stella, V. J. (1986). Estimating the maximal potential for intestinal lymphatic transport of lipophilic drug molecules. *Int. J. Pharm.*, **34**, 175–178.
244. Gershkovich, P., Fanous, J., Qadri, B., Yacovan, A., Amselem, S. and Hoffman, A. (2009). The role of molecular physicochemical properties and apolipoproteins in association of drugs with triglyceride-rich lipoproteins: *In-silico* prediction of uptake by chylomicrons. *J. Pharm. Pharmacol.*, **61**, 31–39.
245. Gershkovich, P. and Hoffman, A. (2005). Uptake of lipophilic drugs by plasma derived isolated chylomicrons: Linear correlation with intestinal lymphatic bioavailability. *Eur. J. Pharm. Sci.*, **26**, 394–404.
246. Ipp, H. and Zemlin, A. (2013). The paradox of the immune response in HIV infection: When inflammation becomes harmful. *Clinica. Chimica. Acta.*, **416**, 96–99.
247. Glechner, A., Wöckel, A., Gartlehner, G., Thaler, K., Strobelberger, M., Griebler, U. and Kreienberg, R. (2013). Sentinel lymph node dissection only versus complete axillary lymph node dissection in early invasive breast cancer: A systematic review and meta-analysis. *Eur. J. Cancer*, **49**, 812–825.
248. Shackleford, D. M., Faassen, W. A., Houwing, N., Lass, H., Edwards, G. A., Porter, C. J. H. and Charman, W. N. (2003). Contribution of lymphatically transported testosterone undecanoate to the systemic exposure of testosterone after oral administration of two andriol formulations in conscious lymph duct-cannulated dogs. *J. Pharmacol. Exp. Ther.*, **306**, 925–933.
249. Caliph, S. M., Charman, W. N. A. and Porter, C. J. H. (2000). Effect of short-, medium-, and long-chain fatty acid-based vehicles on the absolute oral bioavailability and intestinal lymphatic transport of halofantrine and assessment of mass balance in lymph-cannulated and non-cannulated rats. *J. Pharm. Sci.*, **89**, 1073–1084.
250. Sheehe, D. M., Green, J. B. and Green, M. H. (1980). Influence of dietary fat saturation on lipid absorption in the rat. *Atherosclerosis*, **37**, 301–310.
251. Gershkovich, P., Qadri, B., Yacovan, A., Amselem, S. and Hoffman, A. (2007). Different impacts of intestinal lymphatic transport on the oral bioavailability of structurally similar synthetic lipophilic cannabinoids: Dexanabinol and PRS-211,220. *European J. Pharm. Sci.*, **31**, 298–305.

252. Khoo, S.-M., Shackleford, D., Porter, C. H., Edwards, G. and Charman, W. (2003). Intestinal lymphatic transport of halofantrine occurs after oral administration of a unit-dose lipid-based formulation to fasted dogs. *Pharm. Res.*, **20**, 1460–1465.
253. Palin, K. J. and Wilson, C. G. (1984). The effect of different oils on the absorption of probucol in the rat. *J. Pharm. Pharm.*, **36**, 641–643.
254. Noguchi, T., Charman, W. N. A. and Stella, V. J. (1985). Lymphatic appearance of DDT in thoracic or mesenteric lymph duct cannulated rats. *Int. J. Pharm.*, **24**, 185–192.
255. Gursoy, R. N. and Benita, S. (2004). Self-emulsifying drug delivery systems (SEDDS) for improved oral delivery of lipophilic drugs. *Biomed. Pharmacother.*, **58**, 173–182.
256. Elgart, A., Cherniakov, I., Aldouby, Y., Domb, A. and Hoffman, A. (2013). Improved oral bioavailability of bcs class 2 compounds by self nano-emulsifying drug delivery systems (SNEDDS): The underlying mechanisms for amiodarone and talinolol. *Pharm. Res.*, 1–16.
257. Sachs-Barrable, K., Lee, S. D., Wasan, E. K., Thornton, S. J. and Wasan, K. M. (2008). Enhancing drug absorption using lipids: A case study presenting the development and pharmacological evaluation of a novel lipid-based oral amphotericin B formulation for the treatment of systemic fungal infections. *Adv. Drug Deliv. Rev.*, **60**, 692–701.
258. Lawrence, M. J. and Rees, G. D. (2012). Microemulsion-based media as novel drug delivery systems. *Adv. Drug Deliv. Rev.*, **64,** 175–193.
259. Gershkovich, P., Wasan, E. K., Lin, M., Sivak, O., Leon, C. G., Clement, J. G. and Wasan, K. M. (2009). Pharmacokinetics and biodistribution of amphotericin B in rats following oral administration in a novel lipid-based formulation. *J. Antimicrob, Chemother.*, **64**, 101–108.
260. Wasan, E. K., Bartlett, K., Gershkovich, P., Sivak, O., Banno, B., Wong, Z., Gagnon, J., Gates, B., Leon, C. G. and Wasan K. M. (2009). Development and characterization of oral lipid-based Amphotericin B formulations with enhanced drug solubility, stability and antifungal activity in rats infected with *Aspergillus fumigatus* or Candida albicans. *Int. J. Pharm.*, **372**, 76–84.
261. Thornton, S. J. and Wasan, K. M. (2009). The reformulation of Amphotericin B for oral administration to treat systemic fungal infections and visceral leishmaniasis. *Expert Opin. Drug Deliv.*, **6**, 271–284.
262. Wasan, E. K., Gershkovich, P., Zhao, J., Zhu, X., Werbovetz, K., Tidwell, R. R., Clement, J. G., Thornton, S. J. and Wasan, K. M. (2010). A novel tropically stable oral amphotericin B formulation (iCo-010) exhibits efficacy against visceral leishmaniasis in a murine model. *PLoS Negl. Trop. Dis.*, **4**, e913.
263. Thornton, S. J., Wasan, K. M., Piecuch, A., Lynd, L. L. D. and Wasan, E. K. (2010). Barriers to treatment for visceral leishmaniasis in hyperendemic areas: India, Bangladesh, Nepal, Brazil and Sudan. *Drug Dev. Ind. Pharm.*, **36**, 1312–1319.
264. Kydonieus, A. F., Willie, J. J. and Murphy, G. F. (2000). 'Fundamental Concepts in transdermal delivery of drugs', in Kydonieus, A. F. (ed.), *Biochemical Modulation of Skin Reactions. Transdermal, Topicals, Cosmetics.* CRC Press, pp. 1–14.
265. Brown, M. B., Martin, G. P., Jones, S. A. and Akomeah, F. K. (2006). Dermal and transdermal drug delivery systems: Current and future prospects. *Drug Deliv.*, **13**, 175–187.

266. Jakubovic, H. R. and Ackerman, A. B. (1992). 'Structure and function of skin: Development, morphology, and physiology', in Moschella, S. L. and Hurley, H. J. (eds.), *Dermatology*, W. B. Saunders Company, Philadelphia, pp. 3–87.
267. El Maghraby, G. M., Barry, B. W. and Williams, A. C. (2008). Liposomes and skin: From drug delivery to model membranes. *Eur. J. Pharm. Sci.*, **34**, 203–222.
268. Elsayed, M. M. A., Abdallah, O. Y., Naggar, V. F. and Khalafallah, N. M. (2007). Lipid vesicles for skin delivery of drugs: Reviewing three decades of research. *Int. J. Pharm.*, **332**, 1–16.
269. Cevc, G., Schätzlein, A. G., Richardsen, H. and Vierl, U. (2003). Overcoming semipermeable barriers, such as the skin, with ultradeformable mixed lipid vesicles, transfersomes, liposomes, or mixed lipid micelles. *Langmuir*, **19**, 10753–10763.
270. Cevc, G., Schätzlein, A. and Richardsen, H. (2002). Ultradeformable lipid vesicles can penetrate the skin and other semi-permeable barriers unfragmented. Evidence from double label CLSM experiments and direct size measurements. *Biochim. Biophys. Acta (BBA) — Biomembranes*, **1564**, 21–30.
271. Cevc, G. and Gebauer, D. (2003). Hydration-driven transport of deformable lipid vesicles through fine pores and the skin barrier. *Biophysical Journal*, **84**, 1010–1024.
272. Cevc, G. (1996). Transfersomes, liposomes and other lipid suspensions on the skin: Permeation enhancement, vesicle penetration, and transdermal drug delivery. *Crit. Rev. Ther. Drug Carrier Syst.*, **13**, 257–388.
273. Touitou, E., Dayan, N., Bergelson, L., Godin, B. and Eliaz, M. (2000). Ethosomes — novel vesicular carriers for enhanced delivery: Characterization and skin penetration properties. *J. Control. Release*, **65**(3), 403–418.
274. Touitou, E., Godin, B. and Weiss, C. (2000). Enhanced delivery of drugs into and across the skin by ethosomal carriers. *Drug Develop. Res.*, **50**, 406–415.
275. Dayan, N. and Touitou, E. (2000). Carriers for skin delivery of trihexyphenidyl HCl: Ethosomes vs. liposomes. *Biomater.*, **21**, 1879–1885.
276. Horwitz, E., Pisanty, S., Czerninski, R., Helser, M., Eliav, E. and Touitou, E. (1999). A clinical evaluation of a novel liposomal carrier for acyclovir in the topical treatment of recurrent herpes labialis. *Oral Surg., Oral Med., Oral Pathol., Oral Radiol., Endod.*, **87**, 700–705.
277. Paolino, D., Lucania, G., Mardente, D., Alhaique, F. and Fresta, M. (2005). Ethosomes for skin delivery of ammonium glycyrrhizinate: *In vitro* percutaneous permeation through human skin and *in vivo* anti-inflammatory activity on human volunteers. *J. Control. Release*, **106**, 99–110.
278. Norlen, L. and Al-Amoudi, A. (2004). Stratum corneum keratin structure, function, and formation: The cubic rod-packing and membrane templating model. *J. Investig. Dermatol.*, **123**, 715–732.
279. Esposito, E., Cortesi, R., Drechsler, M., Paccamiccio, L., Mariani, P., Contado, C., Stellin, E., Menegatti, E., Bonina, F. and Puglia, C. (2005). Cubosome dispersions as delivery systems for percutaneous administration of indomethacin. *Pharm. Res.*, **22**, 2163–2173.
280. Nielsen, L. S., Schubert, L. and Hansen, J. (1998). Bioadhesive drug delivery systems: I. Characterisation of mucoadhesive properties of systems based on glyceryl mono-oleate and glyceryl monolinoleate. *Eur. J. Pharm. Sci.*, **6**, 231–239.

281. Geraghty, P. B., Attwood, D., Collett, J. H. and Dandiker, Y. (1996). The *in vitro* release of some antimuscarinic drugs from monoolein/water lyotropic liquid crystalline gels. *Pharm. Res.*, **13**, 1265–1271.
282. Norling, T., Lading, P., Engström, S., Larsson, K., Krog, N. and Nissen, S. S. (1992). Formulation of a drug delivery system based on a mixture of monoglycerides and triglycerides for use in the treatment of periodontal disease. *J. Clin. Periodontol.*, **19**, 687–692.
283. Müller, R. H., Radtke, M. and Wissing, S. A. (2002). Solid lipid nanoparticles (SLN) and nanostructured lipid carriers (NLC) in cosmetic and dermatological preparations. *Adv. Drug Deliv. Rev.*, **54**, S131–S155.
284. Wissing, S. A. and Müller, R. H. (2002). Solid lipid nanoparticles as carrier for sunscreens: *In vitro* release and *in vivo* skin penetration. *J. Control. Release*, **81**, 225–233.
285. Ricci, M., Puglia, C., Bonina, F., Giovanni, C. D., Giovagnoli, S. and Rossi, C. (2005). Evaluation of indomethacin percutaneous absorption from nanostructured lipid carriers (NLC): *In vitro* and *in vivo* studies. *J. Pharm. Sci.*, **94**, 1149–1159.
286. Mehnert, W. and Mäder, K. (2001). Solid lipid nanoparticles: Production, characterization and applications. *Adv. Drug Deliv. Rev.*, **47**, 165–196.
287. Diebold, Y., Jarrín, M., Sáez, V., Carvalho, E. L. S., Orea, M., Calonge, M., Seijo, B. and Alonso, M. J. (2007). Ocular drug delivery by liposome–chitosan nanoparticle complexes (LCS-NP). *Biomaterials*, **28**, 1553–1564.
288. Xu, Q., Kambhampati, S. P. and Kannan, R. M. (2013). Nanotechnology approaches for ocular drug delivery. *Middle East Afr. J. Ophthalmol.*, **20**, 26–37.
289. Alghadyan, A., Peyman, G., Khoobehi, B. and Liu, K.-R. (1988). Liposome-bound cyclosporine: Retinal toxicity after intravitreal injection. *Int. Ophthalmol.*, **12**, 105–107.
290. Liu, K. R., Peyman, G. A. and Khoobehi, B. (1989). Efficacy of liposome-bound amphotericin B for the treatment of experimental fungal endophthalmitis in rabbits. *Investigative Ophthalmology & Visual Science*, **30**, 1527–1534.
291. Ahmed, I. and Patton, T. F. (1987). Disposition of timolol and inulin in the rabbit eye following corneal versus non-corneal absorption. *Int. J. Pharm.*, **38**, 9–21.
292. Kaur, I. P., Garg, A., Singla, A. K. and Aggarwal, D. (2004). Vesicular systems in ocular drug delivery: An overview. *Int. J. Pharm.*, **269**, 1–14.
293. El-Gazayerly, O. N. and Hikal, A. H. (1997). Preparation and evaluation of acetazolamide liposomes as an ocular delivery system. *Int. J. Pharm.*, **158**, 121–127.
294. Shaikh, R., Singh, T. R. R., Garland, M. J., Woolfson, A. D. and Donnelly, R. F. (2011). Mucoadhesive drug delivery systems. *J. Pharm. Bioallied Sci.*, **3**, 89–100.
295. Li, N., Zhuang, C., Wang, M., Sun, X., Nie, S. and Pan, W. (2009). Liposome coated with low molecular weight chitosan and its potential use in ocular drug delivery. *Int. J. Pharm.*, **379**, 131–138.
296. Yerushalmi, N. and Margalit, R. (1998). Hyaluronic acid-modified bioadhesive liposomes as local drug depots: Effects of cellular and fluid dynamics on liposome retention at target sites. *Arch. Biochem. Biophys.*, **349**, 21–26.
297. Kaufman, H. E., Steinemann, T. L., Lehman, E., Thompson, H. W., Varnell, E. D., Jacob-LaBarre, J. T. and Gebhardt, B. M. (1994). Collagen-based drug delivery and artificial tears. *J. Ocul. Pharmacol.*, **10**, 17–27.

298. Yerushalmi, N. and Margalit, R. (1994). Bioadhesive, collagen-modified liposomes: Molecular and cellular level studies on the kinetics of drug release and on binding to cell monolayers. *Biochim. Biophys. Acta (BBA) — Biomembranes*, **1189**, 13–20.
299. Norley, S. G., Huang, L. and Rouse, B. T. (1986). Targeting of drug loaded immunoliposomes to herpes simplex virus infected corneal cells: An effective means of inhibiting virus replication *in vitro*. *J. Immunol.*, **136**, 681–685.
300. Norley, S. G., Sendele, D., Huang, L. and Rouse, B. T. (1987). Inhibition of herpes simplex virus replication in the mouse cornea by drug containing immunoliposomes. *Invest. Ophthalmol. Vis. Sci.*, **28**, 591–595.
301. Thakur, A., Fitzpatrick, S., Zaman, A., Kugathasan, K., Muirhead, B., Hortelano, G. and Sheardown, H. (2012). Strategies for ocular siRNA delivery: Potential and limitations of non-viral nanocarriers. *J. Biol. Eng.*, **6**, 7.
302. Liu, H.-A., Liu, Y.-L., Ma, Z.-Z., Wang, J.-C. and Zhang, Q. (2011). A lipid nanoparticle system improves siRNA efficacy in RPE cells and a laser-induced murine CNV model. *Invest. Ophthalmol. Vis. Sci.*, **52**, 4789–4794.
303. Li, X., Nie, S.-F., Kong, J., Li, N., Ju, C.-Y. and Pan, W.-S. (2008). A controlled-release ocular delivery system for ibuprofen based on nanostructured lipid carriers. *Int. J. Pharm*, **363**, 177–182.
304. Shen, J., Sun, M., Ping, Q., Ying, Z. and Liu, W. (2010). Incorporation of liquid lipid in lipid nanoparticles for ocular drug delivery enhancement. *Nanotechnol.*, **21**, 1–10.
305. Cavalli, R., Gasco, M. R., Chetoni, P., Burgalassi, S. and Saettone, M. F. (2002). Solid lipid nanoparticles (SLN) as ocular delivery system for tobramycin. *Int. J. Pharm.*, **238**, 241–245.
306. Shen, J., Deng, Y., Jin, X., Ping, Q., Su, Z. and Li, L. (2010). Thiolated nanostructured lipid carriers as a potential ocular drug delivery system for cyclosporine A: Improving *in vivo* ocular distribution. *Int. J. Pharm.*, **402**, 248–253.
307. Luo, Q., Zhao, J., Zhang, X. and Pan, W. (2011). Nanostructured lipid carrier (NLC) coated with Chitosan Oligosaccharides and its potential use in ocular drug delivery system. *Int. J. Pharm.*, **403**, 185–191.
308. del Pozo-Rodríguez, A., Delgado, D., Solinís, M. A., Gascón, A. R. and Pedraz, J.L. (2008). Solid lipid nanoparticles for retinal gene therapy: Transfection and intracellular trafficking in RPE cells. *Int. J. Pharm.*, **360**, 177–183.
309. Olbrich, C., Bakowsky, U., Lehr, C.-M., Müller, R. H. and Kneuer, C. (2001). Cationic solid-lipid nanoparticles can efficiently bind and transfect plasmid DNA. *J. Control. Release*, **77**, 345–355.
310. Zuckerman, J. N. (2000). The importance of injecting vaccines into muscle, *BMJ*, **321**, 1237–1238.
311. Morein, B. and Simons, K. (1985). Subunit vaccines against enveloped viruses: Virosomes, micelles and other protein complexes, *Vaccine*, **3**, 83–93.
312. Felnerova, D., Viret, J.-F., Glück, R. and Moser, C. (2004). Liposomes and virosomes as delivery systems for antigens, nucleic acids and drugs, *Curr. Opin. Biotechnol.*, **15**, 518–529.
313. The College of Physicians of Philadelphia (2014). The History of Vaccines, available at: http://www.historyofvaccines.org/ (Accessed February 2014).
314. Huckriede, A., Bungener, L., Daemen, T. and Wilschut, J. (2003). 'Influenza Virosomes in Vaccine Development', in Nejat, D. (ed.), *Methods in Enzymology*, Academic Press, pp. 74–91.

315. Almeida, J., Edwards, D. C., Brand, C. and Heath, T. (1975). Formation of virosomes from influenza subunits and liposomes, *Lancet*, **306**, 899–901.
316. Carr, C. M. and Kim, P. S. (1993). A spring-loaded mechanism for the conformational change of influenza hemagglutinin, *Cell*, **73**, 823–832.
317. Bullough, P. A., Hughson, F. M., Skehel, J. J. and Wiley, D. C. (1994). Structure of influenza haemagglutinin at the pH of membrane fusion, *Nature*, **371**, 37–43.
318. Glück, R., Mischler, R., Brantschen, S., Just, M., Althaus, B. andCryz, S. J., Jr. (1992). Immunopotentiating reconstituted influenza virus virosome vaccine delivery system for immunization against hepatitis A, *J. Clin. Invest.*, **90**, 2491–2495.
319. Waelti, E., Wegmann, N., Schwaninger, R., Wetterwald, A., Wingenfeld, C., Rothen-Rutishauser, B. and Gimmi, C. D. (2002). Targeting HER-2/neu with Antirat Neu Virosomes for Cancer Therapy. *Cancer Res.*, **62**, 437–444.
320. Daemen, T., de Mare, A. Bungener, L., de Jonge, J., Huckriede, A. and Wilschut, J. (2005). Virosomes for antigen and DNA delivery. *Adv. Drug Deliv. Rev.*, **57**, 451–463.

Chapter 11

Smart Polymer-Based Nanomedicines

Jonathan M. Hartley and Jindřich Kopeček

Departments of Bioengineering, and of Pharmaceutics and Pharmaceutical Chemistry, University of Utah, Salt Lake City, UT, USA

11.1 Polymer Therapeutics Design and Synthesis

The treatment of deadly pathologies is a high priority among medical doctors and scientists. The pathologies are complicated and treating them can be even more so. This complexity has prompted the development of different classes of therapeutics, such as nucleic acids, proteins, and carbohydrates [1]. However, these molecules suffer several limitations to their efficacy, namely: poor stability in plasma, poor biodistribution, poor solubility, and/or adverse side effects. To overcome these limitations, smart therapeutic systems have been designed to deliver these promising molecules to the site of action. Smart polymer-based therapeutics have been studied extensively in delivering therapeutics [2–5]. Synthetic polymers have been designed in an effort to mimic the functionality of natural polymers; they have been given names such as stimuli sensitive, smart, and intelligent polymers. Of course, these materials do not have an intelligence quotient (IQ), but they undergo rapid physical or chemical changes in response to stimuli [6]. Smart polymers have application in many biomedical fields, such as tissue engineering, cell–sheet engineering, hydrogel systems, and pharmaceuticals. Therapeutics with multiple functionalities and architectures are now possible due to improved polymerization and bioconjugate chemistry techniques.

In this chapter, we will review polymer-based intelligent systems in the context of the architecture of the nanomedicine. The nanomedicine architectures we will review are linear polymers, micelles, polymersomes, nanogels, dendritic polymers, polyplexes, and polymer–protein conjugates. We will review smart therapeutics that respond to stimuli such as pH, temperature, biological factors (enzyme, ligand), ultrasound, magnetism, and light. Nanomedicines often employ sensitivities to multiple stimuli in order to overcome the major barriers to successful delivery [7–9]. In order for a therapeutic to arrive at the desired location and exert its desired function, it must avoid being captured in the circulation by the reticuloendothelial (RES) system, extravasate into the target tissue, enter the cells (if the target is intracellular), traverse the nuclear membrane (if the target is deoxyribonucleic acid, DNA, or other nuclear enzyme), and release the drug. Yet even still, a successful design that overcomes one barrier may still render a therapeutic ineffective due to other challenges, such as immunogenicity, toxicity, or manufacturing complexities [6]. The design of polymer therapeutics must also consider regulatory barriers to clinical approval. The clinical success of smart polymer-based therapeutics will be discussed briefly at the end of this chapter with suggestions on future design and characterization of polymer nanomedicines.

11.1.1 *Polymer principles*

The word polymer originates from the Greek word "polymeres" meaning "having many parts" [10]. The many parts of a polymer are the monomer units that make up the simplest repeating units in the polymer chain. The number of repeating monomers is referred to as the degree of polymerization. A molecule with a degree of polymerization between two and 20 is usually referred to as an oligomer, while larger molecules are called polymers [10].

Two major polymer classifications are relevant to our discussion of smart polymer based nanomedicines: molecular structure and chemical structure [10]. Molecular structure aims to classify polymers based on configurational differences, such as linear, branched, graft, and star (Fig. 11.1). Terms such as homopolymer, copolymer, and terpolymer are used to describe polymers with identical monomers or mixtures of monomers. Chemical structure refers to certain chemical groups present in the polymer and the method of polymer synthesis.

Depending on the molecular and chemical structures of the polymers they may self-assemble into micelles or polymer vesicles (polymersomes) (Fig. 11.1) [11]. Micellar self-assemblies can be cross-linked in the core to provide greater stability in circulation. These cross-linked structures are called nanogels or shell cross-linked micelles [12].

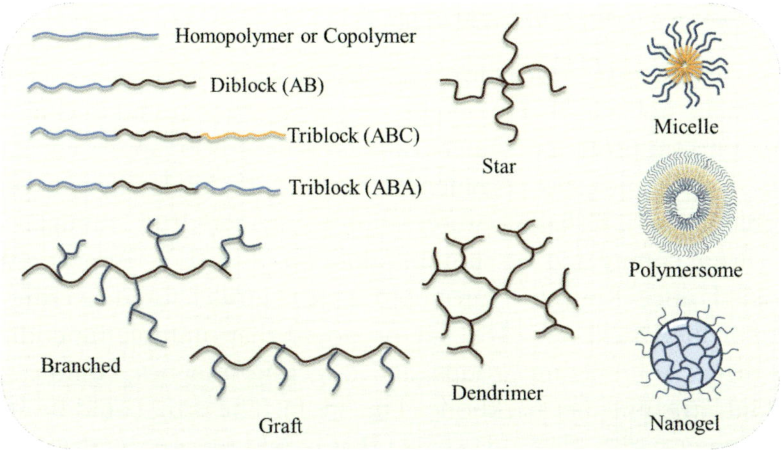

Figure 11.1. A selection of common polymer architectures and self-assembled structures used to deliver low molecular weight (MW) drugs, peptides, proteins, or genetic material. Homopolymers, copolymers, and diblock and triblock copolymers are linear polymers. Branched, graft, star, and dendritic polymer structures are higher order structures. Micelles, polymersomes, and nanogels comprise self-assembled linear polymers often of diblock or triblock nature.

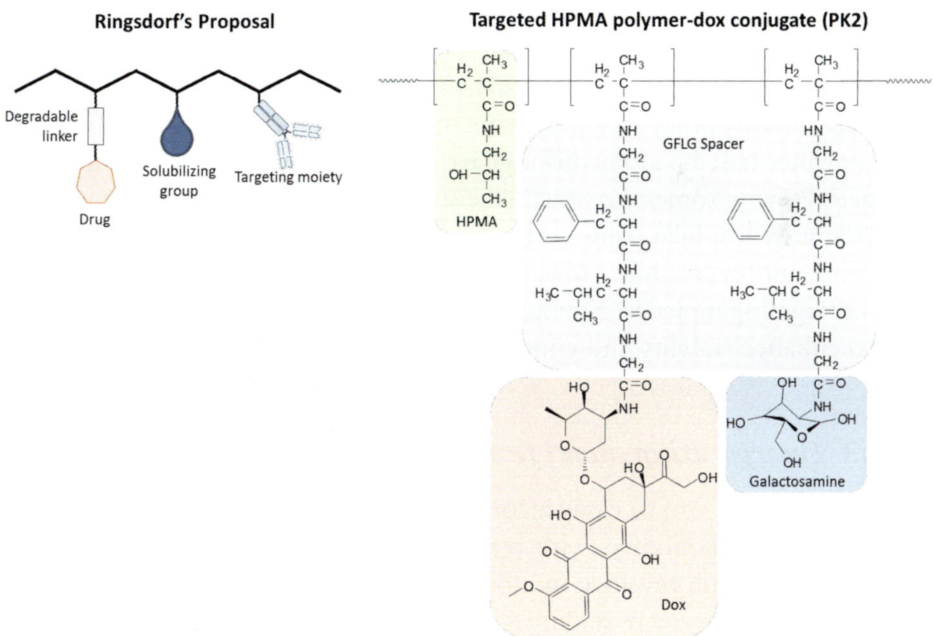

Figure 11.2. Representation of the drug delivery concept presented by Ringsdorf and the clinically tested HPMA copolymer-doxorubicin (Dox) conjugate PK2 comprising a drug (Dox) tethered to a polymer backbone via the enzyme degradable sequence GFLG and a liver-targeting moiety, *N*-acylated galactosamine. HPMA = *N*-(2-hydroxypropyl)methacrylamide.

11.1.2 *Early polymer therapeutics*

Polymer therapeutics have shown considerable promise in reaching the clinic, as many polymer therapeutics have already been tested in clinical trials and over a dozen protein polymer conjugates have been approved. Work on polymer–drug conjugates commenced with work by Jatzkewitz (enzyme degradable linkers), Ushakov (water soluble polymer–drug conjugates), and Mathé (drug conjugated to immunoglobulin) [13–17]. Almost 40 years have passed since Ringsdorf proposed a new model for delivering drugs using a polymer backbone [18]. He proposed that multiple functionalities, such as drugs, solubilizing groups, and targeting moieties, could be incorporated into the polymer backbone (Fig. 11.1). The drug could be attached to the polymer with a cleavable spacer that would release the drug in a specific location. Targeting moieties could also be incorporated to direct the polymer therapeutic to the desired location. Kopeček and coworkers have studied polymer therapeutics based on *N*-(2-hydroxypropyl)methacrylamide (HPMA) extensively, inspired by Ringsdorf's model [19]. The monomer HPMA was designed and synthesized in the early 1970s for use in drug delivery [19–23]. Early work on macromolecular therapeutics has been reviewed [24, 25].

Based on a stringent biological rationale, new polymer-based nanomedicine designs have emerged: polymeric drugs, polymer–drug conjugates, polymer–protein conjugates, polymeric micelles, and polyplexes [1]. Recently, other polymer-based nanomedicines have appeared, such as polymersomes and nanogels. Polymersomes are vesicle-like structures comprising an aqueous core for loading hydrophilic drugs, a hydrophobic shell for loading hydrophobic molecules, and hydrophilic corona for prolonging circulation and/or attachment of targeting moieties. Nanogels are essentially micelles with a cross-linked core to enhance stability and control size.

11.1.3 *Incorporating smart functionality*

Smart functionality can be easily incorporated into synthetic polymers by simply co-polymerizing monomers with reactive functionality, pH-sensitive properties, or other stimuli-sensitive monomers [2]. Stimuli-sensitive block copolymers can be synthesized by adding a temperature/pH-sensitive polymer block. Figure 11.4 depicts various chemical and physical stimuli used in the design of nanomedicines. Internal or chemical stimuli are present in the body, such

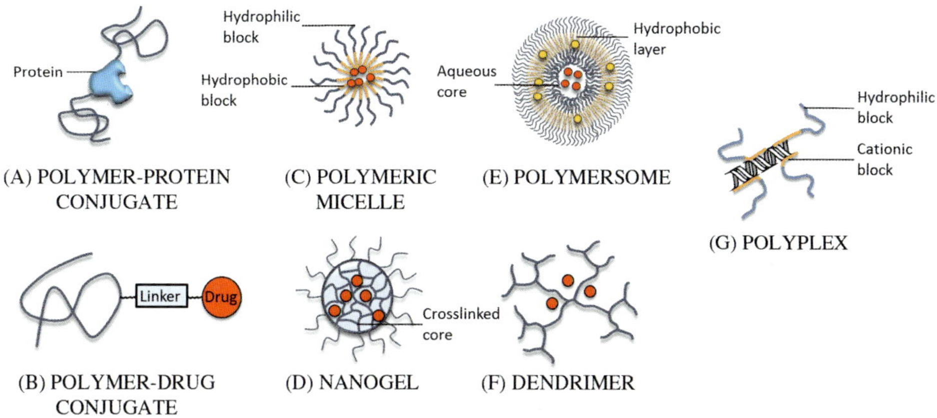

Figure 11.3. Representation of polymer-based nanomedicines being investigated for treatment of cancer and other diseases. (A) synthetic polymers such as polyethylene glycol (PEG) are attached to proteins to make a polymer–protein conjugate. (B) polymeric–drug conjugates comprise a hydrophilic polymer backbone with active agents conjugated via degradable linkers to the backbone. (C) micelles self-assemble from diblock or triblock copolymers comprising hydrophilic and hydrophobic polymer blocks. (D) nanogels are core cross-linked micelles where the core can swell or contract depending on the polymer used. (E) polymersomes comprise diblock or triblock polymers that form vesicle-like structures. (F) dendrimers can sequester drugs in the core or the drugs can be covalently attached. (G) polyplexes comprise a cationic block (for deoxyribonucleic acid, DNA, short interfering ribonucleic acid, siRNA, micro ribonucleic acid, miRNA, complexation) and hydrophilic block.

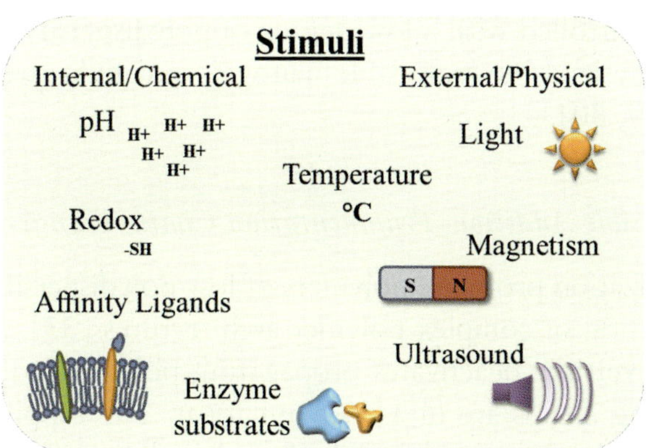

Figure 11.4. Stimuli used to control the targeting and release of active molecules from polymer-based nanomedicines. Internal and external are two classes of stimuli used in therapeutic design.

as pH and redox gradients. Biorecognition is needed in some nanomedicine designs to direct the therapeutic to the desired tissue/cell/organelle [4]. External or physical stimuli are often applied outside the body.

11.1.4 *Synthesis strategies*

The development of controlled/living radical polymerization (CRP) techniques expanded the available design possibilities of smart polymer therapeutics. Additionally, CRP techniques produce polymers with low polydispersities, which is critical for proper pre-clinical characterization and evaluation, as the MW determines the route of elimination and the *in vivo* half-life of the polymer [26]. These new polymerization techniques make it technologically possible to blend polymers that have different stimuli-sensitive properties to produce or tune new functionalities [27]. Bioresponsive functionality and biodegradability can be incorporated into polymers by incorporating natural polymer grafts or blocks into the polymer.

11.1.4.1 *Atom Transfer Radical Polymerization (ATRP)*

In ATRP, a transition metal catalyst reacts with a halogen atom on an organic halide initiator to produce an oxidized metal–halide and a carbon-centered radical. Alkene monomers react with the carbon-centered radical to produce carbon–carbon bonds. The metal–halide complex can react with the propagating chain and suppress bimolecular termination [28]. In this way, polymer MW can be highly controlled with a low degree of polydispersity. Matyjaszewski's group has synthesized biodegradable nanogels and miktoarm star polymers using ATRP [29, 30].

11.1.4.2 *Reversible Addition–Fragmentation Chain-Transfer (RAFT)*

RAFT polymerization produces polymers of low polydispersity and can facilitate the production of complex polymer architectures [33]. A chain transfer agent (CTA) reversibly deactivates propagating polymer radicals so that the majority of living chains are in a dormant form. The capping reactions are rapid and equilibrium is quickly reached, which allows all chains to grow at essentially the same rate, producing narrow MW distributions [33, 34]. RAFT polymerization allows for a large degree of freedom in architectural design and the incorporation of appropriate functionalities. For applications in drug delivery, RAFT polymerization is more versatile in that a greater variety of vinyl monomers are available for reaction in water and functionality can be easily incorporated into the α and ω ends of the polymer for later

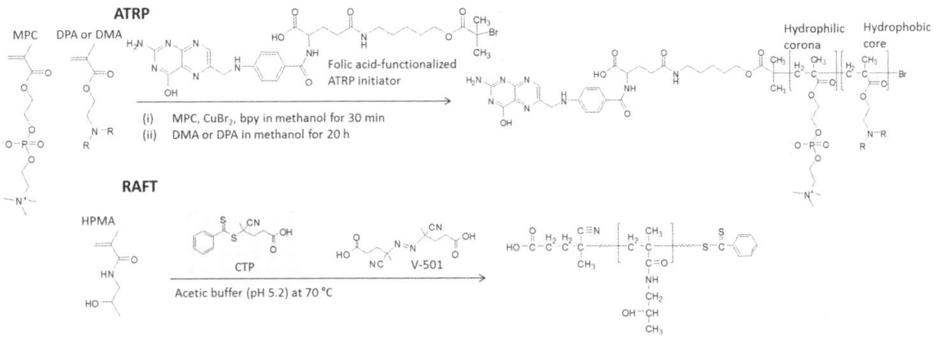

Figure 11.5. Reaction schemes for the RAFT synthesis of polyHPMA and the ATRP synthesis of a folic acid functionalized diblock copolymer [31, 32]. In RAFT synthesis, the monomers add between the sulfur and tertiary carbon. The CTA in this RAFT synthesis is 4-cyanopentanoic acid dithiobenzoate (CTP); the initiator is 4,4′-azobis(4-cyanopentanoic acid) (V-501). In the ATRP synthesis, the monomers add between the bromine and the tertiary carbon. The scheme depicts the synthesis of a diblock copolymer of 2-(methacryloyloxy) ethyl phosphorylcholine (MPC) and the pH sensitive 2-(diisopropylamino)ethyl methacrylate (DPA). ATRP = Atom Transfer Radical Polymerization, RAFT = Reversible Addition–Fragmentation Chain-Transfer.

bioconjugation [35]. CTA can be designed with enzyme-degradable or pH-sensitive structures for smart delivery of therapeutics. RAFT polymerization allows for the creative design of nanopolymers with multiple blocks of differing properties or functional groups.

RAFT polymerization holds several advantages over ATRP, namely: the absence of toxic metal species in polymerization, versatility in CTA or macroCTA use for incorporating biopolymers and functional groups, and the availability of a wide range of monomers for RAFT polymerization [35]. While the thiocarbonylthio moiety of the CTA in RAFT has been found to be toxic, the group can be easily removed by post-polymerization chain end modification [36, 37].

RAFT polymerization has led to further development of poly(*N*-(2-hydroxypropyl)methacrylamide) (pHPMA). In 2005, pHPMA was successfully synthesized using RAFT polymerization [31]. As a result, a second generation of polymer therapeutics has emerged, which will be discussed later.

11.2 Stimuli

11.2.1 *Temperature*

In 1967, Scarpa *et al.* reported the thermal transition behavior of poly(*N*-isopropylacrylamide) (PNIPAM) [38]. This polymer exhibits a lower critical solution temperature (LCST), below which the pendant groups form hydrogen bonds with surrounding water molecules. As the temperature rises, the hydrogen bonding between the water molecules and the pendant groups on the polymer become disrupted and the polymer becomes insoluble. At lower temperatures, water molecules form clathrates around the pendant groups on PNIPAM and as the temperature rises, the polymer chains aggregate, leading to the release of the highly ordered water molecules and, therefore, an increase in entropy. The transition from soluble to insoluble was found to be independent of polymer MW [39]. The polymer PNIPAM has a transition temperature near 37°C, making it an attractive material to use in biomedical applications. Additionally, the transition temperature can be tuned by incorporating either hydrophilic or hydrophobic co-monomers into polymerization [40]. Incorporating hydrophobic monomers into polymerization yields a polymer with a lower LCST; conversely, adding hydrophilic monomers raises the LCST. The rate at which the smart polymer systems constructed from PNIPAM respond to temperature changes depends partially upon the size of the system; for example, nanosized systems respond faster to temperature changes than micro-sized systems [41]. PNIPAM has been heavily studied because the LCST is near body temperature. Thermo-sensitive polymers have been used for applications in topical gels, injectable scaffolds for tissue engineering, and cell–sheet engineering [2].

Other temperature sensitive polymers have been studied besides PNIPAM, such as poly(*N*-vinyl caprolactam) (PVCa), which also has an LCST near body temperature and is biocompatible [42].

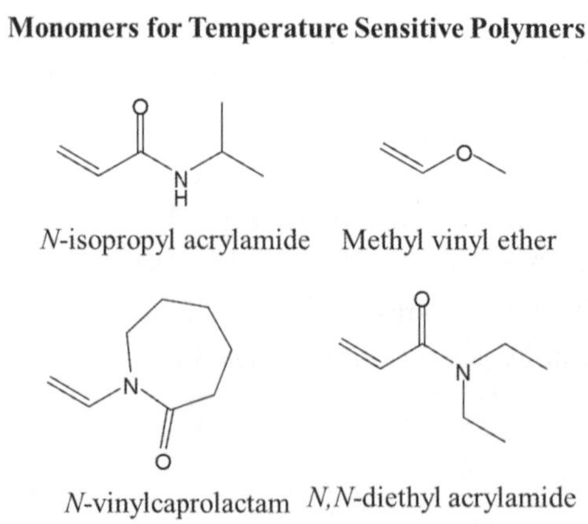

Figure 11.6. Chemical structures of common monomers used to synthesize temperature sensitive polymers.

11.2.2 *pH*

The body consists of a complex arrangement of tissue and sub-cellular compartments between which the biological conditions can be drastically different. One of those conditions is the fluid pH, which can range from a pH of 2.0 in the stomach to a pH of 5.0–8.0 within the small intestine. Not only are there normal pH gradients in the body, but pH changes can arise in pathological situations. For example, tumor tissue has a slightly acidic extracellular environment with a pH of 6.5–7.2 [43, 44]. pH gradients are also encountered in sub-cellular compartments, such as the early endosomal (pH 6.0–6.5), late endosomal (pH 5.0–6.0), and lysosomal (pH 4.5– 5.0) compartments (Fig. 11.7) [45, 46]. After endocytosis, the pH decreases from 7.4 to as low as 5.0 in the late endosome. This results in the disruption of ligand–receptor interactions so that receptors can be recycled. The lowering of pH in the endosomal and lysosomal compartments and the pH changes in pathological tissue can be exploited for smart delivery of therapeutic molecules.

Polymers can be synthesized to incorporate pendant acidic or basic chemical groups that have acid dissociation constant (pK_a) values between 3 and 10 [2]. Small changes in pH result in an altered fraction of ionized groups on the polymer backbone, which affects polymer solubility and conformation. The side chains become protonated, and the interaction between the polymer chain and itself becomes more favorable than the interaction between the backbone and the surrounding solvent.

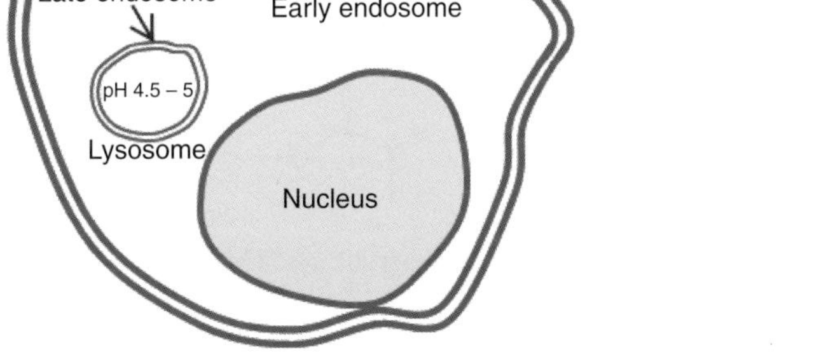

Figure 11.7. pH changes from outside the cell to the lysosome.

The selection of a pH-sensitive linker depends on the ultimate target of the drug. Some drugs need to be released extracellularly to act on membrane proteins, while others need to be released inside the cell to bypass multi-drug resistant pathways. Linkers that are pH-sensitive, such as Schiff's base, hydrazone, esters, and *cis*-aconityl, have been used to conjugate drugs to polymer backbones and facilitate drug release through pH-controlled hydrolysis [2, 47]. Because hydrolysis is the mechanism of degradation of pH-sensitive linkers, the drugs may be released from the carrier prematurely into the blood stream. Choi *et al.* conjugated Dox to a polyHPMA-*co*-APMA polymer via a pH-sensitive *cis*-aconityl bond [48]. The strongest release of Dox occurred at pH 5.0, while minimal release was observed at pH 7.0.

Cationic polymers respond to a low pH environment by absorbing protons in a process termed the "proton sponge effect". These polymers absorb protons, while also acting as a sink for other counter ions. The increase in ions on the endosome results in an increased osmotic pressure that results in membrane rupture and the delivery of cargo to the cytosol of the cell. Figure 11.8 shows some common cationic polymers used in gene delivery,

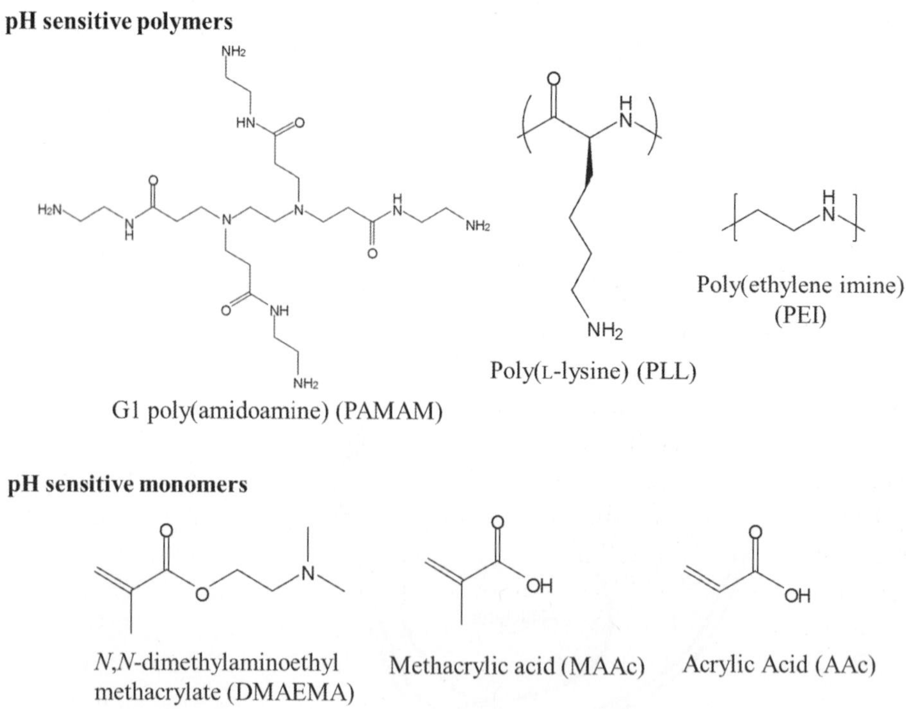

Figure 11.8. Common polymers and monomers used to produce pH-sensitive polymers for delivering genes or drugs.

such as poly(amidoamine) (PAMAM), poly(L-lysine) (PLL), and poly(ethylene imine) (PEI). These polymers can effectively complex with negatively-charged genetic material.

11.2.3 *Light*

Light-sensitive systems can be synthesized by incorporating photo-convertible or photo-labile groups into pendant groups of a polymer chain. A common light-sensitive group used in light-sensitive polymers is azobenzene, which changes between a *cis* and *trans* conformation upon irradiation [49, 50]. The *trans* form is more hydrophobic than the *cis* form, and this difference has been exploited to produce light-sensitive micelles.

Photo-labile groups, such as pyrene and 2-nitrobenzyl moieties, can be cleaved to remove hydrophobic pendant groups to alter the solubility of polymer chains [51]. Andrus *et al.* synthesized a block copolymer containing polyethylene oxide (PEO) as the hydrophilic block and poly(2-nitrobenzyl methacrylate) as the hydrophobic block. Photolysis of 2-nitrobenzyl occurs via either one-photon ultraviolet (UV) or two-photon near infrared at 365 and 700 nm, respectively. Upon irradiation, the 2-nitrobenzyl groups are cleaved off, thereby converting the polymer block to hydrophilic poly(methyl) methacrylate. Once the 2-nitrobenzyl groups are cleaved off, the micelles dissociate irreversibly.

11.2.4 *Reduction*

Interest in redox-sensitive materials has grown because of the ability to use them in the delivery of biomolecules to intracellular compartments. The disulfide is the most common redox-sensitive bond that is incorporated into the polymer therapeutics. The disulfide bond is relatively stable in the blood, but is quickly cleaved (minutes to hours) in a thiol–disulfide exchange reaction [52]. Unlike hydrolytically degraded polymers such as polyesters — which could take many days to degrade — disulfide bonds degrade very quickly in the cell. Glutathione tripeptide (γ-glutamyl-cysteinyl-glycine, GSH) and disulfide GSH (GSSG) exist in non-equilibrium potentials in various cellular compartments [53]. Glutathione reductase and nicotinamide adenine dinucleotide phosphate (NADPH) maintain concentrations of 0.5–10 mM GSH within the cell. The difference in concentration of reducing agents across the cell membrane makes it an attractive stimulus, which can be built into polymeric delivery systems for triggered release or biodegradability. Disulfide bonds were used to attach photo-sensitizer mesochlorin e_6 to HPMA copolymer [54] or short interfering ribonucleic acid (siRNA) to poly(vinyl ether) copolymers [55].

11.2.5 *Ultrasound*

Ultrasound waves are most often produced by piezoelectric transducers and have frequencies greater than 20 kHz [56]. Ultrasound waves can be refracted, reflected, and focused through a medium, much like optical and sound waves . Ultrasound is a pressure wave that moves through a medium, and in the case of drug delivery, through human tissue. The pressure wave causes an expansion and compression of the transmitting medium, which can stress cells and tissues, yet leave cell membranes unharmed unless gas bubbles are present [60, 61]. Ultrasound's impact on tissues can be divided into two categories: thermal effects and non-thermal effects. The absorption of ultrasound energy by cells and tissues can increase the local temperature in the area of application. Non-thermal effects include cavitating bubbles, radiation pressure, acoustic streaming, and radiation torque. Not only can ultrasound be used to disrupt a nanomedicine to cause release, but ultrasound can also facilitate the transport of drugs across cell membranes [62, 63].

11.3 Biorecognition/Biological

The complexity of the tumor environment has made it necessary to develop more smart targeting strategies with polymer therapeutics. Of particular interest are therapeutics that can limit recurrence and metastasis of tumors. The tumor comprises heterogeneous cell populations, some of which could be cancer stem cells with self-renewing ability, which would then cause recurrence of the tumor [64–67]. Selective treatment with a combination of therapeutics is necessary to effectively treat a tumor with heterogeneous cell populations.

It is important to consider several factors in designing an enzyme-sensitive delivery system, such as charge of the carrier, expression of the enzyme, and accessibility of the enzyme degradable sequence. In the case of carrier charge, a negatively-charged dextran carrier has slower degradation kinetics compared to a neutral dextran molecule degraded by similarly charged matrix metalloproteinases (MMPs) [68]. Early polymer–drug conjugates were synthesized containing a Gly–Phe–Leu–Gly (glycine, Gly; phenylalanine, Phe; leucine, Leu) peptide sequence that is a substrate for the thiol-dependent cysteine protease cathepsin B in the lysosomal compartment [19, 69]. Once the drug is internalized via pinocytosis, the conjugate is trafficked to the lysosomes where the drug is cleaved from the polymer.

11.3.1 *Passive targeting*

Passive targeting refers to the accumulation of macromolecules into tissue regions with leaky capillary vasculature — otherwise known as the enhanced permeation and retention (EPR) effect. This effect was discovered by Matsumura and Maeda in 1986 [70]. The vasculature of tumors is especially chaotic and unorganized, which results in gaps between endothelial cells through which nanomedicines may extravasate. Once extravasted into the tumor or inflamed area, the macromolecules tend to become trapped or retained in the tumor space due to poor lymphatic drainage. Nanomedicines designed to treat solid tumors rely on the EPR effect for accumulation at the tumor site [71]. Particles between ~10 and 500 nm in size or ~40 kDa in mass tend to accumulate in the tumor space [71, 72]. The pore cut-off size in the tumor vasculature depends upon the tumor type [73]. Nanomedicines need to circulate for long periods of time to effectively accumulate into the tumor tissue. Consequently, increased MW of the carrier results in higher accumulation in solid tumors [74]. More detailed analyses and reviews of the EPR effect have been published [71, 72, 75].

11.3.2 *Active targeting*

A major goal in polymer therapeutic design is to minimize adverse effects on healthy tissue. One way to minimize interaction with healthy tissue is to place ligands on therapeutics that bind to a specific tissue or cell. A targeting moiety often comes from three different categories: angiogenesis-targeting, cell proliferation-targeting, or tumor cell (or other target cell)-targeting [76].

The type of targeting moiety affects its efficacy. For example, liposomes conjugated to whole monoclonal antibodies (mAbs) accumulated to a lesser extent than liposomes conjugated to antibody fragments (Fab') due to lower circulation time of the whole mAb targeted particle [77]. In addition to the type of targeting molecule, the affinity is important as well. Adams *et al.* found that high affinity single-chain variable fragment (scFV) molecules with a high affinity for HER-2/*neu* could not penetrate tumors as well as lower affinity molecules [78]. This result was initially postulated by Weinstein *et al.*, who termed it the "binding site barrier effect" [79].

A case where active targeting is desirable is in the eradication of cancerous B cells. Branched [80, 81] and linear [82] HPMA copolymers grafted with multiple Fab' fragments of the anti-CD20 antibody 1F5 to the polymer have been evaluated [80]. The conjugates bind to the CD20 antigen on the B cell surface and cross-link the antigen. CD20 cross-linking induces caspase-dependent apoptosis in B cells. Polymer conjugates with a varying number of attached Fab' were synthesized. Conjugates with a higher number of Fab' induced higher levels of apoptosis in Raji B cells.

11.3.3 *Selective targeting*

Because the tumor environment is heterogeneous with multiple cell types, the targeting of different cells within a single cancer or disease is becoming more important, so as to make the treatment more effective. Tumor masses may contain malignant stem cells that are resistant to traditional treatments and are responsible for the recurrence of the cancer. Therefore, it is necessary to target multiple cell types using a combination of drugs.

A selective approach was taken in targeting cancer stem cells present in prostate tumors without using targeting ligands [83, 84]. More specifically, a Hedgehog (Hh) pathway inhibitor was used, since the Hh pathway is critical for stem cell development and plays a role in the recurrence of prostate cancer [85, 86]. Two conjugates were synthesized, each bearing either cyclopamine (CYP) or docetaxel (DTX). RAFT polymerization was used to produce narrow MW distribution copolymers of HPMA and MA-GFLG-TT (TT = thiazolidine-2-thione). A polymer P-GFLG-TT of 50 kDa with a PDI of 1.07 was synthesized, and CYP was later attached to the polymer. The conjugate showed a selective inhibitory effect on cancer stem cells in an RC-92a/hTERT cell line [83]. An *in vivo* model of prostate cancer was used to test the efficacy of combination treatment with two polymer–drug conjugates with enzyme-degradable linkers [84]. One conjugate was directed towards the bulk tumor cells (P-DTX) and the other was directed towards cancer stem cells (P-CYP). The combination treatment showed enhanced anti-tumor effects in PC-3 prostate cancer xenografts compared to single drug treatment. The treatment showed a durable anticancer effect even after treatment was stopped. Accounting for tumor heterogeneity in the conjugate design resulted in an improved treatment for prostate cancer in mice.

11.4 Structures of Polymer-Based Nanomedicines

11.4.1 *Linear polymers*

Smart linear polymer nanomedicines usually employ stimuli-sensitive linkers to release a drug at a specific location. Smart polymer nanomedicines have been studied for the delivery of genetic material, where the polymer backbone contains pH-sensitive pendant groups for endosomal escape. Endosomal escape can also be achieved by incorporating reduction-sensitive moieties that are cleaved once inside the cell, thereby exposing membrane destabilizing groups or simply raising the osmotic pressure. Smart polymer therapeutics are especially needed for the delivery of siRNA, miRNA, or DNA, since multiple barriers need to be traversed before arriving at their target location. Water-soluble polymers used to delivery therapeutics can be generally classified into either biologically-inert or biologically-active.

For biologically inert polymers such as HPMA to be effective, the polymers should incorporate biorecognizable functionalities. Biorecognition at the plasma membrane and in the target sub-cellular compartment are essential for reaching target cells and releasing the drug, respectively. The clinically-tested polymer conjugates PK1 and PK2 each consisted of the enzyme-degradable linker GFLG between the polymer and the drug, and the PK2 conjugate employed the liver hepatocyte (asialoglycoprotein receptor) targeting moiety, *N*-acylated galactosamine. Results from clinical trials showed that the attachment of low MW drugs to polymers decreases the drugs' adverse effects, and that the conjugates were biocompatible [87, 88]. However, the efficacy of these conjugates did not match data from preclinical animal experiments. In order to increase the efficacy of the treatment, more conjugate accumulation is needed at the tumor site via the EPR effect. The amount of polymer that can accumulate in the tumor depends largely on the blood circulation time. Larger MW polymers circulate for longer times in the blood. The PK1 and PK2 conjugates had an HPMA Mw below the renal threshold to ensure elimination from the body.

To improve the intravascular half-life of HPMA conjugates, a second generation HPMA conjugate was designed that incorporates the enzyme-degradable spacer GFLG into the polymer backbone [89–91]. Backbone-degradable polymers were synthesized following the scheme shown in Fig. 11.9. The synthetic process consists of two major steps: first, a telechelic HPMA copolymer is synthesized by RAFT polymerization, followed in the second step by chain extension using alkyne–azide [89, 90] or thiol–ene [91] click reactions. In addition, we synthesized a new RAFT CTA, N^{α},N^{ε}-bis(4-cyano-4-(phenylcarbonothioyl thio)pentanoylglycylphenylalanylleucylglycyl)lysine (Peptide2CTA), containing an enzymatically-degradable (GFLGKGLFG) spacer capped at both ends with 4-cyano-4-(phenylcarbonothioylthio)pentanoate [91]. During RAFT polymerization, the HPMA monomers incorporate at both dithiobenzoate groups of the Peptide2CTA with identical efficiency. Thus, it is possible to prepare degradable diblock copolymers with narrow MW distribution in one step, eliminating the chain extension reaction [91].

Multiblock polyHPMAs with MW as high as 300 kDa and containing degradable oligopeptide sequences were obtained by chain extension followed

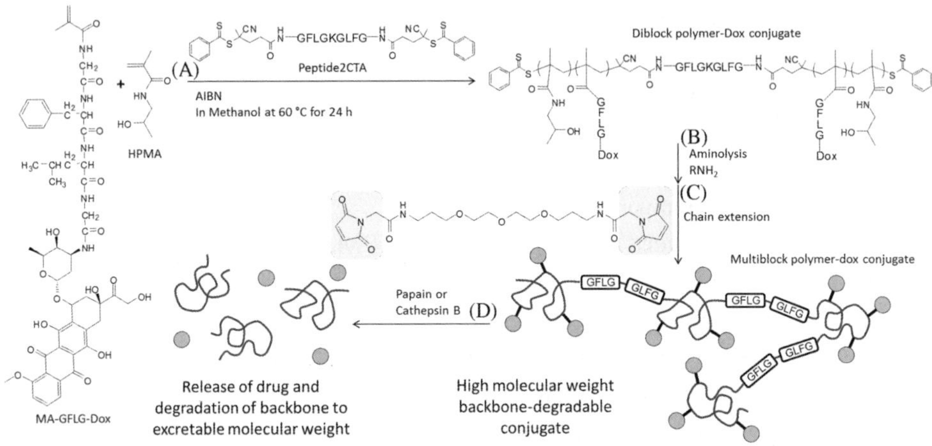

Figure 11.9. Scheme for the synthesis of diblock and multiblock polyHPMA–Dox conjugates. (A) MA-GFLG-Dox and HPMA copolymerized with the CTA Peptide2CTA to produce a degradable diblock conjugate. (B) aminolysis to remove the dithiobenzoate groups. (C) chain extension using a bifunctional chain extender to produce multiblock polymer conjugates. (D) degradation of the multiblock conjugate releasing free drug and degrading the backbone into fragments with MW below the renal threshold. HPMA = *N*-(2-hydroxypropyl) methacrylamide, AIBN = azobisisobutyronitrile, Peptide2CTA = N^{α},N^{ε}-bis(4-cyano-4-(phenylcarbonothioylthio)pentanoylglycylphenylalanylleucylglycyl) lysine, Dox = doxorubicin.

by fractionation using size exclusion chromatography (SEC) [89–91]. These second generation polymer–drug conjugates have shown enhanced antitumor efficacy against human ovarian xenografts in mice compared to first generation conjugates [92, 93].

The release of drug from conjugates using enzyme-degradable linkers is dependent upon the expression of the enzyme in the tissue. Low enzyme expression may prevent therapeutic efficacy of the drug. To overcome this, researchers have utilized other degradable linkers between the polymer and the drug. Etrych *et al.* conjugated Dox to an HPMA copolymer backbone via a pH-sensitive hydrazone bond [94]. At physiological pH, less than 10% of the drug was released, and at a pH of 5.0 after 5 h, nearly 50% of the drug was released. These conjugates were synthesized with two stimuli-sensitive areas on the single linkage to Dox. The linkage comprised a GFLG sequence, which can be cleaved by thiol-dependent cathepsin B in the lysosomes and a hydrazone linkage. Incubation at low pH and with cathepsin B increased Dox release *in vitro* compared to incubation at low pH without enzyme. Hrubý *et al.* synthesized a thermally sensitive and pH-degradable polymer for the delivery of radiotherapeutics [95]. They co-polymerized *N*-isopropylmethacrylamide with a co-monomer bearing hydrophobic pendant groups attached via the pH-sensitive hydrazone bond. The resulting polymer had a transition temperature between room and body temperature. Upon injection to an injured site, the polymer precipitated and released the therapeutic at a rate proportional to the rate of pH-induced hydrolysis of the hydrazone bond.

Linear pH-responsive polymers have been used to develop synthetic non-viral vectors for gene delivery. Non-viral vectors are desirable in that the potential toxicity is reduced and manufacturing is easier. However, the efficacy has traditionally been much lower than viral vectors. Pack *et al.* outlined the design constraints for non-viral vectors, some of which include: protection of genetic material, serum stability, endolysosomal escape, non-toxicity, and non-immunogenicity [96].

The polymers PLL and PEI have been studied extensively for their potential in gene delivery [96]. Polymers of PLL have been modified to target specific cells by conjugation of transferrin, sugars, folate, peptides, and antibodies [96]. PEI has been modified to incorporate targeting ligands as well, and has been found to be effective in *in vivo* gene delivery, yet because PEI has high cytotoxicity, its progression has been limited.

Stayton, Hoffman, and coworkers have investigated pH-sensitive polymers capable of escaping the endosome for gene delivery [41, 96]. Polymers that incorporate pH-sensitive groups, such as ethyl acrylic acid (EAAc) and propyl acrylic acid (PAAc), have the ability to disrupt cellular membranes. Polymers made from EAAc and PAAc undergo sharp transitions from hydrophilic to hydrophobic at low pH values. As the pH decreases in the endosome, the polymers transition to hydrophobic and can disrupt the endosomal membrane. Drugs or biomolecules attached via degradable linkers (disulfide, hydrazone, enzyme-degradable, etc.) to the polymer can then be released into the cytosol [97–99]. Murthy *et al.* prepared graft copolymers with a membrane-disruptive polymer backbone with polyethylene glycol (PEG) grafts that were covalently attached via a pH-sensitive linker (Fig. 11.10) [97]. They attached a targeting moiety to induce endocytosis of the polymer conjugate, after which the PEG chains could be cleaved off in the low pH environment of the endosomes and lysosomes. The low pH caused shedding of the PEG shield, thereby exposing the membrane-disruptive polymer to the organelle membrane. The disruptive polymer consisted of three random co-monomers: butyl methacrylate (BMA), *N,N*-dimethylaminoethyl methacrylate (DMAEMA), and a styrene-acetal monomer. Bae and coworkers used oligomeric sulfonamides (OSAs) to complex DNA and also to promote endosomal escape through a proton-buffering effect [100].

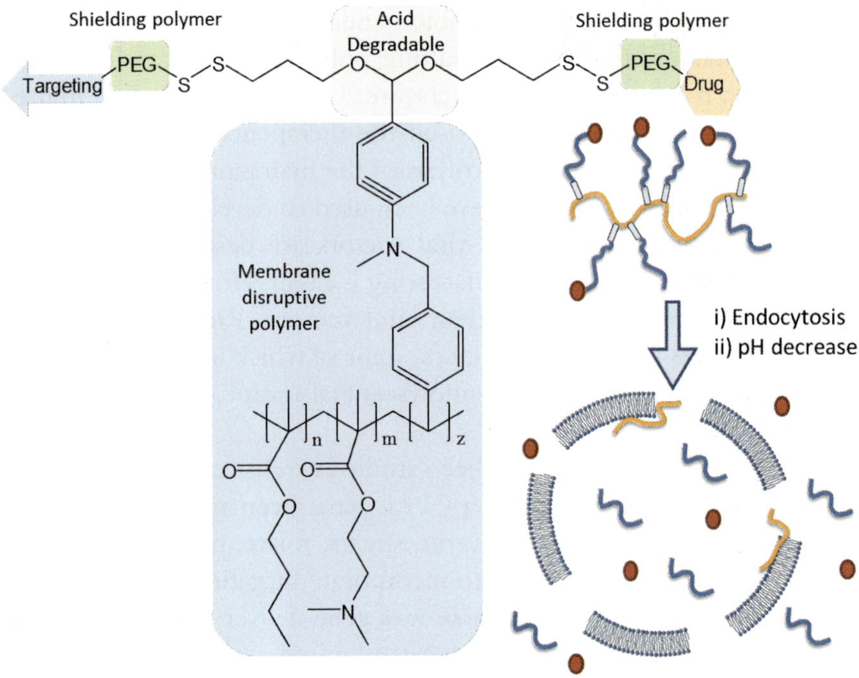

Figure 11.10. Chemical structure and scheme for the endsomolytic polymer shielded with PEG chains attached via acid sensitive linkers. PEG = polyethylene glycol.

11.4.1.1 *In vivo self-assembling polymer therapeutics*

The emergence of stimuli-sensitive polymers resulted from efforts to mimic biological materials. Biological materials are possibly immunogenic, difficult to synthesize, and unstable in biological environments. Synthetic materials have enhanced stability or controllable degradability, are usually immunocompatible, and can be reproducibly synthesized. However, synthetic materials still have limitations compared to biological materials. Some of these limitations can be overcome by combining synthetic and biological materials to produce hybrid constructs with the specificity and nano-scale organization of biology and the immunocompatibility of synthetic polymers.

Several novel stimuli-sensitive nanomedicines have been inspired by the design of materials used in stimuli-sensitive hydrogels. Hybrid hydrogels made up of HPMA copolymer with peptide grafts that form a coiled-coil were synthesized [101, 102]. Using the principles of self-assembly and mulivalency, a new class of smart polymer-based therapeutics emerged, termed drug-free macromolecular therapeutics [103, 104]. These therapeutics are termed drug-free because no low-MW drugs are used in the design. There are three major components to drug-free macromolecular therapeutics: (1) an HPMA copolymer backbone, (2) a specific biorecognition pair of molecules (oligos, morpholinos, peptides), and (3) a targeting moiety. From these three components, two conjugates are produced: a polymer containing multiple grafts of motif1 (P-motif1) and Fab′–motif2 conjugate (Fab′–motif2) (Fig. 11.11). Figure 11.11 shows the therapeutic mechanism where the Fab′–motif2 conjugate binds to CD20 on the surface of B cells thus decorating the surface with motif2. Subsequent treatment with P-motif1 leads to biorecognition at the cell surface and CD20 cross-linking and clustering, which results in apoptosis induction.

Recently, our lab designed a hybrid nanoconjugate using morpholino oligonucleotides as biorecognition motifs, instead of coiled-coil forming peptides [105]. The morpholinos are neutral, as opposed to the oppositely charged

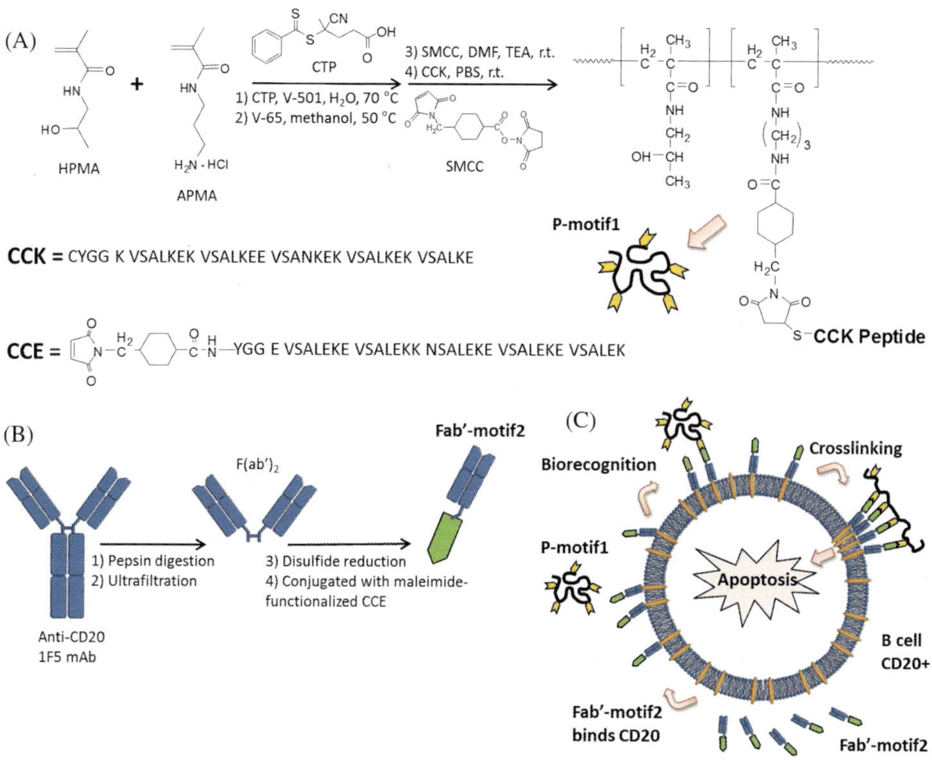

Figure 11.11. Scheme for the therapeutic mechanism of drug-free macromolecular therapeutics. (A) RAFT synthesis of a polymer with reactive amines that were converted to maleimide groups; (B) anti-CD20 antibody 1F5 was digested and reduced to produce a Fab' fragment that is reacted with a maleimide functionalized peptide; (C) scheme showing the binding of Fab′–motif2 to the B cell surface followed by biorecognition and cross-linking after treatment with P-motif1.

coiled-coil peptides. Two complementary 25-bp morpholinos were designed and each was conjugated to an HPMA copolymer or anti-CD20 Fab' fragment. This conjugate system successfully induced apoptosis *in vitro* of Raji B cells through self-assembly at the cell surface and consequent CD20 cross-linking. Furthermore, this system was tested in a non-Hodgkin's lymphoma mouse model. The nanoconjugate prevented lymphoma cell growth and dissemination in the mice compared to controls [105].

11.4.2 *Micelles*

The increased versatility of new polymerization techniques allows for the synthesis of well-defined block copolymers that assemble into micelles at or above the critical micelle concentration (CMC). Below the CMC, the polymers exist as unimers in solution. The transition from unassociated unimers to aggregates or micelles occurs due to competition between entropy and energy [106, 107]. To assemble into micelles, one block consists predominantly of hydrophobic monomers, while the other block consists of hydrophilic monomers. Upon contact with water, these amphiphilic molecules self-assemble into larger structures (lamellas, globules, cylinders, vesicles, or micelles) with a hydrophobic interior and hydrophilic corona [108]. Assembly into certain structures is governed by the nature and length of the hydrophobic region of the polymer [109]. Micelle size ranges between 10 to 100 nm. Micellar systems are particularly useful for drug delivery in that hydrophobic drugs can be non-covalently sequestered in the core of micelles, and the micelle can be triggered to release the contents at the desired location.

Geng *et al.* synthesized copolymers of poly(ethyleneglycol) and poly(capro lactone) (PCL) that self-assembled into cylindrical micelles, or filomicelles [110]. These micelles circulated in the blood stream for as long as one week. Acid-catalyzed hydrolysis of the filomicelles caused them to fragment into smaller micelles that could extravasate into tumor areas and enter cells. Experiments conducted on animal models showed that Paclitaxel-loaded filomicelles significantly reduced the tumor volume relative to treatment with only free drug.

Micelles composed of poly(ethylene glycol)-*b*-poly(amino acids) (PEG-PAA) utilize the biocompatible water-soluble polymer PEG and non-toxic PAA [111, 112]. Kataoka's laboratory synthesized a pH-sensitive, folate-targeted micelle, folate-poly(ethylene glycol)-poly(aspartate hydrazone adriamycin) [Fol-PEG-P(Asp-Hyd-ADR)] [112]. Folate was attached to the PEG chain that extends into solution for binding folate receptors that are over-expressed on cancer cells [113]. Dox was attached to the polyaspartate with an acid-sensitive hydrazone bond. These micelles showed enhanced anti-tumor activity in a mouse model compared to free drug and untargeted micelles [114].

Interestingly the accumulation in the tumor was about equal for targeted and untargeted micelles, but the anti-tumor activity of the folate-targeted micelle was higher than the untargeted micelle.

Du *et al.* synthesized a polyphosphoester–Dox conjugate with dual pH-sensitivity [115]. The polymer nanoparticle surface switches from a negative potential to positive at pH ~6.0 in the tumor environment. The positive charge facilitated cellular uptake.

Klaikherd *et al.* combined three stimuli-sensitive elements into a single micelle [8]. The micelle comprised a temperature-sensitive PNIPAM block, a pH-sensitive poly(2-hydroxyethyl methacrylate) (HEMA)-based block, and a thiol-reductive linker between the two polymer blocks. The hydroxyl on the HEMA monomer units was protected with a tetrahydropyran (THP) derivative, which is cleaved off in mild acidic conditions, thereby converting the HEMA block from hydrophobic to hydrophilic in nature. In the presence of dithiothreitol, the block copolymer is reduced to its respective homopolymers. The respective blocks were synthesized via ATRP, and the end groups were functionalized such that they could be coupled together via a disulfide exchange reaction. At pH 5.0 after 50 h, 30% of entrapped Nile red was released from the micelles, but in the presence of glutathione at pH 5.0, 80% of the dye was released after 24 h. This combination of multiple stimuli into a micelle could release drug at a faster rate than other single-stimuli sensitive systems.

Husseini *et al.* produced micelles from Pluronic P-105, which are ABA type copolymers of PEO (A) and PPO (B) loaded with Dox [116]. The hydrophilic PEO extends into solution, while the hydrophobic PPO aggregates in a core, excluding water [117]. Upon sonication at low frequencies, the Dox was released. After sonication ceased, the drug was re-encapsulated as the micelles reformed. The mechanism of release using low frequency ultrasound is inertial cavitation. However, this type of micelle is unstable in the blood stream, so a micelle with an interpenetrating network was synthesized to prevent disassembly of the micelle *in vivo* [118]. The core consisted of an interpenetrating network of PPO and PNIPAAm or PDMAAm. The cores were swollen at room temperature to allow for drug-loading, and then the volume of the core decreased above the thermal transition temperature [109]. The micelle showed significant inhibition of tumor growth in a bi-lateral solid tumor rat model [119]. However, despite success *in vitro* and promising results *in vivo*, these micelles were found to induce cell dehydration.

11.4.3 *Polymersomes*

Discher *et al.* reported the first polymersome in 1999 made from an amphiphilic block copolymer of polyethyleneoxide-polyethylethylene EO_{40}–EE_{37} [11].

The average MW of the polymer was 3,900 g/mol with a contour length about ten times greater than a phospholipid. The block copolymer self-assembled into liposome-like vesicles with a core thickness of the membrane approximately 8 nm (liposome = 3 nm) surrounding an aqueous core. Interestingly, the elastic behavior of these polymer vesicles was similar to that of phospholipid vesicles; however, the polymersome membranes were ten times less permeable to water compared to phospholipid membranes. They also confirmed that proteins could be encapsulated inside the polymersomes and could be a potential nanocarrier. Hydrophilic drugs can be encapsulated in the aqueous core and hydrophobic drugs can be sequestered in the membrane. A more detailed review of stimuli responsive polymersomes was completed by Meng *et al.* [120].

Polymersomes can be designed to circulate for long periods of time in the blood stream. A PEG corona increases blood circulation time of polymersomes to a greater extent than PEGylated liposomes [121]. Polymersomes can be made from diblock or triblock polymers (AB, ABA, BAB, or ABC).

pH-sensitive polymersomes are the most highly studied due to the pH gradients present in the body, especially in the tumor microenvironment [120]. Polybutadiene-*b*-poly(L-glutamic acid) (PBD-PGA) and poly(L-glutamic acid)-*b*-poly(L-lysine) (PGA-PLL) are polypeptide-based block copolymers that form vesicle-like structures in water [122, 123]. The PBD-PGA structures were between 100 and 150 nm, and maintained their pH responsiveness, even in high salt concentration media.

Poly(lactic acid) (PLA) and poly(ε-caprolactone) (PCL) have been used to create polymersomes that exhibit pH-dependent hydrolysis. Dox was loaded into PEG-PLA polymersomes with different amounts of PEG-PBD mixed in. The release rate increased with increasing amounts of PEG-PLA. These polymersomes slowed tumor growth in an *in vivo* model [124]. However, the drug release rate is slow from polymersomes that are sensitive to pH-induced hydrolysis. To increase the release rate polymers that have a pH-dependent solubility have been used such as poly(2-vinylpyridine) (P2VP) and polyleucine-*b*-polylysine (PLE-PLL) [125, 126]. Incorporating pH-sensitive linkers into the polymer backbone, such as orthoester, hydrazone, and acetal, can increase the sensitivity to changes in pH for more rapid drug release.

Thermo-sensitive polymersomes most often use poly(*N*-isopropylacrylamide) (PNIPAAm) for the hydrophobic portion of a block copolymer. Above the LCST, the PNIPAAm-based block copolymers self-assemble into polymersomes or micelles. Quin *et al.* synthesized the block copolymer poly(ethylene oxide)-*block*-poly(*N*-isopropylacrylamide) that forms polymersomes in solution [127]. Above 32°C, stable polymer vesicles are formed that can encapsulate the hydrophilic drug Dox. Triggered release of Dox was accomplished by lowering the temperature to below 32°C. At room temperature, 70% of Dox was released within 2 h, whereas at 37°C, no Dox was released after 2 h. This construct could release its contents in an area of the body where a cold pack or cryo-probe is applied.

Redox-sensitive polymersomes were formed by incorporating a disulfide bond between the hydrophilic and hydrophobic blocks of a polymer chain PEG-SS-PPS [128]. These polymer chains spontaneously form after hydration of a thin film with water. Within 10 minutes, the release of the polymersome contents was observed *in vitro*. This type of polymersome would be suitable for delivering biomolecules to the cytoplasm of cells, where the concentration of glutathione is higher. It should be noted that a small concentration of cysteine and glutathione are present in the blood and could trigger release of the polymersome contents prematurely.

Light-sensitive polymersomes have been synthesized by incorporating photo-convertible groups into the polymer backbones. Upon illumination, sections of the polymer are cleaved off or altered in a way that changes the thermodynamic balance between the hydrophobicity and hydrophilicity of the polymer. Zhao *et al.* synthesized a copolymer of polymethacrylate and poly(acrylic acid) (PAzoMAb-PAA) containing azobenzene side chains [129]. The UV-induced isomerization of the azobenzene group resulted in disruption of the vesicles, and illumination with visible light caused reaggregation.

Other external stimuli such as magnetism and ultrasound have been considered in polymersome design. Magnetic-sensitive polymersomes were made by encapsulation of γ-Fe_2O_3 nanoparticles in vesicles formed from PBD-PGA [130]. Ultrasound-sensitive polymersomes of PEG-PLA were prepared by encapsulating air inside the core [131].

11.4.4 *Nanogels or shell cross-linked micelles*

Nanogels are polymeric systems with cross-links in the core or shell of a nano-sized particle. Upon hydration or application of a stimulus, the nanogels can swell or collapse. Nanogels have an enhanced ability in preventing drug leakage during circulation as compared to uncross-linked Pluronic micelles [132]. Drugs can be loaded into nanogels in two general ways: chemically (covalent attachment to the polymer) or physically (supramolecular interactions) [12]. Chemically bound drugs exhibit greater stability in circulation; however, physically entrapping drugs may be more versatile, as "cocktails" of drugs could be loaded into a nanogel. Figure 11.12 conceptually shows the function of a stimuli-sensitive nanogel. Loaded drugs can be released upon encountering a

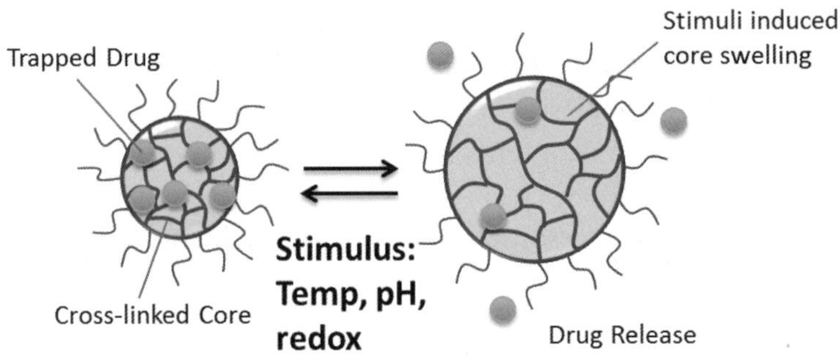

Figure 11.12. Representation of drug release from a nanogel upon stimulus application.

stimulus such as temperature and pH. Depending on the chemical structure of the polymer in the core, the polymer network will either swell or collapse.

Li *et al.* synthesized disulphide cross-linked nanogels by the miniemulsion polymerization of oligo(ethylene glycol) acrylate (OEGA) and 2-(5,5-dimethyl-1,3-dioxan-2-yloxy) ethyl acrylate (DMDEA) [133]. These nanogels were thermo-, redox-, and pH-sensitive. Temperature- and pH- sensitive polymer blocks (OEGA and DMDEA, respectively) caused a swelling or collapse of the gel, while upon exposure to free thiol and low pH, the nanogel dissolved. The gels were capable of encapsulating hydrophobic drugs, and the unloaded nanogels exhibited no toxicity to MCF-7 cells. With reducing agent dithiothreitol (DTT) and at pH 5.0, the nanogels released 80% of Nile red after 5 h of incubation.

Blackburn and Lyon used free-radical precipitation polymerization to create nanogels made of *N*-isopropylacrylamide or *N*-isopropylmethacrylamide, acrylic acid or 4-acrylamidofluorescein, and *N,N′*-methylene-bis-acrylamide as a cross-linker [134]. They were able to tune the core size of the nanogels by varying the surfactant and initiator concentrations.

A disadvantage of micelles is that below their CMC, they disassemble into single unimers. Premature disassembly is highly undesirable in drug delivery applications, so researchers have sought to form cross-links in the micelle to improve stability. Often, triblock copolymers of type ABC are used to form shell cross-linked micelles, where the A block is a hydrophilic polymer (e.g. PEO), the B block is cross-linkable, and the C block is stimuli-responsive [135]. The hydrophilic PEO prevents cross-linking between neighboring micelles.

Li *et al.* synthesized the thermo-sensitive triblock copolymer poly(ethylene oxide)-block-[(*N,N*-dimethylacrylamide)-*stat*-(N-acryloxysuccinimide)]-*block*-(*N*-isopropylacrylamide), [PEO-*b*-P(DMA-*stat*-NAS)-*b*-NIPAM] using

RAFT polymerization [136]. At temperatures above the LCST, micelles of 38 nm were formed. The micelles were cross-linked using cystamine to react with the NAS group, and disulfide bonds were formed between polymer chains. The size of the micelle was not changed after cross-linking, but at temperatures below the LCST, the micelle size increased to 57 nm. The addition of tris(2-carboxyethyl)phosphine (TCEP) triggered the dissolution of the micelle within 30 min.

11.4.5 *Dendritic polymers*

Dendritic polymers are an attractive architectural choice to produce polymeric nanomedicines due to their multifunctional and well defined structures [137]. Dendrimers may have many generations of branches from the core of the molecule, and as the number of generations increases, more void space is available for loading agents and more functional groups are available at the surface for modification with targeting ligands. Bioactive agents may be covalently attached to the dendrimer or non-covalently sequestered in the dendrimer core. Stimuli-sensitive linkers used to conjugate drugs or shielding polymers comprise hydrazone, carbamate, ester, amide, carbazate, disulfide, imine, and enzyme-sensitive peptide sequences [137]. A more complete review of stimuli-sensitive dendrimers has been covered [138].

Haag *et al.* used polyglycerol (PG) as a dendritic core, then functionalized the outside with PEG [139]. The PEG was covalently tethered to the dendritic core via a pH-sensitive imine linker that cleaves at pH 5.0. Dox was encapsulated in the dendritic core to make a chemotherapeutic nanoparticle. Using the same dendritic PG core, they functionalized the corona of the dendrimer with amine groups to produce hyperbranched polyglycerolamine dendrimers [140]. Spermine was conjugated to the PG core via a carbonate bond. siRNA was complexed to the PG-NH_2 carrier for gene silencing. Significant silencing of a variety of genes was achieved *in vitro* in a HeLaS3 cell line. Other researchers have explored polycation dendrimers for the delivery of DNA and siRNA [141–144].

Etrych *et al.* synthesized pH-sensitive HPMA star polymers with a poly(amido amine) PAMAM dendritic core [145]. Semi-telechelic HPMA copolymers were conjugated to the PAMAM core via an enzyme-degradable or -reducible bond. Dox was attached to the HPMA backbone using a pH-sensitive hydrazone bond or the enzyme-degradable sequence GFLG. Dox was released upon lowering of the pH or cleavage by enzymes in the cell. Constructs that employed a thiol reductive tether between HPMA and the

PAMAM core showed faster degradation than the enzyme degradable tether GFLG (4 h versus. 75 h). This difference is likely due to accessibility to the core of the construct by enzymes and glutathione. The architecture of polymers influences efficacy of the intended treatment. They found that star polymers circulated in the blood longer and accumulated in tumors to a greater extent than linear polymers [145]. The enhanced accumulation in the tumor can be explained by the EPR effect. An increased concentration gradient between the blood and the tumor will result in higher accumulation.

11.4.6 *Elastin-like polypeptides*

Polypeptide polymers are an attractive material for use in drug delivery due to their biodegradability and biocompatibility. Similar to PNIPAM, natural elastin (Val-Pro-Gly-Val-Gly)x (valine, Val; proline, Pro; glycine, Gly) exhibits LCST-like behavior. Above a certain critical temperature, the elastin-like polypeptides (ELPs) reversibly coacervate. ELPs have been investigated, where substitutions have been made to the second Val in the pentapeptide repeat [146]. This class of polymers is especially attractive for use in biomedical applications because the ELPs are stimuli-responsive, monodisperse (genetically encoded), and biocompatible [147, 148]. ELPs can be precisely designed using genetic encoding to produce monodisperse polymers in *Escherichia coli*. The temperature transition point can be tuned precisely anywhere in the range of 0–100°C.

Temperature-responsive ELPs have been used to treat solid tumors by inducing local hyperthermia to the target location. Chilkoti's group designed an ELP with a transition temperature a few degrees above body temperature. Heating of the tumor to 42°C resulted in a two-fold increase in accumulation of the ELP compared to unheated tumor [146]. In the same work, thermally-responsive ELPs were compared to a copolymer of *N*-isopropylacrylamide (NIPAAm) and acrylamide (AAm). The poly(NIPAAM–*co*-AAm) also accumulated in the tumor, but to a slightly lesser extent than the ELP. Also of note is that heat not only facilitated accumulation of the ELP into the tumor, but also facilitated cellular uptake *in vitro* [149]. In other work, the group attached a hydrazone functionalized Dox molecule to a C-terminal cysteine residue of an ELP [150]. At pH 4.0, almost 80% of the Dox was released after 72 h.

ELPs are not as versatile as synthetic polymers at incorporating pendant reactive groups. However, new techniques are continually emerging for incorporating non-natural amino acids into the design of ELPs. The immunogenicity should be extensively studied in these constructs, as their potential for response is higher than synthetic polymers.

11.5 Challenges for Translation

Smart polymer nanomedicines have had limited success in reaching the clinic. Synthetic polymer constructs that have reached the clinic are limited to polymer–protein conjugates. The first approval was in the 1990s and many have followed since. Most of these polymer–protein conjugates have PEG as the polymeric component. These conjugates include a protein which has a specific function in a biological system. For example, Adagen® is a PEG–adenosine deaminase conjugate used to treat severe combined immunodeficiency (SCID). Adenosine deaminase is involved in the metabolism of purine and has important roles in maintaining and developing the immune system.

Table 11.1. United States Food and Drug Association (FDA) approved PEG–protein conjugates [151].

Conjugate	Company	Indication	Ref.
PEG-adenosine deaminase (Adagen®)	Enzon	SCID	[152]
PEG-L-asparagase (Oncaspar®)	Enzon	Acute lymphoblastic leukemia	[153]
PEG-interferon α -2b (Pegintron®)	Schering-Plough	Hepatitis C	[154]
PEG-interferon α -2b (Sylatron®)	Merk	Melanoma	[155]
PEG-interferon α -2a (Pegasys®)	Hoffman-La Roche	Hepatitis C	[156]
PEG-growth hormone receptor antagonist (Somavert®)	Pfizer	Acromegaly	[157]
PEG-anti-VEGF aptamer (Macugen®)	OSI	Age-related macular degeneration (AMD)	[158]
PEG-Granulocyte macrophage colony-stimulating factor (Neulasta®)	Amgen	Stimulation of neutrophils during chemotherapy	[159]
PEG-epoetin beta (Mircera®)	Hoffman-La Roche	Anemia	[160]
PEG-anti-tumor necrosis factor α Fab′ (Cimzia®)	UCB Pharma	Crohn's disease, rheumatoid arthritis	[161]
PEG-Uricase (Krystexxa®)	Savient Pharmaceuticals	Gout	[162]
PEG-synthetic erythropoietin Analogue (Omontys®)	Affymax	Anemia	[163]

Limited clinical translation of polymer therapeutics can be attributed partially to fears of potential toxicity of the polymers, poor *in vitro* and *in vivo* experimental models, and complex design [6, 41]. Many smart polymers are not easily degradable in the body and may not be easily excreted by the kidneys. To be excreted by the kidneys, the whole MW distribution of a random hydrophilic neutral polymer (such as polyHPMA) needs to be below approximately 50 kDa; however, to enhance passive targeting through the EPR effect, polymers need to have larger MW.

The lack of standardized and reliable *in vitro* and *in vivo* models is also likely to blame for the failure of smart medicines to reach the clinic [6]. Standardized methods for testing toxicity, efficacy and biocompatibility of nanomedicines are needed for the field to progress.

In a discussion relating to clinical translation, it is important to distinguish between two types of polymers used in polymer therapeutics: biologically active and biologically inert [164]. HPMA polymers are biologically inert and are often coupled to drugs/proteins via a stimuli-sensitive linker to release the active component at the appropriate locations. Biologically active polymers have had little success in clinical translation due to toxicity. However, PEG is considered biologically inert, but researchers have found that in some cases, an anti-PEG immunoglobulin M (IgM) response was elicited [165]. The phenomenon of accelerated blood clearance (ABC) of PEGylated proteins and liposomes in animals has added to fears of PEG immunogenicity. Schellekens *et al.* recently reviewed reports of anti-PEG antibodies; they hypothesized that the ABC effect likely results from a complement mechanism that is activated due to non-specifically bound immunoglobulin G (IgG) and IgM to PEGylated particles [166].

Gaspar and Duncan state in a review of polymer therapeutic development, "As modern polymer chemistry is rapidly evolving to produce ever more elaborate structures; multivalent polymers, branched polymers, graft polymers, dendrimers, dendronised polymers, block copolymers, stars and hybrid glyco- and peptide derivatives (and such structures are often grafted to the surface of other nanosized particles) it is clear that each (potentially heterogeneous) structure will present unique challenges during pharmaceutical development" [164]. These unique challenges extend from basic physico-chemical characterization to industrial scale-up of these complex pharmaceuticals.

The increase in complexity presents a challenge in characterizing the structures to ensure reproducibility for clinical translation. For example, no polymer therapeutics employing endosomal escape technology have reached the clinic despite success *in vitro*. New analytical tools and techniques are needed to address these challenges. In the case of shell cross-linked micelles, it is difficult to determine the degree of cross-linking in the micelle [135]. Without a detailed knowledge of the therapeutic's physical and chemical characteristics then, it is difficult to ensure reproducibility.

11.5.1 *Necessary steps towards safe smart polymer therapeutics*

The step-by-step approach used to bring PK1 and PK2 to clinical trials can be used to bring other polymer-based systems to the clinic (Fig. 11.13) [164]. Characterization of the polymer therapeutic is essential to understand the interactions of the polymer with cells *in vitro* and with the host *in vivo*. Polymers must be non-toxic and produce acceptable immunogenicity responses in their proposed administration route. Polyelectrolyte polymers should be screened early on for antigenicity and cellular immunogenicity. All polymers should be tested *in vitro* and *in vivo* without conjugated drugs or linkers to ensure safety. Then, the polymers can be optimized and a library of conjugates synthesized for biological tests. It is necessary to perform *in vitro* analyses of cytotoxicity, hemocompatibility, complement activation, and rate of pH-dependent hydrolysis or enzyme degradation. These tests are highly sensitive to polymer Mw, cell type, and impurities in the polymer; therefore, it is critical to properly characterize the polymer therapeutic.

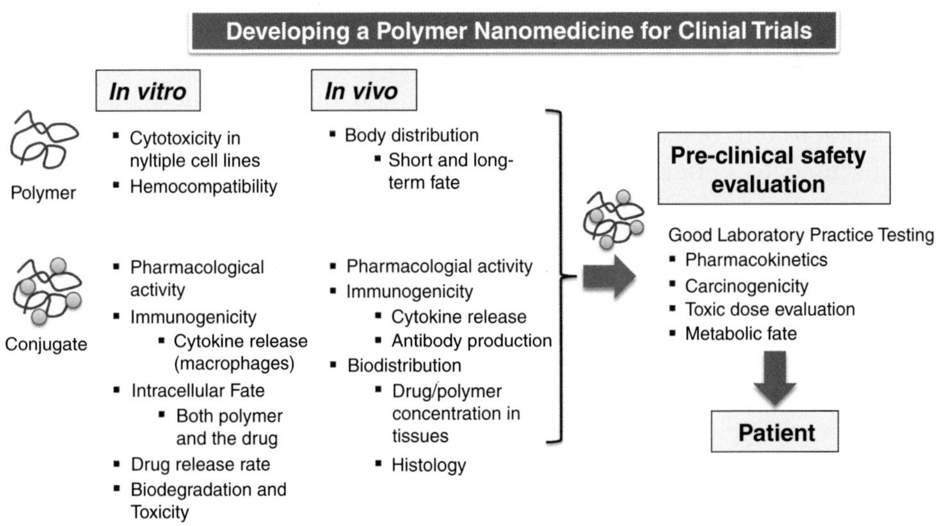

Figure 11.13. Outline for the progress of polymer-based therapeutics from early *in vitro* testing to the clinic.

11.5.1.1 *Biocompatibility*

Polymer biocompatibility is of supreme importance. Two categories of polymer therapeutics result in two different strategies to achieve biocompatibility. There are non-degradable polymers, such as HPMA and PEG, and degradable polymers, such as poly(L-glutamic acid). Non-degradable polymers have a stable backbone chemistry that is not hydrolyzed in the blood. However, non-degradable polymers need to be eliminated from the body after delivery of the payload. Non-degradable polymers first employed designs with MW below 50 kDa to ensure excretion in the kidneys. Non-degradable polymers above this MW will accumulate in the body and could cause adverse side effects. Degradable polymers, on the other hand, are built using chemistries that allow for degradation in the body. For example, PLGA has a polymer backbone consisting of hydrolytic ester bonds that facilitate the degradation of the polymer acids products found naturally in the body. New polymerization techniques have enabled the design of polymers that have degradable and non-degradable portions.

11.5.1.2 *Characterization of smart therapeutics*

The development of nanomedicines and nanomaterials has increased dramatically in the last decade; however, our understanding of the interactions at the interface between these materials and biology has not kept pace. The nano–bio interface comprises dynamic forces, thermodynamic exchanges, and kinetics (Fig. 11.14) [167]. Recent advances in super-resolution imaging make it possible to break the diffraction barrier and increase optical microscopy resolution to tens of nanometers [168, 169]. These high resolution techniques can be used to better understand the interactions of polymer therapeutics with their targets, and also aid in improving the design of smart polymer nanomedicines [170].

In addition to a more careful study of the interactions of medicines with individual cells, improved analytical methods are needed to evaluate nanomedicine distribution in the whole organism. Sensitive optical imaging technologies, such as fluorescence tomography (FMT), allow for facile characterization of nanomedicine biodistribution without sacrificing the animal. However, these methods are not adequate for tracking nanomedicines in humans due to poor light penetration to deeper tissues.

The promise of polymer-based nanomedicines to treat diseases more effectively has become a reality for polymer–protein conjugates; however, for other smart systems, more rigorous and standardized pre-clinical testing is

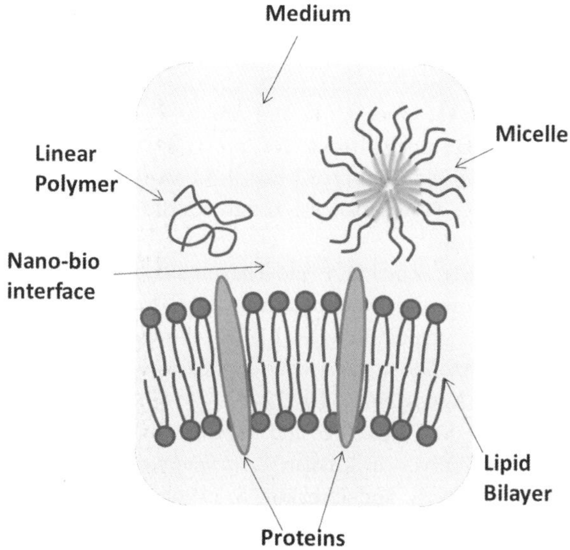

Figure 11.14. Representation of the nano–bio interface at the cell membrane between polymer-based nanomedicines.

required, along with more detailed characterization of the cell–material interactions and the whole body interactions. For successful clinical translation, an appreciation for the biological complexity of the disease is essential, along with an engineering perspective for polymer therapeutic design that can easily be scaled up.

References

1. Duncan, R. (2003). The dawning era of polymer therapeutics. *Nat. Rev. Drug Discov.*, **2**, 347–360.
2. Schmaljohann, D. (2006). Thermo- and pH-responsive polymers in drug delivery. *Adv. Drug Delivery Rev.*, **58**, 1655–1670.
3. Bawa, P., Pillay, V., Choonara, Y. E. and Toit, L. C. D. (2009). Stimuli-responsive polymers and their applications in drug delivery. *Biomed. Mater.*, **4**, 022001.
4. Kopeček, J. (2003). Smart and genetically engineered biomaterials and drug delivery systems. *Eur. J. Pharm. Sci.*, **20**, 1–16.
5. Kumar, A., Srivastava, A., Galaev, I. Y. and Mattiasson, B. (2007). Smart polymers: Physical forms and bioengineering applications. *Prog. Polym. Sci. (Oxford)*, **32**, 1205–1237.
6. Grainger, D. W. (2013). Connecting drug delivery reality to smart materials design. *Int. J. Pharm.*, **454**, 21–24.
7. Cabane, E., Zhang, X., Langowska, K., Palivan, C. G. and Meier, W. (2012). Stimuli-responsive polymers and their applications in nanomedicine. *Biointerphases*, 7, 1–27.
8. Klaikherd, A., Nagamani, C. and Thayumanavan, S. (2009). Multi-stimuli sensitive amphiphilic block copolymer assemblies. *J. Am. Chem. Soc.*, **131**, 4830–4838.

9. Cheng, R., Meng, F., Deng, C., Klok, H.-A. and Zhong, Z. (2013). Dual and multi-stimuli responsive polymeric nanoparticles for programmed site-specific drug delivery. *Biomaterials*, **34**, 3647–3657.
10. Rodriguez, F., Cohen, C., Ober, C. K. and Archer, L. A. (2003). *Principles of Polymer Systems*, 5 ed. Taylor & Francis, Ithaca, NY, pp. 1, 2.
11. Discher, B. M., Won, Y.-Y., Ege, D. S., Lee, J. C.-M., Bates, F. S., Discher, D. E. and Hammer, D. A. (1999). Polymersomes: Tough vesicles made from diblock copolymers. *Science*, **284**, 1143–1146.
12. Chacko, R. T., Ventura, J., Zhuang, J. and Thayumanavan, S. (2012). Polymer nanogels: A versatile nanoscopic drug delivery platform. *Adv. Drug Delivery Rev.*, **64**, 836–851.
13. Jatzkewitz, H. (1955). Peptamin (glycyl-L-leucyl-mescaline) bound to blood plasma expander (polyvinylpyrrolidone) as a new depot form of a biologically active primary amine (mescaline). *Z. Naturforsch.*, **10**, 27–31.
14. Givental, N., Ushakov, S., Panarin, E. and Popova, G. (1965). Experimental studies on penicillin polymer derivatives (in Russian). *Antibiotiki.*, **10**, 701.
15. Shumikhina, K., Panarin, E. and Ushakov, S. (1966). Experimental study of polymer salts of penicillins (in Russian). *Antibiotiki.*, **11**, 767.
16. Panarin, E. and Ushakov, S. (1968). Synthesis of polymer salts and amidopenicillines (in Russian). *Khim. Pharm. Zhur.*, **2**, 28–31.
17. Mathé, G., Loc, T. B. and Bernard, J. (1958). Effect sur la leucémie L1210 de la souris d'une combinaison par diazotation d'a méthoptérine et de γglobulines de hamsters porteurs de cette leucémie par hétérogreffe. *C. R. Acad. Sci.*, **246**, 1626–1268.
18. Ringsdorf, H. (1975). Structure and properties of pharmacologically active polymers. *J. Polym. Sci. Polym. Symp.*, **51**, 135–153.
19. Kopeček, J. and Kopečková, P. (2010). HPMA copolymers: Origins, early developments, present, and future. *Adv. Drug Delivery Rev.*, **62**, 122–149.
20. Kopeček, J. and Bažilová, H. (1973). Poly[*N*-(2-hydroxypropyl)methacrylamide] — I. Radical polymerization and copolymerization. *Eur. Polym. J.*, **9**, 7–14.
21. Říhová, B., Bilej, M., Vetvicka, V., Ulbrich, K., Strohalm, J., Kopeček, J. and Duncan, R. (1989). Biocompatibility of *N*-(2-hydroxypropyl)methacrylamide copolymers containing Adriamycin — immunogenicity, and effect of hematopoietic stem-cells in bone-marrow in vivo and mouse splenocytes and human peripheral-blood lymphocytes in vitro. *Biomaterials*, **10**, 335–342.
22. Říhová, B., Kopeček, J., Ulbrich, K., Pospisil, M. and Mancal, P. (1984). Effect of the chemical-structure of *N*-(2-hydroxypropyl)methacrylamide copolymers on their ability to induce antibody-formation inbred strains of mice. *Biomaterials*, **5**, 143–148.
23. Říhová, B., Veres, K., Fornusek, L., Ulbrich, K., Strohalm, J., Vetvicka, V., Bilej, M. and Kopeček, J. (1989). Action of polymeric prodrugs based on *N*-(2-hydroxypropyl)methacrylamide copolymers. 2. Body distribution and T-cell accumulation of free and polymer bound [125I]Daunomycin. *J. Controlled Release*, **10**, 37–49.
24. Kopeček, J. (1977). Soluble biomedical polymers. *Polim. Med.*, 7, 191.
25. Kopeček, J. (1981). *Soluble Polymers in Medicine*. Williams, D. F. (ed.), CRC Press, Boca Raton, pp. 159–180.
26. Ghandehari, H. and Cappello, J. (1998). Genetic engineering of protein-based polymers: Potential in controlled drug delivery. *Pharm. Res.*, **15**, 813–815.
27. Bajpai, A. K., Shukla, S. K., Bhanu, S. and Kankane, S. (2008). Responsive polymers in controlled drug delivery. *Prog. Polym. Sci.*, **33**, 1088–1118.

28. Wang, J. S. and Matyjaszewski, K. (1995). Controlled/"living" radical polymerization. atom transfer radical polymerization in the presence of transition-metal complexes. *J. Am. Chem. Soc.*, **117**, 5614–5615.
29. Oh, J. K., Siegwart, D. J., Lee, H. I., Sherwood, G., Peteanu, L., Hollinger, J. O., Kataoka, K. and Matyjaszewski, K. (2007). Biodegradable nanogels prepared by atom transfer radical polymerization as potential drug delivery carriers: Synthesis, biodegradation, in vitro release, and bioconjugation. *J. Am. Chem. Soc.*, **129**, 5939–5945.
30. Gao, H. and Matyjaszewski, K. (2007). Arm-first method as a simple and general method for synthesis of miktoarm star copolymers. *J. Am. Chem. Soc.*, **129**, 11828–11834.
31. Scales, C. W., Vasilieva, Y. A., Convertine, A. J., Lowe, A. B. and McCormick, C. L. (2005). Direct, controlled synthesis of the nonimmunogenic, hydrophilic polymer, poly(*N*-(2-hydroxypropyl)methacrylamide) via RAFT in aqueous media. *Biomacromolecules*, **6**, 1846–1850.
32. Licciardi, M., Tang, Y., Billingham, N. C., Armes, S. P. and Lewis, A. L. (2005). Synthesis of novel folic acid-functionalized biocompatible block copolymers by atom transfer radical polymerization for gene delivery and encapsulation of hydrophobic drugs. *Biomacromolecules*, **6**, 1085–1096.
33. Moad, G., Mayadunne, R. T. A., Rizzardo, E., Skidmore, M. and Thang, S. H. (2003). Synthesis of novel architectures by radical polymerization with reversible addition fragmentation chain transfer (RAFT polymerization). *Macromolecular Symposia*, **192**, 1–12.
34. Shipp, D. A. (2005). Living radical polymerization: Controlling molecular size and chemical functionality in vinyl polymers. *J. Macromol. Sci. Polymer Rev.*, **45**, 171–194.
35. York, A. W., Kirkland, S. E. and McCormick, C. L. (2008). Advances in the synthesis of amphiphilic block copolymers via RAFT polymerization: Stimuli-responsive drug and gene delivery. *Adv. Drug Delivery Rev.*, **60**, 1018–1036.
36. Stenzel, M. H., Barner-Kowollik, C., Davis, T. P. and Dalton, H. M. (2004). Amphiphilic block copolymers based on poly(2-acryloyloxyethyl phosphorylcholine) prepared via RAFT polymerisation as biocompatible nanocontainers. *Macromol. Biosci.*, **4**, 445–453.
37. Perrier, S. and Takolpuckdee, P. (2005). Macromolecular design via reversible addition–fragmentation chain transfer (RAFT)/xanthates (MADIX) polymerization. *J. Polym. Sci., Part A: Polym. Chem.*, **43**, 5347–5393.
38. Scarpa, J. S., Mueller, D. D. and Klotz, I. M. (1967). Slow hydrogen-deuterium exchange in a non-.alpha.-helical polyamide. *J. Am. Chem. Soc.*, **89**, 6024–6030.
39. Fujishige, S., Kubota, K. and Ando, I. (1989). Phase transition of aqueous solutions of poly(N-isopropylacrylamide) and poly(N-isopropylmethacrylamide). *J. Phys. Chem.*, **93**, 3311–3313.
40. Chen, G. and Hoffman, A. S. (1995). Graft copolymers that exhibit temperature-induced phase transitions over a wide range of pH. *Nature*, **373**, 49–52.
41. Hoffman, A. S. (2013). Stimuli-responsive polymers: Biomedical applications and challenges for clinical translation. *Adv. Drug Delivery Rev.*, **65**, 10–16.
42. Makhaeva, E. E., Tenhu, H. and Khokhlov, A. R. (1998). Conformational changes of poly (vinylcaprolactam) macromolecules and their complexes with ionic surfactants in aqueous solution. *Macromolecules*, **31**, 6112–6118.
43. Vaupel, P., Kallinowski, F. and Okunieff, P. (1989). Blood flow, oxygen and nutrient supply, and metabolic microenvironment of human tumors: A review. *Cancer Res.*, **49**, 6449–6465.

44. Rofstad, E. K., Mathiesen, B., Kindem, K. and Galappathi, K. (2006). Acidic extracellular pH promotes experimental metastasis of human melanoma cells in athymic nude mice. *Cancer Res.*, **66**, 6699–6707.
45. Grabe, M. and Oster, G. (2001). Regulation of organelle acidity. *J. Gen. Physiol.*, **117**, 329–344.
46. Watson, P., Jones, A. T. and Stephens, D. J. (2005). Intracellular trafficking pathways and drug delivery: Fluorescence imaging of living and fixed cells. *Adv. Drug Deliver Rev.*, **57**, 43–61.
47. Ulbrich, K. and Šubr, Vr. (2004). Polymeric anticancer drugs with pH-controlled activation. *Adv. Drug Deliver Rev.*, **56**, 1023–1050.
48. Choi, W. M., Kopečková, P., Minko, T. and Kopeček, J. (1999). Synthesis of HPMA copolymer containing adriamycin bound via an acid-labile spacer and its activity toward human ovarian carcinoma cells. *J. Bioact. Compat. Polym.*, **14**, 447–456.
49. Alvarez-Lorenzo, C., Bromberg, L. and Concheiro, A. (2009). Light-sensitive intelligent drug delivery systems†. *Photochem. Photobiol.*, **85**, 848–860.
50. Koňák, Č., Kopečková, P. and Kopeček, J. (1992). Photoregulated association of *N*-(2-hydroxypropyl) methacrylamide copolymers with azobenzene-containing side chains. *Macromolecules*, **25**, 5451–5456.
51. Jiang, J., Tong, X., Morris, D. and Zhao, Y. (2006). Toward photocontrolled release using light-dissociable block copolymer micelles. *Macromolecules*, **39**, 4633–4640.
52. Gilbert, H. F. (1995). '[2] Thiol/disulfide exchange equilibria and disulfidebond stability', in Lester, P. (ed.), *Methods Enzymol.*, Academic Press, pp. 8–28.
53. Meng, F., Hennink, W. E. and Zhong, Z. (2009). Reduction-sensitive polymers and bioconjugates for biomedical applications. *Biomaterials*, **30**, 2180–2198.
54. Cuchelkar, V., Kopečková, P. and Kopeček, J. (2008). Synthesis and biological evaluation of disulfide-linked HPMA copolymer-mesochlorin e6 conjugates. *Macromol. Biosci.*, **8**, 375–383.
55. Rozema, D. B., Lewis, D. L., Wakefield, D. H., Wong, S. C., Klein, J. J., Roesch, P. L., Bertin, S. L., Reppen, T. W., Chu, Q. and Blokhin, A. V. (2007). Dynamic PolyConjugates for targeted *in vivo* delivery of siRNA to hepatocytes. *Proc. Natl. Acad. Sci.*, **104**, 12982–12987.
56. Husseini, G. A. and Pitt, W. G. (2008). Micelles and nanoparticles for ultrasonic drug and gene delivery. *Adv. Drug Deliver Rev.*, **60**, 1137–1152.
57. Brennen, C. E. (1995). *Cavitation and bubble dynamics*: Oxford Engineering Series.
58. Wu, J. and Nyborg, W. L. (2008). Ultrasound, cavitation bubbles and their interaction with cells. *Adv. Drug Deliver Rev.*, **60**, 1103–1116.
59. Leighton, T. G. (2007). What is ultrasound? *Prog. Biophys. Mol. Biol.*, **93**, 3–83.
60. Nyborg, W. L. (2001). Biological effects of ultrasound: Development of safety guidelines. Part II: General review. *Ultrasound Med. Biol.*, **27**, 301–333.
61. Rooney, J. A. (1970). Hemolysis near an ultrasonically pulsating gas bubble. *Science*, **169**, 869–871.
62. Schlicher, R. K., Radhakrishna, H., Tolentino, T. P., Apkarian, R. P., Zarnitsyn, V. and Prausnitz, M. R. (2006). Mechanism of intracellular delivery by acoustic cavitation. *Ultrasound Med. Biol.*, **32**, 915–924.
63. Tachibana, K., Uchida, T., Ogawa, K., Yamashita, N. and Tamura, K. (1999). Induction of cell-membrane porosity by ultrasound. *The Lancet*, **353**, 1409.

64. Shackleton, M., Quintana, E., Fearon, E. R. and Morrison, S. J. (2009). Heterogeneity in cancer: Cancer stem cells versus clonal evolution. *Cell*, **138**, 822–829.
65. Dick, J. E. (2009). Looking ahead in cancer stem cell research. *Nat. Biotech.*, **27**, 44–46.
67. Rosen, J. M. and Jordan, C. T. (2009). The increasing complexity of the cancer stem cell paradigm. *Science*, **324**, 1670–1673.
68. Clevers, H. (2011). The cancer stem cell: Premises, promises and challenges. *Nat. Med.*, **17**, 313–319.
69. Chau, Y., Tan, F. E. and Langer, R. (2004). Synthesis and characterization of dextran–peptide–methotrexate conjugates for tumor targeting via mediation by matrix metalloproteinase II and matrix metalloproteinase IX. *Bioconjug. Chem.*, **15**, 931–941.
70. Kopeček, J. and Kopečková, P. (2011). 'Design of Polymer-Drug Conjugates', in Kratz, F., Senter, P. D., Steinhagen, H., (eds.), *Drug Delivery in Oncology*. Wiley-VCH, Weinheim, pp. 483–512.
71. Maeda, H., Seymour, L. W. and Miyamoto, Y. (1992). Conjugates of anticancer agents and polymers — advantagels of macromolecular therapeutics *in vivo. Bioconjug. Chem.*, **3**, 351–362.
72. Fang, J., Nakamura, H. and Maeda, H. (2011). The EPR effect: Unique features of tumor blood vessels for drug delivery, factors involved, and limitations and augmentation of the effect. *Adv. Drug Deliver Rev.*, **63**, 136–151.
73. Torchilin, V. (2011). Tumor delivery of macromolecular drugs based on the EPR effect. *Adv. Drug Deliver Rev.*, **63**, 131–135.
74. Hobbs, S. K., Monsky, W. L., Yuan, F., Roberts, W. G., Griffith, L., Torchilin, V. P. and Jain, R. K. (1998). Regulation of transport pathways in tumor vessels: Role of tumor type and microenvironment. *Proc. Natl. Acad. Sci.*, **95**, 4607–4612.
75. Shiah, J. G., Dvořák, M., Kopečková, P., Sun, Y., Peterson, C. and Kopeček, J. (2001). Biodistribution and antitumour efficacy of long-circulating *N*-(2-hydroxypropyl) methacrylamide copolymer–doxorubicin conjugates in nude mice. *Eur. J. Cancer*, **37**, 131–139.
76. Maeda, H., Nakamura, H. and Fang, J. (2013). The EPR effect for macromolecular drug delivery to solid tumors: Improvement of tumor uptake, lowering of systemic toxicity, and distinct tumor imaging *in vivo. Adv. Drug Deliver Rev.*, **65**, 71–79.
77. Byrne, J. D., Betancourt, T. and Brannon-Peppas, L. (2008). Active targeting schemes for nanoparticle systems in cancer therapeutics. *Adv. Drug Deliver Rev.*, **60**, 1615–1626.
78. Maruyama, K., Takahashi, N., Tagawa, T., Nagaike, K. and Iwatsuru, M. (1997). Immunoliposomes bearing polyethyleneglycol-coupled Fab′ fragment show prolonged circulation time and high extravasation into targeted solid tumors in vivo. *FEBS Lett.*, **413**, 177–180.
79. Fujimori, K., Covell, D. G., Fletcher, J. E. and Weinstein, J. N. (1989). Modeling analysis of the global and microscopic distribution of immunoglobulin G, F(ab′)2, and Fab in tumors. *Cancer Res.*, **49**, 5656–5663.
80. Johnson, R. N., Kopečková, P. and Kopeček, J. (2009). Synthesis and evaluation of multivalent branched HPMA copolymer-Fab′ conjugates targeted to the B-Cell antigen CD20. *Bioconjug. Chem.*, **20**, 129–137.
81. Johnson, R. N., Kopečková, P. and Kopeček, J. (2012). Biological activity of anti-CD20 multivalent HPMA copolymer-Fab′ conjugates. *Biomacromolecules*, **13**, 727–735.
82. Chu, T. W., Yang, J. and Kopeček, J. (2012). Anti-CD20 multivalent HPMA copolymer-Fab′ conjugates for the direct induction of apoptosis. *Biomaterials*, **33**, 7174–7181.

83. Zhou, Y., Yang, J. and Kopeček, J. (2012). Selective inhibitory effect of HPMA copolymer-cyclopamine conjugate on prostate cancer stem cells. *Biomaterials*, **33**, 1863–1872.
84. Zhou, Y., Yang, J., Rhim, J. S. and Kopeček, J. (2013). HPMA copolymer-based combination therapy toxic to both prostate cancer stem/progenitor cells and differentiated cells induces durable anti-tumor effects. *J. Controlled Release*, **172**, 946–953.
85. Karhadkar, S. S., Bova, G. S., Abdallah, N., Dhara, S., Gardner, D., Maitra, A., Isaacs, J. T., Berman, D. M. and Beachy, P. A. (2004). Hedgehog signalling in prostate regeneration, neoplasia and metastasis. *Nature*, **431**, 707–712.
86. Sanchez, P., Hernández, A. M., Stecca, B., Kahler, A. J., DeGueme, A. M., Barrett, A., Beyna, M., Datta, M. W., Datta, S. and i Altaba, A. R. (2004). Inhibition of prostate cancer proliferation by interference with SONIC HEDGEHOG-GLI1 signaling. *Proc. Natl. Acad. Sci. USA.*, **101**, 12561–12566.
87. Vasey, P. A., Kaye, S. B., Morrison, R., Twelves, C., Wilson, P., Duncan, R., Thomson, A. H., Murray, L. S., Hilditch, T. E., Murray, T., Burtles, S., Fraier, D., Frigerio, E. and Cassidy, J. (1999). Canc Res Campaign Phase IIIC. Phase I clinical and pharmacokinetic study of PK1 *N*-(2-hydroxypropyl)methacrylamide copolymer doxorubicin : First member of a new class of chemotherapeutic agents — Drug-polymer conjugates. *Clin. Cancer Res.*, **5**, 83–94.
88. Seymour, L. W., Ferry, D. R., Anderson, D., Hesslewood, S., Julyan, P. J., Poyner, R., Doran, J., Young, A. M., Burtles, S. and Kerr, D. J. (2002). Hepatic drug targeting: Phase I evaluation of polymer-bound doxorubicin. *J. Clin. Oncol.*, **20**, 1668–1676.
89. Yang, J., Luo, K., Pan, H., Kopečková, P. and Kopeček, J. (2011). Synthesis of biodegradable multiblock copolymers by click coupling of RAFT-generated heterotelechelic polyHPMA conjugates. *React. Funct. Polym.*, **71**, 294–302.
90. Luo, K., Yang, J. Y., Kopečková, P. and Kopeček, J. (2011). Biodegradable multiblock poly *N*-(2-hydroxypropyl)methacrylamide via reversible addition-fragmentation chain transfer polymerization and click chemistry. *Macromolecules*, **44**, 2481–2488.
91. Pan, H. Z., Yang, J. Y., Kopečková, P. and Kopeček, J. (2011). Backbone degradable multiblock *N*-(2-Hydroxypropyl)methacrylamide copolymer conjugates via reversible addition-fragmentation chain transfer polymerization and thiol-ene coupling reaction. *Biomacromolecules*, **12**, 247–252.
92. Pan, H., Sima, M., Yang, J. and Kopeček, J. (2013). Synthesis of long-circulating, backbone degradable HPMA copolymer–doxorubicin conjugates and evaluation of molecular-weight-dependent antitumor efficacy. *Macromol. Biosci.*, **13**, 155–160.
93. Zhang, R., Luo, K., Yang, J., Sima, M., Sun, Y., Janát-Amsbury, M. M. and Kopeček, J. (2013). Synthesis and evaluation of a backbone biodegradable multiblock HPMA copolymer nanocarrier for the systemic delivery of paclitaxel. *J. Controlled Release*, **166**, 66–74.
94. Etrych, T., Jelínková, M., Říhová, B. and Ulbrich, K. (2001). New HPMA copolymers containing doxorubicin bound via pH-sensitive linkage: Synthesis and preliminary in vitro and in vivo biological properties. *J. Controlled Release*, **73**, 89–102.
95. Hrubý, M., Kučka, J., Lebeda, O., Macková, H., Babič, M., Konák, C., Studenovský, M., Sikora, A., Kozempel, J. and Ulbrich, K. (2007). New bioerodable thermoresponsive polymers for possible radiotherapeutic applications. *J. Controlled Release*, **119**, 25–33.
96. Pack, D. W., Hoffman, A. S., Pun, S. and Stayton, P. S. (2005). Design and development of polymers for gene delivery. *Nat. Rev. Drug Discov.*, **4**, 581–593.

97. Murthy, N., Campbell, J., Fausto, N., Hoffman, A. S. and Stayton, P. S. (2003). Design and synthesis of pH-responsive polymeric carriers that target uptake and enhance the intracellular delivery of oligonucleotides. *J. Controlled Release*, **89**, 365–374.
98. Murthy, N., Campbell, J., Fausto, N., Hoffman, A. S. and Stayton, P. S. (2003). Bioinspired pH-responsive polymers for the intracellular delivery of biomolecular drugs. *Bioconjug. Chem.*, **14**, 412–419.
99. Murthy, N., Robichaud, J. R., Tirrell, D. A., Stayton, P. S. and Hoffman, A. S. (1999). The design and synthesis of polymers for eukaryotic membrane disruption. *J. Controlled Release*, **61**, 137–143.
100. Kang, H. C. and Bae, Y. H. (2007). pH-tunable endosomolytic oligomers for enhanced nucleic acid delivery. *Adv. Funct. Mater.*, **17**, 1263–1272.
101. Wang, C., Stewart, R. J. and Kopeček, J. (1999). Hybrid hydrogels assembled from synthetic polymers and coiled-coil protein domains. *Nature*, **397**, 417–420.
102. Yang, J. Y., Xu, C. Y., Wang, C. and Kopeček, J. (2006). Refolding hydrogels self-assembled from *N*-(2-hydroxypropyl)methacrylamide graft copolymers by antiparallel coiled-coil formation. *Biomacromolecules*, 7, 1187–1195.
103. Wu, K., Liu, J. H., Johnson, R. N., Yang, J. Y. and Kopeček, J. (2010). Drug-free macromolecular therapeutics: Induction of apoptosis by coiled-coil-mediated cross-linking of antigens on the cell surface. *Angew. Chem. Int. Ed.*, **49**, 1451–1455.
104. Wu, K., Yang, J., Liu, J. and Kopeček, J. (2012). Coiled-coil based drug-free macromolecular therapeutics: In vivo efficacy. *J. Controlled Release*, **157**, 126–131.
105. Chu, T. W., Yang, J., Zhang, R., Sima, M. and Kopeček, J. (2013). Cell surface self-assembly of hybrid nanoconjugates via oligonucleotide hybridization induces apoptosis. *ACS Nano.*, **8**, 719–730.
106. Israelachvili, J. N., Mitchell, D. J. and Ninham, B. W. (1976). Theory of self-assembly of hydrocarbon amphiphiles into micelles and bilayers. *J. Chem. Soc., Faraday Trans.*, **72**, 1525–1568.
107. Masayuki, Y., Kwon, G. S., Teruo, O., Yasuhisa, S., Mayumi, N. and Kazunori, K. (1994). Influencing factors on in vitro micelle stability of adriamycin-block copolymer conjugates. *J. Controlled Release*, **28**, 59–65.
108. Gao, Z. and Eisenberg, A. (1993). A model of micellization for block copolymers in solutions. *Macromolecules*, **26**, 7353–7360.
109. Rapoport, N. (2007). Physical stimuli-responsive polymeric micelles for anti-cancer drug delivery. *Prog. Polym. Sci.*, **32**, 962–990.
110. Geng, Y., Dalhaimer, P., Cai, S., Tsai, R., Tewari, M., Minko, T. and Discher, D. E. (2007). Shape effects of filaments versus spherical particles in flow and drug delivery. *Nat. Nano.*, **2**, 249–255.
111. Lavasanifar, A., Samuel, J. and Kwon, G. S. (2002). Poly(ethylene oxide)-block-poly(l-amino acid) micelles for drug delivery. *Adv. Drug Deliver Rev.*, **54**, 169–190.
112. Bae, Y. and Kataoka, K. (2009). Intelligent polymeric micelles from functional poly(ethylene glycol)-poly(amino acid) block copolymers. *Adv. Drug Deliver Rev.*, **61**, 768–784.
113. Lu, Y. and Low, P. S. (2012). Folate-mediated delivery of macromolecular anticancer therapeutic agents. *Adv. Drug Deliver Rev.*, **64**, 342–352.
114. Bae, Y., Nishiyama, N. and Kataoka, K. (2007). *In vivo* antitumor activity of the folate-conjugated pH-sensitive polymeric micelle selectively releasing adriamycin in the intracellular acidic compartments. *Bioconjug. Chem.*, **18**, 1131–1139.

115. Du, J. Z., Du, X. J., Mao, C. Q. and Wang, J. (2011). Tailor-made dual pH-sensitive polymer–doxorubicin nanoparticles for efficient anticancer drug delivery. *J. Am. Chem. Soc.*, **133**, 17560–17563.
116. Husseini, G. A., Myrup, G. D., Pitt, W. G., Christensen, D. A. and Rapoport, N. Y. (2000). Factors affecting acoustically triggered release of drugs from polymeric micelles. *J. Controlled Release*, **69**, 43–52.
117. Kabanov, A. V. and Alakhov, V. Y. (2002). Pluronic® block copolymers in drug delivery: From micellar nanocontainers to biological response modifiers. *Crit. Rev. Ther. Drug Carrier Syst.*, **19**, 72.
118. Pruitt, J. D., Husseini, G., Rapoport, N. and Pitt, W. G. (2000). Stabilization of pluronic P-105 micelles with an interpenetrating network of *N,N*-diethylacrylamide. *Macromolecules*, **33**, 9306–9309.
119. Nelson, J. L., Roeder, B. L., Carmen, J. C., Roloff, F. and Pitt, W. G. (2002). Ultrasonically activated chemotherapeutic drug delivery in a rat model. *Cancer Res.*, **62**, 7280–7283.
120. Meng, F., Zhong, Z. and Feijen, J. (2009). Stimuli-responsive polymersomes for programmed drug delivery. *Biomacromolecules*, **10**, 197–209.
121. Photos, P. J., Bacakova, L., Discher, B., Bates, F. S. and Discher, D. E. (2003). Polymer vesicles in vivo: Correlations with PEG molecular weight. *J. Controlled Release*, **90**, 323–334.
122. Chécot, F., Lecommandoux, S., Klok, H. A. and Gnanou, Y. (2003). From supramolecular polymersomes to stimuli-responsive nano-capsules based on poly(diene-b-peptide) diblock copolymers. *Eur. Phys. J. E*, **10**, 25–35.
123. Chécot, F., Rodriguez-Hernandez, J., Gnanou, Y. and Lecommandoux, S. (2007). pH-responsive micelles and vesicles nanocapsules based on polypeptide diblock copolymers. *Biomol. Eng.*, **24**, 81–85.
124. Ahmed, F., Pakunlu, R. I., Srinivas, G., Brannan, A., Bates, F., Klein, M. L., Minko, T. and Discher, D. E. (2006). Shrinkage of a rapidly growing tumor by drug-loaded polymersomes: pH-triggered release through copolymer degradation. *Mol. Pharmaceutics*, **3**, 340–350.
125. Bellomo, E. G., Wyrsta, M. D., Pakstis, L., Pochan, D. J. and Deming, T. J. (2004). Stimuli-responsive polypeptide vesicles by conformation-specific assembly. *Nat Mater.*, **3**, 244–248.
126. Borchert, U., Lipprandt, U., Bilang, M., Kimpfler, A., Rank, A., Peschka-Süss, R., Schubert, R., Lindner, P. and Förster, S. (2006). pH-induced release from P2VP–PEO block copolymer vesicles. *Langmuir*, 22, 5843–5847.
127. Qin, S., Geng, Y., Discher, D. E. and Yang, S. (2006). Temperature-controlled assembly and release from polymer vesicles of poly(ethylene oxide)-block-poly(N-isopropylacrylamide). *Adv. Mater.*, **18**, 2905–2909.
128. Cerritelli, S., Velluto, D. and Hubbell, J. A. (2007). PEG-SS-PPS: Reduction-sensitive disulfide block copolymer vesicles for intracellular drug delivery. *Biomacromolecules*, **8**, 1966–1972.
129. Tong, X., Wang, G., Soldera, A. and Zhao, Y. (2005). How can azobenzene block copolymer vesicles be dissociated and reformed by light? *J. Phys. Chem. B*, **109**, 20281–20287.
130. Lecommandoux, S., Sandre, O., Chécot, F., Rodriguez-Hernandez, J. and Perzynski, R. (2005). Magnetic nanocomposite micelles and vesicles. *Adv. Mater.*, **17**, 712–718.

131. Zhou, W., Meng, F., Engbers, G. H. M. and Feijen, J. (2006). Biodegradable polymersomes for targeted ultrasound imaging. *J. Controlled Release*, **116**, e62–e64.
132. Jiwpanich, S., Ryu, J. H., Bickerton, S. and Thayumanavan, S. (2010). Noncovalent encapsulation stabilities in supramolecular nanoassemblies. *J. Am. Chem. Soc.*, **132**, 10683–10685.
133. Qiao, Z. Y., Zhang, R., Du, F. S., Liang, D. H. and Li, Z. C. (2011). Multi-responsive nanogels containing motifs of ortho ester, oligo(ethylene glycol) and disulfide linkage as carriers of hydrophobic anti-cancer drugs. *J. Controlled Release*, **152**, 57–66.
134. Blackburn, W. and Lyon, L. A. (2008). Size-controlled synthesis of monodisperse core/shell nanogels. *Colloid. Polym. Sci.*, **286**, 563–569.
135. Read, E. S. and Armes, S. P. (2007). Recent advances in shell cross-linked micelles. *Chem. Commun.*, 3021–3035.
136. Li, Y., Lokitz, B. S., Armes, S. P. and McCormick, C. L. (2006). Synthesis of reversible shell cross-linked micelles for controlled release of bioactive agents. *Macromolecules*, **39**, 2726–2728.
137. Quadir, M. A. and Haag, R. (2012). Biofunctional nanosystems based on dendritic polymers. *J. Controlled Release*, **161**, 484–495.
138. Calderón, M., Quadir, M. A., Strumia, M. and Haag, R. (2010). Functional dendritic polymer architectures as stimuli-responsive nanocarriers. *Biochimie.*, **92**, 1242–1251.
139. Xu, S., Luo, Y., Graeser, R., Warnecke, A., Kratz, F., Hauff, P., Licha, K. and Haag, R. (2009). Development of pH-responsive core–shell nanocarriers for delivery of therapeutic and diagnostic agents. *Bioorg. Med. Chem. Lett.*, **19**, 1030–1034.
140. Fischer, W., Calderón, M., Schulz, A., Andreou, I., Weber, M. and Haag, R. (2010). Dendritic polyglycerols with oligoamine shells show low toxicity and high siRNA transfection efficiency in vitro. *Bioconjug. Chem.*, **21**, 1744–1752.
141. Beck-Broichsitter, M., Schmehl, T., Gessler, T., Seeger, W. and Kissel, T. (2012). Development of a biodegradable nanoparticle platform for sildenafil: Formulation optimization by factorial design analysis combined with application of charge-modified branched polyesters. *J. Controlled Release*, **157**, 469–477.
142. Reul, R., Nguyen, J. and Kissel, T. (2009). Amine-modified hyperbranched polyesters as non-toxic, biodegradable gene delivery systems. *Biomaterials*, **30**, 5815–5824.
143. Hardy, J. G., Kostiainen, M. A., Smith, D. K., Gabrielson, N. P. and Pack, D. W. (2005). Dendrons with spermine surface groups as potential building blocks for nonviral vectors in gene therapy. *Bioconjug. Chem.*, **17**, 172–178.
144. Jones, S. P., Gabrielson, N. P., Wong, C. H., Chow, H. F., Pack, D. W., Posocco, P., Fermeglia, M., Pricl, S. and Smith, D. K. (2011). Hydrophobically modified dendrons: Developing structure–activity relationships for DNA binding and gene transfection. *Mol. Pharmaceutics*, **8**, 416–429.
145. Etrych, T., Strohalm, J., Chytil, P., Černoch, P., Starovoytova, L., Pechar, M. and Ulbrich, K. (2011). Biodegradable star HPMA polymer conjugates of doxorubicin for passive tumor targeting. *Eur. J. Pharm. Sci.*, **42**, 527–539.
146. Meyer, D. E., Shin, B. C., Kong, G. A., Dewhirst, M. W. and Chilkoti, A. (2001). Drug targeting using thermally responsive polymers and local hyperthermia. *J. Controlled Release*, **74**, 213–224.
147. Chilkoti, A., Christensen, T., MacKay, J. A. (2006). Stimulus responsive elastin biopolymers: applications in medicine and biotechnology. *Curr. Opin. Chem. Biol.*, **10**, 652–657.

148. Urry, D. W., Parker, T. M., Reid, M. C. and Gowda, D. C. (1991). Biocompatibility of the bioelastic materials, poly(GVGVP) and its γ-irradiation cross-linked matrix: Summary of generic biological test results. *J. Bioact. Compat. Polym.*, **6**, 263–282.
149. Raucher, D. and Chilkoti, A. (2001). Enhanced uptake of a thermally responsive polypeptide by tumor cells in response to its hyperthermia-mediated phase transition. *Cancer Res.*, **61**, 7163–7170.
150. Furgeson, D. Y., Dreher, M. R. and Chilkoti, A. (2006). Structural optimization of a "smart" doxorubicin–polypeptide conjugate for thermally targeted delivery to solid tumors. *J. Controlled Release*, **110**, 362–369.
151. Borchmann, D. E., Carberry, T. P. and Weck, M. (2014). "Bio"-macromolecules: Polymer-protein conjugates as emerging scaffolds for therapeutics. *Macromol. Rapid Commun.*, **35**, 27–43.
152. Levy, Y., Hershfield, M. S., Fernandez-Mejia, C., Polmar, S. H., Scudiery, D., Berger, M. and Sorensen, R. U. (1988). Adenosine deaminase deficiency with late onset of recurrent infections: Response to treatment with polyethylene glycol-modified adenosine deaminase. *J. Pediatr.*, **113**, 312–317.
153. Graham, M. L. (2003). Pegaspargase: A review of clinical studies. *Adv. Drug Deliv. Rev.*, **55**, 1293–1302.
154. Wang, Y. S., Youngster, S., Grace, M., Bausch, J., Bordens, R. and Wyss, D. F. (2002). Structural and biological characterization of pegylated recombinant interferon alpha-2b and its therapeutic implications. *Adv. Drug Deliv. Rev.*, **54**, 547–570.
155. Daud, A., Soon, C., Dummer, R., Eggermont, A. M., Hwu, W. J., Grob, J. J., Garbe, C. and Hauschild, A. (2012). Management of pegylated interferon alpha toxicity in adjuvant therapy of melanoma. *Expert Opin. Biol. Ther.*, **12**, 1087–1099.
156. Bailon, P., Palleroni, A., Schaffer, C. A., Spence, C. L., Fung, W. J., Porter, J. E., Ehrlich, G. K., Pan, W., Xu, Z. X. and Modi, M. W. (2001). Rational design of a potent, long-lasting form of interferon: A 40 kDa branched polyethylene glycol-conjugated interferon α-2a for the treatment of hepatitis C. *Bioconjug. Chem.*, **12**, 195–202.
157. Kopchick, J., Parkinson, C., Stevens, E. and Trainer, P. J. (2002). Growth hormone receptor antagonists: Discovery, development, and use in patients with acromegaly. *Endocr. Rev.*, **23**, 623–646.
158. Noël, A., Jost, M., Lambert, V., Lecomte, J. and Rakic, J. M. (2007). Anti-angiogenic therapy of exudative age-related macular degeneration: Current progress and emerging concepts. *Trends Mol. Med.*, **13**, 345–352.
159. Kinstler, O., Molineux, G., Treuheit, M., Ladd, D. and Gegg, C. (2002). Mono-*N*-terminal poly(ethylene glycol)–protein conjugates. *Adv. Drug Deliv. Rev.*, **54**, 477–485.
160. Macdougall, I. C. and Eckardt, K. U. (2006). Novel strategies for stimulating erythropoiesis and potential new treatments for anaemia. *The Lancet*, **368**, 947–953.
161. Sandborn, W. J., Feagan, B. G., Stoinov, S., Honiball, P. J., Rutgeerts, P., Mason, D., Bloomfield, R. and Schreiber, S. (2007). Certolizumab pegol for the treatment of Crohn's disease. *N. Engl. J. Med.*, **357**, 228–238.
162. Biggers, K. and Scheinfeld, N. (2008). Pegloticase, a polyethylene glycol conjugate of uricase for the potential intravenous treatment of gout. *Curr. Opin. Investig. Drugs*, **9**, 422–429.
163. Stead, R. B., Lambert, J., Wessels, D., Iwashita, J. S., Leuther, K. K., Woodburn, K. W., Schatz, P. J., Okamoto, D. M., Naso, R. and Duliege, A. M. (2006). Evaluation of the

safety and pharmacodynamics of Hematide, a novel erythropoietic agent, in a phase 1, double-blind, placebo-controlled, dose-escalation study in healthy volunteers. *Blood*, **108**, 1830–1834.

164. Gaspar, R. and Duncan, R. (2009). Polymeric carriers: Preclinical safety and the regulatory implications for design and development of polymer therapeutics. *Adv. Drug Deliv. Rev.*, **61**, 1220–1231.
165. Ishida, T., Wang, X., Shimizu, T., Nawata, K. and Kiwada, H. (2007). PEGylated liposomes elicit an anti-PEG IgM response in a T cell-independent manner. *J. Controlled Release*, **122**, 349–355.
166. Schellekens, H., Hennink, W. and Brinks, V. (2013). The immunogenicity of polyethylene glycol: Facts and fiction. *Pharm. Res.*, **30**, 1729–1734.
167. Nel, A. E., Madler, L., Velegol, D., Xia, T., Hoek, E. M. V., Somasundaran, P., Klaessig, F., Castranova, V. and Thompson, M. (2009). Understanding biophysicochemical interactions at the nano-bio interface. *Nat. Mater.*, **8**, 543–557.
168. Rust, M. J., Bates, M. and Zhuang, X. (2006). Sub-diffraction-limit imaging by stochastic optical reconstruction microscopy (STORM). *Nat. Meth.*, **3**, 793–796.
169. Betzig, E., Patterson, G. H., Sougrat, R., Lindwasser, O. W., Olenych, S., Bonifacino, J. S., Davidson, M. W., Lippincott-Schwartz, J. and Hess, H. F. (2006). Imaging intracellular fluorescent proteins at nanometer resolution. *Science*, **313**, 1642–1645.
170. Kiick, K. L. (2007). Polymer Therapeutics. *Science*, **317**, 1182–1183.

Chapter 12

Inorganic Nanoparticle-Based Smart Drug Delivery Systems

Jutaek Nam, Sungwook Jung, Sekyu Hwang, Jaejung Song, and Sungjee Kim

Department of Chemistry, Pohang University of Science and Technology, Pohang, Republic of Korea

12.1 Introduction

Inorganic nanoparticles (NPs) have offered many opportunities in biological research fields, providing a novel platform for diagnostics, therapy, and biosensing. The application of inorganic NPs for the delivery of therapeutic agents is a rapidly expanding field with prominent progress because of their inherent promising properties. Inorganic NPs are generally more stable and inert in biological environments than conventional organic material-based drug delivery platforms, such as polymers, dendrimers, micelles, and liposomes. Their large surface area from a high surface-to-volume ratio renders drug reservoirs for effective loading of a variety of therapeutics, from small molecules to large biomolecules, like protein, ribonucleic acid (RNA), or deoxyribonucleic acid (DNA). The pharmacological properties of payload drugs can be greatly improved on the NP surfaces because of the positive attribute of NPs to the solubility, *in vivo* stability, and biodistribution. Inorganic NPs can enhance the accumulation of therapeutics at the site of interest when administered *in vivo*. For example, the unique size scale of inorganic NPs with much larger sizes than small molecules, yet comparable to biological units, such as proteins, can enable their preferential accumulations in tumors through the enhanced permeability and retention effect (EPR) [1, 2]. The dimension is also well suited to penetrate various biological barriers for drug delivery *in vivo,* including gastrointestinal tracts with oral/intraperitoneal injection, skin penetration with transdermal delivery, and air/blood barrier of lungs with intratracheal instillation. In addition, site-selective active delivery can be achieved by conjugating NPs with targeting molecules that recognize specific cell types [3–5]. Contrary to conventional molecule agents, which typically have a single conjugation site, NPs can usually have a number of targeting molecules on the surface because they are usually much larger than targeting molecules and have multiple conjugation sites. The multiple displays of targeting molecules on NPs can enhance the target binding affinity and subsequent cellular uptake as the result of multivalent interactions [6].

An ideal drug delivery system should allow controlled delivery of therapeutic agents in a target site, and subsequent controlled release is also important to fully activate therapeutic actions. Inorganic NPs possess great possibility in their use as versatile drug delivery platforms by taking advantages of facile physicochemical tunabilities upon on-demand design by exploiting well-developed synthetic protocols and flexible surface functionalities. So far, many kinds of inorganic NPs, including semiconductor, metal, metal oxide, and lanthanide-doped NPs have been developed and demonstrated for biological applications [7]. From previous cases, useful perspectives can be gathered on the proper NP design for the fabrication of effective drug delivery platforms. We discuss important NP design parameters that should be considered for efficient payload delivery. Sophisticated fabrication of NPs can also allow the development of stimuli-responsive "smart" drug delivery systems. Finally, some examples of "smart" delivery systems that exploit internal or external stimuli for spatiotemporally controlled payload release are to be discussed.

12.2 Design Consideration of Inorganic NPs for Delivery Application

One of major goals in drug delivery applications is to increase the efficacy of selective delivery into the site of interest by proper design of the delivery platform. The fate of inorganic NPs in biological environments is largely dependent on their physicochemical properties, such as hydrodynamic (HD) size and surface charge. Conjugation of targeting molecules that strongly interact with desired tissues or cells allows target-selective delivery of NPs with less concern over side effects caused by nonspecific distribution of drugs. The following factors play major roles in the selective drug delivery design, which will be discussed in more detail.

12.2.1 *HD size*

Unlike small molecules, NPs and macromolecules larger than a few nanometers in size are generally not able to diffusively pass through the membranes and should be mediated by the membrane-bound vesicles to enter the cells. Inorganic NPs with broad size ranges from a few to several hundred nanometers can enter cells via a series of processes termed endocytosis, of which efficiency is largely dependent on the NPs' HD size. Chithrani *et al.* reported 50 nm as the optimal size for maximal gold NP uptake into various cells among particles with 10–100 nm in diameters (Fig. 12.1) [8, 9]. It was argued that the largest cellular uptake for 50-nm NPs is the result of appropriate balance between the fastest endocytosis and moderate exocytosis rate by demonstrating that the cellular uptake rate is highest for 50-nm NPs, whereas the removal rate is faster for smaller NPs. Similar conclusions were drawn from investigations by other groups: Osaki *et al.* demonstrated that 50-nm-sized artificial glycovirus composed by plasmid DNA and micellar glycocluster particles entered Henrietta Lacks (HeLa) cells more efficiently than 5-nm-sized micellar

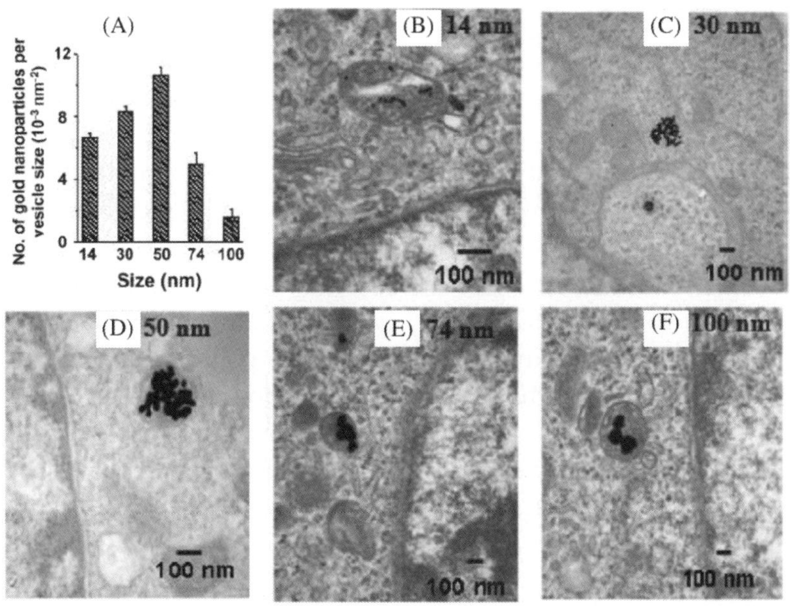

Figure 12.1. Uptake of gold NPs into HeLa cells. (A) Number of NPs by the vesicle diameter versus the NP size. (B–F) TEM images of gold NPs with 14, 30, 50, 74, and 100 nm diameter trapped in endocytotic vesicles [8]. NP = nanoparticle, TEM = transmission electron microscopy.

glycocluster compartments or 15-nm-sized quantum dot (QD) conjugated sugar balls [10], and Wang *et al.* observed the maximal cellular uptake for 45-nm gold NP with the uptake level becoming smaller as the size of gold NPs increased from 45 to 110 nm [11]. The investigations have demonstrated that the maximal cellular uptake of NPs can take place in the size of 40–60 nm, among a range of several to hundred nanometers.

Although many factors, including NP ligand density, membrane adhesion area, membrane stretching, and bending, may be involved in the size-dependent uptake of NPs, the underlying mechanism can be simply explained by the balance between thermodynamic driving forces for wrapping and receptor diffusion kinetics [8, 9]. Membrane-mediated wrapping of NPs at the first stage of endocytosis is a key step that is driven by chemical energy released through ligand–receptor interactions, where receptor depletion limits the total available interaction sites. Small NPs have fewer ligands for interaction with membrane receptors, so adsorption of NPs may not produce enough energy to overcome unfavorable elastic deformation of the membrane, which acts as a thermodynamic barrier for endocytosis [12]. They may dissociate from the receptors before being engulfed by the membrane due to the low binding affinity. In contrast, large NPs can present multivalent binding with membrane receptors by larger numbers of ligands on the surface, which can produce sufficient energy for complete membrane wrapping. However, as large NPs use a large number of membrane receptors, receptor depletion in nearby areas precludes endocytosis of forthcoming NPs by the limited receptor diffusion. The intermediate size range of 40–60 nm seems optimal in terms of multivalent ligand–receptor interactions, yet without inducing significant membrane receptor depletion.

HD size also greatly influences delivery and distribution of NPs *in vivo*. The four major routes for NP delivery *in vivo* include intravenous administration, oral/intraperitoneal injection, transdermal pathway, and intratracheal instillation. Intravenous administration into blood circulation is the most widely used route, which can deliver NPs into many organs through the endothelium of blood vessels. Generally, the majority of intravenously administrated NPs are found in the reticuloendothelial system (RES), such as the liver and spleen, which are usually responsible for clearing foreign particles from blood circulation [13, 14]. The RES can readily eliminate NPs with an HD larger than 10 nm, and this typically results in reduced circulation half-lives for large NPs. When the HD size is reduced to sub-10 nm, smaller accumulation in RESs and more broad distribution over many organs are observed, as the result of being capable of penetrating various organ barriers of specific pore sizes [15]. For example, the HD size limit for the renal barrier was reported to be ~5.5 nm [16] and ~34 nm for the penetration into lung [17], in which only NPs smaller than these limits can enter the organs. Reduction of the HD size below the limit of the renal barrier can achieve fast *in vivo* clearance of NPs through renal filtration and urinary excretions, which is an important aspect of NP design, especially when long-term toxicity of NPs is concerned [16]. Pore size of the endothelial walls is the primary barrier for NP delivery to desired target tissues with intravenous administration. Normal endothelial cells are tightly bound to each other, limiting permeation of NPs, whereas tumor endothelial cells are loosely ordered, comprising leaky blood vasculatures with large-sized pores. As a result, EPR of NPs can be achieved in tumors over normal organs (Fig. 12.2) [2]. NPs smaller than the cutoff size of tumor transvascular pores, the smallest reported to be 100–200 nm [18, 19], can passively accumulate in tumors by the EPR effect, which becomes the most popular strategy for tumor targeting of NPs. Popović *et al.* showed that small NPs can extravasate easily and diffuse away from the leaky blood vessels, whereas extravasation of large NPs was limited by the transvascular pore barriers in the vessel wall [20].

Tumor accumulation levels by the EPR effect is dependent on the diffusion rate into the tumor and subsequent clearance rate out of the tumor, and both are affected by the HD size of NPs. Perrault *et al.* examined the effect of the HD size of NPs on tumor accumulation levels using gold NPs with HD size in the 20–100 nm ranges (Fig. 12.3) [21]. The highest tumor accumulation

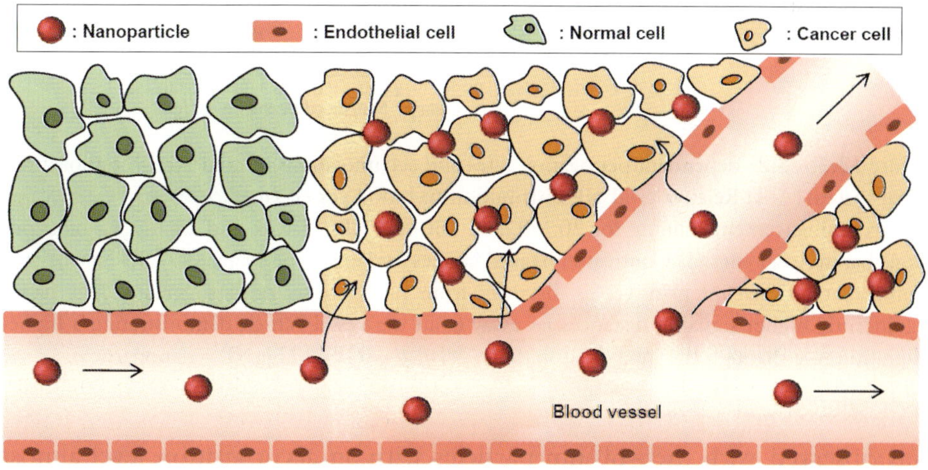

Figure 12.2. Schematic representation of passive tumor targeting by EPR effect. NPs can passively extravasate into tumor through leaky blood vasculature and can result in significant accumulations. NP = nanoparticle.

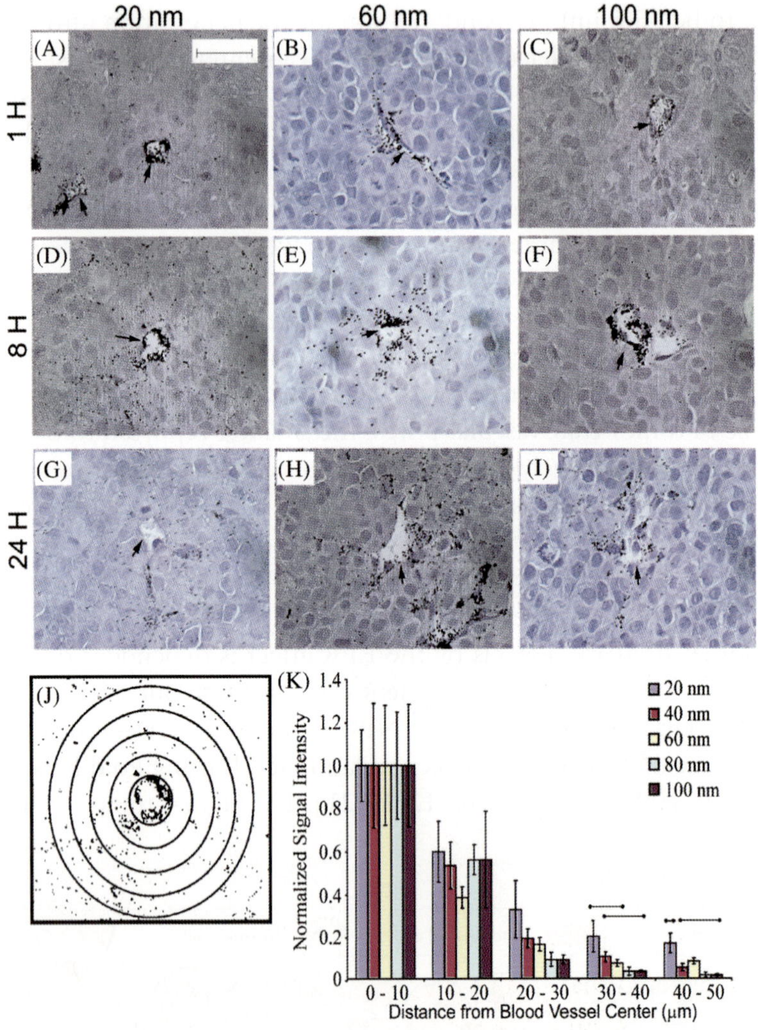

Figure 12.3. Size-dependent permeation rates and accumulations of gold NPs in tumor interstitial spaces. (A–I) Relative gold NP distribution to blood vessels (marked by arrows) revealed by silver enhancement staining for 20, 60, and 100 nm NPs at 1, 8, and 24 h post-injections (scale bar in A: 40 μm). (J) Contrast-enhanced images for densitometry analysis. (K) Normalized densitometry signal intensity versus distances from blood vessel centers. Smaller NPs had a significantly higher signal at distances far away from the vessel center, demonstrating a higher migration rate through the interstitial space ($n = 10$, $p < 0.05$) [21]. NP = nanoparticle.

was observed with the ~60-nm particles with an accumulation level almost two orders of magnitude larger when compared with that of ~20-nm particles, which resulted from the enhanced diffusion into the tumor by the longer blood circulation time along with the reduced clearance from the tumor by the slow permeation rate in tumor interstitial space. For a maximal tumor accumulation, NPs should have long blood circulation time and should be large enough to prevent rapid penetration/clearance through tumor interstitial space as long as they can extravasate into tumor transvascular pores.

NPs can also be administered through oral/intraperitoneal injection, the transdermal pathway, and intratracheal instillation. Similar to cases of intravenous administration, *in vivo* distribution of NPs is largely determined by their HD size. For oral administration, Hillyer *et al.* studied size dependence of biodistribution of 4–58-nm gold NPs and found that penetration from the gastrointestinal tract into other organs typically decreased with increasing particle size [22]. In the gastrointestinal tract, NPs can attach to mucus and transport across the mucus mesh that allows entry to the underlying epithelia. The NPs larger than 10 nm were mostly found in the small intestine and stomach because of their inefficient mucus penetrations, whereas smaller particles broadly distributed to many organs, with the largest accumulation found in the kidney, among others. In contrast, when 13-nm gold NPs were administered into mice via intraperitoneal injection, the majority of particles accumulated in the liver and spleen followed by stomach, small intestine, and kidney, suggesting a different uptake mechanism and easier penetration through epithelia [23]. Upon the transdermal delivery, a significantly larger amount of gold NPs was observed in the deeper region of skin with decreasing particle size within the range of 15–198 nm, indicating a higher permeation rate for smaller particles [24]. NPs are supposed to be delivered by follicular penetration through skin appendages, such as hair follicles and sweat ducts. Large particles such as 198-nm gold NPs were mostly confined in the epidermis region because of the size limit of follicular penetration. Small NPs are also preferred for NP delivery through intratracheal instillation. Semmler-Behnke *et al.* demonstrated that small gold NPs (1.4 nm) can effectively pass the air/blood barrier of the lung, resulting in significant distribution to blood, liver, skin, and carcass after 24 h, whereas 18-nm particles are almost completely trapped in the lungs [15]. Similarly, Sadauskas *et al.* reported that 2-nm gold NPs showed higher translocation into systemic circulation than 40- and 100-nm gold NPs when administered by inhalation [25]. Once particles are introduced into systemic circulation through the air/blood barrier of the lung, they are supposed to follow the fate of intravenously administered NPs. For the fabrication of drug delivery systems, NP HD size should be carefully controlled to achieve efficient accumulation in the target tissue, while reducing non-specific uptakes by other organs, such as the RES. When considering safety, NPs should also allow facile removal from the body to prevent toxic side effects for non-targeted tissues. However, the above requirements are usually contradictory; efficient clearance from the body to avoid adverse side effects is accompanied by the reduction of the *in vivo* residence time required for sufficient accumulation to the target tissue and also results in fast clearance from the target tissue. One way to address this dilemma is to develop a flexible platform that can change its HD size at physiological conditions. To increase *in vivo* residence time, for example, sub-100-nm large clusters consisted of sub-5-nm gold NPs and biodegradable polymers have been developed that can provide sufficient blood residence time for accumulation in a target tissue, while effective renal clearance can be achieved once the polymer component is degraded, releasing small gold NPs [26]. For a longer retention in the target tissue, NPs can be designed to modulate their size in response to selective stimuli; small NPs that can be selectively clustered and form large aggregates in a target tissue would be effectively accumulated after internalization by the increased size, while most untargeted NPs can be rapidly cleared from the body through the renal pathway. Smart NP design that can meet the conflicting demands can open a new avenue of NP applications as a versatile drug delivery vehicle.

12.2.2 *Surface charge*

Surface charge is primarily important for the stable dispersion of NPs in aqueous solution by electrostatic stabilization, where highly charged NP surfaces are often necessary to guarantee the colloidal stability. Besides, surface charge also has a vital role in successful biological application of NPs, since it greatly affects interactions with a variety of biomolecules. The cellular interaction is the first concern for drug delivery. NPs should make contact with and attach to cell surfaces before they can enter the cells. NPs with positively charged surfaces are favored for contact/adsorptions on the membranes because cellular membranes largely consist of negatively charged domains, such as phospholipids with negatively charged head groups and polysulfated proteoglycans [27]. Cho *et al.* have examined the role of surface charge in cellular adsorption and internalization of gold NPs [28]. They showed that positively charged NPs can be more adsorbed on the cell membranes, resulting in a higher level of internalizations when compared with the negatively charged or neutral NPs (Fig. 12.4). It was revealed that the membrane adsorption process is the rate-limiting step for the cellular uptake of NPs because it was much slower than the internalization process, where positively charged NPs can show larger cellular uptake by facile adsorption on the negatively charged cell surface. Uptake of negatively charged particles was also evident and more efficient when compared with the uptake of neutral NPs, presumably due to the interaction with positively charged membrane domains on the cellular surface, though the positive domains should be much scarcer than the negative domains.

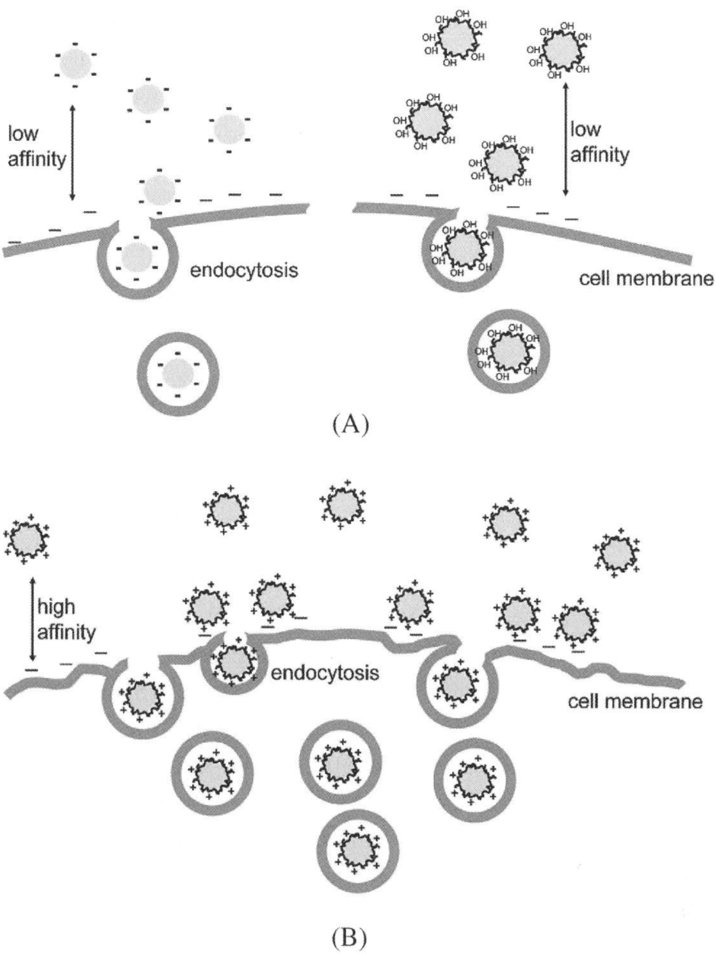

Figure 12.4. Schematics illustrating the interactions between gold NPs and SK-BR-3 cells. (A) Citrate-coated (negatively charged) and poly(vinyl alcohol)-coated (neutral) gold NPs show low affinity with cell membranes. (B) Poly(allyamine hydrochloride)-coated (positively charged) gold NPs display high cell membrane binding affinity [28]. NP = nanoparticle.

Neutral NPs showed the least adsorption on the cell surfaces, which resulted in the lowest level of uptake.

It seems that charged NPs can take advantage of electrostatic attractions with cellular membranes to enhance membrane adsorption and subsequent cellular internalization. In addition, charged NPs can induce surface reconstruction at the points they adsorb on the lipid membrane bilayers; binding of negatively charged NPs causes local gelation in otherwise fluid bilayers, while positively charged NPs induce local fluidization of otherwise gelled bilayers [29]. The structural change into a more fluidic phase may disrupt bilayer integrity and cause direct cell penetration of positively charged NPs without being mediated by endocytotic pathways. Lerueil *et al.* have demonstrated that many types of cationic materials, including cell-penetrating peptide, cationic dendrimers, polymers, and positively charged inorganic NPs, induced membrane-disrupting hole formation in lipid bilayers, regardless of their physical properties, such as size, shape, chemical composition, and deformability [30]. Since the transport across the membrane hole is an energy-independent diffusive process, it should be more favorable than energy-dependent endocytotic pathways. This may be another reason why positively charged NPs show enhanced cellular uptake when compared with negatively charged or neutral NPs.

Surface charge also plays a crucial role in *in vivo* biodistribution and clearance of NPs. Like the adsorption on cell surfaces, charged NPs can also non-specifically bind to other biomolecules *in vivo* via electrostatic interactions. Non-specific interactions with the extracellular matrix and non-targeted cell membrane possibly hamper the efficacy of drug delivery into the target tissue. When NPs are administrated via intravenous injections, adsorption of abundant plasma proteins may promote undesired recognition and clearance by the RES cells [31]. Aggregations of NPs can also be induced during non-specific interactions with such biomolecules, which complicates the *in vivo* applications of NPs. To overcome these aforementioned issues and increase the efficacy of target delivery, NPs are suggested to have a neutral surface charge and high colloidal stability at physiological conditions. Neutral surface ligands with hydroxyl groups, especially poly(ethyleneglycol) (PEG), have been widely used to reduce the non-specific interactions of NPs, while maintaining high colloidal stability. The low non-specific binding of PEGylated NPs with plasma proteins reduces their clearance by the RES, which results in a longer blood circulation time and efficient accumulation in tumors [32, 33]. The molecular weight of PEG for the generation of anti-fouling NP surface is generally reported to be 5–10 kDa [34, 21, 35–37]. However, it is also dependent on the NP core size, in which the combination of smaller NPs and longer PEG chains results in the most improved half-life in blood circulation (Fig. 12.5) [21].

PEG can be replaced by small zwitterionic molecules that act the same way to PEG by net neutral surface charge with great aqueous solubility in various conditions. Park *et al.* have systemically studied the effects of zwitterionic QD surfaces compared with QDs that are positively or negatively surface charged [38]. The zwitterionic QDs were colloidally stable with nearly neutral surface charge under broad pH ranges and in high salt concentrations. The zwitterionic QDs showed minimal non-specific adsorptions to various surfaces including polystyrene and bovine serum albumin, and to cell surfaces (Fig. 12.6).

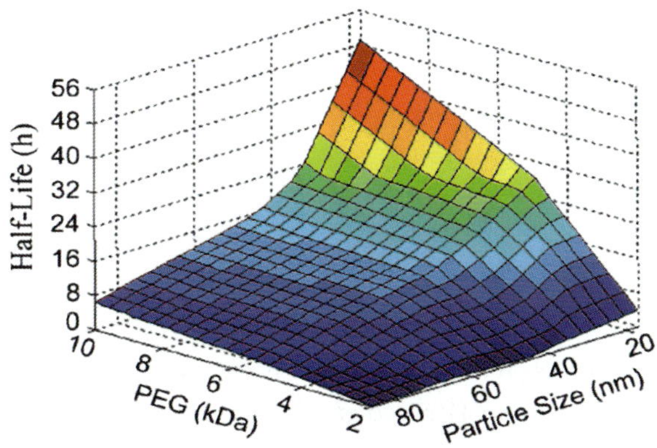

Figure 12.5. Pharmacokinetics mapping of blood half-life against the design parameters of core size and methoxy-PEG (mPEG) molecular weight. Five gold NPs of different core diameters (18–87 nm) and three different mPEG molecular weight brush layers (2, 5, 10 kDa) were used to determine the blood half-life (hours) as the function of particle size (nm) and mPEG molecular weight (kDa) [21].

An ideal NP platform for drug delivery is desired to have good colloidal stability, controllable HD size, and minimal non-specific adsorption levels. The zwitterionic NP surface coating by small surface ligands can be a promising approach to meet the challenges for NP probes. Neutral NPs with PEG surface coating have also been widely used under other administration pathways other than intravenous injection. For oral injection, NPs should adhere to mucosal surfaces to penetrate the gastrointestinal tract and subsequently translocate into systemic circulation to reach the target organ. However, adhesion that is too strong causes mucociliary clearance, inhibiting efficient penetration across the mucus layers. PEG chains provide the necessary adhesive interactions with the mucus network, while effectively avoiding undesirably strong mucus adhesion [39], and as a result, greatly improving the NP transports through diffusion in undiluted human mucus [40]. Neutral surface charge and negligible protein adsorption resulted in rapid penetration across mucus barriers. The high surface coverage of PEG improved the transport rate, whereas large increases in the molecular weight of PEG resulted in the slower transport, suggesting that dense surface coverage of low molecular weight PEG is desired for PEG-coated NPs to rapidly penetrate mucus layers [41]. Passing the air/blood barrier of the lungs after intratracheal instillation is less affected by the surface charge of NPs and mainly dependent on their HD size. However, PEG coating can improve the blood circulation time once NPs are translocated into the systemic circulation [42]. Transdermal penetration of NPs by skin permeation and capillary bed perfusion is dependent on the surface

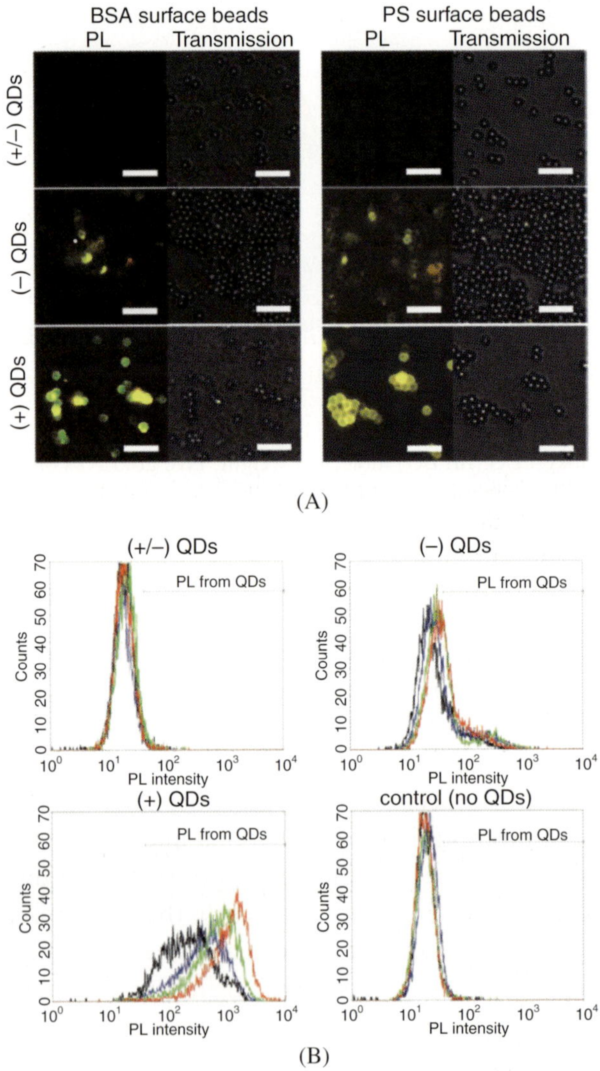

Figure 12.6. Evaluation of non-specific adsorption levels QDs with different surface coating. (A) Non-specific adsorptions of zwitterionic charged (+/−) (top), negatively charged (−) (middle), and positively charged (+) (bottom) QDs on bovine serum albumin surface beads (left) and on polystyrene beads (right); (scale bar: 5 μm). (B) Histogram data of the photoluminescence intensity of HeLa cells co-incubated with (+/−) (top left), (−) (top right), (+) (bottom left) QDs, and a control sample with no QDs (bottom right); ([QD] = 30 nM); incubation times: 30 min: black; 60 min: blue; 90 min: green; 120 min: blue [38]. QD = quantum dot, BSA = bovine serum albumin, PS = polystyrene, PL = photoluminescence.

charge of NPs. Lee *et al.* showed that QDs with carboxylic acid surface coatings had greater tissue deposition than PEG-coated QDs, which is supposed to be the result of agglomeration, tissue binding, or pharmacological activity of NPs after entering the capillary endothelial cells [43].

12.2.3. *Targeting*

When conjugated, targeting molecules can help NPs in selectively recognizing specific membrane receptors or antigens on target cells, resulting in facilitated cell adsorption and internalization. There have been many kinds of targeting molecules developed so far, including proteins (mainly antibodies and their fragments), nucleic acids (aptamers), and other receptor ligands (proteins, peptides, small molecules, etc.) (Table 12.1) [44]. Protein antibodies with known cell membrane-bound receptors can be used to target specific cell types. For example, NPs coated with anti-human epidermal growth factor receptor-2 (anti-HER-2) antibody, such as Herceptin, can target breast cancer cells with high specificity because many breast cancer cells over-express HER-2. Similarly, epidermal growth factor receptor (EGFR) can be targeted by anti-EGFR antibodies, such as Erbitux. Endogenous proteins have been also used for targeted NP delivery to cancer cells; the iron-binding protein transferrin has been widely used to deliver NPs to cells that over-express transferrin receptors. Proteins are typically large in size, for example, whole antibodies such as Herceptin and Cetuximab are 15–20 nm in diameter [44], and the fragment crystallizable (Fc) domain of the antibody can bind to Fc receptors on normal cells and macrophages, leading to increased immunogenicity [2]. Alternatively, small antibody fragments can be used after proper engineering to retain the targeting affinity and specificity from the original whole antibody while eliminating Fc domain. For example, 3–5 nm-sized single-chain variable fragments (scFv), which consist of antibody heavy- and light-chain variable domains connected with a flexible peptide linker, can be used to direct NPs to target cells with high binding affinities and specificities [45, 46]. The targeted NP can have a smaller HD size and reduced immunogenicity by the engineered antibody fragment. Aptamers are short single strand RNA or DNA oligonucleotides that are capable of binding to target antigens with high affinity and specificity. Specific aptamers for targets can be selected from a large number of random sequences by the Systematic Evolution of Ligands by Exponential Enrichment (SELEX) process [47]. Aptamers can have diverse functional groups because SELEX is a chemical process that allows a wide variety of modifications. In addition, they generally have less immunogenicity, which can lead to improved biodistribution [48]. NP surfaces can be easily conjugated with modified aptamers, showing efficient internalization to the target cells [49, 50]. Small peptides or molecules have also been exploited for targeting. Small peptide cyclic arginine-glycine-aspartic acid (cRGD) can bind to cell adhesion interin $\alpha_v\beta_3$ that is abundant on the surface of endothelial cells and various cancer cells. Folic acid (FA) is a well-known small molecule ligand that binds to folate receptors with high affinity; folate conjugation on

Table 12.1. Type and property of various targeting molecules.

Type	MW (kDa)	Diameter (nm)	Features
Monoclonal antibodies			
Whole antibodies	150	15–20	High affinity, divalent, many clinically approved examples, contains biologically active constant (Fc) region, long circulation
Engineered fragments (monovalent)			
ScFv	25	3–5	Lowered affinity, rapid clearance from circulation, renal retention, reduced stability, reduced immunogenicity
Fab′	50	5–10	Can be produced genetically or enzymatically by cleavage of monoclonal antibodies
Nanobody	15	2–3	Smallest antigen-binding fragment, single domain, can bind cryptic epitopes
Engineered fragments (divalent)			
$F(ab')_2$	100	10–15	Improved affinity, can be engineered to a variety of sizes and arrangements of protein domains
Diabodies	50–80	5–10	Mono-specific or bi-specific dimer of ScFv
Minibodies	80	10	Can be produced genetically
Aptamers			
RNA/DNA	10–30	2–3	Rapid clearance, automated chemical synthesis, susceptible to nucleases without chemical modification
Receptor-ligands			
Whole proteins	30–150	Variable	Produced using recombinant DNA technologies, can be biologically active, susceptible to proteases
Peptides	0.5–10	Variable	Facile synthesis and modification, diverse libraries and screening technologies, susceptible to peptidases, renal retention
Small molecules	0.1–1.0	0.5–2.0	Chemical synthesis, simple modification and coupling chemistries, can be biologically active, highly variable affinities

NPs can significantly promote their targeted delivery into cancer cells that over-express folate receptors [51, 52]. Small molecules allow simple conjugation to NPs by means of chemical reactions and the inherent small size promises better control of overall HD size.

The property and targeting efficacy of the conjugates is affected not only by the type and surface density of targeting molecules, but also by the overall physicochemical properties of NPs, such as HD size and surface charge. Chan *et al.* demonstrated that the internalization of Herceptin-coated gold NPs by the breast cancer cell was most efficient for 40 and 50-nm gold NPs among a size range of 2–100 nm [6]. This optimal size for particle uptake is argued due to the accurate balance between multivalent cross-linking of membrane receptors and the process of membrane wrapping involved in receptor-mediated endocytosis. Similarly, an optimal effective size of 44 nm was reported for cellular uptake when gold nanocages conjugated with anti-EGFR antibody were incubated with human glioblastoma cells [53]. The control of NP surface and conjugation reaction is crucial to maintain targeting activity, especially for protein-based targeting molecules, because adsorption on the NP surfaces may cause structural disturbance and denaturation of proteins altering their functions. Zhang *et al.* have shown that gold NPs can cause aggregations of adsorbed proteins, with the partially unfolded structures resulted from the protein-NP assembly [54]. NP surface charge is a key parameter that regulates such protein adsorption and denaturation. Neutral ligands, such as PEG, are preferred for maintaining protein structure, while charged surface ligands are more likely to perturb proteins via strong electrostatic interactions [55]. With highly charged surfaces, even well-fabricated targeted NPs are susceptible of non-specific interactions with biomolecules *in vivo*, which may result in a severe reduction of targeting efficacy by shielding the active sites of targeting molecules or removing the conjugates from circulation through RES uptake before they reach the target site. Because neutral surfaces are preferred to avoid non-specific interactions, NPs are often coated with PEG before/after conjugation with targeting molecules. Small surface ligands with zwitterionic characteristics can also be used for such purposes, as they show a similar property to PEG. Choi *et al.* have shown that zwitterionic QDs with a small HD size (<6 nm) can target tumors with high specificity after conjugation with small targeting molecules (Fig. 12.7) [56]. Most of the remaining QDs are rapidly cleared from the body through renal filtration without being trapped in RES cells, which was ascribed to the small HD size and suppression of non-specific interactions with biomolecules and non-targeted tissues. Zwitterionic surface ligand molecules can be used with other surface molecules that contain

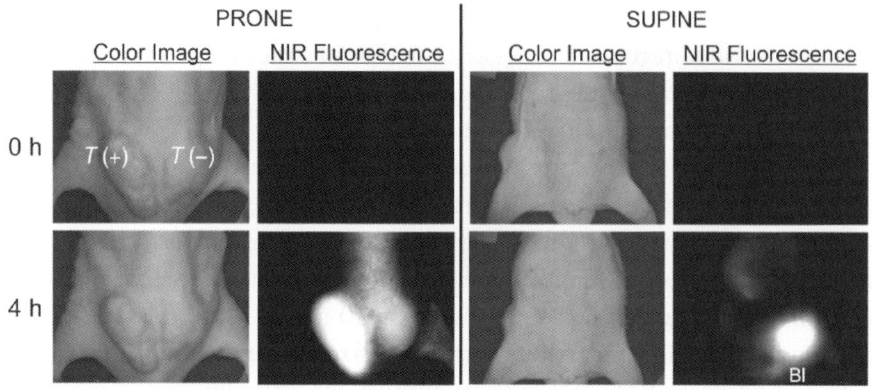

Figure 12.7. Targeting of QDs to human melanoma xenograft tumors. Small zwitterionic QDs conjugated with cRGD (< 6 nm in HD size) preferentially accumulate into the $\alpha_v\beta_3$-positive T(+) tumor when compared with the $\alpha_v\beta_3$-negative T(−) tumor with the positive-to-negative tumor ratio of 5.1 at 4 h post-intravenous injection. Most of the remaining QDs were excreted through renal clearance and were found primarily in the bladder due to their small size and the low non-specific adsorption level [56]. QD = quantum dot, cRGD = cyclic arginine-glycine-asparagine, NIR = near-infrared.

functional groups for bioconjugation chemistry. By carefully controlling the ratio of two surface ligand molecules, the mixed surface ligand system can permit specific targeting of biomarkers, while maintaining the advantages of zwitterions of minimal non-specific adsorptions and high colloidal stability [38].

12.3 Smart Delivery System with Strategic Payload Release

Once drug delivery systems have entered the target site, efficient release of the payload drugs is a prerequisite to fully retain their activity and maximize therapeutic efficacies. Sophisticated designs of NP drug delivery systems allows the exploitation of internal (e.g. pH, intracellular glutathione (GSH)) or external (e.g. light, magnetic field) stimuli to initiate payload release. Some examples of inorganic NP delivery system with the ability of strategic payload release are discussed based on the specific triggers for the drug release.

12.3.1 *pH*

It is known that the extracellular microenvironment of solid tumors is mildly acidic by hypoxia and NPs are internalized into cells via the endocytic pathway, which is exposed to these acidic environments, as endocytic vesicles change from early and late endosomes to lysosomes [57]. Drug delivery systems based on pH-sensitive NPs that can respond to acidic conditions can release and activate payload drugs when accumulated in tumor regions or after being taken up by tumor cells. Nam *et al.* reported a "smart" gold nanoparticle (SAN) consisting of a 10-nm gold sphere and a pH-sensitive surface molecule that contains a hydrolysis-susceptible citraconic amide unit [58]. The amide bonds are stable at neutral or basic conditions, but they tend to hydrolyze abruptly at a pH lower than 7.0. The acidic hydrolysis of surface molecules results in mixed-charge NP surfaces with both negative and positive charges, triggering rapid aggregation among NPs via electrostatic interactions. The aggregation shifts surface plasmon absorbance to near-infrared (NIR) regions by the appearance of coupled surface plasmon modes, and this shift was successively utilized for deep-tissue penetrating photothermal therapy. The SAN-based pH-responsive drug delivery system was further developed by covalently conjugating the anticancer drug doxorubicin at the terminal of surface molecules via a carbodiimide coupling reaction [59]. The SAN conjugate is designed to release payload therapeutics by a pH-triggered linker cleavage under the mild acidic conditions in tumors and simultaneously form aggregates of gold NPs for photothermal therapy (Fig. 12.8). Spatio-temporally concerted release of chemo- and thermotherapeutic agents triggered by the pH change in the tumor environment can take advantage of the high synergistic effect of chemo- and photothermal combination cancer therapy at both the *in vitro* cell level and *in vivo* animal level. In addition to the highly synergistic therapeutic effect, the huge size increase by pH-triggered aggregation of NPs can improve delivery efficacy to a tumor tissue by enhanced retention as well as into tumor cells by blocked exocytosis. Liu *et al.* reported a similar enhanced retention and cellular uptake of NPs by pH-induced aggregation in tumors [60]. It was demonstrated that 16-nm gold NPs with mixed-charge zwitterionic self-assembled monolayers selectively form aggregates in response to the acidic extracellular pH of solid tumors, and as a result, have a considerable blood half-life with much higher tumor accumulation, retention, and cellular internalization than pH-insensitive PEGylated gold NPs.

Such pH-responsive selective size modulation in a tumor can be very promising for drug delivery, especially when using NPs smaller than the renal filtration limit; effective tumor accumulation can be achieved by increased size, while most untargeted NPs can be rapidly cleared from the body through renal pathways, alleviating potential toxic side effects. As for the effort, Hwang *et al.* developed pH-responsive SANs with HD sizes below 6 nm using one phase synthetic method [61]. The small SANs were able to form aggregates in response to mild acidic conditions similar to large SANs, and efficiently accumulated in cancer cells with significant red-shift absorption,

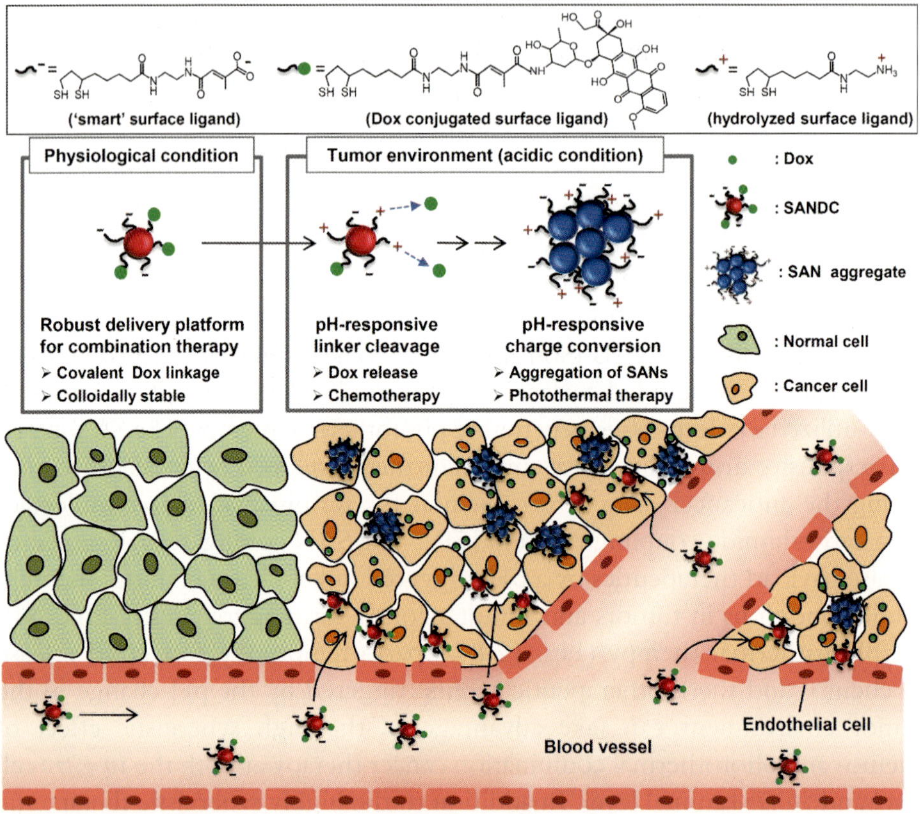

Figure 12.8. Schematic illustration of the working mechanism of "smart" gold nanoparticle doxorubicin conjugate (SANDC). SANDC consists of "smart" gold nanoparticles (SANs) and covalently conjugated Dox. The SANDC is designed to release the Dox by pH-triggered linker cleavage under the mild acidic conditions in tumor. Simultaneously, the SANDC is designed to convert the surface charge from negative to a mixture of negative and positive charges, which induces rapid aggregation among the NPs *via* electrostatic interactions. This spatio-temporally concerted release from SANDCs is exploited for chemo and photothermal combination cancer therapy [59].

enabling deep-tissue penetrating photothermal therapy. Besides the size, the major barrier that limits *in vivo* application of SANs is a large degree of uptake/clearance by RES cells before NPs reach a target site. Direct surface modification using anti-fouling molecules, such as PEG, is not allowed because it alters the physicochemical properties of NPs hampering pH-responsive ability. Encapsulating NPs in vehicles can be an alternative strategy to direct anti-fouling surface modification, which can preserve the initial NP surface as is, while the delivery property is controlled by the vehicle characteristics.

Nam *et al.* reported gold NPs-in-liposome hybrid nanostructures by encapsulating SANs in PEG-grafted liposomal vehicles, in which the PEG-grafted liposome acts as a non-associated surface passivation layer, providing anti-fouling coating for encapsulated SANs (Fig. 12.9) [62]. The pH-responsive function of SANs was well preserved after liposomal encapsulation, whereas their biological fate was determined by the liposomal passivation layer, showing reduced non-specific interactions with cancer cells and RES organs due to surface-grafted PEG. As a result, tumor accumulation of SANs increased by 1.1–2.5-fold, accompanied by a prolonged blood circulation and significantly decreased RES uptake. The pH-triggered aggregation of SANs has also been successfully adopted for cancer diagnosis using gold NP-based bioimaging techniques, such as surface-enhanced Raman scattering imaging [63], and photothermal optical coherent tomography [64].

12.3.2. *Intracellular GSH*

GSH is the most abundant thiol species in the cytoplasm, with its intracellular concentration being substantially higher (1–10 mM) than extracellular thiol concentrations (cysteine 8 μM and GSH 2 μM) [65]. Since some kinds of inorganic NPs, especially gold NPs, have strong binding affinities towards thiol groups, disulfide exchange or place-exchange reactions with intracellular GSH can release payloads bound to the NP surfaces. Hong *et al.* demonstrated GSH-mediated payload release from 2-nm gold NPs with the surface composed of mixed monolayers of a thiolated tetra(ethylene glycol) cationic ligand and a thiolated Bodipy dye (Fig. 12.10) [66]. Reaction with 10-mM GSH significantly released monolayer contents, which was clearly demonstrated by the translocation of hydrophobic Bodipy dyes into the toluene phase after GSH treatment. The release of Bodipy dyes was also observed in living cells by the strong fluorescence signal that appeared after NP incubation. Intracellular GSH-mediated payload release was further verified by pretreating cells with varying levels of glutathione monoester (GSH-OEt) that can be rapidly internalized into cells and hydrolyzed to GSH manipulating cellular GSH levels. Increasing the concentration of GSH-OEt resulted in a dose-dependent increase of Bodipy fluorescence in the cells.

The GSH-mediated payload release was further examined in three-dimensional *in vitro* tumor models using gold NPs carrying either fluorescein or doxorubicin molecules via anchoring of thiol groups [67]. Payload release upon GSH treatment was effective regardless of surface charges and payload types; however, NPs diffuse/localize differently in an *in vitro* three-dimensional tumor model depending on whether they are positively or negatively charged.

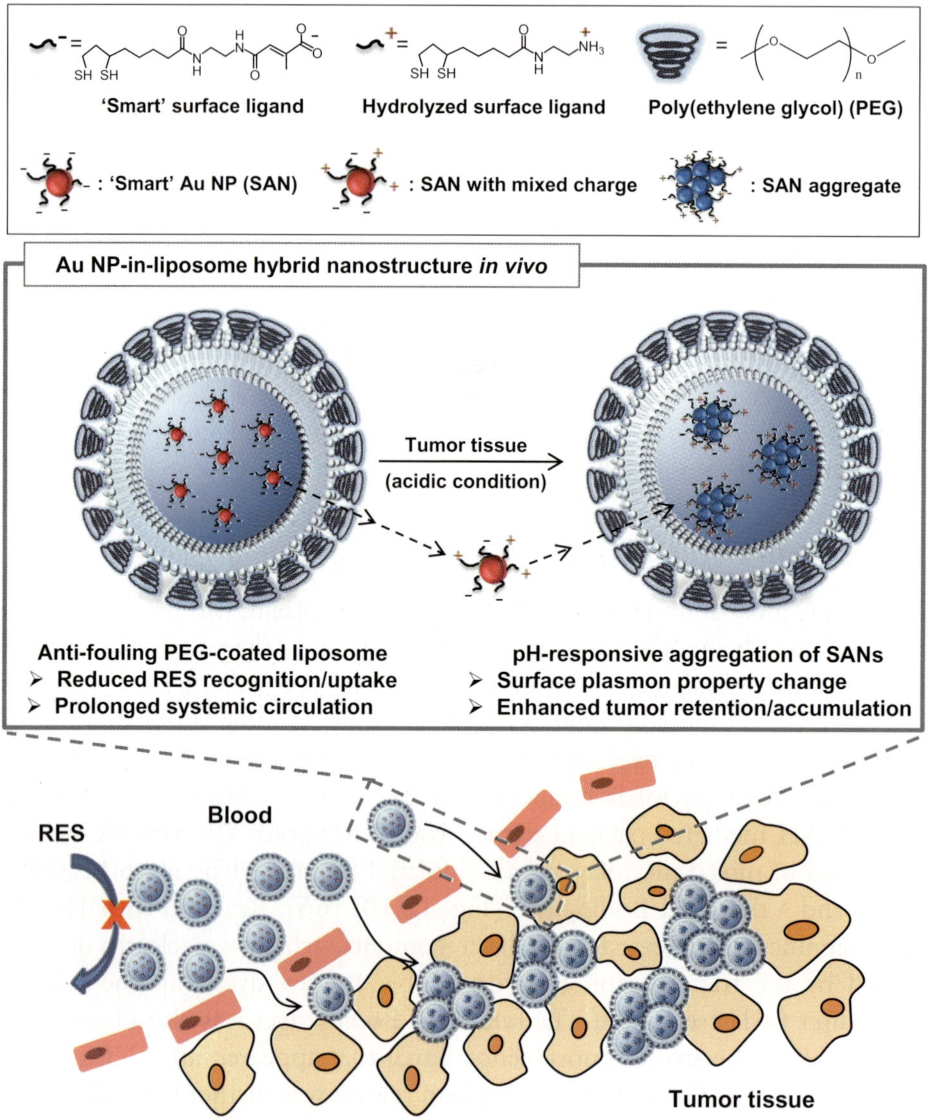

Figure 12.9. Schematic illustration of systemic tumor delivery of Au NPs-in-liposome hybrid nanostructure. The hybrid consists of PEG-grafted liposome and encapsulated SANs. PEG-grafted liposome provides non-associated anti-fouling coating for SANs enhancing tumor accumulation by reduced RES uptake and prolonged blood circulation. Upon the acidic condition in tumor, SANs convert their surface charge from negative to a mixture of negative and positive charges by the change in hydrolysis-susceptible 'smart' surface ligand. This induces rapid aggregation among the NPs *via* electrostatic interactions, resulting in surface plasmon property changes by the appearance of coupled surface plasmon modes [62]. Au = gold, NP = nanoparticle.

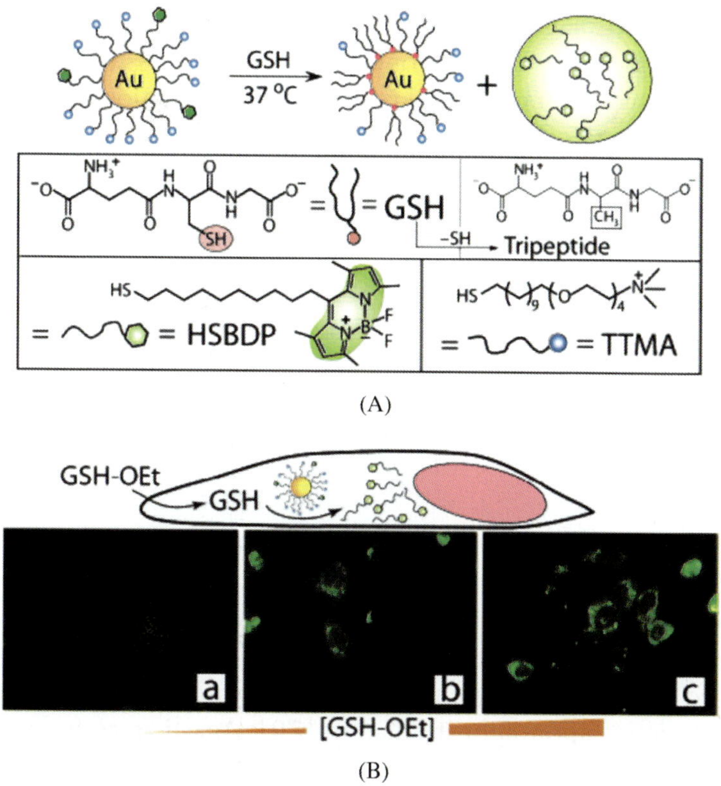

Figure 12.10. GSH-mediated delivery and release using gold NP carriers. (A) Structure of the gold NP carrier and schematic depiction of the GSH-mediated surface monolayer exchange reaction which releases the thiolated hydrophobic Bodipy dye, HSBDP. (B) Schematic representation and fluorescence images when using GSH-OEt as an external stimulus to release HSBDP from gold NPs. The GSH-OEt concentrations were varied from 0, 5, and 20 mM in panels a, b, and c, respectively [66]. GSH = glutathione, GSH-OEt = glutathione monoethyl ester, NP = nanoparticle, Au = gold, HSBDP = thiolated hydrophobic BODIPY dye, TTMA = tetra(ethylene glycol)-lyated cationic ligand.

It was revealed that positive particles had a significantly higher uptake by the cells, showing improved delivery of payload to the majority of cells in tumors, while negative particles diffuse at a faster rate, delivering payload deeper into tissues, and mathematical modeling further supported this finding. A variety of thiolated drugs can be loaded on gold NPs for intracellular delivery and GSH-mediated release. Wang *et al.* developed polyamidoamine dendrimer-encapsulated gold NPs (DEGNPs) as a carrier of thiol-containing anticancer drugs, such as captopril (CPP), 6-mercaptopurine (6-MP), thiolated doxorubicin and cisplatin [68]. DEGNPs are capable of significant loading of thiol-containing drugs with prolonged release kinetics due to the strong anchoring of thiol groups on gold surfaces. The addition of GSH rapidly released the majority of drugs and, likewise, intracellular drug release was efficient, accompanied by the improved cytotoxicity. Non-thiolated payloads can also be effectively released by GSH-triggered place exchange reactions. Kim *et al.* directly monitored GSH-triggered release of the non-specifically adsorbed anticancer drug topotecan (TOPO) from gold NPs using a label-free fluorescence live-cell imaging technique that detects fluorescence from released drugs [69]. The release of a nanomolar amount of TOPO upon *in vitro* GSH treatment was successfully monitored, and *in vivo* fluorescence images of TOPO were also obtained from a subcutaneous injection site in living mice after GSH treatment.

12.3.3 *Light*

Light is one of the most widely used external triggers that can achieve controlled drug release by controlling the site of irradiation. Light-responsive NPs can be obtained using caged compounds that can release covalently linked photo-cleavable moieties upon light irradiation [70]. Agasti *et al.* have developed a gold NP platform for photocontrolled release of a caged anticancer drug (5-fluorouracil, 5-FU) by conjugating the drug to the particle surface through a photo-responsive o-nitrobenzyl (ONB) linkage (Fig. 12.11) [71]. The ONB group could be efficiently cleaved when exposed to ultraviolet (UV)-A (365 nm) irradiation, allowing photocontrolled uncaging and release of the covalently conjugated 5-FU moieties from gold NPs. Upon incubation with cancer cells, the initially non-toxic conjugates induced significant cell death after UV irradiation depending on the NP concentration and light dose, which should be correlated with the amount of the liberated drugs.

The photocleavable ONB moiety has also been exploited to manipulate surface properties of NPs by photocontrolled uncaging processes. Han *et al.* reported photo-responsive gold NPs that can change surface charge from positive to negative upon UV irradiation by releasing a quaternary ammonium and leaving behind a carboxylic acid [72]. Initially, positively charged NPs were strongly associated with DNA by electrostatic attraction; however, the reversal in electrostatics upon UV irradiation led to the release of DNA. DNA transcription assay using T7 RNA polymerase demonstrated that DNA was effectively liberated from gold NPs, while retaining its function. The gold NP–DNA complex was efficiently taken up by cells, in which DNA could be released and localized in the nucleus upon irradiation. The nuclear localized

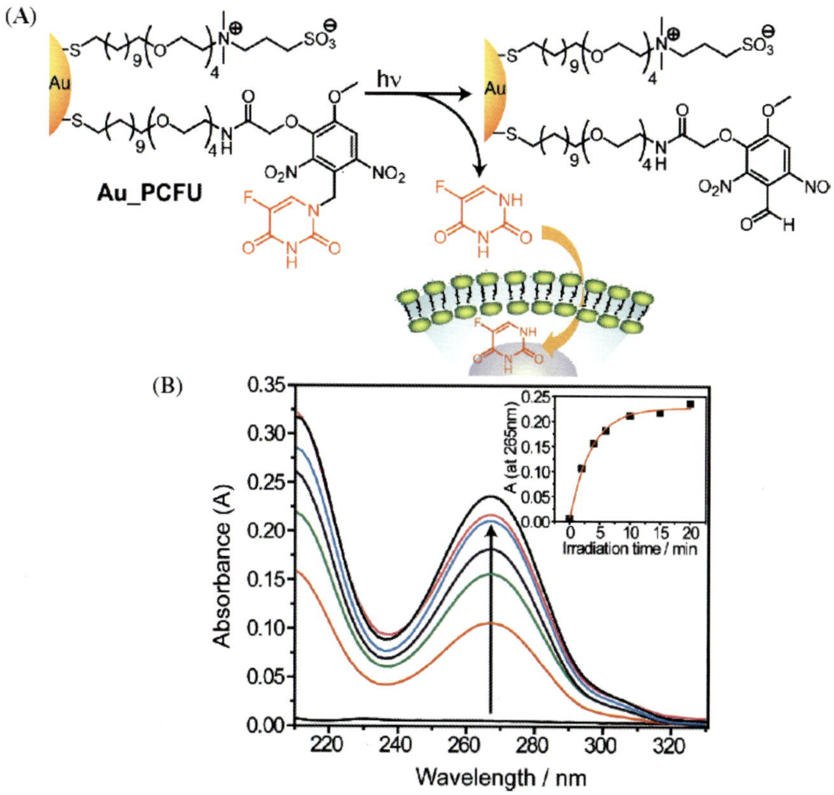

Figure 12.11. Photoregulated release of caged anticancer drugs from gold NPs. (A) Photochemical reaction of 5-fluorouracil (5-FU) conjugated gold NPs (Au_PCFU) and delivery of payload to cell. (B) Overlaid UV-vis spectral changes showing light dose dependent increase of 5-FU release. Inset: The plot of absorbance at 265 nm against irradiation time [71]. NP = nanoparticle.

DNA could act as a genetic therapeutic agent, providing macromolecule drug delivery systems with light-regulated intracellular release.

These photoactivable systems have demonstrated effective photocontrolled payload release; however, this is not suitable for *in vivo* applications because the UV light that is necessary for the uncaging process is very harmful to cells and has a poor tissue penetration depth. NIR light provides longer tissue penetrations by the reduced photon scatterings, and causes minimal photo damage to the biological specimen. To overcome the limitation of UV irradiation and take advantage of NIR light, upconversion NPs (UCNPs) that transduce NIR light to UV light by a multi-photon process have been developed for photocontrolled uncaging and release of payloads [73]. NIR-to-UV UCNPs can be typically obtained by the co-doping of several kinds of lanthanide ions, such as ytterbium (Yb^{3+}) and thulium (Tm^{3+}) on host NPs like sodium yttrium fluoride ($NaYF_4$). Jayakumar *et al.* prepared lanthanide-doped UCNPs coated with thin layers of mesoporous silica that could carry caged macromolecule drugs, such as plasmid DNA and siRNA [74]. Upconverted UV light emitted from the NIR irradiated UCNPs activated caged nucleic acid, rendering it fully functional for exerting its effect in the host cells. Activation/knockdown of green fluorescence protein (GFP) by uncaged plasmid DNA/siRNA was performed in both solution and cells, and patterned activation of GFP clearly demonstrated a photo-controlled mechanism. In-depth photoactivation through tissue phantoms and *in vivo* activation of caged nucleic acids were also successfully demonstrated.

Not only therapeutic effects, but also targeting efficiency can be regulated using the photocontrolled uncaging approach. One of the problems in the use of cancer targeting molecules is although corresponding receptors are overexpressed in many cancer cells, significant expression also occurs in normal cells, limiting tumor delivery capacity. Chien *et al.* developed silica-coated UCNP-based drug delivery systems composed of photo-responsive FA targeting ligands and the anticancer drug doxorubicin (Fig. 12.12) [75]. FA was masked using the photo-cleavable ONB moiety, which suppresses targeting activity. Upon irradiation of a NIR light on UCNPs, the emitted UV light cleaved the photo-labile ONB linkage, recovering the targeting effect of FA. NIR-triggered improvement in cancer targeting was demonstrated both *in vitro* and *in vivo* with effective tumor suppression by enhanced doxorubicin delivery. UNCPs with caged targeting molecules provide a platform for the improvement of selective targeting and possible reduction of adverse side effects from chemotherapy.

12.3.4. *Magnetic field*

When ferromagnetic NPs are reduced to below a critical size, ambient thermal energy can be sufficient to switch and randomize the magnetic spin direction, resulting in a net magnetization of zero; this behavior is called superparamagnetism [76]. Metal NPs or metal oxide NPs are capable of superparamagnetism when the size is reduced to typically less than 20 nm [76]. In this size regime, each NP becomes a single magnetic domain and shows superparamagnetic behavior at the ambient temperature above the so-called blocking temperature. Such NPs have large magnetic moments and behave like giant paramagnetic atoms, showing a fast response against applied magnetic fields with negligible remanence (residual magnetism) and coercitivity (the field required to bring the magnetization to zero). Iron oxide NPs (IONPs) are one of the most well studied magnetic NPs showing strong superparamagnetism at room temperature. The application of an alternating magnetic field (AMF)

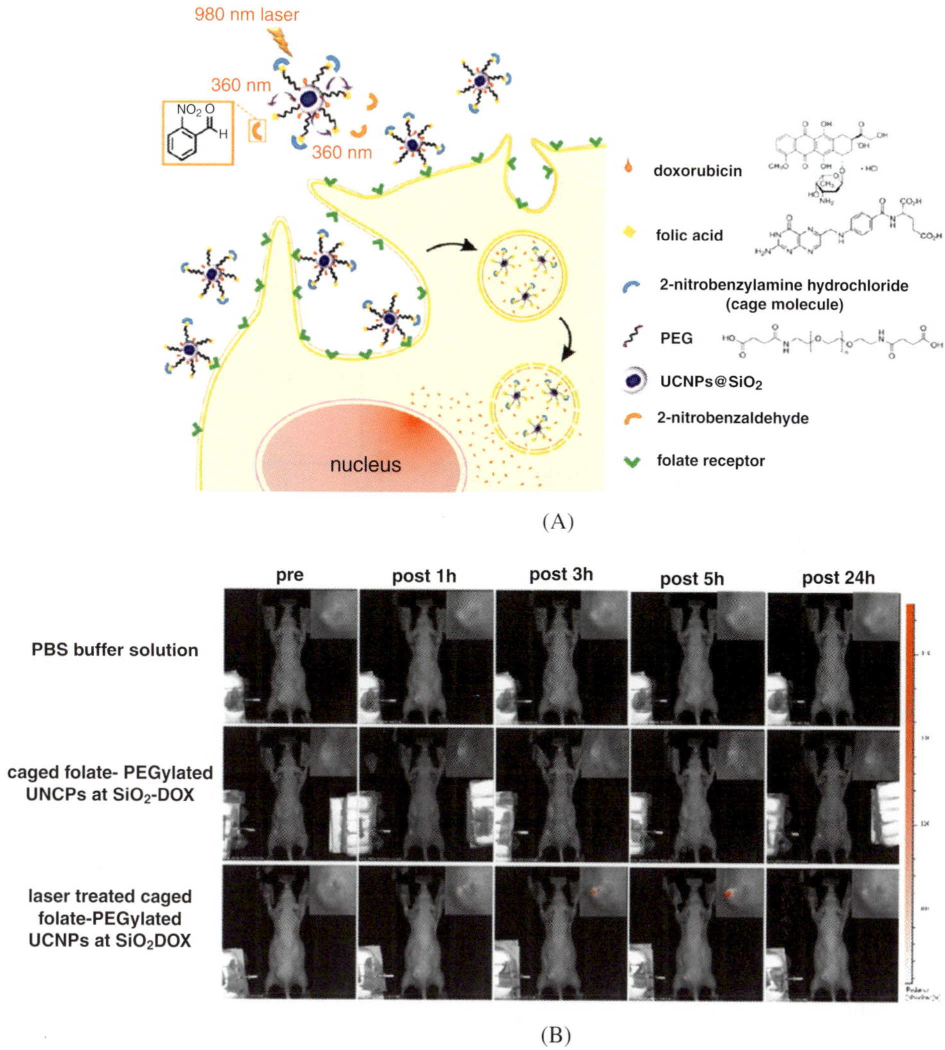

Figure 12.12. (A) Illustration of photocaged UCNPs following NIR laser activation to remove cage molecules from FA and subsequent targeting of cancer cells. (B) Time course upconversion NIR luminescence (emission 800 nm) images taken after intravenous injection of caged folate-PEGylated UCNPs at SiO_2-DOX NPs and with/without NIR laser irradiation to the tumors for 1 hr. Insets show the enlarged tumor region. The bright NIR emission is only observed in the tumor of mouse that received both caged folate-PEGylated UCNPs at SiO_2-DOX and NIR irradiation, indicating enhanced targeting effect by photo-uncaging of folic acid [75]. PEG = polyethylene glycol.

to IONPs induces the production of thermal energy by the magnetic hysteresis loss and Neel/Brownian relaxation, generating high local heating [77].

IONPs can be used as a platform for controlled delivery systems, in which remote release of payload therapeutic agents is stimulated by the heat generated

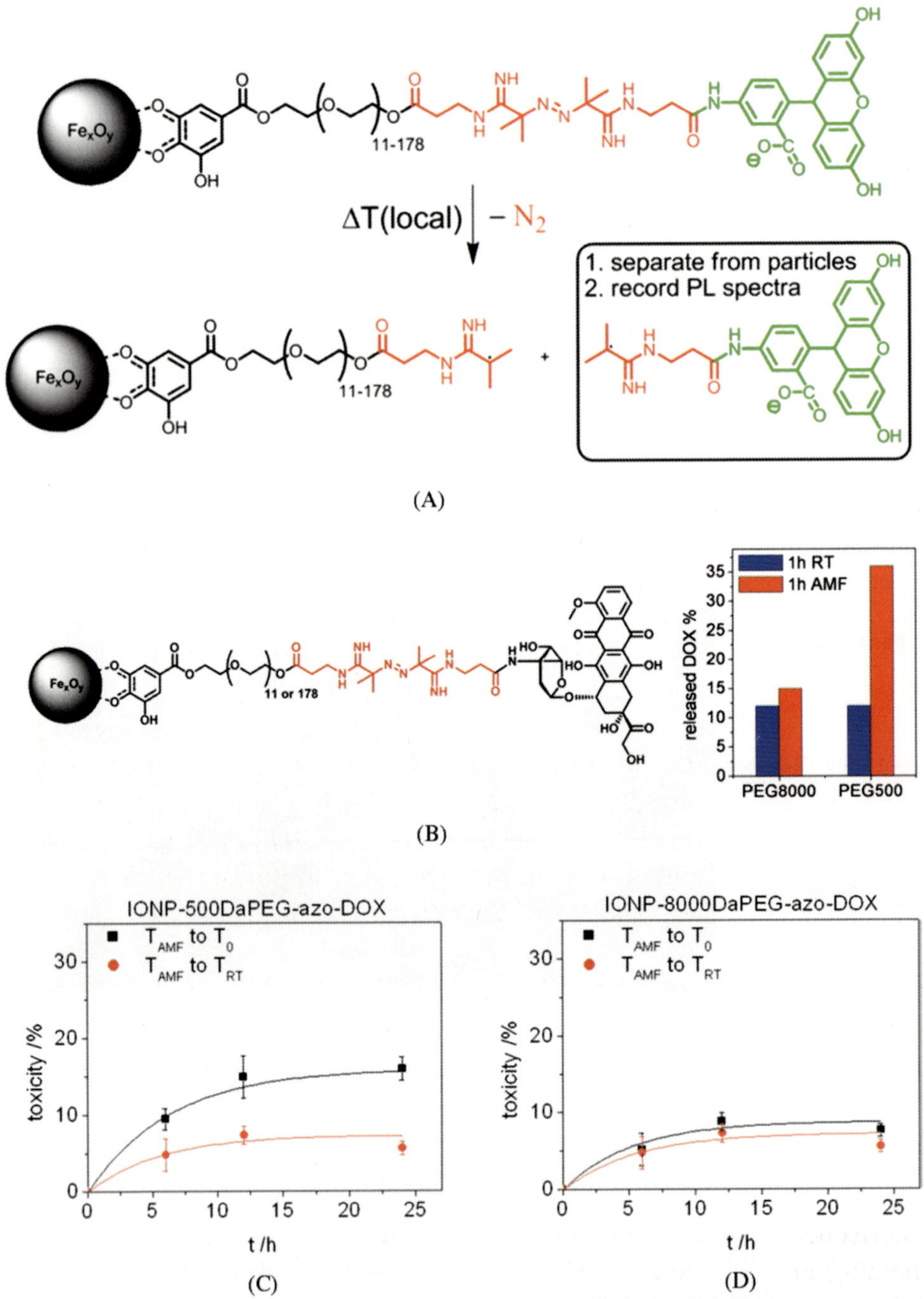

Figure 12.13. (A) Sketch of the readily functionalized IONPs bearing fluoresceineamine dye connected through thermo-labile azo molecule to the tails of PEG spacers. An increase in temperature by AMF exposure results in increased cleavage of the azo group and release of the dye from the particles. (B) Sketch of the readily functionalized IONPs bearing DOX,

Figure 12.13. (*Continue on facing page*) which is covalently conjugated through azo linkage to the PEG spacers of different two different MW (500 and 8000 Da) together with the release profiles for samples exposed for 1 h at RT and under AMF (334.5 kHz, 17 mT), respectively. (C) Cytotoxicity within 24 h of DOX released from IONPs-500 Da PEG after AFM treatment compared to a sample incubated 1 h at RT (T_{RT}) and to a blank kept for 1 h at RT (T_0). (D) Cytotoxicity of DOX released from IONPs-8000 Da PEG after AFM treatment compared to a sample incubated 1 h at RT (T_{RT}) and to a blank (T_0) [78]. PL = photoluminescence.

upon AMF exposure. Riedinger *et al.* developed a IONP platform bound by PEG spacers of distinct molecular weight connected to fluoresceineamine dye through a thermo-labile azo linker (Fig. 12.13) [78]. They monitored the increase in dye fluorescence under AMF exposure to measure remotely controlled payload release by cleavage of the azo group. The local heating with a temperature increase up to 45 °C was found at distances below 0.5 nm from the surface of IONPs, accompanied by significant dye release, which decayed exponentially with increasing distance from IONPs by using higher molecular weight PEG spacers. Controlled drug delivery was demonstrated by replacing fluoresceineamine with the anticancer drug doxorubicin in the same nanoparticle arrangement; under actuation by AMF, distance-dependent control over the release rate of doxorubicin was achieved with triggered *in vitro* cytotoxicity for cancer cell treatments. Similarly, N'Guyen *et al.* prepared functional IONP with magnetically induced drug release ability using thermo-reversible Diels-Alder (DA) reactions between diene and alkene derivatives [79]. Cargo molecules loaded using the DA reaction could be released through retro-DA (rDA) reactions at an elevated temperature (90–110°C). Upon AMF exposure, IONPs can generate sufficient thermal energy in close proximity to the DA cycloadduct to initiate the rDA reaction. Using tetramethylrhodamine-5-C2-maleimide (Rhd-M) as a fluorescence probe and model drug, significant drug release through the rDA reaction was demonstrated under AFM-induced excitation of IONPs without any significant heating of the medium.

12.4 Conclusion

Inorganic NPs have offered many opportunities in biological research fields, holding great promises as a versatile drug delivery platform. The inherent advantages of inorganic NPs, including long-term stability in biological environments, high surface-to-volume ratios with the diversity of surface functionalization, and facile physicochemical tunability upon on-demand design, provide a scaffold for therapeutic agents with the capability of greatly improving pharmacological properties. Fully exploiting the advantages and versatilities of inorganic NPs for a drug delivery platform requires judicious control over their physicochemical properties that crucially regulate NPs' interaction with biological systems. The HD size and surface charge are the most important NP design parameters to be considered for efficient delivery of therapeutic agents in a target site because they can largely affect the cellular interaction, *in vivo* circulation, and tissue permeation. Typically, maximal cellular uptake can take place in the intermediate NP size range of 40–60 nm. Clearance of NPs from blood circulation is mainly caused by non-specific uptake by RES cells when they are larger than 10 nm. Small NPs with sub-10-nm HD size show broad distribution over many organs as they become capable of penetrating various organ barriers, for example, the renal barrier of ~5.5-nm size limit. Reduction of the HD size below the limit of the renal barrier can achieve fast clearance of NPs through urinary excretions. For maximal tumor accumulation, NPs should have long blood circulation time and should be large enough to prevent rapid penetration within tumor interstitial space. NP surface charge and hydrophobicity have a vital role in interactions with cells and *in vivo* compartments. In general, positively charged NPs adsorb more on cell membranes and consequently show a higher level of internalizations than negatively charged or neutral NPs. NPs encounter a large variety of biomolecules *in vivo*, where in many cases, non-specific adsorptions hamper targeted delivery by large non-specific uptake/clearance by RES cells. For optimal performance, NPs are suggested to have neutral surface charge at physiological conditions. PEG and small zwitterionic molecules have been attractive candidates to fabricate neutral NPs with reduced non-specific adsorption of biomolecules. Conjugation of targeting molecules that strongly interact with desired tissues or cells can allow target-specific delivery of NPs with an alleviated concern of side effects caused by the nonspecific release/distribution of therapeutic agents. Many kinds of targeting molecules (protein, peptide, aptamer, small molecule ligand) have been used for targeted NP delivery, where HD size and surface charge should also be carefully considered to achieve proper targeting effects. Once drug delivery systems have entered the target site, the efficient release of cargo therapeutics is a prerequisite to fully retain their activity and maximize therapeutic efficacies. Sophisticated fabrication of NP surfaces can allow smart drug delivery systems that initiate cargo release upon triggering by internal (pH, glutathione) or external (light, magnetic field) stimuli. Drug delivery systems based on acidic pH-sensitive NPs can release and activate cargo drugs when accumulated in a tumor region or after being taken up by tumor cells because extracellular/intracellular tumor environments are acidic. GSH is the most abundant thiol species in the cytoplasm that can trigger cargo release by disulfide exchange or place-exchange reactions. Gold NPs have been widely used for such GSH-mediated intracellular release approaches because of a strong binding affinity towards thiol groups and facile concentration-dependent place exchange reactions. Light-responsive cargo release can be achieved using caged compounds that can release covalently linked photo-cleavable moieties upon light irradiation. The majority of photo-cleavable linkages developed so far respond to UV light. The inherent limit of such UV-responsive systems for *in vivo* applications, light-induced damage to the cells and poor tissue penetration depth, can be overcome by using UCNP platforms that transduce NIR light to UV light by multi-photon processes. Superparamagnetic NPs can generate strong local heating in response to magnetic fields, providing smart drug delivery systems in which remote release of cargo therapeutic agents is stimulated by the heat generated upon AMF exposure via cleavage of the thermo-labile linkage. IONPs are the most popular platforms for such applications due to the efficient AMF-triggered local heating by their strong superparamagnetism.

Gaining full control over the *in vivo* distribution/targeting and release of cargo drugs is the ultimate goal for developments of drug delivery systems. Although there are several hurdles that still need to be addressed and solved for *in vivo* applications, including a long-term toxicity and a sophisticated control on biodistribution/clearance, the successes and the promises that have been demonstrated so far clearly suggest inorganic NPs as a valuable drug delivery platform for future clinical translations.

References

1. Matsumura, Y. and Maeda, H. (1986). A new concept for macromolecular therapeutics in cancer chemotherapy: Mechanism of tumoritropic accumulation of proteins and the anti-tumor agent smancs. *Cancer Res.,* **46**, 6387–6392.
2. Peer, D., Karp, J. M., Hong, S., Farokhzad, O. C., Margalit, R. and Langer, R. (2007). Nanocarriers as an emerging platform for cancer therapy. *Nat. Nanotech.,* **2**, 751–760.
3. Medintz, I. L., Uyeda, H. T., Goldman, E. R. and Mattoussi, H. (2005). Quantum dot bioconjugates for imaging, labelling and sensing. *Nat. Mater.,* **4**, 435–446.
4. Chou, L. Y. T., Ming, K. and Chan, W. C. W. (2011). Strategies for the intracellular delivery of nanoparticles. *Chem. Soc. Rev.,* **40**, 233–245.
5. Smith, A. M., Duan, H., Mohs, A. M. and Nie, S. (2008). Bioconjugated quantum dots for in vivo molecular and cellular imaging. *Adv. Drug Deliv. Rev.,* **60**, 1226–1240.
6. Jiang, W., Kimbetty, Y. S., Rutka, J. T. and Chan, W. C. W. (2008). Nanoparticle-mediated cellular response is size-dependent. *Nat. Nanotech.,* **3**, 145–150.
7. Nam, J., Won, N., Bang, J., Jin, H., Park, J., Jung, S., Jung, S., Park, Y. and Kim, S. (2013). Surface engineering of inorganic nanoparticles for imaging and therapy. *Adv. Drug Deliv. Rev.,* **65**, 622–648.
8. Chithrani, B. D., Ghazani, A. A. and Chan, W. C. W. (2006). Determining the size and shape dependence of gold nanoparticle uptake into mammalian cells. *Nano Lett.,* **6**, 662–668.

9. Chithrani, B. D. and Chan, W. C. W. (2007). Elucidating the mechanism of cellular uptake and removal of protein-coated gold nanoparticles of different sizes and shapes. *Nano Lett.*, 7, 1542–1550.
10. Osaki, F., Kanamori, T., Sando, S., Sera, T. and Aoyama, Y. (2004). A quantum dot conjugated sugar ball and its cellular uptake. On the size effects of endocytosis in the subviral region. *J. Am. Chem. Soc.*, **126**, 6520–6521.
11. Wang, S. H., Lee, C. W., Chiou, A. and Wei, P. K. (2010). Size-dependent endocytosis of gold nanoparticles studied by three-dimensional mapping of plasmonic scattering images. *J. Nanobiotech.*, **8**, 33.
12. Zhang, S., Li, J., Lykotrafitis, G., Bao, G. and Suresh, S. (2009). Size-dependent endocytosis of nanoparticles. *Adv. Mater.*, **21**, 419–424.
13. De Jong, W. H., Hagens, W. I., Krystek, P., Burger, M. C., Sips, A. J. A. M. and Geertsma, R. E. (2008). Particle size-dependent organ distribution of gold nanoparticles after intravenous administration. *Biomaterials*, **29**, 1912–1919.
14. Sonavane, G., Tomoda, K. and Makino, K. (2008). Biodistribution of colloidal gold nanoparticles after intravenous administration: Effect of particle size. *Coll. Surf. B Biointerfaces*, **66**, 274–280.
15. Semmler-Behnke, M., Kreyling, W. G., Lipka, J., Fertsch, S., Wenk, A., Takenaka, S., Schmid, G. and Brandau, W. (2008). Biodistribution of 1.4-and 18-nm gold particles in rats. *Small*, **4**, 2108–2111.
16. Choi, H. S., Liu, W., Misra, P., Tanaka, E., Zimmer, J. P., Itty Ipe, B., Bawendi, M. G. and Frangioni, J. V. (2007). Renal clearance of quantum dots. *Nat. Biotech.*, **25**, 1165–1170.
17. Choi, H. S., Ashitate, Y., Lee, J. H., Kim, S. H., Matsui, A., Insin, N., Bawendi, M. G., Semmler-Behnke, M., Frangioni, J. V. and Tsuda, A. (2010). Rapid translocation of nanoparticles from the lung airspaces to the body. *Nat. Biotech.*, **28**, 1300–1303.
18. Hobbs, S. K., Monsky, W. L., Yuan, F., Roberts, W. G., Griffith, L., Torchilin, V. P. and Jain, R. K. (1998). Regulation of transport pathways in tumor vessels: Role of tumor type and microenvironment. *P. Natl. Acad. Sci. USA*, **95**, 4607–4612.
19. Yuan, F., Dellian, M., Fukumura, D., Leunig, M., Berk, D. A., Torchilin, V. P. and Jain, R. K. (1995). Vascular permeability in a human tumor xenograft: Molecular size dependence and cutoff size. *Cancer Res.*, **55**, 3752–3756.
20. Popović, Z., Liu, W., Chauhan, V. P., Lee, J., Wong, C., Greytak, A. B., Insin, N., Nocera, D. G., Fukumura, D., Jain, R. K. and Bawendi, M. G. (2010). A nanoparticle size series for in vivo fluorescence imaging. *Angew. Chem. Int. Ed.*, **49**, 8649–8652.
21. Perrault, S. D., Walkey, C., Jennings, T., Fischer, H. C. and Chan, W. C. W. (2009). Mediating tumor targeting efficiency of nanoparticles through design. *Nano Lett.*, **9**, 1909–1915.
22. Hillyer, J. F. and Albrecht, R. M. (2001). Gastrointestinal persorption and tissue distribution of differently sized colloidal gold nanoparticles. *J. Pharm. Sci.*, **90**, 1927–1936.
23. Hillyer, J. F. and Albrecht, R. M. (1998). Correlative instrumental neutron activation analysis, light microscopy, transmission electron microscopy, and X-ray microanalysis for qualitative and quantitative detection of colloidal gold spheres in biological specimens. *Microsc. Microanal.*, **4**, 481–490.
24. Sonavane, G., Tomoda, K., Sano, A., Ohshima, H., Terada, H. and Makino, K. (2008). In vitro permeation of gold nanoparticles through rat skin and rat intestine: Effect of particle size. *Coll. Surf. B*, **65**, 1–10.

25. Sadauskas, E., Jacobsen, N. R., Danscher, G., Stoltenberg, M., Vogel, U., Larsen, A., Kreyling, W. and Wallin, H. (2009). Biodistribution of gold nanoparticles in mouse lung following intratracheal instillation. *Chem. Cent. J.,* **3**, 16.
26. Tam, J. M., Tam, J. O., Murthy, A., Ingram, D. R., Ma, L. L., Travis, K., Johnston, K. P. and Sokolov, K. V. (2010). Controlled assembly of biodegradable plasmonic nanoclusters for near-infrared imaging and therapeutic applications. *ACS Nano.,* **4**, 2178–2184.
27. Pujals, S., Fernández-Carneado, J., López-Iglesias, C., Kogan, M. J. and Giralt, E. (2006). Mechanistic aspects of CPP-mediated intracellular drug delivery: Relevance of CPP self-assembly. *Biochimica et Biophysica Acta (BBA) — Biomembranes,* **1758**, 264–279.
28. Cho, E. C., Xie, J., Wurm, P. A. and Xia, Y. (2009). Understanding the role of surface charges in cellular adsorption versus internalization by selectively removing gold nanoparticles on the cell surface with a I2/KI etchant. *Nano Lett.,* **9**, 1080–1084.
29. Wang, B., Zhang, L., Bae, S. C. and Granick, S. (2008). Nanoparticle-induced surface reconstruction of phospholipid membranes. *P. Natl. Acad. Sci. USA,* **105**, 18171–18175.
30. Leroueil, P. R., Berry, S. A., Duthie, K., Han, G., Rotello, V. M., Mcnerny, D. Q., Baker, J. R., Orr, B. G. and Banaszak Holl, M. M. (2008). Wide varieties of cationic nanoparticles induce defects in supported lipid bilayers. *Nano Lett.,* **8**, 420–424.
31. Nel, A. E., Madler, L., Velegol, D., Xia, T., Hoek, E. M. V., Somasundaran, P., Klaessig, F., Castranova, V. and Thompson, M. (2009). Understanding biophysicochemical interactions at the nano-bio interface. *Nat. Mater.,* **8**, 543–557.
32. Cai, W., Shin, D.-W., Chen, K., Gheysens, O., Cao, Q., Wang, S. X., Gambhir, S. S. and Chen, X. (2006). Peptide-labeled near-infrared quantum dots for imaging tumor vasculature in living subjects. *Nano Lett.,* **6**, 669–676.
33. Gao, X., Cui, Y., Levenson, R. M., Chung, L. W. K. and Nie, S. (2004). In vivo cancer targeting and imaging with semiconductor quantum dots. *Nat. Biotech.,* **22**, 969–976.
34. Uchida, K., Otsuka, H., Kaneko, M., Kataoka, K. and Nagasaki, Y. (2005). A reactive poly(ethylene glycol) layer to achieve specific surface plasmon resonance sensing with a high S/N ratio: The substantial role of a short underbrushed PEG layer in minimizing nonspecific adsorption. *Anal. Chem.,* **77**, 1075–1080.
35. Bentzen, E. L., Tomlinson, I. D., Mason, J., Gresch, P., Warnement, M. R., Wright, D., Sanders-Bush, E., Blakely, R. and Rosenthal, S. J. (2005). Surface modification to reduce nonspecific binding of quantum dots in live cell assays. *Bioconjug. Chem.,* **16**, 1488–1494.
36. Ballou, B., Lagerholm, B. C., Ernst, L. A., Bruchez, M. P. and Waggoner, A. S. (2004). Noninvasive imaging of quantum dots in mice. *Bioconjug. Chem.,* **15**, 79–86.
37. Daou, T. J., Li, L., Reiss, P., Josserand, V. and Texier, I. (2009). Effect of poly(ethylene glycol) length on the in vivo behavior of coated quantum dots. *Langmuir,* **25**, 3040–3044.
38. Park, J., Nam, J., Won, N., Jin, H., Jung, S., Jung, S., Cho, S. H. and Kim, S. (2011). Compact and stable quantum dots with positive, negative, or zwitterionic surface: Specific cell interactions and non-specific adsorptions by the surface charges. *Adv. Funct. Mater.,* **21**, 1558–1566.
39. Lai, S. K., Wang, Y. Y. and Hanes, J. (2009). Mucus-penetrating nanoparticles for drug and gene delivery to mucosal tissues. *Adv. Drug Deliv. Rev.,* **61**, 158–171.
40. Lai, S. K., O'hanlon, D. E., Harrold, S., Man, S. T., Wang, Y. Y., Cone, R. and Hanes, J. (2007). Rapid transport of large polymeric nanoparticles in fresh undiluted human mucus. *P. Natl. Acad. Sci. USA,* **104**, 1482–1487.

41. Wang, Y. Y., Lai, S. K., Suk, J. S., Pace, A., Cone, R. and Hanes, J. (2008). Addressing the PEG mucoadhesivity paradox to engineer nanoparticles that "slip" through the human mucus barrier. *Angew. Chem. Int. Ed.*, **47**, 9726–9729.
42. Lipka, J., Semmler-Behnke, M., Sperling, R. A., Wenk, A., Takenaka, S., Schleh, C., Kissel, T., Parak, W. J. and Kreyling, W. G. (2010). Biodistribution of PEG-modified gold nanoparticles following intratracheal instillation and intravenous injection. *Biomaterials*, **31**, 6574–6581.
43. Lee, H. A., Imran, M., Monteiro-Riviere, N. A., Colvin, V. L., Yu, W. W. and Riviere, J. E. (2007). Biodistribution of quantum dot nanoparticles in perfused skin: Evidence of coating dependency and periodicity in arterial extraction. *Nano Lett.*, 7, 2865–2870.
44. Scheinberg, D. A., Villa, C. H., Escorcia, F. E. and Mcdevitt, M. R. (2010). Conscripts of the infinite armada: Systemic cancer therapy using nanomaterials. *Nat. Rev. Clin. Oncol.*, 7, 266–276.
45. Yang, L., Mao, H., Wang, Y. A., Cao, Z., Peng, X., Wang, X., Duan, H., Ni, C., Yuan, Q., Adams, G., Smith, M. Q., Wood, W. C., Gao, X. and Nie, S. (2009). Single chain epidermal growth factor receptor antibody conjugated nanoparticles for *in vivo* tumor targeting and imaging. *Small*, **5**, 235–243.
46. Kirui, D. K. Rey, D. A. and Batt, C. A. (2010). Gold hybrid nanoparticles for targeted phototherapy and cancer imaging. *Nanotech.*, **21**, 105105.
47. Nimjee, S. M., Rusconi, C. P. and Sullenger, B. A. (2005). Aptamers: An emerging class of therapeutics. *Annu. Rev. Med.*, **56**, 555–583.
48. Wang, A. Z., Gu, F., Zhang, L., Chan, J. M., Radovic-Moreno, A., Shaikh, M. R. and Farokhzad, O. C. (2008). Biofunctionalized targeted nanoparticles for therapeutic applications. *Expert Opin. Biol. Ther.*, **8**, 1063–1070.
49. Javier, D. J., Nitin, N., Levy, M., Ellington, A. and Richards-Kortum, R. (2008). Aptamer-targeted gold nanoparticles as molecular-specific contrast agents for reflectance imaging. *Bioconjug. Chem.*, **19**, 1309–1312.
50. Huang, Y. F., Lin, Y. W., Lin, Z. H. and Chang, H. T. (2009). Aptamer-modified gold nanoparticles for targeting breast cancer cells through light scattering. *J. Nanopart. Res.*, **11**, 775–783.
51. Bharali, D. J., Lucey, D. W., Jayakumar, H., Pudavar, H. E. and Prasad, P. N. (2005). Folate-receptor-mediated delivery of InP quantum dots for bioimaging using confocal and two-photon microscopy. *J. Am. Chem. Soc.*, **127**, 11364–11371.
52. Dixit, V., Van Den Bossche, J., Sherman, D. M., Thompson, D. H. and Andres, R. P. (2006). Synthesis and grafting of thioctic acid–PEG–folate conjugates onto Au nanoparticles for selective targeting of folate receptor-positive tumor cells. *Bioconjug. Chem.*, **17**, 603–609.
53. Au, L., Zhang, Q., Cobley, C. M., Gidding, M., Schwartz, A. G., Chen, J. and Xia, Y. (2009). Quantifying the cellular uptake of antibody-conjugated Au nanocages by two-photon microscopy and inductively coupled plasma mass spectrometry. *ACS Nano.*, **4**, 35–42.
54. Zhang, D., Neumann, O., Wang, H., Yuwono, V. M., Barhoumi, A., Perham, M., Hartgerink, J. D., Wittung-Stafshede, P. and Halas, N. J. (2009). Gold nanoparticles can induce the formation of protein-based aggregates at physiological pH. *Nano Lett.*, **9**, 666–671.
55. Aubin-Tam, M. E. and Hamad-Schifferli, K. (2005). Gold nanoparticle–cytochrome c complexes: The effect of nanoparticle ligand charge on protein structure. *Langmuir*, **21**, 12080–12084.

56. Choi, H. S., Liu, W., Liu, F., Nasr, K., Misra, P., Bawendi, M. G. and Frangioni, J. V. (2010). Design considerations for tumour-targeted nanoparticles. *Nat. Nanotech.,* **5**, 42–47.
57. Schmaljohann, D. (2006). Thermo- and pH-responsive polymers in drug delivery. *Adv. Drug Deliv. Rev.,* **58**, 1655–1670.
58. Nam, J., Won, N., Jin, H., Chung, H. and Kim, S. (2009). pH-induced aggregation of gold nanoparticles for photothermal cancer therapy. *J. Am. Chem. Soc.,* **131**, 13639–13645.
59. Nam, J., La, W. G., Hwang, S., Ha, Y. S., Park, N., Won, N., Jung, S., Bhang, S. H., Ma, Y. J., Cho, Y. M., Jin, M., Han, J., Shin, J. Y., Wang, E. K., Kim, S. G., Cho, S. H., Yoo, J., Kim, B. S. and Kim, S. (2013). pH-responsive assembly of gold nanoparticles and "spatiotemporally concerted" drug release for synergistic cancer therapy. *ACS Nano.,* 7, 3388–3402.
60. Liu, X., Chen, Y., Li, H., Huang, N., Jin, Q., Ren, K. and Ji, J. (2013). Enhanced retention and cellular uptake of nanoparticles in tumors by controlling their aggregation behavior. *ACS Nano.,* 7, 6244–6257.
61. Hwang, S., Nam, J., Song, J., Jung, S., Hur, J., Im, K., Park, N. and Kim, S. (2014). A sub 6 nanometer plasmonic gold nanoparticle for pH-responsive near-infrared photothermal cancer therapy. *New J. Chem.,* **38**, 918–922.
62. Nam, J., Ha, Y. S., Hwang, S., Lee, W., Song, J., Yoo, J. and Kim, S. (2013). pH-responsive gold nanoparticles-in-liposome hybrid nanostructures for enhanced systemic tumor delivery. *Nanoscale,* **5**, 10175–10178.
63. Jung, S., Nam, J., Hwang, S., Park, J., Hur, J., Im, K., Park, N. and Kim, S. (2013). Theragnostic pH-sensitive gold nanoparticles for the selective surface enhanced raman scattering and photothermal cancer therapy. *Anal. Chem.,* **85**, 7674–7681.
64. Xiao, P., Li, Q., Joo, Y., Nam, J., Hwang, S., Song, J., Kim, S., Joo, C. and Kim, K. H. (2013). Detection of pH-induced aggregation of "smart" gold nanoparticles with photothermal optical coherence tomography. *Opt. Lett.,* **38**, 4429–4432.
65. Jones, D. P., Carlson, J. L., Samiec, P. S., Sternberg Jr., P., Mody Jr, V. C., Reed, R. L. and Brown, L. A. S. (1998). Glutathione measurement in human plasma: Evaluation of sample collection, storage and derivatization conditions for analysis of dansyl derivatives by HPLC. *Clinica. Chimica. Acta.,* **275**, 175–184.
66. Hong, R., Han, G., Fernández, J. M., Kim, B.-J., Forbes, N. S. and Rotello, V. M. (2006). Glutathione-mediated delivery and release using monolayer protected nanoparticle carriers. *J. Am. Chem. Soc.,* **128**, 1078–1079.
67. Kim, B., Han, G., Toley, B. J., Kim, C.-K., Rotello, V. M. and Forbes, N. S. (2010). Tuning payload delivery in tumour cylindroids using gold nanoparticles. *Nat. Nanotech.,* **5**, 465–472.
68. Wang, X., Cai, X., Hu, J., Shao, N., Wang, F., Zhang, Q., Xiao, J. and Cheng, Y. (2013). Glutathione-triggered "off–on" release of anticancer drugs from dendrimer-encapsulated gold nanoparticles. *J. Am. Chem. Soc.,* **135**, 9805–9810.
69. Kim, M., Ock, K., Cho, K., Joo, S. W. and Lee, S. Y. (2012). Live-cell monitoring of the glutathione-triggered release of the anticancer drug topotecan on gold nanoparticles in serum-containing media. *Chem. Commun.,* **48**, 4205–4207.
70. Nakanishi, J., Nakayama, H., Shimizu, T., Ishida, H., Kikuchi, Y., Yamaguchi, K. and Horiike, Y. (2009). Light-regulated activation of cellular signaling by gold nanoparticles that capture and release amines. *J. Am. Chem. Soc.,* **131**, 3822–3823.
71. Agasti, S. S., Chompoosor, A., You, C. C., Ghosh, P., Kim, C. K. and Rotello, V. M. (2009). Photoregulated release of caged anticancer drugs from gold nanoparticles. *J. Am. Chem. Soc.,* **131**, 5728–5729.

72. Han, G., You, C. C., Kim, B. J., Turingan, R. S., Forbes, N. S., Martin, C. T. and Rotello, V. M. (2006). Light-regulated release of DNA and its delivery to nuclei by means of photolabile gold nanoparticles. *Angew. Chem. Int. Ed.,* **45**, 3165–3169.
73. Carling, C. J., Nourmohammadian, F., Boyer, J. C. and Branda, N. R. (2010). Remote-control photorelease of caged compounds using near-infrared light and upconverting nanoparticles. *Angew. Chem. Int. Ed.,* **49**, 3782–3785.
74. Jayakumar, M. K. G., Idris, N. M. and Zhang, Y. (2012). Remote activation of biomolecules in deep tissues using near-infrared-to-UV upconversion nanotransducers. *P. Natl. Acad. Sci., USA,* **109**, 8483–8488.
75. Chien, Y. H., Chou, Y. L., Wang, S. W., Hung, S. T., Liau, M. C., Chao, Y. J., Su, C. H. and Yeh, C. S. (2013). Near-infrared light photocontrolled targeting, bioimaging, and chemotherapy with caged upconversion nanoparticles *in vitro* and *in vivo*. *ACS Nano.,* **7**, 8516–8528.
76. Jeong, U., Teng, X., Wang, Y., Yang, H. and Xia, Y. (2007). Superparamagnetic colloids: Controlled synthesis and niche applications. *Adv. Mater.,* **19**, 33–60.
77. Shankar, S. S. and Deka, S. (2011). Metal nanocrystals and their applications in biomedical systems. *Sci. Adv. Mater.,* **3**, 169–195.
78. Riedinger, A., Guardia, P., Curcio, A., Garcia, M. A., Cingolani, R., Manna, L. and Pellegrino, T. (2013). Subnanometer local temperature probing and remotely controlled drug release based on azo-functionalized iron oxide nanoparticles. *Nano Lett.,* **13**, 2399–2406.
79. N'guyen, T. T. T., Duong, H. T. T., Basuki, J., Montembault, V., Pascual, S., Guibert, C., Fresnais, J., Boyer, C., Whittaker, M. R., Davis, T. P. and Fontaine, L. (2013). Functional iron oxide magnetic nanoparticles with hyperthermia-induced drug release ability by using a combination of orthogonal click reactions. *Angew. Chem. Int. Ed.,* **125**, 14402–14406.

Chapter 13

Smart Nanopreparations for Cancer

Shaheen Sultana[1] *and Shahnaz Sultana*[2]

[1]*Northern Border University, Rafha Campus, Kingdom of Saudi Arabia*
[2]*Jazan University, Jazan, Kingdom of Saudi Arabia*

13.1 Introduction

The emergence of smart nanocarriers opens up potential opportunities to improve the accumulation of chemotherapy within solid tumors. These systems have the potential to be activated only in cancerous environments and then kill just the cancer cells. Their targeting ability to cancer cells only spare the health of surrounding cells from being harmed, reducing the drug dose, and thus enhancing patient survivability.

13.1.1 *pH-sensitive nanoformulation*

Tumor extracellular pH (pHe, average value: 5.0–6.0) is lower than in normal tissues (pH: 7.4), which is due to the high rate of glycolysis in cancer cells, both in aerobic and anaerobic conditions [1]. pH-sensitive nanoformulations, such as polymeric micelles and nanogels, have been targeted to slightly acidic extracellular pH environments of solid tumors. These systems utilize pH-sensitive polymers, such as poly(L-histidine) (polyHis) or polysulfonamide. The imidazole ring of a polyHis has lone pairs of electrons on the unsaturated nitrogen that endow amphoteric properties by protonation and deprotonation [2]. Doxorubicin (Dox), loaded in pH-sensitive micelles, poly(l-histidine) (M(n):5K)-b-PEG (M(n):5K), showed a 5.8- and 5.2-fold increase in the area under concentration (AUC) curve and half-life time (t(1/2)) (plasma half life) than free DOX in phosphate buffered saline (PBS), respectively. Tumor growth was significantly retarded in mice xenografted with human ovarian carcinoma, which showed preferential accumulation of the anticancer agent in tumor cells [3]. A copolymer of N-isopropylacrylamide (NIPAAm) and chitosan (4:1, m/m) with loading efficiencies 85.7% and 9.6%, respectively, was synthesized. The particles showed a highly pH-dependent *in vitro* release profile, in which the cumulative release rate was significantly enhanced below pH 6.8 and decreased rapidly above pH 6.9 at 36.5 ± 0.5°C. *In vivo* studies in mice revealed significant tumor regression with complete regression in more than 50% of the mice [4]. A newly synthesized biodegradable nanoparticulate using pH-sensitive polymeric systems, such as poly(ethylene oxide) (PEO)-modified poly(beta-amino ester) (PbAE) effectively delivered paclitaxel into the tumor mass [5]. Similarly, researchers have synthesized poly(ethylene glycol) methacrylate-*co*-2-(diisopropylamino)ethyl methacrylate (PEGMA-*co*-DPA) copolymers capable of self-assembly to form pH-sensitive nanoparticles in aqueous environments, as a means of encapsulating the water-insoluble photo-sensitizer, meso-tetra(hydroxyphenyl)chlorin (m-THPC), which exhibited obvious phototoxicity in HT-29 colon cancer cells after light irradiation [6]. Recently, smart nanoparticle systems with activatable cell-penetrating peptide (designated as *dt*ACPP) that is dual-triggered by the lowered pHe and matrix metalloproteinase 2 (MMP2) was constructed. Upon reaching the tumor microenvironment, the linker was cleaved by MMP2, and *dt*ACPP was activated to expose CPP to drive the nanoparticles' internalization into the intratumoral cells (Fig. 13.1) [7, 8]. Recently, a mixed polymeric micelle system developed on three grafted copolymers, including 1,2-distearoyl-sn-glycero-3-phosphoethanolamine-polyethyleneglycol-2000(DSPE-PEG2000), antinucleosome antibody (mAb 2C5)-modified DSPE-PEG3400 (DSPE-PEG3400-2C5), and poly(ethylene glycol)-coupled poly(L-histidine) (PHIS-PEG2000), showed 75–95% drug released within 2 h at pH 5.5. Further modification of the mixed micelles with anti-cancer nucleosome-specific monoclonal antibody 2C5 significantly increased the tumor cell-specific internalization and cytotoxicity [2].

Recently, a novel pH-sensitive and liver-targeted glycyrrhetinic acid-modified alginate/doxorubicin-modified alginate complex nanoparticle (GA-ALG/DOX-ALG NP) was synthesized. As shown in Fig. 13.2, the area under the curve ($AUC_{0-\infty}$) and half-life ($t_{1/2}$) in the liver of GA-ALG/DOX-ALG NPs were 1,156.7 μg h/g and 34.3 h, respectively, which was 11.8- and 3.2-fold higher than that of the DOX·HCl group. Furthermore, the mice treated with GA-ALG/DOX-ALG NPs showed 79.3% inhibition in tumor growth rate, which was much higher than that of the DOX·HCl (48.5%) and DOX-ALG NPs groups (62.7%) [9].

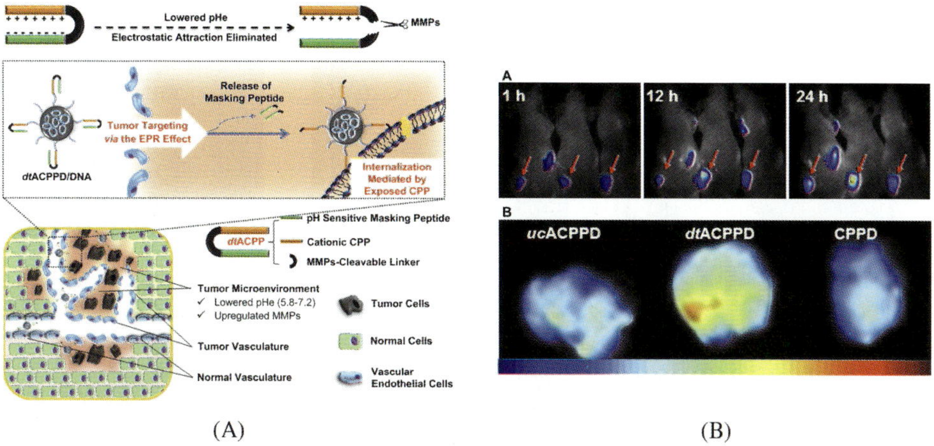

Figure 13.1. a. The targeting and internalization strategy of *dt*ACPP modifiednanoparticles, *dt*ACPPD/DNA. b. The real-time fluorescence imaging shows the *in vivo* distribution of nanocarriers in the subcutaneous tumor-bearing mice. (A) *In vivo* imaging of tumor-bearing mice treated with Red BODIPY-labeled nanocarriers, *uc*ACPPD (left in every image), *dt*ACPPD (middle in every image), or CPPD (right in every image). Images were taken at 1, 12, and 24 h after systemic administration. Red arrow indicates tumor site. (B) Imaging of the excised tumors. Intensity of the signal: dark red is the strongest while dark blue is the weakest, as shown by the bar [7]. pHe = extracellular pH; MMP = matrix metalloproteinase, EPR = enhanced permeability and retention; CPP = cell-penetrating peptide.

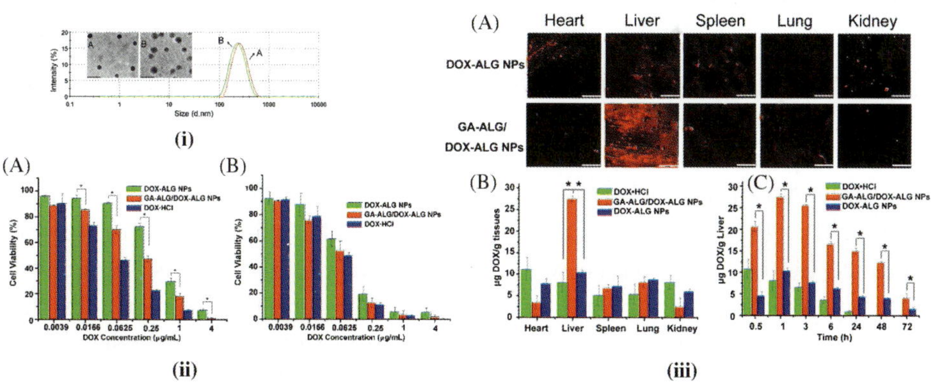

Figure 13.2. (i) The size distribution and transmission electron microscopy (TEM) image of DOX-ALG NPs (A) and GA-ALG/DOX-ALG NPs (B). All scale bars are 200 nm. (ii) Cytotoxicity of free DOX·HCl, DOX-ALG NPs and GA-ALG/DOX-ALG NPs against HepG2 cells for (A) 48 h and (B) 72 h; Means ± SD ($n = 3$). $*p < 0.05$. (iii). A. Micrographs of different tissue homogenates after intravenous injection DOX-ALG NPs and GA-ALG/DOX-ALG NPs at a single dose of 7 mg/kg at 1 h after injection (scale bars correspond to 200 m in all images). (B) Tissue distribution of DOX after intravenous injection of DOX·HCl, DOX-ALG NPs and GA-ALG/DOX-ALG NPs at 1 h after injection and (C) DOX concentration in the liver after intravenous injection of DOX·HCl, DOX-ALG NPs and GA-ALG/DOX-ALG NPs in Kunming mice at a single dose of 7 mg/kg.

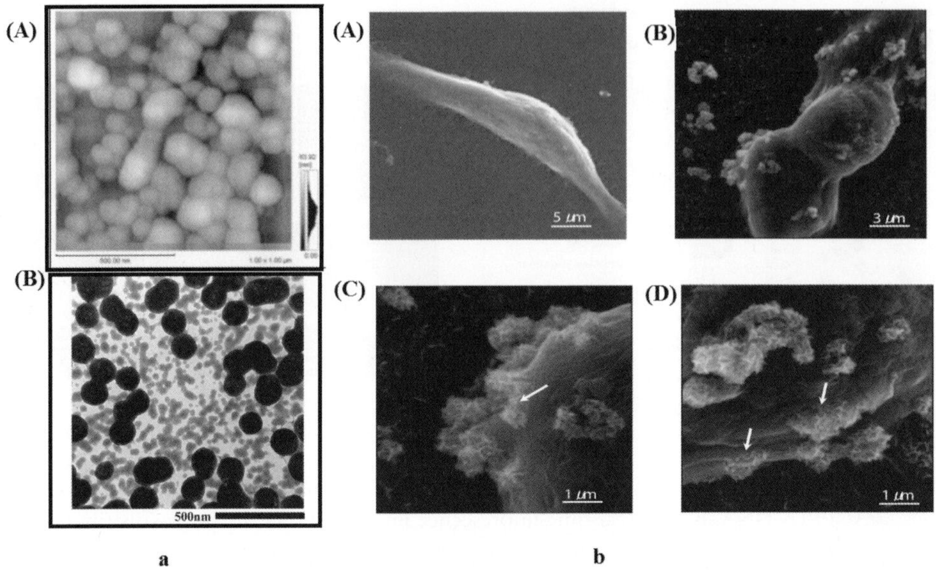

Figure 13.3. a. Characterization of carbonate apatite nanocomposites by A. atomic force microscopy (AFM) and B. TEM pictures b. Scanning electron microscopy revealed a process of uptake behavior into HCT116 colon cancer cells. (A) The control HCT116 devoid of apatite particles. (B) The particles attached on the cell. (C) The particles were getting internalized at as early as 45 min, and (D) further going into the cells at 90 min.

A novel pH-sensitive carbonate apatite demonstrated significant cellular uptake of drug (92%) with high cytotoxicity (Fig. 13.3) [10]. Soppimath and co-workers synthesized core/shell nanoparticles that displayed a pH-sensitive thermal response, which were self-assembled from the amphiphilic tercopolymer, poly(*N*-isopropylacrylamide-*co*-*N*,*N*-dimethylacrylamide-*co*-10-undecenoic acid) (P(NIPAAm-*co*-DMAAm-*co*-UA). The particles were further modified by folic acid, which was conjugated to the hydrophilic segment of the polymer through the free amine group (for targeting cancer cells that overexpress folate receptors), and cholesterol was grafted to the hydrophobic segment of the polymer. The nanoparticles showed greater cellular uptake because of the folate-receptor-mediated endocytosis process when compared with nanoparticles without folate [11].

A multifunctional gold nanorod (GNR)-based nanopreparation was synthesized, in which the chemotherapeutic agent was covalently conjugated onto PEGylated (polyethylene glycol, PEG) nanorods via a hydrazone bond with tumor-targeting ligands onto the distal ends of the PEG arms (Fig. 13.4). The results revealed that conjugated nanocarriers exhibited a higher cellular uptake and cytotoxicity than non-targeted ones *in vitro*.

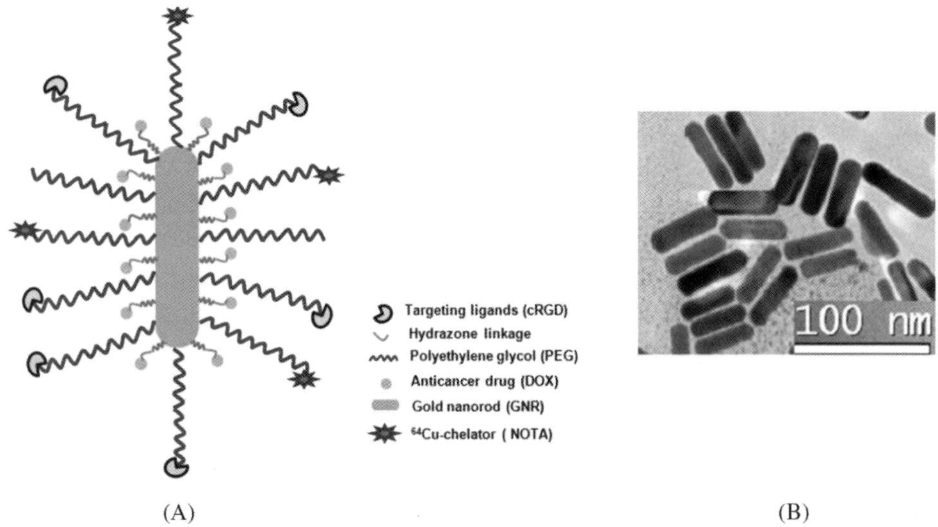

Figure 13.4. (A) A schematic illustration of the multifunctional GNR-DOX-cRGD nanocarriers for tumor-targeted drug delivery and PET imaging. (B) TEM photomicrograph of nanorods. cRGD = cyclic arginine–glycine–aspartic acid, DOX = doxorubicin.

Novel pH-sensitive membranolytic and lysosomolytic nanocarriers (BSA-AL-NLCs) were developed by inserting L-arginine lauril ester (AL) into nanostructure lipid carriers (NLCs) and coated with bovine serum albumin (BSA). The particles displayed pH-dependent release and internalization into human breast cancer cells via the endosome–lysosome pathway *in vitro* [12]. Polyamidoamine dendrimer complexes synthesized from pH-sensitive diblock copolymers of poly(methacryloyl sulfadimethoxine) (PSD) and PEG modified by lactose exhibit selective targeting and higher antitumor efficacy against HepG2 cells [13]. The RGD-modified PEGylated polyamidoamine (PAMAM) dendrimer with Dox conjugated by an acid-sensitive *cis*-aconityl linkage (RGD-PPCD) showed a significantly prolonged half-life, increased AUC, and greater accumulation in brain tumors than in normal brain tissue, when compared with the drug solution [14].

The introduction of a multifunctional strategy, in which folate-PEGylation was applied to modify the surface of polymethacrylate (PMA)-grafted poly(amidoamine) (PAMAM) nanoparticles, significantly enhanced tumor selectivity via folate receptor targeting, long circulation, and the EPR effect [15]. Multifunctional hollow nanoparticles conjugated with folic acid, self-assembled from poly(N-vinylimidazole-co-N-vinylpyrrolidone)-g-poly(D,L-lactide) graft copolymers and methoxyl/functionalized-PEG-PLA diblock copolymers, exhibited a specific on–off switch drug release behavior, owing to the pH-sensitive structure of imidazole and excellent antitumor activity [16].

pH-sensitive liposomes, sterically stabilized by PEG molecules (stealth liposomes) with a targeted moiety, PR_b, which specifically recognizes and binds to integrin $\alpha(5)\beta(1)$-expressing cells, has been synthesized. The system binds to CT26.WT colon carcinoma cells that express integrin $\alpha(5)\beta(1)$, undergo cellular internalization, and deliver a payload directly to cancer cells in a short period of time when compared to other formulations [17].

In another study, plasmid DNA was complexed with polyethyleneimine (PEI) and further with a pH-sensitive diblock copolymer, poly(methacryloyl sulfadimethoxine) (PSD)-block-PEG (PSD-b-PEG) to obtain polymeric nanoparticles. The nanoparticles effectively target the acidic pH of extracellular matrix of tumors (pH:6.6), interact with the cells and demonstrated high cytotoxicity and transfection ability [18]. Chitosan nanoparticles are widely employed as carriers to co-deliver chemotherapeutic agents to achieve targeted drug delivery, with a pH-responsive drug release, and strong cytotoxicity against the cells [19, 20].

13.1.2 *Gold nanoparticles*

Gold nanoparticles (AuNPs) absorb energy from near-infrared (NIR) light and emit it as heat to destroy tumors with minimal side effects. AuNPs are sub-micrometer-sized particles of gold dispersed in a fluid, usually water. The large surface area-to-volume ratio of AuNPs enables their surface to be coated with hundreds of molecules (including therapeutics, targeting agents, and anti-fouling polymers). AuNPs have unique optical and physical properties, making them suitable for biomedical applications with better biocompatibility in disease diagnosis and therapeutics. AuNPs can be prepared and conjugated with many functionalizing agents, such as polymers, surfactants, ligands, dendrimers, drugs, DNA, RNA, proteins, peptides, and oligonucleotides to expand and improve them for photothermal therapy with reduced cytotoxic effects in various cancers, gene therapies, and many other diseases [21]. Injection of the functionalized AuNPs into tumor-bearing mice causes the over-expressed protease of MMP-2 in tumor tissue and intracellular GSH to lead to the rapid release of the anti-tumor drug (doxorubicin) from the functionalized AuNPs to inhibit tumor growth, while being fluorescently imaged simultaneously. The functionalized AuNPs with tumor-triggered drug release properties can further improve the efficacy and reduce side effects significantly [22]. A new technique, "living" PEGylation, that allows for control over the density and composition of heterobifunctional PEG (thiol-terminated poly(ethylene glycol), HS-PEG-R) on AuNPs when they are incubated with U-87 malignant glioblastoma (MG) and non-target cells, demonstrating that targeting ligand density is critical to maximizing the efficiency of the targeting of AuNPs to cancer cells. This also sequentially controls the HS–PEG–R density to develop multifunctional nanoparticles, conjugating positively charged HS–PEG–NH2 at increasing ratios to AuNPs containing negatively charged HS–PEG–COOH to reduce uptake by macrophage cells. The ability to minimize non-specific binding/uptake by healthy cells could further improve targeted nanoparticle efficacy [23]. A folate-conjugated AuNP is an effective agent for targeted nanophotothermolysis and the application of an optical fiber to transport the laser light from the source to the target tissue inside the body. Thus, a new method of nanosurgery, in which a surgeon is able to perform surgery at the cellular or even at the subcellular level, may be possible [24]. AuNPs may have potential as chemotherapeutic agents if conjugated to nuclear targeting ligands. Considering their additional radiotherapeutic properties, targeted AuNPs could also become highly effective in the treatment of cancer with concomitant chemoradiation therapy [25]. The folic acid functionalized single-stranded DNA was designed to target the folate receptor on the cancer cell surface, and the molecular beacon was employed as a drug carrier for activated release associated with intracellular tumor messenger RNA (mRNA). The nanocarrier could reduce the dosage and greatly improve the therapeutic effect of drugs in cancer cells [26].

Dendrimer encapsulated gold nanoparticles (DEGNPs) were synthesized and used as carriers of thiolated anticancer drugs. Thiol-containing drugs, such as captopril and 6-mercaptopurine, loaded within DEGNPs showed an "Off–On" release behavior in the presence of thiol-reducing agents, such as glutathione and dithiothreitol. Much more reduced cytotoxicity as compared to the free anticancer compounds was observed when thiolated doxorubicin and cisplatin were loaded within the nanoparticle [27]. Controlled aggregation of NPs sensitive to the tumor microenvironment can serve as a universal strategy to enhance the retention and cellular uptake of inorganic NPs in tumors, and modifying NPs with a mixed-charge zwitterionic surface can provide an easy way to obtain stealth properties and pH-sensitivity at the same time [28].

Multimodal treatments with thermal therapies at the perfusion nadir in the sub- and supraphysiological temperature regimes increases tumor volumetric destruction by over 60%. Nanoparticle-conjugated vascular disrupting agents (VDAs) significantly improve cancer therapy by preconditioning the tumor vasculature to a secondary insult in a targeted manner [29]. The cytotoxic activity of an anti-cancer drug Dox conjugated to the P-AuNS (α-helix peptides of 17-amino acids conjugated to gold nanospheres) showed different cytotoxicity against the cell lines, depending on the peptide sequence, with a higher efficiency than free Dox at the same concentration. The cytotoxicity by Dox was correlated with the cell penetrating (CP) activity of the peptides against the three cell lines. These results demonstrated that P-AuNSs would be useful tools for the development of new cell-selective drug delivery systems [30]. The multilayer NPs showed excellent tumor targetability and antitumor efficacy in tumor-bearing mice, caused by the enhanced permeation and retention (EPR) effect [31]. Two gold nanoparticle–antibody conjugates, gold–BerH2 antibody (anti-CD30 receptor) and gold–ACT1 antibody (anti-CD25-receptor), were synthesized. Under laser irradiation at suitable power, the high killing efficiency of gold-targeted L-428 Hodgkin's cells was achieved, with little damage to non-targeted cancer cells [32]. FGF1-coated AuNPs are feasible for targeted cancer therapy, as fibroblast growth factor receptors (FGFRs) are over-expressed in a wide variety of tumors [33]. The combination of trastuzumab conjugated Human Epidermal Growth Factor Receptor-2 (HER-2)-targeted AuNPs (Au-T) and X-radiation resulted in the regression of MDA-MB-361 tumors by 46%, as compared to treatment with X-radiation (16.0% increase in tumor volume). No significant normal tissue toxicity was observed. Radiosensitization of breast cancer to X-radiation with AuNPs was successfully achieved with an optimized therapeutic strategy of molecular targeting of HER-2 and intratumoral administration [34]. The conjugated heparin (heparin–deoxycholate sodium) retained its ability to bind with angiogenic factors, showing a significant decrease in endothelial tubular formation. Second, targeting ligands conjugated heparin derivatives have introduced a receptor-mediated delivery of anticancer drugs. Heparin-folic acid–retinoic acid (HFR) bioconjugates for treating cancer cells showed a three-fold higher efficacy than heparin–retinoic acid (HR) [35]. The uniqueness of inorganic nanoparticles, especially GNPs, as drug delivery vehicles and moreover, will present a widespread scenario of GNPs that can be used for the treatment of life-threatening diseases, like cancer [36]. Ultra-small gold nanoclusters (AuNCs), the RNase-A-encapsulated AuNCs (RNase-A-AuNCs), display an exceptionally large Stokes shift of ~210 nm, as well as a single dominant fluorescence lifetime of ~1.5 s, about three orders of magnitude longer than biological autofluorescence. Furthermore, by coupling vitamin B(12) (VB(12)) to the RNase-A-AuNC, a multifunctional nanoplatform is developed that is suitable for simultaneous targeting and imaging of cancer at the cellular level using Caco-2 cell lines as an *in vitro* model. Since VB(12) has effective uptake pathways in the digestive system, this nanoplatform may have the potential for targeted oral drug delivery *in vivo* [37]. The conjugation of streptavidin-coated AuNPs with a biotin-linked prostate-specific membrane antigen (PSMA) inhibitor (CTT54) was used to generate PSMA-targeted AuNPs. *In vitro* evaluations of these targeted AuNPs were conducted to determine PSMA-mediated and time-dependent binding to PSMA-positive androgen-sensitive human prostate adenocarcinoma cells (LNCaP) cells. The PSMA-targeted AuNPs exhibited significantly greater and selective binding to LNCaP cells compared to control non-targeted AuNPs, thus demonstrating the feasibility of this approach [38]. An enhanced capacity for protoporphyrin IX (PpIX) synthesis through 5-aminolevulinic acid (ALA) administration has been reported in cancer cells. ALA–AuNP combinations induced cell death via reactive oxygen species (ROS)-mediated apoptosis after photodynamic therapy (PDT) [39]. RGD gold nanoparticle-mediated delivery of short interfering RNA (siRNA) by intratracheal instillation in mice leads to successful suppression of tumor cell proliferation and respective tumor size reduction. Functionalized GNPs are capable of the targeted delivery of siRNA to cancer cells towards effective silencing of the specific target oncogene. The gold-nanoconjugates trigger a complex inflammatory and immune response that might promote the therapeutic effect of RNA interference (RNAi) to reduce tumor size with low doses of siRNA [40]. 10-nm pH-responsive spherical GNPs and conjugated doxorubicins are designed to form aggregates in mild acidic environments, such as those of tumors. The conjugates accumulate in cancer cells by the rapid phagocytic actions and effective blockage of exocytosis by the increased aggregate size. They also effectively accumulate in tumors up to 17 times over the control because of the enhanced permeation and retention. The synergistic effect is demonstrated by the remarkable reductions in both the therapeutically effective drug dosage and the photothermal laser threshold [41]. Octreotide was conjugated to AuNPs (~20 nm) by a spontaneous reaction with the thiol groups. The AuNP–octreotide system exhibited properties suitable for plasmonic photothermal therapy in the treatment of cervical cancer [42].

NPs in crystalline and amorphous forms received much attention worldwide for their use in many commercial applications, and fueled many research centers to devote to developing and expanding various nano-applications [43–45]. Many nanomaterial applications came out with high expectations from diverse fields, which were grown and diversified into medical areas, including field of clinical trials for their unique optical and physical properties. Recently, the applications of Au-NPs expanded into various biomedical fields, such as biosensors, clinical chemistry, immunoassays, genomics, photothermolysis of cancer cell, microorganisms detection and control, targeted drug delivery, optical imaging and monitoring of biological cells and tissues by exploiting resonance scattering, or *in vivo* photo acoustic techniques [46–48]. AuNPs conjugated with drugs created an interesting theme, which fuelled many interests, especially for their binding property with a vast range of organic and biological molecules, their low toxicity and strong absorption spectrum. Hence,

AuNPs played a major role as carriers for drugs and vaccines to the targeted cells or tissues. At present, the focus is on the structure divergent of synthesized AuNPs, starting from general colloidal Au nanospheres, nanorods, or silica-coated Au nanoshells (NSs), nanocubes, nanorice, nanostars, to nanocages [49, 50]. Finally, surface functionalization protocols were developed to be continuously used in various biomedical applications, such as in the conjugation between AuNPs and molecular probes, including polyclonal and monoclonal antibodies (Abs), DNA oligonucleotides, enzymes, and drugs [51]. A synergistic effect of photothermal therapy and chemotherapy was demonstrated by gold nanostar conjugates [52] (Fig. 13.5).

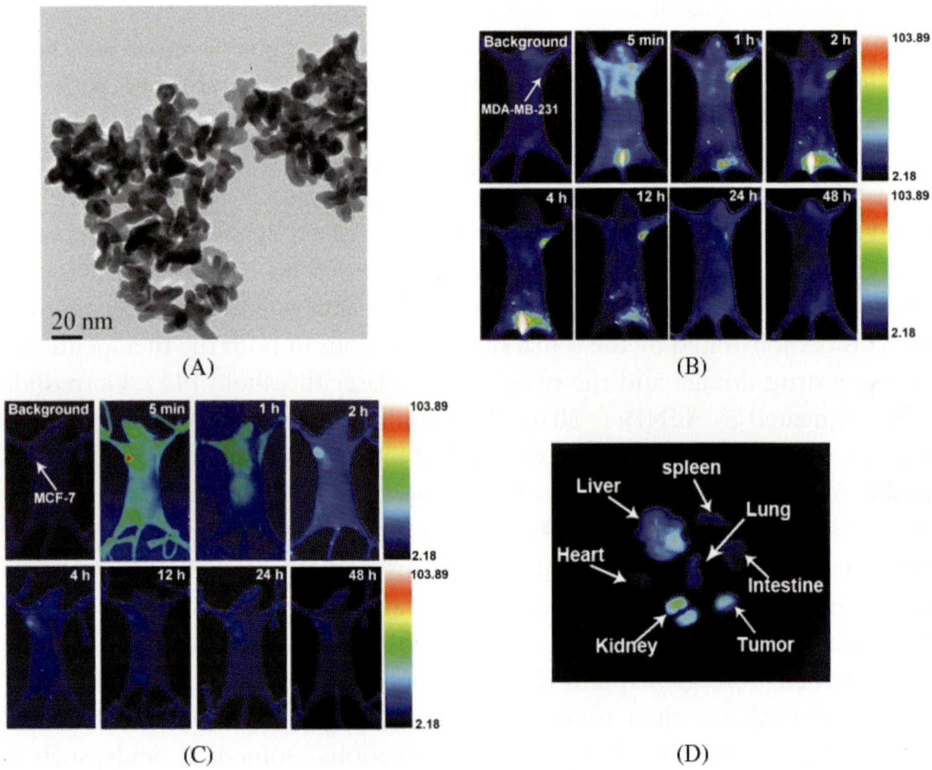

Figure 13.5. (A). Gold nanostar conjugate. The biodistribution of Au-cRGD-MPA showing *in vivo* fluorescence images of Au-cRGD-MPA in (B) MDA-MB-231 (positive $\alpha v\beta 3$ receptor expression), (C) MCF-7 (negative $\alpha v\beta 3$ receptor expression) tumor-bearing mice models at different time intervals (5 min, 1 h, 2 h, 4 h, 12 h, 24 h, and 48 h). (D) The *ex vivo* fluorescence images of isolated organs (heart, liver, spleen, lung, kidney, intestine, and tumor) from the MDA-MB-231 tumor-bearing mice at 48 h post-injection; the *ex vivo* fluorescence image of Au-cRGD-MPA in major organs (heart, liver, spleen, lung, kidney, intestine, and tumor) detected at 48 h post-injection (MDA-MB-231 tumor-bearing mouse).

Tumor necrosis factor-alpha (TNF-α), a cytokine, was proven as an excellent anticancer agent for therapeutic application due to its toxic effects against cancer cells [53]. Later, a nanoparticle drug delivery system was designed with TNF-α conjugated with PEG-coated AuNPs, which efficiently increased damages to tumor cells and decreased the toxicity from TNF-α [54]. TNF-α conjugated with PEG coated AuNPs, given intravenously with a proper dose and time, delayed the tumor growth. It also suppressed the blood flow to the tumor and killed tumor cells by the anti-angiogenesis approach. Even though the particles were administrated intravenously, no accumulation was observed in the organs of healthy animals. Another mechanism assaulted cryoinjury by employing vascular-targeting molecules in ice balls to kill tumor cells completely without any toxic effects. Another Au based conjugate, named as CYT-6091, was created and distributed into the blood stream in delivering TNF-α to solid tumors successfully [55]. The theragnostic gold NPs are cancer-specific because they aggregate rapidly and accumulate selectively in cancerous cells. As a result, both Raman imaging and photothermal efficacy were turned on under a cancerous local environment. In addition, the relatively small hydrodynamic size can have the potential for better access to targeted delivery *in vivo* and facilitated excretion after therapy [56]. Methotrexate (MTX), an inhibitor to dihydrofolatereductase, was used as a chemotherapeutic agent for treating various cancer types. Surface modified AuNPs with the optimal size of PEG chain increased the half life to a few hours, which would be most efficient for cancer therapy [57]. Therefore, AuNPs in drug delivery should be specific, reliable, well targeted, and directed to the site of interest without any hindrance. Anti-epidermal growth factor receptor (EGFR) antibody, called cetuximab (C225), was used as a targeting agent: the anti-cancer drug was gemcitabine and the drug delivery vehicle was AuNPs in Fig. 13.5. This delivery system inhibited the growth of pancreatic tumor cells *in vitro* and orthotropic pancreatic tumor growth *in vivo*. In the same manner, gemcitabine could be used as an anti-cancer drug for other cancers, such as neck, head, breast, and ovarian cancers [58].

13.1.2.1 *AuNPs in drug delivery of lung cancer*

The drug cisplatin completely cured testicular cancer. Cisplatin, Carboplatin, and Oxaliplatin could bind to DNA and prevent the transcription and replication, thereby inducing cellular apoptosis [59, 60]. The reduction in the uptake of drug, tolerance of drug DNA adducts, and increases of tripeptide glutathione in intracellular levels could inactivate cisplatin [61]. AuNPs were used as scaffolds for the platinum-based drug delivery, since they would

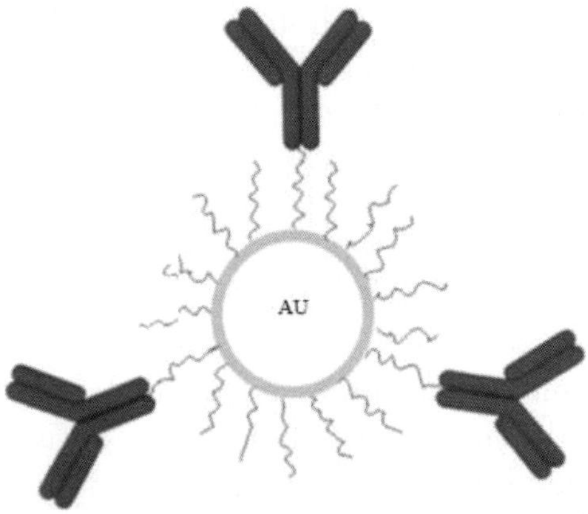

Figure 13.6. Schematic representation of Antibody conjugated AuNPs. Au = gold.

be nontoxic and nonimmunogenic [62–64]. AuNPs were modified with several linkers, like thiols, dicarboxylates, or amine groups, to be conjugated with several drugs, such as cisplatin or BBR3464. These AuNPs with platinum-based drugs were attached with targeting groups, such as aptamers or peptides, in order to increase the specificity towards leukemia and lung cancer, respectively. This approach could suggest a new direction in the future of platinum-based cancer treatments. The platinum-based AuNPs produced better results than oxaloplatin, where they were specifically able to penetrate into the nucleus of lung cancer cells, causing cytotoxicity [65].

Researchers have attempted to treat lung cancer with a variety of types of nanoparticle matrices including lipid, polylactide-co-glycolide, albumin, poly (ω-pentadecalactone-co-butylene-co-succinate), cerium oxide, gold, ultra-small superparamagnetic iron oxide nanoparticles, super paramagnetic iron oxide, lipid-polycation-DNA, N-[1-(2,3-dioleoyloxyl)propyl]-NNN-trimethylammoniummethylsulfate, silica-overcoated magnetic cores, and polyethyleneglycol phosphatidylethanolamine. There are various ways in which nanoparticles enhance drug delivery, and these include encapsulation against immune response, tissue penetration, target selectivity and specificity, delivery monitoring, promoting apoptosis, and blocking pathways for cancer initiation and progression [66].

13.1.2.2 *AuNPs for cytotoxicity studies*

The main reason for cytotoxicity depends on the size of AuNPs. The recent *in vitro* studies with very small AuNPs around 1.4 nm revealed necrosis of the cells from the induction of oxidative stress and mitochondrial damage [67].

Endocytosis of AuNPs would be influenced by their size, surface properties, and aspect ratio. Spherical AuNPs (50 nm) were taken up easily by cancer cell lines, whereas particles less than 50 nm needed to be functionalized with cell penetrating peptides. PEG-coated NPs seemed to avoid phagocytosis and macrophage recognitions [68].

13.1.2.3 *Smart GNPs enhance killing effect on cancer cells*

Glutamic acid (Glu)–AuNPs synthesized by the seeding method technique may have a bright future in tumor-targeted diagnosis and treatment. The effects on cellular uptake and radiosensitizing were investigated using two human cancer cell lines (Henrietta Lacks, HeLa, and MCF-7). Cancer cells took up more Glu–AuNPs than naked AuNPs and lower doses of AuNPs and Glu–AuNPs still enhanced the killing effect using X-ray irradiation, although the apoptotic rate was not altered [69]. Conde *et al.* constructed an antisense gold-nanobeacon consisting of a stem-looped oligonucleotide double-labelled with 3′-Cy3 and 5′-Thiol-C6 and tested for the effective blocking of gene expression in colorectal cancer cells [70].

Unmodified AuNPs inhibit the proliferation of cancer cells in a size- and concentration-dependent manner by abrogating MAP kinase (MAPK) signaling. In addition, these AuNPs reverse epithelial–mesenchymal transitions (EMT) in cancer cells by reducing the secretion of a number of proteins involved in EMT, up-regulating E-Cadherin, and down-regulating Snail, N-Cadherin, and Vimentin. Inhibition of MAPK signaling and reversal of EMT upon AuNP treatment inhibits tumor growth and metastasis in two separate orthotopic models of ovarian cancer [71].

Temozolomide-loaded gold nanostructures are efficient in reducing chemoresistance and destroy 82.7% of cancer stem cells compared with a 42% destruction rate using temozolomide alone. Measurements of *in vitro* cytotoxicity and apoptosis indicate that combination with gold facilitated the ability of temozolomide, an alkylating drug, to alter the resistance of these cancer stem cells, suggesting a new chemotherapy strategy for patients diagnosed with inoperable recurrent malignant glioma [72].

Polysaccharide PST001, which is isolated from the seed kernels of Tamarindus indica (Ti), is an antitumor agent. PST–AuNPs were prepared such that PST001 acted both as a reducing agent and as a capping agent. PST–AuNPs not only retained the antitumor effect of PST001, but also showed an enhanced effect even at a low concentration. It was also found that the nanoparticles exerted their antitumor effects through the induction of apoptosis [73]. In one study, mice received an intravenous injection of GNRs plus NIR laser treatment and showed disappearance of tumors within 15 days with no evidence of recurrence, until the end of the study (3 months) Choline-bound AuNP (choline–AuNP) stabilized with PEI showed increased intra-tumoral uptake as compared to unconjugated AuNPs [74].

13.1.2.4 *Photodynamic therapy*

NIR absorbing GNPs produce heat when excited by light at wavelengths from 700 to 800 nm. This enables these NPs to eradicate targeted tumors. When light is applied to a tumor containing GNPs, the particles rapidly heat up, killing tumor cells in a treatment also known as hyperthermia therapy. A bioclearance study demonstrated that the novel nanocomposite (Au-SNPs≅GA) containing small-size gold nanoparticles (Au-SNPs) and gum arabic (GA) can be gradually excreted by the renal and hepatobiliary system, which might be due to the breakdown and oxidation of GA under irradiating synchrotron X-rays. Thus, the novel gold nanocomposites can be promising photothermal agents for cancer treatment at the therapeutic level, minimizing toxicity concerns regarding long-term accumulation *in vivo* [75]. AuNP exposure to a less powerful laser resulted in small disruptions of the cell membrane due to localized hyperthermia. This did not lead to cell death, but provided a mechanism for killing cancer cells by providing enhanced uptake of drug molecules, thus leading to a new avenue for hyperthermia-anticancer drug combined cancer therapeutics [76]. Multi-functional fluorescent nanodiamond–gold/silver nanoparticles formed by the surface passivation of poly(ethylene glycol) bis(3-aminopropyl) when conjugated with transferrin was more therapeutic than the bare nanodiamond–gold/silver materials, even when exposed to a less energetic laser source. Ultimately, this multi-functional material has great potential for application in simple synthesis. It is non-cytotoxic, supports long-term tracing, and can be used in highly efficient photothermal therapy against cancer cells [77]. Photothermal therapy has been introduced as a selective and noninvasive treatment for cancer, in which exogenous photothermal agents are exploited to achieve the selective destruction of cancer cells. The effect of increasing concentrations of these Barium titanate core-gold NSs (0–100 μg/mL) on human neuroblastoma SH-SY5Y cells, testing the internalization and intrinsic toxicity and validating the hyperthermic functionality of the particles through NIR laser-induced thermoablation experiments [78]. Generation of singlet oxygen by the direct irradiation of naked GNPs is observed using either continuous wave or pulsed laser sources. The underlying mechanism involves plasmon- and hot-electron-mediated reaction pathways and oxygen (O_2) seems to significantly amplify the overall death rates during photothermal treatment (PTT) of cancer cell lines *in vitro* [79]. The heating mechanism is only evident for AuNPs of diameters ≤10 nm, by the electrophoretic oscillation of charged AuNPs across finite length scales in response to a time-varying electric field. These results will assist the future development of nanoparticle-assisted heat production by radiofrequency (RF) fields for applications such as targeted cancer hyperthermia [80]. Thermally induced injury and death of cancer cells were found to begin at 44–45°C, which was achieved using the PTT effect with gold NSs and irradiation with an NIR laser at a power of 2 W for 5 min. The peritoneal macrophage efficiently functioned as a cellular vector for the NS, and the cancer cells surrounding the NS-loaded macrophages selectively lost their cellular viability after being irradiated with the NIR laser [81].

HeLa cells with and without 50-nm AuNPs were irradiated at depths of 1.5, 5, 10, 15, and 20 cm. Conventional beams with square aperture sizes 5, 10, and 15 cm at the isocenter, and flattening filter free (FFF) beams were used. Relative DNA damage enhancement with AuNPs was evaluated by γ-H2AX staining. This opened new possibilities for the clinical development of gold nanoparticle-aided radiation therapy [82]. AuNPs were encapsulated within the amphiphilic diblock copolymer poly(ethylene glycol)-b-poly(ε-capralactone) micelles, which showed a 1.7-fold improvement in the median survival time, compared with mice receiving radiation alone [83] (Fig. 13.7).

13.1.2.5. *Diagnostics*

AuNPs are also used to detect biomarkers in the diagnosis of heart diseases, cancers, and infectious agents. They are also common in lateral flow immunoassays, a common household example being the home pregnancy test.

AuNPs were used as an efficient platform for assembling a number of aptamers for high-efficiency cell capture. Up to 95 aptamers were attached onto each AuNP, resulting in an enhanced molecular recognition capability. Combining DNA nanostructures with microfluidics has great potential for the sensitive isolation of circulating tumor cells (CTCs) and is promising for cancer diagnosis and prognosis [84]. Cancerous tissue and cancer cells took up morechitosan-coated gold/gold sulfide (CS-GGS) nanoparticles and were completely ablated after exposure to NIR light, while benign tissue and noncancerous cells showed less uptake of these nanoparticles, and remained viable after exposure to NIR light [85]. Theranostic microcapsules were successfully fabricated by introducing GNPs into poly(lactic acid) microcapsules through

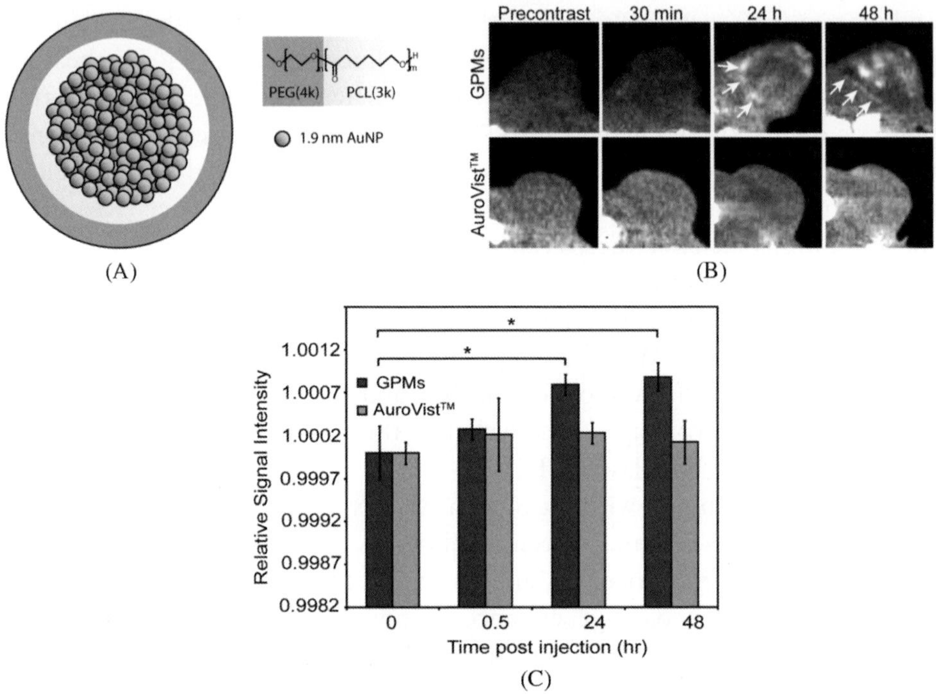

Figure 13.7. A. GNPs encapsulated within amphiphilic diblock copolymer poly(ethylene glycol)-b-poly(ε-capralactone) micelles. B. *In vivo* computerized tomography (CT) images and intensity analysis of nu/nu nude mice with HT1080 flank tumors. (A) Representative CT images in the axial plane prior to injection (precontrast) and 30 min, 24 h, and 48 h post-injection of GPMs (n = 3) or AuroVist (n = 3). Tumor boundaries are indicated by white arrows. (C) Quantitative analysis of CT images. Snnnignal intensity of each tumor was normalized to adjacent paraspinal muscle. For contrast measurement, the relative signal intensity was calculated as the quotient of the postcontrast to precontrast normalized tumor intensity. The asterisk indicates statistical significance ($p < 0.05$).

a double-microemulsion method, followed by depositing graphene oxide onto the microcapsule surface via the electrostatic layer-by-layer self-assembly technique. It was proved that the obtained microcapsules could serve as a contrast agent to simultaneously enhance US imaging and X-ray CT imaging greatly both *in vitro* and *in vivo*. In addition, the *in vivo* therapeutic examinations showed that the microcapsule was an effective agent for photothermal therapy of cancer [86]. The AuNPs have potential applications in cancer diagnosis and therapy. The study demonstrates that fluorescence peak intensity is particle size- and exposure duration-dependent, suggesting that fluorescence intensity can be used as a useful tool for pointing to bioaccumulation and toxicity induced by AuNPs in the different rat organs [87]. The gold vesicles

(GVs) composed of a monolayer of assembled AuNPs show strong absorbance in the NIR range of 650–800 nm, as a result of the plasmonic coupling effect between neighboring AuNPs in the vesicular membranes. The strong NIR absorption and the capability of encapsulating photosensitizer Ce6 in GVs enable trimodality NIR fluorescence/thermal/photoacoustic imaging-guided synergistic photothermal/photodynamic therapy (PTT/PDT) with improved efficacy [88]. Various malignancies using GNPs and ultrashort pulses may be more effective and versatile than previously anticipated, allowing for a diverse and highly specific set of tools for local cancer therapy. Antibody-coated gold nanospheres irradiated by a few resonant femtosecond pulses could elevate ROS formed within the irradiated cells during photothermal cell damage [89]. The phospholipid-based polymeric nanomicelles (PGRFe) formulation efficiently generates plasmonic nanobubbles within cancer cells, as visualized by confocal microscopy, causing cell destruction. The combined magnetic and plasmonic functionalities of the nanoplatform enable magnetic field-directed, imaging-guided, enhanced photo-induced cancer therapy [90].

13.1.3 *Magnetic nanoparticles*

These nanoparticles consist of iron, nickel, and/or cobalt, which display magnetic behavior in the presence of a magnetic field. The particles offer several potential advantages such as non-invasive treatment, tumor targeting efficiency, and quantitative determination through magnetic resonance imaging (MRI) [91]. Magnetic nanoparticles have become important tools for the imaging of prevalent diseases, such as cancer, atherosclerosis, diabetes, and others. While first generation nanoparticles were fairly nonspecific, newer generations have been targeted to specific cell types and molecular targets via affinity ligands. Commonly, these ligands emerge from phage or small molecule screens, or are based on antibodies or aptamers. Secondary reporters and combined therapeutic molecules have further opened potential clinical applications of these materials [92]. A magnetic nanoparticle with a well-tolerated intravenous dose achieved a tumor concentration of 1.9 mg Fe/g tumor in a subcutaneous squamous cell carcinoma mouse model, with a tumor to non-tumor ratio >16. With an applied field of 38 kA/m at 980 kHz, tumors could be heated to 60°C in 2 minutes, durably ablating them with millimeter (mm) precision, leaving surrounding tissue intact [93].

Iron oxide nanoparticle-based therapy represents an important alternative to conventional chemotherapy, radiation, or surgery. Iron oxide nanoparticles are usually composed of three main components: an iron core, a polymer coating, and functional moieties. The biodegradable iron core can be designed to be superparamagnetic. This is particularly important, if the nanoparticles are to be used as a contrast agent for noninvasive MRI. Surrounding the iron core is generally a polymer coating, which not only serves as a protective layer, but is also a very important component for transforming nanoparticles into biomedical nanotools for *in vivo* applications. Finally, different moieties attached to the coating serve as targeting macromolecules, therapeutic payloads, or additional imaging tags [94].

Docetaxel (DTX) is one of the most important anticancer drugs; however, the severity of its adverse effects detracts from its practical use in the clinic. Magnetic nanoparticles of Fe_3O_4 (MgNPs–Fe_3O_4) can enhance the delivery and efficacy of anticancer drugs. Combined treatment suppressed nuclear transcription factor κB expression. These findings offer the possibility that MgNPs–Fe3O4-low dose DTX combination therapy may be effective in treating prostate cancer with limited adverse effects [95].

A novel magnetic directed enzyme/prodrug therapy (MDEPT) set up by conjugating β-glucosidase (β-Glu) to aminated, starch-coated, magnetic iron oxide nanoparticles (MNPs) using glutaraldehyde as the cross-linker showed selective delivery into a subcutaneous tumor of a glioma-bearing mouse [96]. A spray dried suspension of lactose, doxorubicin, and Fe_3O_4 superparamagnetic iron oxide nanoparticles (SPIONs) demonstrates more than twice the spatial deposition and retention compared to a liquid suspension, in regions under the influence of a strong magnetic gradient [97]. Dox-loaded folate-coated Fe_3O_4 MNPs prepared by the co-precipitation and emulsification cross-linking method showed a high degree (81.8%) of apoptosis in human ovarian cancer cell lines when compared to free drug [98].

The effects of MgNPs–Fe_3O_4 or DTX alone, and in combination with prostate cancer (PC-3) cell growth *in vitro,* were studied. MgNPs–Fe_3O_4 alone reduced the viability of LNCaP and PC-3 cells. However, MgNPs–Fe_3O_4 enhanced the cytotoxic effect of a low dose of DTX. Combined treatment suppressed nuclear transcription factor κB expression, whereas MgNPs–Fe_3O_4 or DTX alone was found to be ineffective [95]. The cytotoxic potential of T-cell receptor-transfected human immunodeficiency virus (HIV)-specific T-lymphocyte (CTLs) was improved using the SPION FeraSpin R. The particles demonstrated efficient uptake and cell surface attachment [99].

Nowadays, combination therapy for cancer treatment is becoming more popular and has developed an emerging path for new treatment strategies. Magnetic nanoparticles were developed for co-delivering potent chemotherapeutic drugs, such as Curcumin (herbal drug) and Temozolomide (DNA methylating agent). The dual drug-loaded MNP formulations demonstrated higher cytotoxic effects than single drug-loaded MNP formulations and, thus, demonstrated the synergistic potential of the formulation [100]. Similarly, dual-targeted drug delivery systems represent a potential and promising platform for targeting the cytotoxic agents to tumor sites. A folate- and magnetic-targeted nanocarriers, in which folic acid (FA) was chemically conjugated to one end of diamino-PEG, showed preferential internalization of FA-MNPs into tumor cells, rather than into normal cells [101]. Polymeric vesicles (Pluronic® L-121) dual-loaded with the anti-cancer drug camptothecin and MNPs reduced the cell viability of PC-3 cells measured after 72 h significantly, when compared with drug-free polymersomes, which showed no cytotoxic effects [102]. Magnetoplasmonic nanoplatforms combining GNRs and iron-oxide nanoparticles within PGRFe enabled magnetic field-directed, imaging-guided, enhanced photo-induced cancer therapy [90]. EGFR-targeted superparamagnetic inhalable iron oxide (SPIO) nanoparticles for lung cancer demonstrated enhanced tumor retention with significant inhibition of *in vivo* lung tumor growth [103]. Aptamer$_{\alpha v\beta 3}$-conjugated magnetic nanoparticles (Apt$_{\alpha v\beta 3}$-MNPs) used to enable precise detection of integrin expression exhibited not only cytocompatibility, but also an efficient targeting ability with high magnetic sensitivity [104]. A significantly enhanced photothermal cancer ablation effect was observed when graphene oxide (GO) was decorated by both iron oxide nanoparticles (IONPs) and gold, forming a multi-functional magnetic & plasmonic GO-IONP-Au nanocomposite with strong superparamagnetism and then further coated with PEG [105]. Recently, amine groups was introduced into MNPs using 3-aminopropyltriethoxysilane and further attached with dual-responsive poly(N-isopropylacrylamide)-block-poly(acrylic acid) copolymer, synthesized by reversible addition-fragmentation chain transfer (RAFT) polymerization. Further, to accomplish targeting and fluorescent properties, folic acid and rhodamine B isothiocyanate was tethered to the surface of the nanoparticles. Intracellular uptake studies revealed a preferential uptake of these nanoparticles into cancer cells (HeLa cells) compared to normal fibroblast cells (L929 cells) and significant death of the HeLa cells [106]. Mouse esophageal tumor growth was found to be significantly inhibited by the photothermal effect of Fe_3O_4 nanoparticles coated with carboxyl-terminated poly (ethylene glycol)-phospholipid [107]. Bifunctional mitoxantrone (MTX)-grafted MNPs modified by dopamine–polyethylene glycol–folic acid (DPA–PEG–FA) were able to specifically bind to and significantly inhibit the folate receptor (FR)-positive MCF-7 cells, but not the FR-negative A549 cells [108]. Superparamagnetic nanoparticles grafted with hyperbranched polyglycerol (HPG) and conjugated with MTX MNP-g-HPG-MTX showed eight times more uptake by human head and neck cancer (KB) cells, and can be eight times or more higher than those by 3T3 fibroblasts and RAW macrophages [109]. The poly(lactic-co-glycolic acid) (PLGA)-nanocapsules loaded with maghemite magnetic nanoparticles (MNPs) and selol showed an enhanced cytotoxic effect on tumor cells than normal cells [110]. The magnetic nanocarriers were synthesized based on SPIONs, biocompatible Pluronic F127, and poly(dl-lactic acid) (F127-PLA) copolymer chemically conjugated with tumor-targeting ligand–FA via a facile chemical conjugation method that could be targeted into tumor sites more efficiently and further represented significant therapeutic efficiency to solid tumors [111].

Further advancement involved the development of temperature-sensitive poly(N-isopropylacrylamide-acrylamide-allylamine)-coated iron oxide magnetic nanoparticles. The designed system significantly killed advanced thyroid cancer (ATC) cells when compared to free doxorubicin [112]. Similarly, porphyrin-dextran iron oxide magnetic nanoparticles (PDMN) were found to be effective in inhibiting proliferation and inducing apoptosis of BIU-87 cells [113].

Magnetic iron oxide nanoparticles (MIONPs) were coated with covalently cross-linked biofunctional PEG hydrogel and functionalized with the fibronectin-derived arginine-glycine-aspartic acid-serine (RGDS) sequence to enhance specific cell targeting in biomedical imaging and cancer therapy [114].

MNPs were prepared by the chemical precipitation method and loaded with curcumin (CUR) using diffusion methods, and showed preferential uptake in MDA-MB-231 breast cancer cells in a concentration-dependent manner, superior MRI characteristics, and significantly increased targeting capability of CUR [115].

Chitosan oligosaccharide-stabilized ferrimagnetic iron oxide nanocubes (Chito-FIONs) successfully eradicated target cancer cells through caspase-mediated apoptosis without any severe toxicity [116]. Researchers synthesized hydrotropic micelles (HMs) to encapsulate SPIONs as the MRI agent and PTX as the hydrophobic anticancer drug. The designed particles showed dose-dependent cytotoxicity that was lower than free PTX. HM-SPION-PTX exhibited an 8.1-fold higher T(2) relaxivity than HM-SPION, implying the potential of HM-SPION-PTX as the contrast agent for MRI. When systemically administered to tumor-bearing mice, HM-SPION-PTX was effectively accumulated at the tumor site, allowing its detection using MRI and effective therapy [117].

PTX-loaded thermo-sensitive magnetoliposomes containing 1,2-dipalmitoyl-sn-glycero-3-phosphocholine (DPPC) and 1-palmitoyl-2-oleoyl-sn-glycero-3-phospho-rac-glycerol (PG) were prepared by the thin film hydration method. 89% of cells were found to be killed when the magnetoliposomes with

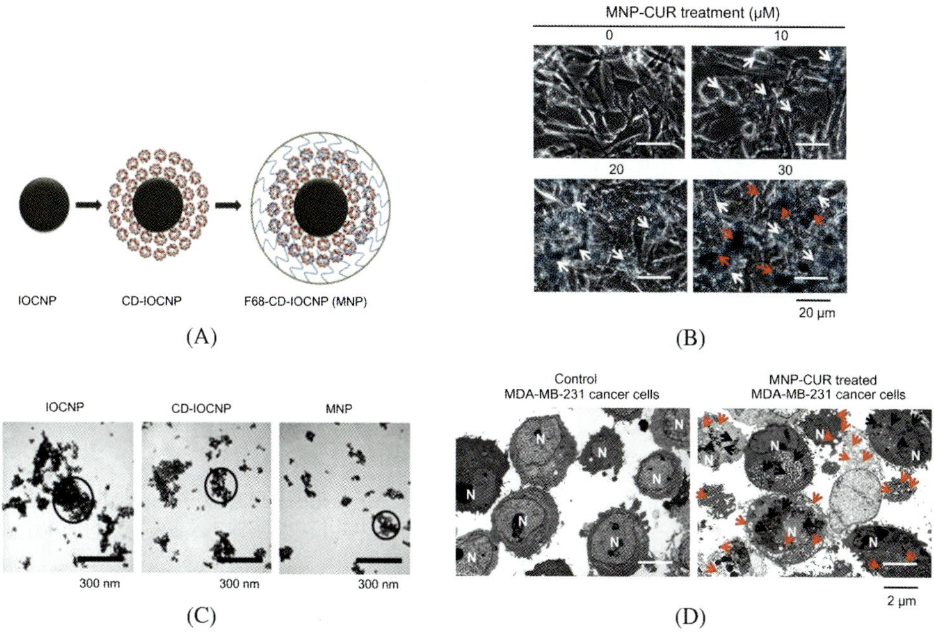

Figure 13.8. (A) Schematic representation of the synthesis route of a multifunctional magnetic nanoparticle. Iron oxide (magnetic) nanoparticles prepared by chemical precipitation reaction in presence of β-cyclodextrin and subsequent F68 polymer coating leads to magnetic nanoparticle nanoformulation. (B) Transmission electron microscopic image of pure iron oxide core nanoparticles (no coating with β-cyclodextrin/F68 polymer), iron oxide nanoparticles coated with 100 mg of β-cyclodextrin, and iron oxide nanoparticles coated with 100 mg of β-cyclodextrin and 200 mg of F68 polymer. (C) Curcumin-loaded magnetic nanoparticle formulation improves curcumin cellular uptake. Phase contrast images of MDA-MB-231 cancer cells treated with curcumin-loaded magnetic nanoparticle formulation (0–30 μM). The intensity of blue stain produced by Prussian blue from the iron of curcumin-loaded magnetic nanoparticle formulation helps improve visualization of internalized nanoparticles of curcumin-loaded magnetic nanoparticle formulation. White arrows highlight less internalization of nanoparticles and red arrows indicate greater accumulation of nanoparticles. (D) Transmission electron microscopic images of free MDA-MB-231 cells and MDA-MB-231 cells incubated with curcumin-loaded magnetic nanoparticle formulation (20 μM) for 6 hours. Black arrows indicate the accumulation and internalized nanoparticles (curcumin-loaded magnetic nanoparticle formulation) and red arrows represent the vacuole and lysosomal activity which induces apoptosis.

100 nM of drug were used to treat HeLa cells in combination with hyperthermia under an alternating current (AC) magnetic field [118]. Polysorbate 80-coated temozolomide-loaded PLGA-based superparamagnetic nanoparticles could be able to promise a multifunctional theragnostic carrier of brain cancer [119].

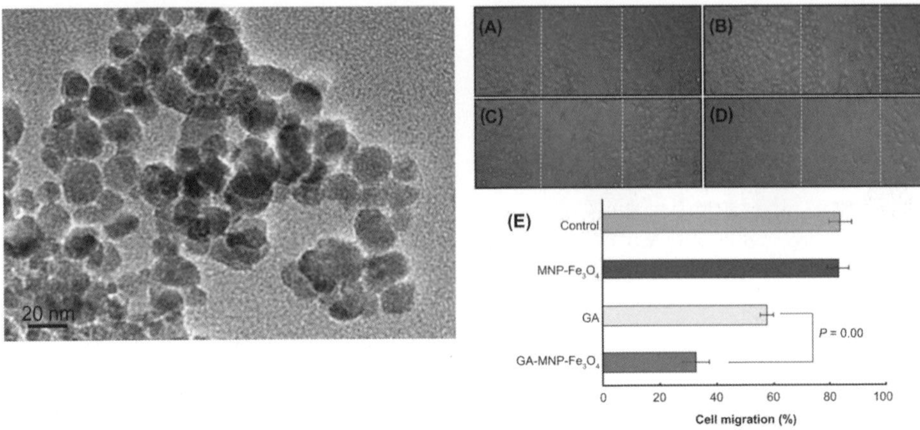

(1) (2)

Figure 13.9. (1). TEM image of magnetic Fe3O4 nanoparticles. (2). Inhibition of Panc-1 pancreatic cancer cell migration after the different treatments for 48 hours. (A) Untreated cells as controls, (B) MNP-Fe_3O_4, (C) GA alone, (D) GA-MNP-Fe_3O_4, and (E) quantified cell motility. GA = gambogic acid, MNP-Fe_3O_4 = magnetic Fe_3O_4 nanoparticle.

Introducing nucleic acids encoding immunostimulatory proteins, such as cytokine interleukin 12 (IL-12), into somatic cells can stimulate an immune response against tumors. Surface modified SPIONs with a combination of polyacrylic acid (PAA) and PEI (SPIONs-PAA-PEI) proved to be safe and effective for the magnetofection of cells and tumors in mice. Magnetofection of cells with plasmid DNA encoding reporter genes using SPIONs-PAA-PEI was superior in transfection efficiency to commercially available SPIONs. Magnetofection of murine mammary adenocarcinoma with plasmid DNA encoding IL-12 using SPIONs-PAA-PEI resulted in a significant antitumor effect [91]. Gambogic acid-loaded MNPs-Fe_3O_4 was able to suppress cancer cell proliferation and prevent cells from migrating effectively. After treatment, Panc-1 pancreatic cancer cells showed a significantly decreased expression of E26 transformation-specific sequence-1 (ETS1), as well as cyclin D1, urokinase plasminogen activator (u-PA), and vascular endothelial growth factor (VEGF) [120].

References

1. Lee, E. S., Gao, Z. and Bae, Y. H. (2008). Recent progress in tumor pH targeting nanotechnology. *J. Control. Release,* **132,** 164–170.
2. Wu, H., Zhu, L. and Torchilin, V. P. (2013). pH-sensitive poly(histidine)-PEG/DSPE-PEG co-polymer micelles for cytosolic drug delivery. *Biomaterials,* **34,** 1213–1222.

3. Gao, Z. G., Lee, D. H., Kim, D. I. and Bae, Y. H. (2005). Doxorubicin loaded pH-sensitive micelle targeting acidic extracellular pH of human ovarian A2780 tumor in mice. *J. Drug Target,* **13,** 391–397.
4. Li, F., Wu, H., Zhang, H., Li, F., Gu, C. H. and Yang, Q. (2009). Antitumor drug Paclitaxel-loaded pH-sensitive nanoparticles targeting tumor extracellular pH. *Carbohydr. Polym.,* 77, 773–778.
5. Shenoy, D., Little, S., Langer, R. and Amiji, M. (2005). Poly(ethylene oxide)-modified poly(beta-amino ester) nanoparticles as a pH-sensitive system for tumor-targeted delivery of hydrophobic drugs: Part 2. In vivo distribution and tumor localization studies. *Pharm. Res.,* **22,** 2107–2114.
6. Peng, C. L., Yang, L. Y., Luo, T. Y., Lai, P. S., Yang, S. J., Lin, W. J. and Shieh, M. J. (2010). Development of pH sensitive 2-(diisopropylamino)ethyl methacrylate based nanoparticles for photodynamic therapy. *Nanotechnology,* **21,** 155103.
7. Huang, S., Shao, K., Kuang, Y., Liu, Y., Li, J., An, S., Guo, Y., Ma, H., He, X. and Jiang, C. (2013). Tumor targeting and microenvironment-responsive nanoparticles for gene delivery. *Biomaterials,* **34,** 5294–5302.
8. Huang, S., Shao, K., Liu, Y., Kuang, Y., Li, J., An, S., Guo, Y., Ma, H. and Jiang, C. (2013). Tumor-targeting and microenvironment-responsive smart nanoparticles for combination therapy of antiangiogenesis and apoptosis. *ACS Nano.,* **7,** 2860–2871.
9. Guo, H., Lai, Q., Wang, W., Wu, Y., Zhang, C., Liu, Y. and Yuan, Z. (2013). Functional alginate nanoparticles for efficient intracellular release of doxorubicin and hepatoma carcinoma cell targeting therapy. *Int. J. Pharm.,* **451,** 1–11.
10. Hossain, S., Yamamoto, H., Chowdhury, E. H., Wu, X., Hirose, H., Haque, A., Doki, Y., Mori, M. and Akaike, T. (2013). Fabrication and intracellular delivery of doxorubicin/carbonate apatite nanocomposites: Effect on growth retardation of established colon tumor. *PLoS One,* **8,** e60428.
11. Soppimath, K. S., Liu, L. H., Seow, W. Y., Liu, S. Q., Powell, R., Chan, P. and Yang, Y. Y. (2007). Multifunctional core/shell nanoparticles self-assembled from pH-induced thermosensitive polymers for targeted intracellular anticancer drug delivery. *Adv. Funct. Mater.,* **17,** 355–362.
12. Li, S., Su, Z., Sun, M., Xiao, Y., Cao, F., Huang, A., Li, H., Ping, Q. and Zhang, C. (2012). An arginine derivative contained nanostructure lipid carriers with pH-sensitive membranolytic capability for lysosomolytic anti-cancer drug delivery. *Int. J. Pharm.,* **436,** 248–257.
13. Liu, D., Hu, H., Zhang, J., Zhao, X., Tang, X. and Chen, D. (2011). Drug pH-sensitive release in vitro and targeting ability of polyamidoamine dendrimer complexes for tumor cells. *Chem. Pharm. Bull. (Tokyo),* **59,** 63–71.
14. Zhang, L., Zhu, S., Qian, L., Pei, Y., Qiu, Y. and Jiang, Y. (2011). RGD-modified PEG-PAMAM-DOX conjugates: *In vitro* and *in vivo* studies for glioma. *Eur. J. Pharm. Biopharm.,* **79,** 232–240.
15. Shen, M., Huang, Y., Han, L., Qin, J., Fang, X., Wang, J. and Yang, V. C. (2012). Multifunctional drug delivery system for targeting tumor and its acidic microenvironment. *J. Control Release,* **161,** 884–892.
16. Lu, P. L., Chen, Y. C., Ou, T. W., Chen, H. H., Tsai, H. C., Wen, C. J., Lo, C. L., Wey, S. P., Lin, K. J., Yen, T. C. and Hsiue, G. H. (2011). Multifunctional hollow nanoparticles based on graft-diblock copolymers for doxorubicin delivery. *Biomaterials,* **32,** 2213–2221.

17. Garg, A. and Kokkoli, E. (2011). pH-Sensitive PEGylated liposomes functionalized with a fibronectin-mimetic peptide show enhanced intracellular delivery to colon cancer cell. *Curr. Pharm. Biotechnol.,* **12,** 1135–1143.
18. Sethuraman, V. A., Na, K. and Bae, Y. H. (2006). pH-responsive sulfonamide/PEI system for tumor specific gene delivery: An in vitro study. *Biomacromolecules,* **7,** 64–70.
19. Wang, Q., Zhang, L., Hu, W., Hu, Z. H., Bei, Y. Y., Xu, J. Y., Wang, W. J., Zhang, X. N. and Zhang, Q. (2010). Norcantharidin-associated galactosylated chitosan nanoparticles for hepatocyte-targeted delivery. *Nanomedicine,* **6,** 371–381.
20. Fan, L., Li, F., Zhang, H., Wang, Y., Cheng, C., Li, X., Gu, C. H., Yang, Q., Wu, H. and Zhang, S. (2010). Co-delivery of PDTC and doxorubicin by multifunctional micellar nanoparticles to achieve active targeted drug delivery and overcome multidrug resistance. *Biomaterials,* **31,** 5634–5642.
21. Das, M., Shim, K. H., An, S. S. and Yi, D. K. (2011). Review on gold nanoparticles and their applications. *J. Toxicol. Inviron. Health Sci.,* **3,** 193–205.
22. Chen, W. H., Xu, X. D., Jia, H. Z., Lei, Q., Luo, G. F., Cheng, S. X., Zhuo, R. X. and Zhang, X. Z. (2013). Therapeutic nanomedicine based on dual-intelligent functionalized gold nanoparticles for cancer imaging and therapy *in vivo*. *Biomaterials,* **34,** 8798–8807.
23. Chen, H., Paholak, H., Ito, M., Sansanaphongpricha, K., Qian, W., Che, Y. and Sun, D. (2013). 'Living' PEGylation on gold nanoparticles to optimize cancer cell uptake by controlling targeting ligand and charge densities. *Nanotechnology,* **24,** 355101.
24. Shakeri-Zadeh, A., Kamrava, S. K., Farhadi, M., Hajikarimi, Z., Maleki, S. and Ahmadi, A. (2014). A scientific paradigm for targeted nanophotothermolysis; the potential for nanosurgery of cancer. *Lasers Med. Sci.,* **29,** 847–853.
25. Huang, C., Bao, Q., Hunting, D., Zheng, Y. and Sanche, L. (2013). Conformation-dependent DNA damage induced by gold nanoparticles. *J. Biomed. Nanotechnol.,* **9,** 856–862.
26. Pan, W., Yang, H., Zhang, T., Li, Y., Li, N. and Tang, B. (2013). Dual-targeted nanocarrier based on cell surface receptor and intracellular mRNA: An effective strategy for cancer cell imaging and therapy. *Anal. Chem.,* **85,** 6930–6935.
27. Wang, X., Cai, X., Hu, J., Shao, N., Wang, F., Zhang, Q., Xiao, J. and Cheng, Y. (2013). Glutathione-triggered "off-on" release of anticancer drugs from dendrimer-encapsulated gold nanoparticles. *J. Am. Chem. Soc.,* **135,** 9805–9810.
28. Liu, X., Chen, Y., Li, H., Huang, N., Jin, Q., Ren, K. and Ji, J. (2013). Enhanced retention and cellular uptake of nanoparticles in tumors by controlling their aggregation behavior. *ACS Nano.,* **7,** 6244–6257.
29. Shenoi, M. M., Iltis, I., Choi, J., Koonce, N. A., Metzger, G. J., Griffin, R. J. and Bischof, J. C. (2013). Nanoparticle delivered vascular disrupting agents (VDAs): Use of TNF-alpha conjugated gold nanoparticles for multimodal cancer therapy. *Mol. Pharm.,* **10,** 1683–1694.
30. Park, H., Tsutsumi, H. and Mihara, H. (2013). Cell penetration and cell-selective drug delivery using alpha-helix peptides conjugated with gold nanoparticles. *Biomaterials,* **34,** 4872–4879.
31. Oh, K. S., Lee, H., Kim, J. Y., Koo, E. J., Lee, E. H., Park, J. H., Kim, S. Y., Kim, K., Kwon, I. C. and Yuk, S. H. (2013). The multilayer nanoparticles formed by layer by layer approach for cancer-targeting therapy. *J. Control Release,* **165,** 9–15.

32. Qu, X., Yao, C., Wang, J., Li, Z. and Zhang, Z. (2012). Anti-CD30-targeted gold nanoparticles for photothermal therapy of L-428 Hodgkin's cell. *Int. J. Nanomed.,* 7, 6095–6103.
33. Szlachcic, A., Pala, K., Zakrzewska, M., Jakimowicz, P., Wiedlocha, A. and Otlewski, J. (2012). FGF1-gold nanoparticle conjugates targeting FGFR efficiently decrease cell viability upon NIR irradiation. *Int. J. Nanomed.,* 7, 5915–5927.
34. Chattopadhyay, N., Cai, Z., Kwon, Y. L., Lechtman, E., Pignol, J. P. and Reilly, R. M. (2013). Molecularly targeted gold nanoparticles enhance the radiation response of breast cancer cells and tumor xenografts to X-radiation. *Breast Cancer Res. Treat.,* **137,** 81–91.
35. Nurunnabi, M., Khatun, Z., Moon, W. C., Lee, G. and Lee, Y. K. (2012). Heparin based nanoparticles for cancer targeting and noninvasive imaging. *Quant. Imaging Med. Surg.,* **2,** 219–226.
36. Khan, M. S., Vishakante, G. D. and Siddaramaiah, H. (2013). Gold nanoparticles: A paradigm shift in biomedical applications. *Adv. Colloid Interface Sci.,* 199–200, 44–58.
37. Kong, Y., Chen, J., Gao, F., Brydson, R., Johnson, B., Heath, G., Zhang, Y., Wu, L. and Zhou, D. (2013). Near-infrared fluorescent ribonuclease-A-encapsulated gold nanoclusters: Preparation, characterization, cancer targeting and imaging. *Nanoscale,* **5,** 1009–1017.
38. Kasten, B. B., Liu, T., Nedrow-Byers, J. R., Benny, P. D. and Berkman, C. E. (2013). Targeting prostate cancer cells with PSMA inhibitor-guided gold nanoparticles. *Bioorg. Med. Chem. Lett.,* **23,** 565–568.
39. Benito, M., Martin, V., Blanco, M. D., Teijon, J. M. and Gomez, C. (2013). Cooperative effect of 5-aminolevulinic acid and gold nanoparticles for photodynamic therapy of cancer. *J. Pharm. Sci.,* **102,** 2760–2769.
40. Conde, J., Tian, F., Hernandez, Y., Bao, C., Cui, D., Janssen, K. P., Ibarra, M. R., Baptista, P. V., Stoeger, T. and De La Fuente, J. M. (2013). In vivo tumor targeting via nanoparticle-mediated therapeutic siRNA coupled to inflammatory response in lung cancer mouse models. *Biomaterials,* **34,** 7744–7753.
41. Nam, J., La, W. G., Hwang, S., Ha, Y. S., Park, N., Won, N., Jung, S., Bhang, S. H., Ma, Y. J., Cho, Y. M., Jin, M., Han, J., Shin, J. Y., Wang, E. K., Kim, S. G., Cho, S. H., Yoo, J., Kim, B. S. and Kim, S. (2013). pH-responsive assembly of gold nanoparticles and "spatiotemporally concerted" drug release for synergistic cancer therapy. *ACS Nano.,* **7,** 3388–3402.
42. Mendoza-Nava, H., Ferro-Flores, G., Ocampo-Garcia, B., Serment-Guerrero, J., Santos-Cuevas, C., Jimenez-Mancilla, N., Luna-Gutierrez, M. and Camacho-Lopez, M. A. (2013). Laser heating of gold nanospheres functionalized with octreotide: *In vitro* effect on HeLa cell viability. *Photomed. Laser Surg.,* **31,** 17–22.
43. Buzea, C., Pacheco, Ii and Robbie, K. (2007). Nanomaterials and nanoparticles: Sources and toxicity. *Biointerphases,* **2,** MR17–MR71.
44. Kawasaki, E. S. and Player, A. (2005). Nanotechnology, nanomedicine, and the development of new, effective therapies for cancer. *Nanomedicine,* **1,** 101–109.
45. Horton, M. A. and Khan, A. (2006). Medical nanotechnology in the UK: A perspective from the London Centre for Nanotechnology. *Nanomedicine,* **2,** 42–48.
46. Stewart, M. E., Anderton, C. R., Thompson, L. B., Maria, J., Gray, S. K., Rogers, J. A. and Nuzzo, R. G. (2008). Nanostructured plasmonic sensors. *Chem. Rev.,* **108,** 494–521.

47. Liu, X., Dai, Q., Austin, L., Coutts, J., Knowles, G., Zou, J., Chen, H. and Huo, Q. (2008). A one-step homogeneous immunoassay for cancer biomarker detection using gold nanoparticle probes coupled with dynamic light scattering. *J. Am. Chem. Soc.*, **130,** 2780–2782.
48. Huang, X., Jain, P. K., El-Sayed, I. H. and El-Sayed, M. A. (2008). Plasmonic photothermal therapy (PPTT) using gold nanoparticles. *Lasers Med. Sci.*, **23,** 217–228.
49. Wang, H., Brandl, D. W., Le, F., Nordlander, P. and Halas, N. J. (2006). Nanorice: A hybrid plasmonic nanostructure. *Nano. Lett.*, **6,** 827–832.
50. Nehl, C. L., Liao, H. and Hafner, J. H. (2006). Optical properties of star-shaped gold nanoparticles. *Nano. Lett.*, **6,** 683–688.
51. Mitamura, K. and Imae, I. (2009). Functionalization of gold nanorods toward their applications. *Plasmonics,* **4,** 23–30.
52. Chen, H., Zhang, X., Dai, S., Ma, Y., Cui, S., Achilefu, S. and Gu, Y. (2013). Multifunctional gold nanostar conjugates for tumor imaging and combined photothermal and chemo-therapy. *Theranostics,* **3,** 633–649.
53. Mocellin, S. and Nitti, D. (2008). TNF and cancer: The two sides of the coin. *Front Biosci.*, **13,** 2774–2783.
54. Visaria, R. K., Griffin, R. J., Williams, B. W., Ebbini, E. S., Paciotti, G. F., Song, C. W. and Bischof, J. C. (2006). Enhancement of tumor thermal therapy using gold nanoparticle-assisted tumor necrosis factor-alpha delivery. *Mol. Cancer Ther.*, **5,** 1014–1020.
55. Visaria, R., Bischof, J. C., Loren, M., Williams, B., Ebbini, E., Paciotti, G. and Griffin, R. (2007). Nanotherapeutics for enhancing thermal therapy of cancer. *Int. J. Hyperthermia,* **23,** 501–511.
56. Jung, S., Nam, J., Hwang, S., Park, J., Hur, J., Im, K., Park, N. and Kim, S. (2013). Theragnostic pH-sensitive gold nanoparticles for the selective surface enhanced Raman scattering and photothermal cancer therapy. *Anal. Chem.*, **85,** 7674–7681.
57. Liu, Y., Shipton, M. K., Ryan, J., Kaufman, E. D., Franzen, S. and Feldheim, D. L. (2007). Synthesis, stability, and cellular internalization of gold nanoparticles containing mixed peptide-poly(ethylene glycol) monolayers. *Anal. Chem.*, **79,** 2221–2229.
58. Pecorelli, S., Pasinetti, B., Tisi, G. and Odicino, F. (2006). Optimizing gemcitabine regimens in ovarian cancer. *Semin. Oncol.*, **33,** S17–S25.
59. Todd, R. C. and Lippard, S. J. (2009). Inhibition of transcription by platinum antitumor compounds. *Metallomics,* **1,** 280–291.
60. Wang, D. and Lippard, S. J. (2005). Cellular processing of platinum anticancer drugs. *Nat. Rev. Drug Discov.*, **4,** 307–320.
61. Kelland, L. (2007). The resurgence of platinum-based cancer chemotherapy. *Nat. Rev. Cancer,* **7,** 573–584.
62. Mckenzie, F., Faulds, K. and Graham, D. (2008). LNA functionalized gold nanoparticles as probes for double stranded DNA through triplex formation. *Chem. Commun. (Camb.)*, 2367–2369.
63. Male, K. B., Lachance, B., Hrapovic, S., Sunahara, G. and Luong, J. H. (2008). Assessment of cytotoxicity of quantum dots and gold nanoparticles using cell-based impedance spectroscopy. *Anal. Chem.*, **80,** 5487–5493.
64. Eck, W., Craig, G., Sigdel, A., Ritter, G., Old, L. J., Tang, L., Brennan, M. F., Allen, P. J. and Mason, M. D. (2008). PEGylated gold nanoparticles conjugated to monoclonal F19 antibodies as targeted labeling agents for human pancreatic carcinoma tissue. *ACS Nano.*, **2,** 2263–2272.

65. Brown, S. D., Nativo, P., Smith, J. A., Stirling, D., Edwards, P. R., Venugopal, B., Flint, D. J., Plumb, J. A., Graham, D. and Wheate, N. J. (2010). Gold nanoparticles for the improved anticancer drug delivery of the active component of oxaliplatin. *J. Am. Chem. Soc.,* **132,** 4678–4684.
66. Chandolu, V. and Dass, C. R. (2013). Treatment of lung cancer using nanoparticle drug delivery systems. *Curr. Drug Discov. Technol.,* **10,** 170–176.
67. Pan, Y., Leifert, A., Ruau, D., Neuss, S., Bornemann, J., Schmid, G., Brandau, W., Simon, U. and Jahnen-Dechent, W. (2009). Gold nanoparticles of diameter 1.4 nm trigger necrosis by oxidative stress and mitochondrial damage. *Small,* **5,** 2067–2076.
68. Akiyama, Y., Mori, T., Katayama, Y. and Niidome, T. (2009). The effects of PEG grafting level and injection dose on gold nanorod biodistribution in the tumor-bearing mice. *J. Control Release,* **139,** 81–84.
69. Song, K., Xu, P., Meng, Y., Geng, F., Li, J., Li, Z., Xing, J., Chen, J. and Kong, B. (2013). Smart gold nanoparticles enhance killing effect on cancer cells. *Int. J. Oncol.,* **42,** 597–608.
70. Conde, J., Larguinho, M., Cordeiro, A., Raposo, L. R., Costa, P. M., Santos, S., Diniz, M. S., Fernandes, A. R. and Baptista, P. V. (2014). Gold-nanobeacons for gene therapy: Evaluation of genotoxicity, cell toxicity and proteome profiling analysis. *Nanotoxicology,* **8,** 521–532.
71. Arvizo, R. R., Saha, S., Wang, E., Robertson, J. D., Bhattacharya, R. and Mukherjee, P. (2013). Inhibition of tumor growth and metastasis by a self-therapeutic nanoparticle. *Proc. Natl. Acad. Sci. USA.,* **110,** 6700–6705.
72. Orza, A., Soritau, O., Tomuleasa, C., Olenic, L., Florea, A., Pana, O., Bratu, I., Pall, E., Florian, S., Casciano, D. and Biris, A. S. (2013). Reversing chemoresistance of malignant glioma stem cells using gold nanoparticles. *Int. J. Nanomedicine,* **8,** 689–702.
73. Joseph, M. M., Aravind, S. R., Varghese, S., Mini, S. and Sreelekha, T. T. (2013). PST-Gold nanoparticle as an effective anticancer agent with immunomodulatory properties. *Colloids Surf. B. Biointerfaces,* **104,** 32–39.
74. Razzak, R., Zhou, J., Yang, X., Pervez, N., Bedard, E. L., Moore, R. B., Shaw, A., Amanie, J. and Roa, W. H. (2013). The biodistribution and pharmacokinetic evaluation of choline-bound gold nanoparticles in a human prostate tumor xenograft model. *Clin. Invest. Med.,* **36,** E133–E142.
75. Liu, C. P., Lin, F. S., Chien, C. T., Tseng, S. Y., Luo, C. W., Chen, C. H., Chen, J. K., Tseng, F. G., Hwu, Y., Lo, L. W., Yang, C. S. and Lin, S. Y. (2013). In-situ formation and assembly of gold nanoparticles by gum arabic as efficient photothermal agent for killing cancer cells. *Macromol. Biosci.,* **13,** 1314–1320.
76. Sun, X., Zhang, G., Keynton, R. S., O'toole, M. G., Patel, D. and Gobin, A. M. (2013). Enhanced drug delivery via hyperthermal membrane disruption using targeted gold nanoparticles with PEGylated Protein-G as a cofactor. *Nanomedicine,* **9,** 1214–1222.
77. Cheng, L. C., Chen, H. M., Lai, T. C., Chan, Y. C., Liu, R. S., Sung, J. C., Hsiao, M., Chen, C. H., Her, L. J. and Tsai, D. P. (2013). Targeting polymeric fluorescent nanodiamond-gold/silver multi-functional nanoparticles as a light-transforming hyperthermia reagent for cancer cells. *Nanoscale,* **5,** 3931–3940.
78. Farrokhtakin, E., Ciofani, G., Puleo, G. L., De Vito, G., Filippeschi, C., Mazzolai, B., Piazza, V. and Mattoli, V. (2013). Barium titanate core — gold shell nanoparticles for hyperthermia treatments. *Int. J. Nanomed.,* **8,** 2319–2331.

79. Pasparakis, G. (2013). Light-induced generation of singlet oxygen by naked gold nanoparticles and its implications to cancer cell phototherapy. *Small,* **9,** 4130–4134.
80. Corr, S. J., Raoof, M., Mackeyev, Y., Phounsavath, S., Cheney, M. A., Cisneros, B. T., Shur, M., Gozin, M., Mcnally, P. J., Wilson, L. J. and Curley, S. A. (2012). Citrate-capped gold nanoparticle electrophoretic heat production in response to a time-varying radiofrequency electric-field. *J. Phys. Chem. C. Nanomater. Interfaces,* **116,** 24380–24389.
81. Yang, T. D., Choi, W., Yoon, T. H., Lee, K. J., Lee, J. S., Han, S. H., Lee, M. G., Yim, H. S., Choi, K. M., Park, M. W., Jung, K. Y. and Baek, S. K. (2012). Real-time phase-contrast imaging of photothermal treatment of head and neck squamous cell carcinoma: An *in vitro* study of macrophages as a vector for the delivery of gold nanoshells. *J. Biomed. Opt.,* **17,** 128003.
82. Berbeco, R. I., Korideck, H., Ngwa, W., Kumar, R., Patel, J., Sridhar, S., Johnson, S., Price, B. D., Kimmelman, A. and Makrigiorgos, G. M. (2012). DNA damage enhancement from gold nanoparticles for clinical MV photon beams. *Radiat. Res.,* **178,** 604–608.
83. Al Zaki, A., Joh, D., Cheng, Z., De Barros, A. L., Kao, G., Dorsey, J. and Tsourkas, A. (2014). Gold-loaded polymeric micelles for computed tomography-guided radiation therapy treatment and radiosensitization. *ACS Nano.,* **8,** 104–112.
84. Sheng, W., Chen, T., Tan, W. and Fan, Z. H. (2013). Multivalent DNA nanospheres for enhanced capture of cancer cells in microfluidic devices. *ACS Nano.,* **7,** 7067–7076.
85. Li, Y., Gobin, A. M., Dryden, G. W., Kang, X., Xiao, D., Li, S. P., Zhang, G. and Martin, R. C. (2013). Infrared light-absorbing gold/gold sulfide nanoparticles induce cell death in esophageal adenocarcinoma. *Int. J. Nanomed.,* **8,** 2153–2161.
86. Jin, Y., Wang, J., Ke, H., Wang, S. and Dai, Z. (2013). Graphene oxide modified PLA microcapsules containing gold nanoparticles for ultrasonic/CT bimodal imaging guided photothermal tumor therapy. *Biomaterials,* **34,** 4794–4802.
87. Abdelhalim, M. A. (2013). Uptake of gold nanoparticles in several rat organs after intraperitoneal administration *in vivo*: A fluorescence study. *Biomed. Res. Int.,* **2013,** 353695.
88. Lin, J., Wang, S., Huang, P., Wang, Z., Chen, S., Niu, G., Li, W., He, J., Cui, D., Lu, G., Chen, X. and Nie, Z. (2013). Photosensitizer-loaded gold vesicles with strong plasmonic coupling effect for imaging-guided photothermal/photodynamic therapy. *ACS Nano.,* **7,** 5320–5329.
89. Minai, L., Yeheskely-Hayon, D. and Yelin, D. (2013). High levels of reactive oxygen species in gold nanoparticle-targeted cancer cells following femtosecond pulse irradiation. *Sci. Rep.,* **3,** 2146.
90. Ohulchanskyy, T. Y., Kopwitthaya, A., Jeon, M., Guo, M., Law, W. C., Furlani, E. P., Kim, C. and Prasad, P. N. (2013). Phospholipid micelle-based magneto-plasmonic nanoformulation for magnetic field-directed, imaging-guided photo-induced cancer therapy. *Nanomedicine,* **9,** 1192–1202.
91. Prijic, S., Prosen, L., Cemazar, M., Scancar, J., Romih, R., Lavrencak, J., Bregar, V. B., Coer, A., Krzan, M., Znidarsic, A. and Sersa, G. (2012). Surface modified magnetic nanoparticles for immuno-gene therapy of murine mammary adenocarcinoma. *Biomaterials,* **33,** 4379–4391.
92. Mccarthy, J. R. and Weissleder, R. (2008). Multifunctional magnetic nanoparticles for targeted imaging and therapy. *Adv. Drug Deliv. Rev.,* **60,** 1241–1251.
93. Huang, H. S. and Hainfeld, J. F. (2013). Intravenous magnetic nanoparticle cancer hyperthermia. *Int. J. Nanomed.,* **8,** 2521–2532.

94. Yigit, M. V., Moore, A. and Medarova, Z. (2012). Magnetic nanoparticles for cancer diagnosis and therapy. *Pharm. Res.,* **29,** 1180–1188.
95. Sato, A., Itcho, N., Ishiguro, H., Okamoto, D., Kobayashi, N., Kawai, K., Kasai, H., Kurioka, D., Uemura, H., Kubota, Y. and Watanabe, M. (2013). Magnetic nanoparticles of Fe_3O_4 enhance docetaxel-induced prostate cancer cell death. *Int. J. Nanomed.,* **8,** 3151–3160.
96. Zhou, J., Zhang, J., David, A. E. and Yang, V. C. (2013). Magnetic tumor targeting of beta-glucosidase immobilized iron oxide nanoparticles. *Nanotechnology,* **24,** 375102.
97. Mcbride, A. A., Price, D. N., Lamoureux, L. R., Elmaoued, A. A., Vargas, J. M., Adolphi, N. L. and Muttil, P. (2013). Preparation and characterization of novel magnetic nano-in-microparticles for site-specific pulmonary drug delivery. *Mol. Pharm.,* **10,** 3574–3581.
98. Fazilati, M. (2014). Folate decorated magnetite nanoparticles: Synthesis and targeted therapy against ovarian cancer. *Cell Biol. Int.,* **38,** 154–163.
99. Williams, J. P., Southern, P., Lissina, A., Christian, H. C., Sewell, A. K., Phillips, R., Pankhurst, Q. and Frater, J. (2013). Application of magnetic field hyperthermia and superparamagnetic iron oxide nanoparticles to HIV-1-specific T-cell cytotoxicity. *Int. J. Nanomedicine,* **8,** 2543–2554.
100. Dilnawaz, F. and Sahoo, S. K. (2013). Enhanced accumulation of curcumin and temozolomide loaded magnetic nanoparticles executes profound cytotoxic effect in glioblastoma spheroid model. *Eur. J. Pharm. Biopharm.,* **85,** 452–462.
101. Licciardi, M., Scialabba, C., Cavallaro, G., Sangregorio, C., Fantechi, E. and Giammona, G. (2013). Cell uptake enhancement of folate targeted polymer coated magnetic nanoparticles. *J. Biomed. Nanotechnol.,* **9,** 949–964.
102. Bleul, R., Thiermann, R., Marten, G. U., House, M. J., St Pierre, T. G., Hafeli, U. O. and Maskos, M. (2013). Continuously manufactured magnetic polymersomes — a versatile tool (not only) for targeted cancer therapy. *Nanoscale,* **5,** 11385–11393.
103. Sadhukha, T., Wiedmann, T. S. and Panyam, J. (2013). Inhalable magnetic nanoparticles for targeted hyperthermia in lung cancer therapy. *Biomaterials,* **34,** 5163–5171.
104. Lim, E. K., Kim, B., Choi, Y., Ro, Y., Cho, E. J., Lee, J. H., Ryu, S. H., Suh, J. S., Haam, S. and Huh, Y. M. (2014). Aptamer-conjugated magnetic nanoparticles enable efficient targeted detection of integrin alphavbeta3 via magnetic resonance imaging. *J. Biomed. Mater. Res. A.,* **102A,** 49–59.
105. Shi, X., Gong, H., Li, Y., Wang, C., Cheng, L. and Liu, Z. (2013). Graphene-based magnetic plasmonic nanocomposite for dual bioimaging and photothermal therapy. *Biomaterials,* **34,** 4786–4793.
106. Sahoo, B., Devi, K. S., Banerjee, R., Maiti, T. K., Pramanik, P. and Dhara, D. (2013). Thermal and pH responsive polymer-tethered multifunctional magnetic nanoparticles for targeted delivery of anticancer drug. *ACS Appl. Mater. Interfaces,* **5,** 3884–3893.
107. Chu, M., Shao, Y., Peng, J., Dai, X., Li, H., Wu, Q. and Shi, D. (2013). Near-infrared laser light mediated cancer therapy by photothermal effect of Fe_3O_4 magnetic nanoparticles. *Biomaterials,* **34,** 4078–4088.
108. Heidari Majd, M., Asgari, D., Barar, J., Valizadeh, H., Kafil, V., Coukos, G. and Omidi, Y. (2013). Specific targeting of cancer cells by multifunctional mitoxantrone-conjugated magnetic nanoparticles. *J. Drug Target.,* **21,** 328–340.
109. Li, M., Neoh, K. G., Wang, R., Zong, B. Y., Tan, J. Y. and Kang, E. T. (2013). Methotrexate-conjugated and hyperbranched polyglycerol-grafted Fe(3)O(4) magnetic nanoparticles for targeted anticancer effects. *Eur. J. Pharm. Sci.,* **48,** 111–120.

110. Estevanato, L. L., Da Silva, J. R., Falqueiro, A. M., Mosiniewicz-Szablewska, E., Suchocki, P., Tedesco, A. C., Morais, P. C. and Lacava, Z. G. (2012). Co-nanoencapsulation of magnetic nanoparticles and selol for breast tumor treatment: In vitro evaluation of cytotoxicity and magnetohyperthermia efficacy. *Int. J. Nanomed.*, **7**, 5287–5299.
111. Huang, C., Tang, Z., Zhou, Y., Zhou, X., Jin, Y., Li, D., Yang, Y. and Zhou, S. (2012). Magnetic micelles as a potential platform for dual targeted drug delivery in cancer therapy. *Int. J. Pharm.*, **429**, 113–122.
112. Koppolu, B., Bhavsar, Z., Wadajkar, A. S., Nattama, S., Rahimi, M., Nwariaku, F. and Nguyen, K. T. (2012). Temperature-sensitive polymer-coated magnetic nanoparticles as a potential drug delivery system for targeted therapy of thyroid cancer. *J. Biomed. Nanotechnol.*, **8**, 983–990.
113. Luo, D. S., Mi, Q. W., Meng, X. J., Gao, Y., Dai, Y. P. and Deng, C. H. (2012). Synergetic killing effects of external magnetic fields combined with porphyrin-dextran magnetic nanoparticles on the human bladder cancer cells. *Beijing Da Xue Xue Bao*, **44**, 524–527.
114. Nazli, C., Ergenc, T. I., Yar, Y., Acar, H. Y. and Kizilel, S. (2012). RGDS-functionalized polyethylene glycol hydrogel-coated magnetic iron oxide nanoparticles enhance specific intracellular uptake by HeLa cells. *Int. J. Nanomedicine*, **7**, 1903–1920.
115. Yallapu, M. M., Othman, S. F., Curtis, E. T., Bauer, N. A., Chauhan, N., Kumar, D., Jaggi, M. and Chauhan, S. C. (2012). Curcumin-loaded magnetic nanoparticles for breast cancer therapeutics and imaging applications. *Int. J. Nanomed.*, **7**, 1761–1779.
116. Bae, K. H., Park, M., Do, M. J., Lee, N., Ryu, J. H., Kim, G. W., Kim, C., Park, T. G. and Hyeon, T. (2012). Chitosan oligosaccharide-stabilized ferrimagnetic iron oxide nanocubes for magnetically modulated cancer hyperthermia. *ACS Nano.*, **6**, 5266–5273.
117. Yoon, H. Y., Saravanakumar, G., Heo, R., Choi, S. H., Song, I. C., Han, M. H., Kim, K., Park, J. H., Choi, K., Kwon, I. C. and Park, K. (2012). Hydrotropic magnetic micelles for combined magnetic resonance imaging and cancer therapy. *J. Control Release*, **160**, 692–698.
118. Kulshrestha, P., Gogoi, M., Bahadur, D. and Banerjee, R. (2012). In vitro application of paclitaxel loaded magnetoliposomes for combined chemotherapy and hyperthermia. *Colloids Surf. B. Biointerfaces*, **96**, 1–7.
119. Ling, Y., Wei, K., Zou, F. and Zhong, S. (2012). Temozolomide loaded PLGA-based superparamagnetic nanoparticles for magnetic resonance imaging and treatment of malignant glioma. *Int. J. Pharm.*, **430**, 266–275.
120. Wang, C., Zhang, H., Chen, Y., Shi, F. and Chen, B. (2012). Gambogic acid-loaded magnetic Fe(3)O(4) nanoparticles inhibit Panc-1 pancreatic cancer cell proliferation and migration by inactivating transcription factor ETS1. *Int. J. Nanomed.*, **7**, 781–787.

Chapter 14

Advances in Smart Nanopreparations for Oral Drug Delivery

German A. Islan, Maximiliano L. Cacicedo, Valeria E. Bosio and Guillermo R. Castro

Department of Chemistry, School of Sciences, Universidad de La Plata, Argentina

14.1 Introduction

Nanotechnology has been growing fast as a new area for oral delivery of molecules with biological interest. Common drugs from the market usually show numerous side effects in the organism and a relatively poor biodisponibility. In this sense, smart nanodevices encapsulating drugs provide an alternative to avoid the barriers that affect the stability of oral delivery formulations, such as the extremely acidic pH of the stomach and the presence of gastrointestinal proteases. These systems are combining the benefits of the controlled release formulations with the advantages of nanotechnology, such as high surface area to volume ratio, capability to access the deepest interstices of many organs (e.g. intestine), or to penetrate the cell membrane. In the present chapter, the development and application of smart nanopreparations, with growing attention in the pharmaceutical industry, are discussed. Among them carriers like liposomes are described, seeing how they evolved to solid lipid nanoparticles (SLNs) and, how these evolved to nanostructures of solid lipid carriers. Nanocrystals of pure drugs and of enzymes, another interesting nanoproduct, have emerged as feasible alternatives to overcome the drawbacks of oral administration drugs. More importantly, the use of biopolymers and compatible synthetic polymers become a new tool for the development of nanosystems. Finally, some inorganic matrices that have been tested in living organisms are considered. Silica compounds, zeolites, and nanoclays, and their role in control release are discussed. Also, silver nanoparticles and their relevant importance in the treatment of multiresistent pathogens are mentioned. However, toxicity of these inorganic systems is now under discussion. Along the manuscript, the most important nanocarriers until now are studied, and the tools for approaching new nanocarriers with potential applications in oral delivery are provided.

14.1.1 *Importance of oral delivery and nanotechnology*

Oral administration is a route that presents some advantages over other administration alternatives in drug delivery. First of all, it is easy to manage by the patients, as they can dose their selves at home following a simple medical prescription. Second, the costs of manufacturing are reduced, because it is not necessary to maintain strictly sterile conditions. A third reason is the lack of specialized facilities (e.g. hospitals) to administer the drug to the patients. Those are the major reasons why about 80% of the drug market is based on oral delivery. However, many factors must be analyzed carefully for the development of oral formulations. The success in the delivery depends on the overpass of major physiological barriers, such as the extremely acidic gastric media, the poor permeability through the intestinal mucosa, and/or the digestion by gastric, pancreatic, and intestinal proteases.

In the last decade, many new oral drug delivery systems have been designed with the purpose of overcoming the physiological barriers, providing a predictable and reproducible controlled release profile, and increasing the biodisponibility of the drugs. These systems maintain steady drug plasma levels for long periods of time and are able to minimize drug side effects while the drug is controllably released. In this sense, nanodevices combine the benefits of the controlled release formulations with the advantages of nanotechnology. The high surface area to volume ratio, the capability to access the deepest interstices of the intestine or to penetrate the cell membrane, ensure a high quality delivery to obtain better therapeutic effects.

14.1.1.1 *Side effects of drugs*

During the last decade, adverse side effects in the gastrointestinal (GI) mucosa were found to be produced by some commercially available drugs. The chronic administration of them combined with high doses to keep the drug in the therapeutic window led to undesirable side effects. Among them, irritation, bleeding, vomiting, nausea, and diarrhea are the most common [1]. For example, ciprofloxacin, an antibiotic which belongs to the fluroquinolones family, tends to form crystals at the high administered doses and produce renal damage and intestinal perforation, in addition to other side effects mentioned before [2]. Typical examples are the non-steroidal anti-inflammatory agents (NSAIDs) aspirin (acetylsalicylic acid) and diclofenac (2-(2,6-dichloranilino)phenylacetic acid), which produce irritation and ulcers in the stomach, and even bleeding in the most serious acute cases.

Thereby, smart nanodevices for oral controlled delivery emerge as a feasible alternative to overcome the side effects associated with conventional drug administration.

14.1.1.2 *Biodisponibility of drugs*

Beyond the mentioned drug side effects, biodisponibility is another major point to take into account for the design of nanopreparations improving drug delivery. The biodisponibility of a drug is directly correlated with the physicochemical and structural properties of the drug molecules. Some drugs show poor solubility in the GI tract, which limits their dissolution in aqueous media, reducing their intestinal absorption. This is the case with taxanes, widely used in cancer treatment (e.g. paclitaxel). Other drugs present a limited oral bioavailability due to a labile structure under physiological conditions (like pH or ionic strength), specially associated with peptide or protein drugs.

There is another phenomenon that affects drug biodisponibility and is known as pre-systemic metabolism (or first-pass metabolism). It denotes the loss of the drug before it reaches general systemic circulation. This type of drug metabolism is very important in oral delivery since it is usually associated with intestinal issues (the lumen and the wall gut, in addition to bacterial metabolism) and hepatic activity.

When drugs are administered within a nanosystem, the probability that they reach the systemic circulation in a non-free state is increased, and it is an alternative to overcome the drug concerns associated with solubility and stability [3].

14.1.2 *Physiological barriers that affect nanopreparations' stability*

14.1.2.1 *Gastric acidity*

The gastric juice produced by the stomach is composed of hydrochloric acid (HCl), potassium chloride (KCl), sodium chloride (NaCl), and the acid protease pepsin. This digestive fluid has an extremely acidic environment with a pH ranging from 1.2 to 3.5. The acidity plays an important role in activating digestive enzymes and in causing protein and peptide proteolysis. Furthermore, the acidic environment inhibits pathogen growth previously ingested via the oral route and possible infections are prevented. However, this is also an obstacle for therapeutic drugs that become inactivated at low pH values.

14.1.2.2 *GI proteases*

Along the GI tract, the different oral formulations (especially the protein-based) are exposed to the proteolytic activity of enzymes. Pepsin is the first enzyme found in the GI tract and is responsible for hydrolyzing proteins into peptides. Its optimum pH is around 2.0, and works properly in the stomach environment. After passing the stomach, the oral intake is passed through different parts of the intestine: the duodenum, jejunum, and ileum. In the duodenum, pancreatic proteases (endopeptidases and exopeptidases) catalyze the breakdown of the remnant peptides. Among them, trypsin and a-chymotrypsin are the major proteases, but the presence of other hydrolases, such as esterases and lipases, can play a relevant role in the protein hydrolytic processes.

14.1.2.3 *Transport mechanisms in the GI tract*

The smart design of nanodevices involves the consideration of the mechanism in which the delivered molecules cross the cell membrane. There are four main mechanisms: paracellular, transcellular, carrier-mediated, and receptor-mediated transport. Parameters of drug molecules, such as hydrophobicity, ionization degree, molecular weight, and pH stability, determine the mechanism of absorption through each pathway. However, the selected carrier can entrap the drug, but show different surface patterns (e.g. antibodies attachment or molecules that are recognized by cell membrane receptors) that target the drug inside the cell in different pathways than in the free molecule [4].

14.1.2.4 *Mucus layer of GI tract*

All orally administered drugs will be in contact within insoluble mucus that is typically coating the entire surface of the gastric mucosa. This barrier plays an important role in determining the absorption and bioavailability of drugs.

This mucosal barrier is mainly composed of glycocalyx (fibrous and fuzzy coat), which possesses mucin glycoprotein. Mucin provides the cohesive and adhesive nature of the mucus, and is mostly a physical barrier against orally administered drugs. But the drugs in contact with mucosal surfaces can be removed by a mucus clearance. However, the strategy to overpass the mucin barrier consists in the design of nanoparticles able to move across the mucus. In this sense, it was found that its size is very important. In fact, it was reported that nanoparticles less than 200 nm in size can effectively diffuse through the mucosal barrier [5].

14.1.2.5 *Residence time*

The time in which a dosage form resides in the stomach is strictly regulated by the mechanism of gastric emptying. This is a valuable issue to consider when a formulation is designed, because the absorption and drug bioavailability depends on the exposed time in the gastric environment. It is clear that if the residence time of the drug is not well established, great fluctuations in plasma drug levels can be produced. Also, some local gastric irritation by the drug itself can appear. In this sense, nanoparticles can be developed to pass across the stomach as quickly as possible, or to show gastro-retentive floating properties to increase their residence time [3].

14.1.3 *Biopharmaceutical considerations*

14.1.3.1 *Mononuclear phagocyte system (MPS)—PEGylation*

Working with nanoparticulate systems led to the obligation of considering some special issues about the interactions between them and the circumvent environment. The human immune system triggers the rapid clearance of nanoparticles in about 5 minutes after intravenous administration, thanks to the action of phagocytic cells. These specialized cells are part of the mononuclear phagocyte system (MPS), and their residence zones are the lymph nodes, the spleen, and the liver. On the other side, the importance of surface chemistry and the structure of nanoparticles in their response to the immune mechanism have been reported. In this sense, surface modification of the nanosystems takes a relevant role for the increment of blood circulation time. Some surfactants and polymers are able to form an aqueous film that covers particle surfaces and helps to avoid the binding of plasma proteins, such as opsonins, and the consequent road to the reticulo-endothelial system (RES). One example is the polyethylene glycol (PEG), a polymer extensively used because of their good biocompatibility properties; PEGs with different molecular weights are already commercially available. PEGylated systems have been widely studied as drug carriers with very positive results. However, some reports have shown that *in vivo* subsequent administration of PEGylated particles leads to the production of anti-PEG that "wakes up" the immune system and an exacerbated response takes place.

14.1.3.2 *The enhanced permeation and retention (EPR) effect*

Cancer cells express abnormal phenotypes, implying many modified biological functions. Consequently, this behavior affects the tissue, turning it defective. In this sense, the tumor endothelium has modified intercellular junctions, making it leaky. Additionally, solid tumors have very few lymphatic vessels, so there is also a lack of drainage and the accumulation of extravased macromolecules is observed. This phenomenon is called the enhanced permeability and retention (EPR) effect. Since an increment in the limit of molecular size for extravasations exists, this phenomenon is significant for molecules larger than 40 kDa. [6]. The EPR effect is also related to the degree of vascularization of the tumor and can be used to increase antitumor efficacy and reduce systemic side effects, thanks to the strategy of passive tumor targeting with, for example, polymer–drug conjugates, proteins, liposomes, micelles, and nanoparticles [7].

14.1.3.3 *Active targeting*

Molecular advances help to deeply understand cell functions and the connectivity between themselves, rather than the transportation of metabolites and nutrients. This comprehension extrapolates to the study of relevant sicknesses, such as cancer. As mentioned before, cancer cells act differently in comparison with normal cells; they over-express proteins in the membrane that are not present in normal cells, at least at that level. This example remarks very well how the knowledge of molecular mechanisms could help in the development of new therapies for certain illnesses. Folic acid receptors are present at low levels in normal cells, but in transformed cells, folic acid receptors, which are membrane proteins, are over-expressed. The attachment of folic acid to nanoparticles helps them to bind specifically with the receptor that is in cancer cells; once the particle–protein interacts with its receptor, an endocytosis process begins. This strategy is called active targeting and has gained relevance in the last several years, mostly because of the promising results. Moreover, this therapy is being developed for brain targeting in order to circumvent the blood–brain barrier (BBB), which avoids the penetration of systemic drugs. Specific receptors of the BBB were used to favor receptor-mediated endocytosis (RME) to the brain. Examples are the transferrin receptor, liporprotein related protein (LPR) receptor, and insulin receptor. Later, a more complex system for double targeting was designed in order to pass through the BBB and also to target tumor cells having as blank only one protein, angiopep-2, which was over-expressed in glioma and BBB cells [4].

14.2 Nanopreparations

The so-called "smart nanopreparations", which are taking advantage of the nanometer size of the systems, such as the high surface area to volume ratio and the capability to reach the deepest interstices of the intestinal epithelium, are clearly improving the oral biodisponibility of encapsulated drugs. In combination, nanopreparations imply the development of systems that expose active surfaces for targeting particles to specific regions of the organism. An important concern is the toxicity and side effects of the nanocarriers after administration. Also, the natural capability of the organism to excrete or reabsorb them once they have been used is a special consideration.

In this section, the most important nanopreparations in oral delivery are classified considering their physicochemical state in: nanocrystals, polymeric nanocarriers, emulsions, and inorganic systems.

14.2.1. *Nanocrystals*

14.2.1.1 *Drug nanocrystals*

14.2.1.1.1 Description

Drug nanocrystals are nanoparticles with a crystalline character [8]. An important characteristic is that the drug nanocrystal is totally composed of drug and no carrier material is required in the formulation (Fig. 14.1a). When drug nanocrystals are dispersed in liquid media, surfactants or polymeric stabilizers are needed to stabilize the particle, obtaining the so-called "nanosuspensions" (Fig. 14.1b).

Depending on the preparation method (Fig. 14.2), drug nanocrystals can be obtained in either crystalline or an amorphous product. In the last case, they are referred to as "nanocrystals in the amorphous state" [11].

14.2.1.1.2 Applications

Nowadays, the pharmaceutical industry is dealing with the limited solubility and dissolution rate of poorly soluble drugs. In this scenario, oral administration of pure drug nanocrystals, prepared by "bottom up" or "top down" technologies, significantly enhances the dissolution behavior of poorly soluble drugs [12]. The improvement of drug absorption as nanocrystals is shown in

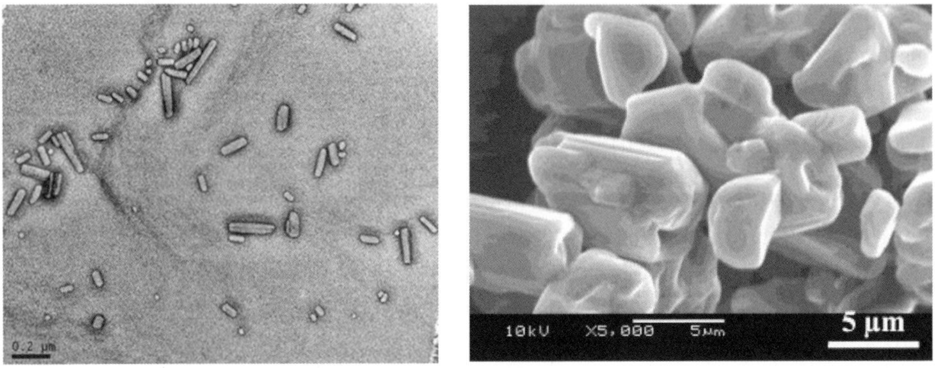

(A) (B)

Figure 14.1. (A) Transmission electron microscope (TEM) image of paclitaxel nanocrystals ([9] with permission); (B) Scanning electron microphotographs of phenytoin particles in an oral nanosuspension (from [10] with permission).

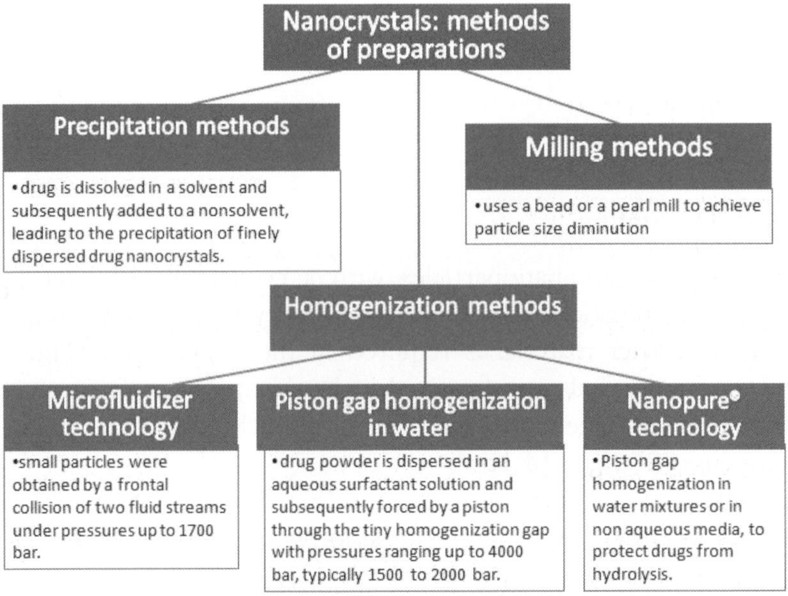

Figure 14.2. Preparation methods of drug nanocrystals.

Fig. 14.3. Some nanocrystal formulations are now available in the market, such as Rapamune® (contains sirolimus, a macrocyclic immunosuppressive drug), Emend® (the drug is aprepitant, used for the treatment of emesis), Tricor® (fenofibrate nanocrystals), or Megace ES® (a progestin antitumor activity drug) [11]. Nanosizing of drugs increases their specific surface area,

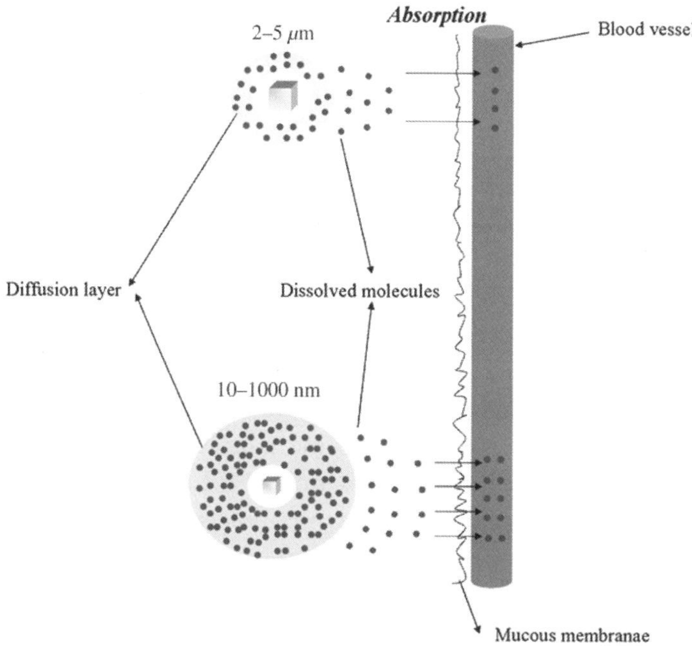

Figure 14.3. Drug nanocrystals form a high drug concentration gradient between the GI tract and blood vessel due to the increased saturation solubility and dissolution velocity in digestive juice, and lead to a significant improvement on absorption (from [17] with permission).

which directly impact their oral absorption and bioavailability. Nanocrystals can be administered as liquid dosage forms (nanosuspensions) or dry dosage forms (tablets or gelatin capsules containing the pellets). An oral matrix pellet containing nanocrystalline ketoprofen (a nonsteroidal anti-inflammatory drug) was successfully developed. The formulation released the drug over a period of 24 hours [13]. In another work, Amphotericin B was prepared as a nanocrystalline suspension to improve the antibiotic oral bioavailability. Results indicated that oral administration of amphotericin B nanosuspension facilitated a significant absorption improvement than the commercial formulations [14]. Similar results for oral fenofibrate nanosuspensions were found, which showed an enhancement in bioavailability compared to micronized suspensions of the drug [15]. In a recent investigation, Nimodipine (a drug used for treatment of high blood pressure) was prepared as nanocrystals by combining microprecipitation and high pressure homogenization techniques. Those nanoparticles showed size-dependence related to the aqueous solubility. The developed nanocrystals exhibited a 2.6-fold higher oral bioavailability than the commercial drug form [16]. Also, a deep study on the structure and stability of Paclitaxel nanocrystals was carried out. Results indicated that

Paclitaxel nanocrystal formulation improved the antitumoral activity of the drug in comparison with Taxol (the commercial formulation) without incidence of toxicity [9].

14.2.1.2 *Cross-linked enzyme: Crystals (CLECs) and aggregates (CLEAs)*

14.2.1.2.1 Description

The cross-linked enzyme crystals (CLECs) were prepared by crystallization of the enzymes, followed by cross-linking with a bifunctional reagent, generally glutaraldehyde [18]. However, CLEC's drawbacks were associated with the need to purify and crystallize the pure protein. For these reasons, the appearance of the so-called cross-linked enzyme aggregates (CLEAs) offered an alternative to CLEC's problems. Formation of CLEAs results from the precipitation of physical aggregates of enzyme molecules using non-denaturing methods and later cross-linked. The reaction leads to the formation of insoluble aggregates (from nano- to micro-meter size) displaying catalytical activity, stable against common denaturing and aggressive environmental conditions, and reusability [19].

14.2.1.2.2 Applications

CLECs and CLEAs have been reported using a wide variety of enzymes such as lipases, laccases, fitases, galactose oxidases, trypsin, glucose oxidases, β-galactosidases, alcohol dehydrogenases, and other enzymes commonly used in oral therapies [18]. In all cases, the enzyme activity was not only retained, but also protected from inactivation at extreme conditions, such as at the acidic gastric pH [20, 21]. An application of CLEAs procedure was performed for subtilisin, which lost its protease activity under gastric conditions. This enzyme was first aggregated using ammonium sulfate and PEG. In the second step, the protein was cross-linked with glutaraldehyde to produce the CLEAs form. The effect of pH, temperature, kinetic parameter, thermal stability, and stability in organic solvents were studied. Results indicate that CLEA of subtilisin presented an enhanced stability in organic solvents and it showed thermal stability up to 60°C. In order to increase the pH stability of the CLEA and to be useful for the oral drug delivery, the developed enzyme nanoparticulates were entrapped in alginate: guar gum beads. This process is considered to be a feasible strategy to deliver protein and peptide drugs, which are usually degraded in the stomach before reaching the jejunum [22]. Recent work from our laboratory describes the development of cross-linked

Figure 14.4. Nanosized cross-linked enzyme aggregates (CLEAs) of Alginate Lyase.

enzyme aggregates of alginate lyase (AL-CLEAs) for oral cystic fibrosis treatment (Fig. 14.4). Meanwhile, free enzyme showed a rapid inactivation under simulated gastric conditions; the AL-CLEAs retained more than 70% of enzymatic activity. Moreover, AL-CLEAs showed thermal stability at temperatures over 37 °C and stability against organic solvents. The developed AL-CLEA system becomes a feasible alternative for cystic fibrosis treatment, as the enzyme reaches the intestine in catalytic conditions to reduce the viscoelasticity of the mucoid alginate produced by opportunistic pathogens [23].

14.2.2 *Polymeric nanocarriers*

The most important reason about why to use polymer carriers for drug delivery is because of the capability to modulate pharmacokinetics and bioavailability properties (Fig. 14.5a) [24]. Polymers could also protect the drug from the strong gastrointestinal environment and these nanocarriers are designed with a site-specific targeting capacity, so the carrier releases the drug in certain conditions. This site-specific targeting is achieved by the conjugation of molecules, like protein (Fig. 14.5b). Also, it is important to define precisely the main characteristics of the nanosystems, such as size, surface properties, and polymer chemistry. For example, the hydrophobicity and hydrophilicity within the polymeric system can be manipulated to accommodate a wide variety of molecules [25]. It has been reported that nanoparticles have a

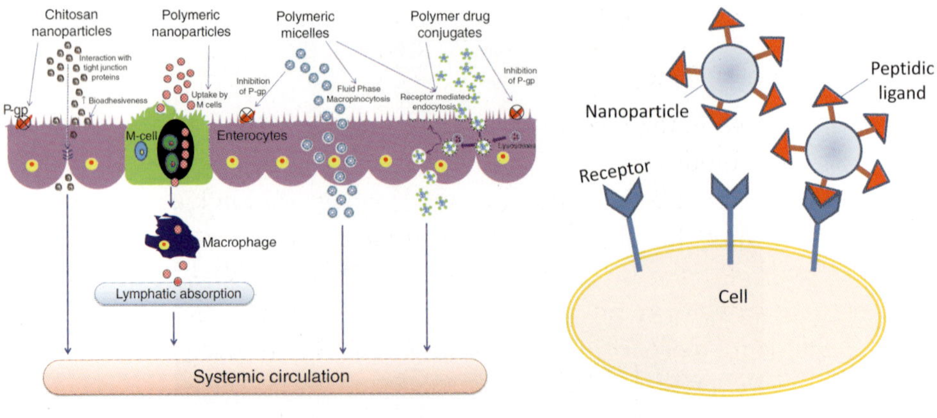

Figure 14.5. (A) Potential absorption mechanisms implemented by polymeric nanocarriers for increasing oral bioavailability of drug substances (from [25] with permission). B) Scheme of targeted nanoparticles with peptidic ligands.

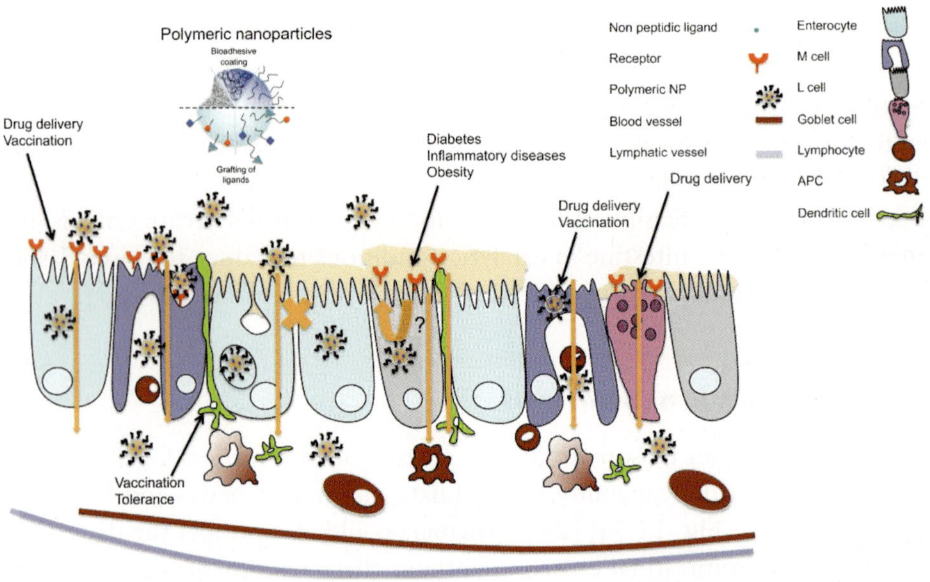

Figure 14.6. Targeting of oral delivery nanoparticles for biomedical applications (from [42] with permission). NP = nanoparticle, APC = antigen-presenting cell.

passive accumulation in inflamed intestinal areas. These nanoparticles remain for several days in the site making the drug delivery system properties very important, since reservoirs of these nanocarriers could be formed in damaged areas [26].

14.2.2.1. *Polymeric nanoparticles*

14.2.2.1.1 Description

Polymers involved in the formation of nanoparticles can be classified into two different types: biodegradable and non-biodegradable. Biodegradable polymers are metabolized by the organism and the degradation kinetics could be described and modeled. However, there is an important issue to take into account, which are the degradation products. Lots of polymers are biodegradable, but some of them release toxic degradation residues. Some examples of biodegradable polymers are poly(lactide-co-glycolide) (PLGA), poly(lactide) (PL), poly(glycolide) (PGA), poly(caprolactone) (PCL) and the so-called biopolymers, which come from natural sources, like pectins, chitosan, alginate, and dextrans. The last two have been reported to inhibit P-glycoproten (P-gp) efflux pumps, thus helping drug resistance in cancer therapies [27]. Another example of a polymer with this property is PEG, which inhibits the efflux of some drugs, like doxorubicin and paclitaxel. PEGylated paclitaxel showed better bioavailability than paclitaxel alone. Other polymers, such as acacia, also possess multidrug resistance (MDR) reversal capability (e.g. intracellular accumulation of epirubicin was significantly increased when tested against MDR Caco-2 cell lines in the presence of acacia) [25]. The other types are the non-biodegradable polymers, like polyethylene imine (PEI) and poly(methyl methacrylate) (PMMA) [28–30].

Nanoparticles present two main structures: nanocapsules and nanospheres. Nanocapsules are polymer sheets, where the drug is confined inside them. On the other hand, a nanosphere is a matrix, where the drug is entrapped within the polymer chains to form a solid structure [31].

One of the most important issues to be considered is the drug burst release from the nanoparticle because of the high surface area and the molecule adsorption produced in the first 60 to 120 min of drug administration. If the drug is applied orally, this initial drug release from nanosystems will occur during the gastric or small intestinal passage and the drug will either be degraded in the gastric environment or taken up into systemic circulation. In such cases, the consequence is the reduction of the available drug amount required at the target site (e.g. distal ileum and colon) and increasing adverse side effects [26]. In this sense, polymeric chemistry is very important in order to solve this problem. Some biopolymer properties help make the matrix entrapment more "sealed", so that the environment effect of the nanosystem is reduced. A good example of that are pectins, especially those with a high degree of methoxylation [32]. Another good strategy is to incorporate these nanoparticles inside microparticulate polymeric carriers, whose job would be to protect drug delivery systems, the nanoparticles, from the media. When the microcarrier arrives at the intestinal tract, the erosion process will release the nanoparticles. This was applied with Tacrolimus-loaded PLGA nanoparticles, where these nanocarriers where encapsulated into Eudragit® P4135F microspheres [33]. Once in the intestinal tract, polymeric nanoparticles will interact with the environment and this interaction will be responsible for the drug pharmacokinetics and bioavailability. One of the mechanisms involved in the last concept is mucoadhesion. The capacity of the nanocarrier to interact with the lumen is very important and will affect residence time in the intestinal tract. For example, charged polymers, like chitosan, could interact with the negatively charged mucin: this surface charge property would be another parameter of polymer modifications, which will collaborate with polymeric nanoparticles design [34]. PEGylation is the other well-known polymer modification applied in nanoparticle synthesis, since promising results were published. PEG helped to circumvent problems associated with nanoparticles, such as particle aggregation, stability, and clearance from body. PEG chains are reported to impart stealth characteristics to the nanoparticles, improve bioadhesion in the gastrointestinal tract, and provide surface hydrophilicity [25, 35, 36].

The nanocarrier capacity to interact through ligand–receptor junctions is a relevant point in drug delivery systems development (Fig. 14.6). The strategy consists of the presence of a specific ligand on the surface of the polymeric nanoparticle. This ligand will interact directly with its receptor, which is situated on the lumen cells of the intestinal tract. However, there are very few receptor-mediated transport systems in the gastrointestinal tract. Some examples of ligands include folic acid [37]; bioadhesins, such as lectins and pectins [38]; peptidic ligands, such as arginine–glycine–aspartate (RGD) [39–41]; and bile acids, such as deoxycholic acid. Lectins have been widely studied for this reason. Nanoparticles containing lectin on the surface are taken up by antigen sampling cells, M-cells of the Peyer's patches. Once internalized, these carriers pave their way into systemic circulation via the lymphatic system. Also, RGD peptides are found to interact with the $\beta 1$ integrin receptors localized at the apical pole on the M-cells of the Peyer›s patches along the GI tract. The Peyer's patches are covered by cells called microfold cells (M-cells). The luminal anti-gens are transported from the mucosal surface of Peyer's patches through the specialized epithelial M-cells, a direct route into the lymphatic system. Advantageously, this avoids first-pass metabolism and decreases the chances of cellular rejection, as M-cells do not express P-glycoprotein efflux pumps. Furthermore, the Payer's patches also have, comparatively (to the rest of the GT), low mucosal coating [31].

14.2.2.2 *Polymeric micelles*

14.2.2.2.1 Description

These systems are conformed by two different regions: one is hydrophobic and the other one is hydrophilic. The hydrophobic core is the site for lipophilic drug encapsulation. Meanwhile, the surface is the hydrophilic region of the particle, which interacts with the aqueous environment. Lipophilic drugs have a lot of problems with its solubility and, therefore, its bioavailability, so any drug carrier that helps with those problems would be taken into account. One typical example is paclitaxel, which is a very powerful molecule for cancer therapies and is widely used, but its low solubility forces the formulation of excipients that are mainly responsible for the toxic effects. Many efforts have been spent in order to encapsulate this drug inside these types of nanocarriers.

Another advantage is that polymeric micelles could be modified by the attachment of targeting molecules, so the carrier would be absorbed by receptor-mediated endocytosis [43]. Also, micelles may either be functionalized with certain polymers to impart different properties, such as temperature- or pH-sensitivity, so the drug is released in these conditions.

Two main strategies are described for the preparation of polymeric micelles (PMs). The first one is called "direct dissolution" and consists of the dissolution of the polymer and drug in an aqueous solvent. This method is adequate when the polymer is moderately hydrophobic. The second strategy involves the usage of organic solvents and is adequate when the polymer and drug are mostly hydrophobic. Once the drug and polymer are dissolved in the organic solvent, this one is lately removed by different methods, like dialysis against water if the solvent is water-miscible; if not, an oil-in-water emulsion method could be done in order to retain the drug. Later, as dry dosage forms are preferred for oral administration, these drug-loaded micelles should be spray dried.

It becomes very relevant to describe the critical micelle concentration (CMC), in order to utilize proper synthesis parameters. The CMC refers to the concentration of copolymer above which micelle formation is favored. Taking into account the fluid volumes of the GI tract (stomach and small intestine) in the fed (740 mL) and fasted (130 mL) states, polymer concentrations may vary anywhere from 135 to 7,700 mg/L following the oral administration of a micelle dose of 100–1,000 mg. Thus, as an indicative guide, a CMC of less than 135 mg/L should be sufficient to resist rapid dissociation upon oral administration.

The PM systems are still in the experimental stage since only a few *in vivo* studies were carried out and only assays on cell culture models were performed.

However, PMs are promising systems, since interesting results were collected recently with paclitaxel. The formulation was administrated intravenously or orally to rats. But, only a limited number of rats were studied and without taking into account the effect of stomach and bile salts [44].

Another interesting point about these systems is their design with the capacity of pH-dependence released of the drug in order to achieve a control delivery mostly in the intestinal environment. Also, this is important to avoid drug precipitation upon release from the micelles. To release their cargo content in a wanted pH environment, the inner core of PMs should be composed of weakly acidic groups able to be ionized at the intestinal pH (pH ≥ 5). The polymer deprotonation increases the core's polarity, while lowering its affinity for the encapsulated drug; in consequence, it is released from the micelle [43].

14.2.2.3 *Polymer–drug conjugates*

14.2.2.3.1 Description

Covalent junctions between drug molecules and polymers serve as drug delivery systems. Additional functionalities could also be attached so that specific properties are reached. Furthermore, upon exploitation of the structure–activity relationship of the drug molecules, various benefits such as increased solubilization, enhanced plasma half-life, bioavailability enhancement due to the increase in hydrodynamic volume and reduced excretion by kidneys, protection towards degrading enzymes, prevention or reduction of aggregation or immune responses, reduction in toxic side effects, and targeting capability can be achieved via polymer–drug conjugates [25].

The most important issue about these types of conjugation is to guarantee the pharmacological activity of the drug once conjugated. In this sense, the election of the functional group, which is going to react is extremely relevant. The linkage chemistry is also important to achieve the desired objective; for example, enzymatically degradable linkages are preferred, since the drug should be released under certain conditions. As mentioned before, once the drug is released, its therapeutic activity must be intact. Polymers such as polyacrylates, chitosans, poly(amidoamine), and thiolated polymers are widely used as oral absorption enhancers for drug–polymer conjugates. A good example of that is chitosan, which gives a mucoadhesive property and high capacity of capturing a large volume of water molecules. Furthermore, chitosan consists of several functional groups (hydroxyl, –OH, and amino, –NH2) that allow for distinct chemical modifications [45]. Chitosan was conjugated to paclitaxel and has been studied and is known to offer several advantages, such as improved water solubility due to increased hydrophilicity, prolonged retention of the conjugate in the gastrointestinal system due to the mucoadhesive properties of chitosan, resulting in overall increased uptake, opening of tight junctions, and enabling the paracellular route for absorption and ability to bypass the CYP-450-dependent metabolism, as the conjugated paclitaxel would no longer be the substrate of these enzymes [25]. An interesting explanation about these permeation enhancer polymers is the capacity to selectively open the specific water channels of the tight junctions, allowing for the passage of hydrophilic compounds through enterocytes. Additionally, the large size of polymer conjugates bypasses the possible systemic toxicity that can be reached by entering into blood circulation. It is also relevant to mention that thiolated polymers demonstrated a better enhancing effect in oral delivery. These polymers are obtained by the incorporation of free thiol groups into polymers like chitosan, with the result of enhanced mucoadhesiveness [25]. These polymeric compounds are capable of increasing local concentrations of glutathione (GT) by reducing oxidized glutathione in the intestine epithelia. GT, in turn, is able to inhibit protein tyrosine phosphate (pTP) by forming disulfide bonds with an active enzyme. Generally, pTP is involved in the closure of tight junctions by dephosphorylation of occluding; thus, pTP inhibition results in prolonging the opening of tight junctions and disruption of cell–cell adhesions, which are accompanied by enhanced permeability for hydrophilic compounds [46].

14.2.3 *Emulsions*

In recent years, lipid-based nanoparticles have emerged as one of the most interesting vehicles throughout the pharmaceutical field, due to their innovative properties. They are able to entrap hydrophobic drugs (e.g. antithumorals or antiparasitaries) and provide a controlled release of them, enhancing the oral absorption of drugs with extensive presystemic metabolism [47]. Considering their low toxicity for clinical applications [48] and the increasing development of new drugs with hydrophobic nature [49], they become a potent tool for oral treatments.

14.2.3.1 *Liposomes*

14.2.3.1.1 Description

They are closed spherical vesicles defined by a lipidic bilayer membrane, mainly composed of natural or synthetic amphipathic lipids (phospholipids, in particular the most common is phosphatidylcholine, PC). The bilayers are separated from one another by aqueous domains and enclose an inner aqueous space, in which aqueous soluble drugs can be loaded. Also, taking advantage of the bilayers membrane nature, lipophilic drugs can be encapsulated inside it (Fig. 14.7).

Liposomes can be classified considering the number of lipid bilayers surrounding the aqueous phase [51] in the following structures:

* Multilamellar vesicles (MLV): several lipid layers separated from one another by a layer of aqueous solution.
* Small unilamellar vesicles (SUV): surrounded by a single lipid bilayer. They are in the range of 20–60 nm.
* Large unilamellar vesicles (LUV): also surrounded by a single lipid layer, but in the size range from nanometers to micrometers.

The liposomes surface can be easy tailored by chemical modification of the lipids, in order to attach molecules with biological functions, such as polymers (e.g. PEG) or antibodies for targeting.

There are several methods and techniques to produce liposomes [52], but the most commonly used at laboratory scale are sonication, extrusion, reverse-phase evaporation, and solvent injection (Fig. 14.8).

14.2.3.1.2 Applications

Liposomes are one of the most extensively studied advanced drug vehicles. In the last several years, liposomes have been used as an oral dosage form, since they protect the entrapped drugs from digestive degradation or increase the biodisponibility of poorly absorbed drugs at the GI tract [53]. It is already possible to find liposome formulations in clinical uses (e.g. AmBisome®, liposomal amphotericin B, and Doxil®, PEGylated liposomes loaded with doxorubicin) and others are in clinical trials. Some liposomal formulations containing different drugs for oral clinical treatments are listed in Table 14.1.

Variants of liposome formulations include a surface modification to target the drug to a specific cell or the inclusion of polymeric cores. A liposomal formulation of paclitaxel (an antithumoral drug) targeting the folate receptor (FR) was designed to overcome the toxicity of Cremophor®EL, the traditional vehicle, and to increase the drug effectiveness. FR-targeted liposomes encapsulating paclitaxel resulted in a four-fold greater cytotoxicity compared to non-targeted liposomes in KB oral carcinoma cells [54]. Other work reported the encapsulation of various anticancer drugs, such as 5-fluorouracil and methotrexate into liposomes with polymeric cores, resulting in an increase of their bioavailability by oral delivery [55].

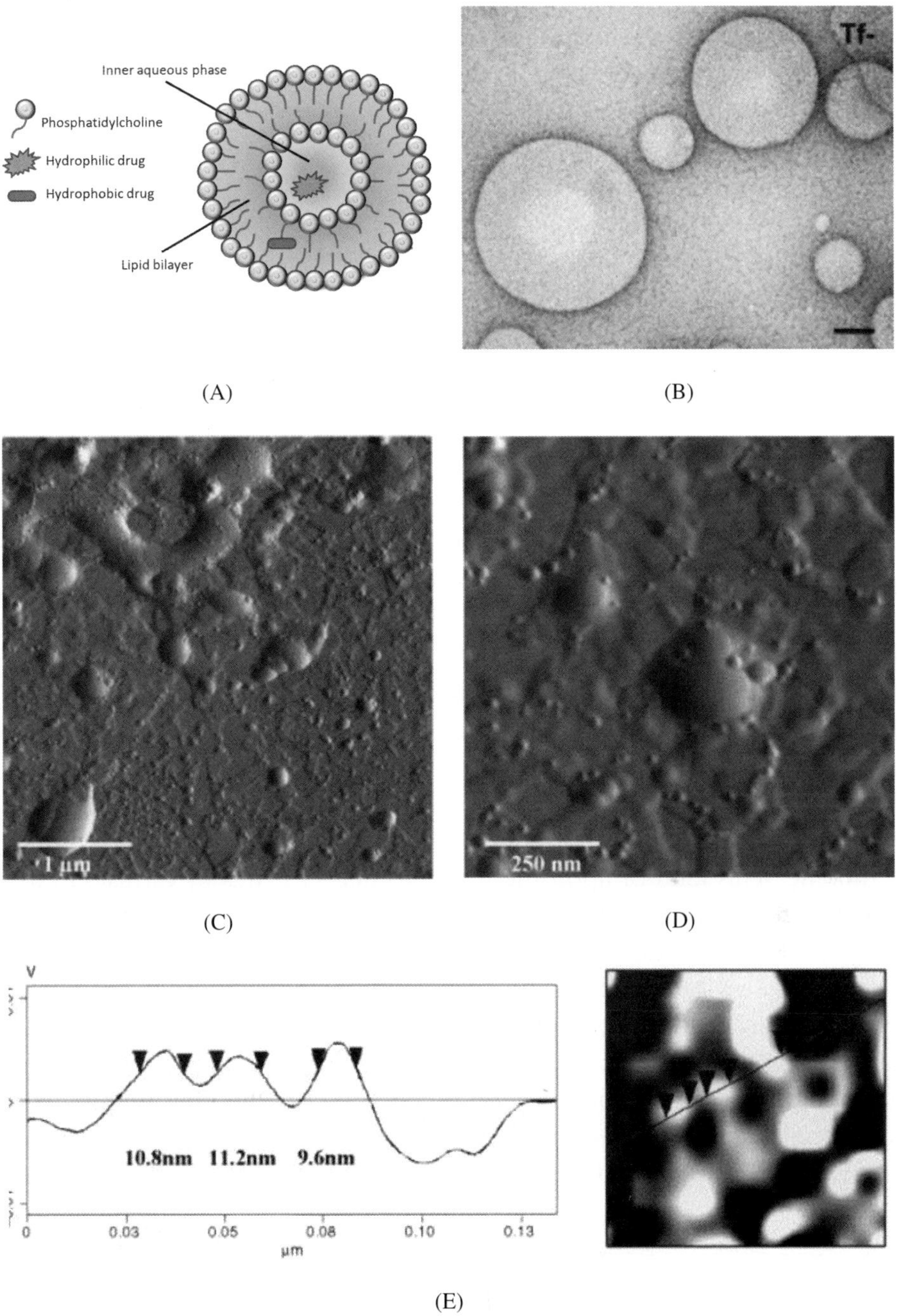

Figure 14.7. (a) Scheme of Liposome structure; (b) TEM image of liposomes prepared with DSPE-PEG2000-COOH, after negative staining with uranyl acetate. Bar=50 nm. (c) Up: Liposomes covalently modified with Transferrin (Tf). Small globular structures are visible at the surface. Below: Line scan of liposome surface. The globular structures were measured to be 10.53 ± 1.45 nm in diameter (images (B) and (C) from [50] with permission).

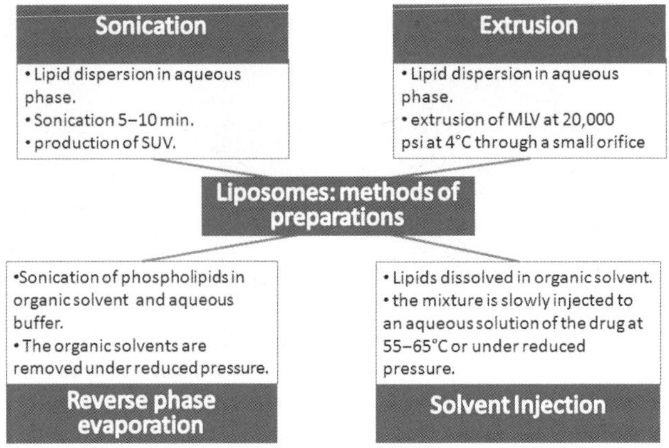

Figure 14.8. Preparation methods of liposomes. MLV = multilamellar vesicles.

Surface-coated liposomes were developed for oral drug delivery of insulin (a peptide drug involved in diabetes control). DPCC-chol liposomes were coated with PEG and showed slower release profiles for insulin than conventional liposomes [56].

Another interesting property of liposomes is their ability to act as pH-sensitive systems. Liposomes that possess low (or high) pH can be prepared containing the encapsulated aqueous drugs with charge in solution. When the pH is neutralized within the liposome (protons go through the membrane), the drug will also be neutralized and will freely pass through the liposomal membrane. This kind of liposome works to deliver drugs by diffusion rather than by direct cell fusion [52].

However, liposomes present some disadvantages and limitations. First of all, they possess low encapsulation percentages of lipophilic drugs and the release profile of drugs cannot be predicted, as they show poor batch-to-batch reproducibility. With respect to the synthesis, they are manufactured through processes in the presence of organic solvents and they are difficult to sterilize. Finally, liposomes are extremely sensitive to the environmental conditions and they are unstable in biological fluids [57].

14.2.3.2 *Solid lipid nanoparticles* (*SLNs*)

14.2.3.2.1 Description

SLNs are devices composed of lipids (triglycerides, glyceride mixtures, or waxes) that are solid at room temperature (Fig. 14.9a). Different lipids are commonly used to produce SLNs, such as stearic acid, cetyl alcohol, cetylpalmitate,

Table 14.1. Drugs encapsulated in liposomes for oral therapies.

Drug	Liposome composition	Treatment	Reference
Amphotericin B	ComercialLipome: AmBisome®	Fungi infection	[58]
Doxorubicin	Doxil® PEGylated liposomes	Cancer	[59]
	EPC, Distearoyl-PC (DSPC)		
Paclitaxel	EPC	Cancer	[60]
	Folate Receptor-targeted liposomes		[54]
5-fluorouracil and methotrexate	liposomes with polymeric cores	Cancer	[55]
Indomethacin	Monophasic vesicles of egg PC	Anti-inflammatory drug	[61]
Insulin	DPPC-chol coated with PEG	Diabetes	[56]
Ampicillin	PC/chol	Bacterial Infection	[62]
Vancomycin, Teicoplain	PC	Bacterial Infection	[63]

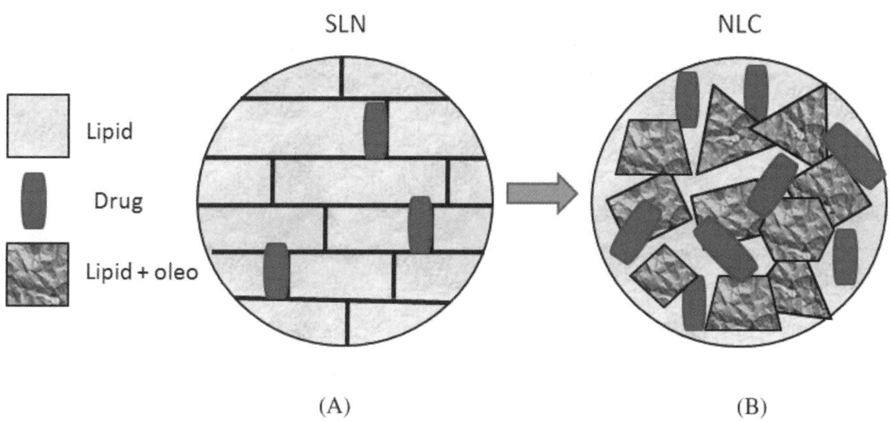

Figure 14.9. Schematic illustration of: (A) Solid Lipid Nanoparticle (SLN) and (B) Nano-structured Lipid Carrier (NLC).

tripalmitin/Dynasan® 116, Compritol® 888 ATO, Glycerylmonostearate, trimyristin/Dynasan® 114, Precirol® ATO5, tristearin/Dynasan® 118, Imwitor® 900, or myristylmyristate. These lipids used during preparation are physiologically compatible, making SLN a safe and effective alternative in comparison to conventional delivery systems [64] Moreover, when administered to living organisms by the oral route, they are eroded and degraded by bile salts and pancreatic lipase [65]. They have become very attractive candidates as carriers for oral drug delivery, due to the unique physiochemical properties of lipids and diversity, their biocompatibility and their ability to enhance oral bioavailability of poorly water-soluble drugs [66]. As controlled release systems, they offered prolonged release profiles, since the mobility of the drug molecules into a solid matrix is considerably lower than an oily phase [49].

In comparison with polymeric nanoparticles, scaling-up of SLNs for oral delivery purposes is possible using procedures like high-pressure homogenization and others (Fig. 14.10) in the absence of toxic solvents [48, 49].

On the contrary, SLN showed some limitations related to their inherent low loading capacity due to the highly ordered crystalline structure of the solid lipid matrix [67]. Even after preparation, particle growth and the dynamics of polymorphic transitions can be produced with the tendency to expulse the encapsulated drug during storage [48].

14.2.3.2.2 Applications

Both hydrophobic and hydrophilic drugs have been incorporated into SLN formulations for oral administration. Different drugs, with their therapeutic application and preparation method are listed in Table 14.2. An improvement

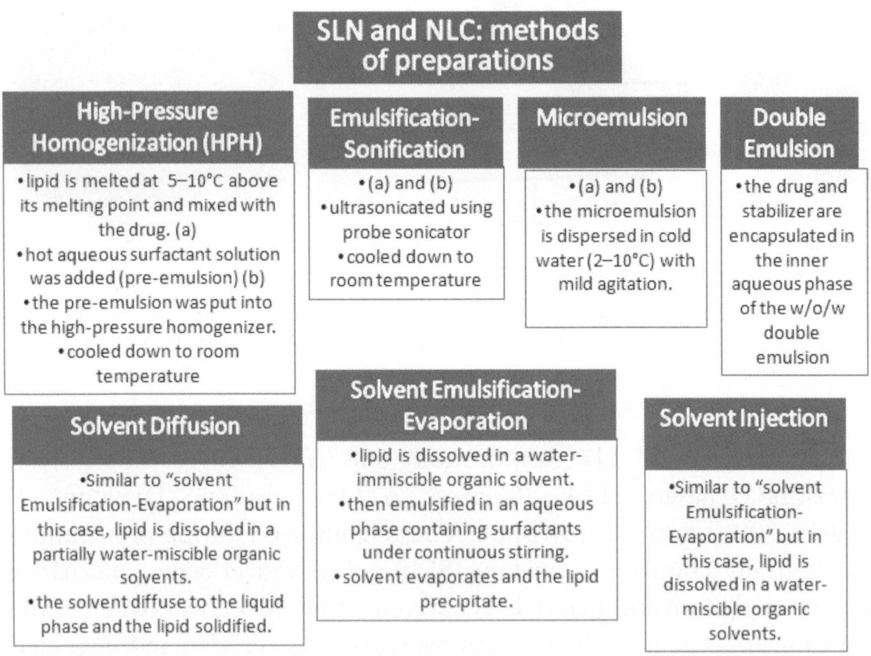

Figure 14.10. Methods of synthesis of SLN and NLC. SLN = solid lipid nanoparticle, NLC – nanostructured lipid carrier.

Table 14.2 Use and preparation methods of SLNs.

Drug	Method of preparation	Treatment	References
Apomorphine	Dispersion-ultrasonication	Parkinson's disease	[72]
Paclitaxel	Phase inversion-based process	Cancer	[69, 70]
Buspirone	Emulsification-evaporation followed by ultrasonification	Anxiolytic psychoactive drug	[75]
Carvedilol	Microemulsion	Anti hypertension	[78]
Clozapine	Hot homogenization followed by ultrasonication	Psychotic disorders (including schizophrenia	[79]
Cyclosporine A	HPH	Immunosuppressive effect	[71]
Digoxin	Ultrasonic and HPH method	Cardiac applications	[80]
Insulin	Emulsion solvent diffusion	Diabetes	[74]
Methotrexate	Solvent diffusion	Cancer, autoimmune diseases, ectopic pregnancy, and for the induction of medical abortions.	[81]
N3-O-toluyl-fluorouracil	Film dispersion–ultrasonication	Cancer	[73]
Pentoxifylline	Homogenization followed by the ultrasonication	Intermittent claudication	[76]
Raloxifene	Solvent emulsification/evaporation method	Osteoporosis	[82]
Rifampicin, Isoniazid, Pyrazinamide	Emulsion solvent diffusion	Tuberculosis	[77]
Salmon calcitonin	Double emulsion	Osteoporosis	[83]

in oral bioavailability was observed, enhancing the GI absorption, bypassing the first-pass metabolism (responsible for the reduction in concentration of the administrated drug before it reaches the systemic circulation), and providing a sustainable release of the drug encapsulated into SLN [68]. The increase in bioavailavility of Raloxifene encapsulated into SLN compared with free drug is shown in Fig. 14.11a. Between the hydrophobic drugs, paclitaxel (antithumoral drug) was loaded into lipid nanocapsules with an efficiency of 99.9 ± 1.0% and a three-fold increase in the oral bioavailability of paclitaxel compared with

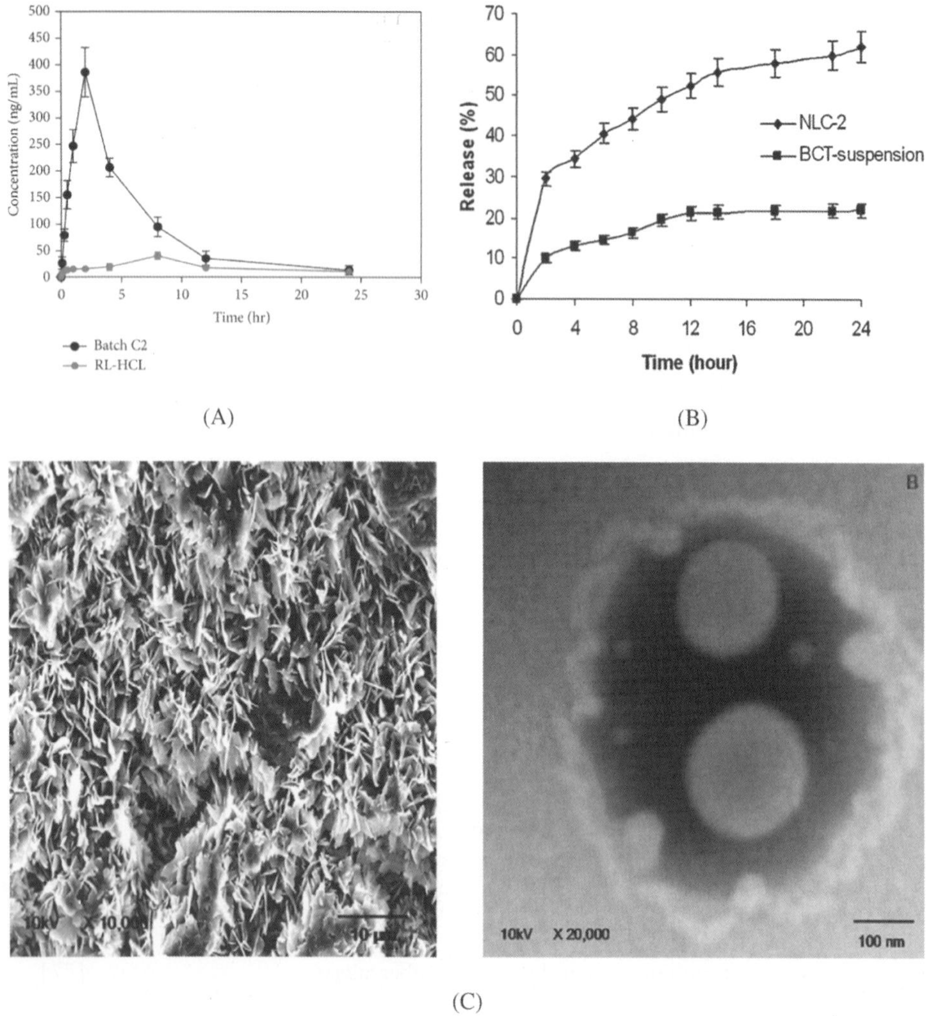

Figure 14.11. (A) Plasma concentration versus time profile of Raloxifene free and into SLN (from [82] with permission). (B) Comparative *in vitro* drug release profiles from optimized batch (NLC-2) and BCT-suspension using 1% SLS as dissolution medium. (C) SEM photomicrographs of drug–lipid melt (A) and lyophilized NLC-2 (B) bicalutamide (from [87] with permission).

Taxol® (the commercial dosage form) orally administered [69, 70]. In the same trend, SLNs with a mean size of 157 nm containing cyclosporine A showing an encapsulation yield of 96.1% were developed. The pharmacokinetic parameters of the developed formulations were established after oral administration in pigs. The study demonstrated that cyclosporine was released in a controlled manner from SLN formulation. Meanwhile, the marketed microemulsion formulation (Sandimmun®) showed a high initial plasma peak; this phenomenon was absent in the case of SLN formulation [71]. In another work, encapsulation of Apomorphine into SLN was carried out, as the drug showed poor oral bioavailability due to the first-pass effect. The formulations were prepared using two emulsifiers, glyceryl monostearate (GMS) and polyethylene glycol monostearate (PMS), revealing a significant effect of emulsifiers on the properties of SLNs. In addition, an improvement in oral bioavailability and brain regional distribution was observed [72].

An interesting formulation was developed by film dispersion–ultrasonication method using hexadecyl trimethyl ammonium bromide (a cationic tenside) to prepare cationic SLNs containing N3-O-toluyl-fluorouracil (TFu). Spherical shape nanoparticles with a mean size of 178.8 ± 9.99 nm showed an encapsulation percentage of 71.03 ± 1.19%. The oral bioavailability of TFu–SLNs was highly enhanced in comparison with TFu suspensions, when both formulations were orally administrated to mice. The study demonstrated that the bio-adhesion through electrostatic interaction between the positively charged SLNs and the negatively charged mucosal surface was responsible for the enhanced absorption [73].

Other studies were focused on the encapsulation of proteins into SLNs. Insulin, a peptide involved in the treatment of diabetes, was encapsulated into lectin-modified SLNs. The formulation was prepared by a modified double dispersion method showing a high drug-entrapment efficiency (>60%). Insulin into SLNs was protected against *in vitro* degradation of digestive enzymes. Pharmacokinetic studies demonstrated that bioavailability of insulin was almost similar when it was administered into SLNs by the oral route, in comparison with subcutaneous injection of insulin in rats [74].

With respect to water-soluble drugs, some formulations were developed with the lowest encapsulation percentages (less than 50%). However, an enhancement of the insulin bioavailability was observed. SLNs of buspirone–HCl were produced by emulsification–evaporation, followed by the sonication method. The particles of 345.7 nm showed a loading efficiency of 32.8%. *In vitro* experiments revealed that 90% of buspirone was released in 4.5 hours. Pharmacokinetic studies of 15 mg·kg−1 buspirone orally administered in the form of free drug and SLNs showed an increase of bioavailability of the drug in male Wistar rats [75]. Similar results were found for Pentoxifylline encapsulated into SLN prepared by homogenization, followed by the sonication method, demonstrating that SLN formulation is a promising approach to improve the oral bioavailability of drugs, which are directly affected by a high first-pass effect [76]. Finally, SLN encapsulating the anti-tuberculosis drugs rifampicin, isoniazid, and pyrazinamide have been developed and studies in animal models resulted in a successful management of tuberculosis when they were administrated every ten days [77].

14.2.3.3 *Nano-structured lipid carrier*

14.2.3.3.1 Description

The highly ordered crystalline arrangement of solid lipids directly affects the loading and release properties of drugs encapsulated into SLNs. When polymorphic transitions take place during storage, the incorporated drug tends to be expelled from the lipid matrix [84]. To overcome these drawbacks, nanostructure lipid carriers (NLCs) were developed as second-generation lipid nanocarriers. They are composed of solid lipid matrices that are incorporated with liquid oils at defined ratios (preferably 70:30 to 99:9:0.1), which decreases the highly ordered crystallinity of the solid matrix (Fig. 14.9b). In addition, the oil droplets into the solid matrix increase the drug loading capacity in comparison with SLNs.

14.2.3.3.2 Applications

Many works developed NLC systems to obtain absorption enhancing effects of low water-soluble drugs, promoted by the absorption promoting effect of the lipid in the route of administration. Vinpocetine (a vasodilator and anti-inflammatory agent) was encapsulated into NLCs by a high-pressure homogenization method. Pharmacokinetic studies in Wistar rats showed a relative increase in bioavailability of 322% of vinpocetine-loaded NLCs than vinpocetine suspension [85]. In other investigations, Domperidone-loaded NLCs were prepared by hot homogenization followed by an ultrasonication technique, using trimyristin as solid lipid, cetylrecinoleate as liquid lipid, and a mixture of soy phosphatidylcholine (99%) and tween 80 as the surfactant. The *in vitro* release kinetics showed that NLC provides a controlled release for more than 24 hours [86]. Another lipophilic drug, progesterone, shows poor bioavailability by oral administration due to its intensive hepatic metabolism. In this sense, progesterone-loaded NLCs were prepared to improve the drug oral bioavailability, in order to maintain an effective concentration in the blood serum in reduced doses. Mono stearin and stearic acid were used as solid lipids, while oleic acid was used as the oil lipid [41]. An investigation on phase transition behavior during the NLC processing was carried out with NLCs entrapping Bicalutamide, another poorly water-soluble drug commonly used for prostate cancer treatment. NLCs were prepared by hot high-pressure homogenization (HPH), showing a Bicalutamide controlled release profile [87] (Figs. 14.11b and 14.11c). Lovastatin (a cholesterol-lowering drug) also showed poor oral bioavailability due to a rapid metabolism in the gut and liver. An NLC formulation containing Lovastatin was developed and showed desirable size, zeta potential, drug-loading capacity, and release properties. The oral drug bioavailability was increased from 4% to 13% and 24% when the drug was administered from NLCs containing soybean phosphatidyl choline (SPC) and Myverol (lipophilic emulsifier), respectively. In addition, an *in vivo* real-time bioluminescence imaging experiment revealed a high stability of the Myverol system over the SPC system in the stomachal environment [88]. Furthermore, NLCs have the property to protect the entrapped pH-sensitive molecules in the presence of simulated gastric fluid, and slowly released the drug in simulated intestinal fluid. In an *in vitro* study, Lutein was entrapped in NLC to improve its oral bioavailability via the enhancement of the drug's solubility in the GI tract. It was found that the ultrasonication duration and drug load significantly impacted the NLC size [89]. Finally, some works reported that the combination of surfactants produced more stable systems with high encapsulation efficiency and desired particle size. Hence, Glipizide (anti-diabetic agent) was entrapped into NLC prepared with three different types of surfactant (Poloxamer 188, Soya lecithin, Sodium taurocholate) and their mixtures. The formulation prepared by combination of the three mentioned surfactants resulted in particles with smallest mean size and narrow distribution. Also, NLC exhibited a biphasic release pattern of Glipizide with an initial burst release followed by a slow release profile for a long period of time [90].

14.2.4. *Inorganic and hybrid systems*

In the last decade, inorganic systems have provided a new platform to "use the nature to emulate the nature" with different biomedical applications. Silica compounds and zeolites, are the most widely used in the nano field. In addition, other metallic systems like gold (Au), silver (Ag), or magnetite minerals are increasing their importance, but toxicity concerns are now under discussion for oral delivery. Nanoparticles, nanotubes, nanomembranes, and nanoporous microparticles are some examples of different kinds of typical structures of these materials [91–94]. Of particular interest, is the combination of inorganic and organic compounds, such as biopolymers, to produce new and versatile drug delivery devices to enhance their potential applications.

14.2.4.1 *Silica*

14.2.4.1.1 Description

Porous silica materials refer to those with nanostructured pores or capillaries. Some properties, like large surface areas and pore volume, make them special for different applications. Their surface could be modified and there is a good capability to control pore size through the synthesis process. Also, they are classified as a biocompatible and inert material.

These particular properties help to resolve a problem in drug delivery systems for poorly water-soluble drugs. Thanks to high surface, the system reaches a high free energy state, which could be lowered with the adsorption of molecules; in this case, pharmacophores. Once they are located inside nanopores, adsorbed molecules remain in a non-crystalline state, called the amorphous state, which is physically unstable because of its high internal energy. However, when amorphous aggregates exist inside these structures, a stabilization process takes place [95]. Amorphous drug conformation is advantageous for drug delivery because there is an increment in solubility and dissolution rate, which leads to the conclusion of a much higher bioavailability. Molecules confined in porous matrices behave differently, since the change in its dynamics and thermodynamics play a role in system stability. For example, at –50°C (223.15 K), water confined in porous silicon dioxide (SiO_2) has mobility on an order of magnitude higher than water molecules in ice crystals at the same temperature. Furthermore, some mesoporous silica materials have been developed specifically for drug delivery. For example mobil composition of matter (MCM-41) is frequently used for slow drug release and was synthesized for the first time in 1992 with a quaternary ammonium surfactant, which served as a structure modifier (see "mesoporous silica synthesis"). Some years later in 1998, Santa Barbara Amorphous (SBA-15) was prepared and it employs amphiphilic poly(alkylene oxide) triblock copolymers as templates in order to direct pore size and achieve faster drug release kinetics [96].

As told, MCM-41 and SBA-15 were prepared with structure direction compounds; this leads us to talk about a relevant advantage of mesoporous silica materials, which is its synthesis flexibility. Since its synthesis is based on a sol–gel process, particle size can be controlled during the preparation including surfactants micelles or polymers, which act as templates for silica. After synthesis, the template is removed and the appearance of pores from 2 nm to several tens of nanometers can be observed. This capability allows for the control of drug kinetics because a decrease in pore size makes it more difficult for a drug to escape from the system [97]. It is also important to remember that, as mentioned before, nanopores lead drug molecules to be in a non-crystalline form modifying their behavior and allowing a drug super saturation state. Additionally, many modifications help to achieve new applications for mesoporous silica: as an example, amino and carboxyl groups can form bifunctional modifiers with biopolymers like PEG and chitosan, which can be attached to material surfaces [96, 98, 99].

14.2.4.1.2 Preparation

The sol– process is the traditional method used for mesoporous silica synthesis. This process involves a series of reactions where the starting material is usually an alkoxysilane (e.g., tetraethoxysilane (TEOS), $Si(OCH_2CH_3)_4$). Hydrolysis is the first step of the reaction with the formation of silanol groups (Si-OH), followed by condensation, and the subsequent formation of siloxane bridges (Si–O–Si). Once condensation reaches a critical size, particles aggregate and form a continuous sponge-like framework. The sol–gel reaction is influenced by temperature, pH, type of catalyst, solvent nature, and electrolytes. In addition, some silanol groups, called terminal silanol groups, always remain without forming bridges. They are responsible for drug–matrix interactions.

Moreover, there are three different methods for drug-loading into these matrices: the organic solvent immersion method, incipient wetness impregnation, or the melt method. The most used protocol for poorly aqueous soluble drugs is solvent immersion. There are two processes that take place during loading: first, immersing the mesoporous material into a concentrated drug solution in order to get the drug inside by capillary action, and second, a diffusion process to get adsorbed in pore walls. On the other hand, in the incipient wetness impregnation method, a very concentrated drug solution is used, close to the solubility limit. The volume of drug solution must be equal to the volume of the material pores. This method is used when small quantities of drug are available. It must be mentioned that the loading method affects the loading rate of a matrix and also the release kinetics; the more disordered structure that is obtained, the faster the release [98].

14.2.4.1.3 Applications

Naturally, these mesoporous materials are applied to overcome some particular problems of poorly water-soluble drugs, like paclitaxel. To name some of them: poor dissolution rate, slow intestinal absorption, and low bioavailability. In consequence, when a drug of this kind is administered, it cannot complete its clinical action mostly because of being eliminated from the GI tract before being fully absorbed. In addition to the previously described facts, these drugs have many negative side effects, which the patient has to suffer during treatment. The main cause is that drug dosage must be increased to

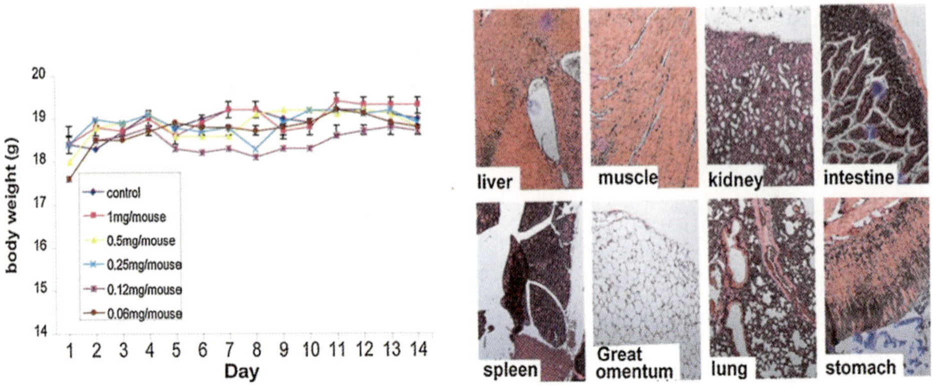

(A) (B)

Figure 14.12. Toxicity evaluation of mesoporous silica nanoparticles. 12 female mice i.v. injected with indicated doses of mesoporous silica nanoparticles, twice per week for 14 days: (A) Average body weights of mice. (B) Representative tissue sections of mice stained with hematoxylin and eosin (form [73] with permission).

toxic limits in order to reach therapeutic drug levels in the blood. In conclusion, mesoporous materials can improve drug dissolution and also the permeation/absorption rate, since modifications could be done to reach mucoadhesive properties [100–102] and no toxicity effects were observed after being tested in living organisms (Fig. 14.12).

14.2.4.2 *Zeolites and Nanoclays*

14.2.4.2.1 Description

14.2.4.2.1.1 *Zeolites*

The zeolites are a natural crystalline alumino–silicate material that is conformed with micropores, and present a regularity of architecture and composition, with different and unique characteristics related with structural properties like pore size, inter-void space, and cages dimensionality inside the structures (that permit the running of different channel systems through them). This material is composed by alkaline metal ions and water molecules. The tetrahedral three-dimensional (3D) structures of aluminate (AlO_4) and silicate (SiO_4) generates a system of channels and pores, where the AlO_4 is responsible for the negative charge in the crystalline structure and the presence of alkaline ions keeps a general neutral material [103].

Related to the composition, the properties of the extra cations (like sites, types, and numbers) can give extra properties to this material, useful in terms

Figure 14.13. The SEM images of the zeolite particles. (Insets) histograms representing statistical size distribution of the zeolite particles were generated from SEM image analysis for (A) BEA and (B) NaX-FAU zeolites particles (from [104] with permission).

of accommodating different kinds of molecules (like drugs) inside these channels, cages, and pores. In addition, the different dimensions of these cavities can control the release rate of the drug diffusion through them (Fig. 14.13) [104].

These materials can be synthesized too (only about 40 known zeolites have a natural origin, while around 150 were synthesized in the last time), like the case of beta zeolite, which has a smaller pore size than general mesoporous materials (0.7 nm versus 50 nm on average), that awards different properties, and can be used like nanocapsules for drug delivery systems. These systems appear like almost ideal materials to transport and release nanosized particles: large amounts of drugs (and macromolecules like enzymes) in different phases can be added and integrated in this kind of structures by different methods because of an ion-exchange mechanism [105].

14.2.4.2.1.2 *Nanoclays*

The clays have shown very advantageous properties in different forms in the industry arena, and now they are starting to participate in the biomedical field. There are many kinds of this material in nature showing different properties based on the quantity of water that this kind of geologic mineral is composed of. The size of these particles, to be considered like clays, normally have to be less than 2 microns, but it depends on the field of work. The "nanoclays" are starting to be studied and used for biomedical applications, such as montmorillonite, which was the first used for drug delivery systems [106]. For this system, the drug is loaded between the inter-sheets spaces by an ion-exchange mechanism, in which the stacked layer is not involved. With this loading procedure, the drug is bound ionically to the framework surface of the montomorillonite (a cationic nanoclay) made by sheets. In nature, these kinds of cationic nanoclays are very common, and the layers of aluminum and silicates are charged negatively, while anionic clays (like hydrocalcite) are not so common in nature, but can be synthesized to load anionic molecules.

Another example of nanoclays are the nanotubes of halloysite, rolled sheets of alumino–silicates to form tubes of 50 up to 80 nm of external diameter and only 10 to 15 nm of internal diameter, which can reach the length of microns. This material is formed in volcanic zones ($Al_2Si_2O_5(OH)_4 \times nH_2O$), similar to kaolinite, but with more water located in the free space of the adjacent layers. The aluminum is disposed to form the internal surface of the nanotubes, while the silicate groups are on the external surfaces (Fig. 14.14). The dispersability in polymers of these systems is much higher compared with the normal platy clays that need an exfoliation procedure [107].

These nanotubes present advantages too, in order to prolong the release time of molecules. The normal procedure of drug-loading is by vacuum cycles from drug solution, where the drug can be controlled by the number of cycles, solution concentration, release profiles by the halloysite nature, drug size, and solubility; all parameters related with the drug–halloysite ionic interactions. Moreover, by combining these nanotubes with a polymeric matrix, release profiles can be extended for weeks [106].

14.2.4.2.2 Applications

14.2.4.2.2.1 *Zeolites*

Different types of synthetic and natural zeolites were studied for gastrointestinal models of drug delivery to delay the release profile for a controlled therapy. Anticancer therapies are one target for these kinds of systems. The administration of the anticancer drug 5-fluorouracil, like a model molecule, was recently studied in different kinds of synthetic microporous zeolite particles (particle size and pore diameter were compared for BEA and FAU, two different zeolites: beta polymorph A, disordered framework zeolite in the first case, and a zeolite classified by the International Zeolite Association like a faujasite structure with a cubic space group crystallization in the second case) [104]. In the work, an improvement of the interaction between the matrix within the drug was found in the case of a more hydrophobic ambience with the BEA. The hypothesis is that in BEA, the van der Waals interactions are propitiated compared with electrostatic interactions. This difference in terms of interactions between the drug and the framework would be responsible for differences in the drug diffusion rates, where BEA systems show lower release profiles compared to FAU systems (Fig. 14.15).

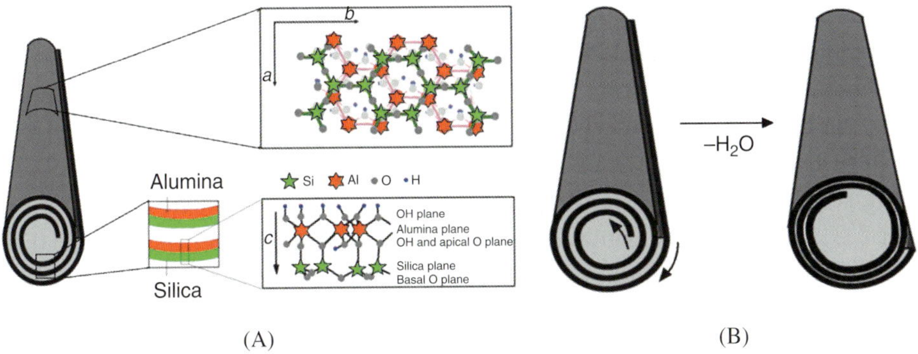

Figure 14.14. (A) Halloysite crystal structure; tubular morphology, side view and top view of the halloysite layer; a, b, c are crystalline direct ions. (B) Scheme of thehalloysite (from [107] with permission).

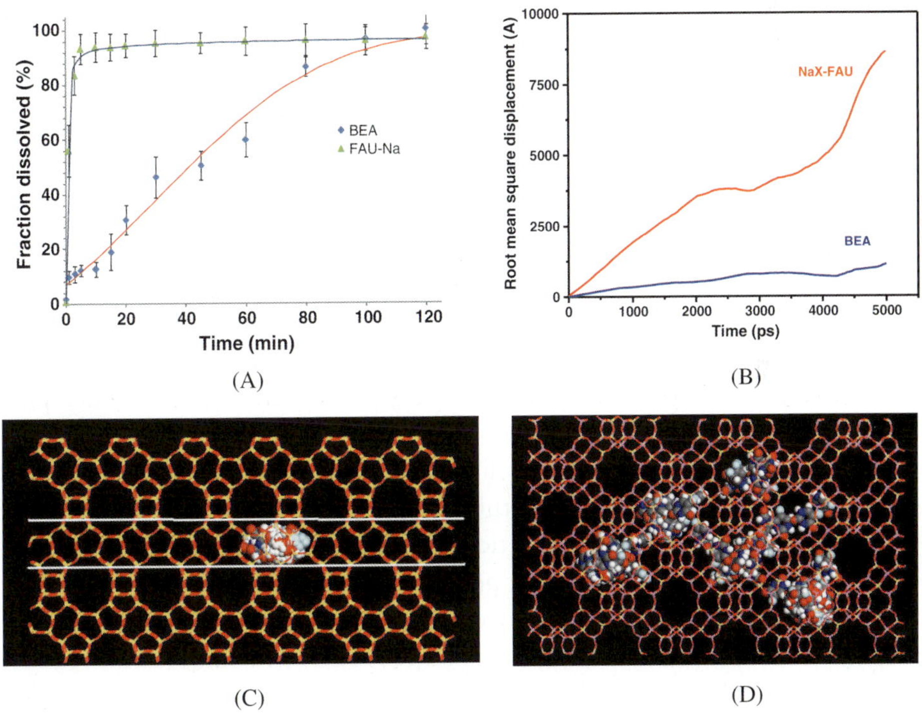

Figure 14.15. (A) Release profiles of 5-FU from BEA and NaX-FAU particles in 0.1 N HCl. (B) Root-mean displacement graphs for 5-FU inside the BEA framework and NaX-FAU. Diffusion of 5-FU through the channel system of (C) BEA and (D) NaX-FAU. The location of the drug molecule is shown every 10 ps during the first 1,000 ps of the simulations (from [104] with permission).

In another example, zeolite X and a mixture between zeolites A and X were studied for ketoprofen delivery. These two types of zeolites are chemically produced, and the mixture was made by physical procedures. By a simple soaking method, the ketoprofen was included in the matrix, with a charge yield of about 30% (w/w). Moreover, the release profiles showed less than 10% drug release at the gastric level and the bulk release begins when zeolites are at pH 5.0, showing the advantage of this material for GI delivery therapies [103].

Other common GI administration drugs were tested with zeolites matrices in different works, like ibuprofen. In a recent work, the load of ibuprofen in micro- and mesoporous zeolites showed similar results [106]. These results are indicating that the loading is not related to pore and crystal size, superficial area and volume of pores, as well as the content of aluminum in the composition of each zeolites tested. However, the release profile showed strict dependency on these parameters. Mesoporous zeolites showed a lower diffusion rate of ibuprofen, probably because of the van der Waals forces with the silanol functional groups in the host framework. At the same time, zeolites with a high content of aluminum showed a release of only 60% of the loaded ibuprofen after 72 h, which are evidencing the high interactions between ibuprofenate spices and aluminum. Meanwhile, the ibuprofen release was 100% from zeolites with low aluminum content in the first 24 h [105].

14.2.4.2.2.2 *Nanoclays*

Tetracycline–HCl was the first model drug for halloysite as a vehicle in the controlled release of molecules. This hydrophobic antibiotic and khellin like a model of hydrophobic vasodilator helped to study this material like a possible matrix in drug delivery processes, showing completely different release profiles. While the antibiotic was liberated after 10 hours, the vasodilator could be maintained in the nanotube lumen for a week. The admixing of these systems in polymers (synthetic polymers like epoxi, polyvinyl alcohol, polyacrylates and biopolymers like pectin, chitosan, and gelatin) for coatings with layer-by-layer processes improved the systems with a retard release profile, reduced the finals rates by about 75% (months) and eliminated the burst effect (fast initial release) [106]. In other works, halloysite lumen was loaded with macromolecules, like proteins and DNA [107].

14.2.4.3 *Silver nanoparticles*

14.2.4.3.1 Description

Among the metallic compounds, silver nanoparticles (AgNPs) are discussed in this section, due to their interesting antibacterial properties and potential uses in oral delivery. Silver nanoparticles are metal materials in the nanometer size, and their properties strongly depend on size and surface configuration.

Different methods of preparing silver nanoparticles were found. Physical approaches are based on evaporation–condensation methods, by the use of a tube furnace at atmospheric pressure. This procedure, however, is not suitable because with the large volume of the tube furnace, there is considerable energy consumption and a lot of time is required to reach thermal stability [108]. Chemical methods are the most common for AgNP preparation and involve the reduction of the silver salt ($AgNO_3$) with different reducing agents. The resulting AgNPs are obtained as colloidal dispersions in water or organic solvents. Their stability strictly depends on the use of protective agents that are absorbed or bound onto the nanoparticle surface, to avoid their agglomeration and prevent oxidation [108]. In recent years, new green biosynthetic methods have emerged as a simple and eco-friendly alternative to overcome the environmental toxicity associated with conventional methods to obtain AgNPs. Natural reducing agents, such as saccharides or microorganisms like bacteria, fungus, or plant extracts, are employed in the green synthesis strategy [109]. However, strict control of reaction conditions and the microorganism growth are required to obtain nanoparticles with defined and reproducible properties [110].

14.2.4.3.2 Applications

Metal nanoparticles have become a powerful device for the treatment of infections, and in particular, of multi-resistant pathogens. AgNPs are capable of inhibiting the microbial growth from the first contact between the bacteria and the nanoparticle [111]. No observable clinically relevant toxicity markers were found when human voluntaries were orally administered with AgNPs in a period of 14 days [112]. The fact could be explained considering the remarkable human cell resistance to the toxic effects of silver nanoparticles compared to other living organisms. In addition, silver nanoparticles exhibit a lower toxicity than silver ions and it decreases when silver is encapsulated in the form of nanoparticles [113]. However, high concentrations of AgNPs orally administrated become toxic in living organisms, which is the reason why their doses need to be carefully adjusted [114]. The coating, synthesis procedure, and nature of the AgNPs finally determine their toxicity [115].

Beside their toxicity, AgNPs were proposed as an alternative for oral treatment of severe intestinal infections, which were not able to be eradicated with conventional antibiotics [116]. Through an established mechanism, AgNPs are directly attached to bacterial cell walls, thus disturbing cell–wall permeability and cellular respiration [117]. It was demonstrated that after 12 hours of

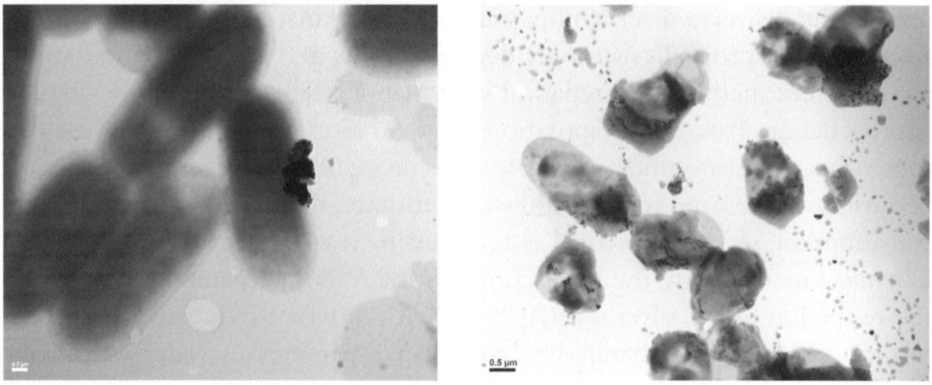

(A) (B)

Figure 14.16: Ag-NPs killing mechanism against *E.Coli.* (A) interaction and aggregation of Ag-NPs on cell wall; (B) bacterial lysis produced by membrane damage (**Islan**] unpublished).

incubation, AgNPs induced the membrane damage of opportunistic pathogens like *Pseudomonas aeruginosa* or *Escherichia coli* with subsequent bacterial death [116] (Fig. 14.16).

14.3 Conclusions

Nanotechnology has been growing as a new area for oral delivery of molecules with biological interest. In this sense, smart nanoparticles encapsulating drugs, provide an alternative to overcome side effects in the organism and poor biodisponibility.

The high surface area to volume ratio to improve drug adsorption, the protection against extreme gastric acidity and protease digestion to maintain drug stability, the bypassing of the first-pass metabolism to increase the drug biodisponibility in the target organ and the enhancement of penetration through gastro-intestinal mucosal are some advantages that smart nanosystems provide to improve drug dissolution and absorption in physiological conditions, with a reduction of side effects.

Besides, the toxicity of these systems is now under discussion, and the perspectives in the field of oral delivery in food and medicine are very promising.

Acknowledgments and Conflict of Interest Statement

The argentine grants from CONICET (National Council for Science and Technology, PIP 0214), The National Agency of Scientific and Technological Promotion (ANPCyT, PICT 2011–2116), Fundación Argentina de

Nanotecnología and UNLP (National University of La Plata, 11/X545 and PRH 5.2) are gratefully acknowledged.

The authors have no conflict of interest.

References

1. Lanas, A. (2010). A review of the gastrointestinal safety data: A gastroenterologist's perspective. *Rheumatology,* **48**, ii3–ii10.
2. Patterson, D. R. (1991). Quinolone toxicity: Methods of assessment. *Am. J. Med.*, **91**, 35S.
3. Talevi, A., Blanch, B. L., Castro, G. R. and Rosen, Y. (2014). 'An Introduction to Pharmacokinetics: From Conventional to Advanced Systemic Drug Delivery Systems', in *Advancing Drug Delivery: An Integrated Clinical and Engineering Approach*, CRC Press., Miami, FL, pp. 1–51.
4. Ren, J., Shen, S., Wang, D., Xi, Z., Guo, L., Pang, Z., Qian, Y., Sun, X. and Jiang, X. (2012). The targeted delivery of anticancer drugs to brain glioma by PEGylated oxidized multi-walled carbon nanotubes modified with angiopep-2. *Biomaterials,* **33**, 3324–3333.
5. Yun, Y., Cho, Y. W. and Park, K. (2013). Nanoparticles for oral delivery: Targeted nanoparticles with peptidic ligands for oral protein delivery. *Adv. Drug Deliv. Rev.,* **65**, 822–832.
6. Marcucci, F. and Corti, A. (2012). How to improve exposure of tumor cells to drugs — promoter drugs increase tumor uptake and penetration of effector drugs. *Adv. Drug Deliv. Rev.,* **64**, 53–68.
7. Narang, A. S. and Varia, S. (2011). Role of tumor vascular architecture in drug delivery. *Adv. Drug Deliv. Rev.,* **63**, 640–658.
8. Patel, A. P., Patel, J. K., Patel, K. S., Deshmukh, A. B. and Mishra, B. (2011). A review on drug nanocrystal a carrier free drug delivery. *IJRAP*, **2**, 448–458.
9. Deng, J., Huang, L. and Liu, F. (2010). Understanding the Structure and Stability of Paclitaxel Nanocrystals. *Int. J. Pharm.,* **390**, 242–249.
10. Niwa, T., Miura, S. and Danjo, K. (2011). Universal wet-milling technique to prepare oral nanosuspension focused on discovery and preclinical animal studies — Development of particle design method. *Int. J. Pharm.*, **405**, 218–227.
11. Junghanns, J. U. A. H. and Müller, R. H. (2008). Nanocrystal technology, drug delivery and clinical applications. *Int. J. Nanomed.,* **3**, 295–309.
12. Gao, L., Liu, G., Ma, J., Wang, X., Zhou, L., Li X. and Wang, F. (2013). Application of drug nanocrystal technologies on oral drug delivery of poorly soluble drugs. *Pharm. Res.,* **2**, 307–324.
13. Vergote, G. J., Vervaet, C., Driessche, I., Hoste, S., Smedt, S., Jain, R. A., Demeester, J., Ruddy S. and Remon, J. P. (2001). An oral controlled release matrix pellet formulation containing nanocrystalline ketoprofen. *Int. J. Pharm.,* **219**, 81–87.
14. Kayser, O., Olbrich, C., Yardley, V., Kiderlen, A. F. and Croft, S. L. (2003). Formulation of amphotericin B as nanosuspension for oral administration. *Int. J. Pharm.*, **254**, 73–75.
15. Hanafy, A., Spahn-Langguth, H., Vergnault, G., Grenier, P., Grozdanis, M. T. and Lenhardt, T. (2007). Pharmacokinetic evaluation of oral fenofibrate nanosuspensions and SLN in comparison to conventional suspensions of micronized drug. *Adv. Drug Deliv. Rev.,* **59**, 419–426.

16. Lu, J., Liong, M., Li, Z., Zink, J. I. and Tamanoi, F. (2013). Biocompatibility, biodistribution, and drug-delivery efficiency of mesoporous silica nanoparticles for cancer therapy in animals. *Small,* **6**, 1794–1805.
17. Gao, L., Liu, G., Ma, J., Wang, X., Zhou, L. and Li, X. (2012). Drug nanocrystals: In vivo performances. *J. Control. Release,* **160**, 418–430.
18. Sheldon, R. A. (2007). Cross-linked enzyme aggregates (CLEA®s): Stable and recyclable Biocatalysts. *Biochem. Soc. Trans.,* **3**, 1583–1587.
19. Schoevaart, R., Wolbers, M., Golubovic, M., Ottens, M., Kieboom, A., Van Rantwijk, F., Van der Wielen, L. and Sheldon, R. (2004). Preparation, optimization, and structures of cross-linked enzyme aggregates (CLEAs). *Biotechnol. Bioeng.*, **87**, 754–762.
20. Wilson, L., Fernández-Lorente, G., Fernández-Lafuente, R., Illanes, A., Guisán, J. M. and Palomo, J. M. (2006). CLEAs of lipases and poly-ionic polymers: A simple way of preparing stable biocatalysts with improved properties. *Enzyme Microb. Technol.*, **39**, 750–755.
21. Ribeiro, M. H. L. and Rabaça, M. (2011). Cross-linked enzyme aggregates of naringinase: Novel biocatalysts for naringin hydrolysis. *Enz. Res.*, **2011**, 1–8.
22. Sangeetha, K. and Abraham, T. E. (2008). Preparation and characterization of cross-linked enzyme aggregates (CLEA) of subtilisin for controlled release applications. *Int. J. Biol. Macromol.*, **43**, 314–319.
23. Islan, G. A., Martinez, Y. N., Illanes, A. and Castro, G. R. (2014). Development of novel alginate lyase cross-linked aggregates for the oral treatment of cystic fibrosis. *RSC Adv.,* **4**, 11758–11765.
24. Li, S. D. and Huang, L. (2008). Pharmacokinetics and biodistribution of nanoparticles. *Mol. Pharm.,* **5**, 496–504.
25. Thanki, K., Gangwal, R. P., Sangamwar, A. T. and Jain, S. (2013). Oral delivery of anticancer drugs: Challenges and opportunities. *J. Control. Release,* **170**, 15–40.
26. Collnot, E. M., Ali, H. and Lehr, C. M. (2012). Nano- and microparticulate drug carriers for targeting of the inflamed intestinal mucosa. *J. Control. Rel.,* **161**, 235–246.
27. Werle, M. (2008). Natural and Synthetic Polymers as Inhibitors of Drug Efflux Pumps. *Pharm. Res.,* **25**, 500–511.
28. Torchilin, V. P. (2012). Multifunctional nanocarriers. *Adv. Drug Deliv. Rev.,* **64**, 302–315.
29. Shen, J., Sun, H., Xu, P., Yin, Q., Zhang, Z., Wang, S., Yu, H. and Li, Y. (2013). Simultaneous inhibition of metastasis and growth of breast cancer by co-delivery of twist shRNA and paclitaxel using pluronic P85-PEI/TPGS complex nanoparticles. *Biomaterials,* **34**, 1581–1590.
30. Karaman, D. Ş., Gulin-Sarfraz, T., Hedström, G., Duchanoy, A., Eklund, P. and Rosenholm, J. M. (2014). Rational evaluation of the utilization of PEG-PEI copolymers for the facilitation of silica nanoparticulate systems in biomedical applications. *J. Colloid Interf.,* **418**, 300–310.
31. Hunter, A. C., Elsomb, J., Wibroec, P. P. and Moghimic, S. M. (2012). Polymeric particulate technologies for oral drug delivery and targeting: A pathophysiological perspective. *Maturitas,* **73**, 5–18.
32. Islan, G. A., de Verti, I. P., Marchetti, S. G. and Castro, G. R. (2012). Studies of ciprofloxacin encapsulation on alginate/pectin matrixes and its relationship with biodisponibility. *Appl. Biochem. Biotech.* **167**, 1408–1420.
33. Lamprecht, A., Yamamoto, H., Takeuchi, H. and Kawashima, Y. (2005). A pH-sensitive microsphere system for the colon delivery of tacrolimus containing nanoparticles. *J. Control. Release,* **104**, 337–346.

34. Lai, S. K., Wang, Y. Y. and Hanes, J. (2009). Mucus-penetrating nanoparticles for drug and gene delivery to mucosal tissues. *Adv. Drug Deliv. Rev.,* **61**, 158–171.
35. Jokerst J. V., Lobovkina, T., Zare, R. N. and Gambhir, S. S. (2011). Nanoparticle PEGylation for imaging and therapy. *Nanomedicine (Lond.),* **6**, 715–728.
36. Knop, K., Hoogenboom, R., Fischer, D. and Schubert, U. S. (2010). Poly(ethylene glycol) in drug delivery: Pros and cons as well as potential alternatives. *Angew. Chem. Int. Ed. Engl.,* **49**, 6288–6308.
37. Jain, S., Rathi, V. V., Jain, A. K., Das, M. and Godugu, C. (2012). Folate-decorated PLGA nanoparticles as a rationally designed vehicle for the oral delivery of insulin. *Nanomedicine (Lond.),* 7, 1311–1337.
38. Mishra, N., Tiwari, S., Vaidya, B., Agrawal, G. P. and Vyas, S. P. (2011). Lectin anchored PLGA nanoparticles for oral mucosal immunization against hepatitis B. *J. Drug Target.,* **19**, 67–78.
39. Plapied, L., Duhem, N., des Rieux, A. and Préat, V. (2011). Fate of polymeric nanocarriers for oral drug delivery. *Curr. Opin. Colloid Interf. Sci.,* **16**, 228–237.
40. Zhang, F., Huang, X., Zhu, L., Guo, N., Niu, G., Swierczewska, M., Lee S., Xu, H., Wang, A. Y., Mohamedali, K. A., Rosenblum, M. G., Lu, G. and Chen, X. (2012). Noninvasive monitoring of orthotopic glioblastoma therapy response using RGD-conjugated iron oxide nanoparticles. *Biomaterials,* **33**, 5414–5422.
41. Yuan, H., Wang, L. L., Du, Y. Z., You, J., Hu, F. Q. and Zeng, S. (2007). Preparation and characteristics of nanostructured lipid carriers for control-releasing progesterone by melt-emulsification. *Colloids Surf. B,* **60**, 174–179.
42. Des Rieux, A., Pourcelle, V., Cani, P. D., Marchand-Brynaert, J. and Préat, V. (2013). Targeted nanoparticles with novel non-peptidic ligands for oral delivery. *Adv. Drug Deliv. Rev.,* **65**, 833–844.
43. Gaucher, G., Satturwar, P., Jones, M. C., Furtos, A. and Leroux, J. C. (2010). Polymeric micelles for oral drug delivery. *Eur. J. Pharm. Biopharm.,* **76**, 147–158.
44. Lee, S. C., Huh, K. M., Lee, J. H., Cho, Y. W., Galinsky, R. E. and Park, K. (2007). Hydrotropic polymeric micelles for enhanced paclitaxel solubility: In vitro and in vivo characterization. *Biomolecules,* **8**, 202–208.
45. Al-Hilal T. A., Alam F. and Byun Y. (2012). Oral drug delivery systems using chemical conjugates or physical complexes. *Adv. Drug Deliv. Rev.,* **65**, 845–864.
46. Clausen A. E., Kast C. E. and Bernkop-Schnurch A. (2002). The Role of Glutathione in the Permeation Enhancing Effect of Thiolated Polymers. *Pharm. Res.,* **19**, 602–608.
47. Zhang, Z., Gao, F., Jiang, S., Ma, L. and Li, Y. (2012). Nano-based drug delivery system enhances the oral absorption of lipophilic drugs with extensive presystemic metabolism. *Curr. Drug Metab.,* **13**, 1110–1118.
48. Puri, A., Loomis, K., Smith, B., Lee, J. H., Yavlovich, A., Heldman, E. and Blumenthal, R. (2009). Lipid-based nanoparticles as pharmaceutical drug carriers: From concepts to clinic. *Crit. Rev. Ther. Drug Carrier Syst.,* **26**, 523–580.
49. Martins, S., Sarmento, B., Ferreira, D. C. and Souto, E. B. (2007). Lipid-based colloidal carriers for peptide and protein delivery — liposomes versus lipid nanoparticles. *Int. J. Nanomed.,* **2**, 595–607.
50. Anabousi, S., Laue, M., Lehr, C. M., Bakowsky, U. and Carsten Ehrhardt, C. (2005). Assessing transferrin modification of liposomes by atomic force microscopy and transmission electron microscopy. *Eur. J. Pharm. Biopharm.,* **60**, 295–303.
51. Giddi, H. S., Arunagirinathan, M. A. and Bellare, J. R. (2007). Self-assembled surfactant nano-structures important in drug delivery: a review. *Indian J. Exp. Biol.,* **45**, 133–159.

52. Dua, J. S., Rana, A. C. and Bhandari, A. K. (2012). Liposome: Methods of preparation and applications. *Int. J. Pharm. Stud. Res.*, **3**, 14–20.
53. Chiang, C. M. and Weiner, N. (1987). Gastrointestinal uptake of liposomes II. In vivo studies. *Int. J. Pharm.*, **40**, 143–150.
54. Wu, J., Liu, Q. and Lee, R. J. (2006). A folate receptor-targeted liposomal formulation for paclitaxel. *Int. J. Pharm.*, **316**, 148–153.
55. Moutardier, V., Tosini, F., Vlieghe, P., Cara, L., Delpero, J. R. and Clerc, T. (2003). Colloidal anticancer drugs bioavailabilities in oral administration models. *Int. J. Pharm.*, **260**, 23–38.
56. Iwanaga, K., Ono, S., Narioka, K., Morimoto, K., Kakemi, M., Mamoru, S. Y. and Oku, N. (1997). Oral delivery of insulin by using surface coating liposomes: Improvement of stability of insulin in GI tract. *Int. J. Pharm.*, **157**, 73–80.
57. Beija, M., Salvayre, R., Lauth-de Viguerie, N. and Marty, J. D. (2012). Colloidal systems for drug delivery: From design to therapy. *Trends Biotechnol.*, **30**, 485–496.
58. Moen, M. D., Lyseng-Williamson, K. A. and Scott, L. J. (2009). Liposomal amphotericin B: A review of its use as empirical therapy in febrile neutropenia and in the treatment of invasive fungal infections. *Drugs,* **69**, 361–392.
59. Krishna, R., McIntosh, N., Riggs, K. W. and Mayer, L. D. (1999). Doxorubicin encapsulated in sterically stabilized liposomes exhibits renal and biliary clearance properties that are independent of valspodar (PSC 833) under conditions that significantly inhibit nonencapsulated drug excretion. *Clin. Cancer Res.,* **5**, 2939–2947.
60. Ceruti, M., Crosasso, P., Brusa, P., Arpicco, S., Dosio, F. and Cattel, L. (2000). Preparation, characterization, cytotoxicity and pharmacokinetics of liposomes containing water-soluble prodrugs of paclitaxel. *J. Control. Release,* **63**, 141–153.
61. Soehngen, E. C. Godin-Ostro, E., Fielder, F. G., Ginsberg, R. S., Slusher, M. A. and Weiner, A. L. (1988). Encapsulation of indomethacin in liposomes provides protection against both gastric and intestinal ulceration when orally administered to rats. *Arthritis Rheum.*, **31**, 414–422.
62. Bakker-Woudenberg, A. J. M., Lokerse, A. F., Vinkvan den Berg, J. C. and Roerdink, F. H. (1988). Liposome-encapsulated ampicillin against Listeria monocytogenes *in vivo* and *in vitro. Infection,* **16**, 165–168.
63. Onyeji, C. O., Nightingale, C. G. and Marangos, M. N. (1994). Enhanced killing of methicillin-resistant Staphylococcus aureus in human macrophages by liposome-entrapped vancomycin and teicoplanin. *Infection,* **22**, 338–342.
64. Muller, R. H., Ruhl, D., Runge, S., Schulze-Forster, K. and Wolfgang, M. (1997). Cytotoxicity of solid lipid nanoparticles as a function of the lipid matrix and the surfactant. *Pharm. Res.,* **14**, 458–462.
65. Roger, E., Lagarce, F., Garcion, E. and Benoit, J. P. (2010). Biopharmaceutical parameters to consider in order to alter the fate of nanocarriers after oral delivery. *Nanomedicine,* **5**, 287–306.
66. Mei, L., Zhang, Z., Zhao, L., Huang, L., Yang, X. L., Tang, J. and Feng, S. S. (2013). Pharmaceutical nanotechnology for oral delivery of anticancer drugs. *Adv. Drug Deliv. Rev.,* **65**, 880–890.
67. Wissing, S. A., Kayzer, O. and Müller, R. H. (2004). Solid lipid nanoparticles for parenteral drug delivery. *Adv. Drug. Deliv. Rev.,* **56**, 1257–1272.
68. Das, S. and Chaudhury, A. (2011). Recent advances in lipid nanoparticle formulations with solid matrix for oral drug delivery. *AAPS Pharm. Sci. Tech.,* **12**, 62–76.

69. Peltier, S., Oger, J. M., Lagarce, F., Couet, W. and Benoit, J. P. (2006). Enhanced oral paclitaxel bioavailability after administration of paclitaxel-loaded lipid nanocapsules. *Pharm. Res.*, **23**, 1243–1250.
70. Roger, E., Lagarce, F., Garcion, E. and Benoit, J. P. (2010). Reciprocal competition between lipid nanocapsules and P-gp for paclitaxel transport across Caco-2 cells. *Eur. J. Pharm. Sci.*, **40**, 422–429.
71. Muller, R. H., Runge, S., Ravelli, V., Mehnert, W., Thunemann, A. F. and Souto, E. B. (2006). Oral bioavailability of cyclosporine: Solid lipid nanoparticles (SLN®) versus drug nanocrystals. *Int. J. Pharm.*, **317**, 82–89.
72. Tsai, M. J., Huang, Y. B., Wu, P. C., Fu, Y. S., Kao, Y. R., Fang, J. Y. and Tsai, Y. H. (2011). Oral apomorphine delivery from solid lipid nanoparticleswith different monostearate emulsifiers: pharmacokinetic and behavioral evaluations. *J. Pharm. Sci.*, **100**, 547–557.
73. Liu, D., Liu, C., Zou, W. and Zhang, N. (2010). Enhanced gastrointestinal absorption of N3-O-toluyl-fluorouracil by cationic solid lipid nanoparticles. *J. Nanopart. Res.*, **12**, 975–984.
74. Zhang, Z., Lv, H. and Zhou, J. (2009). Novel solid lipid nanoparticles as carriers for oral administration of insulin. *Pharmazie*, **64**, 574–578.
75. Varshosaz, J., Minayian, M., and Moazen, E. (2010). Enhancement of oral bioavailability of pentoxifylline by solid lipid nanoparticles. *J. Liposome Res.*, **20**, 115–123.
76. Varshosaz, J., Tabbakhian, M. and Mohammadi, M. Y. (2010). Formulation and optimization of solid lipid nanoparticles of buspirone HCl for enhancement of its oral bioavailability. *J. Liposome Res.*, **20**, 286–296.
77. Pandey, R., Sharma, S. and Khuller, G. K. (2005). Oral solid lipid nanoparticle-based antitubercular chemotherapy. *Tuberculosis*, **85**, 415–420.
78. Sanjula, B., Shah, F. M., Javed, A. and Alka, A. (2009). Effect of poloxamer 188 on lymphatic uptake of carvedilol-loaded solid lipid nanoparticles for bioavailability enhancement. *Drug Target*, **17**, 249–256.
79. Manjunath, K. and Venkateswarlu, V. (2005). Pharmacokinetics, tissue distribution and bioavailability of clozapine solid lipid nanoparticles after intravenous and intraduodenal administration. *J. Control. Release*, **107**, 215–228.
80. Hu, L., Jia, H., Luo, Z., Liu, C. and Xing, Q. (2010). Improvement of digoxin oral absorption in rabbits by incorporation into solid lipid nanoparticles. *Pharmazie*, **65**, 110–113.
81. Paliwal, R., Rai, S., Vaidya, B., Khatri, K., Goyal, A. K., Mishra, N., Mehta, A. and Vyas, S. P. (2009). Effect of lipid core material on characteristics of solid lipid nanoparticles designed for oral lymphatic delivery. *Nanomedicine*, **5**, 184–191.
82. Kushwaha, A. K., Vuddanda, P. R., Karunanidhi, P., Singh, S. K. and Singh, S. (2013). Development and Evaluation of Solid Lipid Nanoparticles of Raloxifene Hydrochloride for Enhanced Bioavailability. *BioMed. Res. Int.*, **2013**, 1–9.
83. Martins, S., Silva, A. C., Ferreira, D. C. and Souto, E. B. (2009). Improving oral absorption of samon calcitonin by trimyristin lipid nanoparticles. *J. Biomed. Nanotechnol.*, **5**, 76–83.
84. Iqbal, M. A., Md, S., Sahni, J. K., Baboota, S., Dang, S. and and Ali, J. (2012). Nanostructured lipid carriers system: Recent advances in drug delivery. *J. Drug Target.*, **20**, 813–830.
85. Zhuang, C. Y., Li, N., Wang, M., Zhang, X. N., Pan, W. S., Peng, J. J., Pan, Y. S. and Tang, X. (2010). Preparation and characterization of vinpocetine loaded nanostructured lipid carriers (NLC) for improved oral bioavailability. *Int. J. Pharm.*, **394**, 179–185.

86. Thatipamula, R., Palem, C., Gannu, R., Mudragada, S., and Yamsani, M. (2011). Formulation and in vitro characterization of domperidone loaded solid lipid nanoparticles and nanostructured lipid carriers. *Daru.*, **19**, 23–32.
87. Kumbhar, D. D. and Pokharkar, V. B. (2013). Engineering of a nanostructured lipid carrier for the poorly water-soluble drug, bicalutamide: Physicochemical investigations. *Colloids Surf. A.*, **416**, 32–42.
88. Chen, C. C., Tsai, T. H., Huang, Z. R. and Fang, J. Y. (2010). Effects of lipophilic emulsifiers on the oral administration of lovastatin from nanostructured lipid carriers: Physicochemical characterization and pharmacokinetics. *Eur. J. Pharm. Biopharm.*, **74**, 474–482.
89. Liu, C. H. and Wu, C. T. (2010). Optimization of nanostructured lipid carriers for lutein delivery. *Colloids Surf. B.*, **353**, 149–156.
90. Jain, N. K. and Ram, A. (2011). Development and characterization of nanostructured lipid carriers of oral hypoglycemic agent: Selection of surfactants. *Int. J. Pharm. Sci. Rev. Res.*, 7, 125–130.
91. Saha, D., Warren, K. E. and Naskar, A. K. (2014). Soft-templated mesoporous carbons as potential materials for oral drug delivery. *Carbon,* **71**, 47–57.
92. Seabra, A. and Durán, N. (2013). Biological applications of peptides nanotubes: An overview. *Peptides,* **39**, 47–54.
93. Allaker, R. P. and Memarzadeh, K. (2014). Nanoparticles and the control of oral infections. *Int. J. Antimicrob. Ag.*, **43**, 95–104.
94. Bosio, V. E., Cacicedo, M., Calvignac, B., León, I., Beuvier, T., Boury, F. and Castro, G. R. (2014). Synthesis and characterization of $CaCO_3$-biopolymer hybrid nanoporous microparticles for controlled release of doxorubicin. *Colloids Surf. B*, in press. doi:10.1016/j.colsurfb.2014.09.011.
95. Qian K. K. and Bogner R. H. (2012). Application of mesoporous silicon dioxide and silicate in oral amorphous drug delivery systems. *J. Pharm. Sci.*, **101**, 444–463.
96. Van Speybroeck, M., Mellaerts, R., Martens, J. A., Annaert, P., Van den Mooter, G., and Augustijns, P. (2011). 'Ordered mesoporous silica for the delivery of poorly soluble drugs', in Wilson, C. G. and Crowley, P. J. (eds.), *Controlled Release in Oral Drug Delivery, in Advances in Delivery Science and Technology,* Springer, New York, pp. 203–219.
97. Van Speybroeck, M., Mols, R., Mellaerts, R., Thi, T. D., Martens, J. A., Van Humbeeck, J., Annaert, P., Van den Mooter, G. and Augustijns, P. (2010). Enhanced absorption of the poorly soluble drug fenofibrate by tuning its release rate from ordered mesoporous silica. *Eur. J. Pharm. Sci.*, **41**, 623–630.
98. Xu, W., Riikonen, J. and Lehto, V. P. (2013). Mesoporous systems for poorly soluble drugs. *Int. J. Pharm.,* **453**, 181–197.
99. He, Q., Zhang, J., Shi, J., Zhu, Z., Zhang, L., Bu, W., Guo, L. and Chen, Y. (2010). The effect of PEGylation of mesoporous silica nanoparticles on nonspecific binding of serum proteins and cellular responses. *Biomaterials,* **31**, 1085–1092.
100. Bikiaris, D. N. (2011). Solid dispersions, part I: Recent evolutions and future opportunities in manufacturing methods for dissolution rate enhancement of poorly water-soluble drugs. *Expert Opin. Drug Deliv.*, **8**, 1501–1519.
101. Kawabata, Y., Wada, K., Nakatani, M., Yamada, S. and Onoue, S. (2011). Formulation design for poorly water-soluble drugs based on biopharmaceutics classification system: Basic approaches and practical applications. *Int. J. Pharm.*, **420**, 1–10.

102. Kinnari, P., Makila, E., Heikkila, T., Salonen, J., Hirvonen, J. and Santos, H. A. (2011). Comparison of mesoporous silicon and non-ordered mesoporous silica materials as drug carriers for itraconazole. *Int. J. Pharm.,* **414**, 148–156.
103. Rimoli, M. G., Rabaioli, M. R., Melisi, D., Curcio, A., Mondello, S., Mirabelli, R. and Abignente, E. (2008). Synthetic zeolites as a new tool for drug delivery. *J. Biomed. Mater. Res. A.,* **87**, 156–164.
104. Spanakis, M., Bouropoulos, N., Theodoropoulos, D., Sygellou, L., Ewart, S., Moschovi, A. M., Siokou, A., Niopas, I., Kachrimanis, K., Nikolakis, V., Cox, P. A., Vizirianakis, I. S. and Fatouros, D. G. (2014). Controlled release of 5-fluorouracil from microporous zeolites. *Nanomedicine: Nanotechnol. Biol. & Med.,* **10**, 197–205.
105. Gonzalez, G., Sagarzazu, A. and Zoltan, T. (2013). Influence of Microstructure in Drug Release Behavior of Silica Nanocapsules. *J. Drug Deliv.,* **2013**, 1–8.
106. Ward, C. J., DeWitt, M. and Davis, E. W. (2012). Halloysite Nanoclay for Controlled Release Applications. *Nanomaterials for Biomedicine,* **10**, 209–238.
107. Lvov, Y. and Abdullayev, E. (2013). Functional polymer–clay nanotube composites with sustained release of chemical agents. *Progr. Polym. Sci.,* **38**, 1690–1719.
108. El-Nour, K. M. M. A., Eftaiha, A., Al-Warthan, A. and Ammar, R. A. A. (2010). Synthesis and applications of silver nanoparticles. *Arab. J. Chem.,* **3**, 135–140.
109. Sharma, V. K., Yngard, R. A. and Lin, Y. (2009). Silver nanoparticles: Green synthesis and their antimicrobial activities. *Adv. Colloid & Interface Sci.,* **145**, 83–96.
110. Cauerhff, A. and Castro, G. R. (2013). Bionanoparticles, a green nanochemistry approach. *Electron. J. Biotechnol.,* **16**, 1–10.
111. Durán, N., Marcato, P. D., De Conti, R., Alves, O. L., Costa, F. T. M. and Brocchi, M. (2010). Potential use of silver nanoparticles on pathogenic bacteria, their toxicity and possible mechanisms of action. *J. Braz. Chem. Soc.,* **21**, 949–959.
112. Munger, M. A., Radwanski, P., Hadlock, G. C., Stoddard, G., Shaaban, A., Falconer, J., Grainger, D. W. and Deering-Rice, C. E. (2014). *In vivo* human time-exposure study of orally dosed commercial silver nanoparticles. *Nanomedicine: Nanotechnol, Biol. Med.,* **10**, 1–9.
113. De Lima, R., Seabra, A. B. and Durán, N. (2012). Silver nanoparticles: A brief review of cytotoxicity and genotoxicity of chemically and biogenically synthesized nanoparticles. *J. Appl. Toxicol.,* **32**, 867–879.
114. Kim, Y. S., Song, M. Y., Park, J. D., Song, K. S., Ryu, H. R., Chung, Y. H., Chang, H. K., Lee, J. H., Oh, K. H., Kelman, B. J., Hwang, I. K. and Yu, I. J. (2010). Subchronic oral toxicity of silver nanoparticles. *Particle and Fibre Toxicol.,* 7, 20-30.
115. Tejamaya, M., Römer, I., Merrifield, R. C. and Lead, J. R. (2012). Stability of citrate, PVP, and PEG coated silver nanoparticles in ecotoxicology media. *Environm. Scie. Technol.,* **46**, 7011–7017.
116. Islan, G. A., Mukherjee, A. and Castro, G. R. (2015). Development of biopolymer nanocomposite for silver nanoparticles and Ciprofloxacin controlled release. *Int. J. Biol. Macromol.,* **72**, 740–750.
117. Hayden, S. C., Zhao, G., Saha, K., Phillips, R. L., Li, X., Miranda, O. R., Rotello, V. M., El-Sayed, M. A., Schmidt-Krey, I. and Bunz, U. H. (2012). Aggregation and interaction of cationic nanoparticles on bacterial surfaces. *J. Am. Chem. Soc.,* **134**, 6920–6923.

Chapter 15

Smart Theranostic Nanosystems

Michael Li, Ratul Mukerji, and Kam W. Leong

*Department of Biomedical Engineering,
Duke University, Durham, NC, USA*

15.1 Introduction

Nanomedicine in the form of nanoparticles continues to impact healthcare with many benefits realized through therapeutic and imaging modalities. The potential to specifically target cells in therapy or to provide early diagnoses makes nanomedicine an attractive solution to many of today's health concerns, with many applications seen in cancer treatment. Numerous designs of imaging and therapeutic nanoparticles differing in surface chemistry, size, shape, and composition has been proposed and evaluated. The coupling of therapeutic and imaging (theranostic) capabilities for a nanosystem provides a unique platform and opportunity for which researchers can deliver a multifunctional tool that can respond to certain environmental cues, described as "smart" systems. This review details the progress in the development of "smart", theranostic nanosystems, organized by their environmental cues, including pH, light, and magnetism. It highlights the opportunities and challenges when designing and applying this technology and discusses what implications they may hold for the field of nanomedicine in the future.

15.2 pH-Sensitive Theranostic Nanosystems

Many smart nanosystems target different pH levels in tissues to achieve specificity. Nanosystems with sensitivity to pH levels are particularly useful for and most widely seen in cancer diagnosis and therapy. The reason for this is a significant decrease in extracellular and intracellular pH in tumor cells, with pH reaching levels as low as 5.7, compared to a normal tissue pH of 7.4. Acidic tumor microenvironments arise from the altered tumor vasculature causing insufficient delivery of metabolites, such as oxygen, to many areas within the tumor. This causes an increase in production of lactic acid under anaerobic conditions and hydrolysis of adenosine triphosphate (ATP), contributing to the increased acidity in tumor microenvironments [1]. Furthermore, intracellular vesicles (where drugs and drug carriers are internalized) in tumors are acidified, reaching pH levels as low as 4.5 [2]. This inherent difference in pH between healthy tissue and tumor tissue provides a way to differentially target therapy to tumor tissue, where recent pH-sensitive theranostics have shown immense promise.

A great deal of diversity is seen in pH-sensitive nanosystems. The basic design usually entails a polymeric backbone of the nanoparticle or micelle system that is biodegradable, nontoxic, and biocompatible. Polymeric carriers, such as chitosan and poly(lactic acid) (PLA), are used to encapsulate or bind a therapeutic agent and an imaging agent [1, 3]. Therapeutic agents, specific to cancer treatment, include doxorubicin (Dox), paclitaxel, and sirolimus, all effective anti-cancer drugs. Imaging agents, most commonly relying on magnetic resonance imaging (MRI), include iron oxide or manganese components in the core of the nanoparticles. Alternatively, gold nanoparticles (AuNPs) are used as the core of many pH-sensitive theranostic systems. AuNPs can be functionalized with a variety of different therapeutic surface groups via acid-sensitive linkers, and have inherent fluorescent imaging properties due to their absorbance and scattering over several ranges in the wavelength spectrum depending on size and shape.

15.2.1 *Mechanisms for pH sensitivity*

There are a handful of different and innovative mechanisms for responsiveness to pH seen in recent smart theranostic systems. The first common mechanism, especially in cancer treatment involving Dox, is the disruption of the electrostatic interactions between positively charged Dox and the negatively charged groups on the polymeric backbone (Fig. 15.1). At physiological and near-neutral pH, the electrostatic interactions remain strong, keeping the Dox from being released. This often entails the degradation of the polymeric backbone as well, to further stimulate drug release by a matrix-degradation release mechanism. Jing *et al.* reported this design in their study with PLA nanoparticles conjugated with Dox and manganese-porphyrin. In acidic pH, PLA degraded faster, the re-protonated Dox became hydrophilic, and the uncharged carboxylic acid weakened the complexation. All conspired to promote Dox release [4]. Guo *et al.* utilized the same mechanism for

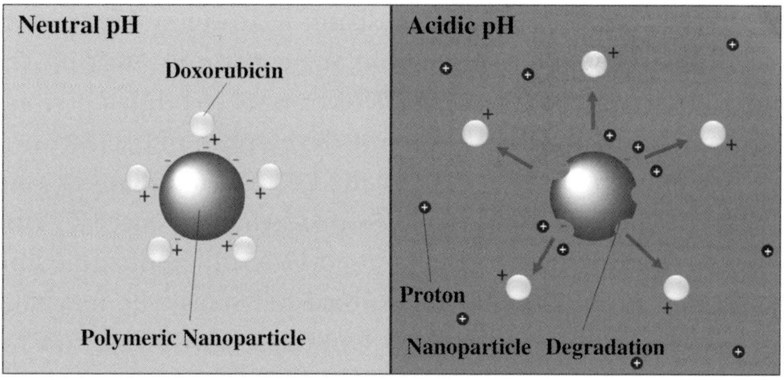

Figure 15.1. Schematic diagram of protonation mechanism of pH-responsive release of Dox. At neutral pH (left), the electrostatic interactions between positively charged Dox and the negatively charged groups of the polymeric nanoparticle are intact, keeping the Dox from being released. At acidic pH (right), the higher proton concentration results in disruption of that interaction and accelerated degradation of the polymer, resulting in Dox release.

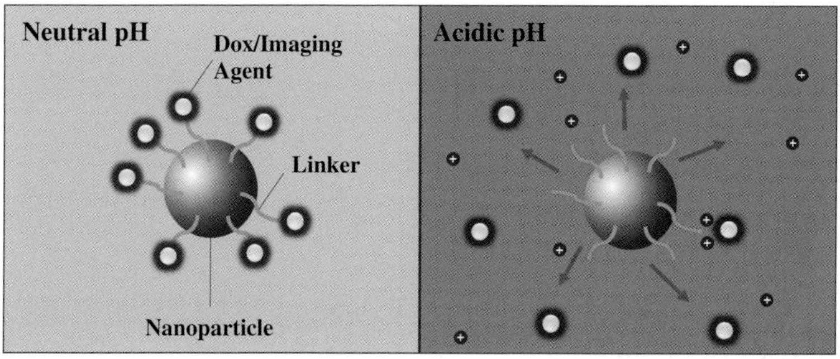

Figure 15.2. Schematic diagram of acid-sensitive linker release of payload. At neutral pH (left), the linker is unaffected and keeps the payload bound to the nanoparticle. At acidic pH (right), the linker is cleaved and the payload is released. The nanoparticle can be polymeric or gold-based, and the payload can be drugs (not restricted to protonated drugs such as Dox) and/or imaging agents.

pH-responsiveness in their study with micelles made of poly(amidoamine)–poly(l-lactide)-*b*-poly(ethylene glycol) [5]. The negatively charged polyester groups in the polymeric backbone associated with the positively charged Dox, and this interaction was disrupted by increased acidic conditions, resulting in increased DOX release. This mechanism seems to be specific to treatments with protonated drugs like DOX, and allows for an inherent pH-sensitivity that does not require the conjugation of an additional moiety for pH-sensitivity.

Another mechanism is the use of an acid-sensitive linker (Fig. 15.2). Examples of linkers used successfully in targeted drug delivery include hydrazone, esters, acetal bonds, and disulfide bridges. Hydrazone linkers hydrolyze at low pH while being stable at neutral pH, making them ideal for targeted delivery in the lowered pH of tumors [6]. Huang *et al.* demonstrated success with hydrazone linkers in developing pH-sensitive theranostics, where Adriamycin (ADR) was attached to the micelles via a hydrazone bond and magnetic nanoparticles were embedded in the core [7]. An interesting study performed by Xiao *et al.* also used Dox and hydrazone linkers, taking aspects from both of the aforementioned pH-sensitivity mechanisms [2]. This study used gold nanorods GNRs conjugated with Dox via hydrazone and cyclic arginine–glycine–aspartic acid (cRGD) peptides (for tumor targeting) for a smart theranostic system for cancer therapy. Figure 15.3 shows the pH-sensitive release of payloads *in vitro*.

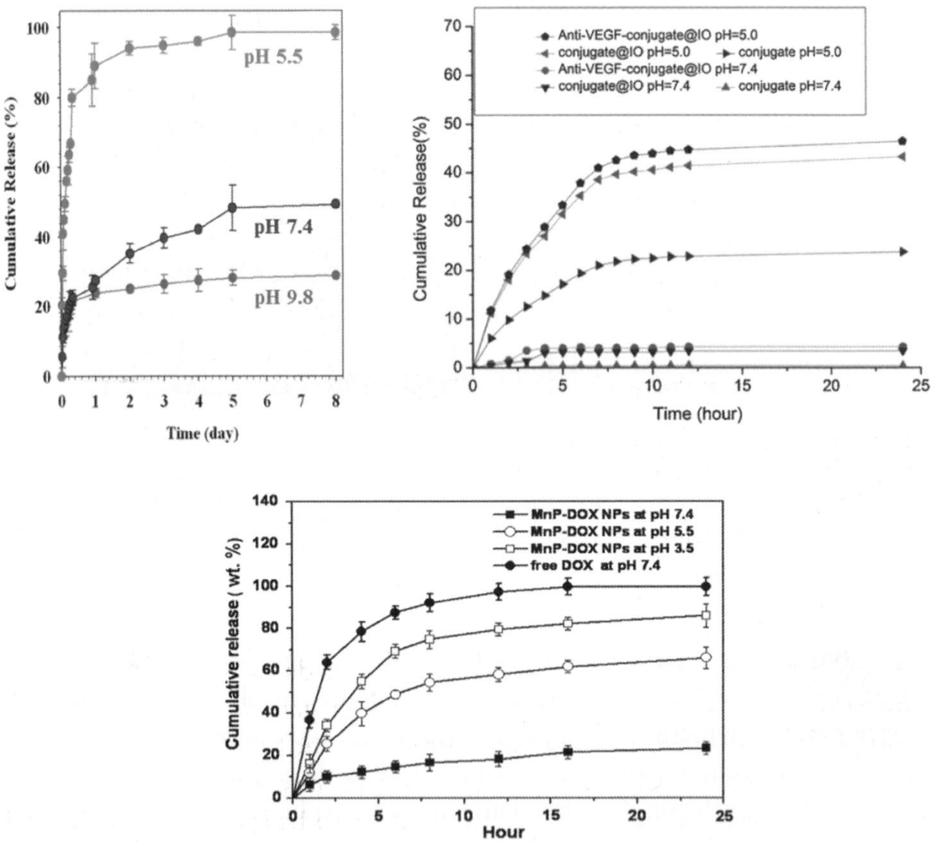

Figure 15.3. Examples of *in vitro* pH-sensitive release profiles of different theranostic systems: significant differences are seen in release levels at lowered pH compared to neutral and high pH. Top-left: Lim *et al.* [8], Top-right: Huang *et al.* [7], Bottom: Jing *et al.* [4].

15.2.2 *Strategies for targeting*

The development of pH-sensitive nanosystems for targeted drug delivery faces several challenges, including biocompatibility, stability, controlled release, and effective targeting. The former three design criteria are considered with great detail when choosing the backbone and therapeutic agent of the nanosystem. The different pH-sensitive release mechanisms have been discussed; targeting mechanisms are also of great importance. Nanosystems can be designed to release their payloads at decreased pH levels, but they need to be physically present at the sites of lowered pH before release can occur. For this to happen, effective tissue or cell targeting is paramount.

Indeed, the diversity of nanosystems has resulted in a number of innovative targeting mechanisms seen in the literature of pH-sensitive theranostics. Antibodies, peptides, and aptamers are the most commonly used targeting agents. Cancer cells over-express numerous membrane proteins and antigens, which could be the binding targets of engineered peptides and antibodies, respectively (Fig. 15.4). The previously mentioned study performed by Xiao *et al.* used GNRs surface-conjugated with cRGD peptides for targeting, along with an anticancer drug and imaging agent [2]. RGD peptides are known for their importance in cell attachment, and this particular cRGD peptide was designed to bind to the $\alpha_v\beta_3$ integrin, which is over-expressed on tumor cells. Xiao *et al.* confirmed the targeting ability of the cRGD peptide via flow

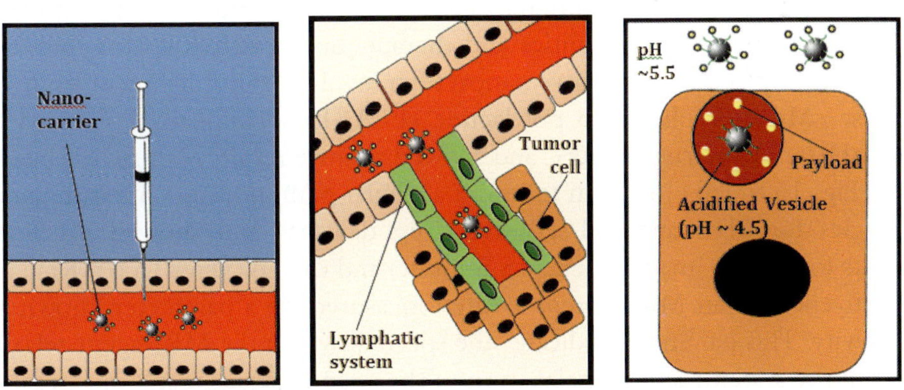

Figure 15.4. Schematic of nanocarrier pathway to tumor. (Left) Nanocarriers injected into bloodstream. (Middle) Nanocarriers transported to tumor area via lymphatic system. Targeting system, such as peptide or antibody, makes this process faster and more selective. (Right) Extracellular tumor environment, with a slight decrease in pH, results in initial release of payload. Acidified vesicles (sharp decrease in pH) uptake nanocarriers via endocytosis result in full release of payload within tumor cells.

cytometry analysis, as cells treated with non-cRGD nanoparticles showed a much lower uptake compared to the positive group.

Antibodies have also been effectively used for targeting cancer therapy. Huang *et al.* used anti-vascular endothelial growth factor (VEGF) antibodies to target the over-expressed VEGF present on many cancer cells, including cells in hepatoma [7]. *In vivo* imaging confirmed the superior targeting ability of anti-VEGF nanoparticles compared to plain nanoparticles. Guo *et al.* conjugated an anti-CD105 antibody to their polymeric nanosystem; CD105 is expressed almost exclusively on tumor endothelial cells and serves as an excellent target for treating proliferating tumors [5].

It is difficult to say which targeting mechanism works the best in these smart theranostic systems, because the studies performed are different and target different types of cancer. Some nanosystems, like Koo's chitosan-based system simply leverages on the enhanced permeation and retention (EPR) effect of nanoparticles for enhanced efficacy [9]. The addition of a targeting antibody or peptide would give further specificity on top of the EPR effect, although the extra level of complexity would add to the cost and energy of development.

15.2.3 *Application of pH-sensitive theranostics*

As mentioned previously, the majority of the applications seen with pH-sensitive theranostics are within the realm of cancer treatment, albeit in diverse ways. The combination of different nanosystem backgrounds, pH-sensitivity mechanisms, imaging modalities and therapeutic agents allows for this vast diversity. Jing *et al.*'s innovative nanosystem involved the attachment of manganese (Mn)-poryphyrin to polylactide (PLA)-Dox nanoparticles for MRI and pH-sensitive drug delivery, and showed positive results with *in vivo* imaging (Fig. 15.5, left) [4]. Lim followed a similar path in using chitosan nanoparticles for MRI and pH-sensitive drug delivery, and showed excellent results for *in vivo* imaging (Fig. 15.5, right) and therapy (Fig. 15.6) [8]. The effectiveness of an MRI system is often measured via a parameter known as relaxivity. This parameter indicates the speed at which the MRI agent relaxes within the system; quicker relaxation of the agent's atoms result in a greater relaxivity and greater contrast in imaging, which is desired. Nanocarriers enclosing magnetic agents has resulted in the improvement of half-lives and, thus, relaxivities of these agents for MRI [4]. Smart theranostic systems have shown relaxivity values comparable to many MRI agents used commercially, which will be discussed in-depth later in this review [4]. Though MRI is the most common imaging modality seen with pH-sensitive nanosystems, alternate

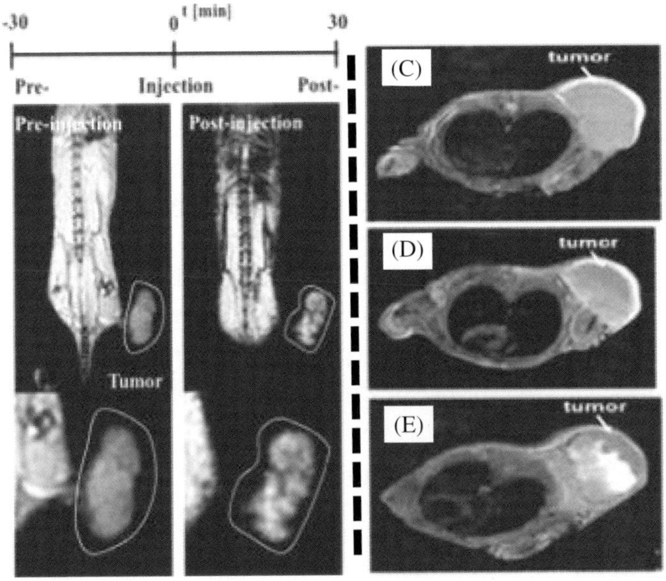

Figure 15.5. *In vivo* MRI imaging of sample pH-sensitive theranostic systems. (Left) Tail-vein injection of chitosan-DMNPs; accumulation of magnetic nano particles in tumor over time allows for improved imaging [3]. (Right) Intravenous injection of MnP-Dox nanoparticles shows improved contrast in MRI over time (C)–(E) [4].

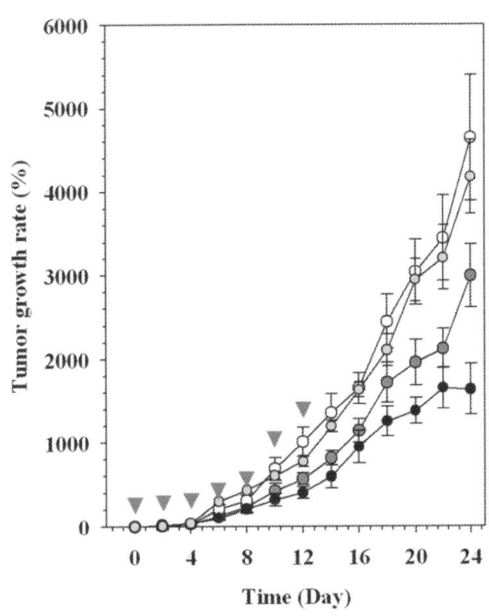

Figure 15.6. *In vivo* therapeutic efficacy trial of chitosan-DMNPs shows effectiveness in delaying tumor regression [3]. Black — C-DMNPs, Gray — Dox, White — Saline.

imaging modalities are seen as well. Jung *et al.* used pH-sensitive AuNPs for selective surface enhanced Raman scattering (imaging) and photothermal cancer therapy, with positive results *in vivo* [10]. The internalized AuNPs are made to aggregate and form "hot spots" at individual nanoparticle interfaces, allowing for enhanced Raman scattering and photothermal ablation of cancer cells. Photo-sensitive nanosystems can be considered a separate category of smart theranostic nanosystems and will be covered more in-depth in the following section, but this brings up an important point. The diversity afforded by these smart nanosystems often means there is overlap in the mechanisms of sensitivity of smart nanosystems (pH, photothermal, magnetic, etc.); this only improves their effectiveness.

15.3 Photo-Responsive Systems

Photo-responsive (PR) systems differ from other theranostic nanosystems because their response mechanism does not rely on cues such as enzymatic degradation, pH, and magnetic actuation. Instead, laser irradiation provides an exact, highly tunable stimulus that strikes a localized area of imaging and therapeutic interest. Primarily, irradiation has demonstrated capability not only in image-enhancement, but also in therapeutic amplification in aspects such as PR drug release and photothermal therapy (PTT). Like the other classes of nanosystems here within this account, the design parameters are numerous and often will overlap with design or innovative schematics of other smart nanosystems. Nonetheless, the PR theranostic systems detailed in this review demonstrate novelty and hold promising potential in the field of nanomedicine. These systems can be constructed to possess a blend of targeting moieties, imaging agents, and therapeutics. The following sections will provide an overview of photo nanosystems' design, applications, challenges, and successes.

15.3.1 *Design and synthesis of PR theranostic nanoplatforms*

In the past decade, numerous classes of nanomaterials with strong near-infrared (NIR) absorbance have been widely used as photothermal agents for cancer PTT treatment, including various gold nanostructures, carbon nanomaterials, palladium (Pd) nanosheets, copper sulfide (CuS) nanoparticles, and even organic nanoparticles [11]. NIR absorbance is desirable because the part of the region ranging from 700–900 nm is known as the "transparent window". In this window, light can penetrate soft tissue due to reduced interference from water and hemoglobin within the blood, as well as scattering from tissue [12].

AuNPs (H_2O) have been researched extensively, due to their biocompatibility, bio-inertness, and light scattering ability at resonant wavelengths from the excitation of plasmon oscillations. This phenomenon is known as localized surface plasmon resonance (LSPR). These nanoparticles come in many shapes and sizes such as spherical, rod-like, and even cage morphologies [12, 13, 14]. Gold and other PR nanomaterials can provide enhanced imaging, enhanced therapy, or a combination of these, depending on design and material (Fig. 15.7). Enhanced therapy typically results from hyperthermia, which is the heating of tissues within the range of 42–46 °C [15]. Irreversible necrosis is achieved through heating at 43 °C for 240 min or an equivalent 54 °C for 1 second [15]. Overall, the systems and their designs described have their respective advantages, as well as disadvantages. From the examples provided

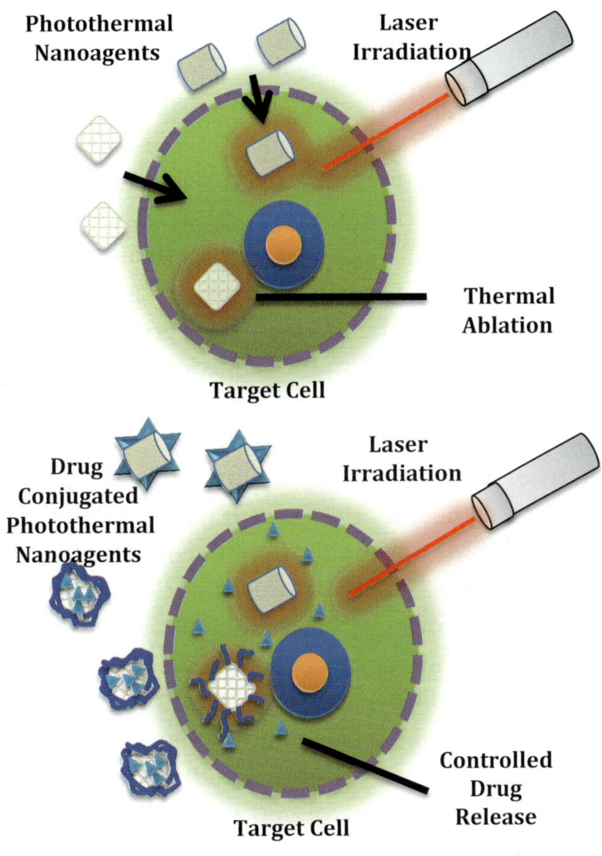

Figure 15.7. (left). Schematic of photothermal nanoagents to targeted regions. There are primarily two modes of therapy using laser irradiation: photothermal ablation for the necrosis of tumor tissue (top) and heating for the controlled release of drug molecules (bottom).

in the following section, there are certain parameters that are essential. The characteristics of a successful PR nanosystem include: (i) delivery of enough nanoparticles to a region of interest to mediate photothermal ablation while limiting systemic side effects; (ii) accurate dosage for sustained effects within the therapeutic window; and (iii) monitoring of the spatiotemporal heat profile and response to the therapeutic nanosystem [15]. Due to the diverse range of materials PR nanosystems encompass, the examples in the following paragraph will cover design overviews of each system, primarily categorized by material.

15.3.2 *Gold-based PR theranostic tanosystems*

AuNPs are perhaps the most popular in light-reactive theranostic nanosystems. The plasmon excitation phenomenon, known as LSPR, allows for not only enhanced scattering for imaging purposes, but also hyperthermic responses as well [13]. The inherent duality of laser irradiation makes gold an attractive material, although it does not always make for an effective theranostic. For example, solid spherical gold morphologies do not possess LSPR within the "transparent window", making tissue penetration rather limited. On the other hand, non-spherical/solid morphologies such as rod, shell, and cage possess LSPR in the NIR region. Xia *et al.* demonstrated the power of NIR LSPR and its numerous modalities through the use of gold nanocages (GNCs). A novel class of nanostructures, GNCs possess a cubic morphology with thin and porous walls. Scale-up procedures have been found to be simple, using a galvanic replacement between gold nanocubes and aqueous hydrochloric acid [12]. The multiple modalities experimented in Xia's work demonstrate the true, all-encompassing ability of PR theranostics. For hyperthermic-controlled drug release, the surface of the GNCs was coated with poly(N-isopropylacrylamide) through thiol-chemistry, which is used in a variety of gold conjugations. Upon exposure to NIR laser irradiation or even high-intensity focused ultrasound, these polymer coatings reach above their lower critical solution temperature and collapse, releasing encaged drugs within the polymer–gold construct. Aside from drug release, the hyperthermic potential was demonstrated using a murine model (Fig. 15.8). A positron emission tomography (PET)/computerized tomography (CT) scan done before and after treatment of control and polyethylene glycol (PEG)ylated GNC treated tumor-bearing mice showed the effectiveness of ablation as well as NIR absorbance. A highly popular imaging modality realized within the realm of gold nanomaterials is the use of two-photon imaging, with GNCs

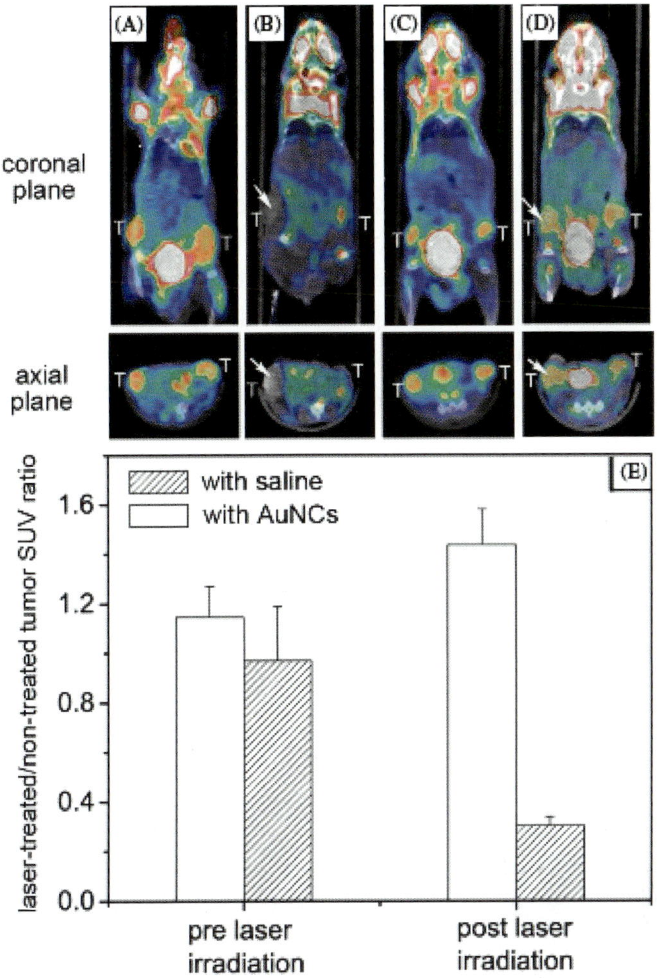

Figure 15.8. *In vivo* F-FDG PET/CT results of the mice after IV administration of PEGylated GNCs before (A) and after treatment (B) and saline control before (C) and after (D). The white arrows highlight regions of laser exposure (0.7 W/cm^2 at 10 min) and the graph (bottom) shows the standardized uptake values of laser treated tumors to non-treated tumors [12].

being no exception. Xia demonstrated the ability to monitor the cellular uptake of GNCs, while also causing photothermal damage, due to an overlap in the excitation laser and LSPR peaks. This type of concurrent imaging and ablation is not always desirable, depending on its intended purpose, as it affects the integrity of the sample; background scattering at this wavelength can also lower imaging sensitivity [14]. While doping the gold construct with silver or other noble metals is an alternative solution, other researchers have recently exploited the use of AuNPs and their lower plasmon resonance wavelengths as an advantage in imaging and therapy [12, 14]. Lukianova *et al.* demonstrated the generation of plasmonic nanobubbles (PNBs) of aggregating gold, spherical nanoclusters with an LSPR of ~530 nm [14]. These unique PNBs were able to generate enhanced optical imaging, mechanical cell disruption with negligible photothermal effects, and real-time image-guided therapy. A laser pulse creates the transient phenomenon of PNBs, with larger laser fluences generating larger bubbles. Optical imaging bubbles have shown enhancement of increased contrast of 10–100-fold versus AuNPs alone. Mechanical disruption of these bubbles could be achieved through laser energy of 1 J cm^{-2} while other PR nanosystems would require energy on the scale of 30–1,000 J cm^{-2}. To demonstrate this model *in vivo*, Lukianova used a zebrafish embryo model, transplanted with 60-nm AuNPs conjugated with C225 anti-epithelial growth factor receptor (EGFR) and DiI fluorescent-labeled C4-2B prostate cells (Fig. 15.9). The DiI-tagged prostate cells allow for tracking of cell viability. Each cancer cell was scanned, located, and centrally positioned under the laser. Under a laser with fluence of 125 mJ cm^{-2}, a

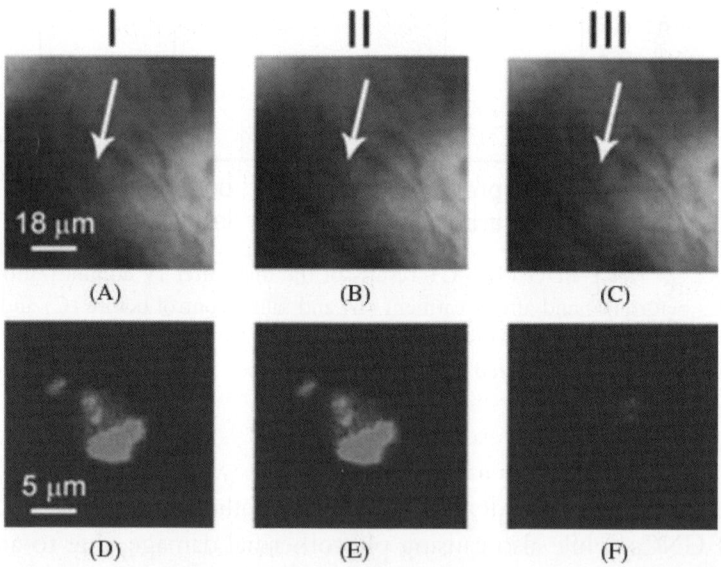

Figure 15.9. Brightfield images (A–C) of the zebrafish embryo with prostate cancer cells incubated with GNPs and DiI dye before PNB generation (I), after the 1st PNB at a pulse of 125 mJ/cm^2 (II) and after a 2nd PNB at a pulse of 175 mJ/cm^2 (III). Fluorescence of cancer cells (arrows in A–C) is shown in D–F, as brightfield imaging cannot distinguish cancer from host cells. Significant loss of cell viability was observed after the 2nd PNB generation [16].

PNB generated enhanced optical contrast, while a second fluence of 175 mJ cm^{-2} demonstrated even higher contrast, followed shortly by a significant loss in DiI fluorescence, indicating an effective PNB-based therapy [16]. The use of lower wavelengths would require the use of transparent media for access to deeper tissue. Optical fiber probes can circumvent this issue of low penetration but in the present, these PNBs can see their applications within subcutaneous tumor imaging and extracorporeal methods of processing tumor tissues.

15.3.3 *Silica-based and other PR theranostic nanosystems*

The PR nanosystem realm is not completely dominated by gold nanomaterials. Composite nanosystems have seen their popularity rise in theranostics. Mesoporous silica nanoparticles possess unique properties, such as large surface areas, tunable pore sizes, and well-defined surface properties, that make them an attractive composite material [17]. For example, Zhang *et al.* developed mesoporous silica-coated GNRs as a theranostic agent. Using two-photon imaging, Zhang found that one can monitor the cellular uptake of Au≅SiO2 conjugated with DOX (Fig. 15.10). In addition, he developed two therapeutic modalities using one construct. At lower laser intensity, Dox-loaded GNRs show enhanced drug release while at a higher intensity, photothermal effects aid in destroying cancerous cells. The use of silica helped prevent the aggregation of GNRs, which can shift LSPR to the visible wavelength spectrum and lead to a decrease in tissue penetration. It also helped increase the specific drug area, allowing for a higher payload of Dox while the transparency of the surrounding silica shell helped minimize laser interference. The use of nanocomposites is not limited to just silica-gold. Other "smarter" nanosystems have been developed, such as Ma and company's iron oxide and graphene oxide nanocomposite. The use of superparamagnetic iron oxide nanoparticles (SPIONs) allowed for T_2-weighted MRI, as well as magnetic drug targeting. Tumor-bearing Balb/c mice were injected with PEGylated graphene oxide-SPIONs to examine whole body *in vivo* imaging capability (Fig. 15.11). The images displayed 24 hours after injection showed a remarkably darker contrast in areas of interest, such as the liver and the tumor. The increased tumor contrast was most likely due to the enhanced permeability and retention (EPR) effect. The system was conjugated with Dox via π–π stacking, and subsequent drug release was enhanced due to the protonation effects of a lower pH in the tumor microenvironment [18]. The capabilities did not end there, as graphene oxide could absorb light at a wavelength of 808 nm, rendering the nanocomposite suitable for thermal ablation of cancer cells. Active targeting through an applied magnetic field also allowed for localization of

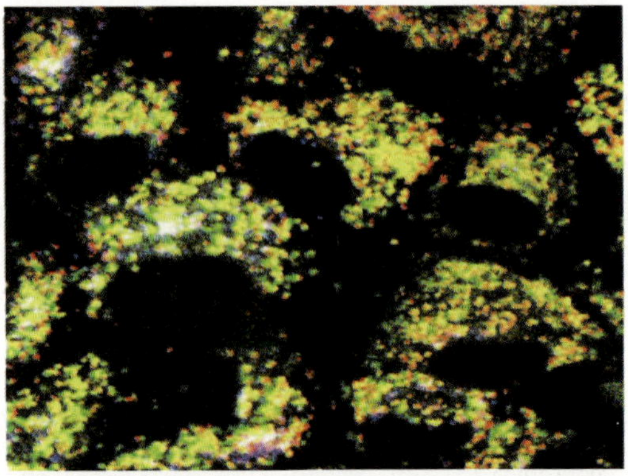

(A)

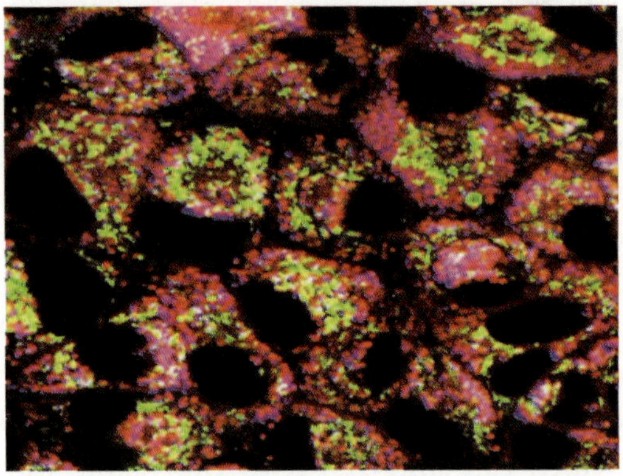

(B)

Figure 15.10. Two-photon confocal microscopic images show the intracellular localization of DOX (red) and gold/silica constructs (blue) with Lysa-Tracker (green) and Mito-Tacker (green) as organelle-specific probes [13].

the nanocomposite to regions of interest, demonstrating a potential for image analysis, therapeutic release, and optimal biodistribution. External laser irradiation of different powers on different concentrations of the nanocomposite proved to be effective at 50 mg/L on 4T1 cancer cells. This type of theranostic tool not only demonstrates the photothermal potential of theranostics, but also successfully hybridizes other types of materials and concepts prevalent in other smart nanosystems to create a powerful apparatus.

15.4 Magnetically Responsive Systems

Similar to the other systems described in this chapter, magnetic actuation provides spatial and temporal control, while affording the ability to be operated

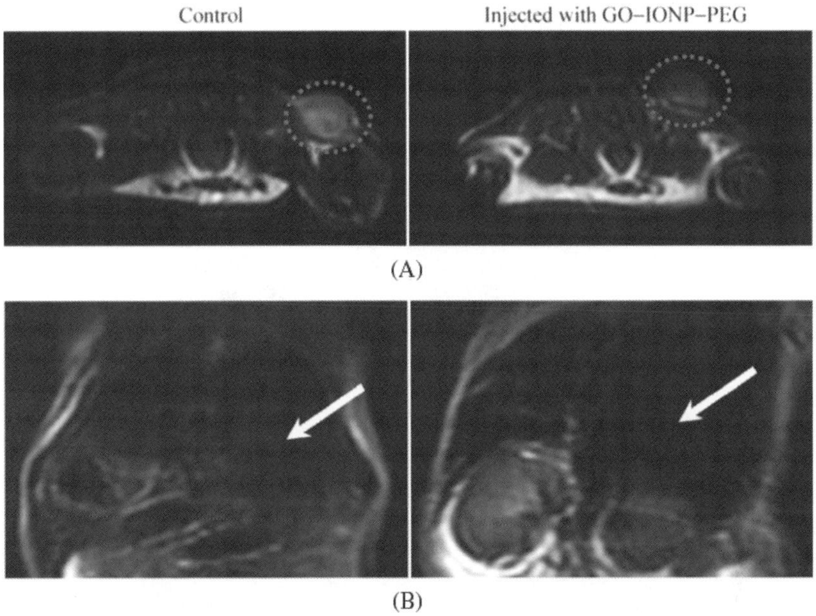

Figure 15.11. *In vivo* MRI of tumor-bearing mice injected with and without graphene-oxide SPIONs. Transversal (a) and longitudinal (b) show significant darkening of nanoparticle injected mice. Liver (arrows) and tumors (red circles) are marked [18].

outside of the body, providing a non-invasive approach to theranostics [19]. Utilizing the power of magnetization, these nanosystems are typically categorized by the responding mechanisms: magnetically responsive targeting and magnetically stimulated drug release (Fig. 15.12). Within these two broadly classified "smart" nanosystems, the design parameters of these two categories are boundless. The separation of stimuli is merely nominal and the dividing line is blurred, meaning the design innovations between these two smart nanosystems share many features in common. The appeal of each design is unique — due to the myriad of designs — and deciding which features are most suitable is a case-to-case evaluation. This account will provide an overview of these design parameters, successful applications, and opportunities that lie for future directions. Since the advent of magnetic nanoparticles, there have been many new designs and innovations based on magnetic properties for imaging and diagnostics. Magnetic nanoparticles exhibit not only unique material properties, but also an integrated design capability for cell targeting, imaging, and therapy, which render them as ideal platform materials for theranostics [20].

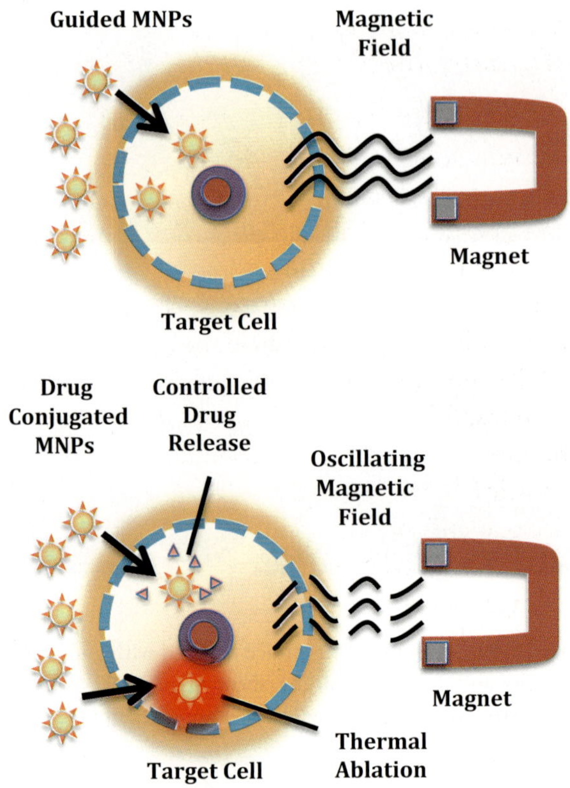

Figure 15.12. The schematic diagram shows the methods in therapy administration using magnetic fields. These techniques include the guidance of MNPs under the influence of a magnetic field (top), the release of drug after exposure to an oscillating field (bottom) and thermal ablation of MNPs after exposure to an oscillating field (bottom).

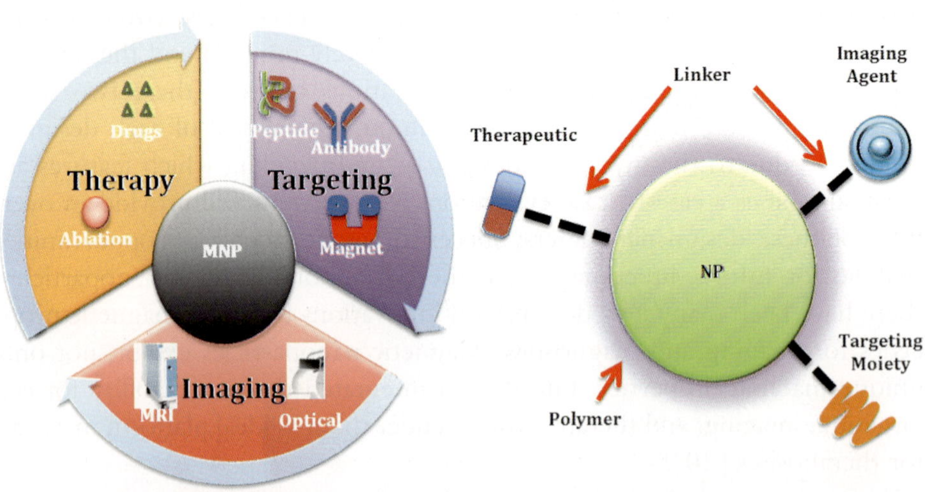

Figure 15.13. Schematic diagram shows the typical parameters examined when designing these magnetic nanoparticles (left). A typical nanoparticle "template" is shown in greater detail taking into account these parameters (right).

15.4.1 *Design and synthesis of theranostic magnetic nanoplatforms*

The challenges of producing these theranostic nanosystems have given rise to many novel designs, each tailored for its unique purpose (Fig. 15.13). Among various nanosystems studied thus far, magnetic nanoparticles containing ferromagnetic iron (Fe) and cobalt (Co), as well as their alloys and oxides, have been proven to be the most promising probes for theranostics [21]. The core material chosen must display properties at strong enough magnetic sensitivity in order to achieve maximal efficiency for imaging and therapeutic purposes [21]. While the majority of this section covers the use of magnetic iron-based nanosystems, ferromagnetic compounds such as iron are not limited to the therapeutic modalities described herein. Iron-doped zinc oxide (ZnO) has shown theranostic potential due to its therapeutic potential through the creation of reactive oxidative species, as well as enhanced MRI sensitivity [22]. It serves as a point that useful magnetic properties can be instilled within nanosystems to complement other nanosystems classified within pH-sensitive or photo-responsive realms. For example, Wang *et al.* used an acid-sensitive mesoporous ZnO to serve as a nanocarrier for doxorubicin. $MnFe_2O_4$ was conjugated as a T2 relaxation enhancement for MRI (nanoporous platform). Nonetheless, iron oxide remains a staple in the design of many magnetic nanoplatforms due to its biocompatibility as well as its low toxicity. The most common oxidized forms of iron manifested in theranostics today are magnetite (Fe_3O_4) and maghemite (γ-Fe_2O_3) [23]. These nanosystems generally rely upon inorganic materials, although some possess organic modifications, resulting in high-contrast imaging and magnetic properties for targeting, drug release, and therapy.

15.4.2. *Design parameters of magnetically responsive nanosystems*

Smart nanosystems possess many tunable properties such as size, surface area, surface chemistry, shape, and zeta potential, all of which can drastically affect a nanoparticle's theranostic potential. The surface chemistry (a coating material) of a magnetic nanoparticle plays a large role in conjugations, such as the attachment of targeting moieties, drugs, and contrasting agents, and may even indirectly impact diffusivity. Without a coating, magnetic nanoparticles display hydrophobicity and a tendency to aggregate [23]. Propensity to agglomerate is in part due to the charge of proteins and other polyelectrolytes prevalent in the serum. A suitable coating material is typically one that shows good biocompatibility, such as polysaccharides, synthetic polymers, lipids, and proteins [24]. For nanoplatforms that forgo the use of magnetic force, active targeting can still be employed through a multitude of motifs such as peptides, antibodies, and aptamers, a feasible process in part due to surface chemistry. These site-directed conjugates allow for accumulation of these theranostics in areas of need without the need for a strong magnetic field. In addition to this active targeting, the surface chemistry of a magnetic nanoparticle can allow for the loading of drugs and imaging agents, if an imaging modality is not inherent or strong enough within the platform already. Other coatings employing hydrophilic natural and synthetic polymers have been used due to their ability to prevent particle aggregation, increase solubility, and improve the stability of the particulates [26]. Hydrophilicity is an important factor because it not only increases the circulation time of nanoparticle, but also prevents macrophage clearance and plasma protein adsorption [26]. Dextran coating is a prime example, gaining popularity due to good biocompatibility and high stability [24].

Physical properties, such as size and magnetic potential, also play a large role within the design to create an effective theranostic nanoplatform. Aside from surface character, size will determine the magnetic properties, as well as the platform's ability to evade reticuloendothelial system clearance. While the exact parameters vary, sizes should be within the range of 10–300 nm, as any particles larger will be subject to rapid liver clearance and any particles smaller have been shown to be susceptible to renal filtration. Super paramagnetism typically arises in iron oxide particles of less than 20 nm in diameter. This property translates into a higher susceptibility to external magnetic field changes, a critical consideration in targeting and drug release [21]. The issue of magnetization and optimization of magnetic properties falls under the design parameter of magnetic potential. There are three properties within magnetism that are taken into consideration when tuning magnetic nanoplatforms: magnetization (M), coercivity (H_c), and magnetocrystalline anisotropy constant (K) [20]. These properties each have a corresponding mechanism of importance imparted to them. The figure below shows the relationship of these values to theranostic aims (Fig. 15.14).

Taking into account all these factors, there are several techniques available for the preparation of iron oxide nanoparticles. Precipitation of iron salts with *in situ* and post-addition of coating materials is a relatively popular synthesis technique. It allows for a more homogeneous yield of magnetic nanoparticles albeit in smaller amounts [25]. Thermal decomposition of organometallic compounds also offers good control over final particle size, shape, and crystal

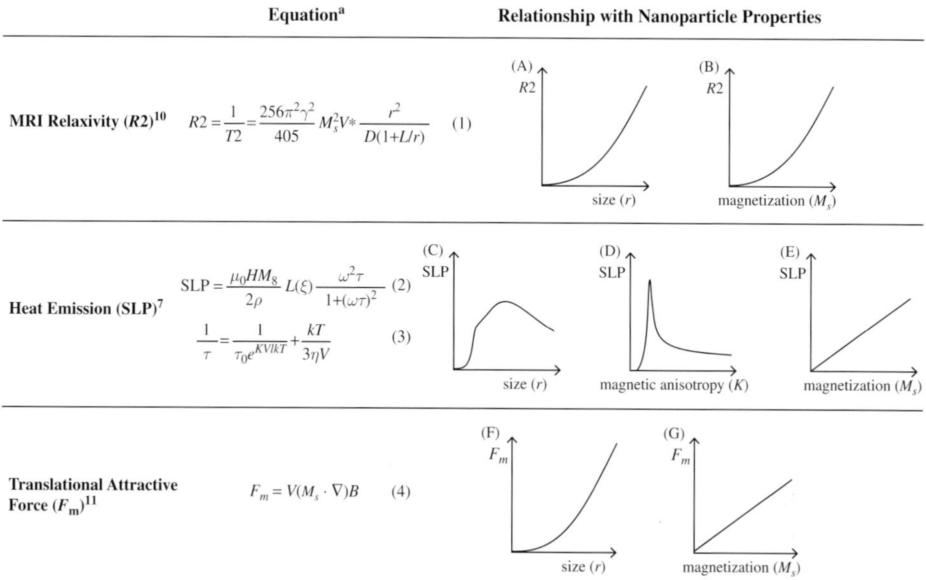

Figure 15.14. These equations represent the functional properties of magnetic nanoparticles. The importance of magnetization, coercivity, and magnetocrystalline anisotropy constant are realized through the roles they play in regard to MNP design [20].

structure. In this method, iron oxide nanoparticles are synthesized in organic solvent, meaning the final product typically needs surface modifications to impart aqueous stability [25].

15.4.3 *Application of magnetic nanosystems*

As one can imagine, a stimulus that can be externally triggered and fine-tuned will find attractive applications in imaging, cell tracking, drug delivery, and gene delivery. It is not without its challenges, however. This section will detail the challenges faced and innovations developed to compensate for limitations encountered.

In the realm of imaging, MRI is the most extensively used modality. As iron is the most used material, prominent relaxation time is measured in relation to T2 (negative contrast MRI imaging) and most magnetic imaging for the studies described is realized through this relaxation. The technique is based on the difference in nuclear magnetic relaxations of the H_2O protons in biological solutions and around solid tissues [21]. A contrast agent modifies the nuclear magnetic relaxation rate of its surrounding protons and changes the signal contrast. Its contrast enhancement effect is measured by the relaxation rate, R = 1/T (s^{-1}) and the relaxivity, r = R/concentration ($mM^{-1}\bullet s^{-1}$). A higher relaxivity corresponds to a better contrast effect [21]. When all these elements are combined, what results is a powerful, multifunctional device that cannot only treat, but also images effectively. Due to many complications and variables within a chaotic biological environment on a patient–patient basis, MRI using magnetic nanoparticles is not without its limitations. Most MRI inaccuracies arise from pre-existing physiological conditions, such as incomplete fat suppression, blood stream air bubbles, and calcification [20]. In order to combat this, Wang *et al.* devised a T1–T2 dual imaging system. An "error-free" drug system, gadolinium-embedded iron oxide nanoparticles were designed to show the benefits of having two imaging modalities. Observed minimal interference proved promising and perhaps provides the foundation for future studies to be done using an error-free, enhanced diagnostic approach in theranostic nanosystems.

Drug release for these smart nanosystems can be designed to respond to magnetic field oscillation. The magnetic field oscillation leads to a change from magnetic energy into thermal energy through the frictions caused by the physical rotation of a nanoparticle (Brownian relaxation), and the magnetization reversal within the nanoparticle (Neel relaxation). The use of alternating magnetic fields at high frequencies enhances the permeability of encapsulated contents. Lu *et al.* showed that an oscillating magnetic field could also affect the permeability of microcapsules by acting on aggregates of the nanoparticles, expanding the applications of magnetic oscillation. Additionally, Sanson *et al.* showed the benefits of polymer encapsulation with enhanced permeability in a novel manner using proof-of-concept *in vitro* studies. He used a radio-frequency magnetic field for smart drug release, while also exploring magnetic guidance potential *in vitro.* A hydrophobically-coated IONP was loaded within the membrane of poly(trimethylene carbonate)-b-poly(L-glutamic acid) (PTMC-b-PGA) block copolymer vesicles using a nanoprecipitation process, allowing for a low polydispersity. Doxorubicin was also loaded using a similar process, with a loading content of up to 70% maghemite. The magneto-liposomes were initially in a gel state and upon heating, reached the transition temperature that turned the gel into a fluid. Guided nanoparticles into targeted tissue could then exhibit controlled drug release through radio-frequency magnetic hyperthermia, demonstrating viability of this therapy and concept. The magnetophoretic ability of these magnetic polymersomes was able to reach a speed of 1 μm/s, a value similar to other studies citing enhanced uptake from magnetic targeting (Fig. 15.15). A further example of magnetic hyperthermic potential is seen through Hugo De

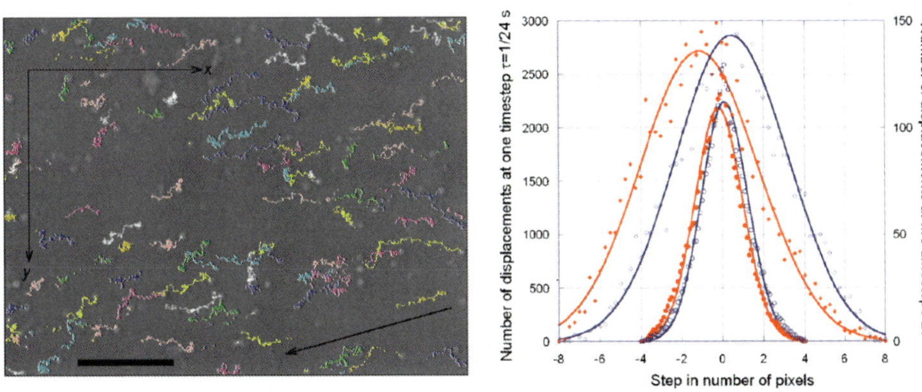

(A) (B)

Figure 15.15. Video microscopy (inverted brightfield image) superposed with trajectories of 50% weight loading of IONPs under a magnetic field gradient of 18.5 T/m (arrow) over a period of 10s (A). The scale bar is 20 μm and the displacement is visualized with the x (red) and y (blue) shifts in the histograms (B) [27].

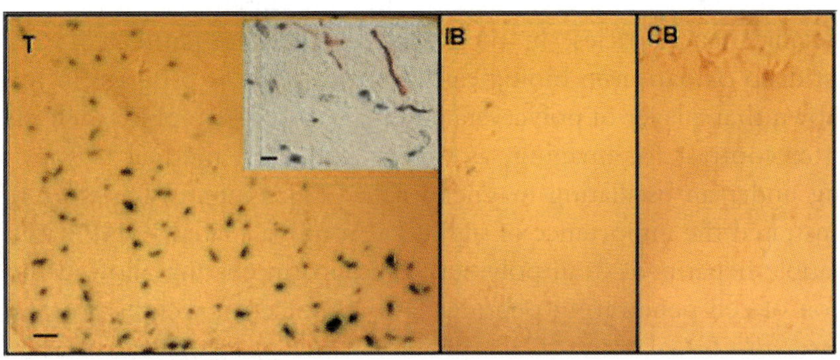

Figure 15.17. Histochemical analysis of tumor (T), ipsilateral brain (IB), and contralateral brain (CB) tissue. Staining was done with X-Gal and the enlarged inset shows capillaries stained with Burrstone's Red [28].

Olveira's work, which developed polymersomes loaded with both maghemite nanoparticles (γ–Fe_2O_3) and doxorubicin. A high frequency magnetic field at 500 kHz demonstrated enhanced drug release kinetics. No significant heating was measured during the dispersion, leading the researchers to believe that minimal thermal heating results in increased permeation of the polymersome membrane and subsequent drug release, thus verifying decreased Henrietta Lacks (HeLa) cell viability due to Dox release and not thermal effects (Fig. 15.16).

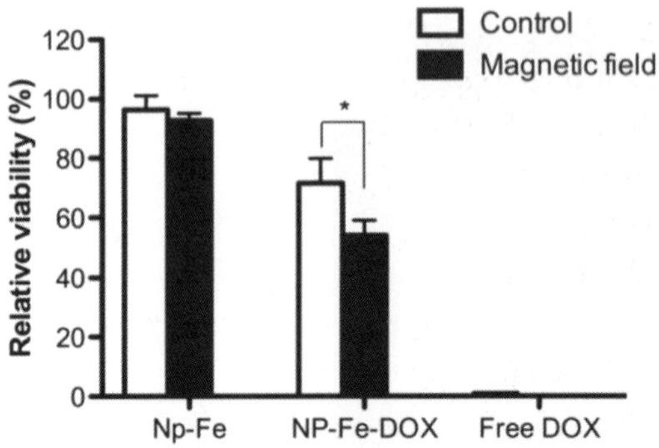

Figure 15.16. MTT assay of HeLa cells after 72-hour exposure to magnetic polymersomes with and without magnetic field (48 hours following a 24h incubation period) to demonstrate effective controlled drug release [23].

Polymersomes have shown themselves to be more than just controlled drug releasers. In addition to controlled release under AC magnetic fields, these magnetic polymersomes can be functionalized to increase half-life *in vivo,* targeting specificity, and even increase shelf-life [27].

Metallic nanoparticles, in their natural state, are difficult to stabilize against deep oxidation in biological situations. Recent progress in this area has shown that a layer of polycrystalline magnetite or graphite shell can lead to better contrast for imaging, as well as enhanced thermal therapy when heating under an oscillating magnetic field [21]. Huining He *et al.* further demonstrated the importance of stability through a coated IONP with a biocompatible heparin–dextran polymer. The heparin coating allowed the conjugation of cell-penetrating peptides, a low molecular weight protamine in this case, and ß-Galactose (Gal) (proof of concept in this case). The study demonstrated >350-fold brain tumor retention through the use of topographic magnetic field optimization versus passive, conventional permeable and retention targeting. As with most cases involving the blood–brain barrier, the ability to attain efficient capture and retention in a localized region within the brain while minimizing brain toxicity is a difficult feat. The staining in Fig. 15.17 shows a histochemical analysis and a significant accumulation of ß-Gal, 465 kDa within the tumoral region, while minimizing ipsi- and contra-lateral brain tissue exposure (Fig. 15.17). With these promising results, optimized smart nanosystems will find their use in regions that have been considered inaccessible to effective drug therapy.

Super paramagnetism is generally expressed at sizes less than 20 nm, which can be tricky for the desired biodistribution. This size range has demonstrated a propensity to leak out of capillary tissue, resulting in low tumor retention. To redress this, Hayashi *et al.* created a theranostic SPION by the conjugation of PEG and folic acid to help tumor retention and prevent capillary leakage. The SPIONs used were less than 10 nm in diameter, but coatings of PEG and FA in combination with a nanocluster assembly, 100 nm in size, counteracted unwanted leakage, as well as enhanced relaxivity for contrast. As a result, this study showed sufficient tumor accumulation and after exposure to an alternating current (AC) magnetic field, significantly stunted tumor growth. For imaging capability, Hayashi's nanocluster assembly also proved markedly well against commercial contrast agents, such as Resovist (Fig. 15.18). With a 20-fold higher saturation magnetization compared to Resovist, the clustered SPIONs showed a much higher magnetic moment as compared with individual SPIONs with a much smaller concentration of iron. In addition, the therapeutic potential was observed to be effective using a murine tumor model. After verification of thermal increase in H_2O, folic acid and PEG-conjugated SPION NCs were injected and allowed to accumulate for 24 h before an AC magnetic field of frequency 230 kHz was applied for 20 minutes. Thermal images taken immediately afterward showed localized heating within tumoral regions. Survival rate of the mice using the targeted construct was 100% 12 weeks after treatment, while all other treatment groups failed after 8 weeks. After 5 weeks of treatment, tumor volume was suppressed to 10% of that found in the control group (Fig. 15.18). As a result, it shows the importance of not just immunogenic evasion through PEGylation, but also the need for specific targeting to reach a critical threshold of therapeutic accumulation within the tumoral region in order for effective treatment.

15.5 Conclusion

The promise of smart theranostic systems is clear upon examination of several interesting studies done in the past five years. In this review, we have discussed three categories of such nanosystems: pH-sensitive, photo-responsive, and magnetically responsive. A diverse array of pH-sensitive systems rely on a handful of sensitivity mechanisms, the most prominent of which include acid-sensitive linkers and disruption of the electrostatic interaction between positively charged anti-cancer drugs and negatively charged groups on the polymer. Strong *in vitro* results of controlled release and cellular uptake are seen with these nanosystems, and *in vivo* results are headed in a positive direction as well. The majority of these systems utilizes MRI as their imaging modality, and compare well to commercial MRI agents. Generally promising results are also seen in nanosystems of the other two categories discussed. However, there are a few points to keep in mind.

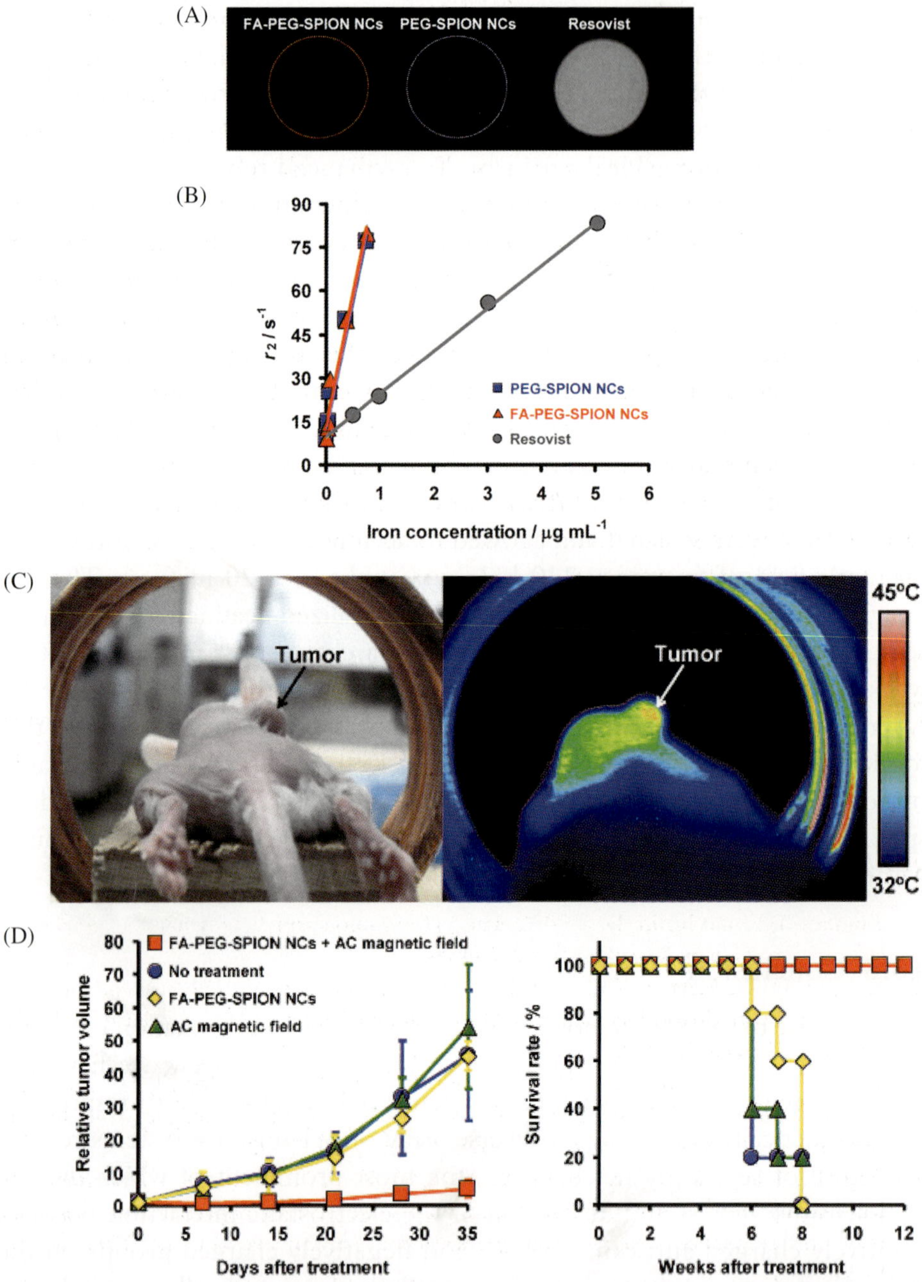

Figure 15.18. MRI of phantoms containing targeted SPION nanoclusters, untargeted SPION nanoclusters, and commercial imaging agent Resovist (A). The graph below (B) shows relaxivity dependence on iron concentration. *In vivo* results show hyperthermia after subsequent exposure to an AC magnetic field (C) and the graphs below (D) show tumor growth suppression and survival rate of mice in various treatment groups [29].